AF676321

Sustainable Functional Food Processing

Sustainable Functional Food Processing

Editors

Danijela Bursać Kovačević
Predrag Putnik

MDPI • Basel • Beijing • Wuhan • Barcelona • Belgrade • Manchester • Tokyo • Cluj • Tianjin

Editors
Danijela Bursać Kovačević
University of Zagreb
Croatia

Predrag Putnik
University North
Croatia

Editorial Office
MDPI
St. Alban-Anlage 66
4052 Basel, Switzerland

This is a reprint of articles from the Special Issue published online in the open access journal *Foods* (ISSN 2304-8158) (available at: https://www.mdpi.com/journal/foods/special_issues/sustainable_functional_food_processing).

For citation purposes, cite each article independently as indicated on the article page online and as indicated below:

LastName, A.A.; LastName, B.B.; LastName, C.C. Article Title. *Journal Name* **Year**, *Volume Number*, Page Range.

ISBN 978-3-0365-5301-6 (Hbk)
ISBN 978-3-0365-5302-3 (PDF)

© 2022 by the authors. Articles in this book are Open Access and distributed under the Creative Commons Attribution (CC BY) license, which allows users to download, copy and build upon published articles, as long as the author and publisher are properly credited, which ensures maximum dissemination and a wider impact of our publications.
The book as a whole is distributed by MDPI under the terms and conditions of the Creative Commons license CC BY-NC-ND.

Contents

About the Editors

Danijela Bursać Kovačević

Danijela Bursać Kovačević, PhD, is a distinguished Croatian researcher and associate professor with international recognition and influence in the field of food science and technology. She is a Professor of Chemistry and Technology of Fruits and Vegetables and related disciplines at the Faculty of Food Technology and Biotechnology, University of Zagreb (Zagreb, Croatia). She is currently a member of the editorial boards of several indexed journals and author of more than 200 articles published in international journals and at national and international symposia and conferences. In addition, she is currently the project leader of the scientific project "Hurdle technology and 3D printing for sustainable fruit juice processing and preservation". In 2021, she was awarded two valuable prizes: the first is the Tanner Award for co-author of the most cited article published in *Comprehensive Reviews in Food Science and Food Safety* in 2018 (https://ift.onlinelibrary.wiley.com/doi/10.1111/1750-3841.15207), and the second is the Clarivate™ Award for highly cited researchers who rank in the top 1% of citations for agricultural science in Web of Science™ (https://recognition.webofscience.com/awards/highly-cited/2021).

Predrag Putnik

Predrag Putnik, PhD, is senior researcher born in Zagreb, graduated from the University in Zagreb, Biotechnical Sciences (field of Food Engineering). He has a research background in applied laboratory work (chemistry, biology, bioinformatics, food science, nutrition, molecular biology, and industrial food engineering), information technology, and teaching. He has +100 peer-reviewed scientific papers and book chapters, where most of the journals have the highest impact factor (Q1) for food science and technology. With two other colleagues, he edited a book titled "Agri-Food Industry Strategies for Healthy Diets and Sustainability", Academic Press, Elsevier. He is an associate and guest editor for a number of food science journals with high Impact Factors on Web of Science™. During his scientific research work, he collaborates(ed) with 266 scientists from 108 different institutions from a total of 29 countries from all continents around the world. In addition, there were +30 international conference presentations and invited talks. In 2021, he received the Tanner Award for co-authoring the most-cited paper published in Comprehensive Reviews in Food Science and Food Safety, Wiley, in 2018. In 2021, he was elected as a Highly Cited Researcher ranked in the top 1% by citations for Agricultural Sciences by Web of Science™.

foods

MDPI

Editorial

Sustainable Functional Food Processing

Predrag Putnik [1,*] and Danijela Bursać Kovačević [2,*]

[1] Department of Food Technology, University North, Trg dr. Žarka Dolinara 1, 48000 Koprivnica, Croatia
[2] Department of Food Engineering, Faculty of Food Technology and Biotechnology, University of Zagreb, Pierottijeva 6, 10000 Zagreb, Croatia
* Correspondence: pputnik@alumni.uconn.edu (P.P.); dbursac@pbf.hr (D.B.K.)

Citation: Putnik, P.; Kovačević, D.B. Sustainable Functional Food Processing. *Foods* **2021**, *10*, 1438. https://doi.org/10.3390/foods10071438

Received: 3 June 2021
Accepted: 9 June 2021
Published: 22 June 2021

Publisher's Note: MDPI stays neutral with regard to jurisdictional claims in published maps and institutional affiliations.

Copyright: © 2021 by the authors. Licensee MDPI, Basel, Switzerland. This article is an open access article distributed under the terms and conditions of the Creative Commons Attribution (CC BY) license (https://creativecommons.org/licenses/by/4.0/).

Functional nutrition has become one of the main directions for a healthy lifestyle and sustainable food production due to its promising positive influence on health and its association with the use of raw materials of natural origin [1–3]. Therefore, it has attracted great interest from both consumers and manufacturers concerned about human well-being and sustainable economic growth [4]. Not surprisingly, new socio-demographic trends (e.g., longer life expectancy, promotion of healthy lifestyles, better healthcare, etc.) supported functional sector to become an increasingly lucrative segment of the food industry with a rapidly growing market [5].

Functional foods are industrially processed or unprocessed natural foods that have beneficial health effects beyond their basic nutritional value when consumed regularly [6]. Consumers today are increasingly looking for products that are safe, natural, have Generally Recognized as Safe (GRAS) status, and are produced using sustainable and/or ecologically sound technologies. For this reason, "functional food" is an increasingly popular term in the social and scientific spheres, so the industry is continuously investing in the development of a sector that can offer products with additional benefits for consumer health. Here, it is important to emphasize that clinical trials (randomized, double-blind, and placebo-controlled) should be conducted for the foods that are to be labeled as "functional" in order to draw conclusions about the health benefits of the products [6].

For instance, functional juices and other beverages produced from indigenous fruits (which are economically poorly explored) represent an interesting niche, with all the above characteristics to satisfy the interests of different food markets [7,8]. Accordingly, recent trends in the juice and beverage industry are aimed at producing functional juices and beverages with various raw materials such as vitamins and their precursors; minerals; fiber; unsaturated fatty acids; BACs, including polyphenols, carotenoids, chlorophylls, tannins, etc.; various antioxidants; probiotics; and prebiotics [9,10]. Due to their considerable nutritional value, fruit juices have been found to be excellent carriers or transport vehicles for probiotic bacteria.

Here, the focus is on functional ingredients such as BACs [11] and probiotics [12], which are responsible for numerous beneficial effects of functional foods on health. Unfortunately, the majority of BACs are thermolabile, which is particularly important for food production, where classical heat treatments (e.g., pasteurization) are still used. As thoroughly documented, this leads to food degradation and affects the quality of the final product. In order to prevent such negative effects of the production process, scientists and engineers have focused on developing economical and environmentally friendly technologies capable of maintaining the nutritional and sensory quality of the food as well as microbiological stability during functional food processing [13]. Such approaches are based on low energy consumption and on the use of low-impact processing and "hurdle technology," combining advanced (e.g., high power ultrasound (HPU), pulsed electric field (PEF), high pressure processing (HPP), etc.) and conventional food technologies, e.g., pasteurization [12,14,15]. Moreover, aside from food quality and safety, food design is also important for sensory appeal for consumers and economic success. Hence, technologies such as 3D food printing can be particularly useful in functional food production [16]. In the food industry, 3D printing is being explored in many areas, such as personalized and digitized nutrition, supply chain simplification, and expanding food offerings.

Aside from above mentioned, production of functional foods with application of innovative technologies is becoming increasingly popular (e.g., juices, dairy etc.) due to growing governmental requirements and support to decrease high expenditure and disposal of toxic chemicals and energy [12]. Since raw materials used for production of functional foods commonly employ suitable raw materials (fruits, vegetables, legumes etc.) they are mixed well with probiotics and/or biologically active compounds from plant and/or animal origin [10,16–22]. General direction for production of such foods is to have adherence to local diets (e.g., Mediterranean diets) while supplying local food markets with healthy alternatives for consumers. To that end it is encouraging existence, at (inter-)national levels (e.g., Research Executive Agency from European Commission; Croatian Science Foundation etc.), of initiatives to financially support projects able to develop nutritious foods with sustainable processing while overcoming limitations associated with upscaling of advanced technology in manufacturing. For instance, good example represents European SFS-funding program or 3D-SustJuice project funded by the Croatian Science Foundation (Hurdle technology and 3D printing for sustainable fruit juice processing and preservation) which is strongly related to this special issue and serves as a vehicle for dissemination of important data to scientific community [23]. Consequently, this Special Issue of *Foods* collected data relevant to sustainable functional food production [1,10,12,17,22,24–26], hurdle technology [12], advanced food processing [1,2,12,17,24–32], functional beverages [1,12,17,18,25,33], probiotics and BACs [1,2,12,17,20–22,24–32], and authentic fruits [7,25,30].

For this valuable collection of data, editors would like to thank all the authors who submitted their papers to this special issue and congratulate them for publishing their articles with *FOODS*. This would not be possible without the strong support of all devoted reviewers and their constructive comments, for which we are profoundly grateful. We would also like to acknowledge the 3D-SustJuice Project funded by the Croatian Science Foundation. Last but not least, we would like to thank Prof. Dr. Christopher John Smith, Editor-in-chief of *FOODS*, and entire MDPI team for their support of this special issue.

Author Contributions: Conceptualization, P.P. and D.B.K.; writing—original draft preparation, P.P. and D.B.K.; writing—review and editing, P.P. and D.B.K.; project administration, D.B.K. All authors have read and agreed to the published version of the manuscript.

Funding: This research was funded by the Croatian Science Foundation through the funding of the "Hurdle Technology and 3D Printing for Sustainable Fruit Juice Processing and Preservation" project, number "IP-2019-04-2105".

Conflicts of Interest: The authors declare no conflict of interest. The funders had no role in the design of the study; in the collection, analyses, or interpretation of data; in the writing of the manuscript, or in the decision to publish the results.

References

1. Ogori, A.F.; Amove, J.; Aduloju, P.; Sardo, G.; Okpala, C.O.R.; Bono, G.; Korzeniowska, M. Functional and quality characteristics of ginger, pineapple, and turmeric juice mix as influenced by blend variations. *Foods* **2021**, *10*, 525. [CrossRef]
2. Gallotti, F.; Lavelli, V. The effect of uv irradiation on vitamin d2 content and antioxidant and antiglycation activities of mushrooms. *Foods* **2020**, *9*, 1087. [CrossRef] [PubMed]
3. Šojić, B.; Tomović, V.; Kocić-Tanackov, S.; Bursać Kovačević, D.; Putnik, P.; Mrkonjić, Ž.; Đurović, S.; Jokanović, M.; Ivić, M.; Škaljac, S.; et al. Supercritical extracts of wild thyme (*Thymus serpyllum* L.) by-product as natural antioxidants in ground pork patties. *LWT* **2020**, *130*, 109661. [CrossRef]
4. Putnik, P.; Bezuk, I.; Barba, F.J.; Lorenzo, J.M.; Polunić, I.; Bursać Kovačević, D. Sugar reduction: Stevia rebaudiana Bertoni as a natural sweetener. In *Agri-Food Industry Strategies for Healthy Diets and Sustainability*; Academic Press: London, UK, 2020; pp. 123–152.
5. Musina, O.; Rashidinejad, A.; Putnik, P.; Barba, F.J.; Abbaspourrad, A.; Greiner, R.; Roohinejad, S. The use of whey protein extract for manufacture of a whipped frozen dairy dessert. *Mljekarstvo* **2018**, *68*, 254–271. [CrossRef]

6. Granato, D.; Barba, F.J.; Bursać Kovačević, D.; Lorenzo, J.M.; Cruz, A.G.; Putnik, P. Functional Foods: Product Development, Technological Trends, Efficacy Testing, and Safety. *Annu. Rev. Food Sci. Technol.* **2020**, *11*, 93–118. [CrossRef] [PubMed]
7. Duralija, B.; Putnik, P.; Brdar, D.; Bebek Markovinović, A.; Zavadlav, S.; Pateiro, M.; Domínguez, R.; Lorenzo, J.M.; Bursać Kovačević, D. The Perspective of Croatian Old Apple Cultivars in Extensive Farming for the Production of Functional Foods. *Foods* **2021**, *10*, 708. [CrossRef] [PubMed]
8. Roselló-Soto, E.; Barba, F.; Putnik, P.; Bursać Kovačević, D.; Lorenzo, J.; Cantavella-Ferrero, Y. Enhancing Bioactive Antioxidants' Extraction from "Horchata de Chufa" By-Products. *Foods* **2018**, *7*, 161. [CrossRef] [PubMed]
9. Kandylis, P. Grapes and Their Derivatives in Functional Foods. *Foods* **2021**, *10*, 672. [CrossRef] [PubMed]
10. Hinestroza-Córdoba, L.I.; Duarte Serna, S.; Seguí, L.; Barrera, C.; Betoret, N. Characterization of Powdered Lulo (*Solanum quitoense*) Bagasse as a Functional Food Ingredient. *Foods* **2020**, *9*, 723. [CrossRef] [PubMed]
11. Lorenzo, J.M.; Munekata, P.E.; Putnik, P.; Bursać Kovačević, D.; Muchenje, V.; Barba, F.J. *Sources, Chemistry, and Biological Potential of Ellagitannins and Ellagic Acid Derivatives*; Studies in Natural Products Chemistry; Elsevier: Amsterdam, The Netherlands, 2019; pp. 189–221.
12. Putnik, P.; Pavlić, B.; Šojić, B.; Zavadlav, S.; Žuntar, I.; Kao, L.; Kitonić, D.; Kovačević, D.B. Innovative Hurdle Technologies for the Preservation of Functional Fruit Juices. *Foods* **2020**, *9*, 699. [CrossRef]
13. Stamenković, Z.; Pavkov, I.; Radojčin, M.; Tepić Horecki, A.; Kešelj, K.; Bursać Kovačević, D.; Putnik, P. Convective Drying of Fresh and Frozen Raspberries and Change of Their Physical and Nutritive Properties. *Foods* **2019**, *8*, 251. [CrossRef] [PubMed]
14. Škegro, M.; Putnik, P.; Bursać Kovačević, D.; Kovač, A.P.; Salkić, L.; Čanak, I.; Frece, J.; Zavadlav, S.; Ježek, D. Chemometric Comparison of High-Pressure Processing and Thermal Pasteurization: The Nutritive, Sensory, and Microbial Quality of Smoothies. *Foods* **2021**, *10*, 1167. [CrossRef]
15. Galanakis, C.M. Functionality of Food Components and Emerging Technologies. *Foods* **2021**, *10*, 128. [CrossRef]
16. Tomašević, I.; Putnik, P.; Valjak, F.; Pavlić, B.; Šojić, B.; Bebek Markovinović, A.; Bursać Kovačević, D. 3D printing as novel tool for fruit-based functional food production. *Curr. Opin. Food Sci.* **2021**, *41*, 138–145. [CrossRef]
17. Žuntar, I.; Petric, Z.; Bursać Kovačević, D.; Putnik, P. Safety of Probiotics: Functional Fruit Beverages and Nutraceuticals. *Foods* **2020**, *9*, 947. [CrossRef]
18. Petric, Z.; Žuntar, I.; Putnik, P.; Bursać Kovačević, D. Food–Drug Interactions with Fruit Juices. *Foods* **2021**, *10*, 33. [CrossRef] [PubMed]
19. Zavadlav, S.; Blažić, M.; Van de Velde, F.; Vignatti, C.; Fenoglio, C.; Piagentini, A.M.; Pirovani, M.E.; Perotti, C.M.; Bursać Kovačević, D.; Putnik, P. *Sous-Vide* as a Technique for Preparing Healthy and High-Quality Vegetable and Seafood Products. *Foods* **2020**, *9*, 1537. [CrossRef] [PubMed]
20. Wi, G.; Bae, J.; Kim, H.; Cho, Y.; Choi, M.-J. Evaluation of the Physicochemical and Structural Properties and the Sensory Characteristics of Meat Analogues Prepared with Various Non-Animal Based Liquid Additives. *Foods* **2020**, *9*, 461. [CrossRef]
21. Małecki, J.; Tomasevic, I.; Djekic, I.; Sołowiej, B.G. The Effect of Protein Source on the Physicochemical, Nutritional Properties and Microstructure of High-Protein Bars Intended for Physically Active People. *Foods* **2020**, *9*, 1467. [CrossRef]
22. Lopes, M.; Pierrepont, C.; Duarte, C.M.; Filipe, A.; Medronho, B.; Sousa, I. Legume Beverages from Chickpea and Lupin, as New Milk Alternatives. *Foods* **2020**, *9*, 1458. [CrossRef]
23. Bursać Kovačević, D.; Putnik, P. *Hurdle Technology and 3D Printing for Sustainable Fruit Juice Processing and Preservation (IP-2019-04-2105)*; Croatian Science Foundation: Zagreb, Croatia, 2020; Volume 1000000 HRK, Available online: http://sustjuice.pbf.hr/ (accessed on 3 June 2021).
24. Barišić, V.; Flanjak, I.; Kopjar, M.; Benšić, M.; Jozinović, A.; Babić, J.; Šubarić, D.; Miličević, B.; Doko, K.; Jašić, M.; et al. Does High Voltage Electrical Discharge Treatment Induce Changes in Tannin and Fiber Properties of Cocoa Shell? *Foods* **2020**, *9*, 810. [CrossRef]
25. Adetoro, A.O.; Opara, U.L.; Fawole, O.A. Effect of Carrier Agents on the Physicochemical and Technofunctional Properties and Antioxidant Capacity of Freeze-Dried Pomegranate Juice (*Punica granatum*) Powder. *Foods* **2020**, *9*, 1388. [CrossRef]
26. Mocanu, G.-D.; Chirilă, A.C.; Vasile, A.M.; Andronoiu, D.G.; Nistor, O.-V.; Barbu, V.; Stănciuc, N. Tailoring the Functional Potential of Red Beet Purées by Inoculation with Lactic Acid Bacteria and Drying. *Foods* **2020**, *9*, 1611. [CrossRef]
27. Nguyen, T.T.; Rosello, C.; Bélanger, R.; Ratti, C. Fate of Residual Pesticides in Fruit and Vegetable Waste (FVW) Processing. *Foods* **2020**, *9*, 1468. [CrossRef]
28. González-Toxqui, C.; González-Ángeles, Á.; López-Avitia, R.; González-Balvaneda, D. DryingHabaneroPepper (*Capsicum chinense*) by Modified Freeze Drying Process. *Foods* **2020**, *9*, 437. [CrossRef] [PubMed]
29. Liu, Y.; Ragaee, S.; Marcone, M.F.; Abdel-Aal, E.-S.M. Composition of Phenolic Acids and Antioxidant Properties of Selected Pulses Cooked with Different Heating Conditions. *Foods* **2020**, *9*, 908. [CrossRef]
30. González, C.M.; Gil, R.; Moraga, G.; Salvador, A. Natural Drying of Astringent and Non-Astringent Persimmon "Rojo Brillante". Drying Kinetics and Physico-Chemical Properties. *Foods* **2021**, *10*, 647. [CrossRef] [PubMed]
31. Kumla, J.; Suwannarach, N.; Tanruean, K.; Lumyong, S. Comparative Evaluation of Chemical Composition, Phenolic Compounds, and Antioxidant and Antimicrobial Activities of Tropical Black Bolete Mushroom Using Different Preservation Methods. *Foods* **2021**, *10*, 781. [CrossRef] [PubMed]

32. Zhang, W.; Chen, C.; Pan, Z.; Zheng, Z. Vacuum and Infrared-Assisted Hot Air Impingement Drying for Improving the Processing Performance and Quality of *Poria cocos* (Schw.) Wolf Cubes. *Foods* **2021**, *10*, 992. [CrossRef]
33. Wiesner, A.; Szuta, M.; Galanty, A.; Paśko, P. Optimal Dosing Regimen of Osteoporosis Drugs in Relation to Food Intake as the Key for the Enhancement of the Treatment Effectiveness—A Concise Literature Review. *Foods* **2021**, *10*, 720. [CrossRef]

foods

Review

The Perspective of Croatian Old Apple Cultivars in Extensive Farming for the Production of Functional Foods

Boris Duralija [1], Predrag Putnik [2,*], Dora Brdar [3], Anica Bebek Markovinović [3], Sandra Zavadlav [4], Mirian Pateiro [5], Rubén Domínguez [5], José M. Lorenzo [5,6] and Danijela Bursać Kovačević [3]

1 Department of Pomology, Faculty of Agriculture, University of Zagreb, Svetošimunska Cesta 25, 10000 Zagreb, Croatia; bduralija@agr.hr
2 Department of Food Technology, University North, Trg dr. Žarka Dolinara 1, 48000 Koprivnica, Croatia
3 Faculty of Food Technology and Biotechnology, University of Zagreb, Pierottijeva 6, 10000 Zagreb, Croatia; brdar.dora@gmail.com (D.B.); abebekmarkovinovic@pbf.hr (A.B.M.); dbursac@pbf.hr (D.B.K.)
4 Department of Food Technology, Karlovac University of Applied Sciences, Trg J. J. Strossmayera 9, 47000 Karlovac, Croatia; sandra.zavadlav@vuka.hr
5 Centro Tecnológico de la Carne de Galicia, Rúa Galicia N° 4, Parque Tecnológico de Galicia, San Cibrao das Viñas, 32900 Ourense, Spain; mirianpateiro@ceteca.net (M.P.); rubendominguez@ceteca.net (R.D.); jmlorenzo@ceteca.net (J.M.L.)
6 Área de Tecnología de los Alimentos, Facultad de Ciencias de Ourense, Universidad de Vigo, 32004 Ourense, Spain
* Correspondence: pputnik@alumni.uconn.edu

Citation: Duralija, B.; Putnik, P.; Brdar, D.; Bebek Markovinović, A.; Zavadlav, S.; Pateiro, M.; Domínguez, R.; Lorenzo, J.M.; Bursać Kovačević, D. The Perspective of Croatian Old Apple Cultivars in Extensive Farming for the Production of Functional Foods. *Foods* **2021**, *10*, 708. https://doi.org/10.3390/foods10040708

Academic Editor: Urszula Tylewicz

Received: 8 February 2021
Accepted: 15 March 2021
Published: 26 March 2021

Publisher's Note: MDPI stays neutral with regard to jurisdictional claims in published maps and institutional affiliations.

Copyright: © 2021 by the authors. Licensee MDPI, Basel, Switzerland. This article is an open access article distributed under the terms and conditions of the Creative Commons Attribution (CC BY) license (https://creativecommons.org/licenses/by/4.0/).

Abstract: The Republic of Croatia has a long tradition of fruit growing due to its geographical location, climatic conditions, and high quality of fruit crops, especially apple fruits. Apples can be used for the formulation of functional foods either in processed form (e.g., juice), or as a by-product (e.g., apple pomace). However, there is a growing demand for functional foods derived from ancient and traditional plant sources as they are recognized as a very valuable source of health-promoting bioactive ingredients. Similarly, old apple cultivars (*Malus domestica* Borkh.) are characterized by good morphological and pomological properties, less need for chemicals during cultivation and the higher share of biologically active compounds (BACs) with better sensory acceptability compared to commercial cultivars. However, their nutritional and biological potential is underestimated, as is their ability to be processed into functional food. The importance in preserving old apple cultivars can also be seen in their significance for improving the nutritional composition of other apple cultivars through innovative cultivation strategies, and therefore old local apple cultivars could be of great importance in future breeding programs.

Keywords: old apple cultivar; biologically active compounds; functional food; agriculture; extensive farming

1. Introduction

The Republic of Croatia (RC) is a country with a long tradition of fruit production and processing, and the cultivation of old fruit cultivars (e.g., apples) in extensive farming occupies an important place for the economy, agronomy, and public health. Old fruit cultivars in Croatia are a valuable asset and natural heritage, which does not receive enough attention for popularization and processing. This is because the old apple cultivars were mainly grown locally in numerous small orchards and were not involved in scientific breeding programs. Therefore, only the cultivars obtained through systematic breeding have gained more importance on the market. Apple (*Malus domestica* Borkh.) is a fruit species belonging to the rose family (Rosaceae), and is the most commonly grown and consumed fruit in RC. The annual average consumption of apples per household member was 11.5 kg in 2017 [1]. However, there are no accurate data on how many old apple cultivars are grown in RC and how many are consumed fresh and/or for processing.

Apple is a commercially important fruit in RC and the extension of commercial life and reduction of postharvest losses of fruits is mainly based on storage at low temperatures alone or in combination with modified atmospheres (MAs) and controlled atmospheres (CA), which are primarily aimed at reducing total metabolism and thus delaying ripening and senescence. In the last decade, the traditional CA storage regimes (oxygen concentrations above 1 kPa) have been replaced in apples by the use of ultra-low oxygen (ULO) concentrations (<1 kPa). The latest dynamic CA (DCA) allows the use of much lower oxygen concentrations [2]. In addition, the activity of the enzyme polyphenoloxidase, which is the key factor of the enzymatic browning, can be a problem for post-harvesting apples, which is why various pre-treatments such as dipping solutions and non-thermal technologies are suggested [3,4].

Apple fruit quality includes a large group of external and internal characteristics. External fruit quality includes color, shape, size, and absence of defects, while internal quality (which determines eating quality) consists of taste, texture, aroma, nutritional value, sweetness, acidity (contributes to flavor), shelf life, and absence of defects [5]. For instance, after harvesting and before transportation to market or storage, fruits are calibrated by hand or sorting machine based on weight, fruit size, quality, and color characteristics. In addition, apple is a climacteric fruit and requires an increase in respiration rate and ethylene production to trigger the ripening process in an autocatalytic reaction [6]. Ripening is a complex process involving many factors, including hormonal control, which regulates biochemical and physiological changes that determine the final organoleptic and nutritional characteristics of the fruit [7]. Fruit ripening and senescence led to quality degradation and decrease in fruit firmness due to the degradation of pectin, cellulose, and hemicellulose under the action of enzymes such as pectin methylesterase (PME), polygalacturonase (PG), pectin lyase (PL), cellulase, and β-glucosidase (β-Glu) [8].

Looking at the nutritional aspects, apple fruits contain sugars, acids, vitamins, minerals, pectins, and water [9,10]. Additional to their nutritional value, they are also a valuable and easily accessible source of various biologically active compounds (BACs), especially polyphenols such as chlorogenic acid, (+)-catechin, (−)-epicatechin, phloretin, quercetin, and phloridzin [11]. In a recent study, old Croatian apple cultivars were found to have higher overall quality compared to conventional ones [12]. Therefore, old apple cultivars offer great potential for the production of functional products with higher yield of BACs. BACs have been found to contribute to food functionality due to their antioxidant activity that neutralizes free radicals, prevents the formation of new ones in the body, and repairs the cellular damages caused by them [13]. Increased free radicals lead to oxidative stress resulting in oxidative damages, cell death, tissue damages, and various diseases [14,15].

Dietary fiber, found in apples, is also a plant substance with many benefits for human health such as reduction in fat and cholesterol absorption, normalization of digestion, maintenance of intestinal health, and involvement in diabetes control [16,17]. There are also reports of apple pomace, a by-product of the fruit juice industry, being used as raw material for fiber-enriched functional foods [18].

Recently, consumer demand for traditional 'old' apples and their products is increasing, as they follow the trend of consuming natural foods, without added pesticides and additives. Unfortunately, old orchards are decaying recently, and therefore, more and older apple cultivars are being lost. As a result, there are a large number of valuable sources of genetic material with desirable fruit morphological-pomological characteristics, for which, some studies have demonstrated better sensory acceptability compared to commercial cultivars [19]. In addition, many old apple cultivars are particularly important due to genes for pests and diseases resistance, drought tolerance, winter hardiness, and unique fruit quality [20].

New knowledge and research on old apple cultivars would strengthen the market for these fruits and their products, prevent the loss of this valuable genetic material, and contribute to greater biodiversity to promote health and overall well-being. Although there is little literature on old apple cultivars in Croatia, the main objective of this article is to give an overview of the perspective of growing old apple cultivars in the Republic of Croatia, with an overview of their biological values and potential in the production of functional foods.

2. Apple Production in the Republic of Croatia

In the total fruit production of the Republic of Croatia in 2018, apple stands out as the most economically important fruit species [1]. According to FAOSTAT data (The Food and Agriculture Organization Corporate Statistical Database), apple cultivation in the Republic of Croatia shows an increase from 2016 to 2018, although in the same period, the area of apple orchards decreased [21]. According to Croatian Bureau of Statistics, the total production of apples in 2017 was 56,570 tons on 4838 ha of cultivated area, and the production increased by 63.47% in 2018 [1]. Data with distribution of apple production in the Republic of Croatia for 2017 and 2018 are given in Supplementary Table S1.

In addition, data from Croatian Bureau of Statistics (2020) also show an overall decrease in the area under apple trees in the Republic of Croatia in the period from 2010 to 2019 [1]. At the same time, apple production decreased (Supplementary Table S2). It seems as agricultural production of apples becomes more effective in Croatia, however, overall decrease of areas for cultivation of this cultivar seems discouraging.

According to the age of apple trees, four basic classes can be distinguished in the Republic of Croatia (Supplementary Figure S1). In 2012, apple trees aged 5 to 14 years were predominant (46.30%), followed by young apple trees aged less than 5 years (24.86%), trees aged 15 to 24 years (16.43%), and trees older than 25 years with the lowest share (12.41%). Five years later (2017), the proportion of trees younger than 5 years decreased almost threefold, while apple trees aged 5 to 14 years are still the most represented. The structure of orchards and apple orchards changes significantly over time, which of course also affects fruit quality.

The Croatian Bureau of Statistics also has data on the area of apples by density of plantations in hectares. The density of plantations with less than 400 apple trees/ha occupied an area of 3.84 ha in 2012, and in 2017, there was a significant increase in the area below 361.32 ha. The density of 400 to 1599 apple trees/ha in 2012 was found to be 1207.95 ha of arable land, while in 2017, the area decreased to 1148.15 ha. A drastic decrease in the area planted with a planting density of 1600 to 3199 trees/ha can be seen from 2012 to 2017. In 2012, the area planted at the indicated density was 3541.01 ha and in 2017, it decreased to 1964.04 ha. The planting density of 3200 and more apple trees/ha recorded the largest increase in the area from 45.84 ha in 2012 to 985.58 ha in 2017 [1]. So, data imply that Croatian growers prefer smaller orchards for production of apples.

Various apple cultivars are grown in the Republic of Croatia, and their areas of cultivation change significantly over time. The most common apple cultivars in Europe are 'Golden Delicious', 'Gala', 'Jonagold', 'Red Delicious', 'Idared', 'Elstar', 'Granny Smith', 'Braeburn', 'Fuji', 'Jonathan', and 'Pink Lady' and many of these cultivars are grown in Croatia [22]. The total area of apple tree plantations decreased from 2012 to 2017. In 2017, the largest area was cultivated with Idared (36%), Golden Delicious (16%), and Jonagold (10%), where only Golden Delicious increased plantation for given period. The largest jump in the increase of planted area was recorded by 'Fuji' (+137%), followed by 59% increase for 'Gala' (Supplementary Table S3). These cultivars are valued by the consumers and the demand for them has increased, and so has the area under cultivation. On the other hand, for Elstar and Florina, the largest decrease in cultivation was noticed. Interestingly, it was noticed a drastic decrease in the area planted with 'Idared' in 2017. However, as it was earlier planted on the largest area as compared to other cultivars, the relative decrease in cultivation was only around 19%.

Old apple cultivars in extensive farming are becoming more popular and are increasingly in demand on the market. A very important branch of production in the Republic of Croatia is the supply of apples to its own market. Consumers in developing markets demand information about the origin of food and the impacts of production on the environment and food safety. With consumer preference for domestic apples that are produced and processed in a sustainable manner, a further increase in the production and processing of old apple cultivars is expected.

2.1. Old Apple Cultivars in Extensive Farming in the Republic of Croatia

Old apple cultivars are the natural and cultural heritage of the Republic of Croatia, as well as a valuable source of genetic diversity. Cultivars that have adapted to local agro-ecological growing conditions and have grown there for a long period of time are recognized as the basis for market-oriented, organic apple cultivation [23]. The main requirement of traders and consumers is high fruit quality [24]. Sensory characteristics (appearance, texture, taste, and aroma), nutritional value, chemical composition, mechanical properties, and functional ingredients define fruit quality [25]. Harvesting timing and post-harvest treatments such as storage, handling methods, and transportation to markets, as well as handling in retail also determine apple quality [23]. According to Westwood [26], the primary quality-related parameters are sugar and acid content, color, firmness, texture, juiciness, taste, nutritional value, absence of disease or insects, and general appearance [26]. Old cultivars show good resistance to biotic and abiotic stress factors [27], and are characterized by different morphological and pomological characteristics compared to commercial apple cultivars.

Skendrović Babojelić et al. [28] stated that fruits do not always look as the first-class fruit [29], but they are distinguished by their different fullness of taste and especially expressed aroma. Janjić [23] describes the morphological and physical characteristics (mass, height, and width of apple fruit) of the following old cultivars: 'Roter Pogatscher' ('Božićnica') (Figure 1A), 'Blumen Calvill' Grafenštajnka') (Figure 1B), 'Großer Rheinischer Bohnapfel' ('Bobovec') (Figure 1C), 'Grüner Stettiner' ('Zeleni štetinec') (Figure 1D), 'Weißer Winterkalvill' ('Bijeli zimski kalvil'), 'London Pippin' ('London peping'), 'Yellow Bellflower' ('Lijepocvjetka'), and 'Reinette de Champagne' ('Šampanjka').

Figure 1. Old apple cultivars 'Roter Pogatscher' ('Božićnica') (**A**), 'Blumen Calvill' ('Grafenštajnka') (**B**), 'Großer Rheinischer Bohnapfel' ('Bobovec') (**C**), and 'Grüner Stettiner' ('Zeleni štetinec') (**D**) (This is a painting by a Croatian artist Greta Turković).

'Weißer Winterkalvill' ('Bijeli zimski kalvil') is characterized by medium-thick to thick fruits of irregular asymmetrical shape, delicate pale green to yellow peel, and greenish to white flesh with a slightly sour taste. This cultivar has the largest mass and the largest average width, compared to other analyzed cultivars, but it is susceptible to diseases and pests and more demanding for cultivation. 'Großer Rheinischer Bohnapfel' ('Bobovec') shows abundant yield with medium-large to large fruits of smooth green peel with red streaks, flat in shape, and resistant to shocks [30]. This cultivar is suitable for processing due to a juicy flesh with good resistant to browning.

'London Pippin' is described as a heavy fruit of medium size with the highest height, compared to the other analyzed cultivars, with a medium-thick straw-yellow peel [23]. The flesh is also yellow and has a desirable texture. The fruits of this cultivar are sensitive to transport and many losses are documented due to rot of the fruit during storage [30]. 'Yellow Bellflower' ('Lijepocvjetka') is an old cultivar of medium-sized to large cone-shaped fruit with a straw-yellow peel on the sunny side. The fruit is firm at harvest time, later becoming gentle and sensitive, and brown spots appear on the otherwise white-yellowish flesh.

'Reinette de Champagne' ('Šampanjka') is characterized by medium-sized flattened fruits with the smallest height with a very delicate peel of greenish-yellow to white-yellowish color. The flesh of this cultivar is white and has a very fine texture that is not degraded by storage. 'Grüner Stettiner' ('Zeleni štetinec') also has large and flattened fruits, a thin green peel, and a light-yellow flesh with a sweet-sour taste without a special aroma. The plant cells of flesh have a profound capacity to regenerate therefore the fruits of this cultivar may be stored late spring.

When considering the overall fruit quality, it can be assumed that the desirable fruit quality might be obtained by the cultivar 'Großer Rheinischer Bohnapfel' ('Bobovec') with the highest share of healthy fruits (92%) after storage for 160 days in a refrigerator at normal atmosphere. It is followed by 'Yellow Bellflower' ('Lijepocvjetka') cultivar (77%), 'Grüner Stettiner' ('Zeleni štetinec') (71%), 'Reinette de Champagne' ('Šampanjka') (63%), 'London Pippin' ('London peping') (60%), and 'Weißer Winterkalvill' ('Bijeli zimski kalvil') (43%), while 'Roter Pogatscher' ('Božićnica') cultivar counts only 40% of healthy fruits, thus it is the cultivar with the highest proportion of rotten fruit [23]. For cultivars with a higher percentage of rotten fruits, it is recommended to perform storage in the refrigerator for a shorter time in order to prevent spoilage and degradation of bioactive compounds [23].

Vujević described the morphological characteristics of two other old cultivars from the Bjelovar-Bilogora County: 'Reinette du Canada' ('Kanada') and 'Goldparmäne' ('Zlatna zimska parmenka') [31]. The fruits of 'Reinette du Canada' are characterized by a moderately flattened shape of reddish peel with a basic green color. The fruits are sensitive to storage, but have a high tolerance to transport. The firm flesh of the fruit is yellowish-white in color and has a sweet-sour taste. 'Goldparmäne' is another old cultivar with a moderately flattened shape of the fruit with firm consistency. The average weight, as well as the height and width of the fruit of this cultivar are higher than reported values for 'Reinette du Canada'. The peel is dark red, while the flesh is yellowish white and sweet in taste.

Skenderović Babojelić et al. [28] conducted a physico-chemical analysis of the old apple cultivars ('Großer Rheinischer Bohnapfel'—Bobovec, 'Roter Pogatscher'—Božićnica, 'Yellow Bellflower'—Lijepocvjetka, and 'Goldparmäne'—Zlatna zimska parmenka) grown in the territory of the Topusko, the municipality in Sisak-Moslavina County (continental Croatia). Authors analyzed the color parameters via colorimeter, the hardness of the fruit with a manual penetrometer, the proportion of soluble dry matter with a refractometer, and the proportion of total acids by titration method. The highest fruit hardness was observed in the cultivars 'Großer Rheinischer Bohnapfel' and 'Citronka', slightly lower in the cultivar 'Goldparmäne', and the lowest in the cultivars 'Roter Pogatscher' and 'Yellow Bellflower'. 'Citronka' cultivar exhibited the highest starch degradation index while the lowest value was found in cultivar 'Großer Rheinischer Bohnapfel'. During the process of fruit ripening, the starch is intensively decomposed into simple sugars [32]. In cultivars 'Goldparmäne' and 'Citronka', the highest values of soluble dry matter were determined, and the cultivar 'Roter Pogatscher' was the fruit with the lowest values. The content of soluble dry matter increases during fruit ripening and storage, and it is a good indicator of the sugar content in the apple fruit [33]. The highest number of total acids was determined in the cultivars: 'Yellow Bellflower' and 'Goldparmäne', which were identified as the most acid apples, followed by 'Roter Pogatscher', 'Großer Rheinischer Bohnapfel', and 'Citronka'. Acids have an important role in fruits, as they can slow down the harmful effects of bacteria, degradation of ingredients, and spoilage. In the process of ripening, sugar accumulates, and total acids are broken down, resulting in fruits that become more and more harmonious in taste. In the conclusion, all investigated cultivars were different due to all evaluated parameters, although all cultivars were of good quality and acceptable physiochemical properties. Authors suggested that it is important to preserve old apple cultivars as due to the withering of old trees and a possibility of losing an important source of genetic material, and consequently reduce assortment of apple cultivars on the markets [28].

Hoehn et al. [28] found that consumers are primarily concerned on the fruit size and color when choosing an apple, and that other properties are less important to them. Old cultivars are more resistant to diseases and pests, do not require a large number of pesticide application, as well as intensive care, so they are easily adaptable to organic farming, although they can rarely be purchased in stores. Despite some uniform pomological properties have been observed in old cultivars as compared to commercial apple cultivars, they possess great potential for organic and ecological fruit growing that is becoming increasingly popular. This is well aligned with increasing consumers' awareness about old cultivars with their specific morphological properties and valuable bioactive compounds, vitamins and minerals without the risks of harmful effects of applied pesticides.

2.2. *Sustainable Technologies for Cultivation, Selection, and Preservation of Old Apple Cultivars*

Mainly old apple cultivars grown in Croatia are found in rural areas in small orchards, and they are very well adapted to local environmental conditions. The old cultivars are grafted on the more vigorous seedling rootstocks and develop a much larger root system and a higher tree crown compared to the modern cultivars grafted on the less vigorous rootstocks. This improves resistance to climatic damage such as higher resistance to drought, longer shelf life, and better anchorage under windy conditions [34]. Compared to intensive industrial apple production in Croatia, where a small number (less than 20) of cultivars exist in plantations (Table 1), more than 50 different genotypes are present in cultivation on small farms. The displacement of old and locally well-adapted cultivars by a few widely used modern cultivars has led to a dramatic loss of genetic diversity in orchards [35]. Many of the well-known, international dominant apple cultivars are closely related, whereas old cultivars were collected over a longer period of time and are more diverse [36]. One of the main objectives of apple breeding networks is the re-diversification of cultivar use and the widespread application of genetic analysis to produce separate cultivar lists for each region containing those best suited for environmentally friendly production [37].

Table 1 lists some factors that have an important influence on apple fruit traits. Understanding the factors that can influence fruit characteristics is critical to obtain a high-quality raw material for processing. Global apple production faces many challenges such as reduced biodiversity in orchards, climate change, water scarcity and pollution, harmful chemical residues in fruits, use of non-renewable resources, less nutrients in modern apple fruits, etc. In the future, all attained knowledge should be taken into consideration and the new apple orchard must be planned for more sustainable farming practices. Some old cultivars have many advantages for sustainable cultivation, as they are better adapted on the environmental factors in the growing area and require less use of energy and chemicals.

Table 1. Factors affecting apple fruit quality in sustainable cultivation.

Factor	Effect on Fruit Quality Characteristics	Ref.
Cultivar	Mean fruit mass, shape, firmness, SSC, TA, color	[38]
	Polyphenols (flavonols, dihydrochalcones, flavanols, phenolic acids, anthocyanins)	[39]
	Antioxidant capacity, mineral content	[40]
Rootstock	Mean fruit mass, firmness, SSC	[41]
Interstock	Firmness, SSC, starch content	[42]
Tree age	Firmness, flavor, color	[43]
Environment	Mean fruit mass, firmness, SSC, TA, ascorbic acid	[44]
	Color, anthocyanins	[45]
Plant densities	Soluble solids content, organic acids, sugars	[46]
Training and pruning	SSC, TA, mineral contents	[47]
Production system	Mean fruit mass, mineral content	[48]
Yield	Mean fruit mass, SSC, color	[49]
Agro-techniques	Mean fruit mass, firmness, sugars	[50]
	Mean fruit mass, firmness, SSC, TA, antioxidant capacity	[51]
Harvest time	Mean fruit mass, firmness, ethylene concentration	[52]

The most representative and widely accepted criteria and objectives of sustainable agriculture were adopted in the Den Bosch Declaration [53]. Briefly, sustainable agriculture can be defined as the efficient production of safe, high quality agricultural products in a manner that protects the natural environment, improves the social and economic conditions of farmers, their employees, and local communities, and ensures the health and welfare of all managed species [54]. The guiding principle of sustainable agricultural production is to make the most efficient use of available resources and production potential while minimizing adverse impacts on soils, water, air, and biota [27].

FAO (The Food and Agriculture Organization) sets out the five key principles that balance the social, economic, and environmental dimensions of sustainability: (i) improving the efficiency of resource use; (ii) conserving, protecting, and enhancing natural ecosystems; (iii) protecting and enhancing rural livelihoods and social well-being; (iv) strengthening the resilience of people, communities, and ecosystems; and (v) promoting good stewardship of both natural and human systems [27].

In the temperate zone, cultivated apple is an important fruit crop with an annual production of over 80 million tons on nearly 5 million hectares [55]. The old apple cultivars have a long history of cultivation and grow in different locations in Croatia. Some of the most important conditions for successful growth and development are temperature, water, light, suitable soil, and proper management system. Compared to modern orchard cultivation practices with old cultivars, there are many differences (Supplementary Table S4).

In apple production, there is increasing interest in developing more environmentally friendly production, either through integrated production or organic management [56]. To ensure the sustainability of fruit production, farmers are being pushed to adopt farming practices that may reduce yield and profitability [57]. Management improvements and technologies adopted should be site-specific in terms of soil, landscape, and climate, and should take into account the specific types of apple orchards and farms [58].

3. Apple as a Functional Food

3.1. Apples and Apple Products as a Source of Functional Ingredients

Consumers today demand value-added foods (e.g., functional foods) that are sustainably produced and processed, considered safe, fresh, natural, and have important nutritional value [59]. Therefore, as the most widely consumed temperate fruit species in the world, apple has great potential for the production of functional foods. The chemical composition of apple fruit is extremely complex as it is rich in many nutrients and thus considered as a major source of phytochemicals in the human diet [60]. To that end, Scalbert et al. [61] discovered several phenolic compounds in apple, being (+)-catechin and (−)-

epicatechin (flavan-3-ols or flavanols), phloridzin (dihydrocalconglycosides), quercetins (flavonols), cyanidin-3-O-galactosides (anthocyanins), hydroxycinnamic acids, and hydroxycinnamic acid (chlorogenic acid and *p*-coumaroylquinic acid). The phenolic profile and antioxidant capacity for different parts of the apple in different cultivars are given in Table 2. The presence of these and other bioactive components has also been reported for apple products and by-products (Table 3).

Phenolic compounds are not evenly distributed in apple fruit, i.e., certain components of phenolic compounds are present in certain parts of the fruit while very few or none are present in remaining parts. For this reason, many researchers propose to use apple peels or apple cores as a by-product or waste for the subsequent extraction of phenolic compounds. Tsao et al. [62] reported five times more polyphenols in apple peel than in apple flesh. The authors attributed this difference to the defensive role of the peel in protecting the fruit from harmful UV light and environmental pathogens. McGhie et al. [63] found that about 46% of apple polyphenols are found in the peel, i.e., all flavonols (quercetin derivatives) are determined only in the peel, which would explain the greater antioxidant capacity of the peel (Table 2). Chinnici et al. [27] concluded that flavonols, flavanols, and procyanidins are most contribute to the antioxidant activity of the apple peel, about 90% of the total activity calculated.

Among the processed apple products, apple juice as well as apple cider are the most popular [64]. Due to the different localization of polyphenols in the pulp, apple juice and fresh apple differ in their composition. During juice production, only some of the phenolic compounds are extracted into the juice, while most of the polyphenols remain in the solid residue after the juice is pressed [65]. The solid residue consists of peels and cores, and therefore phenolic compounds such as quercetin glycosides and dihydrocalcones are present in small amounts in apple juice [27].

Regarding juice quality, consumers prefer clear apple juices that do not lose many valuable components with high antioxidant potential during the production process [66]. During the clarification process, mainly (−)-epicatechin and procyanidins are mostly removed. Lee et al. [67] observed a significant contribution of flavonoids such as quercetin, (−)-epicatechin, and procyanidin B2 to the total antioxidant activity of apples, in contrast to the antioxidant contribution of vitamin C. The antioxidant activity (AOA) of juices obtained by pressing was 10% of that of fresh 'Joanagold' apples, while the AOA of clear juices showed only 3% of that of fresh apples, with a 50% decrease in chlorogenic acid and a 3% decrease in catechins. After processing and subsequent storage of apple juice at room temperature (25 °C) for nine months, a significant loss of antioxidant polyphenols was observed. Despite the fact that polyphenols in apple juices were more stable than vitamin C, significant losses of quercetin (60%) and procyanidins (100%) were observed [27,68]. Van der Sluis et al. [69] showed that elevated temperature during apple juice storage negatively affected the stability of polyphenolic antioxidants, with quercetin glycosides and epicatechins being the most heat sensitive, and phloridzin and chlorogenic acid the most stable.

Apple fruit pigmentation is controlled by the relative amounts of anthocyanins, chlorophylls, and carotenoids. Anthocyanins and carotenoids have been shown to have potent antioxidant and anticancer properties [70]. Since they have potential beneficial effects on human health, new cultivars with improved pigmentation are being developed, such as red-fleshed apples with increased anthocyanin concentration.

Apples, apple peels, and apple flesh are also an important source of triterpene compounds (1.635–3.173 mg g^{-1} dry weight), being ursolic acid the most significant constituent (72.1–81.2%), followed by oleanolic acid, corosolic acid, and betulinic acid [71]. Dashbaldan et al. [72] determined the profile of neutral triterpenoids, triterpenic acids, steroids, and esters in apple cultivar 'Antonovka', and observed significant changes in triterpenoid contents during fruit growth and ripening. The importance of triterpenoids found in apples was demonstrated by their high potential of anti-proliferative and anticancer activities [72].

The by-products of apple fruit processing contain considerable amounts of fibers, and therefore have attracted the attention of scientists and industry due to the increasing waste utilization [73]. There are about 2.21 g of dietary fiber in 100 g of apple fruit. Insoluble dietary fiber (cellulose and hemicellulose) accounts for 70%, while soluble dietary fiber (pectin) accounts for 30% [74]. Sun-Waterhouse et al. [75] isolated dietary fibers with high pectin polysaccharide content and phenolic antioxidants from 'Granny Smith', which could be a new type of functional ingredients. Phenolic compounds can bind to cellulose and pectin through covalent bonds via esters or carbon from the cell wall to form insoluble bound phenols.

Table 2. The phenolic profile and antioxidant capacity for different parts of the apple in different cultivars.

Cultivar (Sample)	Phenolic Compund	Concentration (mg 100 g^{-1})	Antioxidant Capacity *	Conclusion Remarks	Reference
Fuji cl. Kiku8 (peel)	chlorogenic acid	8.96 ± 1.22	EC_{50} = 17.34	- The content of phenolic compounds was influenced by the clone and the part of fruit (peel vs. mesocarp). - With no impact of cultivar, higher antioxidant capacity and total phenols were found in peel samples as compared to mesocarp samples. Therefore, peels accounted for the increased index of antiradical capacity in comparison with mesocarp. - The highest antioxidant capacity was determined in peels of Fuji and mesocarp of Golden cl. B. - Quercetin-glucoside was determined only in apple mesocarp, while in peels was not present. - Golden cl. B was found as the sample with highest content of phenolic compounds in peels and mesocarp. - The highest content of (+)-catechin was determined in peels of Fuji cl. Kiku8 and Golden cl. B. - The highest content of (−)-epicatechin was detected in peels of Fuji cl. Kiku8.	[76]
	(+)-catechin	12.91 ± 0.5			
	(−)-epicatechin	9.94 ± 0.76			
	phloretin glucoside	1.04 ± 0.25			
	quercetin glucoside	14.92 ± 2.32			
Braeburn. Hillwell (peel)	chlorogenic acid	2.82 ± 0.38	EC_{50} = 22.67		
	(+)-catechin	9.81 ± 0.73			
	(−)-epicatechin	5.12 ± 0.65			
	phloretin glucoside	0.77 ± 0.4			
	quercetin glucoside	11.54 ± 5.27			
Golden cl. B (peel)	chlorogenic acid	8.64 ± 2.12	EC_{50} = 18.667		
	(+)-catechin	12.18 ± 1.11			
	(−)-epicatechin	6.03 ± 1.11			
	phloretin glucoside	1.05 ± 0.72			
	quercetin glucoside	23.51 ± 2 78			
Fuji cl. Kiku8 (mesocarp)	chlorogenic acid	5.76 ± 1.22	EC_{50} = 39.236		
	(+)-catechin	0.98 ± 0.31			
	(−)-epicatechin	1.75 ± 0.64			
	phloretin glucoside	1.09 ± 0.06			
Braeburn. Hillwell (mesocarp)	chlorogenic acid	5.32 ± 1.33	EC_{50} = 58.48		
	(+)-catechin	0.91 ± 0.18			
	(−)-epicatechin	1.91 ± 0.33			
	phloretin glucoside	0.75 ± 0.09			
Golden cl. B (mesocarp)	chlorogenic acid	7.52 ± 0.9	EC_{50} = 26.596		
	(+)-catechin	2.20 ± 1.34			
	(−)-epicatechin	1.46 ± 0.17			
	phloretin glucoside	1.00 ± 0.11			
Rome Beauty (mesocarp)	Total phenols	93.0 ± 4.1	ND	- The total phenols of the peels were significantly higher than the mesocarp and mesocarp + peel values within all cultivars, while the total phenols of the mesocarp were not significantly lower than the mesocarp + peels contents. - Within all apple cultivars, the peels contained the highest content of flavonoids, followed by the mesocarp + peel and the mesocarp. - The total antioxidant capacity of the peels was higher than that of the flesh or flesh + peel for all cultivars.	[77]
Idared (mesocarp)		75.7 ± 4.0	ND		
Cortland (mesocarp)		103.2 ± 12.3	EC_{50} = 103.9 ± 16.5		
Golden Delicious (mesocarp)		97.7 ± 8.9	EC_{50} = 155.3 ± 11.7		
Rome Beauty (peel and mesocarp)		159.0 ± 15.1	EC_{50}= 26.5 ± 0.3		
Idared (peel and mesocarp)		120.1 ± 15.0	EC_{50}= 125.1 ± 58.8		
Cortland (peel and mesocarp)		119.0 ± 14.9	EC_{50}= 74.1 ± 4.0		
Golden Delicious (peel and mesocarp)		129.7 ± 9.7	EC_{50}= 107.7 ± 22.7		
Rome Beauty (peel)		500.2 ± 13.7	EC_{50}= 12.4 ± 0.4		
Idared (peel)		588.9 ± 83.02	EC_{50}= 13.6 ± 0.2		
Cortland (peel)		388.5 ± 82.4	EC_{50}= 15.7 ± 0.3		
Golden Delicious (peel)		309.1 ± 32.1	EC_{50}= 20.2 ± 0.7		

Table 2. *Cont.*

Cultivar (Sample)	Phenolic Compund	Concentration (mg 100 g^{-1})	Antioxidant Capacity *	Conclusion Remarks	Reference
Idared (peel)	procyanidin B2	23.47 ± 0.01	6.02 ± 0.104 mmol trolox 100 mL^{-1}	- In both apple varieties, total phenolic content was greater in the peel, followed by the mesocarp + peel and the mesocarp. - 'Idared' apple peel had a higher TPC than 'Fuji'. - Quercetin was only determined in the peel samples, mostly in the form of glycosides, galactosides, xylosides, arabinosides, rhamnosides, being the rutinosides as the most common. - The highest antioxidant capacity was determined in the peels, followed by the samples of peels and mesocarp, while the lowest values were detected in the mesocarp samples. - Antioxidant capacity in both ABTS and DPPH assays was - positively correlated with total phenolic compounds found in the peel, mesocarp + peel, and mesocarp.	[78]
	phloridzin	4.32 ± 0.13			
	(−)-epicatechin	3.33 ± 0.04			
	chlorogenic acid	2.54 ± 0.06			
	quercetin glucoside	0.56 ± 0.05			
	rutin	3.26 ± 0.06			
	qercetin	0.08 ± 0.001			
Fuji (peel)	procyanidin B2	31.41 ± 0.13	5.13 ± 0.23 mmol trolox 100 mL^{-1}		
	phloridzin	ND			
	(−)-epicatechin	6.39 ± 0.09			
	chlorogenic acid	ND			
	quercetin glucoside	3.88 ± 0.004			
	rutin	8.33 ± 0.14			
	qercetin	ND			
Idared (peel and mesocarp)	procyanidin B2	12.41 ± 0.43	2.46 ± 0.06 mmol trolox 100 mL^{-1}		
	phloridzin	1.54 ± 0.09			
	(−)-epicatechin	1.12 ± 0.02			
	chlorogenic acid	9.16 ± 0.13			
	quercetin glucoside	0.3 ± 0.02			
	rutin	2.93 ± 0.05			
Fuji (peel and mesocarp)	procyanidin B2	9.48 ± 0.35	2.69 ± 0.08 mmol trolox 100 mL^{-1}		
	phloridzin	0.87 ± 0.03			
	(−)-epicatechin	1.70 ± 0.05			
	chlorogenic acid	4.69 ± 0.03			
	quercetin glucoside	0.13 ± 0.01			
	rutin	0.95 ± 0.02			
Idared (mesocarp)	procyanidin B2	3.13 ± 0.027	1.67 ± 0.04 mmol trolox 100 mL^{-1}		
	phloridzin	0.72 ± 0.02			
	(−)-epicatechin	0.45 ± 0.01			
	chlorogenic acid	8.05 ± 0.08			
	quercetin glucoside	ND			
Fuji (mesocarp)	procyanidin B2	0.57 ± 0.01	2.09 ± 0.04 mmol trolox 100 mL^{-1}		
	phloridzin	1.16 ± 0.08			
	(−)-epicatechin	3.82 ± 0.03			
	chlorogenic acid	3.13 ± 0.027			
	quercetin glucoside	0.72 ± 0.02			

Table 2. *Cont.*

Cultivar (Sample)	Phenolic Compund	Concentration (mg 100 g^{-1})	Antioxidant Capacity *	Conclusion Remarks	Reference
104 apple cultivars (whole fruit)	catechin epicatechin procyanidin B1 procyanidin B2 chlorogenic acid coumaroylquinic acid phloridzin phloretin-xyloglucoside quercetin-galactoside and quercetin-glucoside rutin quercetin-rhamnoside	1.66 7.72 2.71 8.58 17.44 2.18 2.38 3.63 1.45 0.48 1.45	-	- There are large differences between apple cultivars with respect to polyphenol content and profile. - Two major subclasses of polyphenols, flavan-3-ols and phenolic acids, were found as predominant in the apple polyphenol profile. - By calculating the flavan-3-ol to phenolic acid ratio, apple cultivars can be classified into flavan-3-ol rich or phenolic - acid rich.	[79]
Renetta *Red Delicious* *Granny Smith* *Morgendulf* *Golden Delicious* *Royal Gala* *Braeburn* *Fuji*	Total phenols	211.9 131.1 121.0 105.8 86.3 83.9 75.4 66.2	-	- Flavanols (catechin and proanthocyanidins) were reported as the major class of polyphenols (71–90%) in red apples, followed by hydroxycinnamates (4–18%), flavonols (1–11%), dihydrochalcones (2–6%), and anthocyanins (1–3%).	[80]

* TPC—Total phenolic content; ABTS—The 2,2-azino-bis(3-ethylbenzothiazoline-6-sulfonic acid) (ABTS•+) radical cation-based assay; DPPH (2,2-diphenyl-1-picryl-hydrazyl-hydrate) scavenging activity; EC50 = mg of tissue on fresh weight basis required to obtain 50% DPPH scavenging; ND = not detected.

Table 3. The phenolic profile and antioxidant capacity in apple products and apple by-products.

Sample	Phenolic Compund	Concentration (mg 100 g^{-1})	Antioxidant Capacity *	Conclusion Remarks	Reference
Industrial apple pomace consisted of 3 cultivars: *Fuji, Qinguan*, *Granny Smith* (six fractions gradually eluted with aqueous alcohol (20%, 40%, 60%, 80% and 100%)	chlorogenic acid syrigin procyanidins B2 caffeic acid cinnamic acid phloridzin quercetin hyperin (−)-epicatechin	1.30 0.44 32.31 0.15 0.40 0.18 23.93 5.20 ND	43.45 ± 3.45 51.65 ± 9.57 90.96 ± 10.23 71.54 ± 2.41 10.12 ± 2.31 9.68 ± 4.55	- Fraction 3 had the highest phenolic content, while the lowest contents was determined in fractions 5 and 6. - The capacity of scavenging free radicals varied in the order: fraction 3 > fraction 4 > fraction 2 > fraction 1 > fraction 5 > fraction 6. - Fraction 3 had the highest contents of chlorogenic acid, syrigin, procyanidins B2, and quercetin. - None of the nine phenolic compounds detected was found in fractions 5 and 6.	[81]
Limón Montés (single-cultivar pomaces)	chlorogenic acid (−)-epicatechin phloridzin quercetin	681.5 161.1 587.2 252.0	12.4		
De la Riega (single-cultivar pomaces)	chlorogenic acid (−)-epicatechin phloridzin quercetin	1415.5 314.6 730.2 96.0	13.5		
M1, M3—48 h M2—10 h M4—36 h (Mixture of Asturian apples, hydraulic press, and different degrees of exposure to air during the pressing process)	chlorogenic acid (−)-epicatechin phloridzin quercetin chlorogenic acid (−)-epicatechin phloridzin quercetin chlorogenic acid (−)-epicatechin phloridzin qercetin	586.7 ND 302.5 144.2 602.4 287.1 594.7 109.9 375.3 ND 292.5 87.1	7.6 12.5 10.3	- Eleven different cider apple pomaces (six single-cultivar and five from the cider-making industry) were investigated for phenolic profiles and antioxidant capacity. - The group of single-cultivar pomaces showed higher contents of chlorogenic acid, (−)-epicatechin, procyanidin B2 and dihydrochalcones, whereas the industrial samples revealed higher amounts of up to four unknown compounds, with absorption maxima between 256 and 284 nm. - 'Meana' was the cultivar with the lowest amount of trimers and tetramers, and the highest in other flavanols, followed by 'De la Riega and Carrió'. - Phloridzin was determined as the main dihydrochalcone present in the apple pomaces, followed by phloretin-2′-xyloglucoside. - Chlorogenic acid was the major phenolic acid in all the samples. - Asturian cultivars present higher concentrations of (−)-epicatechin, chlorogenic acid and phloridzin than those observed in the Basque region.	[82]
G—1.5 h (Mixture of Asturian and foreign apples Pneumatic, Bucher–Guyer type press)	chlorogenic acid (−)-epicatechin phloridzin quercetin	259.8 167.5 451.6 186.3	8.2		

Table 3. *Cont.*

Sample	Phenolic Compund	Concentration (mg 100 g^{-1})	Antioxidant Capacity *	Conclusion Remarks	Reference
Cripps Pink *Golden Delicious* (Minimally processed apples: treated with antibrowning solution (calcium ascorbate)	chlorogenic acid (−)-epicatechin phloridzin quercetin-3-galactoside chlorogenic acid (−)-epicatechin phloridzin quercetin-3-galactoside	1.13 0.05 0.03 0.03 0.60 0.07 ND 0.03	12.68 ± 0.26 9.50 ± 0.38 (mmol trolox kg^{-1})	- The application of anti-browning agents did not affect the amount of phenolic compounds, but showed improved antioxidant capacity compared to control samples. - Phenolic compounds were stable while the values of antioxidant capacity decreased during storage.	[83]
Cloudy juice from '*Golden Delicious*' treated by High Power Ultrasound (HPU) and stored at 4 °C	Total phenols Total flavan-3-ols (0 day)	1.86 ± 0.09 0.53 ± 0.03	5.93 ± 0.2 (mmol trolox g d. $w.^{-1}$)	- HPU significantly decreased phenolic compounds and antioxidant capacity in the samples. - Storage had a significant effect on total phenols, flavan-3-ols and DPPH values, decreasing the values by 89.21%, 82.80%, and 79.51%, respectively.	[84]
	Total phenols Total flavan-3-ols (7th day)	0.40 ± 0.09 0.08 ± 0.03	1.60 ± 0.2 (mmol trolox g d. $w.^{-1}$)		

* DPPH scavenging rate (%); ND = not detected.

3.2. Bioactive Potential of Old Apple Cultivars from Extensive Farming

Old apple cultivars were more appreciated by consumers according to the tested sensory properties and nutritional characteristics as compared to the commercial 'Golden Delicious' [85]. Old Portuguese apples have been shown to have higher bioactive potential, containing more fiber, protein, sugar, β-carotene, vitamin E, magnesium, and polyphenolic compounds compared to the commercial vs. such as 'Fuji', 'Gala Galaxy', 'Golden', 'Reineta Parda', and 'Starking' [27].

Old apple cultivars have qualitatively identified the same subgroups of polyphenols as the commercial cultivars [39,86,87]. Further results of other studies indicated that the proportions of phenolic compounds were equal in old and new apple cultivars [88]. Iacopini et al. [89] investigated an antiradical potential of old vs. commercial Italian apple cultivars. It was found, that the higher total phenolic content as well as total flavonoids were in two of the four old varieties analyzed, as compared with commercial cultivars. When observing individual phenolic compounds, the HPLC qualitative pattern was similar in all the examined cultivars although higher values was confirmed for old cultivars. Moreover, old cultivars exhibited a higher antioxidant activity compared to the commercial cultivars.

Preti and Tarola [40] recently evaluated fourteen ancient apple cultivars grown in northeast Italy to investigate their nutraceutical properties comprising polyphenols, antioxidant capacity, as well as four major minerals (Na, K, Mg, Ca) in comparison to commercial cultivars. All the analyses were performed on apple peel and pulp, separately. Peel samples showed a significantly higher contents in phenolic compounds with respect to pulp, almost threefold than in pulp for ancient cultivars and quadrupled for commercial cultivars. Dihydroclacones phloridzin and phloretin were mostly found in peel, with higher proportions in ancient apples as compared to commercial counterparts. The importance of phloridzin was described with influence on lower susceptibility of apple fruit to the most important diseases, thus providing resistance to the common apple pathogens such as *Venturia inaequalis* and *Erwinia amylovora* [90].

Though old apple cultivars have shown higher polyphenol content, it is important to note that environmental conditions may have an important impact on the amount of polyphenols [12]. The polyphenol profile could also be affected by different farming methods, i.e., agricultural practice such as conventional, integrated, or organic [91]. The color of apples, and therefore the proportion of pigments, can be influenced by the geographical location of orchards [92]. Volz and McGhie [93] concluded that differences in polyphenol content between cultivars might be as the result of genetic variability. Carbone et al. [76] reported that the genotype of old apple cultivars could have a positive impacts on the content of bioactive compounds, and for this reason, old cultivars were presented as an important source of genes for future breeding programs. Due to the favorable differences in the characteristics of old cultivars, which have developed through their long-term growth in Croatia, these cultivars should be preserved for the future and further popularization among growers, producers, and consumers [27].

Jakobek and Barron [39] analyzed the peel and flesh of the old apple cultivars from the area of Slavonia (Continental Croatia) with the aim of highlighting and preserving the biological diversity of apple cultivars with the greater bioactive potential. Using the high-performance liquid chromatography (HPLC) with a diode-array detector (DAD, they determined the content and composition of polyphenolic compounds, and the results indicated that the following cultivars were the richest in polyphenols: 'Zimnjara', 'Lještarka', and 'Adamova Zvijezda'. The authors noted that the cultivars differ according to their content of individual polyphenolic groups, some of them are richer in the phenolic acids, and others in flavan-3-ols, i.e., cultivars in which phenolic acids occupy a larger proportion, while others that contain a smaller amount of flavan-3-ols, and vice versa. Lastly, it can be concluded that it is possible to classify old apple cultivars based on the predominant proportion of flavan-3-ols or phenolic acids [79].

According to the results for the investigated commercial apple cultivars, the content and composition of polyphenols in old apple cultivars also significantly varies depending on the parts of the fruit. Jakobek et al. [87] showed that the concentration of phenolic acids was significantly higher in the peel than in the flesh for all samples of old apple cultivars, which is consistent with research that was already available [63,87,94–96]. Authors investigated thirteen old apple cultivars 'Ljubeničarka', 'Astrahan', 'Crvenka', 'Kardinal', 'Kraljevina', 'Ružica', 'Pisanica', 'Petrovka', 'Slavonska Srčika', 'Bjeličnik', 'Ledenara', 'Štegerova', and 'Jaje' grown in the region Slavonia (Mihaljevci, near Požega) in Continental Croatia.

Proanthocyanidins are proven as the most abundant analyzed bioactive compounds that made up between 70–90% of total polyphenolic compounds in apples [62,80,88,95,97]. Jakobek et al. [87] firstly analyzed oligomeric pronatocyanidins in the old local apple cultivars from Southeastern European region after acid hydrolysis in the presence of the organic compound fluoroglucinol ($C_6H_6O_3$). In this way, the authors were able to obtain information about the constituent units of proanthocyanidins, as well as their locations within the complex molecular structure, using HPLC coupled to UV–vis detection and ultra-high performance liquid chromatography with quadrupole time-of-flight (UPLC-Q-TOF). In the majority of previous studies, the proportion of oligomeric proanthocyanidins is often neglected because there was no conversion to subunits, but only monomers, dimers, and trimers are considered [62,98–101]. For this reason, Jakobek et al. [87] could not correctly compare the obtained results with the results found in the previous studies. They proved that this method is effective for the characterization and quantification of proanthocyanidins in apple fruits.

Phenolic acids were the second most abundant subgroup in apples, with proportions of 6–25% in the flesh and 1–10% in the peel. The highest values were detected in the samples of the cv. 'Slavonska srčika', followed by 'Kardinal', 'Astrahan', 'Kraljevina' (red apples), and 'Bjeličnik' (green or yellow apples). Flavonols were found mainly in all peel samples, in a proportion of 1–13% of the total content of phenolic compounds in the peel. Much smaller amounts were found in the flesh, which is consistent with the results of research conducted for other apple cultivars [62,95,97]. 'Astrahan' and 'Slavonska srčika' showed the highest proportions of flavonols in their composition, while the lowest values were found in the cv. 'Štegerova' [87]. Dihydrochalcones were observed in peel samples, from 1–10% of the total polyphenolic content, and a smaller share were found in the flesh of the fruit. The cultivars with the highest concentration of dihydrochalcones in the peel were 'Petrovka' and 'Slavonska srčika', and the smallest amount was detected in the cultivar 'Štegerova'. Anthocyanins were found only in the samples of red-peeled apples, in a proportion of 1–7% of the total polyphenol content. The lowest value was detected in the samples of the cultivar 'Slavonska srčika', which was found to be the richest in the remaining phenolic compounds. The highest amounts of anthocyanins were determined in the samples of the cultivar 'Ljubeničarka'. In conclusion, 'Slavonska Srčika' was highlighted as the cultivar with the largest proportion of all phenolic compounds in total phenolic content, except anthocyanins. The cultivar 'Ljubeničarka' had a reddish-colored flesh, which is unusual for apples, because the cultivar contained anthocyanins in the flesh, as well as in the peel. Scientists predict that 'Ljubeničarka' could be important in future apple growing programs due to its attractive red color and high bioactive potential [27].

Jakobek et al. [12] reported that the most of the analyzed old cultivars from the towns of Donji Miholjac and Gornji Tkalec and the village of Rude contained higher amounts of polyphenols in the flesh and in the peel as compared to the commercial apple cultivars. This is precisely the quality characteristic with bioactive potential that should be emphasized for old cultivars. Some cultivars can be distinguished by a higher proportion of polyphenols in the peel, as is the case with 'Pisanike', 'Adamove zvijezde', 'Zelenike', and 'Kanada'. Cultivars that can be characterized by the higher amounts of polyphenols identified in the flesh were 'Božićnica 2', 'Boskop', 'Zimnjara', and 'Crveni boskop'. All identified polyphenols from five different subgroups, have already been reported in the literature [74,102,103]. By comparing the total amounts of identified subgroups of polyphenols in flesh samples, it was noticed that phenolic acids were the most prevalent. Moreover, flavonols gave the main contribution to the total amounts of polyphenols in apple peel samples.

Apples are also a natural source of important dihydrochalcones [104,105]. Dihydrochalcones are a specific subgroup of polyphenols, found mainly in apples with the potential to lower blood glucose levels, which may be useful in diabetes management [106,107]. Almost all old cultivars were found to had a higher proportion of dihydrochalcones compared to the commercial cultivars which is also one of the important reasons for considering more extensive apple growth [87].

In further analysis of polyphenols in old cultivars, it is important to point out that the flesh of the fruit contributes more to the absorption of polyphenols in the body of a person who consumes an apple. The peel only makes up about 10% of the entire fruit and is not always consumed with it. For this reason, cultivars that contain more polyphenols in the flesh might be a better source of polyphenols. Although the peel does not contribute to polyphenolic intake to the same extent as the pulp, it contains important polyphenolic groups such as quercetin derivatives and a high concentration of other polyphenols [87]. Differences between genotypes accounted for 46–97% of the total difference in the concentration of total polyphenols and polyphenolic groups in flesh and peel [93]. Flavonols in the peel protect the fruit from UV radiation, although they are more sensitive to environmental changes due to their sensitivity to light and temperature changes [93]. It is suggested that the high content of flavonols (quercetin derivatives) in the peel of the analyzed old apple cultivars is due to environmental conditions and is very likely to change depending on climatic conditions [87]. The content of other identified polyphenols, especially in the pulp, could result from genetic variability [87]. Table 4 provides an overview of research results on the main polyphenolic subgroups of old apple cultivars grown in Croatia.

Lanzerstorfer et al. [108] investigated the content of minerals, phosphates, and trace elements, as well as the content of polyphenols in apple juices from old apple cultivars. They found large differences between the investigated cultivars regarding the content of the mentioned elements. The authors have concluded that the old apple cultivars can serve as functional apple products with emphasis on desirable health effects.

Apples were shown to have the potential to cause allergic reactions [109]. In this regard, existing studies have shown that old cultivars are better tolerated by individuals with apple intolerance than new cultivars due to their high polyphenol contents [27]. Vegro et al. [110] demonstrated that the genetic material of old cultivars is less allergenic [110]. Barreira et al. [111] suggested in their study that phenolic compounds from old apple cultivars, which are more than those present in the commercial cultivars, can be used in dermal formulations due to the many useful properties such as antioxidant or antimicrobial activity. Ikumi et al. [112] also proposed oral antidiabetic agents based on phloridzin conjugates.

Table 4. The main polyphenolic subgroups of old apple cultivars grown in Croatia (mg kg^{-1} fresh sample weight).

Cultivar	Sample	Total flavan-3-ols	Total Dihydrochalcones	Total Phenolic Acids	Total Flavonols	Total Anthocyanins	Total Phenols	Reference
Crvenka	peel	1179	212	319	964	200	2874	[12]
Crveni boskop		316	168	178	644	41	1347	
Pisanika		653	195	396	2513	44	3801	
Lještarka		542	169	42	1994	64	2811	
Božićnica 1		400	222	176	639	42	1479	
Božićnica 2		493	267	224	1342	93	2319	
Coxs orange		332	32	83	583	46	1076	
Ivanlija		287	47	517	1532	17	2400	
Boskop		268	287	138	240	12	945	
Bobovac		484	80	260	1104	12	1940	
Slavonska srčika		102	53	18	359	4	536	
Kolerova srčika		1077	133	519	1038	8	2775	
Batulenka		305	28	42	552	3	930	
Gravenstein		287	20	23	266	9	605	
Mašanka		280	45	105	1006	5	1441	
Kanada		364	231	161	2316	3	3075	
Kandil Sinap		438	79	27	119	/	663	
Citronka		99	51	221	458	/	829	
Zimnjara		256	232	308	312	/	1108	
Zlatica		242	19	51	388	/	700	
Gloria Mundi		231	26	51	1376	/	1584	
Zelenika		550	54	68	2451	/	3123	
Krastavka		225	103	35	209	/	573	
Adamova zvijezda		1358	151	547	1486	5	3547	
Crvenka	mesocarp	33	9	145	8	/	195	
Crveni boskop		56	30	598	20	/	704	
Pisanika		56	13	134	18	/	221	
Lještarka		23	10	141	17	/	191	
Božićnica 1		42	42	423	20	/	527	
Božićnica 2		112	48	1058	26	/	1294	
Coxs orange		47	18	457	20	/	542	
Ivanlija		42	18	400	22	/	482	

Table 4. *Cont.*

Cultivar	Sample	Total flavan-3-ols	Total Dihydrochalcones	Total Phenolic Acids	Total Flavonols	Total Anthocyanins	Total Phenols	Reference
Boskop		152	68	840	27	/	1087	
Bobovac		48	7	134	16	/	205	
Slavonska srčika		24	2	56	13	/	95	
Kolerova srčika		93	17	325	19	/	454	
Batulenka		10	2	55	12	/	79	
Gravenstein		48	11	258	14	/	331	
Kandil Sinap		140	14	118	13	/	285	
Citronka		31	18	226	11	/	286	
Zimnjara		75	47	603	16	/	741	
Zlatica		9	5	64	13	/	91	
Mašanka		9	7	155	17	/	188	
Kanada		17	12	128	18	/	175	
Gloria Mundi		23	4	107	20	/	154	
Zelenika		15	6	72	16	/	109	
Krastavka		66	12	137	18	/	233	
Adamova zvijezda		48	7	205	16	/	276	
Ljubeničarka	mesocarp of red and light red apples	3892.0	53.1	534.4	/	318.8	4798.3	[87]
Astrahan		3342.0	67.0	663.6	3.8	/	4076.4	
Crvenika		3804.0	138.1	259.0	/	/	4201.1	
Kardinal		5482.0	113.3	1011.1	/	/	6606.4	
Kraljevina		2978.0	82.0	750.9	2.4	/	3813.3	
Ružica		4412.0	60.8	381.1	/	/	4853.9	
Pisanica		3228.0	95.8	606.1	2.8	/	3932.7	
Slavonska Srčika		5326.0	149.5	1381.3	3.4	/	6860.2	
Ljubeničarka	peel of red and light red apples	9984.0	124.0	212.3	199.4	761.0	11,280.7	
Astrahan		8386.0	269.4	64.2	1455.8	556.8	11,318.2	
Crvenika		7538.0	486.8	34.9	399.4	437.2	8896.3	
Kardinal		9694.0	290.2	739.2	427.0	251.8	11402.2	
Kraljevina		11,788.0	207.7	393.1	294.4	410.4	13,093.6	
Ružica		8948.0	472.0	98.2	279.6	402.2	10,200.0	
Pisanica		7122.0	424.2	479.6	397.4	403.0	8826.2	
Slavonska srčika		11,062.0	707.2	1190.0	963.6	79.4	14,002.2	

Table 4. *Cont.*

Cultivar	Sample	Total flavan-3-ols	Total Dihydrochalcones	Total Phenolic Acids	Total Flavonols	Total Anthocyanins	Total Phenols	Reference
Petrovka	mesocarp of green or yellow apples	1844.0	168.2	639.4	/	/	2651.6	
Bjeličnik		2450.0	102.1	842.8	/	/	3394.9	
Ledenara		3884.0	105.4	376.9	4.4	/	4370.7	
Štegerova		3300.0	45.1	508.0	/	/	3853.1	
Jaje		3448.0	1.6	579.9	3.0	/	4142.5	
Petrovka	peel of green or yellow apples	6124.0	769.3	248.2	251.4	/	7392.9	
Bjeličnik		4532.0	418.9	611.7	322.8	/	5885.4	
Ledenara		8862.0	388.2	116.0	419.4	/	9785.6	
Štegerova		5922.0	74.5	386.2	85.2	/	6467.9	
Jaje		5474.0	338.6	199.0	471.4	/	6483.0	

4. The Influence of Apples on Health

The health effects of apples are influenced by the availability of the bioactive compounds contained in apple and by their absorption and metabolism in the human body. The bioavailability of polyphenols depends on the amount of phenolic compounds that are released from solid foods in the body and can pass the intestinal barrier [113]. Fruits and vegetables are naturally composed of hydrated cells with phenolic compounds in cell vacuoles that are only weakly bound to the cell wall [114]. The mechanical action of digestion causes these cells to burst and allows the release of the phenolic compounds, while at the same time, the acidic environment of the stomach and the alkaline environment of the intestine facilitate the release of polyphenols close to the cell wall. Wruss et al. [115] explained the reason for the inconsistent results of clinical studies on the health benefits of apple and individual flavonoids by the specific pharmacokinetics that occur due to differences in small intestinal length, intestinal microbiota, or genetic factors of each individual. These variations in each human organism have significant effects on polyphenol metabolism. The current research available on the health benefits of apples is summarized below (Supplementary Figure S2).

For instance, a synergistic interaction of polyphenol rich foods and the gut microbiota has been demonstrated. Microorganisms in the colon can release polyphenols from the fibrous environment and break them down into phenolic acids, while polyphenols are able to stimulate the growth of beneficial bacterial species and inhibit the growth of pathogenic species [116]. The effect of polyphenols in the human body also depends on the genetic characteristics of the individual. In addition, there is growing evidence that the bioavailability and bioefficiency of polyphenols are influenced by the environment in which the polyphenols are found, i.e., the other bioactive components of the apple, as well as by the dose of polyphenols ingested. Jakobek [117] demonstrated a synergistic interaction between dietary fiber and flavonoids, which has a positive effect on human health, as well as the interaction of lipids and polyphenols which reduces the process of fat absorption, and thus has positive effects on health. Dietary fiber possesses a protective role in the treatment and prevention of certain diseases such as intestinal diseases [118]. The beneficial effects of apples consumption on vascular function and blood pressure prevention have been demonstrated, lowering blood lipids, reducing inflammation and preventing hyperglycemia [27].

The protective effect of apples and other fruits against cardiovascular diseases are attributed to the high content of polyphenols and their specific composition. Apples make an important contribution to the intake of macronutrients associated with the prevention of cardiovascular disease. There is a significant association between intake of more dietary flavonoids and the reduction of mortality, especially coronary mortality in women [119]. Knekt et al. [119] observed an association of increased quercetin intake with lower coronary heart disease mortality. Arts et al. [120] showed in a prospective study of postmenopausal Iowa women that reduced mortality from coronary heart disease has favored the intakes of the flavonoids (+)-catechin and (−)-epicatechin. Recent studies indicated a 46% reduction risk of cardiovascular disease mortality in elderly men by (−)-epicatechin intake, with 28% of the total (−)-epicatechin intake coming from apples. An increasing number of studies reported a lower incidence of coronary heart disease and cardiovascular disease in subjects consuming large amounts of dietary fibers [121]. Consumption of 120 g of apple flesh with 80 g of peels provided a higher intake of quercetin, (−)-epicatechin, and other flavonoids, compared to the control group that consumed only the flesh of the apple. Higher flavonoid consumption resulted in lower systolic blood pressure and pulse pressure in a randomized controlled trial of 30 healthy men and women [122]. The study showed an acute increase in nitric oxide, which causes smooth muscle relaxation, leading to dilatation of blood vessels and a lowering of a blood pressure, i.e., vasodilation. When studying the effect of apples on cholesterol levels, consumption of three apples per day resulted in a 5–8% reduction in total cholesterol, while consumption of apple juice (375–720 mL) had no effect on plasma cholesterol levels and had a deleterious effect on plasma triglyceride levels due to its high fructose content [123]. In postmenopausal women, significantly lower total serum cholesterol levels were found after 6 months of consumption of dried apples compared to 6 months of consumption of prunes [124]. In addition, a study of apple polyphenol consumption (1500 mg daily for four weeks) reflected that total cholesterol was reduced by 4.5% in 48 men and women with high cholesterol levels [125]. Phenolic compounds from apples have also been shown to increase lipoprotein lipase activity, thereby lowering blood cholesterol levels [126]. In human digestion, pectin can potentially lower plasma lipid levels by binding to cholesterol in the gastrointestinal tract [127]. Although the effect of pectin on lowering cholesterol has been reported, the relatively low pectin content in apples suggests the presence of other apple components that may have an effect.

Inflammatory processes are present in a variety of human diseases, and there is evidence that polyphenols have anti-inflammatory effects [128]. In a study of 8335 adults in the United States, apple consumption was inversely related to C-reactive protein (CRP) levels. CRP serves as an inflammatory marker and is therefore an important indicator of inflammation [129]. Chai et al. [124] observed that consumption of dried apples over a 12-month period reduced CRP levels by 32%, but this did not reach a statistically significant difference compared to the control sample. In a meta-analysis of studies on increasing fiber consumption in humans, six out of seven studies reported significant reductions in CRP levels. Prebiotic fibers have been shown to affect intestinal permeability, reducing the absorption of lipopolysaccharide, an endotoxin that releases Gram-negative bacteria that elicit a strong immune response in humans [9].

Western dietary habits are considered to cause disease, while a diet rich in fruits and vegetables is associated with risk reduction. The increase in type 2 diabetes worldwide is a cause for public health concern, since type 2 diabetes may increase the risks of cardiovascular disease [130]. Apples have been highlighted as an important dietary component that has the potential to reduce the growing prevalence of type 2 diabetes. Consumption of more than one apple per day is associated with a significant reduction in type 2 diabetes risk (by 28%) compared with non-apple consumption [131]. The evidence that certain types of polyphenols may reduce the risk of type 2 diabetes comes from a study of 2915 participants. This study found that each 2.5-fold increase in flavonol intake was associated with a 26% lower incidence of type 2 diabetes [132].

Additionally, higher intake of soluble fibers has a beneficial effect on reducing the development of type 2 diabetes. Dietary fiber has been attributed to a beneficial effect on weight loss, and thus on the treatment of obesity [16]. Consumption of two apples per day for two weeks resulted in a significant increase in human intestinal bifidobacteria and fecal acetic acid, i.e., a positive effect on the health of the gut microbiota [133].

Dysbiosis or disturbance of the intestinal microflora is a term used to describe the disturbance of the natural balance of microorganisms in the digestive tract. The disturbance results from a decrease in the proportion of beneficial microorganisms, an overgrowth of potentially harmful microorganisms or a decrease in overall microbiological diversity [134]. The effect of quercetin supplementation on the suppression of dysbiosis of the intestinal microbiota, caused by a diet high in sucrose and fat has been demonstrated. Quercetin reduced the ratio of *Firmicutes* and *Bacteroidetes* (markers of intestinal health) and inhibited the growth of *Erysipelotrichaceus*, *Bacillus*, and *Eubacterium cylindroides* bacteria (bacterial species associated with diet-induced obesity) [135]. Phloretin, found in apples, has been shown to act as an inhibitor of pathogenic biofilm by *Escherichia coli* production and as an anti-inflammatory agent in inflammatory bowel disease [136]. A recent study confirmed a protective effect of phloridzin on antioxidant stress, DNA damage, and apoptosis in $H_2O_2^-$ induced HepG2 cells, therefore many studies suggested that phloridzin can be used in the production of functional foods [137]. In conclusion, the health effects of apples are diverse and have been shown to be beneficial, so the recorded evidence points to the potential of apples for the production of functional products.

5. The Perspective and Development of Apple-Based Functional Products Behind Functional Ingredients

Functional foods are defined as industrially processed foods or unprocessed natural foods that have been shown to have beneficial health effects beyond basic nutritional value when consumed regularly as part of a varied diet [138]. For this reason, functional food is an increasingly popular term in social and scientific circles. Food manufactures are also investing in the development of industrially processed foods that may have additional health benefits for consumers. It is very important that clinical (randomized, double-blind, and placebo-controlled) trials are conducted before drawing conclusions about the health benefits of functional foods [139]. Clinical studies and experimental evidence will provide an answer as to which food ingredient is functionally effective for the human body and to what extent. In addition, for a food to be labeled functional, it must be confirmed in intervention tests that it complies with the regulations of the country. Some of the regulations that functional foods must comply with are the European Food Safety Authority (EFSA) in European Union and the Food and Drug Administration (FDA) in the United States [140–142]. Functional foods must be safe and freely available to consumers without the need for a prescription from a physician [143]. It is necessary to emphasize that functional foods are not medicine; they will not prevent or cure diseases, as various internal and external factors play a crucial role in the occurrence of the disease [144,145]. To some extent, consumers regulate the trends in the food industry and their growing preference for safe, fresh, and natural foods that have health benefits may explain the importance of the adaptability of the food industry and investment in the production of functional products. Therefore, the production of functional products is expected to increase rapidly across the globe.

Apple juice is commonly mixed with another liquid ingredients (extract, tea, beverage, etc.) to formulate an ideal functional food formulation with improved physicochemical properties, nutritional characteristics, and sensory acceptability [146]. A recent study aimed to formulate a functional cake based on apple pomace flour (powder) as a substitute for wheat and rice flour to produce a gluten free product suitable for celiac patients [147]. Apple pomace flour had lower protein content (1.25%) and higher fiber content (56%) compared to wheat and rice flours. The proximate composition of the flour showed that the content of total phenols in apple pomace flour was 4 times higher than in wheat and 7 times higher than in rice flour. The authors concluded that by replacing 100% of the flour with apple pomace, the physical and sensory properties could be satisfactory according to consumer expectations. Although cakes with apple pomace flour had a harder texture and lower specific volume, their general acceptance in terms of good smell and taste made them highly desirable products [27].

The whole unripe apples 'Golden Delicious' without non-edible parts (seeds and peduncle) were used to obtain apple flour by convection drying at 50 °C for 4 h [148]. Grounded dried pieces were used for making spaghetti-type pasta by replacing 50% of durum wheat semolina with oat bran in order to monitor cooking quality, digestibility, antioxidant, nutritional, and texture characteristics. The cooked pasta with apple flour showed the higher content of total phenols and scavenging capacity in comparison with the control samples. Based on the good retention of bioactive compounds and antioxidant capacity after the cooking process of the noodles, the authors concluded that apple flour could be considered as a sustainable food ingredient for the development of a functional food [148]. In addition, apple pomace water extract could also be considered as a very valuable ingredient for the supplementation or development of fortified foods such as functional yogurt formulations [149]. Yogurt fortified with apple pomace extract showed improved fiber content and antioxidant properties compared to plain yogurt. In this regard, probiotic yogurt inoculated with *Lactobacillus acidophilus*, *Streptococcus thermophilus*, and *Bifidobacterium bifidum* to which 3% apple pomace flour was added showed optimal functional properties in terms bioactivity and sensory characteristics compared to control samples [27].

The use of innovative technologies such as high hydrostatic pressure processing (HHPP) can also help to improve the functional properties of the by-products of the apple 'Golden Delicious' [150]. De la Peña Armada et al. [150] found that HHPP under 200 MPa for 15 min was sufficient to improve the solubilization of cell wall components such as pectins, increase the content of soluble carbohydrates that could act as prebiotics, and maintain the content of total phenols compared to untreated samples. The use of HHPP to modify the chemical properties of apple by-products even makes them more suitable for industrial application in the production of functional foods.

Pulsed electric fields (PEF) is another innovative approach used to improve the functional quality of apples without affecting their physicochemical properties [151]. Important changes in phenolic profile and quality properties of raw apples (firmness, color properties, soluble solids content, pH, titratable acidity) were induced by PEF treatment as a function of specific energy and time after treatment. Lower energy (0.01 kJ kg^{-1}) had a positive effect on bioactive compound content, while higher energy (1.8 and 7.3 kJ kg^{-1}) resulted in irreversible quality changes (texture and color). This PEF effect was explained as a response of apple tissues to oxidative stress, which led to accumulation of bioactive compounds after treatment. Another possible explanation is related to matrix changes and increased extractability of bioactive compounds during their determination [27].

6. Conclusions

Food origins and safety are increasingly becoming critical factors in the selection of fruit cultivars for direct consumption, as well as for processing into products. Old apple cultivars with a long history of cultivation on the territory in the Republic of Croatia have been shown to generally require fewer agricultural inputs compared to commercial apple cultivars, but are nevertheless unsuitable for large-scale distribution.

However, their growth, yield, and desirable fruit quality characteristics make them more attractive for both, as an important gene source for apple breeding programs and as commercial cultivars for the domestic supermarket channel and touristic markets.

Their potential for processing into functional foods is supported by evidence of health benefits due to higher levels of polyphenolic compounds and antioxidant capacity as compared to commercial cultivars. Improving yields and efficiency, developing market products, improving knowledge, experience and entrepreneurship, strengthening cooperation, improving product safety and quality e.g., by implementing certification programs, and agro-logistics are the main priorities for improving the competitiveness of old apple cultivars and their functional products on the Croatian and EU markets.

Supplementary Materials: The following are available online at https://www.mdpi.com/article/10.3390/foods10040708/s1, Figure S1: Area of apples by cropping density in hectares (%) in the Republic of Croatia for 2012 and 2017, Figure S2. Health benefits of apple, Table S1: Distribution of apple production in the Republic of Croatia for 2017 and 2018, Table S2: Area under cultivation, production and yield of apples by the main growing regions in the Republic of Croatia, Table S3: Total area of plantations by apple cultivar in hectares in the Republic of Croatia, Table S4: Some differences between modern cultivation and old apple cultivars growing in Croatia.

Author Contributions: Conceptualization, P.P. and D.B.K.; writing—original draft preparation, B.D., D.B., D.B.K.; writing—review and editing, B.D., P.P., A.B.M., S.Z., M.P., R.D. and J.M.L.; project administration, D.B.K. All authors have read and agreed to the published version of the manuscript.

Funding: This research was funded by the Croatian Science Foundation through the funding of the 'Hurdle Technology and 3D Printing for Sustainable Fruit Juice Processing and Preservation' project, number 'IP-2019-04-2105' and by the "Modification of cheese ripening process and development of whey-based products—SIRENA" project, co-financed by the European Union from the European Structural and Investment Funds in the financial period 2014–2020, and the Operational Programme Competitiveness and Cohesion. Contract No: KK.01.1.1.04.0096.

Conflicts of Interest: The authors declare no conflict of interest. The funders had no role in the design of the study; in the collection, analyses, or interpretation of data; in the writing of the manuscript, or in the decision to publish the results.

References

1. Croatian Bureau of Statistics. Available online: https://www.dzs.hr/ (accessed on 5 September 2020).
2. Brizzolara, S.; Manganaris, G.A.; Fotopoulos, V.; Watkins, C.B.; Tonutti, P. Primary metabolism in fresh fruits during storage. *Front. Plant Sci.* **2020**, *11*. [CrossRef] [PubMed]
3. Lante, A.; Tinello, F.; Nicoletto, M. UV-A light treatment for controlling enzymatic browning of fresh-cut fruits. *Innov. Food Sci. Emerg. Technol.* **2016**, *34*, 141–147. [CrossRef]
4. Tinello, F.; Mihaylova, D.; Lante, A. Effect of dipping pre-treatment with unripe grape juice on dried "Golden Delicious" apple slices. *Food Bioprocess Technol.* **2018**, *11*, 2275–2285. [CrossRef]
5. Musacchi, S.; Serra, S. Apple fruit quality: Overview on pre-harvest factors. *Sci. Hortic.* **2018**, *234*, 409–430. [CrossRef]
6. Putnik, P.; Roohinejad, S.; Greiner, R.; Granato, D.; Bekhit, A.E.-D.A.; Bursać Kovačević, D. Prediction and modeling of microbial growth in minimally processed fresh-cut apples packaged in a modified atmosphere: A review. *Food Control* **2017**, *80*, 411–419. [CrossRef]
7. Pérez-Llorca, M.; Muñoz, P.; Müller, M.; Munné-Bosch, S. Biosynthesis, metabolism and function of auxin, salicylic acid and melatonin in climacteric and non-climacteric fruits. *Front. Plant Sci.* **2019**, *10*. [CrossRef] [PubMed]
8. Li, C.; Zhang, J.; Ge, Y.; Li, X.; Wei, M.; Hou, J.; Cheng, Y.; Lv, J. Postharvest acibenzolar-S-methyl treatment maintains storage quality and retards softening of apple fruit. *J. Food Biochem.* **2020**, *44*. [CrossRef]
9. Koutsos, A.; Tuohy, K.; Lovegrove, J. Apples and cardiovascular health—Is the gut microbiota a core consideration? *Nutrients* **2015**, *7*, 3959–3998. [CrossRef]
10. Ploscutanu, G.; Elisei, A.M.; Buzia, O.D. Nutraceutical properties of apples and derived products (pomace, seeds, peels). *Rev. Chim.* **2019**, *70*, 934–939. [CrossRef]
11. Hyun, T.K.; Jang, K.I. Apple as a source of dietary phytonutrients: An update on the potential health benefits of apple. *EXCLI J.* **2016**, *15*, 565–569.
12. Jakobek, L.; Ištuk, J.; Buljeta, I.; Voća, S.; Žlabur, J.Š.; Babojelić, M.S. Traditional, indigenous apple varieties, a fruit with potential for beneficial effects: Their quality traits and bioactive polyphenol contents. *Foods* **2020**, *9*, 52. [CrossRef] [PubMed]
13. Pellegrini, N.; Vitaglione, P.; Granato, D.; Fogliano, V. Twenty-five years of total antioxidant capacity measurement of foods and biological fluids: Merits and limitations. *J. Sci. Food Agric.* **2019**, *100*, 5064–5078. [CrossRef] [PubMed]
14. Sharma, R.; Padwad, Y. Perspectives of the potential implications of polyphenols in influencing the interrelationship between oxi-inflammatory stress, cellular senescence and immunosenescence during aging. *Trends Food Sci. Technol.* **2020**, *98*, 41–52. [CrossRef]
15. Mustafa, S.K.; Oyouni, A.A.W.; Aljohani, M.M.H.; Ahmad, M.A. Polyphenols more than an antioxidant: Role and scope. *J. Pure Appl. Microbiol.* **2020**, *14*, 47–61. [CrossRef]
16. Anderson, J.W.; Baird, P.; Davis, R.H., Jr.; Ferreri, S.; Knudtson, M.; Koraym, A.; Waters, V.; Williams, C.L. Health benefits of dietary fiber. *Nutr. Rev.* **2009**, *67*, 188–205. [CrossRef]
17. Skinner, R.C.; Gigliotti, J.C.; Ku, K.-M.; Tou, J.C. A comprehensive analysis of the composition, health benefits, and safety of apple pomace. *Nutr. Rev.* **2018**, *76*, 893–909. [CrossRef]
18. Issar, K.; Sharma, P.C.; Gupta, A. Utilization of apple pomace in the preparation of fiber-enriched acidophilus yoghurt. *J. Food Process. Preserv.* **2017**, *41*. [CrossRef]
19. Vujević, P.; Milinović, B.; Jelačić, T.; Halapija Kazija, D.; Čiček, D. Kvaliteta Ploda Tradicionalnih Sorti Jabuka u Kolekcijskom Nasadu na Pokušalištu HCPHS Donja Zelina. In *Zbornik Sažetaka 11 Međunarodnog Kongresa "Oplemenjivanje Bilja, Sjemenarstvo i Rasadničarstvo"*; Matotan, Z., Ed.; Hrvatsko Agronomsko Društvo: Zagreb, Croatia, 2018; pp. 74–75.
20. Lončarić, A.; Skendrović Babojelić, M.; Kovač, T.; Šarkan, B. Pomological properties and polyphenol content of conventional and traditional apple cultivars from Croatia. *Food Health Dis. Sci. Prof. J. Nutr. Diet.* **2019**, *8*, 19–24.
21. Food and Agriculture Organization of the United Nations (FAO). Available online: http://www.fao.org/faostat/en/ (accessed on 16 November 2020).
22. Ivković, F. Sortiment jabuka u proizvodnji u EU i Hrvatskoj i koje sorte saditi. *Glas. Zaštite Bilja* **2011**, *34*, 17–19.
23. Janjić, V. *Possibilities of Preservation and Revitalization of Orchard with Old Apple Cultivarsat a Rehabilitation Centre 'Ozalj'*; University of Zagreb: Zagreb, Croatia, 2016.
24. Kouassi, A.B.; Durel, C.-E.; Costa, F.; Tartarini, S.; van de Weg, E.; Evans, K.; Fernandez-Fernandez, F.; Govan, C.; Boudichevskaja, A.; Dunemann, F.; et al. Estimation of genetic parameters and prediction of breeding values for apple fruit-quality traits using pedigreed plant material in Europe. *Tree Genet. Genomes* **2009**, *5*, 659–672. [CrossRef]
25. Abbott, J.A. Quality measurement of fruits and vegetables. *Postharvest Biol. Technol.* **1999**, *15*, 207–225. [CrossRef]
26. Westwood, M.N. *Temperate—Zone Pomology, Physiology and Culture*, 3rd ed.; Timber Press: Portland, OR, USA, 1993; p. 523.
27. Spanos, G.A.; Wrolstad, R.E.; Heatherbell, D.A. Influence of processing and storage on the phenolic composition of apple juice. *J. Agricu. Food Chem.* **1990**, *38*, 1572–1579. [CrossRef]

28. Skendrović Babojelić, M.; Korent, P.; Šindrak, Z.; Jemrić, T. Pomological characteristics and fruit quality traditional apple cultivars. *Glas. Zaštite Bilja* **2014**, *3*, 20–27.
29. Eur-Lex. Commission Regulation (EC) No 85/2004. of 15 January 2004 laying down the marketing standard for apples. *Off. J. Eur. Union* **2004**, l 13/13.
30. Adamič, F.; Bohutinski, O.; Dimitrovski, T.; Gavrilović, M.; Jovančević, R.; Stanković, D.; Vitolović, V. *Jugoslavenska Pomologija—*Jabuka; Proleter-Bečej: Bečej, Serbia, 1963; p. 219.
31. Vujević, B. *Pomological Characteristics of Fruits of Old Apple Cultivars in Bjelovar-Bilogora County*; University of Zagreb: Zagreb, Croatia, 2016.
32. Pašalić, B. *Berba, Pakovanje i Skladištenje Plodova Voćaka*; Poljoprivredni Fakultet: Banja Luka, Bosnia and Herzegovina, 2006.
33. Hoehn, E.; Gasser, F.; Guggenbühl, B.; Künsch, U. Efficacy of instrumental measurements for determination of minimum requirements of firmness, soluble solids, and acidity of several apple varieties in comparison to consumer expectations. *Postharvest Biol. Technol.* **2003**, *27*, 27–37. [CrossRef]
34. Di Vaio, C.; Cirillo, C.; Buccheri, M.; Limongelli, F. Effect of interstock (M.9 and M.27) on vegetative growth and yield of apple trees (cv "Annurca"). *Sci. Hortic.* **2009**, *119*, 270–274. [CrossRef]
35. Urrestarazu, J.; Denancé, C.; Ravon, E.; Guyader, A.; Guisnel, R.; Feugey, L.; Poncet, C.; Lateur, M.; Houben, P.; Ordidge, M.; et al. Analysis of the genetic diversity and structure across a wide range of germplasm reveals prominent gene flow in apple at the European level. *BMC Plant Biol.* **2016**, *16*. [CrossRef]
36. Janick, J.; Moore, J.N. *Fruit Breeding, Volume I: Tree and Tropical Fruits*; Wiley: New York, NY, USA, 1996.
37. Tóth, M.; Ficzek, G.; Király, I.; Honty, K.; Hevesi, M. Evaluation of old Carpathian apple cultivars as genetic resources of resistance to fire blight (*Erwinia amylovora*). *Trees* **2012**, *27*, 597–605. [CrossRef]
38. Jemrić, T.; Babojelić Skendrović, M.; Fruk, G.; Šindrak, Z. Fruit quality of nine old apple cultivars. *Not. Bot. Horti Agrobot. Cluj-Napoca* **2013**, *41*. [CrossRef]
39. Jakobek, L.; Barron, A.R. Ancient apple varieties from Croatia as a source of bioactive polyphenolic compounds. *J. Food Compos. Anal.* **2016**, *45*, 9–15. [CrossRef]
40. Preti, R.; Tarola, A.M. Study of polyphenols, antioxidant capacity and minerals for the valorisation of ancient apple cultivars from Northeast Italy. *Eur. Food Res. Technol.* **2020**, *247*, 273–283. [CrossRef]
41. Kviklys, D.; Kviklienė, N.; Bite, A.; Lepsis, J.; Univer, T.; Univer, N.; Uselis, N.; Lanauskas, J.; Buskienė, L. Baltic fruit rootstock studies: Evaluation of 12 apple rootstocks in North-East Europe. *Hortic. Sci.* **2012**, *39*, 1–7. [CrossRef]
42. Duralija, B. *Determination of Optimum Harvest Timing of Apple Cultivars Alkmene, Prima, Elstar, Hi Early and Golden B in Orchards Donja Zelina and Mrzović*; University of Zagreb: Zagreb, Croatia, 1994.
43. Tahir, I.I.; Johansson, E.; Olsson, M.E. Improvement of quality and storability of apple cv. Aroma by adjustment of some pre-harvest conditions. *Sci. Hortic.* **2007**, *112*, 164–171. [CrossRef]
44. Ilie, A.V.; Petrisor, C.; Hoza, D.; Oltenescu, V. Influence of soil type on yield and quality of different apple cultivars. *Bull. Univ. Agric. Sci. Vet. Med. Cluj Napoca. Hortic.* **2017**, *74*. [CrossRef]
45. Lin-Wang, K.U.I.; Micheletti, D.; Palmer, J.; Volz, R.; Lozano, L.; Espley, R.; Hellens, R.P.; ChagnÈ, D.; Rowan, D.D.; Troggio, M.; et al. High temperature reduces apple fruit colour via modulation of the anthocyanin regulatory complex. *Plant Cell Environ.* **2011**, *34*, 1176–1190. [CrossRef]
46. Hudina, M.; Stampar, F.; Zadravec, P. The influence of planting density on sugar and organic acid content in apple (*Malus Domestica* Borkh.). *Acta Hortic.* **2001**, *557*, 313–320. [CrossRef]
47. Pavičić, N.; Skendrović Babojelić, M.; Jemrić, T.; Šindrak, Z.; Ćosić, T.; Karažija, T.; Ćosić, D. Effects of combined pruning treatments on fruit quality and biennial bearing of 'Elstar' apple (*Malus domestica* Borkh.). *J. FoodAgric. Environ.* **2009**, *7*, 510–515. [CrossRef]
48. Reganold, J.P.; Glover, J.D.; Andrews, P.K.; Hinman, H.R. Sustainability of three apple production systems. *Nature* **2001**, *410*, 926–930. [CrossRef]
49. Meland, M. Effects of different crop loads and thinning times on yield, fruit quality, and return bloom in *Malus* × *domestica* Borkh. 'Elstar'. *J. Hortic. Sci. Biotechnol.* **2015**, *84*, 117–121. [CrossRef]
50. Mašán, V.; Burg, P.; Čížková, A.; Skoupil, J.; Zemánek, P.; Višacki, V. Effects of irrigation and fertigation on yield and quality parameters of 'Gala' and 'Fuji' apple. *Acta Univ. Agric. Et Silvic. Mendel. Brun.* **2018**, *66*, 1183–1190. [CrossRef]
51. MiloŠEviĆ, T.; MiloŠEviĆ, N. Tree vigor, yield, fruit quality, and antioxidant capacity of apple (*Malus* × *domestica* Borkh.) influenced by different fertilization regimes: Preliminary results. *Turk. J. Agric. For.* **2019**, *43*, 48–57. [CrossRef]
52. Ewing, B.L.; Peck, G.M.; Ma, S.; Neilson, A.P.; Stewart, A.C. Management of apple maturity and postharvest storage conditions to increase polyphenols in cider. *HortScience* **2019**, *54*, 143–148. [CrossRef]
53. FAO/Netherlands. Elements for Strategies and Agenda for Action. Strategies and Tools for Sustainable Agriculture and Rural Development. In *FAO/Netherlands Conference on Agriculture and the Environment, held 15–19 April 1991 in S-Hertogenbosch, The Netherlands*; FAO: Rome, Italy, 1991; p. 27.
54. SAI Platform. 2020. Available online: https://saiplatform.org/our-commitment/ (accessed on 5 September 2020).
55. Food and Agriculture Organization (FAO). 2020. Available online: http://www.fao.org/faostat/en/#data/QC/visualize (accessed on 5 September 2020).

56. Samnegård, U.; Alins, G.; Boreux, V.; Bosch, J.; García, D.; Happe, A.K.; Klein, A.M.; Miñarro, M.; Mody, K.; Porcel, M.; et al. Management trade-offs on ecosystem services in apple orchards across Europe: Direct and indirect effects of organic production. *J. Appl. Ecol.* **2018**, *56*, 802–811. [CrossRef]
57. Merwin, I.A.; Pritts, M.P. Are modern fruit production systems sustainable? *HortTechnology* **1993**, *3*, 128–136. [CrossRef]
58. Wang, N.; Wolf, J.; Zhang, F.-s. Towards sustainable intensification of apple production in China—Yield gaps and nutrient use efficiency in apple farming systems. *J. Integr. Agric.* **2016**, *15*, 716–725. [CrossRef]
59. Putnik, P.; Lorenzo, J.; Barba, F.; Roohinejad, S.; Režek Jambrak, A.; Granato, D.; Montesano, D.; Bursać Kovačević, D. Novel food processing and extraction technologies of high-added value compounds from plant materials. *Foods* **2018**, *7*, 106. [CrossRef]
60. Francini, A.; Sebastiani, L. Phenolic compounds in apple (*Malus x domestica Borkh.*): Compounds characterization and stability during postharvest and after processing. *Antioxidants* **2013**, *2*, 181–193. [CrossRef] [PubMed]
61. Scalbert, A.; Williamson, G. Dietary intake and bioavailability of polyphenols. *J. Nutr.* **2000**, *130*, 2073S–2085S. [CrossRef] [PubMed]
62. Tsao, R.; Yang, R.; Xie, S.; Sockovie, E.; Khanizadeh, S. Which polyphenolic compounds contribute to the total antioxidant activities of apple? *J. Agric. Food Chem.* **2005**, *53*, 4989–4995. [CrossRef] [PubMed]
63. McGhie, T.K.; Hunt, M.; Barnett, L.E. Cultivar and growing region determine the antioxidant polyphenolic concentration and composition of apples grown in New Zealand. *J. Agric. Food Chem.* **2005**, *53*, 3065–3070. [CrossRef] [PubMed]
64. Zhang, S.; Hu, C.; Guo, Y.; Wang, X.; Meng, Y. Polyphenols in fermented apple juice: Beneficial effects on human health. *J. Funct. Foods* **2021**, *76*. [CrossRef]
65. Van der Sluis, A.A.; Dekker, M.; Skrede, G.; Jongen, W.M.F. Activity and concentration of polyphenolic antioxidants in apple juice. 1. Effect of existing production methods. *J. Agric. Food Chem.* **2002**, *50*, 7211–7219. [CrossRef] [PubMed]
66. Markowski, J.; Płocharski, W. Determination of phenolic compounds in apples and processed apple products. *J. Fruit Ornam. Plant Res.* **2006**, *14*, 133–142.
67. Lee, K.W.; Kim, Y.J.; Kim, D.-O.; Lee, H.J.; Lee, C.Y. Major phenolics in apple and their contribution to the total antioxidant capacity. *J. Agric. Food Chem.* **2003**, *51*, 6516–6520. [CrossRef] [PubMed]
68. Miller, N.J.; Diplock, A.T.; Rice-Evans, C.A. Evaluation of the total antioxidant activity as a marker of the deterioration of apple juice on storage. *J. Agric. Food Chem.* **1995**, *43*, 1794–1801. [CrossRef]
69. Van der Sluis, A.A.; Dekker, M.; van Boekel, M.A.J.S. Activity and concentration of polyphenolic antioxidants in apple juice. 3. Stability during storage. *J. Agric. Food Chem.* **2005**, *53*, 1073–1080. [CrossRef]
70. López-Pedrouso, M.; Bursać Kovačević, D.; Oliveira, D.; Putnik, P.; Moure, A.; Lorenzo, J.M.; Domínguez, H.; Franco, D. In vitro and in vivo Antioxidant Activity of Anthocyanins. In *Anthocyanins—Aantioxidant Properties, Sources and Health Benefits*; Lorenzo, J.M., Barba, F.J., Munekata, P., Eds.; Nova Science Publishers, Inc.: New York, NY, USA, 2020; pp. 169–204.
71. Butkevičiūtė, A.; Liaudanskas, M.; Kviklys, D.; Zymonė, K.; Raudonis, R.; Viškelis, J.; Uselis, N.; Janulis, V. Detection and analysis of triterpenic compounds in apple extracts. *Int. J. Food Prop.* **2018**, *21*, 1716–1727. [CrossRef]
72. Dashbaldan, S.; Pączkowski, C.; Szakiel, A. Variations in triterpenoid deposition in cuticular Waxes during development and maturation of selected fruits of *Rosaceae* family. *Int. J. Mol. Sci.* **2020**, *21*, 9762. [CrossRef]
73. Zhu, Z.; Gavahian, M.; Barba, F.J.; Roselló-Soto, E.; Bursać Kovačević, D.; Putnik, P.; Denoya, G.I. Valorization of Waste and By-Products from Food Industries Through the Use of Innovative Technologies. In *Agri-Food Industry Strategies for Healthy Diets and Sustainability, New Challenges in Nutrition and Public Health*; Barba, F.J., Putnik, P., Bursać Kovačević, D., Eds.; Academic Press: Oxford, UK, 2020; pp. 249–266.
74. Li, D.; Sun, L.; Yang, Y.; Wang, Z.; Yang, X.; Zhao, T.; Gong, T.; Zou, L.; Guo, Y. Young apple polyphenols postpone starch digestion in vitro and in vivo. *J. Funct. Foods* **2019**, *56*, 127–135. [CrossRef]
75. Sun-Waterhouse, D.; Farr, J.; Wibisono, R.; Saleh, Z. Fruit-based functional foods I: Production of food-grade apple fibre ingredients. *Int. J. Food Sci. Technol.* **2008**, *43*, 2113–2122. [CrossRef]
76. Carbone, K.; Giannini, B.; Picchi, V.; Scalzo, R.L.; Cecchini, F. Phenolic composition and free radical scavenging activity of different apple varieties in relation to the cultivar, tissue type and storage. *Food Chem.* **2011**, *127*, 493–500. [CrossRef]
77. Wolfe, K.; Wu, X.; Liu, R.H. Antioxidant activity of apple peels. *J. Agric. Food Chem.* **2003**, *51*, 609–614. [CrossRef]
78. Loncaric, A.; Dugalic, K.; Mihaljevic, I.; Jakobek, L.; Pilizota, V. Effects of sugar addition on total polyphenol content and antioxidant activity of frozen and freeze-dried apple purée. *J. Agric. Food Chem.* **2014**, *62*, 1674–1682. [CrossRef]
79. Ceymann, M.; Arrigoni, E.; Schärer, H.; Nising, A.B.; Hurrell, R.F. Identification of apples rich in health-promoting flavan-3-ols and phenolic acids by measuring the polyphenol profile. *J. Food Compos. Anal.* **2012**, *26*, 128–135. [CrossRef]
80. Vrhovsek, U.; Rigo, A.; Tonon, D.; Mattivi, F. Quantitation of polyphenols in different apple varieties. *J. Agric. Food Chem.* **2004**, *52*, 6532–6538. [CrossRef]
81. Bai, X.; Zhang, H.; Ren, S. Antioxidant activity and HPLC analysis of polyphenol-enriched extracts from industrial apple pomace. *J. Sci. Food Agric.* **2013**, *93*, 2502–2506. [CrossRef]
82. García, Y.D.; Valles, B.S.; Lobo, A.P. Phenolic and antioxidant composition of by-products from the cider industry: Apple pomace. *Food Chem.* **2009**, *117*, 731–738. [CrossRef]
83. Putnik, P.; Bursać Kovačević, D.; Herceg, K.; Pavkov, I.; Zorić, Z.; Levaj, B. Effects of modified atmosphere, anti-browning treatments and ultrasound on the polyphenolic stability, antioxidant capacity and microbial growth in fresh-cut apples. *J. Food Process. Eng.* **2017**, *40*. [CrossRef]

84. Bursać Kovačević, D.; Bilobrk, J.; Buntić, B.; Bosiljkov, T.; Karlović, S.; Rocchetti, G.; Lucini, L.; Barba, F.J.; Lorenzo, J.M.; Putnik, P. High-power ultrasound altered the polyphenolic content and antioxidant capacity in cloudy apple juice during storage. *J. Food Process. Preserv.* **2019**, *43*. [CrossRef]
85. Donno, D.; Beccaro, G.L.; Mellano, M.G.; Marinoni, D.T.; Cerutti, A.K.; Canterino, S.; Bounous, G. Application of sensory, nutraceutical and genetic techniques to create a quality profile of ancient apple cultivars. *J. Food Qual.* **2012**, *35*, 169–181. [CrossRef]
86. Giomaro, G.; Karioti, A.; Bilia, A.R.; Bucchini, A.; Giamperi, L.; Ricci, D.; Fraternale, D. Polyphenols profile and antioxidant activity of skin and pulp of a rare apple from Marche region (Italy). *Chem. Cent. J.* **2014**, *8*. [CrossRef] [PubMed]
87. Jakobek, L.; García-Villalba, R.; Tomás-Barberán, F.A. Polyphenolic characterisation of old local apple varieties from Southeastern European region. *J. Food Compos. Anal.* **2013**, *31*, 199–211. [CrossRef]
88. Wojdyło, A.; Oszmiański, J.; Laskowski, P. Polyphenolic compounds and antioxidant activity of new and old apple varieties. *J. Agric. Food Chem.* **2008**, *56*, 6520–6530. [CrossRef]
89. Iacopini, P.; Camangi, F.; Stefani, A.; Sebastiani, L. Antiradical potential of ancient Italian apple varieties of *Malus* × *domestica* Borkh. in a peroxynitrite-induced oxidative process. *J. Food Compos. Anal.* **2010**, *23*, 518–524. [CrossRef]
90. Petkovsek, M.; Slatnar, A.; Stampar, F.; Veberic, R. Phenolic compounds in apple leaves after infection with apple scab. *Biol. Plant.* **2011**, *55*. [CrossRef]
91. Santarelli, V.; Neri, L.; Sacchetti, G.; Di Mattia, C.D.; Mastrocola, D.; Pittia, P. Response of organic and conventional apples to freezing and freezing pre-treatments: Focus on polyphenols content and antioxidant activity. *Food Chem.* **2020**, *308*. [CrossRef]
92. Yuri, J.A.; Moggia, C.; Sepulveda, A.; Poblete-Echeverría, C.; Valdés-Gómez, H.; Torres, C.A. Effect of cultivar, rootstock, and growing conditions on fruit maturity and postharvest quality as part of a six-year apple trial in Chile. *Sci. Hortic.* **2019**, *253*, 70–79. [CrossRef]
93. Volz, R.K.; McGhie, T.K. Genetic variability in apple fruit polyphenol composition in *Malus* × *domestica* and *Malus sieversii* germplasm grown in New Zealand. *J. Agric. Food Chem.* **2011**, *59*, 11509–11521. [CrossRef]
94. Karaman, Ş.; Tütem, E.; Başkan, K.S.; Apak, R. Comparison of antioxidant capacity and phenolic composition of peel and flesh of some apple varieties. *J. Sci. Food Agric.* **2013**, *93*, 867–875. [CrossRef] [PubMed]
95. Lamperi, L.; Chiuminatto, U.; Cincinelli, A.; Galvan, P.; Giordani, E.; Lepri, L.; Del Bubba, M. Polyphenol levels and free radical scavenging activities of four apple cultivars from integrated and organic farming in different Italian areas. *J. Agric. Food Chem.* **2008**, *56*, 6536–6546. [CrossRef]
96. Veberic, R.; Trobec, M.; Herbinger, K.; Hofer, M.; Grill, D.; Stampar, F. Phenolic compounds in some apple (*Malus domestica* Borkh) cultivars of organic and integrated production. *J. Sci. Food Agric.* **2005**, *85*, 1687–1694. [CrossRef]
97. Khanizadeh, S.; Tsao, R.; Rekika, D.; Yang, R.; Charles, M.T.; Rupasinghe, H.P.V. Polyphenol composition and total antioxidant capacity of selected apple genotypes for processing. *J. Food Compos. Anal.* **2008**, *21*, 396–401. [CrossRef]
98. Chinnici, F.; Gaiani, A.; Natali, N.; Riponi, C.; Galassi, S. Improved HPLC determination of phenolic compounds in cv. Golden Delicious apples using a monolithic column. *J. Agric. Food Chem.* **2004**, *52*, 3–7. [CrossRef] [PubMed]
99. Escarpa, A.; González, M.C. High-performance liquid chromatography with diode-array detection for the determination of phenolic compounds in peel and pulp from different apple varieties. *J. Chromatogr. A* **1998**, *823*, 331–337. [CrossRef]
100. Napolitano, A.; Cascone, A.; Graziani, G.; Ferracane, R.; Scalfi, L.; Di Vaio, C.; Ritieni, A.; Fogliano, V. Influence of variety and storage on the polyphenol composition of apple flesh. *J. Agric. Food Chem.* **2004**, *52*, 6526–6531. [CrossRef] [PubMed]
101. Valavanidis, A.; Vlachogianni, T.; Psomas, A.; Zovoili, A.; Siatis, V. Polyphenolic profile and antioxidant activity of five apple cultivars grown under organic and conventional agricultural practices. *Int. J. Food Sci. Technol.* **2009**, *44*, 1167–1175. [CrossRef]
102. Piccolo, E.L.; Landi, M.; Massai, R.; Remorini, D.; Conte, G.; Guidi, L. Ancient apple cultivars from Garfagnana (Tuscany, Italy): A potential source for 'nutrafruit' production. *Food Chem.* **2019**, *294*, 518–525. [CrossRef]
103. Kschonsek, J.; Wiegand, C.; Hipler, U.-C.; Böhm, V. Influence of polyphenolic content on the in vitro allergenicity of old and new apple cultivars: A pilot study. *Nutrition* **2019**, *58*, 30–35. [CrossRef]
104. Gosch, C.; Halbwirth, H.; Kuhn, J.; Miosic, S.; Stich, K. Biosynthesis of phloridzin in apple (*Malus domestica* Borkh.). *Plant Sci.* **2009**, *176*, 223–231. [CrossRef]
105. Gosch, C.; Halbwirth, H.; Stich, K. Phloridzin: Biosynthesis, distribution and physiological relevance in plants. *Phytochemistry* **2010**, *71*, 838–843. [CrossRef]
106. Kobori, M.; Masumoto, S.; Akimoto, Y.; Oike, H. Phloridzin reduces blood glucose levels and alters hepatic gene expression in normal BALB/c mice. *Food Chem. Toxicol.* **2012**, *50*, 2547–2553. [CrossRef] [PubMed]
107. Mei, X.; Zhang, X.; Wang, Z.; Gao, Z.; Liu, G.; Hu, H.; Zou, L.; Li, X. Insulin sensitivity-enhancing activity of phlorizin is associated with lipopolysaccharide decrease and gut microbiota changes in obese and type 2 diabetes (db/db) mice. *J. Agric. Food Chem.* **2016**, *64*, 7502–7511. [CrossRef] [PubMed]
108. Lanzerstorfer, P.; Wruss, J.; Huemer, S.; Steininger, A.; Müller, U.; Himmelsbach, M.; Borgmann, D.; Winkler, S.; Höglinger, O.; Weghuber, J. Bioanalytical characterization of apple juice from 88 grafted and nongrafted apple varieties grown in upper Austria. *J. Agric. Food Chem.* **2014**, *62*, 1047–1056. [CrossRef]
109. Paris, R.; Pagliarani, G.; Savazzini, F.; Aloisi, I.; Iorio, R.A.; Tartarini, S.; Ricci, G.; Del Duca, S. Comparative analysis of allergen genes and pro-inflammatory factors in pollen and fruit of apple varieties. *Plant Sci.* **2017**, *264*, 57–68. [CrossRef]

110. Vegro, M.; Eccher, G.; Populin, F.; Sorgato, C.; Savazzini, F.; Pagliarani, G.; Tartarini, S.; Pasini, G.; Curioni, A.; Antico, A.; et al. Old apple (*Malus domestica* L. Borkh) varieties with hypoallergenic properties: An integrated approach for studying apple allergenicity. *J. Agric. Food Chem.* **2016**, *64*, 9224–9236. [CrossRef] [PubMed]
111. Barreira, J.C.M.; Arraibi, A.A.; Ferreira, I.C.F.R. Bioactive and functional compounds in apple pomace from juice and cider manufacturing: Potential use in dermal formulations. *Trends Food Sci. Technol.* **2019**, *90*, 76–87. [CrossRef]
112. Ikumi, Y.; Kida, T.; Sakuma, S.; Yamashita, S.; Akashi, M. Polymer-phloridzin conjugates as an anti-diabetic drug that inhibits glucose absorption through the Na^+/glucose cotransporter (SGLT1) in the small intestine. *J. Control. Release* **2008**, *125*, 42–49. [CrossRef]
113. Hedrén, E.; Diaz, V.; Svanberg, U. Estimation of carotenoid accessibility from carrots determined by an in vitro digestion method. *Eur. J. Clin. Nutr.* **2002**, *56*, 425–430. [CrossRef]
114. Parada, J.; Aguilera, J.M. Food microstructure affects the bioavailability of several nutrients. *J. Food Sci.* **2007**, *72*, R21–R32. [CrossRef]
115. Wruss, J.; Lanzerstorfer, P.; Huemer, S.; Himmelsbach, M.; Mangge, H.; Höglinger, O.; Weghuber, D.; Weghuber, J. Differences in pharmacokinetics of apple polyphenols after standardized oral consumption of unprocessed apple juice. *Nutr. J.* **2015**, *14*. [CrossRef]
116. Hervert-Hernández, D.; Goñi, I. Dietary polyphenols and human gut microbiota: A review. *Food Rev. Int.* **2011**, *27*, 154–169. [CrossRef]
117. Jakobek, L. Interactions of polyphenols with carbohydrates, lipids and proteins. *Food Chem.* **2015**, *175*, 556–567. [CrossRef] [PubMed]
118. Barber, T.M.; Kabisch, S.; Pfeiffer, A.F.H.; Weickert, M.O. The health benefits of dietary fibre. *Nutrients* **2020**, *12*, 3209. [CrossRef]
119. Knekt, P.; Kumpulainen, J.; Järvinen, R.; Rissanen, H.; Heliövaara, M.; Reunanen, A.; Hakulinen, T.; Aromaa, A. Flavonoid intake and risk of chronic diseases. *Am. J. Clin. Nutr.* **2002**, *76*, 560–568. [CrossRef]
120. Arts, I.C.W.; Jacobs, D.R.; Harnack, L.J.; Gross, M.; Folsom, A.R. Dietary catechins in relation to coronary heart disease death among postmenopausal women. *Epidemiology* **2001**, *12*, 668–675. [CrossRef]
121. Threapleton, D.E.; Greenwood, D.C.; Evans, C.E.L.; Cleghorn, C.L.; Nykjaer, C.; Woodhead, C.; Cade, J.E.; Gale, C.P.; Burley, V.J. Dietary fibre intake and risk of cardiovascular disease: Systematic review and meta-analysis. *BMJ* **2013**, *347*, f6879. [CrossRef]
122. Bondonno, C.P.; Yang, X.; Croft, K.D.; Considine, M.J.; Ward, N.C.; Rich, L.; Puddey, I.B.; Swinny, E.; Mubarak, A.; Hodgson, J.M. Flavonoid-rich apples and nitrate-rich spinach augment nitric oxide status and improve endothelial function in healthy men and women: A randomized controlled trial. *Free Radic. Biol. Med.* **2012**, *52*, 95–102. [CrossRef] [PubMed]
123. Jensen, E.N.; Buch-Andersen, T.; Ravn-Haren, G.; Dragsted, L.O. Mini-review: The effects of apples on plasma cholesterol levels and cardiovascular risk—A review of the evidence. *J. Hortic. Sci. Biotechnol.* **2009**, *84*, 34–41. [CrossRef]
124. Chai, S.C.; Hooshmand, S.; Saadat, R.L.; Payton, M.E.; Brummel-Smith, K.; Arjmandi, B.H. Daily apple versus dried plum: Impact on cardiovascular disease risk factors in postmenopausal women. *J. Acad. Nutr. Diet.* **2012**, *112*, 1158–1168. [CrossRef]
125. Nagasako-Akazome, Y.; Kanda, T.; Ikeda, M.; Shimasaki, H. Serum cholesterol-lowering effect of apple polyphenols in healthy subjects. *J. Oleo Sci.* **2005**, *54*, 143–151. [CrossRef]
126. Yao, N.; He, R.-R.; Zeng, X.-H.; Huang, X.-J.; Du, T.-L.; Cui, J.-C.; Hiroshi, K. Hypotriglyceridemic effects of apple polyphenols extract via up-regulation of lipoprotein lipase in triton WR-1339-induced mice. *Chin. J. Integr. Med.* **2012**, *20*, 31–35. [CrossRef] [PubMed]
127. Brouns, F.; Theuwissen, E.; Adam, A.; Bell, M.; Berger, A.; Mensink, R.P. Cholesterol-lowering properties of different pectin types in mildly hyper-cholesterolemic men and women. *Eur. J. Clin. Nutr.* **2011**, *66*, 591–599. [CrossRef] [PubMed]
128. González, R.; Ballester, I.; López-Posadas, R.; Suárez, M.D.; Zarzuelo, A.; Martínez-Augustin, O.; Medina, F.S.D. Effects of flavonoids and other polyphenols on inflammation. *Crit. Rev. Food Sci. Nutr.* **2011**, *51*, 331–362. [CrossRef] [PubMed]
129. Sproston, N.R.; Ashworth, J.J. Role of C-reactive protein at sites of inflammation and infection. *Front. Immunol.* **2018**, *9*, 754. [CrossRef]
130. Shaw, J.E.; Sicree, R.A.; Zimmet, P.Z. Global estimates of the prevalence of diabetes for 2010 and 2030. *Diabetes Res. Clin. Pr.* **2010**, *87*, 4–14. [CrossRef]
131. Song, Y.; Manson, J.E.; Buring, J.E.; Sesso, H.D.; Liu, S. Associations of dietary flavonoids with risk of type 2 diabetes, and markers of insulin resistance and systemic inflammation in women: A prospective study and cross-sectional analysis. *J. Am. Coll. Nutr.* **2005**, *24*, 376–384. [CrossRef]
132. Jacques, P.F.; Cassidy, A.; Rogers, G.; Peterson, J.J.; Meigs, J.B.; Dwyer, J.T. Higher dietary flavonol intake is associated with lower incidence of type 2 diabetes. *J. Nutr.* **2013**, *143*, 1474–1480. [CrossRef] [PubMed]
133. Shinohara, K.; Ohashi, Y.; Kawasumi, K.; Terada, A.; Fujisawa, T. Effect of apple intake on fecal microbiota and metabolites in humans. *Anaerobe* **2010**, *16*, 510–515. [CrossRef]
134. Pickard, J.M.; Zeng, M.Y.; Caruso, R.; Núñez, G. Gut microbiota: Role in pathogen colonization, immune responses, and inflammatory disease. *Immunol. Rev.* **2017**, *279*, 70–89. [CrossRef]
135. Etxeberria, U.; Arias, N.; Boqué, N.; Macarulla, M.T.; Portillo, M.P.; Martínez, J.A.; Milagro, F.I. Reshaping faecal gut microbiota composition by the intake of trans-resveratrol and quercetin in high-fat sucrose diet-fed rats. *J. Nutr. Biochem.* **2015**, *26*, 651–660. [CrossRef]

136. Lee, J.-H.; Regmi, S.C.; Kim, J.-A.; Cho, M.H.; Yun, H.; Lee, C.-S.; Lee, J.; Payne, S.M. Apple flavonoid phloretin inhibits *Escherichia coli* o157:H7 biofilm formation and ameliorates colon inflammation in rats. *Infect. Immun.* **2011**, *79*, 4819–4827. [CrossRef] [PubMed]
137. Wang, H.; Cheng, J.; Wang, H.; Wang, M.; Zhao, J.; Wu, Z. Protective effect of apple phlorizin on hydrogen peroxide-induced cell damage in HepG2 cells. *J. Food Biochem.* **2019**, *43*. [CrossRef] [PubMed]
138. Granato, D.; Barba, F.J.; Bursać Kovačević, D.; Lorenzo, J.M.; Cruz, A.G.; Putnik, P. Functional foods: Product development, technological trends, efficacy testing, and safety. *Annu. Rev. Food Sci. Technol.* **2020**, *11*, 93–118. [CrossRef]
139. Assmann, G.; Buono, P.; Daniele, A.; Della Valle, E.; Farinaro, E.; Ferns, G.; Krogh, V.; Kromhout, D.; Masana, L.; Merino, J.; et al. Functional foods and cardiometabolic diseases. *Nutr. Metab. Cardiovasc. Dis.* **2014**, *24*, 1272–1300. [CrossRef]
140. Brown, L.; Caligiuri, S.P.B.; Brown, D.; Pierce, G.N. Clinical trials using functional foods provide unique challenges. *J. Funct. Foods* **2018**, *45*, 233–238. [CrossRef]
141. Cassidy, Y.M.; McSorley, E.M.; Allsopp, P.J. Effect of soluble dietary fibre on postprandial blood glucose response and its potential as a functional food ingredient. *J. Funct. Foods* **2018**, *46*, 423–439. [CrossRef]
142. Mak, K.-K.; Tan, J.-J.; Marappan, P.; Balijepalli, M.K.; Choudhury, H.; Ramamurthy, S.; Pichika, M.R. Galangin's potential as a functional food ingredient. *J. Funct. Foods* **2018**, *46*, 490–503. [CrossRef]
143. Lenssen, K.G.M.; Bast, A.; de Boer, A. Clarifying the health claim assessment procedure of EFSA will benefit functional food innovation. *J. Funct. Foods* **2018**, *47*, 386–396. [CrossRef]
144. Chibisov, S.; Singh, M.; Singh, R.B.; Halabi, G.; Horiuchi, R.; Takahashi, T. Functional Food Security for Pre-Vention of Obesity and Metabolic Syndrome. In *The Role of Functional Food Security in Global Health*; Singh, R.B., Watson, R.R., Takahashi, T., Eds.; Academic Press: Cambridge, MA, USA, 2019; pp. 145–156.
145. Rao, S.S.; Singh, R.B.; Takahashi, T.; Juneja, L.R.; Fedacko, J.; Shewale, A.R. Economic Burden of Noncommu- Nicable Diseases and Economic Cost of Functional Foods for Prevention. In *The Role of Functional Food Security in Global Health*; Singh, R.B., Watson, R.R., Takahashi, T., Eds.; Academic Press: Cambridge, MA, USA, 2019; pp. 57–68.
146. De Souza, C.C.; Oliveira, C.A.; Pires, J.F.; Pimentel, T.C.; Raices, R.S.L.; Nogueira, L.C. Physicochemical characteristics and sensory acceptance of a mixed beverage based on organic apple juice and cardamom tea (*Elettaria cardamomum*) with allegation of functional properties. *Food Sci. Technol.* **2020**, *40*, 669–676. [CrossRef]
147. Azari, M.; Shojaee-Aliabadi, S.; Hosseini, H.; Mirmoghtadaie, L.; Hosseini, S.M. Optimization of physical properties of new gluten-free cake based on apple pomace powder using starch and xanthan gum. *Food Sci. Technol. Int.* **2020**, *26*, 603–613. [CrossRef]
148. Espinosa-Solis, V.; Zamudio-Flores, P.B.; Tirado-Gallegos, J.M.; Ramírez-Mancinas, S.; Olivas-Orozco, G.I.; Espino-Díaz, M.; Hernández-González, M.; García-Cano, V.G.; Sánchez-Ortíz, O.; Buenrostro-Figueroa, J.J.; et al. Evaluation of cooking quality, nutritional and texture characteristics of pasta added with oat bran and apple flour. *Foods* **2019**, *8*, 299. [CrossRef]
149. Fernandes, P.A.R.; Ferreira, S.S.; Bastos, R.; Ferreira, I.; Cruz, M.T.; Pinto, A.; Coelho, E.; Passos, C.P.; Coimbra, M.A.; Cardoso, S.M.; et al. Apple pomace extract as a sustainable food ingredient. *Antioxidants* **2019**, *8*, 189. [CrossRef]
150. De la Peña Armada, R.; Villanueva-Suárez, M.J.; Mateos-Aparicio, I. High hydrostatic pressure processing enhances pectin solubilisation on apple by-product improving techno-functional properties. *Eur. Food Res. Technol.* **2020**, *246*, 1691–1702. [CrossRef]
151. Ribas-Agustí, A.; Martín-Belloso, O.; Soliva-Fortuny, R.; Elez-Martínez, P. Enhancing hydroxycinnamic acids and flavan-3-ol contents by pulsed electric fields without affecting quality attributes of apple. *Food Res. Int.* **2019**, *121*, 433–440. [CrossRef]

foods

MDPI

Review

Fate of Residual Pesticides in Fruit and Vegetable Waste (FVW) Processing

Tri Thanh Nguyen [1], Carmen Rosello [2,3], Richard Bélanger [4] and Cristina Ratti [1,*]

1 Soils and Agri-Food Engineering Dept, Institute of Nutrition and Functional Foods, Université Laval, Quebec City, QC G1V 0A6, Canada; thanh-tri.nguyen.1@ulaval.ca
2 Chemical Engineering Group, Chemistry Department, Universitat des Iles Balears, Palma, 07122 Mallorca, Spain; carmen.rossello@uib.es
3 Soils and Agri-Food Engineering Dept, Université Laval, Quebec City, QC G1V 0A6, Canada
4 Plant Science Dept, Université Laval, Quebec City, QC G1V 0A6, Canada; Richard.Belanger@fsaa.ulaval.ca
* Correspondence: Cristina.Ratti@fsaa.ulaval.ca

Received: 16 September 2020; Accepted: 12 October 2020; Published: 15 October 2020

Abstract: Plants need to be protected against pests and diseases, so as to assure an adequate production, and therefore to contribute to food security. However, some of the used pesticides are harmful compounds, and thus the right balance between the need to increase food production with the need to ensure the safety of people, food and the environment must be struck. In particular, when dealing with fruit and vegetable wastes, their content in agrochemicals should be monitored, especially in peel and skins, and eventually minimized before or during further processing to separate or concentrate bioactive compounds from it. The general objective of this review is to investigate initial levels of pesticide residues and their potential reduction through further processing for some of the most contaminated fruit and vegetable wastes. Focus will be placed on extraction and drying processes being amid the main processing steps used in the recuperation of bioactive compounds from fruit and vegetable wastes.

Keywords: pesticide; fruit wastes; vegetable wastes; drying; extraction; intensification technologies

1. Introduction

Fruits and vegetables are key elements of a healthy human diet, owing to their high proportion of fibers, vitamins and minerals. In 2010, 6.7 million deaths worldwide were attributed to a low intake of fruits and vegetables causing poor health and a higher risk to develop non-communicable illnesses [1]. In agreement with global trends, fruit and vegetable production rose in 2009 to above 500 and 850 MMT, respectively, and their waste due to primary production, to 40 and 70 MMT [2]. Fruit and vegetable waste arise mainly from tricky processing and inadequate handling of the produce. At the same time, a Swedish study [3] determined that, among the diverse types of produce, apples, tomatoes, peppers and grapes are the ones generating almost 50% of the wastage at the supermarket level. From all this wasted biomass, the development of interesting by-products with applications in food, cosmetic and pharmaceutical industries could be a promising pathway to reach a resource-efficient circular economy [4].

Currently, pesticides are commonly employed to ensure successful fruit and vegetable production. However, their often large-spectrum biocide activity and potential risk to the consumer represent a growing source of concern for the general population and environment [5]. Over the past two decades, many of the most toxic pesticides have been withdrawn from agricultural and/or household practices. Yet others, such as organophosphate insecticides, are still applied to certain crops [6].

A great deal of attention nowadays surrounds the so-called 'dirty dozen', a list of 12 fruits and vegetables with the highest concentration of pesticides. Strawberry, apple, grape, tomato and potato figure prominently among them [7]. Handling after harvesting may markedly decrease the pesticide residues in most fruits and vegetables for human consumption, as a result of the peeling and washing processes [8]. On the other hand, the non-edible parts of fruits and vegetables after processing constitute around 10 to 60% of the total weight of the product, and are composed of peel, skin, seeds, sheaths, etc. [9]. Skin and peel are the main constituents of these wastes, representing more than 50% [2]. Thus, the content of agrochemicals in waste of fruits and vegetables should be monitored, especially if originating from peel and skin, and eventually minimized before or during further processing, targeting the separation or concentration of bioactive compounds from it. It is important to point out that health problems due to pesticide intake are not only related to the toxicity level of the agrochemicals but also to their concentration and exposure time.

The general objective of this work is to review the initial levels of pesticide residues in fruits and vegetables and their potential reduction/increase through further processing for some of the 'dirty dozen' fruit and vegetable wastes. Focus will be placed on extraction and drying processes being among the main processing steps used in the recuperation of bioactive compounds from such wastes.

2. Bibliographic Research Methodology

A literature search on pesticide residues in fruit and vegetable wastes (FVW) and the effect of extraction and drying processes on reducing/increasing pesticide residues was carried out on the ScienceDirect, PubChem, and Google Scholar database. The combination of 'keywords' used for the search includes 'pesticide residues', 'fruit waste', 'vegetable waste' AND 'processing', 'pesticide residues' AND 'extraction' OR 'drying, AND 'pesticide residues' AND ('PEF' OR 'Ultrasound' OR 'Microwaves'). The literature reference sections of the retrieved articles were used to find more studies that might have been missed out during the literature search.

3. Pesticides in Fruits and Vegetables

3.1. Classification and Properties

"Pesticide" is a term for all insecticides, herbicides, fungicides, rodenticides, wood preservatives, garden chemicals and household disinfectants that may be used to kill some pests. Pesticides may be classified based on several parameters, depending on the needs. In any case, the three more popular pesticide classifications are based on (i) the mode of entry to the plant, (ii) the pesticide function and the pest organism they kill, and (iii) the chemical composition of the pesticide [10].

Based on their chemical composition, pesticides can be classified in the groups of organochlorines, organophosphates, carbamates and pyrethroids. Tables 1–4 provide information on the previous four groups of pesticides, including health and environmental hazards, chemical formulas and the main fundamental properties of selected common pesticide compounds.

Table 1. Some examples of organochlorines pesticides, which are organic compounds with five or more chlorine atoms [10–13].

Examples	Physicochemical Characteristics	Health and Environment Risks
Dichlorodiphenyltrichloroethane (DDT)	Melting point: 108.5 °C Solubility in water: 25×10^{-3} mg/L (25 °C) Solubility in ethanol (20×10^{3} mg/L) Solubility in ether (280×10^{3} mg/L) Vapor pressure (20 °C): 2.53×10^{-5} Pa	Probable carcinogen Reproductive effect Liver and kidney problem Eye, nose, skin, throat irritant

Table 1. *Cont.*

Examples	Physicochemical Characteristics	Health and Environment Risks
Captan	Melting point: 178 °C Soluble in water: 3.3 mg/L (25 °C), in acetone: 21 g/L, chloroform: 70 g/L, cyclohexanone: 23 g/L, and in isopropanol: 1.7 g/L. Vapor pressure (20 °C): 13.3 × 10^{-5} Pa.	Probable carcinogen, allergen Induces hyperplasia of the crypt cells Potent eye irritant Mild skin irritant
Lindane	Melting point: 112.5 °C Solubility in water: insoluble. Moderately soluble in ethanol, ether, benzene acetone. Vapor pressure (20 °C): 125.32 × 10^{-5} Pa	Suspected carcinogen Affects central nervous system, and respiratory, reproductive systems.
Endosulfan	Melting point: 70 to 100 °C Solubility in water (22 °C): 0.33 mg/L. Vapor pressure (25 °C): 133.32 × 10^{-5} Pa	Causes DNA damage Potential correlation between endosulfan and leukemia.
Aldrin	Melting point: 104 °C Solubility in water: slightly soluble (0.003%) Vapor Pressure (20 °C): 100 × 10^{-4} Pa	Causes problems with the central nervous system (the brain and spinal cord), and the liver. Eye, skin and mucous membrane irritants.
Dieldrin	Melting point: 176 to 177 °C Solubility in water: 0.186 mg/L at 25–29 °C Vapor Pressure (20 °C): 2.37 × 10^{-5} Pa	Causes problems with the central nervous system (the brain and spinal cord) and the liver. Eye, skin and mucous membrane irritants.
Chlordane	Melting point: 102 to 106 °C Solubility in water: 1 × 10^{-3} mg/L (20 °C) Vapor Pressure (25 °C): 133.32 × 10^{-5} Pa	Chlordane interacts with the human erythrocyte membrane and change its morphology.

Table 2. Some examples of organophosphates pesticides, which are esters of phosphoric acid, containing a phosphate group as their basic structural framework [10,13–15].

Examples	Physicochemical Characteristics	Health and Environment Risks
Parathion	Melting point: 6 °C Solubility in water: 6.54 mg/L at 24 °C. High solubility in xylene and butanol. Vapor Pressure (20 °C): 503.94 × 10^{-5} Pa	Depressed red blood cell cholinesterase activity, nausea, and headache. Affects central nervous system, blood, respiratory systems, eyes and skin.
Methyl parathion	Melting point: 37 °C.Solubility in water: 37.7 mg/L (20 °C)Vapor Pressure (25 °C): 46.7 × 10^{-5} Pa	Causes neuropsychiatric disorders in humans after chronic exposure as well as hematological and ocular alterations. Reduces cholinesterase levels in the brain, erythrocytes, and plasma.
Malathion	Melting point: 156 to 157 °C Solubility in water: 145 mg/L at 20 °C. Soluble in ethanol and acetone; very soluble in ethyl ether. Vapor Pressure (25 °C): 23.73 × 10^{-3} Pa	Inhibition of Acetylcholinesterase activity Affects central nervous system, respiratory systems. Eye, nose, skin irritant
Diazinon	Flash point: 82 °C. Solubility in water: 40 mg/L Vapor Pressure (20 °C): 111.99 × 10^{-4} Pa	Eye and skin irritant Causes gastrointestinal symptoms.
Glyphosate	Melting point: 184.5 °C Solubility in water: 12 × 10^{3} mg/L (25 °C) Vapor Pressure (25 °C): <1 × 10^{-5} Pa	Probable carcinogen Eye and skin irritant

Table 3. Some examples of carbamates pesticides, which are derived from carbamic acid (NH_2COOH) [10,13,16–18].

Examples	Physicochemical Characteristics	Health and Environment Risks
Carbaryl	Melting point: 142 °C Solubility in water: 0.1 mg/L Soluble in most popular organic solvents: dimethylformamide, dimethyl sulfoxide, acetone, cyclohexanone. Vapor Pressure (25 °C): 18.13 × 10^{-5} Pa	Inhibit progesterone biosynthesis of primary human granulose-lutein cells.
Propiconazole	Boiling point: 180 °C Soluble in water: 100 mg/L (25 °C), in hexane: 47 g/L, and in most organic solvents. Completely miscible with ethanol, and acetone. Vapor pressure (25 °C): 13.3 × 10^{-5} Pa.	Skin irritant Liver toxicity and central nervous system effects Adverse changes in erythrocytes
Carbofuran	Melting point: 151 °C Solubility in water: 320 mg/L Highly soluble in N-methyl-2-pyrrolidone, dimethylformamide, dimethyl sulfoxide, acetone, acetonitrile, methylene, chloride, cyclohexanone, benzene, and xylene. Vapor Pressure (33 °C): 2.7 × 10^{-3} Pa	Inhibit acetylcholinesterase. Toxicity affects to vision, growth and predator avoidance skills of fish early life stages.
Propoxur	Melting point: 86 to 92 °C Solubility in water: 1.75 × 10^3 mg/L (20 °C). Vapor Pressure (20 °C): 39.99 × 10^{-5} Pa	Inhibit of Acetylcholinesterase activity. Probable carcinogen after long-term oral or inhalation exposure.
Aminocarb	Melting point: 93 °C Solubility in water: 0.9 × 10^3 mg/L. Soluble in polar organic solvents. Moderately soluble in aromatic solvents. Vapor Pressure (20 °C): 2.3 × 10^{-3} Pa	Reduces in immune responsiveness in exposed animals Decrease in activity of acetylcholinesterase. degenerative changes in both liver and kidney

Table 4. Some examples of pyrethrins and pyrethroids pesticides, which are synthesized by duplicating the structure of natural pyrethrins, components of pyrethrum flowers are the optically active esters derived from (+)-*trans*-chrysanthemic acid and (+)-*trans*-pyrethroic acid [10,13,19,20].

Examples	Physicochemical Characteristics	Health and Environment Risks
Permethrin	Melting point: 34 °C Solubility in water: 5.5 × 10^{-3} mg/L Vapor Pressure (20 °C): 2.87 × 10^{-6} Pa	Eye, skin, and respiratory irritant. Affects central nervous system.
Cypermethrin	Melting point: 81.3 °C Solubility in water (20 °C): 4 × 10^{-3} mg/L Soluble in ethanol: 337 × 10^3 mg/L, hexane 103 × 10^3 mg/L Vapor Pressure (20 °C): 2.27 × 10^{-7} Pa	Causes DNA damage and oxidative stress in gill cells of fish.
Deltamethrin	Melting point: 98 °C Solubility in water: Insoluble Vapor Pressure (25 °C): 2.0 × 10^{-6} Pa	Causes neurotoxicity and liver dysfunction accompanied by elevated reactive oxygen species (ROS) levels.

Table 1 shows the most common examples of pesticides within the organochlorine group, which are organic compounds with five or more chlorines atoms attached. They are widely used as insecticides, such as dichlorodiphenyltrichloroethane (DDT), that is effectively used for the control of malaria in many tropical developing countries [21]. However, owing to the nature of their characteristics (volatile, low polarity, low aqueous solubility, and high lipid solubility), these pesticides have a long-term residual permanence in the environment after application. Moreover, their bioaccumulation and toxicity characteristics may cause hypertension, cardiovascular disorders and other health-related problems in humans, resulting in their ban in many developed countries [22].

Another group of pesticides, the organophosphates (Table 2), includes organic compounds that contain phosphodiester bond in their basic structure. As a result, they easily decompose when applied on plants, and soil, causing reduced environmental pollution. Their activity is mainly directed toward the inhibition of acetylcholinesterase, which controls the functions of the nervous system [23]. The most common examples of organophosphate pesticides shown in Table 2, have higher water solubility than those in the organochlorines group, but they are also more soluble in organic solvents.

Most pesticides belonging to the carbamate group (Table 3) are highly soluble in common organic solvents. Their activity is similar to those of organophosphate pesticides, as they also inhibit the enzyme acetylcholinesterase [23]. Pyrethrin and pyrethroid (Table 4) have low water solubility, while others such as deltamethrin are not water soluble. Nevertheless, they easily decompose when exposed to light, and are only slightly toxic to mammals and birds, so they are generally considered as the safest insecticides for use in food consumption [10].

Pesticide physicochemical properties shown in Tables 1–4, such as solubility and vapor pressure, lead to differences in pesticides plant uptake, environmental distribution, as well as their elimination during fruit and vegetable harvesting and processing. For example, solubility of pesticides in water or organic solvents, plays a key role on their ability to be dissolved in solvents with different polarities during extraction of valuable compounds [13]. Moreover, pesticides with higher vapor pressure are more likely to volatilize i.e., during drying as water evaporates, while low vapor pressure pesticides tend to accumulate in liquid phases, soil or biota. In a study by Sood et al. (2004) [24], the percentage of dimethoate pesticide residue left after the drying step during green tea manufacture was the lowest (23.4%) related to its higher vapor pressure. For pesticides with low water solubility, high vapor pressure contributes markedly in decreasing their content, such as 19% loss of tridemorph residue during black tea drying, which was higher than hexaconazole, propiconazole, and carbendazim residues, less than 7% [25].

3.2. *Toxicity and Maximal Allowed Concentration*

Pesticides are reported to have an impact on human health and have been linked to illnesses, ranging from acute ailments to chronic diseases, such as cancer, reproductive disorders, and endocrine-system dysfunctions [26]. Tables 1–4 summarize some health and environment issues for some selected pesticides. For these reasons, it is widely agreed that the use of pesticides should be carefully monitored to prevent negative effects on health, ground water sources and the environment.

Maximum residue limits (MRLs) (expressed in $\mu g\ kg^{-1}$) are the highest levels of residues expected to be found in food products when the pesticide is used in accordance with its label [27]. The MRLs are systematically set far below levels considered to be unsafe for humans, meaning that food residues containing higher levels than the MRL are not necessarily unsafe for consumption [28].

MRLs were established and recommended by the World Health Organization (WHO) and the Food and Agriculture Organization (FAO). MRLs are also subject to specific legal requirements in most countries, such as those set by the Pest Management Regulatory Agency (PMRA) in Canada, the Food and Drug Administration (FDA) in the United States, and the European Commission (EC) in Europe.

Table 5 shows a comparison of MRLs (pesticides mentioned in Tables 1–4 for fruits and vegetables popular in Canada, i.e., apples, potatoes, tomatoes and strawberries), established by the PMRA, FDA, and EC. Generally, MRLs set by the EC are lower than in North America, where a lower tolerance for pesticide residue limits in fruits and vegetables is applied, i.e., MRLs of cypermethrin on strawberries was set by Canada to be twice as high as the EC, while in the US, MRLs of carbaryl on apples are 1200 times higher than those in Europe. As well, MRLs for malathion and diazinon are 25 to 400 times higher in North America than in Europe. However, in other cases, MRLs from the US and EC are comparable, such as those for DDT and glyphosate in tomato, or permethrin in potato. In rarer instances, lower MRLs are imposed in the US than in Europe, such as in the case of glyphosate in potato and strawberry, cypermethrin in tomato, and deltamethrin in potato.

However, sometimes pesticides that are banned in Canada or the US, are still permitted in Europe. For example, in the case of organochlorines, lindane was banned in Canada from December 2004 [29] because of its toxicity and persistence in the food chain, together with aldrin, dieldrin, and chlordane, while the EC still allows them, albeit in very low MRLs, sometimes three to ten times lower than the accepted values by the FDA, depending on the type of fruits and vegetables. For organophosphates, parathion and methyl parathion are forbidden in Canada and not applied in United States on fruits and vegetables mentioned in Table 5, because of increasing concerns regarding hazards to wildlife and human health, while still being accepted in Europe. Aminocarb in the carbamates group is not allowed in the listed fruits and vegetables of Canada, US and Europe, because of its toxicity for human health and environment. For carbaryl residues, MRLs set by the EC are very low, from 20 to 1200 times lower than US and Canada. Similar situations are described in Table 5 for pyrethrins and members of the pyrethroid group.

3.3. *Fruits and Vegetables with the Highest Presence of Pesticides*

The list of twelve fruits and vegetables with the highest amounts of pesticide residues (named "dirty dozen") is annually published by the Environmental Working Group (EWG), a nonprofit organization. In 2019, the "dirty dozen" ranking was composed, in order of importance by: strawberry, spinach, kale, nectarine, apple, grape, peach, cherry, pear, tomato, celery, and potato. These products were found to have higher levels of pesticides than all other ones over the year [7].

From the data obtained by the United Stated Department of Agriculture (USDA) for their Pesticide Data Program in recent years, strawberry may contain as many as 45 different types of residues. Other fruits and vegetables also present a high number of pesticide residues, such as apples (47), grapes (56), cherries (42), tomatoes (35), potatoes (35), sweet bell peppers (53), etc. Among them, tetrahydrophthalimide (THPI), a metabolite from the non-systemic fungicide-captan, was found in 55% of strawberry samples, while permethrin, an insecticide of the pyrethroid family, dominated in 52% of spinach samples; formetanate hydrochloride, a carbamate pesticide that inhibit cholinesterase, in 53% of nectarines; diphenylamine (DPA), an aromatic amine used as a scald inhibitor for apples, was found in 83% of samples; imidacloprid, a systemic insecticide, in 48% of grapes; fludioxonil, a non-systemic fungicide, in 48% of peaches; boscalid, a non-systemic fungicide, in 65% of cherries; pyrimethanil, an anilinopyrimidine class of fungicides, in 40% of pears; endosulfan, an organochlorine insecticide and acaricide, in 17% of tomatoes; and chlorpropham, a carbamate herbicide, in 80% of potato samples [30].

Table 5. Maximum residue limits (MRLs) for fruits and vegetables in Europe, the United States and Canada.

Types of Pesticide	Examples	MRLs (μg kg^{-1})											
		European Commission [1]				US-FDA [2]				PCPA Canada [3]			
		Apple	Potato	Tomato	Strawberry	Apple	Potato	Tomato	Strawberry	Apple	Potato	Tomato	Strawberry
Organochlorines	DDT	50	50	50	500	100	—	50	100	For fresh vegetable: 500			
	Captan	10^4	30	100	100	25×10^3	50	50	2×10^4	5000	—	5000	5000
	Lindane	10	10	10	10	—	500	—	500	Banned			
	Endosulfan	50	50	50	100	—	—	—	—	2000	—	1000	1000
	Aldrin	10	10	10	10	30	100	50	50	—	—	—	—
	Dieldrin	10	10	10	10	30	100	50	50	—	—	—	—
	Chlordane	10	10	10	20	100	100	100	100	—	—	—	—
Organophosphates	Parathion	50	50	50	100	—	—	—	—	Banned			
	Methyl parathion	10	10	10	50	—	—	—	—	Banned			
	Malathion	20	20	20	20	8000	8000	8000	8000	2000	500	3000	8000
	Diazinon	10	10	10	50	500	100	750	500	750		750	750
	Glyphosate	100	500	100	2000	200	200	100	200	—	—	—	—
Carbamates	Carbaryl	10	10	10	50	12,000	2000	5000	4000	5000	200	5000	7000
	Propiconazole	150	10	300	50	—	—	3000	1300	—	—	3000	1300
	Carbofuran	1	1	2	50	—	—	—	—	—	500		400
	Propoxur	50	50	50	100	—	—	—	—	Banned			
	Aminocarb	—	—	—	—	—	—	—	—	—	—	—	—
Pyrethrins and pyrethroids	Permethrin	50	50	50	100	50	50	2000		1000	50	500	—
	Cypermethrin	1000	50	500	100	—	—	—	—	1000	100	300	200
	Deltamethrin	200	300	70	—	200	40	200	—	400	40	300	200

Note: [1] EU Pesticides database. Retrieved from http://ec.europa.eu/food/plant/pesticides/eu-pesticides-database/public/?event=homepage&language=EN. [2] United States Department of Agriculture. Retrieved from https://www.fas.usda.gov/maximum-residue-limits-mrl-database. [3] MRLs for pesticides regulated under the Pest Control Products Act (PCPA). Retrieved from https://www.canada.ca/en/health-canada/services/consumer-product-safety/pesticides-pest-management/public/protecting-your-health-environment/pesticides-food/maximum-residue-limits-pesticides.html.

3.4. Pesticide Application and Physical Location in Fruits and Vegetables

Pesticides are mainly sprayed on fruits and vegetables and accumulate often on the outer peel or skin, the cuticle [31]. The pesticide could be adsorbed by the plant surface (waxy cuticle and root surfaces) and enter the plant transport system (systemic) to protect it from pests that penetrate the skin; other pesticides may stay on the surface of the plant (contact). While still on the surface of the crop, the pesticide is exposed to environmental factors such as wind and sun, and may be washed off during rainfall. As well, they can undergo volatilization, photolysis, or chemical and microbial degradation [32].

As mentioned above, pesticide residues commonly accumulate on the peel or skin. For instance, thiabendazole and ortho-phenyl-phenol was detected in harvested citrus fruit peels [33], residues of organochlorine pesticides (DDT and its derivatives, lindane, HCB) and organophosphorous pesticides (pirimiphos-methyl, dimethoate, malathion) were detected in potato skins [34]. Difenoconazole was found to be present in tomato skin [35]. In another study, Abou-Arab (1999) [36] reported that hexachlorobenzene, o.p-DDD, p.p-DDD, dimethoate and profenofos were present in the skin of tested tomato, where they were two to seven times higher than in their pulp. In the same study, organophosphate (dimethoate and profenofos) residues were also reported in the seeds of tomato. In the same manner, hexythiazox, a non-systemic acaricide, is applied on the surface of fruits by contact mode, and although it can be easily washed off, it is also absorbed in the pulp of treated strawberry [37].

3.5. Pesticide Analytical Determination

As mentioned previously, pesticides applied in fruits and vegetables are classified based on various criteria, such as mode of entry, mode of action or chemical composition and characteristics [10]. Accordingly, its residues contain not only their main compounds, but also their metabolites and/or degradation products, which have different physicochemical characteristics (vapor pressure, polarity, solubility). This multicompound presence results in difficult and complex methods to isolate pesticide residues in micro-quantities from fruit and vegetable matrices.

Pesticide residues in fruits and vegetables are analyzed through two steps: (a) extraction and clean-up of the target analytes from the matrix, and (b) determination of the target analytes [38]. For the first step, various techniques could be used, such as liquid-liquid extraction (LLE), solid phase extraction (SPE), solid phase micro extraction (SPME), and QuEChERS (quick, easy, cheap, effective, rugged, and safe) extraction.

Liquid-liquid extraction (LLE), also known as partitioning, is a separation process consisting of the transfer of a solute from one solvent to another, the two solvents being immiscible or partially miscible with each other [39]. Organic solvents such as acetonitrile, ethyl acetate, chloroform, hexane, 1,2-dichloromethane, etc. are usually used in LLE methods for the determination of pesticide residues in food and the environment, due to their good solubility in several immiscible liquids, such as in water and organic solvents. For instance, de Pinho et al. (2010) [40] used a mixture of acetonitrile and ethyl acetate (6.5 mL:1.5 mL) as the solvent for extraction of chlorpyrifos, λ-cyhalothrin, cypermethrin and deltamethrin in honey samples. Acetonitrile was also used as an extraction liquid for carbamates (aldicarb, carbofuran and carbaryl) in water samples [41].

Solid-phase extraction (SPE) is one of the most widely used packing column or cartridge extraction methods. Analytes are initially adsorbed onto suitable solids depending on their interaction. Then, a selective organic solvent is used to remove interferences, and then another solvent is selected to elute out the target analytes. Advantages of SPE methods are the reduction of solvents quantities, short concentration time, and improved yield recovery [42]. A study by Torreti et al. (1992) [43] analyzed 15 organochlorine pesticide residues from samples of animal feed, using a C18 SPE column as clean up procedure providing high recovery (70–100%). In another study, a multiresidues method for analysis of 90 pesticide residues with different physicochemical properties in fruits and vegetables was developed, where a polystyrene divinylbenzene column (LiChrolut EN) was used as an effective SPE method for clean-up and pre-concentration procedures of the pesticides from water-diluted acetone extracts [44].

In a recent study, a combination of graphitized carbon black and primary secondary amine (GCB/PSA) was used as SPE method for clean-up process, followed by the injection of fruit and vegetable extracted samples into the UHPLC-TOF/MS to analyze 60 targeted pesticides [45].

Solid phase microextraction (SPME) is a simple, low cost, easily automated and on-site sampling method when compared to SPE. It involves two processes: analytes are separated from the sample by the coating, and the desorption of concentrated analytes are analyzed by an analytical instrument [46]. Because of its advantages, particularly that of being solvent-free, SPME formed by a silica fiber coated with a polyacrylate (PA) film was used in clean-up procedures, followed by GC-MS, for the determination of organophosphate pesticides in wine and fruit juices [47], and of 14 pesticide residues (clofentezine, carbofuran, diazinon, methyl parathion, malathion, fenthion, thiabendazole, imazalil, bifenthrin, permethrin, prochloraz, pyraclostrobin, difenoconazole and azoxystrobin) in mango fruit [48].

The QuEChERS (Quick, easy, cheap, effective, rugged, and safe) sample preparation is a simple, fast, and inexpensive method, originally described by Anastassiades et al. (2003) [49], for the determination of pesticide residues in fruits and vegetables. The QuEChERS technique involves two steps: a liquid-liquid extraction and dispersive solid-phase extraction clean-up. The samples pre-treated using QuEChERS are clean enough to be analyzed using gas or liquid chromatography [50]. Due to the numerous advantages of this method, it was used by many researchers. In a recent study, QuEChERS process provided satisfactory results with high recovery (acceptable ranges) of 72 pesticides in carrot, corn, melon, rice, soy, silage, tobacco, cassava, lettuce and wheat [51], and 11 fungicides, three insecticides in strawberry by-products [52]. In another research on optimization of the clean-up step of QuEChERS method in coffee leaf extracts [53], it was possible to analyze 52 pesticides by LC-MS/MS. For this, the clean-up procedure of QuEChERS method was modified with different combinations of adsorbents, resulting in high recovery (>70%). Recently, the combination of modified QuEChERS method by adding of acetonitrile with 0.1% formic acid, followed by UHPLC-MS/MS determination, was applied by Lee et al. (2018) [54] for a multiresidue analysis of 310 pesticides in brown rice, orange, and spinach, which resulted in 87–89% of the pesticides at spiking level of 10 ng g^{-1} met the acceptability criteria of DG-SANTE guidelines (recovery 70–120%, and RSD ≤ 20%).

For the second step in pesticide analytical determination, i.e., the detection or analysis of target analytes (pesticides) in foods, numerous conventional analytical methods are used such as gas chromatography (GC), high performance liquid chromatography (HPLC), or more delicate including gas chromatography associated with mass spectrometry (GC-MS), liquid chromatography associated with mass spectrometry (LC-MS), and ultra-high performance liquid chromatography tandem mass spectrometry (UHPLC-MS/MS).

For volatile pesticides, which can be easily vaporized, GC is a popular separation method applied in several studies. It is usually coupled with specific detectors, including flame ionization detector (FID), as for the analysis of organophosphorus pesticides in onion, grape and apple juices [55], or pyrethroid pesticides in vegetable oils [56]. Electron capture detector (ECD) has been used as well in the determination of chlorpyrifos-methyl, fenitrothion, procymidone and vinclozolin on peach [57], and the flame photometric detector (FPD), to determine 11 organophosphorus pesticide residues on cabbage, kale and mustard samples [58]. Mass spectrometer detectors (MS and tandem MS) are also popular choices for pesticide determination, as MacLoughlin et al. (2018) [59] analyzed 35 commonly used pesticides by GC-MS, and 381 different types of pesticides in grapes were monitored by GC/MS-MS [60].

On the other hand, for high polarity and non-volatile extracted analytes, HPLC analytical techniques are preferably used as an effective separation method. It can be coupled with detectors such as UV in the case of analysis of pyrethroid residues in fruit and vegetable samples [61] or MS and tandem MS in the determination of malathion, diazinon, imidacloprid and triadimefon in fruit juices (apple, cherry, raspberry, orange and pineapple) [62].

In recent years, UHPLC started the use of smaller stationary-phase particle size (≤2 μm) than those used in classical LC (3–5 μm), for the detection of 21 pesticide residues in tomato and sweet pepper samples, coupled with tandem mass spectrometry (UHPLC-MS/MS) [63]. In another study, time-of-flight mass spectrometry (TOF-MS) was combined with UHPLC to detect 60 pesticides in 286 vegetable and fruit samples [45].

4. Fruit and Vegetable Waste (FVW)

4.1. Common Types of Fruit and Vegetable Wastes

Fruits and vegetables generate at least 10 to 60% of waste materials that are composed of leaves, roots, tubers, skin, pulp, seeds, peel, pomace, etc. [9]. Their percentage and type of waste vary from process to process, and fruit and vegetable types, such as sliced apples, generated 11% of seed and pulp as waste; papaya yielded about 9% of peel, 7% of seeds, 32% unusable pulp as wastage products, and mangos produced 14% of seeds, 11% of peels, 18% unusable pulp [2]; while in the case of apple juice processing, apple pomace as a discarded solid residue dominated 25–30% of the total processed fruits [64]; grape pomace, a wastage of wine production, mainly containing of seeds and peels, contained about 20% of the total weight [65]. Potato peel, a wastage of potato processing, can vary from 15 to 40%, depending on the procedure applied to remove the skin, i.e., steam, abrasion or lye peeling, while in tomato juice processing, only 3–7% of the raw material is lost as waste [66].

4.2. Potential Applications of Fruit and Vegetable Wastes

Since fruit and vegetable wastes are a source of dietary fiber and bioactive compounds such as phenolics or vitamins, their utilization received an increased attention recently for the application as functional ingredients in the food industry [67], or in other industrial applications such as pharmaceutical and/or nutraceuticals, healthcare and chemicals [2].

For instance, by-products extracted from potato peel waste [68], apple peel [69] and passion fruit peel [70] have high antioxidant activity. They can also be used as a base for fermentation reactions, and in healthy and functional food production as a dietary fiber source. Furthermore, in another study, the viscosity and emulsifying properties of pectin extracted from potato peel waste could be improved when treated with high pressure processing [71].

Apple pomace, a by-product of apple processing, has the potential to be incorporated as a natural stabilizer and texturizer in yogurt fermentation by increasing the gelation pH, shortening the fermentation time, and developing a firmer and more consistent yogurt gel during cold storage [72]. Moreover, apple pomace was incorporated into wheat flour as a fiber source to improve the rheological characteristics of cake batter [73]. The characterization of polyphenols, proantocyanidin profiles and antioxidant activity of pomace from different varieties of grapes [74,75] pointed out the possible exploitation of winemaking by-products as inexpensive and easily available sources of bioactive compounds. Different classes of polyphenols and ursolic acid were extracted from grape pomace as well [76], which was also used to enrich yogurt and salad dressing by increasing dietary fiber, total phenolic content, and delaying the lipid oxidation of samples during refrigeration storage [77]. In addition, grape seed flour can be used as an ingredient for cereal bars, pancakes and noodles products [78]. The availability and the potential of wine by-products, grape pomace and stems, obtained from ten different grape (*Vitis vinifera* L.) varieties as raw materials for the production of dietary fiber concentrates, were in addition evaluated with regard to their potential incorporation as dietary fiber concentrates into the food chain [79]. Minjares-Fuentes et al. [80,81] reported that an ultrasound-assisted procedure for the extraction of pectins from grape pomace could be a good option for the extraction of functional pectins and hemicellulosic polysaccharides from grape pomace at the industrial level.

Umaña et al. (2020) [82] investigated the revalorization of mushroom by-product (stalks of A. bisporus) by extracting its components (ergosterol and antioxidant components from mushroom by-products and the attainment of a β-glucan rich residue).

Zhang et al. (2018) [83] studied the effects of dynamic high-pressure micro-fluidization (DHPM) on the physicochemical properties and rheological properties of pectin extracted from black-cherry tomato waste (pomace). In other studies, tomato peel was reported to have amino acids and fatty acids, besides a high content of antioxidants such as flavonoids, phenolic acids, lycopene, ascorbic acid and minerals (Ca, Cu, Mn, Zn, and Se) [84–86].

Sójka et al. (2013) [87] reported that a by-product of strawberry juice production (strawberry press cake), consisting of 40% seeds, 4% sand, and about 55% exhausted strawberry flesh, is an important source of nutrient and polyphenolic composition, including proanthocyanidins, ellagitannins, and especially dimeric agrimoniin. Furthermore, the ingestion of extracts of industrial strawberry pomace showed beneficial health properties with positive changes in the population of intestinal microflora [88].

Moreover, pectin was also extracted from several fruit and vegetable waste sources, such as orange peel [89], melon peel [90], banana peel [91,92], peach pomace [93], potato peel [71], and berry fruit residues [94]. Pectin from berry residues were found to have high quality and purity parameters that make it suitable for commercialization, either as an additive in food or for the elaboration of medicine-related compounds [94].

In addition to valuable bio-functionalities, fruit and vegetable wastes also have increased traditional nutritional values [84–86]. For example, the waste of seven types of underground vegetables (beet, turnip, carrot, sweet potato, radish, potato, ginger) was found to contain vitamin C levels ranging from 44 to 123 mg/100 g, riboflavin from 0.3 to 0.8 mg/100 g, thiamin around 0.4 mg/100 g, and niacin from 0.2 to 1.6 mg/100 g, and a high content of calcium, sodium, magnesium, iron, manganese, zinc, potassium and phosphorus [95].

4.3. FVW by-Products Processing

Among the multiple potential applications shown for FVW, an increased special attention is recently being paid to the extraction of polyphenols and antioxidants from agri-food by-products and converting these extracts into stable powders.

Grape pomace or wine marc is one such agri-food processing waste, which is generated in the range of 5–9 million tons per year worldwide by the wine industry. A recent manuscript deals with possible uses for red wine processing waste, proposed to convert grape pomace into powder with a processing line, during which pomace was separated to be air-dried in an oven at temperatures 45–50 °C for 72 h to remove the remaining mixture of water and alcohol. Then, dried grape pomace was ground after removing seeds and stems and was used to mix with refined wheat (5 to 20%) in cookies to increase their polyphenol content [96]. Another application of grape by-products is the obtention of dietary and phenolic concentrates. For this, the freeze-dried pomace was ground into a homogeneous powder. Acetone (50% aqueous) was used as a solvent for phenolic and dietary fiber extraction. Acetone was evaporated at 60 °C under vacuum. Phenolic compounds were purified with butanone by solvent fractionation process and then freeze-dried again, while dietary fiber concentrate was treated by freeze-drying of the extracted solid residue [97]. In another study, the antioxidant extraction process from grape stalk, in order to quantify the influence of the previous drying operation on extraction kinetics, was evaluated by Garcia-Perez, et al. (2010) [98].

For the extraction/concentration of polyphenols from apple waste, the dried pomace was milled and sieved through a 20-mesh (0.84 mm) sieve into a homogeneous powder. Five (5) g of dried powder was mixed with 200 mL of 70% alcohol (solvent for polyphenol extraction). Alcohol was evaporated under reduced pressure, and then polyphenol compounds were concentrated in the powder by freeze-drying [99]. In another study, apple pomace was used to obtain pectin by extraction [100]. For this, nitric acid solution (pH = 2.5) was employed as an effective solvent for extraction at 80 °C during 1 h, and ethanol was used as agent in pectin precipitation, then pectin was filtered and vacuum dried at 45 °C to constant weight, and finally ground.

Furthermore, potato peel waste was also analyzed to be converted into an antioxidant powder. The phenolic compounds, antioxidant and antiviral activities of extracts were evaluated in dried potato peel (dried at 45 °C for 24 h and then powdered). Absolute ethanol with 5% acetic acid (95:5 ratio) was used as an extraction solvent for 72 h, and then the samples were filtered and concentrated by freeze-drying [101].

As shown in the few previous examples of processing lines for extracting valuable compounds from FVW, but also in many other cases found in the literature [102], extraction and drying processes are crucial operations which are always present in the further processing of plant-based food wastes.

5. Drying and Extraction in FVW by-Products Processing

Drying and extraction processes play an important role in the elimination and separation of contaminants and moisture content, by concentrating bioactive compounds from waste. They also prevent undesirable biochemical changes during storage for further processing. Salim et al. (2017) [102] reviewed conventional and emerging technologies for the conversion of FVW into value added products, where diverse drying and extraction methodologies were thoroughly described and analyzed. Moreover, other techniques, such as enzyme-assisted, subcritical water, microwave-assisted [103], and ultrasound-assisted extraction were proposed by Adetunji et al. (2017) [104], to improve the efficiency of industrial extractions from FVW. In the following paragraphs, a brief description of the principles underlying drying and extraction processes, and intensification methodologies that sometimes assist these processes, will be presented, in order to better understand how these systems could impact in the retention/disposal of pesticide residues.

Drying is a widespread technique in the food industry and a subject of continuous interest in food research. Most food products are dried for improved milling or mixing characteristics in further processing [105]. However, negative changes in food quality may occur during air-drying [106,107]. During the drying of food products, and especially of by-products such FVW, the moisture to be removed does not consist of only one component [108], but of a mixture of two or more components (multicomponent mixture), which could be thermodynamically non-ideal liquid solutions. The material to dry may contain ethanol, acetone, acids such as acetic or nitric, and even several different pesticides, competing with water in the vaporization process. The interaction between these compounds in the mixture through hydrogen bonding, dipolar interaction, or electrostatic interaction, and their different physicochemical characteristics (vapor pressure, boiling point, dissolution in water), will differentiate normal drying (where just water is evaporating) with multicompound drying, where the components can be evaporated, degraded, or co-evaporated together with water. This inter-relationship of compounds in the mixture affects multiple phenomena, such as the sorption behavior, gas-solid mass transfer, and the multicomponent vapor-liquid equilibrium. Furthermore, during the process, the wet bulb temperature changes, the initial composition of moisture, the identity and initial composition of the liquid system, the vapor pressure of different compounds, and the characteristics of the solid can influence the evaporation of selected components [109]. In the multicomponent drying case, drying behavior and its kinetics could not be simply explained by Fick's law [108] and, therefore, other mathematical representations of the phenomenon, such as the Maxwell-Stefan equations, should be used [110]. The diffusion behavior of solvents in a mixture during evaporation differed from when they are alone; it may cause a decrease or increase in the drying time, depending on their solubility, diffusion coefficient, which could be influenced by their concentration and partial pressure [111], or their ratio of gas-side mass transfer coefficient between them in mixture, affecting which component is removed preferentially [112]. The interaction between moisture and solid also affects the drying behavior (drying rate or diffusional paths, composition curves) when other solvent, such as, for instance, isopropyl alcohol is combined with water during drying [113]. Ho and Udell (1995) [114] investigated the influence of different binary mixture systems (toluene with o-xylene, benzene with o-xylene, and toluene with octane) at variable concentration and applied time on reducing of hydrocarbon contaminated soil. In pharmaceutical applications, during the drying of itraconazole, drying kinetics and dried particle morphology were affected by various weight fractions of binary solvent mixture of dichloromethane and ethanol [115]. Moreover, the pervaporation of a multi-compound (binary, ternary) mixture solvent from a membrane or solid state, as described by Heintz and Stephan (1994) [116], could also be explained by the mutual interaction effect between components (water and organic compounds); this frictional interaction force may lead to decrease or increase of component flux in diffusion. The friction coefficient is influenced by the size and shape of molecule of component during its movement through membrane, and this coefficient is accounted for, in part, in the modified Maxwell-Stefan equations [111].

Extraction is an important separation process used in various food processing applications as well. In this process, a desired component in a solid/liquid phase is separated by contact with a suitable solvent. Thus, the compositions of both phases change simultaneously during extraction, until equilibrium is reached. These phases are subsequently separated, and the desired component is recovered from the liquid phase [107]. Extraction constitutes a main processing stage to produce certain food products (oils, sugars), or to isolate desired compounds (antioxidants, vitamins). It could be useful to remove contaminants and other undesirable components and toxins present in food sources. Commonly used solvents in the extraction of food components are water, ethanol (or ethanol-water mixtures), hexane, and carbon dioxide, but the trend is toward the use of natural chemicals [117]. The rate of extraction is influenced by the solid-liquid interface area, the concentration gradient (to ensure a complete extraction, a sufficient gradient must be maintained between the concentration of solute at the surface of the solid and in the solvent), and by the mass transfer coefficient (an increase in temperature increases the rate of solution of the solute in the solvent and also the rate of diffusion of solute through the solution) [107,118]. Concentration changes as a function of extraction time could be represented by empirical-type models, such as Peleg's model [119], Page's or Weibull's model [82,120–122]. Additionally, in liquid-solid extraction, the solubility of compounds in different solvents, and the polarity characteristic of solvents are factors that influence the yield of extracted compounds [123,124].

5.1. Process Intensification (PI)

Different strategies are necessary to transform waste into valuable by-products. As mentioned before, intense work has been carried out focusing on the selection, characterization and stabilization of different agro-food by-products [125,126]. In this sense, drying and extraction are key processes for such valorization, but they could pose various techno-economic and environmental challenges, including low product yields, excessive energy consumption, or valuable compound deterioration such as carotenoids or polyphenols. As well, recent trends in extraction techniques have largely focused on finding solutions that minimize the use of solvents, sometimes causing health and environmental threats. Many of those challenges can be addressed by the application of innovative intensification technologies, such as pulsed electric field [127], ultrasound [128], or microwave [103], among others. The aim of these techniques is the improvement of traditional processes by increasing production yields, reductions in equipment size, energy use and waste, and increasing product quality. In terms of process safety, the reduction of plant size results in a smaller volume of toxic and flammable inventories within processes, thereby reducing the possibility of major explosions [129]. However, if intensification is to be applied to food production in the coming years, an integral analysis of the application of these new technologies need to be explored in detail, so as to offer sustainability in processing and cost-effective production of high-quality extracts.

Over the years, the implementation of PI has evolved into two distinct classifications involving the application of intensification technologies as pretreatments prior to processing, or else, during the process itself. In the following paragraphs, descriptions of some of the most used PI technologies applied to drying and extraction processes and their food applications will be individually presented.

5.1.1. Pulsed Electric Field (PEF)

Pulsed electric field (PEF) is an emerging technology with a wide variety of applications in the food and biotechnology sectors. It has been originally applied as a non-thermal process to the inactivation of bacteria, molds, and yeasts with promising results; other applications include the inactivation and modification of enzymes with negligible or minimum changes to the sensory, physicochemical, and nutritional characteristics of the product [107]. Despite the fact that PEF has been initially used in food processing as a separate and independent process, it can also be utilized as a pretreatment method, in order to enhance the subsequent process kinetics or to modify a quality of final products [130].

In a PEF system, the energy derived from a high voltage power supply is stored in an energy storage capacitor bank and discharged through a food material in a treatment chamber by the supporting of pulse generator to generate the necessary electric field in the food [131]. Important parameters that determine PEF processing impact are the treatment time and the electric field intensity (kV cm^{-1}), which is the ratio between the peak voltage (kV) and the gap distance (cm) between the electrodes in the treatment chamber.

PEF causes an irreversible loss of the membrane function as a semipermeable barrier between the bacterial cell and its environment [132]. Moreover, PEF treatment also leads to a cell membrane disintegration or electroporation phenomenon, intensifying any process based on mass transfer in cellular systems [133], and increasing the vibration and rotation of polar molecules [134].

Regarding drying, interesting technologies related to the application of moderate-continuously (MEF) or high voltage-pulsed (PEF) electric fields were investigated as a pre-treatment of different drying processes, such as osmotic dehydration, vacuum drying convective drying or freeze-drying, significantly increasing the dehydration kinetics [135–137]. These treatments can induce the formation of pores (permanent or reversible) in the cell membrane, facilitating the mass transport, such as in the research by Ostermeier et al. (2018) [138], prior to the convective drying of onion. Cell disintegration and enhanced mass transfer resulted in a 30% reduction in drying time. The effective diffusion coefficient increased from 3.7×10^{-9} to 1.8×10^{-8} m^2/s by increasing field intensity up to 1.07 kV cm^{-1}. Generally, the greater effect of PEF on the drying rate could be observed when drying was carried out at moderate temperatures. In the case of thermal sensitive foods, the enhancing of the convective drying rate only at moderate temperatures is very important for the keeping both of products' quality and energy economy, and PEF treatment showed advantages by an increase of the effective moisture diffusivity, allowing a decrease in the drying temperature from 70 °C to 50 °C during convective drying of potato, at moderate electric field strengths (E = 300–400 V cm^{-1}) [139]. However, the electroporation effect on cell membranes by the PEF treatment results in cell membrane breakdown, and natural compounds such as polyphenols, β-carotene and others may be released and lost by oxidation, or a color change by browning reactions during drying may occur. For solving this problem, oxidation reducing agents such as sodium sulfite can be added to protect natural compounds from oxidation [140]. Regarding PEF pre-treatment influence on antioxidant capacity, no significant changes have been reported for the convective drying of blueberries [141].

PEF has also been applied as a pre-treatment for extraction yield improvement through the electroporation phenomenon, for instance in the case of juice extraction from whole fruits, where the yield increased by 25% for orange, 37% for grapefruit and 59% for lemon [142]. Moreover, PEF treatment improved the polyphenol content after extraction [143], where the yield of total polyphenols extracted from orange peel increased from 20% to 159% for PEF pre-treated at 1 to 7 kV/cm, respectively. The quantity of flavonoids (naringin and hesperin) also increased from 1 to 3.1 mg/100 g and from 1.3 to 4.6 mg/100 g of fresh weight orange peel, without and with PEF pre-treatment, respectively. Conditions of PEF treatment played an important role in the change of cellular disintegration index (Zp), which was used to determine the effect of PEF conditions to permeabilize samples. In the case of lemon peels, Zp values increased to a highest value of 0.55, when the electric field strength and treatment time increased (up to 9 kV/cm, 30 pulses of 3 μs) [144].

Thus, PEF application yields to high-quality and less processed products, but it has a high initial cost for setting up the system. In addition, the requirement of major costs for power supply, the need for a high-speed electrical switch of the pulse generator when operated at a high pulse frequency and large-scale applications, are the main disadvantages of PEF technology [145].

5.1.2. Ultrasound (US)

Ultrasound waves are above the audible range (>20 kHz), with low-intensity ultrasound having frequencies higher than 100 kHz at intensities below 1 W cm^{-2} and high-intensity ultrasound, between 20 and 100 kHz at intensities higher than 1 W cm^{-2} [146]. Low-intensity ultrasound is used to transmit energy through a medium, without or with minimal physical and chemical changes in the material, therefore it can be employed for food analysis and quality control. In contrast, high-intensity ultrasound employs higher power levels for desired physical and chemical properties changes for various bioprocessing applications [147]. Ultrasound waves can be applied both as a pretreatment to the vegetable matrix prior to processing or assisting the actual process, in order to accelerate mass transfer by different mechanisms [148].

In general, applied ultrasound produces alternating compressions and decompressions which affect liquid and solid materials differently. In liquids, the provoked effects are pressure variations and stirring or cavitation. In solid materials, the "sponge effect" is predominant, which produces the release of liquid from the inner part of the particle to the surface and an entry of fluid from outside. Therefore, the forces involved in this mechanical effect could be higher than the surface tension of the water molecules inside the solid, making the mass exchange easier. Moreover, other effects could be occurring, such as changing of viscosity, surface tension, or deformation of solid material [149–151].

In terms of the application of US to drying, high intensity ultrasound produces a series of effects that can enhance heat and mass transfer. In fact, US has been applied to intensify convection drying or atmospheric freeze-drying of different products and by-products [128,152,153], achieving important reductions in operation time and energy consumption [154]. In most cases, ultrasound-assisted processing was used as a method to improve appearance characteristics (color, tastes) of dried fruit products by modifying drying kinetics [133], or to preserve bioactive compounds such as polyphenols, anthocyanins and flavonoids [128]. The effect ultrasound on diffusion and mass transport processes during drying can be quantified by the increase in effective diffusivity values, which can be influenced by sample tissue characteristics such as porosity (ε) and hardness (H). As a result, when 31 kW m^{-3} of ultrasound power was applied to samples being dried at 40 °C and 1 m s^{-1}, a *De* increase of 87% in apple (ε = 0.233, H = 25.92N) was obtained, while only 57% was observed in the case of cassava (ε = 0.029, H = 38.28N) [155].

As ultrasound-assisted drying reduces the drying time, bioactive compounds content and antioxidant activity should be protected from thermal exposure and better maintained during drying. According to Vallespir et al. (2019) [128], losses of total polyphenol, ascorbic acid, and vitamin E contents in kiwifruit dried with ultrasound (20.5 kW m^{-3}) drying at 15 °C were lower than in dried samples without ultrasound application. In another study, when apple was dried by convective drying at temperature 30 °C with ultrasound (18.5 and 30.8 kW m^{-3}), the loss of the total polyphenol was lower (34%) than without ultrasound (39%). However, at higher temperatures (50 and 70 °C), ultrasound assistance promoted a higher degradation of polyphenols; 39% loss compared to 20–27% without treatment [156]. Later, the same researchers found out that US can be effectively used in shortening the drying of apples at temperatures below 10 °C without compromising the quality [157]. In general terms, the application of ultrasound can reduce the drying time and protect bioactive compounds only when applied at low temperatures. However, these compounds can be negatively affected when ultrasound-assisted drying is applied at higher temperatures as a result of an increased temperature and thermal exposure in the sample from ultrasound energy [148]. The reader is encouraged to obtain more detailed information from the interesting review article on food drying enhancement by ultrasound [158,159].

Llavata et al. (2020) [160] have compared the influence of different pre-treatments (ultrasound, pulsed electric fields, high pressure processing or ethanol) on the drying process. For this purpose, researchers reviewed the current findings in some of these alternative pre-treatments, addressing their effectiveness on drying enhancement as well as of their impact on quality parameters, such as the retention of bioactive compounds, the color or the texture of the final product.

Regarding separation and recovery of different biocompounds, acoustically assisted solid-liquid extraction [161] has demonstrated high efficiency, by not only improving the recovery yields, but also accelerating the overall process [162]. It has been applied to improve the extraction of compounds with bioactivity [163] and the separation of trace elements [164]. It could also be an alternative to enhancing the sugar release using milder conditions (temperature, type of acid or acid concentration) during the pretreatment of lignocellulose in the second-generation ethanol production [165]. In a study by Caldas et al. (2018) [166], the yield of phenolic compounds in grape skin by ultrasound assisted extraction was twice as high as that obtained by mechanical agitation extraction (80 mg GAE g^{-1}), while extraction time was reduced three times. The results from another study also showed that the maximum yield of phenolic compounds by the ultrasound-assisted extraction of grape pomace was achieved within 10 min, compared with 20 h of the industrial batch extraction [167]. The ultrasonic degradation of cell tissue is rapid and occurs within the first minute of treatment, therefore, this intensification technology is usually used for air and light sensitive bioactive compound extraction, such as lycopene from tomato waste [168]. Most recently, Umaña et al. (2020) found that ultrasound-assisted extraction from mushroom by-products yielded up to 2 times higher in ergosterol and 46% in phenolic compounds, depending on ethanol concentration and US power density [82]. Furthermore, in a new comprehensive review of ultrasound assisted extraction, Dzah et al. (2020) concluded that ultrasound assistance is considered nowadays a preferred extraction method, due to its versatility and the ability to use less or no organic solvent, although successful results depend largely on the type of plant material, solvent and the micro-environmental extraction parameters [169].

An interesting alternative could be the combination of both US and MEF/PEF treatments. Thus, Mello et al. (2019) [170] found that the PEF pre-treatment of orange peel significantly increased the effects on drying rate of ultrasound application during drying. In this sense, the synergistic effect of US and PEF has also been reported in the extraction of betanin from beetroot [171]. PEF/MEF application may induce changes in the structure, e.g., modifying the porosity. This fact can enhance the effectiveness of US application, because the magnitude of the ultrasound effects is greater when the porosity is higher.

5.1.3. Microwaves (MW)

The microwave-assisted extraction (MAE) process is considered as an emerging technology, particularly for compound extraction from biomaterials by using microwave energy—electromagnetic radiations with a frequency from 0.3 to 300 GHz. Microwaves can penetrate biomaterials and heat them directly by interaction with polar molecules, such as water in the biomaterials [172]. The medium of biomaterials has the ability to absorb and convert microwave energy into thermal energy, dependent on their dielectric properties, which are one of the primary features for its selection as the extracting solvent in the MAE process [103]. Compared with conventional solvent extraction methods, MAE is a novel method providing lower extraction times and solvent consumption [173]. MAE has been widely applied as an effective technique for phenolic compounds extraction from tomatoes [174], grape marc (skins and seeds) [175], pomegranate peels [176] and peanut skins [177], which have high antioxidant activities. Additionally, the MAE process is also considered as a promising method by its effective on receiving higher yield and better quality of extracted fucoidan from brown algae [178], pectin from press residues of berry fruits [94], an acidic polysaccharide from blackberries [179], etc. Arrutia et al. (2020) [180] recently presented a scaled-up continuous flow system for the microwave-assisted extraction of prebiotic hairy pectin from potato waste. Microwave heating is necessary in this process due to its selective and rapid heating, which avoid the deterioration of hairy pectins. Figure 1 shows a schema of this interesting proposal, which represents a concrete step forward towards the implementation of microwave-assisted extraction in the food waste treatment industry. By using the system depicted in Figure 1, it was possible for the authors to recover hairy pectins in the product tank, while starch was concentrated in the feed tank as a sub-product [180].

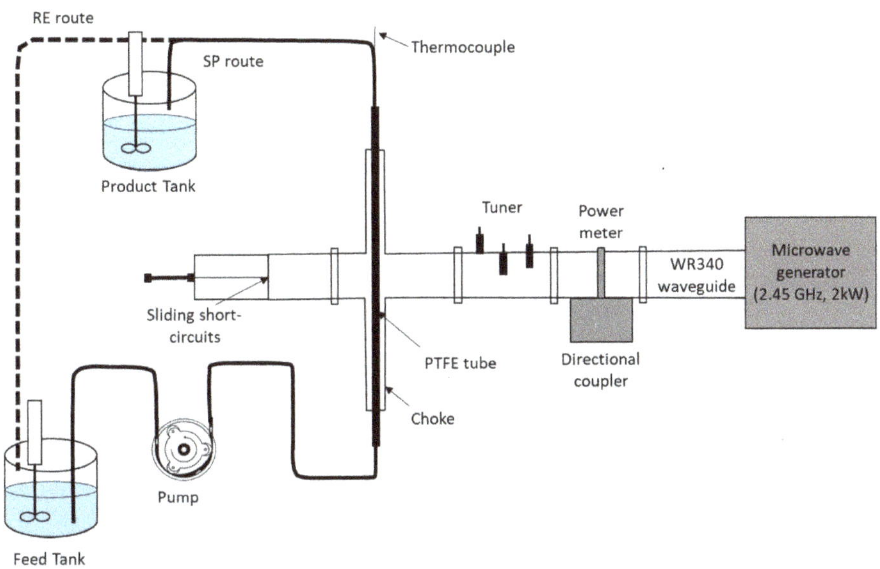

Figure 1. Schema of a continuous-flow MW processing system (adapted from Arrutia et al. 2020) [180].

Microwave has also been applied to food drying to benefit from its quick internal heat generation by the rapid polarization and depolarization of water molecules inside the food material [181], or due to shorter drying times (up to 69%) when compared to hot air drying [182]. In a study by M'hiri et al. (2018) [183], the retention of total phenol and total flavonoids contents was the highest (68.73 and 61.44%, respectively), when combined microwave-air drying (90 W/75 °C) was applied on an industrial lemon by-product. Oil extraction from microwave-dried Hass avocados resulted in high quality and stable avocado oils when compared to those obtained from air-dried avocados [184]. Regarding antioxidant retention, microwave drying was also advantageous for raspberries, for which antioxidant retention was 41.7%, 1.7 times compared to hot air drying alone [185]. The reader is invited to obtain more detailed information about microwave assisted drying in comprehensive reviews found in the literature [186–188].

6. Processing on Pesticide Residues Reduction

6.1. General Food Processing

Processing factors (PF) estimate the effect of processing methods pesticide on residue levels and the disposition of the residues in the processed products, calculated and considered by the Joint FAO/WHO Meeting on Pesticide Residues as follows [189]:

$$\text{PF} = \frac{\text{residues in processed product}\left(\text{mg Kg}^{-1}\right)}{\text{residues in raw agricultural commodity}\left(\text{mg Kg}^{-1}\right)} \quad (1)$$

PF values lower than 1 indicate a reduction in the residue level and higher than 1, a concentration effect [181].

Fruits and vegetables, like other foods, are treated through culinary and food processing before they are consumed. The effects of these handling techniques on pesticide residue levels in fruits and vegetables may be influenced by the physical location of the residues (Section 3.4), as well as the physico-chemical properties, such as solubility, volatility, hydrolytic rate constants, water-octanol partition coefficient and thermal degradation [190]; some shown in Tables 1–4. Among the main food processing techniques applied to fruit and vegetable products before consumption are accounted washing, peeling, juicing, blanching, fermentation, baking, etc.

Washing with ambient temperature water is the most common procedure in both household and commercial preparations. Washing could be reasonably effective in removing residual pesticides of fruits and vegetables, only if the remaining pesticide concentration is low [191]. The reduction of pesticide residues during washing depends on fruit and vegetable types and their characteristics, as reported by Kar et al. (2012) [192]. For instance, the washing of cabbage and cauliflower with tap water removed about 17–40% of chlorantraniliprole residues [192], while in another study, it eliminated 30–50% of phosalone residues in apple [193]. In the case of tomato, washing with tap water yielded a 10%, 15%, 9%, 19%, 23% and 16% loss of HCB, lindane, p,p-DDT, dimethoate, profenofos, and pirimiphos-methyl, respectively [36]. In addition, the effectiveness of residue reduction by washing depends on the time elapsed since the last pesticide application. For example, a study done by Balinova et al. (2006) [57] demonstrated that a reduction in residues through washing decreased with the sampling time (1 to 3 days). Furthermore, no correlation with the solubility and polarity of compound residues could be found, attributed to the penetration of the residues into the cuticle or tissues of the fruit [57].

Peeling is the most effective process to eliminate residual pesticide before consumption. Balinova et al. (2006) [57] reported that the peeling of peaches for baby food was identified as the most effective treatment for decreasing chlorpyrifos-methyl and fenitrothion residues, two organophosphate pesticides. Moreover, potato peeling allowed 71–75% reduction of organochlorine and organophosphate pesticides residues (malathion, lindane, HCB, p,p-DDT) [34], and chlorpropham, a herbicide and sprout suppressant, 91 to 98% [194]. The effective residue reduction by peeling was also reported by Rawn et al. (2008) [8], since captan residue, a non-systemic organophosphate fungicide, contained on post-harvest apple samples (25.8–5100 ng/g) was removed by about 98% through rinsing and peeling, much more than by rinsing alone (50%). The peeling process also showed its effectiveness in the removal of residues when applied to tomato samples, where the concentration of organochlorines (aldrin, dieldrin, endosulfan, endrin, heptachlor, methoxychlor) and organophosphates (chlorpyriphos, dimethoate, malathion, ethyl-parathion, methyl-parathion) significantly decreased by 28 ± 7% three days after harvest, whereas dimethoate and ethyl-parathion were entirely removed ten days after harvest [195]. In another study, chlorpyriphos contained in red pepper was effectively reduced by more than 93% in peeled samples (from 0.064 to 0.004 mg kg^{-1}) [196]. Moreover, in the case of cucumber, the peeling process was the most effective way to reduce carbaryl residues when compared to washing and storage [197].

Juicing is done by pressing fruits or vegetables, sometimes assisted with enzymatic treatment to increase juice yield. During juicing, the combination of supplementary procedures (washing, pressing, sterilization and enzymatic treatment) greatly helps reduce pesticide residues in the matrix. For instance, Li et al. (2015) [198] obtained a 85–95% decrease of β-cypermethrin, chlorpyrifos, tebuconazole, acetamiprid and bendazim in apples matrix during juice processing. In another study, the reduction of HCB, lindane, p,p-DDT, dimethoate, profenofos and pirimiphos-methyl residues ranged from 73 to 78% during tomato juicing [36].

Pesticide residues can also be reduced by other thermal processes such as baking, where undesirable molecules in the tissue co-evaporate with water, or simply degrade [199,200]. For example, detected levels of profenofos residues were reduced from 11.5 ppm to 0.22 ppm in fresh peeled potatoes through microwave-baking, or to 0.19 ppm through oven-baking [201].

Storage of fruits and vegetables prior to processing may represent an important step in degradation of pesticide residues over time, but it varies greatly according to the active ingredient. For instance,

Holland et al. (1994) [190] detected the presence of the fungicide dodin and insecticide phosalone in apples after five months of cold storage at 1–3 °C. Athanasopoulos and Pappas (2000) [202] reported differences in the degradation rate of azinphos methyl between apple and lemon, based on their acidity. In general, storage conditions will impact the fate of residues. This is the case of azinphos-ethyl, where its half-life was measured at 10 days for apples on trees, 83 days for apples stored at ambient conditions (18 ± 5 °C, RH~60%), 91 days for apples in controlled-atmosphere rooms, and 136 days for apples in refrigerated rooms (0 ± 0.5 °C, RH~85%) [203].

6.2. Changes in Pesticide Residues during Drying

In food processing, drying methods may cause an appreciable decline in pesticide residues as a result of evaporation, degradation and/or co-evaporation. However, different drying methods may impose different effects on pesticides. As no information was found on waste of fruits and vegetables, Table 6 resumes some of the research publications dealing with the impact of drying on pesticide residues in fruits and vegetables. For example, the oven-drying of chili pepper at 60 °C for 35 h caused large reductions (37–49%) of clothianidin, diethofencarb, imidacloprid, and tetraconazole, with processing factors (*PF*) in the range of 0.51–0.63. Conversely, moderate reductions (16 and 22%) of methomyl and methoxyfenozide were observed with *PF* of 0.78 and 0.84, while no reduction of chlorfenapyr, folpet, and indoxacarb was present (*PF* of 0.96–0.98) [204]. Oven-drying also caused a high reduction of both dicarboxymides (iprodione and procymidone) residues (57 and 41%, respectively) in grapes as reported by Cabras et al. (1998) [205]. In another study on sun-dried grapes, residues of chlorpyrifos, diazinon, methidathion and dimethoate decreased by 73, 92, 82 and 39%, respectively [206]. However, drying processes can also lead to a higher concentration of pesticides in their by-products, simply from the loss of water in the treated sample. For example, levels of iprodione residues in sun-dried raisin increased 1.6 times, and that of phosalone 2.8 times compared to fresh fruit [205]. In other studies, the level of triadimenol residues, a metabolite of triadimefon, in sun-dried jujube was found to be more than twice as high as that found in fresh jujube, as a result of degradation of triadimefon into triadimenol during sun-drying [181].

Table 6. Effect of drying methods on residual pesticides in fruits and vegetables.

F&V Produce	Pesticide Compounds	Operation	Conditions	Results	Reference
Apple [193], apple pomace [207]	Phosalone [193], kelthane [207]	Rotating 'Hatmacker' drum dryer [193], natural drying [207]	Steam pressure (5 bars), discharge rate (150 L/h), rotation speed (5–76 cm/s) [193]. In the dark, under UV light or sunlight [207].	Phosalone levels were reduced from 22 to 77%. Manufacturers should seek the total elimination of surface residues, i.e., peeling the fruit [193] to improve quality. The loss of kelthane residues was mainly due to volatility rather than photodecomposition [207].	[193,207]
Apricot	Phosalone, iprodione, diazinon, procymidone, bitertanol [208], fenitrothion, dimethoate, omethoate, ziram [209]	Sun drying [208] and ventilated oven [208,209].	Sunlight for 7 days [208] and ventilated oven at 100 °C for 30 min and at 70 °C for 12 h [208,209].	Pesticide residues present in dried fruit were lower than in the fresh fruit (half after sun drying). The exception was phosalone, which increased by 50 (sun-drying) and by 3 times for oven-drying [208]. Omethoate and ziram residues almost doubled after drying, while fenitrothion disappeared and dimethoate remained constant [209].	[208,209]
Chili pepper	Chlorfenapyr, clothianidin, diethofencarb, folpet, imidacloprid, indoxacarb, methomyl, methoxyfenozide and tetraconazole	Oven drying	60 °C for 35 h	Large reductions (37–49%) in clothianidin, diethofencarb, imidacloprid, and tetraconazole. Moderate reductions (16 and 22%) in methomyl and methoxyfenozide, respectively. No effect of drying on chlorfenapyr, folpet, and indoxacarb levels.	[204]

Table 6. *Cont.*

F&V Produce	Pesticide Compounds	Operation	Conditions	Results	Reference
Grape	Iprodione and procymidone [205] Benalaxyl, dimethoate, iprodione, metalaxyl, phosalone, procymidone, vinclozolin [210] Chlorpyrifos, diazinon, methidathion and dimethoate [206] Dimethomorph, famoxadone and cymoxanil [211] Azoxystrobin [212] Pyraclostrobin and metiram [213] Quinoxyfen [214]	Oven drying [205,214] Sun drying & oven drying [206,210] Natural drying [211,213]	70 °C for 24 h [205,214]. No operating conditions [210]. Direct sunlight for 21 days and in an oven at 50 °C for 72 h, at 60 °C for 60 h, at 70 °C for 48 h, at 80°C for 36 h [206]. Shade and outdoors, for 15 days [211] or 25 days [213]. Direct sunlight for 15 days [212]	Iprodione and procymidone decreased by 57 and 41%, respectively [205]. Benalaxyl, phosalone, metalaxyl, and procymidone residues in sun-dried grapes were the same as those on fresh grapes, whereas those of iprodione were higher (1.6 times) and vinclozolin and dimethoate, lower. For the oven-drying process, benalaxyl, metalaxyl, and vinclozolin showed the same residue values in fresh and dried fruits, whereas iprodione and procymidone resides were lower in raisins [210]. Chlorpyrifos, diazinon, methidathion and dimethoate decreased by 73, 92, 82 and 39%, respectively [206]. PF values for raisin processing were 1.03 to 1.14 for dimethomorph, 1.95 to 2.09 for famoxadone, and 1.99 to 1.35 for cymoxanil [211]. Pre-treatment with alkali and sun drying effectively removed a substantial amount of azoxystrobin residues. Commercial production of raisins is, however, carried out with sun drying only [212]. PF values 1.01 to 1.31 for metiram and 1.34 to 1.10 for pyraclostrobin indicated residue concentration after drying [213]. The residue levels in oven dried raisins were comparable to fresh grapes. The lower degradation in the oven-dried sample could be explained by the absence of the degradation effect due to solar radiation [214].	[205,206,210–214]
Honeysuckle (*Lonicera japonica*)	Thiamethoxam and thiacloprid	Three drying methods: sun-, natural (shade)-, and oven drying.	Oven-drying at 30, 40, 50, 60, and 70 °C	59.4–81.0% residue reduction after sun- and oven-drying at 70 °C, higher than for shade- and oven-drying at lower temperatures (at 30 to 60 °C).	[215]
Jujube	Dichlorvos, malathion, chlorpyrifos, triadimefon, hexaconazole, myclobutanil, kresoxim-methyl, tebuconazole, epoxicona-zole, bifenthrin, and cyhalothrin	Microwave drying	Microwave oven (700 W) for 4 min	Degradation rates were from 67% to 93%	[181]
Kumquat candied fruit	Dimethoate, chlorpyrifos, malathion, methidathion and triazophos	Convective drying	60–80 °C	PF of dimethoate, malathion and triazophos after drying were >1, which could be due to the water loss.	[216]

Table 6. *Cont.*

F&V Produce	Pesticide Compounds	Operation	Conditions	Results	Reference
Okra	Malathion, carbaryl [217], endosulfan [217,218], bifenthrin and profenofos [218]	Convective drying [217], sun drying [218]	No specific conditions were found [217,218]	91.8% malathion, 78% carbaryl and 57.4% endosulfan removal [217]. Sun drying helped to decrease endosulfan up to 5.5%, profenos up to 11% and bifenthrin, up to 75%. Bifenthrin was more affected by sun drying because it is hydrolyzed in the presence of UV rays [218].	[217,218]
Pleurotus ostreatus mushroom	Carbendazim	Sun drying and freeze-drying	Direct sunlight (sun drying) and at −86 °C with vacuum of 0.06 mbar (freeze-drying)	Direct sun-drying removed higher carbendazim amounts than freeze-drying, with removal rates ranging between 70 and 97%.	[219]
Plum	Bitertanol, diazinon, iprodione, phosalone, procymidone, and vinclozolin [220] Buprofezin, L-cyhalothrin, spirodiclofen, indoxacarb, acetamiprid, imidacloprid, emamectin benzoate [221]	Oven drying [220] Sunlight drying [221]	Temperature: 30 min at 95 °C, 30 min at 90 °C, 16 h at 85 °C [220] Sunlight drying for 26 days with avg. air temp. 17.6 °C, relative hum. 67.3%, solar radiation 546.3 W m^{-2}; no rain fell [221].	PF factor was around 3, however pesticide residues were lower or similar in dried than in fresh fruits: phosalone showed the same value, while procymidone, iprodione, and bitertanol were lower (0.6, 2.3 and 3.2 times, respectively). [220]. The insecticide residue reductions during sunlight drying was variable and related to the pesticides' physico-chemical properties. The whole industrial prune processing has an important reduction effect on pesticide residues [221].	[220,221]
Red pepper	Chlorpyriphos and fenitrothion	Sun/hot air drying	—	Sun or hot air-drying eliminated a 20–30% of residues.	[218,222]
Shiitake mushroom	Carbendazim, thiabendazole, procymidone, bifenthrin, λ-cyhalothrin, and β-cyfluthri	Drying	Sunlight (26–33 °C, 20 days) and hot-air drying (30–53 °C in the first 10 h, 53–60 °C in the last 10 h)	Removal rate of pesticides by sunlight exposure drying (36.2–94.6%) was higher than that of hot-air drying (26.0–68.1%)	[223]
Spring onion	Etofenprox	Drying	Oven (80 °C for 24 h) and freeze-dried (3 days)	Removal rate by oven dried (85.5%) higher than freeze-dried (66.6%)	[224]

In general, the observed changes during drying are dependent on the type of pesticide compound and may be correlated with the difference in vapor pressure of the mixture. For example, in the article by Noh et al. (2015) [204], the reduction by drying of tetraconazole higher than indoxacarb could be related with its higher vapor pressure (0.18 mPa compared to 2.5×10^{-5} mPa, respectively). As well, in Özbey et al. (2017) [206], the 92% decrease of diazinon in dried grapes could be due to its higher vapor pressure compared to the three other pesticides. Moreover, seven pesticide residues in a group of eleven were higher in dried jujube than in the fresh one [181]; this could also be related to their low vapor pressure. Therefore, from the physicochemical properties of pesticides (Tables 1–4), an estimate of their behavior during drying could be extrapolated (for example, a reduction of chlorpyrifos during drying could be higher than propiconazole, and azoxystrobin).

To end, from the results presented in Table 6, the presence of UV radiation in sun drying seems to enhance the loss of pesticides during drying, such is the case of quinoxyfen in sun-dried grapes compared to oven-dried [214], thiamethoxam and thiacloprid in sun-dried honeysuckle compared to shade natural dried [215], and bifenthrin, which was more affected by sun drying because it is hydrolyzed in the presence of UV rays [218]

6.3. *Change in Pesticide Residues during Extraction*

When extraction methods are applied to foods containing pesticides, the percentage transfer of residues into the solvent will depend on the polarity and solubility of pesticide compounds. Water infusion has been extensively studied regarding to pesticide transfer during tea brewing. For example, the percentage transfer of phosphamidon residue to the tea brew was the highest (33%), followed by dimethoate (26%), monocrotophos (20%), malathion (12%), methyl parathion (10%), quinalphos (8%), and finally chlorpyrifos (3%), as a direct indication of their polarity [225]. Chen et al. (2015) [226] also reported that the transfer rate of nineteen different pesticide residues from tea during brewing was influenced by the octanol–water partition coefficient, and pesticide water solubility. Similarly, Kumar et al. (2005) [227] found that the percentage of propargite residues from manufactured tea to infusion media was in the range of 24–40%, based on the water solubility of residues and their partition coefficient in the solvent.

Other than research on the pesticide transfer during the brewing of tea or other leaves for infusion, there is otherwise little specific investigations on the fate of pesticides during different types of extractions from fruits and vegetables, and none on their waste. Table 7 resumes some of the few publications on the subject. Some of the works listed in Table 7 are just extraction methods developed for analytical determination purposes, but they could indicate the fate of pesticides in similar extraction conditions during the processing of fruit and vegetable wastes. Watanabe et al. (2013) [228] and Iwafune et al. (2014) [229] developed a water-based extraction method for separating pesticides from green pepper, tomato and spinach with high yields. Jaggi et al. (2001) [225] found that most organochlorines, organophosphates, and synthetic pyrethroids residues could be extracted by n-hexane. On the other hand, dimethoate residues were best dissolved in chloroform (96–100%) and those of phosphamidon in dichloromethane (89–95%), two polar solvents. This suggests that the latter residues could possibly be extracted by solvents such as water, whereas less polar pesticides will be best extracted with non-polar solvents, such as n-hexane.

Blanching is a heat treatment for enzyme inactivation, enhancing drying rate and food quality. In addition, it could play an important role in the reduction of polar pesticide residues and toxic constituents in vegetables and fruits. This reduction is explained by the degradation of toxic or pesticide substances that are washed off into the blanching water [230], or by the dissolution of the cuticular waxy layer [231]. Among various processes (washing with tap water, microwave cooking, in-pack sterilization, blanching), hot water blanching was the most effective way to remove deuteratedethylenethiourea, ethylenethiourea, deltamethrin, 3,5-dichloroaniline and boscalid residues in spinach [232].

Table 7. Effect of extraction on residual pesticides in plant-based foods (mainly fruits and vegetables but also including leaves for infusion).

F&V Produce	Pesticide Compounds	Operation	Conditions	Results	Reference
Apples, asparagus, beets, cucumbers, green beans, lettuce, nectarines, peaches, peas, raspberries, spinach, strawberries, tomatoes	Captan, chlorothalonil, iprodione, vinclozolin, endosulfan, permethrin, methoxychlor, malathion, etc.	Water immersion (rinsing)	Ambient temperature for 15–30 s.	A short rinse in tap water reduces pesticide residues on many types of produce. Water solubility of pesticides did not play a significant. The majority of pesticide residue appears to reside on the surface.	[233]
Cabbage, cantaloupe, pear, white potato	Over 33 types of pesticides coming from different families	Accelerated solvent extraction	Dionex ASE 200 extractor; solvent acetone/dichloromethane; 110 °C; 1500 psi; 2 cycles	Accelerated solvent extraction with acetone/dicholoromethane was able to extract a wide range of pesticide residues	[234]
Fruit juice soft drinks (bottles and cans of different brands from 15 European countries)	Over 100 pesticide compounds from 8 different families	Solid-phase extraction (SPE)	HLB cartridges (200 mg) Soft drink samples passed through the cartridges at a flow rate of 3 mL min^{-1}.	Carbendazim, thiabendazole, imazalil, prochloraz, malathion, and iprodione were detected in fruit soft drinks, which are mainly those applied to crops as postharvest treatment. The presence of these pesticides in fruit-based soft drinks could be attributed to the use of the peels in the extracts. Therefore, steps should be taken with the aim of removing any traces of pesticides in these products.	[235]
Green pepper, tomato, spinach	Acetamiprid, clothianidin, dinotefuran, flonicamid, imidacloprid, methomyl, pymetrozine, thiacloprid, and thiamethoxam	Water extraction	Please refer to reference.	Water extraction of downsized samples allow quantitative recovery of hydrophilic pesticides	[228]

Table 7. *Cont.*

F&V Produce	Pesticide Compounds	Operation	Conditions	Results	Reference
Green pepper, tomato	CPMF, dinotefuran, CPMA, Nitenpyram, thiamethoxam, clothianidin, imidacloprid, thiaclopridamide, acetamiprid, thiacloprid	Water extraction	Please refer to reference.	This water-based extraction method is convenient to remove pesticides and could be utilized for regular monitoring of neonicotinoid insecticides and their metabolites in high water content crops.	[229]
Peppermint leaves	Malathion, fenitrothion, dimethoate, chlorpyrifos and pirimiphos-ethyl	Infusion	Boiling water (2 g in 100 mL), in 5, 10, 15 and 20 min.	Residues of dimethoate into the infusion was highest (91%), followed by malathion (62%) and fenitrothion (38%)	[236]
Spinach	Boscalid, mancozeb, iprodione, propamocarb, and deltamethrin	Blanching	Sample immersed in hot water (88 °C) for 5 min.	Decreased residue of propamocarb (70%), iprodione, and others by 10 to 58%.	[232]
Tea leaves	Phosphamidon, dimethoate, monocrotophos, malathion, methyl parathion, quinalphos, and chlorpyrifos [185], propargite [187]	Extraction (infusion)	Boiling water (5 g in 200 mL), for 2, 5 and 10 min [185] and for 2 min [187]	Residue to the tea brew: 33%, 26%, 20%, 12%, 10%, 8%, and 3%, respectively [185]; 24–40% in infusion media [187]	[225,227]
Tomato (cherry), and farm produce i.e., tomatoes, apples, green peppers, peaches, oranges, and lemons	Acephate, malathion, carbaryl, bifenthrin, cypermethrin, cyhalothrin, permethrin, chlorothalonil, and imidacloprid	Water immersion (washing)	Pure water (for all produce) or other washing solutions (for Cherry tomatoes) with added chemical compounds (600 rpm for 1 min, at 10 °C forwashing solutions while for pure water, at 5, 10, and 22 °C, w/wo sonication).	Cherry tomatoes washed at 22 °C presented the highest reduction. Residues in contaminated produce decreased from 40 to 90%. Sonication used with the washing process would increase pesticide removal from produce surfaces.	[237]
Thyme and stinging nettle leaves	Fenitrothion, dimethoate, chlorpyrifos and pirimiphos-ethyl	Infusion	Boiling water (2 g in 100 mL), in 5, 10, 15 and 20 min.	The residues of dimethoate (highest water solubility) transferred into the infusions (89–86%), followed by fenitrothion (27–29%), pirimiphos -ethyl (8–14%) and chlorpyrifos (8–8%) during 5 min infusion.	[238]

6.4. Impact of Intensification Technologies on Pesticide Reduction

PEF treatment has been applied as an effective method for reducing pesticide residues, their degradation level affected by electric field strength and the number of pulses. Ultrasound may also play an important role by itself or combined to other processes in pesticide residue degradation, such as the degradation rate value of diazinon in apple juice when treated at 500 W, or ultrasound power, which was 1.26 and 1.55 times higher than when treated at 300 W and 100 W, respectively [239]. The ultrasound application time also influenced the pesticide residue degradation and thus, the percentage of degradation of phorate, an organophosphorus pesticide, increased about 16% when the ultrasound treatment time increased from 60 to 120 min at 500 W power [240]. In the case of farm produce (i.e., tomatoes, apples, green peppers, peaches, oranges, and lemons), Al-Taher et al. (2013) [237] found out that sonication used with the washing process would increase pesticide removal from produce surfaces, depending on the washing treatment and on the pesticide. Finally, microwave application could also be considered to be an effective method for the removal of pesticide residues on fruits. Pesticide residues were degraded at higher rates (from 67% to 93%) in jujube fruit dried by microwave drying (700 W, 4 min), when compared to just hot air drying, which was highly corelated with the vapor pressure and water solubility of these pesticide compounds [181].

7. Conclusions

The present literature review pointed out that pesticide residues in fruit and vegetable wastes (FVW) processing could pose a problem on human health and environment. The localization of pesticides in foods varies with the nature of molecules, type and portion of plant material and environmental factors, but are usually mostly present in their outer parts of fruits and vegetables. Fruit and vegetable wastes being composed mainly of skin and peel, especially for apple, pepper, tomato, potato, grape and orange wastage, could imply that FVW as raw material for further processing into valuable by-products may be concentrated in agrochemicals.

By-products from FVW are usually bioactive extracts or powders obtained in production lines where air-drying and extraction processes are commonly employed, which may result in increased pesticide/solvent residue concentrations. Drying could concentrate or reduce agrochemicals content depending on the pesticide chemical properties and the type of drying method used. Furthermore, residues in olive oil or apple juice showed great variability upon processing, depending on water solubility of the pesticides and pre-treatments. Thus, the agrochemical content in FVW should be monitored and eventually minimized before or during further by-products processing.

It is surprising how little in-depth research exists on the interaction between pesticide compounds and drying or extraction processes. There are only empirical studies on multi-compound drying of such mixtures and on the application of intensification technologies in by-product processing. In addition, studies on the fate of pesticides during the obtention of extracts from FVW are practically lacking from the literature. This review, being of an exploratory and interpretive nature, thus raised a number of opportunities for future research in the area of the impact of drying and extraction on the fate of pesticide residues in by-products processing from FVW, both in terms of theory development and concept validation.

Author Contributions: Conceptualization, C.R. (Cristina Ratti); methodology, T.T.N. and C.R. (Cristina Ratti); validation, T.T.N., C.R. (Carmen Rosello), R.B. and C.R. (Cristina Ratti); formal analysis, T.T.N., C.R. (Carmen Rosello), R.B. and C.R. (Cristina Ratti); investigation, T.T.N. and C.R. (Cristina Ratti).; resources, C.R. (Cristina Ratti); data curation; writing—original draft preparation, T.T.N. and C.R. (Cristina Ratti); writing—review and editing, C.R. (Carmen Rosello) and R.B.; visualization, T.T.N. and C.R. (Cristina Ratti); supervision, C.R. (Cristina Ratti); project administration, C.R. (Cristina Ratti); funding acquisition, C.R. (Cristina Ratti). All authors have read and agreed to the published version of the manuscript.

Funding: This research was funded by NSERC (National Science and Engineering Research Council of Canada), grant number RGPIN/04774-2017.

Conflicts of Interest: The authors declare no conflict of interest.

References

1. Lim, S.S.; Vos, T.; Flaxman, A.D.; Danaei, G.; Shibuya, K.; Adair-Rohani, H.; Amann, M.; Anderson, H.R.; Andrews, K.G.; Aryee, M.; et al. A comparative risk assessment of burden of disease and injury attributable to 67 risk factors and risk factor clusters in 21 regions, 1990–2010: A systematic analysis for the Global Burden of Disease Study 2010. *Lancet* **2012**, *380*, 2224–2260. [CrossRef]
2. Sagar, N.A.; Pareek, S.; Sharma, S.; Yahia, E.M.; Lobo, M.G. Fruit and Vegetable Waste: Bioactive Compounds, Their Extraction, and Possible Utilization. *Compr. Rev. Food Sci. Food Saf.* **2018**, *17*, 512–531. [CrossRef]
3. Kjørstad, E. These Are the Fruits and Vegetables We Waste the Most. Available online: http://sciencenordic.com/these-are-fruits-and-vegetables-we-waste-most (accessed on 25 September 2018).
4. Lokesh, K.; Ladu, L.; Summerton, L. Bridging the gaps for a "circular" bioeconomy: Selection criteria, bio-based value chain and stakeholder mapping. *Sustainability* **2018**, *10*, 1695. [CrossRef]
5. Shokrzadeh, M.; Saravi, S.S.S. Pesticides in Agricultural Products: Analysis, Reduction, Prevention. In *Pesticides-Formulations, Effects, Fate*; Stoytcheva, M., Ed.; InTech: Rijeka, Croatia, 2012; pp. 225–242.
6. Roberts, J.R.; Reigart, R.J. Organophosphate Insecticides. In *Recognition and Management of Pesticide Poisonings*; Roberts, J.R., Reigart, R.J., Eds.; The Environmental Protection Agency's Office of Pesticide Programs: Washington, DC, USA, 2013; pp. 43–55. ISBN 9780470699010.
7. EWG Dirty DozenTM EWG's 2019 Shopper's Guide to Pesticides in ProduceTM. Available online: https://www.ewg.org/foodnews/dirty-dozen.php (accessed on 30 April 2019).
8. Rawn, D.F.K.; Quade, S.C.; Sun, W.F.; Fouguet, A.; Bélanger, A.; Smith, M. Captan residue reduction in apples as a result of rinsing and peeling. *Food Chem.* **2008**, *109*, 790–796. [CrossRef]
9. Sharma, R.; Oberoi, H.S.; Dhillon, G.S. Fruit and Vegetable Processing Waste: Renewable Feed Stocks for Enzyme Production. In *Agro-Industrial Wastes as Feedstock for Enzyme Production*; Dhillon, G.S., Kaur, S., Eds.; Elsevier Inc.: Amsterdam, The Netherlands, 2016; pp. 23–59. ISBN 9780128026120.
10. Yadav, Y.-C.; Devi, N.-L. Pesticides Classification and Its Impact on Human and Environment. In *Environmental Science and Engineering*; Chandra, R., Gurjar, B.R., Govil, J.N., Eds.; Studium Press LLC: Houston, TX, USA, 2017; Volume 6, pp. 140–158. ISBN 1-62699-094-8.
11. Suwalsky, M.; Rodríguez, C.; Villena, F.; Sotomayor, C.P. Human erythrocytes are affected by the organochloride insecticide chlordane. *Food Chem. Toxicol.* **2005**, *43*, 647–654. [CrossRef]
12. Xu, D.; Liang, D.; Guo, Y.; Sun, Y. Endosulfan causes the alterations of DNA damage response through ATM-p53 signaling pathway in human leukemia cells. *Environ. Pollut.* **2018**, *238*, 1048–1055. [CrossRef]
13. Zacharia, J.T. Identity, Physical and Chemical Properties of Pesticides. In *Pesticides in the Modern World-Trends in Pesticides Analysis*; Stoytcheva, M., Ed.; InTech: Rijeka, Croatia, 2011; pp. 1–20. ISBN 978-953-307-437-5.
14. Edwards, F.L.; Tchounwou, P.B. Environmental toxicology and health effects associated with methyl parathion exposure—A scientific review. *Int. J. Environ. Res. Public Health* **2005**, *2*, 430–441. [CrossRef]
15. Krstic, D.Z.; Colovic, M.; Bavcon Kralj, M.; Franko, M.; Krinulovic, K.; Trebse, P.; Vasic, V. Inhibition of AChE by malathion and some structurally similar compounds. *J. Enzyme Inhib. Med. Chem.* **2008**, *23*, 562–573. [CrossRef]
16. Cheng, S.; Chen, J.; Qiu, Y.; Hong, X.; Xia, Y.; Feng, T.; Liu, J.; Song, L.; Zhang, Z.; Wang, X. Carbaryl inhibits basal and FSH-induced progesterone biosynthesis of primary human granulosa-lutein cells. *Toxicology* **2006**, *220*, 37–45. [CrossRef]
17. Dias, E.; Morais, S.; Ramalheira, E.; Pereira, M.L. Characterization of the toxicological effects of aminocarb on rats: Hematological, biochemical, and histological analyses. *J. Toxicol. Environ. Heal-Part A Curr. Issues* **2014**, *77*, 849–855. [CrossRef]
18. Pessoa, P.C.; Luchmann, K.H.; Ribeiro, A.B.; Veras, M.M.; Correa, J.R.M.B.; Nogueira, A.J.; Bainy, A.C.D.; Carvalho, P.S.M. Cholinesterase inhibition and behavioral toxicity of carbofuran on Oreochromis niloticus early life stages. *Aquat. Toxicol.* **2011**, *105*, 312–320. [CrossRef]
19. Tiwari, S.; Tiwari, R.; Singh, A. Impact of cypermethrin on fingerlings of common edible carp (*Labeo rohita*). *Sci. World J.* **2012**, *2012*, 1–7. [CrossRef] [PubMed]
20. Paravani, E.V.; Simoniello, M.F.; Poletta, G.L.; Casco, V.H. Cypermethrin induction of DNA damage and oxidative stress in zebrafish gill cells. *Ecotoxicol. Environ. Saf.* **2019**, *173*, 1–7. [CrossRef] [PubMed]
21. Devi, R.; Yadav, S.; Nehra, R.; Pundir, C.S. An amperometric hypoxanthine biosensor based on Au@FeNPs for determination of hypoxanthine in meat samples. *Int. J. Biol. Macromol.* **2013**, *62*, 629–635. [CrossRef]

22. Jayaraj, R.; Megha, P.; Sreedev, P. Review Article. Organochlorine pesticides, their toxic effects on living organisms and their fate in the environment. *Interdiscip. Toxicol.* **2016**, *9*, 90–100. [CrossRef]
23. Fenik, J.; Tankiewicz, M.; Biziuk, M. Properties and determination of pesticides in fruits and vegetables. *TrAC-Trends Anal. Chem.* **2011**, *30*, 814–826. [CrossRef]
24. Sood, C.; Jaggi, S.; Kumar, V.; Ravindranath, S.D.; Shanker, A. How manufacturing processes affect the level of pesticide residues in tea. *Sci. Food Agric.* **2004**, *84*, 2123–2127. [CrossRef]
25. Karthika, C.; Muraleedharan, N. Influence of manufacturing process on the residues of certain fungicides used on tea. *Toxicol. Environ. Chem.* **2010**, *92*, 1249–1257. [CrossRef]
26. Weisenburger, D.D. Human health effects of agrichemical use. *Hum. Pathol.* **1993**, *24*, 571–576. [CrossRef]
27. PCPA Maximum Residue Limits for Pesticides. Available online: https://www.canada.ca/en/health-canada/services/consumer-product-safety/pesticides-pest-management/public/protecting-your-health-environment/pesticides-food/maximum-residue-limits-pesticides.html (accessed on 8 April 2019).
28. EFSA. *2008 Annual Report on Pesticide Residues according to Article 32 of Regulation (EC) No 396/2005*; EFSA: Parma, Italy, 2010; Volume 8.
29. *CEC North American Regional Action Plan on Lindane and other Hexachlorocyclohexane Isomers: Final Evaluation Report*; Commission for Environmental Cooperation: Montreal, QC, Canada, 2013.
30. Pesticide Action Network-What's on My Food. Available online: http://www.whatsonmyfood.org/index.jsp (accessed on 15 April 2019).
31. Knoche, M.; Lang, A. Ongoing growth challenges fruit skin integrity. *CRC Crit. Rev. Plant Sci.* **2017**, *36*, 190–215. [CrossRef]
32. Keikotlhaile, B.M.; Spanoghe, P.; Mebdoua, S.; Keikotlhaile, B.M.; Spanoghe, P. Pesticide residues in fruits and vegetables. In *Bioactive Molecules in Food*; Stoytcheva, M., Ed.; InTech: Rijeka, Croatia, 2011; p. 39. ISBN 978-953-307-532-7.
33. Taube, J.; Vorkamp, K.; Förster, M.; Herrmann, R. Pesticide residues in biological waste. *Chemosphere* **2002**, *49*, 1357–1365. [CrossRef]
34. Soliman, K.M. Changes in concentration of pesticide residues in potatoes during washing and home preparation. *Food Chem. Toxicol.* **2001**, *39*, 887–891. [CrossRef]
35. Kong, Z.; Dong, F.; Xu, J.; Liu, X.; Zhang, C.; Li, J.; Li, Y.; Chen, X.; Shan, W.; Zheng, Y. Determination of difenoconazole residue in tomato during home canning by UPLC-MS/MS. *Food Control* **2012**, *23*, 542–546. [CrossRef]
36. Abou-Arab, A.A.K. Behavior of pesticides in tomatoes during commercial and home preparation. *Food Chem.* **1999**, *65*, 509–514. [CrossRef]
37. Saber, A.N.; Malhat, F.M.; Badawy, H.M.A.; Barakat, D.A. Dissipation dynamic, residue distribution and processing factor of hexythiazox in strawberry fruits under open field condition. *Food Chem.* **2015**, *196*, 1108–1116. [CrossRef]
38. Prodhan, M.D.; Alam, S.N.; Uddin, M.J. Analytical Methods in Measuring Pesticides in Foods. In *Pesticide Residue in Foods: Sources, Management, and Control*; Springer International Publishing: Cham, Switzerland, 2017; pp. 135–145. ISBN 9783319526836.
39. Berk, Z. Chapter 11—Extraction. In *Food Science and Technology*; Berk, Z., Ed.; Academic Press: San Diego, CA, USA, 2013; pp. 287–309. ISBN 978-0-12-415923-5.
40. de Pinho, G.P.; Neves, A.A.; de Queiroz, M.E.L.R.; Silvério, F.O. Optimization of the liquid-liquid extraction method and low temperature purification (LLE-LTP) for pesticide residue analysis in honey samples by gas chromatography. *Food Control* **2010**, *21*, 1307–1311. [CrossRef]
41. Goulart, S.M.; Alves, R.D.; Neves, A.A.; de Queiroz, J.H.; de Assis, T.C.; de Queiroz, M.E.L.R. Optimization and validation of liquid-liquid extraction with low temperature partitioning for determination of carbamates in water. *Anal. Chim. Acta* **2010**, *671*, 41–47. [CrossRef]
42. Hennion, M.-C. Solid-phase extraction: Method development, sorbents, and coupling with liquid chromatography. *J. Chromatogr. A* **1999**, *856*, 3–54. [CrossRef]
43. Torreti, L.; Simonella, A.; Dossena, A.; Torreti, E. Determination of organochlorine pesticide residues by solid phase extraction and dual-column HRGC. *J. High. Resolut. Chromatogr.* **1992**, *15*, 99–101. [CrossRef]
44. Štajnbaher, D.; Zupančič-Kralj, L. Multiresidue method for determination of 90 pesticides in fresh fruits and vegetables using solid-phase extraction and gas chromatography-mass spectrometry. *J. Chromatogr. A* **2003**, *1015*, 185–198. [CrossRef]

45. Sivaperumal, P.; Anand, P.; Riddhi, L. Rapid determination of pesticide residues in fruits and vegetables, using ultra-high-performance liquid chromatography/time-of-flight mass spectrometry. *Food Chem.* **2015**, *168*, 356–365. [CrossRef] [PubMed]
46. Zhang, Z.; Yang, M.J.; Pawliszyn, J. Solid-Phase Microextraction: A solvent-free alternative for sample preparation. *Anal. Chem.* **1994**, *66*, 844A–853A. [CrossRef]
47. Zambonin, C.G.; Quinto, M.; De Vietro, N.; Palmisano, F. Solid-phase microextraction—Gas chromatography mass spectrometry: A fast and simple screening method for the assessment of organophosphorus pesticides residues in wine and fruit juices. *Food Chem.* **2004**, *86*, 269–274. [CrossRef]
48. Menezes Filho, A.; dos Santos, F.N.; de Paula Pereira, P.A. Development, validation and application of a methodology based on solid-phase micro extraction followed by gas chromatography coupled to mass spectrometry (SPME/GC-MS) for the determination of pesticide residues in mangoes. *Talanta* **2010**, *81*, 346–354. [CrossRef]
49. Anastassiades, M.; Lehotay, S.J.; Stajnbaher, D.; Schenck, F.J. Fast and easy multiresidue method employing acetonitrile extraction/partitioning and "dispersive solid-phase extraction" for the determination of pesticide residues in produce. *J. AOAC Int.* **2003**, *86*, 412–431. [CrossRef]
50. Iqbal, S.; Iqbal, M.M.; Javed, M.; Bahadur, A.; Yasien, S.; Hurr, A.; Ahmad, N.; Raheel, M.; Liu, G. Modified QuEChERS extraction method followed by simultaneous quantitation of nine multi-class pesticides in human blood and urine by using GC-MS. *J. Chromatogr. B* **2020**, *1152*, 122227. [CrossRef]
51. Viera, M.S.; Rizzetti, T.M.; de Souza, M.P.; Martins, M.L.; Prestes, O.D.; Adaime, M.B.; Zanella, R. Multiresidue determination of pesticides in crop plants by the quick, easy, cheap, effective, rugged, and safe method and ultra-high-performance liquid chromatography tandem mass spectrometry using a calibration based on a single level standard addition in the sample. *J. Chromatogr. A* **2017**, *1526*, 119–127.
52. Sojka, M.; Miszczak, A.; Sikorski, P.; Zagibajlo, K.; Karlinska, E.; Kosmala, M. Pesticide residue levels in strawberry processing by-products that are rich in ellagitannins and an assessment of their dietary risk to consumers. *NFS J.* **2015**, *1*, 31–37. [CrossRef]
53. Trevisan, M.T.S.; Owen, R.W.; Calatayud-Vernich, P.; Breuer, A.; Picó, Y. Pesticide analysis in coffee leaves using a quick, easy, cheap, effective, rugged and safe approach and liquid chromatography tandem mass spectrometry: Optimization of the clean-up step. *J. Chromatogr. A* **2017**, *1512*, 98–106. [CrossRef] [PubMed]
54. Lee, J.; Shin, Y.; Lee, J.; Lee, J.; Kim, B.J.; Kim, J.H. Simultaneous analysis of 310 pesticide multiresidues using UHPLC-MS/MS in brown rice, orange, and spinach. *Chemosphere* **2018**, *207*, 519–526. [CrossRef] [PubMed]
55. Farajzadeh, M.A.; Afshar Mogaddam, M.R.; Rezaee Aghdam, S.; Nouri, N.; Bamorrowat, M. Application of elevated temperature-dispersive liquid-liquid microextraction for determination of organophosphorus pesticides residues in aqueous samples followed by gas chromatography-flame ionization detection. *Food Chem.* **2016**, *212*, 198–204. [CrossRef] [PubMed]
56. Farajzadeh, M.A.; Khoshmaram, L.; Nabil, A.A.A. Determination of pyrethroid pesticides residues in vegetable oils using liquid-liquid extraction and dispersive liquid-liquid microextraction followed by gas chromatography-flame ionization detection. *J. Food Compos. Anal.* **2014**, *34*, 128–135. [CrossRef]
57. Balinova, A.M.; Mladenova, R.I.; Shtereva, D.D. Effects of processing on pesticide residues in peaches intended for baby food. *Food Addit. Contam.* **2006**, *23*, 895–901. [CrossRef] [PubMed]
58. Sapahin, H.A.; Makahleh, A.; Saad, B. Determination of organophosphorus pesticide residues in vegetables using solid phase micro-extraction coupled with gas chromatography-flame photometric detector. *Arab. J. Chem.* **2015**, 1–11. [CrossRef]
59. Mac Loughlin, T.M.; Peluso, M.L.; Etchegoyen, M.A.; Alonso, L.L.; de Castro, M.C.; Percudani, M.C.; Marino, D.J.G. Pesticide residues in fruits and vegetables of the argentine domestic market: Occurrence and quality. *Food Control* **2018**, *93*, 129–138. [CrossRef]
60. Stoytcheva, M. *Pesticides-Formulations, Effects, Fate*; Stoytcheva, M., Ed.; InTech: Rijeka, Croatia, 2011; ISBN 9789533075327.
61. Bagheri, H.; Yamini, Y.; Safari, M.; Asiabi, H.; Karimi, M.; Heydari, A. Simultaneous determination of pyrethroids residues in fruit and vegetable samples via supercritical fluid extraction coupled with magnetic solid phase extraction followed by HPLC-UV. *J. Supercrit. Fluids* **2016**, *107*, 571–580. [CrossRef]
62. Timofeeva, I.; Shishov, A.; Kanashina, D.; Dzema, D.; Bulatov, A. On-line in-syringe sugaring-out liquid-liquid extraction coupled with HPLC-MS/MS for the determination of pesticides in fruit and berry juices. *Talanta* **2017**, *167*, 761–767. [CrossRef]

63. Martins, M.L.; Kemmerich, M.; Prestes, O.D.; Maldaner, L.; Jardim, I.C.S.F.; Zanella, R. Evaluation of an alternative fluorinated sorbent for dispersive solid-phase extraction clean-up of the quick, easy, cheap, effective, rugged, and safe method for pesticide residues analysis. *J. Chromatogr. A* **2017**, *1514*, 36–43. [CrossRef]
64. Bhushan, S.; Kalia, K.; Sharma, M.; Singh, B.; Ahuja, P.S. Processing of apple pomace for bioactive molecules. *Crit. Rev. Biotechnol.* **2008**, *28*, 285–296. [CrossRef]
65. De Souza, V.B.; Thomazini, M.; Balieiro, J.C.D.C.; Fávaro-Trindade, C.S. Effect of spray drying on the physicochemical properties and color stability of the powdered pigment obtained from vinification byproducts of the Bordo grape (*Vitis labrusca*). *Food Bioprod. Process.* **2015**, *93*, 39–50. [CrossRef]
66. Schieber, A.; Stintzing, F.C.; Carle, R. By-products of plant food processing as a source of functional compounds—Recent developments. *Trends Food Sci. Technol.* **2001**, *12*, 401–413. [CrossRef]
67. Balasundram, N.; Sundram, K.; Samman, S. Phenolic compounds in plants and agri-industrial by-products: Antioxidant activity, occurrence, and potential uses. *Food Chem.* **2006**, *99*, 191–203. [CrossRef]
68. Sepelev, I.; Galoburda, R. Industrial potato peel waste application in food production: A Review. *Res. Rural Dev.* **2015**, *1*, 130–136.
69. Martins, M.P.; Cortés, E.J.; Eim, V.; Mulet, A.; Cárcel, J.A. Stabilization of apple peel by drying. Influence of temperature and ultrasound application on drying kinetics and product quality. *Dry. Technol.* **2019**, *37*, 559–568. [CrossRef]
70. Do Nascimento, E.M.G.C.; Mulet, A.; Ascheri, J.L.R.; De Carvalho, C.W.P.; Cárcel, J.A. Effects of high-intensity ultrasound on drying kinetics and antioxidant properties of passion fruit peel. *J. Food Eng.* **2015**, *170*, 108–118. [CrossRef]
71. Xie, F.; Zhang, W.; Lan, X.; Gong, S.; Wu, J.; Wang, Z. Effects of high hydrostatic pressure and high pressure homogenization processing on characteristics of potato peel waste pectin. *Carbohydr. Polym.* **2018**, *196*, 474–482. [CrossRef]
72. Wang, X.; Kristo, E.; LaPointe, G. The effect of apple pomace on the texture, rheology and microstructure of set type yogurt. *Food Hydrocoll.* **2019**, *91*, 83–91. [CrossRef]
73. Sudha, M.L.; Baskaran, V.; Leelavathi, K. Apple pomace as a source of dietary fiber and polyphenols and its effect on the rheological characteristics and cake making. *Food Chem.* **2007**, *104*, 686–692. [CrossRef]
74. González-Centeno, M.R.; Jourdes, M.; Femenia, A.; Simal, S.; Rosselló, C.; Teissedre, P.L. Proanthocyanidin composition and antioxidant potential of the stem winemaking byproducts from 10 different grape varieties (*Vitis vinifera* L.). *J. Agric. Food Chem.* **2012**, *60*, 11850–11858. [CrossRef]
75. Garau, M.C.; González-Centeno, M.R.; Luna, J.M.; Negre, A.; Rosselló, C.; Femenia, A. Potential of landrace winery byproducts (*Vitis vinifera* L.) as a source of phenolic compounds with antioxidant properties. *J. Int. Sci. Vigne Vin* **2015**, *49*, 241–251.
76. Pintać, D.; Majkić, T.; Torović, L.; Orčić, D.; Beara, I.; Simin, N.; Mimica-Dukić, N.; Lesjak, M. Solvent selection for efficient extraction of bioactive compounds from grape pomace. *Ind. Crops Prod.* **2018**, *111*, 379–390. [CrossRef]
77. Tseng, A.; Zhao, Y. Wine grape pomace as antioxidant dietary fibre for enhancing nutritional value and improving storability of yogurt and salad dressing. *Food Chem.* **2013**, *138*, 356–365. [CrossRef] [PubMed]
78. Rosales Soto, M.U.; Brown, K.; Ross, C.F. Antioxidant activity and consumer acceptance of grape seed flour-containing food products. *Int. J. Food Sci. Technol.* **2012**, *47*, 592–602. [CrossRef]
79. González-Centeno, M.R.; Rosselló, C.; Simal, S.; Garau, M.C.; López, F.; Femenia, A. Physico-chemical properties of cell wall materials obtained from ten grape varieties and their byproducts: Grape pomaces and stems. *LWT-Food Sci. Technol.* **2010**, *43*, 1580–1586. [CrossRef]
80. Minjares-Fuentes, R.; Femenia, A.; Garau, M.C.; Meza-Velázquez, J.A.; Simal, S.; Rosselló, C. Ultrasound-assisted extraction of pectins from grape pomace using citric acid: A response surface methodology approach. *Carbohydr. Polym.* **2014**, *106*, 179–189. [CrossRef] [PubMed]
81. Minjares-Fuentes, R.; Femenia, A.; Garau, M.C.; Candelas-Cadillo, M.G.; Simal, S.; Rosselló, C. Ultrasound-assisted extraction of hemicelluloses from grape pomace using response surface methodology. *Carbohydr. Polym.* **2016**, *138*, 180–191. [CrossRef] [PubMed]
82. Umaña, M.; Eim, V.; Garau, C.; Rosselló, C.; Simal, S. Ultrasound-assisted extraction of ergosterol and antioxidant components from mushroom by-products and the attainment of a β-glucan rich residue. *Food Chem.* **2020**, *332*, 127390. [CrossRef]

83. Zhang, W.; Xie, F.; Lan, X.; Gong, S.; Wang, Z. Characteristics of pectin from black cherry tomato waste modified by dynamic high-pressure microfluidization. *J. Food Eng.* **2018**, *216*, 90–97. [CrossRef]
84. Četojević-Simin, D.; Djilas, S.; Stajčić, S.; Ćetković, G.; Čanadanović-Brunet, J.; Mandić, A. Tomato waste: Carotenoids content, antioxidant and cell growth activities. *Food Chem.* **2014**, *172*, 225–232.
85. Elbadrawy, E.; Sello, A. Evaluation of nutritional value and antioxidant activity of tomato peel extracts. *Arab. J. Chem.* **2016**, *9*, S1010–S1018. [CrossRef]
86. Strati, I.F.; Gogou, E.; Oreopoulou, V. Enzyme and high pressure assisted extraction of carotenoids from tomato waste. *Food Bioprod. Process.* **2015**, *94*, 668–674. [CrossRef]
87. Sójka, M.; Klimczak, E.; Macierzynski, J.; Kołodziejczyk, K. Nutrient and polyphenolic composition of industrial strawberry press cake. *Eur Food Res. Technol* **2013**, *237*, 995–1007. [CrossRef]
88. Kosmala, M.; Zduńczyk, Z.; Kołodziejczyk, K.; Klimczak, E.; Juskiewicz, J.; Zduńczyk, P. Chemical composition of polyphenols extracted from strawberry pomace and their effect on physiological properties of diets supplemented with different types of dietary fibre in rats. *Eur. J. Nutr.* **2014**, *53*, 521–532. [CrossRef]
89. Tovar, A.K.; Godínez, L.A.; Espejel, F.; Ramírez-Zamora, R.M.; Robles, I. Optimization of the integral valorization process for orange peel waste using a design of experiments approach: Production of high-quality pectin and activated carbon. *Waste Manag.* **2019**, *85*, 202–213. [CrossRef]
90. Rezaei, K.; Hosseini, S.S.; Raji, Z.; Khodaiyan, F.; Kiani, H. Extraction optimization and physicochemical properties of pectin from melon peel. *Int. J. Biol. Macromol.* **2017**, *98*, 709–716.
91. Khamsucharit, P.; Laohaphatanalert, K.; Gavinlertvatana, P.; Sriroth, K.; Sangseethong, K. Characterization of pectin extracted from banana peels of different varieties. *Food Sci. Biotechnol.* **2018**, *27*, 623–629. [CrossRef]
92. Qiu, L.P.; Zhao, G.L.; Wu, H.; Jiang, L.; Li, X.F.; Liu, J.J. Investigation of combined effects of independent variables on extraction of pectin from banana peel using response surface methodology. *Carbohydr. Polym.* **2010**, *80*, 326–331. [CrossRef]
93. Pagán, J.; Ibarz, A.; Llorca, M.; Pagán, A.; Barbosa-Cánovas, G.V. Extraction and characterization of pectin from stored peach pomace. *Food Res. Int.* **2001**, *34*, 605–612. [CrossRef]
94. Bélafi-Bakó, K.; Cserjési, P.; Beszédes, S.; Csanádi, Z.; Hodúr, C. Berry pectins: Microwave-assisted extraction and rheological properties. *Food Bioprocess. Technol.* **2012**, *5*, 1100–1105. [CrossRef]
95. Khattak, K.F.; Rahman, T.U. Analysis of vegetable's peels as a natural source of vitamins and minerals. *Int. Food Res. J.* **2017**, *24*, 292–297.
96. Maner, S.; Sharma, A.K.; Banerjee, K. Wheat flour replacement by wine grape pomace powder positively affects physical, functional and sensory properties of cookies. *Proc. Natl. Acad. Sci. USA* **2017**, *87*, 109–113. [CrossRef]
97. Zhang, L.; Zhu, M.; Shi, T.; Guo, C.; Huang, Y.; Chen, Y.; Xie, M. Recovery of dietary fiber and polyphenol from grape juice pomace and evaluation of their functional properties and polyphenol compositions. *Food Funct.* **2017**, *8*, 341–351. [CrossRef]
98. Garcia-Perez, J.V.; García-Alvarado, M.A.; Carcel, J.A.; Mulet, A. Extraction kinetics modeling of antioxidants from grape stalk (*Vitis vinifera var. Bobal*): Influence of drying conditions. *J. Food Eng.* **2010**, *101*, 49–58. [CrossRef]
99. Bai, X.; Zhang, H.; Ren, S. Antioxidant activity and HPLC analysis of polyphenol-enriched extracts from industrial apple pomace. *J. Sci. Food Agric.* **2013**, *93*, 2502–2506. [CrossRef]
100. Constenla, D.; Ponce, A.G.; Lozano, J.E. Effect of pomace drying on apple pectin. *LWT-Food Sci. Technol.* **2002**, *35*, 216–221. [CrossRef]
101. Silva-Beltrán, N.P.; Chaidez-Quiroz, C.; López-Cuevas, O.; Ruiz-Cruz, S.; López-Mata, M.A.; Del-Toro-sánchez, C.L.; Marquez-Rios, E.; Ornelas-Paz, J.D.J. Phenolic compounds of potato peel extracts: Their antioxidant activity and protection against human enteric viruses. *J. Microbiol. Biotechnol.* **2017**, *27*, 234–241. [CrossRef]
102. Salim, N.S.M.; Singh, A.; Raghavan, V. Potential utilization of fruit and vegetable wastes for food through drying or extraction techniques. *Nov. Tech. Nutr. Food Sci.* **2017**, *1*, 1–13. [CrossRef]
103. Singh, A.; Nair, G.R.; Liplap, P.; Gariepy, Y.; Orsat, V.; Raghavan, V. Effect of dielectric properties of a solvent-water mixture used in microwave-assisted extraction of antioxidants from potato peels. *Antioxidants* **2014**, *3*, 99–113. [CrossRef]
104. Adetunji, L.R.; Adekunle, A.; Orsat, V.; Raghavan, V. Advances in the pectin production process using novel extraction techniques: A review. *Food Hydrocoll.* **2017**, *62*, 239–250. [CrossRef]

105. Araya-Farias, M.; Ratti, C. Dehydration of Foods: General Concepts. In *Advances in Food Dehydration*; Ratti, C., Ed.; CRC Press: Boca Raton, FL, USA, 2008; pp. 19–54.
106. Ratti, C. Hot air and freeze-drying of high-value foods: A review. *J. Food Eng.* **2001**, *49*, 311–319. [CrossRef]
107. Brennan, J.G. *Food Processing Handbook*; Wiley-VCH Verlag &, Co. KGaA: Weinheim, Germany, 2006; Volume 1, ISBN 9783527307197.
108. Toledo, R.T. Fundamentals of Food Process. In *Engineering*, 3rd ed.; Toledo, R.T., Ed.; Springer: Berlin/Heidelberg, Germany, 1999; ISBN 0878495223.
109. Carrín, M.E.; Crapiste, G.H. Convective Drying of Foods. In *Advances in Food Dehydration*; Ratti, C., Ed.; CRC Press: Boca Raton, FL, USA, 2008; pp. 123–151.
110. Krishna, R.; Wesselingh, J.A. The Maxwell-Stefan approach to mass transfer. *Chem. Eng. Sci.* **1997**, *52*, 861–911. [CrossRef]
111. Izák, P.; Bartovská, L.; Friess, K.; Šípek, M.; Uchytil, P. Comparison of various models for transport of binary mixtures through dense polymer membrane. *Polymer* **2003**, *44*, 2679–2687. [CrossRef]
112. Thurner, F.; Schlünder, E.U. Progress towards understanding the drying of porous materials wetted with binary mixtures. *Chem. Eng. Process.* **1986**, *20*, 33–41. [CrossRef]
113. Wolf, H. Selective drying of hygroscopic materials wetted with binary mixtures. *Dry. Technol.* **1994**, *12*, 1445–1470. [CrossRef]
114. Ho, C.K.; Udell, K.S. Mass transfer limited drying of porous media containing an immobile binary liquid mixture. *Int. J. Heat Mass Transf.* **1995**, *38*, 339–350. [CrossRef]
115. Wulsten, E.; Kiekens, F.; van Dycke, F.; Voorspoels, J.; Lee, G. Levitated single-droplet drying: Case study with itraconazole dried in binary organic solvent mixtures. *Int. J. Pharm.* **2009**, *378*, 116–121. [CrossRef]
116. Heintz, A.; Stephan, W. A generalized solution-diffusion model of the pervaporation process through composite membranes Part II. Concentration polarization, coupled diffusion and the influence of the porous support layer. *J. Memb. Sci.* **1994**, *89*, 153–169. [CrossRef]
117. Aguilera, J.M. Solid-Liquid Extraction. In *Extraction Optimization in Food Engineering*; Tzia, C., Ed.; Marcel Dekker: New York, NY, USA, 2003; p. 438. ISBN 0824747127.
118. Chaiklahan, R.; Chirasuwan, N.; Bunnag, B. Stability of phycocyanin extracted from Spirulina sp.: Influence of temperature, pH and preservatives. *Process. Biochem.* **2012**, *47*, 659–664. [CrossRef]
119. Cheung, Y.C.; Wu, J.Y. Kinetic models and process parameters for ultrasound-assisted extraction of water-soluble components and polysaccharides from a medicinal fungus. *Biochem. Eng. J.* **2013**, *79*, 214–220. [CrossRef]
120. Jokić, S.; Velić, D.; Bilić, M.; Bucić-Kojić, A.; Planinić, M.; Tomas, S. Modelling of the process of solid-liquid extraction of total polyphenols from soybeans. *Czech. J. Food Sci.* **2010**, *28*, 206–212. [CrossRef]
121. Natolino, A.; Da Porto, C. Kinetic models for conventional and ultrasound assistant extraction of polyphenols from defatted fresh and distilled grape marc and its main components skins and seeds. *Chem. Eng. Res. Des.* **2020**, *156*, 1–12. [CrossRef]
122. González-Centeno, M.R.; Comas-Serra, F.; Femenia, A.; Rosselló, C.; Simal, S. Effect of power ultrasound application on aqueous extraction of phenolic compounds and antioxidant capacity from grape pomace (*Vitis vinifera L.*): Experimental kinetics and modeling. *Ultrason. Sonochem.* **2015**, *22*, 506–514. [CrossRef]
123. Do, Q.D.; Angkawijaya, A.E.; Tran-Nguyen, P.L.; Huynh, L.H.; Soetaredjo, F.E.; Ismadji, S.; Ju, Y.H. Effect of extraction solvent on total phenol content, total flavonoid content, and antioxidant activity of Limnophila aromatica. *J. Food Drug Anal.* **2014**, *22*, 296–302. [CrossRef]
124. Van Ngo, T.; Scarlett, C.J.; Bowyer, M.C.; Ngo, P.D.; Van Vuong, Q. Impact of different extraction solvents on bioactive compounds and antioxidant capacity from the root of *Salacia chinensis* L. *J. Food Qual.* **2017**, *2017*, 9305047. [CrossRef]
125. Peanparkdee, M.; Iwamoto, S. Bioactive compounds from by-products of rice cultivation and rice processing: Extraction and application in the food and pharmaceutical industries. *Trends Food Sci. Technol.* **2019**, *86*, 109–117. [CrossRef]
126. Nayak, A.; Bhushan, B. An overview of the recent trends on the waste valorization techniques for food wastes. *J. Environ. Manag.* **2019**, *233*, 352–370. [CrossRef]
127. Dufton, G.; Mikhaylin, S.; Gaaloul, S.; Bazinet, L. Positive impact of pulsed electric field on lactic acid removal, demineralization and membrane scaling during acid whey electrodialysis. *Int. J. Mol. Sci.* **2019**, *20*, 797. [CrossRef] [PubMed]

128. Vallespir, F.; Rodríguez, Ó.; Cárcel, J.A.; Rosselló, C.; Simal, S. Ultrasound assisted low-temperature drying of kiwifruit: Effects on drying kinetics, bioactive compounds and antioxidant activity. *J. Sci. Food Agric.* **2019**, *99*, 2901–2909. [CrossRef] [PubMed]
129. Wang, H.; Mustaffar, A.; Phan, A.N.; Zivkovic, V.; Reay, D.; Law, R.; Boodhoo, K. A review of process intensification applied to solids handling. *Chem. Eng. Process. Process. Intensif.* **2017**, *118*, 78–107. [CrossRef]
130. Wiktor, A.; Witrowa-Rajchert, D. Pulsed Electric Fields as Pretreatment for Subsequent Food Process Operations. In *Handbook of Electroporation*; Miklavcic, D., Ed.; Springer International Publishing: Cham, Switzerland, 2016; pp. 1–16. ISBN 978-3-319-26779-1.
131. Barbosa-Cánovas, G.V.; Góngora-Nieto, M.M.; Pothakamury, U.R.; Swanson, B.G. Fundamentals of high-intensity pulsed electric fields (PEF). In *Preservation of Foods with Pulsed Electric Fields*; Barbosa-Cánovas, G.V., Ed.; Academic Press: San Diego, CA, USA, 1999; pp. 1–19.
132. Wouters, P.C.; Alvarez, I.; Raso, J. Critical factors determining inactivation kinetics by pulsed electric field food processing. *Trends Food Sci. Technol.* **2001**, *12*, 112–121. [CrossRef]
133. Nowacka, M.; Wiktor, A.; Anuszewska, A.; Dadan, M.; Rybak, K.; Witrowa-Rajchert, D. The application of unconventional technologies as pulsed electric field, ultrasound and microwave-vacuum drying in the production of dried cranberry snacks. *Ultrason. Sonochem.* **2019**, *56*, 1–13. [CrossRef]
134. Chen, F.; Zeng, L.; Zhang, Y.; Liao, X.; Ge, Y.; Hu, X.; Jiang, L. Degradation behaviour of methamidophos and chlorpyrifos in apple juice treated with pulsed electric fields. *Food Chem.* **2009**, *112*, 956–961. [CrossRef]
135. Chauhan, O.P.; Sayanfar, S.; Toepfl, S. Effect of pulsed electric field on texture and drying time of apple slices. *J. Food Sci. Technol.* **2018**, *55*, 2251–2258. [CrossRef]
136. Lammerskitten, A.; Wiktor, A.; Siemer, C.; Toepfl, S.; Mykhailyk, V.; Gondek, E.; Rybak, K.; Witrowa-Rajchert, D.; Parniakov, O. The effects of pulsed electric fields on the quality parameters of freeze-dried apples. *J. Food Eng.* **2019**, *252*, 36–43. [CrossRef]
137. Telfser, A.; Gómez Galindo, F. Effect of reversible permeabilization in combination with different drying methods on the structure and sensorial quality of dried basil (*Ocimum basilicum L.*) leaves. *LWT* **2019**, *99*, 148–155. [CrossRef]
138. Ostermeier, R.; Giersemehl, P.; Siemer, C.; Töpfl, S.; Jäger, H. Influence of pulsed electric field (PEF) pre-treatment on the convective drying kinetics of onions. *J. Food Eng.* **2018**, *237*, 110–117. [CrossRef]
139. Lebovka, N.I.; Shynkaryk, N.V.; Vorobiev, E. Pulsed electric field enhanced drying of potato tissue. *J. Food Eng.* **2007**, *78*, 606–613. [CrossRef]
140. Huang, W.; Feng, Z.; Aila, R.; Hou, Y.; Carne, A.; Bekhit, A.E.D.A. Effect of pulsed electric fields (PEF) on physico-chemical properties, β-carotene and antioxidant activity of air-dried apricots. *Food Chem.* **2019**, *291*, 253–262. [CrossRef]
141. Yu, Y.; Jin, T.Z.; Xiao, G. Effects of pulsed electric fields pretreatment and drying method on drying characteristics and nutritive quality of blueberries. *J. Food Process. Preserv.* **2017**, *41*. [CrossRef]
142. El Kantar, S.; Boussetta, N.; Lebovka, N.; Foucart, F.; Rajha, H.N.; Maroun, R.G.; Louka, N.; Vorobiev, E. Pulsed electric field treatment of citrus fruits: Improvement of juice and polyphenols extraction. *Innov. Food Sci. Emerg. Technol.* **2018**, *46*, 153–161. [CrossRef]
143. Luengo, E.; Álvarez, I.; Raso, J. Improving the pressing extraction of polyphenols of orange peel by pulsed electric fields. *Innov. Food Sci. Emerg. Technol.* **2013**, *17*, 79–84. [CrossRef]
144. Peiro, S.; Luengo, E.; Segovia, F.; Raso, J.; Pilar Almajano, M. Improving polyphenol extraction from lemon residues by pulsed electric fields. *Waste Biomass Valor* **2019**, *10*, 889–897. [CrossRef]
145. Jeyakondan, S.; Jayas, D.S.; Holley, R.A. Pulsed electric field processing of foods: A review. *J. Food Prot.* **2016**, *62*, 1088–1096. [CrossRef]
146. Mason, T.J.; Chemat, F.; Vinatoru, M. The extraction of natural products using ultrasound or microwaves. *Curr. Org. Chem.* **2011**, *15*, 237–247. [CrossRef]
147. Soria, A.C.; Villamiel, M. Effect of ultrasound on the technological properties and bioactivity of food: A review. *Trends Food Sci. Technol.* **2010**, *21*, 323–331. [CrossRef]
148. Rodríguez, Ó.; Eim, V.; Rosselló, C.; Femenia, A.; Cárcel, J.A.; Simal, S. Application of power ultrasound on the convective drying of fruits and vegetables: Effects on quality. *J. Sci. Food Agric.* **2018**, *98*, 1660–1673. [CrossRef]
149. Cárcel, J.A.; García-Pérez, J.V.; Benedito, J.; Mulet, A. Food process innovation through new technologies: Use of ultrasound. *J. Food Eng.* **2012**, *110*, 200–207. [CrossRef]

150. Cárcel, J.A.; García-Pérez, J.V.; Riera, E.; Mulet, A. Influence of high-intensity ultrasound on drying kinetics of persimmon. *Dry. Technol.* **2007**, *25*, 185–193. [CrossRef]
151. Ratti, C. Novel Food Dryers and Future Perspectives. In *Advances in Food Dehydration*; Ratti, C., Ed.; CRC Press: Boca Raton, FL, USA, 2008; pp. 447–458. ISBN 9780429147296.
152. Cárcel, J.A.; Castillo, D.; Simal, S.; Mulet, A. Influence of temperature and ultrasound on drying kinetics and antioxidant properties of red pepper. *Dry. Technol.* **2019**, *37*, 486–493. [CrossRef]
153. Carrión, C.; Mulet, A.; García-Pérez, J.V.; Cárcel, J.A. Ultrasonically assisted atmospheric freeze-drying of button mushroom. Drying kinetics and product quality. *Dry. Technol.* **2018**, *36*, 1814–1823. [CrossRef]
154. Brines, C.; Mulet, A.; García-Pérez, J.V.; Riera, E.; Cárcel, J.A. Influence of the ultrasonic power applied on freeze drying kinetics. *Phys. Procedia* **2015**, *70*, 850–853. [CrossRef]
155. Ozuna, C.; Álvarez-Arenas, T.G.; Riera, E.; Cárcel, J.A.; Garcia-Perez, J.V. Influence of material structure on air-borne ultrasonic application in drying. *Ultrason. Sonochem.* **2014**, *21*, 1235–1243. [CrossRef]
156. Rodríguez, Ó.; Santacatalina, J.V.; Simal, S.; Garcia-Perez, J.V.; Femenia, A.; Rosselló, C. Influence of power ultrasound application on drying kinetics of apple and its antioxidant and microstructural properties. *J. Food Eng.* **2014**, *129*, 21–29. [CrossRef]
157. Santacatalina, J.V.; Contreras, M.; Simal, S.; Cárcel, J.A.; Garcia-Perez, J.V. Impact of applied ultrasonic power on the low temperature drying of apple. *Ultrason. Sonochem.* **2016**, *28*, 100–109. [CrossRef]
158. Musielak, G.; Mierzwa, D.; Kroehnke, J. Food drying enhancement by ultrasound – A review. *Trends Food Sci. Technol.* **2016**, *56*, 126–141. [CrossRef]
159. Cárcel, J.A.; García-Pérez, J.V.; Riera, E.; Rosselló, C.; Mulet, A. Drying Assisted by Power Ultrasound. In *Modern Drying Technology*; John Wiley & Sons, Ltd.: Hoboken, NJ, USA, 2014; pp. 237–278. ISBN 9783527631704.
160. Llavata, B.; García-Pérez, J.V.; Simal, S.; Cárcel, J.A. Innovative pre-treatments to enhance food drying: A current review. *Curr. Opin. Food Sci.* **2020**, *35*, 20–26. [CrossRef]
161. Quintin, D.; Garcia-Gomez, P.; Ayuso, M.; Sanmartin, A.M. Active biocompounds to improve food nutritional value. *Trends Food Sci. Technol.* **2019**, *84*, 19–21. [CrossRef]
162. Ribeiro, A.; Ruphuy, G.; Lopes, J.C.; Dias, M.M.; Barros, L.; Barreiro, F.; Ferreira, I.C.F.R. Spray-drying microencapsulation of synergistic antioxidant mushroom extracts and their use as functional food ingredients. *Food Chem.* **2015**, *188*, 612–618. [CrossRef]
163. Galviz-Quezada, A.; Ochoa-Aristizábal, A.M.; Arias Zabala, M.E.; Ochoa, S.; Osorio-Tobón, J.F. Valorization of iraca (Carludovica palmata, Ruiz & Pav.) infructescence by ultrasound-assisted extraction: An economic evaluation. *Food Bioprod. Process.* **2019**, *118*, 91–102.
164. Machado, I.; Faccio, R.; Pistón, M. Characterization of the effects involved in ultrasound-assisted extraction of trace elements from artichoke leaves and soybean seeds. *Ultrason. Sonochem.* **2019**, *59*, 104752. [CrossRef]
165. Polachini, T.C.; Mulet, A.; Telis-Romero, J.; Cárcel, J.A. Influence of high-intensity ultrasound application on the kinetics of sugar release from acid suspensions of artichoke (*Cynara scolymus*) biomass. *Chem. Eng. Process.-Process. Intensif.* **2019**, *145*. [CrossRef]
166. Caldas, T.W.; Mazza, K.E.L.; Teles, A.S.C.; Mattos, G.N.; Brígida, A.I.S.; Conte-Junior, C.A.; Borguini, R.G.; Godoy, R.L.O.; Cabral, L.M.C.; Tonon, R.V. Phenolic compounds recovery from grape skin using conventional and non-conventional extraction methods. *Ind. Crops Prod.* **2018**, *111*, 86–91. [CrossRef]
167. Goula, A.M.; Thymiatis, K.; Kaderides, K. Valorization of grape pomace: Drying behavior and ultrasound extraction of phenolics. *Food Bioprod. Process.* **2016**, *100*, 132–144. [CrossRef]
168. Amiri-Rigi, A.; Abbasi, S.; Scanlon, M.G. Enhanced lycopene extraction from tomato industrial waste using microemulsion technique: Optimization of enzymatic and ultrasound pre-treatments. *Innov. Food Sci. Emerg. Technol.* **2016**, *35*, 160–167. [CrossRef]
169. Dzah, C.S.; Duan, Y.; Zhang, H.; Wen, C.; Zhang, J.; Chen, G.; Ma, H. The effects of ultrasound assisted extraction on yield, antioxidant, anticancer and antimicrobial activity of polyphenol extracts: A review. *Food Biosci.* **2020**, *35*. [CrossRef]
170. Mello Jr, R.E.; Fontana, A.; Corrêa, J.; Karaağaç, E.; García-Pérez, J.V.; Carcel, J.A. Influence of the pretreatment with pulsed electric field on the hot air drying of orange peel. In Proceedings of the Eurodrying'2019, Torino, Italy, 10–12 July 2019.

171. Menor-García, L.; García-Pérez, J.V.; Raso, J.; Benedito, J.; Cárcel, J.A. Enhancement of betanin extraction from red beetroot by ultrasounds and pulsed electric fields. In Proceedings of the 3rd World Congress on Electroporation, Toulouse, France, 3–6 September 2019.
172. Wang, L.; Weller, C.L. Recent advances in extraction of nutraceuticals from plants. *Trends Food Sci. Technol.* **2006**, *17*, 300–312. [CrossRef]
173. Pérez-Serradilla, J.A.; Luque De Castro, M.D. Microwave-assisted extraction of phenolic compounds from wine lees and spray-drying of the extract. *Food Chem.* **2011**, *124*, 1652–1659. [CrossRef]
174. Li, H.; Deng, Z.; Wu, T.; Liu, R.; Loewen, S.; Tsao, R. Microwave-assisted extraction of phenolics with maximal antioxidant activities in tomatoes. *Food Chem.* **2012**, *130*, 928–936. [CrossRef]
175. Casazza, A.A.; Aliakbarian, B.; Mantegna, S.; Cravotto, G.; Perego, P. Extraction of phenolics from *Vitis vinifera* wastes using non-conventional techniques. *J. Food Eng.* **2010**, *100*, 50–55. [CrossRef]
176. Kaderides, K.; Papaoikonomou, L.; Serafim, M.; Goula, A.M. Microwave-assisted extraction of phenolics from pomegranate peels: Optimization, kinetics, and comparison with ultrasounds extraction. *Chem. Eng. Process. Process. Intensifi Cation* **2019**, *137*, 1–11. [CrossRef]
177. Ballard, T.S.; Mallikarjunan, P.; Zhou, K.; Keefe, S.O. Microwave-assisted extraction of phenolic antioxidant compounds from peanut skins. *Food Chem.* **2010**, *120*, 1185–1192. [CrossRef]
178. Hahn, T.; Lang, S.; Ulber, R.; Muffler, K. Novel procedures for the extraction of fucoidan from brown algae. *Process. Biochem.* **2012**, *47*, 1691–1698. [CrossRef]
179. Yang, Y.; Lei, Z.; Zhao, M.; Wu, C.; Wang, L.; Xu, Y. Microwave-assisted extraction of an acidic polysaccharide from Ribes nigrum L.: Structural characteristics and biological activities. *Ind. Crops Prod.* **2020**, *147*, 112249. [CrossRef]
180. Arrutia, F.; Adam, M.; Calvo-Carrascal, M.Á.; Mao, Y.; Binner, E. Development of a continuous-flow system for microwave-assisted extraction of pectin-derived oligosaccharides from food waste. *Chem. Eng. J.* **2020**, *395*, 125056. [CrossRef]
181. Zhao, L.; Liu, F.; Ge, J.; Ma, L.; Wu, L.; Xue, X. Changes in eleven pesticide residues in jujube (*Ziziphus jujuba* Mill.) during drying processing. *Dry. Technol.* **2018**, *36*, 965–972. [CrossRef]
182. Szadzińska, J.; Łechtańska, J.; Pashminehazar, R.; Kharaghani, A.; Tsotsas, E. Microwave- and ultrasound-assisted convective drying of raspberries: Drying kinetics and microstructural changes. *Dry. Technol.* **2019**, *37*, 1–12. [CrossRef]
183. M'hiri, N.; Ghali, R.; Ben Nasr, I.; Boudhrioua, N. Effect of different drying processes on functional properties of industrial lemon byproduct. *Process. Saf. Environ. Prot.* **2018**, *116*, 450–460. [CrossRef]
184. Santana, I.; Castelo-Branco, V.N.; Guimarães, B.M.; de, O. Silva, L.; Peixoto, V.O.D.S.; Cabral, L.M.C.; Freitas, S.P.; Torres, A.G. Hass avocado (*Persea americana Mill.*) oil enriched in phenolic compounds and tocopherols by expeller-pressing the unpeeled microwave dried fruit. *Food Chem.* **2019**, *286*, 354–361. [CrossRef]
185. Rodriguez, A.; Rodriguez, M.M.; Lemoine, M.L.; Mascheroni, R.H. Study and comparison of different drying processes for dehydration of raspberries. *Dry. Technol.* **2017**, *35*, 689–698. [CrossRef]
186. Zhang, M.; Tang, J.; Mujumdar, A.S.; Wang, S. Trends in microwave-related drying of fruits and vegetables. *Trends Food Sci. Technol.* **2006**, *17*, 524–534. [CrossRef]
187. Guo, Q.; Sun, D.W.; Cheng, J.H.; Han, Z. Microwave processing techniques and their recent applications in the food industry. *Trends Food Sci. Technol.* **2017**, *67*, 236–247. [CrossRef]
188. Chizoba Ekezie, F.-G.; Sun, D.-W.; Han, Z.; Cheng, J.-H. Microwave-assisted food processing technologies for enhancing product quality and process efficiency: A review of recent developments. *Trends Food Sci. Technol.* **2017**, *67*, 58–69. [CrossRef]
189. *FAO/WHO Updating the Principles and Methods of Risk Assessment: MRLs for Pesticides and Veterinary Drugs*; FAO: Rome, Italy, 2006.
190. Holland, P.T.; Hamilton, D.; Ohlin, B.; Skidmore, M.W. *Effects of Storage and Processing on Pesticide Residues in Plant Products*; IUPAC: London, UK, 1994; Volume 66.
191. Kaushik, G.; Satya, S.; Naik, S.N. Food processing a tool to pesticide residue dissipation—A review. *Food Res. Int.* **2009**, *42*, 26–40. [CrossRef]
192. Kar, A.; Mandal, K.; Singh, B. Decontamination of chlorantraniliprole residues on cabbage and cauliflower through household processing methods. *Bull. Environ. Contam. Toxicol.* **2012**, *88*, 501–506. [CrossRef]
193. Mergnat, T.; Fritsch, P.; Saint-Joly, C.; Truchot, E.; Saint-Blanquat, G. Reduction in phosalone residue levels during industrial dehydration of apples. *Food Addit. Contam.* **1995**, *12*, 759–767. [CrossRef]

194. Lentza-Rizos, C.; Balokas, A. Residue levels of chlorpropham in individual tubers and composite samples of postharvest-treated potatoes. *J. Agric. Food Chem.* **2001**, *49*, 710–714. [CrossRef]
195. Reiler, E.; Jørs, E.; Bælum, J.; Huici, O.; Alvarez Caero, M.M.; Cedergreen, N. The influence of tomato processing on residues of organochlorine and organophosphate insecticides and their associated dietary risk. *Sci. Total Environ.* **2015**, *527–528*, 262–269. [CrossRef]
196. Chavarri, M.J.; Herrera, A.; Arino, A. Pesticide residues in field-sprayed and processed fruits and vegetables. *J. Sci. Food Agric.* **2004**, *84*, 1253–1259. [CrossRef]
197. Hassanzadeh, N.; Bahramifar, N.; Esmaili-Sari, A. Residue content of carbaryl applied on greenhouse cucumbers and its reduction by duration of a pre-harvest interval and post-harvest household processing. *J. Sci. Food Agric.* **2010**, *90*, 2249–2253. [CrossRef]
198. Li, M.; Liu, Y.; Fan, B.; Lu, J.; He, Y.; Kong, Z.; Zhu, Y.; Jian, Q.; Wang, F. A chemometric processing-factor-based approach to the determination of the fates of five pesticides during apple processing. *LWT-Food Sci. Technol.* **2015**, *63*, 1102–1109. [CrossRef]
199. Sharma, J.; Satya, S.; Kumar, V.; Tewary, D.K. Dissipation of pesticides during bread-making. *Chem. Health Saf.* **2005**, *12*, 17–22. [CrossRef]
200. Uygun, U.; Koksel, H.; Atli, A. Residue levels of malathion and its metabolites and fenitrothion in post-harvest treated wheat during storage, milling and baking. *Food Chem.* **2005**, *92*, 643–647. [CrossRef]
201. Habiba, R.A.; Ismail, S.M.M.; Ali, H.M. Biochemical effects of profenofos residues in potatoes. *J. Agric. Food Chem.* **1992**, *40*, 1852–1855. [CrossRef] [PubMed]
202. Athanasopoulos, P.E.; Pappas, C. Effects of fruit acidity and storage conditions on the rate of degradation of azinphos methyl on apples and lemons. *Food Chem.* **2000**, *69*, 69–72. [CrossRef]
203. Pappas, C.J.; Athanasopoulos, P.E.; Kyriakidis, N.B. Degradation of azinphos-ethyl in apples stored in different conditions. *J. Agric. Food Chem.* **1998**, *46*, 2092–2095. [CrossRef]
204. Noh, H.H.; Kim, D.K.; Lee, E.Y.; Chang, M.I.; Im, M.H.; Lee, Y.D.; Kyung, K.S. Effects of oven drying on pesticide residues in field-grown chili peppers. *J. Korean Soc. Appl. Biol. Chem.* **2015**, *58*, 97–104. [CrossRef]
205. Cabras, P.; Angioni, A.; Garau, V.L.; Melis, M.; Pirisi, F.M.; Cabitza, F.; Pala, M. Pesticide residues in raisin processing. *J. Agric. Food Chem.* **1998**, *46*, 2309–2311. [CrossRef]
206. Özbey, A.; Karagöz, Ş.; Cingöz, A. Effect of drying process on pesticide residues in grapes. *J. Food* **2017**, *42*, 204–209. [CrossRef]
207. Archer, T.E.; Toscano, R.A. Fate of kelthane residues on apple pomace exposed to drying in the dark, sunlight and ultraviolet light irradiation. *Bull. Environ. Contam. Toxicol.* **1972**, *7*, 353–357. [CrossRef]
208. Cabras, P.; Angioni, A.; Garau, V.L.; Melis, M.; Pirisi, F.M.; Cabitza, F.; Cubeddu, M. Pesticide residues on field-sprayed apricots and in apricot drying processes. *J. Agric. Food Chem.* **1998**, *46*, 2306–2308. [CrossRef]
209. Cabras, P.; Angioni, A.; Garau, V.L.; Minelli, E.V.; Cabitza, F.; Cubeddu, M. Residues of some pesticides in fresh and dried apricots. *J. Agric. Food Chem.* **1997**, *45*, 3221–3222. [CrossRef]
210. Cabras, P.; Angioni, A. Pesticide residues in grapes, wine, and their processing products. *J. Agric. Food Chem.* **2000**, *48*, 967–973. [CrossRef]
211. Shabeer, T.P.A.; Banerjee, K.; Jadhav, M.; Girame, R.; Utture, S.; Hingmire, S.; Oulkar, D. Residue dissipation and processing factor for dimethomorph, famoxadone and cymoxanil during raisin preparation. *Food Chem.* **2015**, *170*, 180–185. [CrossRef]
212. Lentza-Rizos, C.; Avramides, E.J.; Kokkinaki, K. Residues of azoxystrobin from grapes to raisins. *J. Agric. Food Chem.* **2006**, *54*, 138–141. [CrossRef]
213. Shabeer, T.P.A.; Girame, R.; Hingmire, S.; Banerjee, K.; Sharma, A.K.; Oulkar, D.; Utture, S.; Jadhav, M. Dissipation pattern, safety evaluation, and generation of processing factor (PF) for pyraclostrobin and metiram residues in grapes during raisin preparation. *Environ. Monit. Assess.* **2015**, *187*. [CrossRef]
214. Cabras, P.; Angioni, A.; Garau, V.L.; Pirisi, F.M.; Cabitza, F.; Pala, M.; Farris, G.A. Fate of quinoxyfen residues in grapes, wine, and their processing products. *J. Agric. Food Chem.* **2000**, *48*, 6128–6131. [CrossRef]
215. Fang, Q.; Shi, Y.; Cao, H.; Tong, Z.; Xiao, J.; Liao, M.; Wu, X.; Hua, R. Degradation dynamics and dietary risk assessments of two neonicotinoid insecticides during *Lonicera japonica* planting, drying, and tea brewing processes. *J. Agric. Food Chem.* **2017**, *65*, 1483–1488. [CrossRef]
216. Chen, W.; Liu, Y.; Jiao, B. Dissipation behavior of five organophosphorus pesticides in kumquat sample during honeyed kumquat candied fruit processing. *Food Control.* **2016**, *66*, 87–92. [CrossRef]

217. Nath, G.; Jat, N.R.; Srivastava, B.P. Effect of washing, cooking and dehydration on the removal of some insecticides from Okra (*Abelmoschus esculentus Moench.*). *J. Food Sci. Technol.* **1975**, *12*, 127–130.
218. Bajwa, U.; Sandhu, K.S. Effect of handling and processing on pesticide residues in food—A review. *J. Food Sci. Technol.* **2014**, *51*, 201–220. [CrossRef]
219. Xia, E.; Tao, W.; Yao, X.; Wang, J.; Tang, F. Effects of processing on carbendazim residue in *Pleurotus ostreatus*. *Food Sci. Nutr.* **2016**, *4*, 645–650. [CrossRef]
220. Cabras, P.; Angioni, A.; Garau, V.L.; Pirisi, F.M.; Brandolini, V.; Cabitza, F.; Cubeddu, M. Pesticide residues in prune processing. *J. Agric. Food Chem.* **1998**, *46*, 3772–3774. [CrossRef]
221. Alister, C.; Araya, M.; Becerra, K.; Volosky, C.; Saavedra, J.; Kogan, M. Industrial prune processing and its effect on pesticide residue concentrations. *Food Chem.* **2018**, *268*, 264–270. [CrossRef]
222. Lee, M.G. Reduction of chlorpyrifos and fenitrothion residues in red pepper peel by washing and drying. *Food Sci. Biotechnol.* **2001**, *10*, 429–432.
223. Liu, T.; Zhang, C.; Peng, J.; Zhang, Z.; Sun, X.; Xiao, H.; Sun, K.; Pan, L.; Liu, X.; Tu, K. Residual behaviors of six pesticides in shiitake from cultivation to postharvest drying process and risk assessment. *J. Agric. Food Chem.* **2016**, *64*, 8977–8985. [CrossRef]
224. Hwang, K.W.; Bang, W.S.; Jo, H.W.; Moon, J.K. Dissipation and removal of the etofenprox residue during processing in spring onion. *J. Agric. Food Chem.* **2015**, *63*, 6675–6680. [CrossRef]
225. Jaggi, S.; Sood, C.; Kumar, V.; Ravindranath, S.D.; Shanker, A. Leaching of pesticides in tea brew. *J. Agric. Food Chem.* **2001**, *49*, 5479–5483. [CrossRef]
226. Chen, H.; Pan, M.; Pan, R.; Zhang, M.; Liu, X.; Lu, C. Transfer rates of 19 typical pesticides and the relationship with their physicochemical property. *J. Agric. Food Chem.* **2015**, *63*, 723–730. [CrossRef]
227. Kumar, V.; Sood, C.; Jaggi, S.; Ravindranath, S.D.; Bhardwaj, S.P.; Shanker, A. Dissipation behavior of propargite—an acaricide residues in soil, apple (*Malus pumila*) and tea (*Camellia sinensis*). *Chemosphere* **2005**, *58*, 837–843. [CrossRef] [PubMed]
228. Watanabe, E.; Kobara, Y.; Baba, K.; Eun, H. Reduction of hazardous organic solvent in sample preparation for hydrophilic pesticide residues in agricultural products with conventional liquid chromatography. *J. Agric. Food Chem.* **2013**, *61*, 4792–4798. [CrossRef]
229. Iwafune, T.; Ogino, T.; Watanabe, E. Water-based extraction and liquid chromatography-tandem mass spectrometry analysis of neonicotinoid insecticides and their metabolites in green pepper/tomato samples. *J. Agric. Food Chem.* **2014**, *62*, 2790–2796. [CrossRef] [PubMed]
230. Xiao, H.-W.; Pan, Z.; Deng, L.-Z.; El-Mashad, H.M.; Yang, X.-H.; Mujumdar, A.S.; Gao, Z.-J.; Zhang, Q. Recent developments and trends in thermal blanching—A comprehensive review. *Inf. Process. Agric.* **2017**, *4*, 101–127. [CrossRef]
231. Ketata, M.; Desjadins, Y.; Ratti, C. Effect of liquid nitrogen pretreatments on osmotic dehydration of blueberries. *J. Food Eng.* **2013**, *116*, 202–212. [CrossRef]
232. Bonnechère, A.; Hanot, V.; Jolie, R.; Hendrickx, M.; Bragard, C.; Bedoret, T.; Loco, J. Van Effect of household and industrial processing on levels of five pesticide residues and two degradation products in spinach. *Food Control.* **2012**, *25*, 397–406. [CrossRef]
233. Krol, W.J.; Arsenault, T.L.; Pylypiw, H.M.; Incorvia Mattina, M.J. Reduction of pesticide residues on produce by rinsing. *J. Agric. Food Chem.* **2000**, *48*, 4666–4670. [CrossRef]
234. Adou, K.; Bontoyan, W.R.; Sweeney, P.J. Multiresidue method for the analysis of pesticide residues in fruits and vegetables by accelerated solvent extraction and capillary gas chromatography. *J. Agric. Food Chem.* **2001**, *49*, 4153–4160. [CrossRef]
235. García-Reyes, J.F.; Gilbert-López, B.; Molina-Díaz, A.; Fernández-Alba, A.R. Determination of pesticide residues in fruit-based soft drinks. *Anal. Chem.* **2008**, *80*, 8966–8974. [CrossRef]
236. Ozbey, A.; Uygun, U. Behaviour of some organophosphorus pesticide residues in peppermint tea during the infusion process. *Food Chem.* **2007**, *104*, 237–241. [CrossRef]
237. Al-Taher, F.; Chen, Y.; Wylie, P.; Cappozzo, J. Reduction of pesticide residues in tomatoes and other produce. *J. Food Prot.* **2013**, *76*, 510–515. [CrossRef]
238. Ozbey, A.; Uygun, U. Behaviour of some organophosphorus pesticide residues in thyme and stinging nettle tea during infusion process. *Int. J. Food Sci. Technol.* **2007**, *42*, 380–383. [CrossRef]

239. Zhang, Y.; Zhang, W.; Liao, X.; Zhang, J.; Hou, Y.; Xiao, Z.; Chen, F.; Hu, X. Degradation of diazinon in apple juice by ultrasonic treatment. *Ultrason. Sonochem.* **2010**, *17*, 662–668. [CrossRef] [PubMed]
240. Zhang, Y.; Zhang, Z.; Chen, F.; Zhang, H.; Hu, X. Effect of sonication on eliminating of phorate in apple juice. *Ultrason. Sonochem.* **2012**, *19*, 43–48. [CrossRef] [PubMed]

Publisher's Note: MDPI stays neutral with regard to jurisdictional claims in published maps and institutional affiliations.

© 2020 by the authors. Licensee MDPI, Basel, Switzerland. This article is an open access article distributed under the terms and conditions of the Creative Commons Attribution (CC BY) license (http://creativecommons.org/licenses/by/4.0/).

foods

MDPI

Article

Functional and Quality Characteristics of Ginger, Pineapple, and Turmeric Juice Mix as Influenced by Blend Variations

Akama Friday Ogori [1], Julius Amove [2], Precious Aduloju [2], Giacomo Sardo [3], Charles Odilichukwu R. Okpala [4,*], Gioacchino Bono [3] and Małgorzata Korzeniowska [4]

1 Department of Home Sciences, Faculty of Agriculture, Federal University Gashua, Gashua P.M.B.1005, Nigeria; ogorifaraday@gmail.com
2 Department of Food Science and Technology, Federal University of Agriculture Makurdi, Makurdi 970211, Nigeria; amovejulius@yahoo.com (J.A.); preshus06@gmail.com (P.A.)
3 Institute for Biological Resources and Marine Biotechnologies (IRBIM), National Research Council of Italy (CNR), 91026 Mazara del Vallo, Italy; giacomosardo88@gmail.com (G.S.); gioacchino.bono@cnr.it (G.B.)
4 Faculty of Biotechnology and Food Science, Wrocław University of Environmental and Life Sciences, 51-630 Wrocław, Poland; malgorzata.korzeniowska@upwr.edu.pl
* Correspondence: charlesokpala@gmail.com

Abstract: In this current work, the functional and quality characteristics of ginger, pineapple, and turmeric juice mix as influenced by blend variations were investigated. Specifically, the blends had constant ginger amounts, decreased pineapple, and increased turmeric proportionally. Additionally, the functional properties involved physicochemical (pH, soluble solids (SS), total titratable acidity (TA) and viscosity), proximate (moisture, protein, fat and ash), minerals (Ca, and Mg) and vitamin C and β-carotene analyses, whereas quality properties involved microbiological and sensory analyses. The results showed that as quantities of pineapple and turmeric respectively decreased and increased, there was significant increases in Ca, Mg, vitamin C, and β-carotene contents ($p < 0.05$). Across the blends, the degree of significant differences ($p < 0.05$) in the protein, fat, and ash seemed more compared to those of moisture contents. Despite the increases in pH and viscosity, and decreases in SS and TA, the increases in turmeric potentially reinforced by ginger most likely decreased the bacterial/fungi counts, as well as inhibition zones. Increasing and decreasing the respective amounts of turmeric and pineapple might not necessarily make the blends more acceptable, given the decreases in appearance, taste, aroma, and mouthfeel scores.

Keywords: ginger; pineapple; turmeric; juice mix; physicochemical properties; microbiological quality; sensory attributes

Citation: Ogori, A.F.; Amove, J.; Aduloju, P.; Sardo, G.; Okpala, C.O.R.; Bono, G.; Korzeniowska, M. Functional and Quality Characteristics of Ginger, Pineapple, and Turmeric Juice Mix as Influenced by Blend Variations. *Foods* **2021**, *10*, 525. https://doi.org/10.3390/foods10030525

Academic Editors: Danijela Bursać Kovačević and Predrag Putnik

Received: 31 January 2021
Accepted: 25 February 2021
Published: 3 March 2021

Publisher's Note: MDPI stays neutral with regard to jurisdictional claims in published maps and institutional affiliations.

Copyright: © 2021 by the authors. Licensee MDPI, Basel, Switzerland. This article is an open access article distributed under the terms and conditions of the Creative Commons Attribution (CC BY) license (https://creativecommons.org/licenses/by/4.0/).

1. Introduction

Broadly, fruits can be grouped into two categories, namely: dry and fleshy/succulent fruits, and this is largely based on the physical ripe condition [1]. When properly harvested, fruits like orange, pineapple, and watermelon are edible, fleshy, and sweet [2]. Processing of fruit involve enzymes, extraction, and evaporation activities. Additionally, the suitability of a fruit juice and its concentrate/extract for an intended application remains dependent on its quality [1]. Fruits endocarps and mesocarps contain various phytochemical compounds resembling vegetables, with higher amounts of free waters, but lower amounts of carbohydrate, fat, and protein [3]. When the natural liquid of freshly harvested fruit like orange is squeezed, a juice drink is produced and is available for immediate consumption [1]. The regular consumption of fruits and its juices, most importantly, helps to make up for diet nutritional losses as well as maintain health and wellbeing [4]. Anticipating how the freshness of fruit (as well as vegetable) quality in the form of juice drink would continually keep remains challenging [1].

The relatively high metabolic activity in fruits like apple, banana, and pineapple, for instance, continues even after harvesting, which makes them highly perishable [5].

Among the above-mentioned fruits, pineapple (*Ananas comosus*) stands unique because it is among the few bromeliads that produce edible fruits, with morphologically fused berries around a central core [6]. Pineapples comprise antioxidants/polyphenolic compounds, natural enzymes, and pro-vitamins [4,7]. Specifically, a ripe and ready-to-harvest pineapple would have (above-mentioned) berries comprise bioactive and phenolics contents/non-toxic compounds, which presents promising therapeutic potentials that help to enhance immune response [6]. Essentially, the fleshy and juicy pulp makes pineapple an excellent blend to obtain new flavours in beverages and juice mixes. Moreover, mixed juice blends produced from various fruits can help combine basic nutrients and provide improved nutritional value [4].

Ginger (*Zingiber officinale* Rosc) is an underground rhizome or stem of herbaceous perennial species of family Zingiberaceae, also considered typically indigenous to many tropical/subtropical countries [8–10]. As a widely established monocotyledon herb, the main products of ginger include dry or fresh rhizome, as well as ground ginger (powder) [9–11]. It can also be used as a whole juice extract and in drink/tea after blending process [12]. The rhizome/stem of ginger, in addition to comprising such proximate components like ash, carbohydrate, fiber, moisture, and protein, has volatile oil of stem that contributes to provide its pleasant aroma [10,13–15]. Additionally, ginger contains ascorbic acid, β-carotene, curcumin, gingerol, linalool, paradol, γ-terpinene, as well as terpinen-4-ol [10,16–19]. The swollen rhizome/stem of ginger has been associated with antimicrobial, anti-inflammatory, and anti-carcinogenic properties [12].

Turmeric (*Curcuma longa* Linn.), equally an underground rhizome like ginger, and within the family of Zingiberaceae, is largely available either in dry or fresh forms [10,20]. Turmeric, largely cultivated across warm climatic regions of the globe, serves as a common food additive mostly in powdered form. Turmeric (powdered), positioned as a colorant, can serve as a flavouring agent in food formulations [21,22]. Commonly grown in many parts of Nigeria, the production of turmeric has made its sales provide economic and regional benefits [23]. To convert turmeric into a stable commodity, there is need for a number of processing operations, which includes boiling, cleaning, slicing, curing, drying, grading, milling, and packaging [24]. For emphasis, turmeric not only fortifies the drinks that it is added to, it is also able to improve the nutritional quality [23]. Besides its role as spice, food preservative, and coloring material, turmeric occupies a space in traditional medicine given the many scientific studies that revealed its many bioactivities like anti-inflammatory, anti-bacterial, anti-carcinogenic, anti-diabetes, and antioxidant capacities [25]. Largely, turmeric comprises 60% turmerone, 25% zingiberene, and 1.5–5% volatile oil. In particular, turmeric comprises three curcuminoids, namely: bisdemethoxycurcumin (0.30–9.10%), curcumin (diferuloylmethane) (71.50–94%), and demethoxy-curcumin (6–19.4%), which cumulates to the curcuminoids (2.5–8%) that bring about the yellow coloration [10,26,27].

Blending spices with fruits to form a juice mix is becoming increasingly popular in Nigeria, with high promise of spreading to the West Africa sub-region. Additionally, there is increasing notion among many that ginger, pineapple, and turmeric juice mix is affordable, nutritionally enriching, as well as filling, and this is yet to be scientifically verified. To our best knowledge, the blend variations of ginger, pineapple, and turmeric juice mix has not been studied. It is anticipated that a juice mix of this type could result in a nourishing composite with promising functional and sensory qualities. To supplement existing information, the aim of this current study was to determine the functional and quality characteristics of ginger, pineapple, and turmeric juice mix as influenced by blend variations. Specifically, the functional properties involved minerals and vitamins, physico-chemical, and proximate components, whereas quality properties involved microbiological and sensory components.

2. Materials and Methods

2.1. Overview of Experimental Program

The schematic overview of the experimental program, depicting the essential stages from the collection of ginger, pineapple and turmeric, and preparation of individual juices, to the formulation to make the mix juice, and then, the functional and quality analyses, is given in Figure 1. For emphasis, this current study specifically targeted determining the functional and quality characteristics of ginger, pineapple, and turmeric juice mix as influenced by blend variations. Specifically, the functional properties involved minerals and vitamins and physicochemical and proximate components, whereas quality properties involved microbiological and sensory components. The end goal is to achieve a juice mix that could bring about a nourishing composite with promising functional and sensorial attributes.

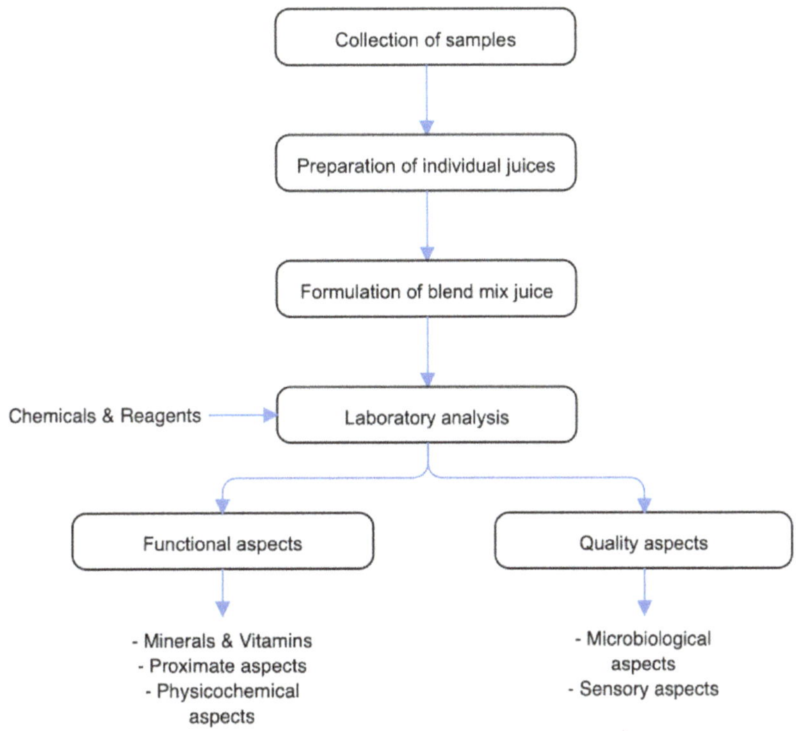

Figure 1. The schematic overview of the experimental program, depicting the essential stages from the collection of ginger, pineapple and turmeric, and preparation of individual juices, to the formulation to make the mix juice, and then, the functional and quality aspects of the laboratory analyses.

2.2. Collection of Samples

The ripe pineapple, matured turmeric, and ginger were purchased from the Railway (7.72732° N, 8.53193° E) and Wadata (7.74527° N, 8.51339° E) markets situated in Makurdi, Benue State, Nigeria. All samples were taken to the laboratory for sample preparation and analysis.

2.3. Chemicals and Reagents

All the chemicals and reagents utilized in this current study were reagent grade standard.

2.4. Preparation of Pineapple Fruit Juice

The preparation of pineapple fruit juice followed the method of Okwori et al. [28] with slight modifications, depicted in Figure 2. Pineapple fruits were selected and washed with 5% HOCl solution and thoroughly rinsed with distilled water before peeling with a sterilized knife. The fruits are cut into sizes of about 3–4 mm thick and juice extraction using a juice extractor. The pineapple juice was filtered using sterile muslin cloth, which was folded into two layers and filtered into a clean transparent bowl. The juice was filled into an air-tight screwed cap, pasteurized, and refrigerated at ~4 °C prior to analysis.

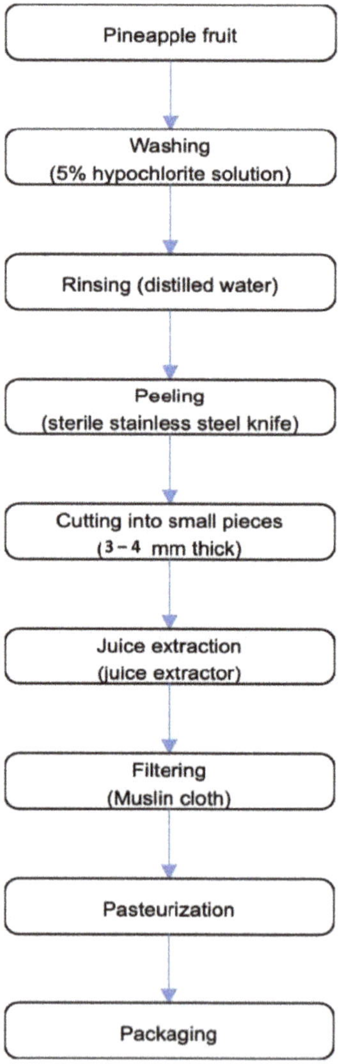

Figure 2. The preparation of pineapple fruit juice (Okwori et al., [28]).

2.5. Preparation of Turmeric and Ginger Juice

Following the method prescribed by the Top 10 Home Remedies Team [29], herein depicted in Figure 3, five fresh turmeric rhizomes were rinsed under clean running tap water to remove the dirts. The turmeric rhizomes were peeled and then cut into pieces and put into the blender, and at the same time, supplemented little equivalents of clean/filtered water were added to ease friction during blending. The juice pulp was then filtered using a sterile muslin cloth to get the juice, which was subsequently refrigerated at ~4 °C, until required.

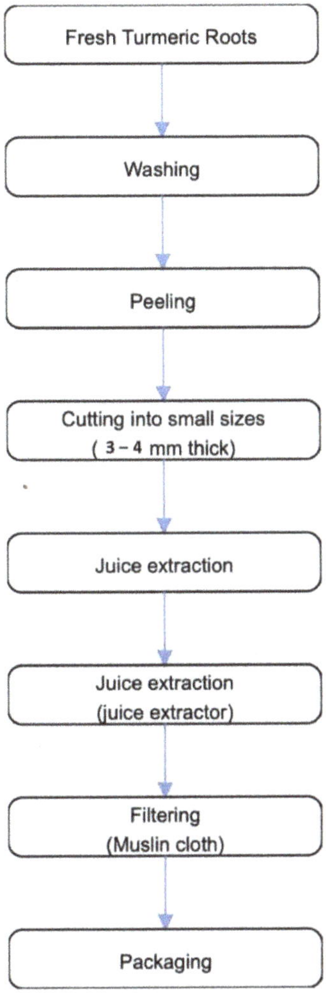

Figure 3. The preparation of turmeric juice (Source: Top 10 Home Remedies Team [29]).

The preparation of ginger juice is similar to that of turmeric juice. Fresh ginger roots were washed under clean running tap water, peeled, and then cut into smaller pieces, thereafter, they were subjected to blending, and at the same time, supplemented with little amounts of clean/filtered water to ease friction during blending. The juice pulp was then filtered using a sterile muslin cloth to get the juice, which was then refrigerated at ~4 °C, until required.

2.6. Formulation of Pineapple, Turmeric and Ginger Blend Juice Mix

The formulation of pineapple, turmeric, and ginger blend juice mix is given in Table 1. The juice from pineapple, turmeric, and ginger juices were blended at varied proportions. This followed the method demonstrated by local artisans, but with slight modifications to enable reproducibility. Specifically, the control sample was pineapple only, that is, PJ:TJ:GJ = 100:0:0. The blends kept the ginger amounts constant, decreased the pineapple and increased the turmeric amounts by proportion. Next, the juice blends were mixed by stirring, bottled with screw caps before the pasteurization at 65 °C for 5 min, in a thermostatically controlled water bath, and thereafter, cooled at ambient temperature of about 27 °C. At the end, the blend juice mix samples were refrigerated at ~4 °C until required for analysis.

Table 1. Formulation of Ginger, Pineapple, and Tumeric Juice Mix by Blends.

Samples (PJ:TJ:GJ)	Pineapple (mL)	Turmeric (mL)	Ginger (mL)	Total (mL)
100:0:0	100	0	0	100
80:10:10	80	10	10	100
70:20:10	70	20	10	100
60:30:10	60	30	10	100
50:40:10	50	40	10	100

PJ = Pineapple juice; TJ = Turmeric juice; GJ = Ginger juice.

2.7. Functional Analysis.

2.7.1. Minerals and Vitamins Measurements

Determination of β-Carotene

The β-carotene of samples was determined using the AOAC method [30]. About 5 g of the sample was transferred into a separating funnel and a solution containing 60 mL of hexane; 40 mL of ethanol were swirled vigorously after adding 2 mL of 2% NaCl. This was then allowed to stand for 30 min after which the lower layer was discarded. The absorbance of the top layer was determined at a wavelength of 460 mm using a spectrophotometer, using the equation below:

$$TC = \frac{absorbance}{100\, specific\ extinction\ \times\ pathlength\ of\ the\ cell}$$

where,

TC: Total carotenoids (mg)
Molar extinction coefficient (∑) = 15×10^{-4}
Specific extinction coefficient (∑) = (∑ × molar mass of β-carotene)
Molar mass of β-carotene = 536.88 g/mol
Path length of cell = 1 cm

Determination of Calcium and Magnesium

The mineral composition (specific to Ca and Mg) of samples were determined by AOAC acid digestion method [31]. Ash obtained after incineration at 600 °C was dissolved in 5 mL HCl solution and transferred into a 50 mL volumetric flask. The resulting solution was made to mark with distilled water. The mineral contents were then measured using atomic absorption spectrophotometer (AAS), and mineral composition results were recorded.

Determination of Vitamin C

The vitamin C of samples was determined using the method described by Ikewuchi and Ikewuchi [32]. The quantities of vitamin C present are measured by the tiny additions of acidified starch (termed "reaction mix"), followed by droplets of iodine until purple color. Any vitamin C will "neutralize" the iodine, to prevent the purple color formation. In line with this, iodine solution (0.1 M) was prepared using 10 g of KI, and starch solution, using 0.25 g of starch powder. In order to actualize the vitamin C, a blank solution (25 mL) was made from the sample, and 10 drops of starch solution were added. The mixture was titrated with iodine solution until the first black blue color, which persisted for ~20 s. Blended juice samples (25 mL) were titrated exactly the same way as the standard solution. The initial and final volume of iodine solution required to produce the color change at the end points were recorded. Subsequently, the vitamin C concentration was determined as follows:

$$\text{vitamin C concentration in the juices (g/100 mg)} = y/b$$

where

b = titre (mL) from the titration of the standard vitamin C solution

y = titre (mL) from the titration of the sample solution.

2.7.2. Proximate Measurements

Determination of Moisture

The moisture of samples is determined by the AOAC method [31]. Cleaned crucible is dried in the oven at 100 °C for 1 h to constant weight and then cooled in the desiccator. Approximately 2 g of the samples were weighed into the crucible and dried at 100 °C to a constant weight, and calculated as below:

$$\%Moisture = \frac{Weightloss \times 100\%}{Weight\ of\ samples}$$

Determination of Crude Protein

The crude protein of samples was determined using the AOAC method [31] with slight modifications. Approximately 1 g of the sample was placed with a selenium catalyst in the micro Kjeldahl digestion flask. The mixture was digested to clean clear solution. The flask was cooled and then diluted with distilled water to the 50 mL mark of a conical flask, 5 mL of the mixture was transferred into distillation apparatus, and 5 mL of 2% boric acid added unto 100 mL conical flask (the receiver flask) with four drops of methyl red indicator. Then, 50% of NaOH was constantly added to the digested sample until the solution turned cloudy, indicating the solution had achieved alkalinity. Distillation was carried out in the boric acid solution at the receiver flask. During the distillation process, the pink color of the solution in the receiver flask turned blue, indicating the presence of ammonia. The resulting solution in the conical flask was then titrated with 0.1 M HCl and the protein content calculated as below:

$$\%Nitrogen \times 6.25\ (1\ \text{mL}\ of\ 0.1\ NHCL = 0.0014\ \text{gN})$$

$$Nitrogen = \frac{Titrevalue - blank \times 0.0014\ N \times 100\% \times 25}{Weight\ of\ sample \times 5\ mlaliquot}$$

Determination of Crude Fat

The crude fat of samples was determined using the AOAC method [31] with slight modifications. The 100 mL beaker used was washed and dried in an oven for 1 h at 105 °C, and thereafter cooled in a desiccator and weighed. Approximately 10 mL of the samples was mixed with hexane in a separating funnel, and the organic layer was transferred into the pre-weighed beaker, subject to water bath, and thereafter weighed. The crude fat was determined using the equation below:

$$\%crudelipid = \frac{Weight\ of\ the\ fat\ \times 100\%}{Weight\ of\ the\ sample}$$

Determination of Ash

The ash of samples was determined from the loss in weight during incineration following the AOAC method [30] with slight modifications. This method allows the entire organic matter to be burnt off, without the appreciable decomposition of the ash constituent. Approximately 5 g of the samples were placed in the incinerator. The ashing was done at a furnace of 600 °C for 6 h and calculated as below:

$$Ash\ Content = \frac{Weight\ of\ ash \times 100\%}{Weight\ of\ the\ sample}$$

2.7.3. Physicochemical Measurements

Determination of pH

The pH of samples was determined using a pH meter, calibrated with buffers standard. The electrode was rinsed with distilled water, the electrode was then dipped into 5 g of the sample, which had been dissolved in 50 mL of water.

Determination of Soluble Solids

The soluble solids of samples were determined using the AOAC method [30]. The prism of the refractometer was cleaned and a drop of the blended juice was placed on the prism and closed. The °Brix was read using the scale of the refractometer when held close to the eyes.

Determination of Titratable Acid (TA)

The titratable acid of samples was determined using the AOAC method [30] with slight modifications. Approximately 10 mL of the juice was pipetted into a conical flask and 25 mL of distilled water added to make a solution. Approximately 200 mL of 0.1 M of NaOH was titrated against the sample using phenolphthalein as an indicator, to achieve color pink as an end point. The corresponding burette reading was taken using the following formula:

$$TA = \frac{Titre \times blank \times Normality\ of\ base \times mlequivalent\ of\ citricacid}{Weight\ of\ Sample}$$

where, TA = titrable acidity (%)

Determination of Viscosity

The juice samples viscosity was determined using a Brookfield viscometer (model Lv-3, Middleboro, MA 02346, USA) with the spindle set at 60 rpm, after which the readings were recorded in millipascal-second (mPa.s).

2.8. Quality Analysis

2.8.1. Microbiological Evaluation

Microbiological analysis of the juice mix was carried out following the method described by Adegoke [33], with slight modifications, following the pour-plate method. This enabled the determinations of total bacteria and fungi counts. Homogenized (~60 s) quantities of blend (~2 g) with 15 mL of diluents was prepared. Serial ten-fold dilution of homogenate involved 0.1 mL of aliquots aseptically introduced into sterile Petri dishes, after which molten agar (~45 °C) was poured unto them, mixed and then allowed to set. The different agar plates were incubated for ~24 h. Nutrient Agar (NA) was used for the enumeration of total bacteria count and was then incubated at 37 °C for 24–48 h. Sabourd Dextrose Agar (SDA) was used for the enumeration of total fungi count then incubated at room temperature (28 ± 2 °C) for 3–5 days. The microbiological analysis were reported in terms of logarithm of colony forming units (log cfu/mL) of the blend sample.

Antibacterial activity of the juice extracts was determined by molten Agar well diffusion technique following the method of Abubakar et al. [34] with slight modifications. The test organism (*Salmonella typhii*) was diluted with Muller Hinton broth to 0.5% McFarland equivalent standard. Approximately 25 mL of Mueller Hinton Agar (HiMedia) plates were checked for sterility and streaked with an overnight broth cultured of bacterial isolate, using sterile cotton buds. A standard sterile cock borer of 6 mm diameter was used to make uniform wells on the surface of the streaked agar media. With the aid of a micropipette, the wells were filled up with 200 µL each of the undiluted blended juice extract (sample A–E). The plates were then allowed to stand for ~1 h in the refrigerator to allow proper diffusion of the extract. Amoxycillin (~25 mg/mL) solution was prepared and served as the control [31]. Following the method of Rahman et al. [35], all the plates were incubated at 37 °C for ~24 h, after which the antibacterial activity was evaluated based on the diameters of zones of inhibition and recorded in millimeter (mm).

2.8.2. Sensory Evaluation

The ginger, pineapple, and turmeric juice mix blends were subjected to sensorial evaluation. This was done with the help of 10 ($N = 10$) panelists, comprising of students and staff of the Food Science and Technology Department, Federal University of Agriculture Makurdi. Specifically, these panelists underwent sensorial training prior to their participation at this study. Importantly, the panelists' participation was voluntary. Additionally and prior to their participation, the verbal consent was taken from all the panelists. To ensure privacy, gender was not indicated. The selection criteria was based on complete participation of sensory training for this study. The samples were presented in a white plastic cup to each panelist. Each sample presented was coded. Essentially, each panelist was provided with adequate space to ensure there was no co-operation during the sampling of juice mix blends. The sensory attributes comprised appearance, taste, aroma, mouthfeel, and general acceptability. Consistent with the method described by Iwe [36], the sensory attributes were individually considered based on a 9-point Hedonic scale, which had the least value (numeric value = 1) designated as 'disliked extremely', and the highest value (numeric value = 9) designated as 'liked extremely'.

2.9. Statistical Analysis

One-way analysis of variance (ANOVA) was used to analyse the emergent data. The results were presented in terms of mean values ± standard deviation (SD) from duplicate measurements. The mean values were resolved with the help of Fisher's Least Significant Difference (LSD). The probability level of statistical significance was set at $p < 0.05$ (95% confidence interval). IBM SPSS software (version 22.0) was used to do the data analysis.

3. Results and Discussion

3.1. Functional Aspects Minerals and Vitamins Variations

The minerals and vitamins variations of ginger, pineapple, and turmeric juice mix as influenced by blends can be seen in Table 2. Clearly, significant differences ($p < 0.05$) in Ca, Mg, vitamin C, and β-carotene contents were found across samples. Specifically, the Ca, Mg, vitamin C, and β-carotene contents increased significantly ($p < 0.05$) as quantities of pineapple and turmeric were respectively decreased and increased. The control PJ:TJ:GJ = 100:0:0 obtained the lowest values for Ca (7.37 ± 0.09 mg/100 mL), Mg (5.37 ± 0.07 mg/100 mL), vitamin C (73.60 ± 0.71 mg/100 mL) and β-carotene (67.92 ± 0.76 mg/100 mL), compared to other samples, which showed varied ranges (Ca = from 8.78 to 18.09 mg/100 mL; Mg = from 6.59 to 8.54 mg/100 mL; vitamin C = from 86.74 to 122.97 mg/100 mL; and β-carotene = from 83.19 to 1454.10 mg/100 mL). Increases in Ca, Mg, vitamin C, and β-carotene would most likely be attributed to the addition of turmeric. The vitamin C in fresh turmeric rhizome/root could show very promising levels [10,37,38]. Additionally, the vitamin C in ginger could also show very promising levels [39]. Moreover, the vitamin C in fresh pineapple juice (control) of this current work appeared higher compared with those reported elsewhere, like ~14.1 mg/100 g reported by Ikewuchi and Ikewuchi [32]; 22.5–33.5 mg/100 g reported by Achinewhu and Hart [40]; ~52 mg/100 g reported by Rodríguez et al. [41]; and ~54 mg/100 g reported by Chakraborty, Rao, and Misra [42]. Besides, both vitamin C and β-carotene might not be responsible for the antioxidant capacity of pineapples [40]. Vitamin C would belong to hydrophilic, whereas carotenoids would belong to lipophilic antioxidants [40]. Moreover, the processing of pineapples into juice could likely be affecting, not only quantities of vitamin C [43], but also those of Ca, Mg, and β-carotene contents obtained at this current study.

Table 2. Minerals and vitamins variations of Ginger, Pineapple, and Tumeric Juice Mix as influenced by Blends.

Samples	Minerals and Vitamin Composition			
PJ:TJ:GJ	Ca (mg/100 mL)	Mg (mg/100 mL)	Vitamin C (mg/100 mL)	β-Carotene (mg/100 mL)
100:0:0	7.37 ^{e} ± 0.09	5.37 ^{e} ± 0.07	73.60 ^{d} ± 0.71	67.92 ^{e} ± 0.76
80:10:10	8.78 ^{d} ± 0.04	6.59 ^{d} ± 0.01	86.74 ^{c} ± 0.23	83.19 ^{d} ± 3.30
70:20:10	12.73 ^{c} ± 0.04	7.67 ^{c} ± 0.02	101.81 ^{b} ± 0.27	199.14 ^{c} ± 1.15
60:30:10	15.63 ^{b} ± 0.19	7.89 ^{b} ± 0.01	103.22 ^{b} ± 0.82	1318.10 ^{b} ± 1.97
50:40:10	18.09 ^{a} ± 0.01	8.54 ^{a} ± 0.01	122.98 ^{a} ± 1.45	1454.10 ^{a} ± 1.69
LSD	0.257	0.081	2.65	5.089

Values are means ± standard deviation (SD) of duplicate determinations. Means in the same column with the same superscript are not significantly different at ($p > 0.05$), Key: PJ = Pineapple juice; TJ = Turmeric juice; GJ = Ginger juice, LSD = Least significant difference.

Proximate and Physicochemical Variations

The proximate variations of ginger, pineapple, and turmeric juice mix as influenced by blends can be seen in Table 3. Clearly, the degree of significant differences ($p < 0.05$) were more in the protein, fat, and ash compared to moisture contents across samples. The control PJ:TJ:GJ = 100:0:0 obtained the lowest values for moisture (95.89 ± 0.00%), crude protein (0.008 ± 0.001%), fat (0.051 ± 0.001%), and ash (0.125 ± 0.004%) contents, compared to the blend samples, which showed varied ranges (moisture = from 96.86 to 98.18%; protein = from 0.013 to 0.261%; fat = from 0.061 to 0.168%; ash = from 0.287 to 0.585%). Akusu, Kiin-Kabari and Ebere [4] reported fresh pineapple juice to have about 88% moisture, 1% crude protein, and 2% ash contents, different from values of this current study. Specifically, the increasing amounts of turmeric appears not to dramatically influence the moisture of the blend juice mix, compared to its noticeable influences on the protein, fat, and ash contents. Increases in crude fat and protein might be because of essential oils in ginger and turmeric [10,11]. To the consumer, increases in ash contents portrays the juice mix as a strong mineral source [44]. Ginger and tumeric generally have competitive proximate components, with ranging amounts of 7–13% moisture, 6–12% protein, 60–72% carbohydrate, and 3–7% ash [10,13–15]. The marginal influence that increasing tumeric amounts had on moisture might strongly impact on the viscosity of the juice mix blend. Potentially, the blends PTG60:30:10 and PTG50:40:10 respectively with moisture contents of ~98%, would proffer higher sensorial implications compared to the others, especially on both appearance and taste attributes.

Table 3. Proximate variations of Ginger, Pineapple, and Tumeric Juice Mix as influenced by Blends.

Samples	Tested Parameters (%)			
P:T:G	Moisture	Protein	Fat	Ash
PTG100:0:0	95.89 ^{b} ± 0.00	0.008 ^{e} ± 0.001	0.051 ^{e} ± 0.001	0.125 ^{d} ± 0.004
PTG80:10:10	96.86 ^{ab} ± 1.37	0.013 ^{d} ± 0.001	0.061 ^{d} ± 0.001	0.287 ^{c} ± 0.006
PTG70:20:10	97.91 ^{a} ± 0.01	0.018 ^{c} ± 0.001	0.072 ^{c} ± 0.001	0.322 ^{c} ± 0.001
PTG60:30:10	98.12 ^{a} ± 0.21	0.087 ^{b} ± 0.001	0.101 ^{b} ± 0.001	0.456 ^{b} ± 0.063
PTG50:40:10	98.18 ^{a} ± 0.07	0.261 ^{a} ± 0.001	0.168 ^{a} ± 0.002	0.585 ^{a} ± 0.049
LSD	1.597	0.003	0.007	0.081

Values are means ± standard deviation (SD) of duplicate determinations. Means in the same column with the same superscript are not significantly different at ($p > 0.05$); Key: PJ = Pineapple juice; TJ = Turmeric juice; GJ = Ginger juice; and LSD = Least significant difference.

The physicochemical variations of ginger, pineapple, and turmeric juice mix as influenced by blends can be seen in Table 4. Clearly, there were significant differences ($p < 0.05$) found in pH, SS, TA, and viscosity across samples. The control PJ:TJ:GJ = 100:0:0 obtained the least values in pH (3.81 ±0.007) and viscosity (300.11 ±0.12 m.Pa.s), but peak values in SS (11.95 ±0.07 °Brix) and TA (0.9005 ±0.07%). Across the blends, noticeable ($p < 0.05$) increases were obtained in pH (from 3.83 to 4.01) and viscosity (from 301.68 to 850.06 m.Pa.s), whereas decreases were obtained in SS (from 9.32 to 4.90 °Brix) and TA (from 0.8425 to 0.5425%).

The physicochemical variations from increases in turmeric and decreases in pineapple amounts appear interesting. Increases in pH and viscosity demonstrates the impact turmeric could have in the blend juice mix [45]. Despite the increases in pH and decreases in TA arising from increasing amounts of turmeric [44], the blend mix juice having a peak pH of ~4 would appear somewhat less susceptible to microbial deterioration particularly to the most familiar neutrophilic microorganisms like *Escherichia coli*, staphylococci, and *Salmonella* spp., which are unable to thrive in acidic pH conditions [46]. The decreases in SS might have happened because both turmeric and ginger constituents hold less sugar content(s) compared to those of pineapple [47]. The blends' viscosity, increasing with quantities of turmeric at this study, might be attributable to its starch [48].

Table 4. Physicochemical variations of Ginger, Pineapple, and Tumeric Juice Mix as influenced by Blends.

Samples	Tested Parameters			
PJ:TJ:GJ	**pH**	**SS (°Brix)**	**TA (%)**	**Viscosity (mPa.s)**
100:0:0	$3.81^{d} \pm 0.007$	$11.95^{a} \pm 0.07$	$0.9005^{a} \pm 0.07$	$300.11^{d} \pm 0.12$
80:10:10	$3.83^{c} \pm 0014$	$9.32^{b} \pm 0.028$	$0.8425^{b} \pm 0.35$	$301.68^{d} \pm 0.78$
70:20:10	$3.85^{c} \pm 0.001$	$8.10^{c} \pm 0.001$	$0.6727^{c} \pm 0.09$	$351.06^{c} \pm 1.35$
60:30:10	$3.89^{b} \pm 0.014$	$7.42^{d} \pm 0.016$	$0.5775^{d} \pm 0.21$	$501.61^{b} \pm 2.27$
50:40:10	$4.01^{a} \pm 0.076$	$4.90^{e} \pm 0.004$	$0.5425^{e} \pm 0.35$	$850.06^{a} \pm 0.06$
LSD	0.024	0.081	0.0640	3.178

Values are means ± standard deviation (SD) of duplicate determinations. Means in the same column with the same superscript are not significantly different at ($p > 0.05$), Key: PJ = Pineapple juice; TJ = Turmeric juice; GJ = Ginger juice; LSD = Least significant difference; SS = Soluble Solids; and TA = Titratable Acid.

3.2. *Quality Aspects Microbiological Variations*

The microbiological variations of ginger, pineapple, and turmeric juice mix as influenced by blends is shown in Table 5. Across the samples, the bacterial count ranged between 5.0×10^3 and 1.6×10^4 log cfu/mL, whereas the fungi count ranged between 5.0×10^3 and 2.8×10^4 log cfu/mL. The blend PJ:TJ:GJ = 60:30:10 obtained the highest bacterial (1.6×10^4 log cfu/mL) and fungi (2.8×10^4 log cfu/mL) counts. The increases in turmeric reduced the bacterial and fungi counts. The control sample (PJ:TJ:GJ = 100:0:0) obtained the lowest bacterial count, but not so for fungi count. Moreover, the control bacterial and fungi counts both resembled one another ($p > 0.05$). In general, both bacterial and fungi counts were below the microbiological limits prescribed by the Food and Agriculture Organization (FAO) of the United Nations for formulated foods, which is 5×10^5 log cfu/mL, which is largely applicable to both aerobic plate counts (APC) and moulds [49]. The increased turmeric amounts are strengthened by the ginger present, which might have probably brought about the decreases in bacterial counts herein, which points to the antimicrobial capacity (of turmeric).

Table 5. Microbiological variations of Ginger, Pineapple, and Tumeric Juice Mix as influenced by Blends.

Samples (PJ:TJ:GJ)	Bacterial Count (log cfu/mL)	Fungi Count (log cfu/mL)
100:0:0	5.0×10^3	5.0×10^3
80:10:10	3.0×10^4	1.0×10^4
70:20:10	1.2×10^4	8.0×10^3
60:30:10	1.6×10^4	2.8×10^4
50:40:10	8.0×10^3	4.2×10^3

Values are means of duplicate determinations. Key: PJ = Pineapple juice; TJ = Turmeric juice; GJ = Ginger juice.

The antimicrobial inhibition of ginger, pineapple, and turmeric juice mix as influenced by blends, can be seen in Table 6. For emphasis, the test organism was *Salmonella typhii*, and the control used was amoxycillin antibiotics. The result shows that inhibition zones ranged between 12.50 mm (100:0:0) and NSI (No Significant Inhibition) (50:40:10) compared with the control that remained at approximately 20 mm. For emphasis, the control helps to show how the inhibition zone faired compared with those of the blends. Clearly, the antimicrobial activity is depicted by the lowering of inhibition zone as the turmeric was increased. Although the ginger amounts were constant, there is high chance that its presence contributed in strengthening the decreases in the inhibition zones at this study. Spices generally demonstrate antimicrobial activity against bacteria, yeast, molds, and viruses, given its diverse phytochemical components (e.g., alcohols, aldehydes, ethers, hydrocarbons, ketones, as well as phenols), which help to lengthen and stabilize food storage shelf time [10]. Nonetheless, this result goes a step further to demonstrate the presence of active compounds like gingerol, shogaols, and zingerone in ginger, and curcuminoids in tumeric, which provides it with the antimicrobial properties against bacteria such as *Bacillus coagulans, B.cereus, B. subtilis, Escherichia coli, Klebsiella pneumoniae, Pseudomonas aeruginosa, Staphylococcus aureus*, and *S. epidermidis* [10,50–53].

Table 6. Antimicrobial inhibition of Ginger, Pineapple, and Tumeric Juice Mix as influenced by Blends.

Samples (PJ:TJ:GJ)	Inhibition Zones of Juice Samples (mm)	Control (mm)
100:0:0	12.50	19.90
80:10:10	12.00	19.90
70:20:10	8.00	19.85
60:30:10	NSI	19.90
50:40:10	NSI	20.00

NSI: No Significant Inhibition. Means of two duplicate determinations. Key: PJ = Pineapple juice; TJ = Turmeric juice; GJ = Ginger juice; The control helps to show how the inhibition zone faired compared with those of the blends.

Sensory Variations

The sensory variations of ginger, pineapple, and turmeric juice mix as influenced by blends, can be seen in Table 7. The appearance scores across samples ranged from 7.87 (PJ:TJ:GJ =100:0:0) to 6.73 (PJ:TJ:GJ = 50:40:10). That of taste ranged from 7.27 (PJ:TJ:GJ = 100:0:0) to 3.80 (PJ:TJ:GJ = 50:40:10). Aroma across samples ranged from 7.33 (PJ:TJ:GJ = 100:0:0) to 5.87 (PJ:TJ:GJ = 50:40:10). Mouth feel across samples ranged from 7.40 (PJ:TJ:GJ = 100:0:0) to 4.20 (PJ:TJ:GJ = 50:40:10). General acceptability across samples ranged from 7.60 (PJ:TJ:GJ = 100:0:0) to 5.00 (PJ:TJ:GJ = 50:40:10). Besides, the sensory evaluation has a crucial role to play in judging the quality of a given food product. Control obtained peak values in aroma, mouth feel, and taste, which might have contributed to its lead overall acceptability. Increasing the turmeric and decreasing the pineapple might not necessarily make the blend juice mix more acceptable, given the decreases obtained in appearance, taste, aroma, and mouthfeel scores. Moreover, the sample blend 80:10:10 might be the more preferred compared to the others. Putting together the functional and quality data obtained thus far, we consider the turmeric, ginger, and pineapple blend juice mix nutritionally rich and consumer safe, yet, it might not be generally preferred specifically at this study.

Table 7. Sensory variations of Ginger, Pineapple, and Tumeric Juice Mix as influenced by Blends.

Samples PJ:TJ:GJ	Appearance	Taste	Aroma	Mouth Feel	General Acceptability
100:0:0	7.87^{ab}	7.27^{a}	7.33^{a}	7.40^{a}	7.60^{a}
80:10:10	8.27^{a}	5.87^{b}	6.67^{a}	6.20^{b}	6.33^{b}
70:20:10	7.73^{ab}	4.80^{bc}	6.47^{a}	5.93^{b}	5.87^{bc}
60:30:10	7.93^{ab}	4.47^{c}	5.93^{a}	5.73^{b}	5.33^{bc}
50:40:10	6.73^{b}	3.80^{c}	5.87^{a}	4.20^{c}	5.00^{c}
LSD	1.27	1.18	1.32	1.15	1.10

Values are means of two duplicate determinations. Means in the same column with the same superscript are not significantly different ($p > 0.05$), Key: PJ = Pineapple juice; TJ = Turmeric juice; GJ = Ginger juice LSD = Least significant difference.

4. Conclusions

The functional and quality characteristics of ginger, pineapple, and turmeric juice mix as influenced by blend variations has been determined. The Ca, Mg, vitamin C, and β-carotene contents increased significantly ($p < 0.05$) as quantities of pineapple and turmeric respectively decreased and increased. The degree of significant differences ($p < 0.05$) across samples appeared more in the protein, fat, and ash compared to moisture contents. Despite reducing the bacterial and fungi counts with the inhibition zone, increasing the turmeric and decreasing the pineapple might not necessarily make the blend juice mix more acceptable.

Given the blend results, the nutritional components of the juice mix blends of the current study require further exploration. For instance, the fruit genotype and climatic/storage conditions, together with different geographical regions, could be an influence on functional and quality outcomes of a given juice mix, and this warrants investigation at a future study. Another future work should target to investigate the antioxidant capacity, bioactive components, and total phenolic content of the same juice mix blends. Given the notion that many in Nigeria who take this juice mix consider it nutritionally enriching, future epidemiological and/or economic studies are warranted, as this could help provide additional information that will help substantiate this (notion). Additionally, a direction of future work could also be focused to determine the physicochemical, rheological, quality, and shelf life attributes of this blend juice mix under varying storage conditions.

Author Contributions: Conceptualization, A.F.O. and J.A.; Data curation, A.F.O., P.A. and G.B.; Formal analysis, A.F.O., J.A., P.A., G.S., C.O.R.O., G.B. and M.K.; Funding acquisition, C.O.R.O. and M.K.; Investigation, A.F.O., J.A., P.A.; Methodology, A.F.O., J.A., P.A., G.S., G.B. and M.K.; Project administration, A.F.O., J.A., P.A. and C.O.R.O.; Software, G.S., G.B. and M.K.; Validation, P.A., G.S., G.B. and M.K.; Visualization, G.S., C.O.R.O., G.B. and M.K.; Writing—original draft, A.F.O. and J.A.; Writing—review & editing, G.S., C.O.R.O., G.B. and M.K. All authors have read and agreed to the published version of the manuscript.

Funding: Publication financed by the project UPWR 2.0: international and interdisciplinary programme of development of Wrocław University of Environmental and Life Sciences, co-financed by the European Social Fund under the Operational Program Knowledge Education Development, under contract No. POWR.03.05.00-00-Z062/18 of 4 June 2019.

Data Availability Statement: Data sharing not applicable.

Acknowledgments: The authors would like to appreciate the laboratory staff support at the Department of Food Science and Technology, Federal University of Agriculture Makurdi—Nigeria. Authors G.S. and G.B. appreciate the funding support from IRBIM-CNR—Mazara del Vallo, Italy. Authors C.O.R.O. and M.K. appreciate the funding support from Wrocław University of Environmental and Life Sciences, Poland.

Conflicts of Interest: The authors have declared no conflict of interest.

References

1. Taylor, B. Fruit and Juice Processing. In *Chemistry and Technology of Soft Drinks and Fruit Juices*, 2nd ed.; Ashurst, R.P., Ed.; Blackwell Publishing Ltd.: Oxford, UK, 2005; pp. 35–67. [CrossRef]
2. Mauseth, J.D. *Botany, an Introduction to Plant Biology*; Jomes and Barlet Publishers: Boston, MA, USA, 2003; p. 285.
3. Agada, R.E. *Quality Evaluation of Juice from Blends of Cucumber and Pineapple Fruits*; Department of Food Science and Technology, Federal University of Agriculture: Makurdi, Benue State, Nigeria, 2018.
4. Akusu, O.M.; Kiin-Kabari, D.B.; Ebere, C.O. Quality Characteristics of Orange/Pineapple Fruit Juice Blends. *Am. J. Food. Sci. Technol.* **2016**, *4*, 43–47. [CrossRef]
5. Offia-Olua, B.I.; Ekwunife, O. Production and evaluation of the physico-chemical and sensory qualities of mixed fruit leather and cakes produced from apple (Musa Pumila), banana (Musa Sapientum), pineapple (Ananas Comosus). *Niger. Food J.* **2015**, *33*, 22–28. [CrossRef]
6. Bhaskar, J.J.; Chilkunda, N.D.; Salimath, P.V. Banana (Musa sp. var. elakki bale) Flower and Pseudostem: Dietary Fiber and Associated Antioxidant Capacity. *J. Agric. Food Chem.* **2012**, *60*, 427–432. [CrossRef]

7. Ogawa, E.M.; Costa, H.B.; Ventura, J.A.; Caetano, L.C.; Pinto, F.E.; Oliveira, B.G.; Barroso, M.E.S.; Scherer, R.; Endringer, D.C.; Romão, W. Chemical profile of pineapple cv. Vitória in different maturation stages using electrospray ionization mass spectrometry. *J. Sci. Food Agric.* **2018**, *98*, 1105–1116. [CrossRef]
8. Ajav, E.; Kolapo, A.; Ogunlade, C.A. Physical Properties of Ginger (Zingiber officinale). *Glob. J. Sci. Front. Res.* **2007**, *14*, 2249–4626.
9. Bailey-Shaw, Y.A.; Williams, L.A.D.; Junor, G.-A.O.; Green, C.E.; Hibbert, S.L.; Salmon, C.N.A.; Smith, A.M. Changes in the Contents of Oleoresin and Pungent Bioactive Principles of Jamaican Ginger (*Zingiber officinale* Roscoe.) during Maturation. *J. Agric. Food Chem.* **2008**, *56*, 5564–5571. [CrossRef]
10. Shahidi, F.; Hossain, A. Bioactives in spices, and spice oleoresins: Phytochemicals and their beneficial effects in food preservation and health promotion. *J. Food Bioact.* **2018**, *3*, 8–75. [CrossRef]
11. Varakumar, S.; Umesh, K.V.; Singhal, R.S. Enhanced extraction of oleoresin from ginger (*Zingiber officinale*) rhizome powder using enzyme-assisted three phase partitioning. *Food Chem.* **2017**, *216*, 27–36. [CrossRef] [PubMed]
12. Samaniego-Sánchez, C.; Inurreta-Salinas, Y.; Quesada-Granados, J.; Blanca-Herrera, R.; Villalón-Mir, M.; De La Serrana, H.L.-G.; Martínez, M.L. The influence of domestic culinary processes on the Trolox Equivalent Antioxidant Capacity of green tea infusions. *J. Food Compos. Anal.* **2011**, *24*, 79–86. [CrossRef]
13. Charles, D.J. *Antioxidant Properties of Spices, Herbs and Other Sources*; Springer Science & Business Media: New York, NY, USA, 2013; ISBN 978-1-4614-4310-0.
14. Parthasarathy, V.A.; Chempakam, B.; Zachariah, T.J. *Chemistry of Spices*; CABI: Oxfordshire, UK, 2008; p. 464; ISBN 978-1-84593-405-7.
15. Peter, K.V. (Ed.) *Handbook of Herbs and Spices*, 2nd ed.; Woodhead Publishing Ltd.: Cambridge, UK, 2016; Volume 2, p. 624; ISBN 9780081016176.
16. Cheng, X.-L.; Liu, Q.; Peng, Y.-B.; Qi, L.-W.; Li, P. Steamed ginger (*Zingiber officinale*): Changed chemical profile and increased anticancer potential. *Food Chem.* **2011**, *129*, 1785–1792. [CrossRef]
17. Embuscado, M.E. Spices and herbs: Natural sources of antioxidants—A mini review. *J. Funct. Foods* **2015**, *18*, 811–819. [CrossRef]
18. Leja, K.B.; Czaczyk, K. The industrial potential of herbs and spices—A mini review. *Acta Sci. Pol. Technol. Aliment.* **2015**, *15*, 353–368. [CrossRef] [PubMed]
19. Yashin, A.; Yashin, Y.; Xia, X.; Nemzer, B. Antioxidant Activity of Spices and Their Impact on Human Health: A Review. *Antioxidants* **2017**, *6*, 70. [CrossRef]
20. Nisar, T.; Iqbal, M.; Raza, A. Turmeric: A promising spice for phytochemical and antimicrobial activities. *Am. Eurasian J. Agric. Environ. Sci.* **2015**, *15*, 1278–1288. [CrossRef]
21. Amalraj, A.; Pius, A.; Gopi, S. Biological activities of curcuminoids, other biomolecules from turmeric and their derivatives—A review. *J. Tradit. Complement. Med.* **2017**, *7*, 205–233. [CrossRef]
22. Sharifi-Rad, J.; El Rayess, Y.; Rizk, A.A.; Sadaka, C.; Zgheib, R.; Zam, W.; Sestito, S.; Rapposelli, S.; Neffe-Skocińska, K.; Zielińska, D.; et al. Turmeric and Its Major Compound Curcumin on Health: Bioactive Effects and Safety Profiles for Food, Pharmaceutical, Biotechnological and Medicinal Applications. *Front. Pharmacol.* **2020**, *11*. [CrossRef]
23. Idowu-Adebayo, F.; Toohey, M.J.; Fogliano, V.; Linnemann, A.R. Enriching street-vended zobo (Hibiscus sabdariffa) drink with turmeric (Curcuma longa) to increase its health-supporting properties. *Food Funct.* **2021**, *12*, 761–770. [CrossRef]
24. Muogbo, P.C.; Gbabo, A.; Nwakuba, N.R.; Onyekwere, I.; Obasa, P.A. Livelihood enhancement through mechanization of turmeric production in Nigeria. In Proceedings of the 3rd NIAE-SE Regional Conference, University of Nigeria Nsukka, Enugu State, Nigeria, 27–30 August 2018.
25. Kalaycıoğlu, Z.; Torlak, E.; Akın-Evingür, G.; Özen, I.; Erim, F.B. Antimicrobial and physical properties of chitosan films incorporated with turmeric extract. *Int. J. Biol. Macromol.* **2017**, *101*, 882–888. [CrossRef] [PubMed]
26. Kumar, N.; Sakhya, K.S. Ethnopharmacological properties of Curcuma longa: A review. *Int. J. Pharm. Sci. Res.* **2013**, *4*, 103–112.
27. Sabale, P.; Modi, A.; Sabale, V. Curcuma longa Linn. A phytochemical and phytopharmacological review. *Res. J. Pharmacogn. Phytochem.* **2013**, *5*, 59–68.
28. Okwori, E.; Onu, R.; Adamu, M.; Chindo, H.; Dikko, H.; Odunze, I.; Baidu, A.; Natala, C.; Eze, P. Production and shelf life determination of fruit/vegetable juices using watermelon, cucumber, pineapple and carrot. *Afr. J. Food Sci. Technol.* **2017**, *8*, 34–39. [CrossRef]
29. Top 10 Home Remedies Team. Available online: https://www.top10homeremedies.com/author/ap (accessed on 21 December 2020).
30. AOAC. *Official Methods of Analysis of AOAC International*, 18th ed.; AOAC International: Gaithersburg, MD, USA, 2005.
31. AOAC. *Official Methods of Analysis of Association of Official Analytical Chemistry International*, 5th ed.; AOAC International: Washington, DC, USA, 1990.
32. Ikewuchi, C.J.; Ikewuchi, C.C. Iodometric determination of the ascorbic acid (vitamin C) content of some fruits con-sumed in a university community in Nigeria. *Global J. Pure Appl. Sci.* **2011**, *17*, 47–49.
33. Iwe, M.O. *Handbook of Sensory Methods and Analysis*; Rejoint Communication Services Ltd.: Enugu, Nigeria, 2002.
34. Adegoke, G.O. *Understanding Food Microbiology*, 2nd ed.; Alleluian Ventures: Ibadan, Nigeria, 2004; p. 216.
35. Abubakar, S.; Akanbi, B.O.; Nasir-Naeem, K.O.; Abdulsalam, Z.N. Phytochemical and In-vitro Antibacterial Activity of the Leaves and Stem Extracts of *Sesbania grandiflora* (L.) against Some Clinical Isolates. *Br. Microbiol. Res. J.* **2015**, *8*, 424–433. [CrossRef]
36. Rahman, M.M.; Sheikh, M.M.I.; Sharmin, S.A.; Islam, S.; Rahman, M.A.; Rahman, M.M.; Alam, M.F. Antibacterial activity of leaf juice and extracts of *Moringa oleifera* Lam. against some human pathogenic bacteria. *CMU J. Nat. Sci.* **2009**, *8*, 219–227.

37. Balasubramanian, K. Molecular Orbital Basis for Yellow Curry Spice Curcumin's Prevention of Alzheimer's Disease. *J. Agric. Food Chem.* **2006**, *54*, 3512–3520. [CrossRef]
38. Vyas, K. The Cure is in the Roots: Turmeric. *J. Nutr. Disord. Ther.* **2015**, *5*, 3. [CrossRef]
39. Oboh, G.; Akinyemi, A.J.; Ademiluyi, A.O. Antioxidant and inhibitory effect of red ginger (*Zingiber officinale* var. Rubra) and white ginger (Zingiber officinale Roscoe) on Fe2+ induced lipid peroxidation in rat brain in vitro. *Exp. Toxicol. Pathol.* **2012**, *64*, 31–36. [CrossRef]
40. Achinewhu, S.C.; Hart, A.D. Effect of processing and storage on the ascorbic acid (vitamin C) content of some pineapple varieties grown in the Rivers State of Nigeria. *Plant Foods Hum. Nutr.* **1994**, *46*, 335–337. [CrossRef]
41. Rodríguez, Ó.; Gomes, W.; Rodrigues, S.; Fernandes, F.A. Effect of acoustically assisted treatments on vitamins, antioxidant activity, organic acids and drying kinetics of pineapple. *Ultrason. Sonochem.* **2017**, *35*, 92–102. [CrossRef] [PubMed]
42. Chakraborty, S.; Rao, P.S.; Mishra, H.N. Effect of combined high pressure–temperature treatments on color and nutritional quality attributes of pineapple (*Ananas comosus* L.) puree. *Innov. Food Sci. Emerg. Technol.* **2015**, *28*, 10–21. [CrossRef]
43. Uckiah, A.; Goburdhun, D.; Ruggoo, A. Vitamin C content during processing and storage of pineapple. *Nutr. Food Sci.* **2006**, *39*, 398–412. [CrossRef]
44. Nwaekpe, J.O.; Anyaegbunam, H.N.; Okoye, B.C.; Asumugha, G.N. Promotion of Turmeric for the Food/Pharmaceutical Industry in Nigeria. *Am. J. Exp. Agric.* **2015**, *8*, 335–341. [CrossRef]
45. Ekeledo, E.N.; Omodamiro, R.M.; Oti, E. Development and evaluation of turmeric: Ginger based pineapple drinks and food flavourings. *Asian J. Plant Sci. Res.* **2013**, *3*, 139–141. Available online: www.pelagiaresearchlibrary.com (accessed on 2 December 2020).
46. Biology—LibreTexts. Module 8.3. The Effects of pH and Temperature on Microbial Growth. Available online: https://bio.libretexts.org/Courses/Manchester_Community_College_(MCC)/Remix_of_Openstax%3AMicrobiology_by_Parker_Schneegurt_et_al/08%3A_Microbial_Growth/8.03%3A_The_Effects_of_pH_on_Microbial_Growth (accessed on 15 February 2021).
47. Ehirim, F.N.; Mokorie, S.U.; Ebiringa, D.C. Effect of pineapple juice dilution of varying levels of yoghurt quality. In Proceedings of the NIFST Annual Conference, Lagos, Nigeria, 23–27 October 2006; pp. 35–36.
48. Kuttigounder, D.; Lingamallu, J.R.; Bhattacharya, S. Turmeric Powder and Starch: Selected Physical, Physicochemical, and Microstructural Properties. *J. Food Sci.* **2011**, *76*, C1284–C1291. [CrossRef]
49. Andrews, W. *Manuals of Food Quality Control. 4. Microbiological Analysis*; FAO Food & Nutrition Paper 1414, Rev. 1; Food and Agriculture Organization of the United Nations: Rome, Italy, 1992; p. 344; ISBN 92-5-103189-4.
50. Gupta, A.; Mahajan, S.; Sharma, R. Evaluation of antimicrobial activity of Curcuma longa rhizome extract against Staphylococcus aureus. *Biotechnol. Rep.* **2015**, *6*, 51–55. [CrossRef]
51. Lourenço, T.; Mendonça, E.; Nalevaiko, P.; Melo, R.; Silva, P.; Rossi, D. Antimicrobial effect of turmeric (Curcuma longa) on chicken breast meat contamination. *Braz. J. Poult. Sci.* **2013**, *15*, 79–82. [CrossRef]
52. Pankaj, S.; Balqees, A.T.; Najat, A.N.; Rahma, A.M.; Sah, P.; Al-Tamimi, B.; Al-Nassri, N.; Al-Mamari, R. Effect of temperature on antibiotic properties of garlic (*Allium sativum* L.) and ginger (*Zingiber officinale* Rosc.). *Afr. J. Biotechnol.* **2012**, *11*, 16192–16195. [CrossRef]
53. Yassen, D.; Ibrahim, E.A. Antibacterial activity of crude extracts of ginger (*Zingiber officinale* Roscoe) on Escherichia coli and *Staphylococcus aureus*: A study in vitro. *Indo Am. J. Pharm. Res.* **2016**, *6*, 5830–5835.

foods

MDPI

Article

Characterization of Powdered Lulo (*Solanum quitoense*) Bagasse as a Functional Food Ingredient

Leidy Indira Hinestroza-Córdoba [1,2], Stevens Duarte Serna [2], Lucía Seguí [2], Cristina Barrera [2] and Noelia Betoret [2,*]

1 Grupo de Valoración y Aprovechamiento de la Biodiversidad, Universidad Tecnológica del Chocó, AA.292, Calle 22 No. 18B-10, Quibdó-Chocó CP. 270001, Colombia; leihicor@doctor.upv.es

2 Institute of Food Engineering for Development, Universitat Politècnica de València, CP 46022 València, Spain; steduase@etsiamn.upv.es (S.D.S.); lusegil@upv.net.upv.es (L.S.); mcbarpu@tal.upv.es (C.B.)

* Correspondence: noebeval@tal.upv.es; Tel.: +34-963877000 (ext. 83624)

Received: 8 May 2020; Accepted: 1 June 2020; Published: 2 June 2020

Abstract: The stabilization of fruit bagasse by drying and milling technology is a valuable processing technology to improve its durability and preserve its valuable biologically active components. The objective of this study was to evaluate the effect of lyophilization and air temperature (60 °C and 70 °C) in hot air-drying as well as grinding conditions (coarse or fine granulometry) on physico-chemical properties; water interaction capacity; antioxidant properties; and carotenoid content of powdered lulo bagasse. Air-drying kinetics at 60 °C and 70 °C and sorption isotherms at 20 °C were also determined. Results showed that drying conditions influence antioxidant properties and carotenoid content while granulometry slightly influenced fiber and water interaction properties. Fiber content was near 50% and carotenoid content was higher than 60 μg/g dry matter in lyophilized powder. This β-carotene content is comparable to that provided by carrot juice. Air-drying at 60 °C only reduced carotenoids content by 10%.

Keywords: fruit by-products; lulo bagasse powder; dehydration; fiber; antioxidant properties; carotenoids

1. Introduction

Lulo (*Solanum quitoense* Lam) is one of the most important tropical fruits in Colombia. According to the report by the Ministry of Agriculture and Rural Development of Colombia (2017), Colombia is the third ranked country in Latin America with the largest number of hectares cultivated with fruit. The lulo harvest from different varieties, covering 10,539 hectares, has increased from 67,473 tons in 2012 to 79,872 tons in 2017 [1].

In recent years, lulo fruit has raised much interest in the global market due to its organoleptic characteristics, pleasant aroma, acidic and refreshing taste [2] and its high content of bioactive components such as fiber, minerals (phosphorus, calcium, iron), vitamins (thiamin, riboflavin, vitamin C), and other specific compounds (carotenes, lutein, zeaxanthin, chlorogenic acid, and bioactive amines) [3]. Moreover, a recent study has demonstrated the antihypertensive effect of compounds present in its juice and responsible for its bitter taste. These compounds are identified as bioactive amines N^1, N^4, N^8-tris-(Dihydrocafeoyl), spermidine, and N^1, N^8-bis-(Dihydrocafeoyl) spermidine [4].

Studies about the uses of the lulo fruit in the food industry are scarce. Lulo fruit is consumed as juice, in desserts, and for jellies; processed to make frozen concentrated juice and tea infusions as well as fermented for the elaboration of alcoholic beverages. In Colombia, it is the main ingredient for making the Lulada fruit cocktail liquor such as Ecuador's Canelazo and Colada Morada cocktail drinks. Juice production is the most frequently industrialization option. However, a large number of by-products are produced with associated environmental problems. According to data from [5], around 9.76 million tons of fruit and vegetable by-products are generated every year. Lulo bagasse includes the fruit skin as well as traces of pulp and seeds. Traditional utilization for such food industry by-products include feed for livestock, fertilizers, or agricultural substrates [6]. However, these uses do not provide adequate added value when considering the valuable active components contained in the by-products.

The stabilization of fruit bagasse by drying and milling technology is a valuable processing technology to improve its durability and preserve its biologically active components. Chemical composition and structural characteristics of raw material largely determine physico-chemical and functional properties of the final powder [7]. Powders obtained can be used as a healthy natural ingredient or as a raw material to extract bioactive compounds for other uses. The effect of drying and milling technology has been studied in fruit pomaces such as apple, grapes, cherry, blackcurrant, strawberry, raspberry, or blackberry [8]. However, as far as we know, no study has been done to date with lulo bagasse. The objective of this study was to evaluate the effect of lyophilization and air temperature (60 °C and 70 °C) in hot air-drying as well as grinding conditions (coarse or fine granulometry) on physico-chemical, water interaction, and antioxidant properties of powdered lulo bagasse. The effect on the content of the three major carotenoids has been also evaluated.

2. Materials and Methods

2.1. Lulo Bagasse Preparation

Two kg of fresh lulo fruits (*Solanum quitoense* Lam.), equivalent to 8–9 pieces, from Colombia were purchased in the Central Market in València (Spain). Whole lulo fruits were washed, blended for 10 min in a domestic blender (Phillips Avance Collection Standmixer, 800 W 2 L), and filtered with a stainless steel 500 μm sieve. After filtering, lulo juice and a solid paste were separated. The solid paste, referred to as the lulo bagasse from now on, was labeled and stored at 4 °C in a freezer until further processing.

2.2. Dehydration and Milling of Lulo Bagasse

Lulo bagasse was dehydrated by hot air-drying and lyophilization. Hot air-drying was carried out in a convective dryer (Pol-eko Aparatura, Katowice, Poland) at 60 °C and 70 °C until aw ≤ 0.3 was achieved. Lyophilization was performed in a lyophilizer (Telstar, Lioalta-g) at 0.05 mbar for 24 h after samples were frozen at −40 °C.

Dehydrated lulo bagasse was milled in a domestic food processor (Thermomix®, Vorwerk, Spain) to obtain two different granulometries (fine and coarse). Fine granulometry resulted from milling at 10,000 rpm for 2 min in 20 s intervals and the coarse one by milling at 4000 rpm for 20 s and, subsequently at 10,000 rpm for 20 s in 5 s intervals. Fine and coarse lulo bagasse powders were stored in opaque glass jars in conditions of controlled relative humidity.

2.3. Analytical Determinations

Water activity was measured with a dew point hygrometer (Aqualab 4TE Decagon devices Inc. Pullman, WA, USA) at the temperature of 20 °C. Moisture content was measured by drying until constant weight was achieved [9]. Total soluble solids content was determined in an ABBE ATAGO 3-T refractometer thermostated at 20 °C. Before the measurement, dried samples were water diluted in the proportion 1:10 (g/mL). Fiber content was determined following the Van Soest method as described by Mertens et al. [10] Neutral detergent fiber, acid detergent fiber, and lignin detergent fiber were analyzed and cellulose, hemicellulose, and lignin were calculated from those results.

2.4. Water Interaction and Emulsifying Properties

Solubility as the mass fraction of dissolved material during powder rehydration was determined following the procedure described by Mimouni et al. [11].

Hygroscopicity, defined as the capability of a product to absorb water, was evaluated according to the Cai and Corke [12] method by weighing 0.5 g of each sample and taking them to an airtight chamber next to a saturated solution of sodium sulfate (Na_2SO_4).

Wettability or the time taken by the powders to get completely wet was assessed by weighing 2 g of sample in a beaker with 20 mL of distilled water at 25 °C [13].

Swelling capacity is defined as the ratio between the volume that a sample occupies after hydration for a certain time and the original weight of the sample [14,15]. One gram of sample was weighed in a graduated conical tube, to which 10 mL of water was added to hydrate the sample for 18 h at 25 °C.

Water holding capacity or water bound by gravity at atmospheric pressure was determined by measuring water content of the precipitate after mixing 0.2 g of sample and 10 mL of distilled water and left to stand for 18 h at 25 °C [14].

Water retention capacity is defined as water content remaining bond after hydration and centrifugation. 1 g of sample was weighed in a conical centrifuge tube and 10 mL of water was added, allowing hydration for 18 h at 25 °C. After that, samples were centrifuged for 30 min at 514× *g*. The precipitate was weighed and lyophilized to obtain the dry weight of the sample [14].

Oil retention capacity was measured following the method described by Garau et al. [16]. A 0.2 g of sample was mixed with 1.5 g of sunflower oil at room temperature. After that, the mixture was centrifuged at 1500× *g* for 5 min, the supernatant was removed, and the precipitate was weighted. Oil retention capacity was expressed in grams of absorbed oil per gram of initial sample.

Emulsifying activity was determined by the method described by Yasumatsu et al. [17]. A 2% (*w*/*v*) aqueous solution was prepared with the sample in a graduated tube. Seven mL of the prepared solution was mixed with 7 mL of sunflower oil and homogenized for 5 min in a vortex at 2400 rpm. After that, the mix was centrifuged at 12,857× *g* for 5 min. The volume of the emulsion was measured and referred to the total fluid volume. The emulsifying stability was determined by a similar procedure explained for emulsifying activity, except that the emulsions were heated at 80 °C for 30 min before centrifugation at 514× *g* for 5 min.

2.5. Particle Size

Particle size was determined on wet and dry dispersions. In both cases, laser diffraction equipment (Mastersizer, Malvern Instruments Limited, Worcerster, Great Britain) was used with a short-wavelength blue light source in conjunction with forward and backscatter detection to enhance sizing performance in the range 0.01–1000 μm. For the wet measurement, a small quantity of each sample was diluted in deionized water until it reached an obscuration of 8–9%. For the dry measurement, a small amount of each sample was put directly into the equipment until it reached an obscuration of 8–9%. Particle size distribution measurements were characterized through average equivalent volume diameter D [3,4].

2.6. Optical Properties

CIE*L*a*b* color coordinates were determined from the surface reflectance spectra obtained between 400 and 700 nm, when measuring on white and black backgrounds, considering standard light source D65 and standard observer 10° (Minolta spectrophotometer CM-3600d, Japan).

2.7. Antioxidant Properties

For antioxidant extraction, the samples were mixed with an 80:20 (*v*/*v*) methanol–water solution in the proportion 1:10 (*w*/*v*) and centrifuged at 10,000 rpm for 5 min at 20 °C (Selecta, "Medrifriger BL-S"). The next analyses were carried out in the supernatant.

2.8. Total Phenols and Flavonoids Content

Total phenols were determined following the Folin–Ciocalteu method [18,19]. A sample of the 0.125 mL of extract, 0.125 mL of Folin–Ciocalteu reagent (Sigma-Aldrich), and 0.5 mL of double-distilled water were mixed and allowed to react for 6 min. After that, 1.25 mL of 7% (*m*/*v*) sodium carbonate solution and 1 mL of double distilled water were added. Absorbance was measured in a spectrophotometer (Thermo Scientific, Helios Zeta U/Vis) at 765 nm. A blank was used as a reference and allowed to react for 90 min. A standard gallic acid curve ranging from 0 to 500 mg/L was obtained to express results in milligrams of gallic acid equivalent (GAE) per gram of dry sample.

Flavonoid content was determined following the method described by Luximon-Ramma et al. [20] A 1.5 mL sample of extract and 1.5 mL of a 2% (*w*/*v*) aluminum chloride solution were mixed and left in the dark for 10 min. Absorbance was measured on a spectrophotometer (Thermo Scientific, Helios Zeta U/Vis) at 368 nm. The resulting data were compared to a standard quercetin curve ranging from 0 to 350 mg/L. The results were expressed in milligrams of quercetin equivalent (EQ) per gram of dry sample.

2.9. DPPH and ABTS Methods

Antioxidant capacity was determined following the DPPH (2,2-diphenyl-1-picrylhydrazyl) method described by Kuskoski et al. [21] and Stratil et al. [22], with some modifications. A total of 0.1 mL of the extract, 0.9 mL of methanol, and 2 mL of the methanol–DPPH solution were mixed and absorbance was measured at 517 nm in a spectrophotometer (Thermo Scientific, Helios Zeta U/Vis). The results were expressed as milligrams of Trolox equivalent (TE) per gram of dry matter, using the Trolox calibration curve within a 0 to 500 mg/L concentration range.

The antioxidant activity was also evaluated following the ABTS (2,2′-azino-bis(3-ethylbenzothiazoline-6-sulfonic acid) radical method described by Re et al. [23]. A solution of 7 mM of ABTS and 2.45 mM of potassium persulfate was prepared and left to stand in the dark at room temperature for 16 h. ABTS was mixed with phosphate buffer to reach an absorbance of 0.70 ± 0.02, read at 734 nm. A 0.1 mL of extract was added to 2.9 mL of ABTS solution and absorbance was measured at 734 nm in a spectrophotometer (Thermo Scientific, Helios Zeta UV/Vis) after 0, 3 and 7 min of reaction time. The results were expressed as mg of Trolox equivalent (TE) per gram of dry matter.

2.10. Carotenoid Content by HPLC (High-Performance Liquid Chromatography)

Carotenoids were extracted according to the procedure described by Rodrigues et al. [24] and Bunea et al. [25], with some modifications. One gram of sample was mixed with methanol/ethyl acetate/petroleum ether (1:1:1, *v*/*v*/*v*) as the extraction solvent. The extract was saponified for 12 h in the dark at room temperature, using a 30% (*v*/*v*) KOH solution in methanol. The sample was washed with saturated saline solution, evaporated in a rotary evaporator (T < 30 °C), and analyzed by HPLC. HPLC analysis was performed in a HPLC Alliance 2995 system, using a separation module (Waters, 2695) made up of a pump and a DAD detector (2996, Waters, Milford, MA, USA). Carotenoids were separated with a YMC C30 column [5 μm, 250 mm × 4.6 mm (internal diameter)], using ternary gradient elution made up of acetonitrile:water (9:1, *v*/*v*) with 0.25% trietylamine (solvent A) and ethyl acetate with 0.25% trietylamine (Solvent B). Carotenoids were quantified at a flow of 1 mL/min. The results were expressed in micrograms per 100 g of dry sample.

2.11. Sorption Isotherms

Sorption isotherms were determined according to the method described by Wolfe et al. [26] with some modifications. This technique involves the use of saturated salt solutions to maintain a known and controlled humidity environment inside a closed jar at a fixed temperature condition. One gram of sample was placed in a closed jar together with one of the next saturated salt solutions: LiCl (aw = 0.1),

CH_3COOK (aw = 0.23, $MgCl_2$ (aw = 0.32), K_2CO_3 (aw = 0.43), Mg $(NO_3)_2$ (aw = 0.52), NaCl (aw = 0.75), KCl (aw = 0.85), and $BaCl_2$ (aw = 0.90) at 20 °C. The samples were weighed every eight days until constant weight was reached. Once the samples reached equilibrium, moisture content was measured.

2.12. *Statistical Analysis*

All determinations were made in triplicate and the statistical analysis of the data was performed in a Statgraphics Centurion XVII software package, making use of a simple or multifactorial analysis of variance (ANOVA) at a 95% confidence level ($p < 0.05$).

3. Results

3.1. *Hot Air-Drying of Lulo Bagasse*

Figure 1 shows the hot air-drying curves at 60 °C and 70 °C and drying speed curves for lulo bagasse.

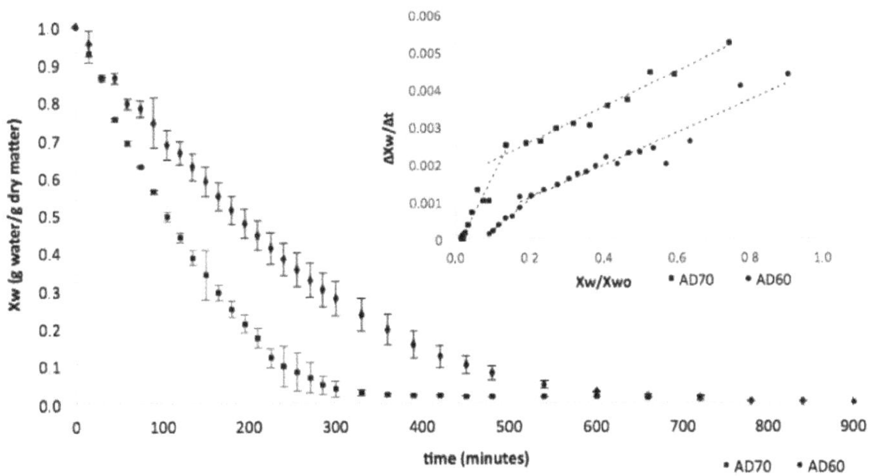

Figure 1. Hot air-drying and drying rate curves at 60 °C and 70 °C of lulo bagasse. AD60 and AD70: hot air-drying at 60 °C and 70 °C, respectively.

For a curve comparison, reduced humidity (X_w/X_{w0}) was used. An effect of air temperature on the kinetics of the process can be observed. Drying with air at 60 °C requires approximately 10–12 h until the product reaches an almost constant weight, which corresponds to a humidity close to 2%. Increasing the temperature to 70 °C decreased the time required to 6–7 h. Although when obtaining a functional food ingredient it is necessary to consider the effect of air-drying temperature on bioactive compounds (antioxidants such as carotenoids in lulo powder), the effect on processing time, production, and energy consumption will be relevant when it comes to setting up an industrial process. In Figure 1, it can be observed that an increase in the air-drying temperature resulted in greater values of drying rates that decreased along all the air-drying process. Experimental values were linearly adjusted to model equation showed at Table 1. Two stages with different kinetics were clearly identified. Kinetic parameters obtained for each stage at both temperatures and correlation coefficients (R^2) are included in Table 1.

Table 1. Kinetics of air drying of lulo bagasse at 60 °C and 70 °C.

	60 °C	70 °C
First stage: $\frac{\Delta Xw}{\Delta t} = k_1 \frac{Xw}{Xw_0} + k_2$	$\frac{Xw}{Xw_0} \in [1, 0.195]$	$\frac{Xw}{Xw_0} \in [1, 0.129]$
k_1	0.0043	0.0048
k_2	0.0002	0.0016
R^2	0.9176	0.9625
Second stage: $\frac{\Delta Xw}{\Delta t} = k'_1 \frac{Xw}{Xw_0} + k'_2$	$\frac{Xw}{Xw_0} \in [0.195, 0.089]$	$\frac{Xw}{Xw_0} \in [0.129, 0.024]$
k'_1	0.0084	0.0195
k'_2	−0.0006	−0.0003
R^2	0.987	0.9319

Xw: moisture content; Xw_0: initial moisture content.

In the first stage, the decline of drying rate was not affected by air temperature. Considering that in the drying chamber the relation between air volume and the mass of lulo bagasse was large enough to avoid changes in the air conditions, the lack of temperature effect on the decline of drying rate revealed an internal control. Structural characteristics and composition determine water molecular transport from the innermost layers to the surface of the solid bed. However, the drying kinetics change when the moisture of the product drops below 40% (*w/w*) at 60 °C and below 30% (*w/w*) at 70 °C (these values are the result of considering a X_{w0} = 3.4 g water/g dry matter and the relation $\frac{Xw}{Xw_0}$ indicated in Table 1). From this humidity, decreasing the drying rate with moisture depends on the air temperature being more pronounced at 70 °C. Phase transitions of some components, together with a greater compaction of the lulo bagasse bed would probably explain these differences.

3.2. Moisture Sorption Isotherms of Lulo Powders

The sorption isotherms at 20 °C of the lulo bagasse powders are shown in Figure 2.

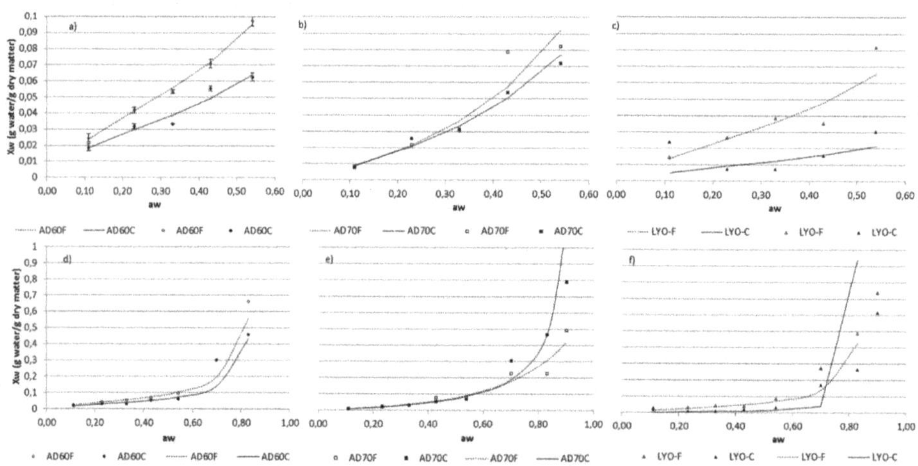

Figure 2. Sorption isotherms: experimental data and BET (Brunauer-Emmett-Teller) (**a–c**) and GAB (Guggenheim-Anderson-de Boer) (**d–f**) adjustment.

The curves demonstrate an important increase in water activity with low increasing equilibrium moisture content, following the type II and III BET classification shape, which is usual for non-structured and non-porous solid foods [27]. It is the typical form of plant products rich in simple sugars such as fructose or glucose and macromolecules such as cellulose or hemicellulose, with low ability to adsorb water molecules. The curves are similar to that for dried persimmon leaves [28]. Figure 2 shows the agreement between the experimental data and predicted isotherms using BET (Brunauer-Emmett-Teller) and GAB (Guggenheim-Anderson-de Boer) models. Model parameters and correlation coefficients (R^2) are included in Table 2. The BET model was adjusted considering aw values below 0.55, which is consistent with the assumptions assumed by the model [29]. GAB model setting has included values up to 0.8.

Table 2. Parameters from BET and GAB adjustment.

	BET [29] $Xw=\frac{W_0Ca_w}{(1-a_w)(1+(C-1)a_w)}$			GAB [26] $Xw=\frac{W_0CKa_w}{(1-Ka_w)(1-Ka_w+CKa_w)}$			
	W_0	C	R^2	W_0	C	K	R^2
AD60F	0.050	5.902	0.992	0.113	−4.103	−0.614	0.929
AD60C	0.033	8.207	0.912	0.022	0.481	−3.208	0.837
AD70F	0.089	0.768	0.058	0.056	0.897	1.299	0.918
AD70C	0.060	1.208	0.061	0.063	0.937	1.108	0.929
LYO-F	0.048	2.726	0.918	0.293	−4.611	−0.095	0.934
LYO-C	0.007	−246.4	0.619	0.056	−4.694	−0.077	0.589

AD60, AD70: hot air drying at 60 °C and 70 °C respectively; LYO: lyophilized; F: fine granulometry; C: coarse granulometry.

Among all sorption isotherm model parameters, the monolayer moisture content is recognized as the most important one. It corresponds to the moisture content affording the longest storage time period with minimum quality loss by deteriorative reactions (except fat oxidation) at a given temperature. From a physico-chemical point of view, it is related to the number of sorption sites available on the material surface. Therefore, the conditions of milling or drying inducing changes in surface structural characteristics are expected to lead to changes in monolayer moisture content. Furthermore, the physical state (crystalline or amorphous) in which the food components are found and the phase transitions along the dehydration process conditioned by temperature and kinetics, strongly influence water retention.

A lower variability and greater consistency in the monolayer moisture values can be observed in Table 2, as obtained from the BET model. The values were slightly higher in the fine powder than in the coarse one, probably because the fine powder had a larger specific surface, having a greater number of accessible active points. Lyophilized powders showed lower values, which can be associated with a greater disruption of cell structure that occurs during this treatment. The greatest difference between the air-dried powders at 60 °C and 70 °C is given by the value of parameter C. It is an empirical parameter directly related to the net heat of sorption being the difference between the heat of sorption of the first layer of molecules of water and the others. It is considered that C values greater than 2 are associated with higher adsorption forces (type II sorption isotherms corresponding to dried powders at 60 °C). The differences obtained between air-dried powders at 60 °C or 70 °C could be related to the effect of temperature on the kinetics of the process and the associated phase transitions.

3.3. *Physico-Chemical Properties*

Table 3 shows the total soluble solids content (x_{ss}), water activity (aw), and moisture content (x_w) of different lulo bagasse powders. All drying conditions allow for reducing the aw below 0.27, which is considered a usual value in food powders such as milk powder or instant coffee [30]. Corresponding moisture contents were also very low, the drying process having contributed to reducing most of the free water content responsible for spoilage reactions. Statistically significant differences among treatments were detected for the three parameters, the differences being lower between fine and coarse powders. The most significant differences appeared between the two temperatures of hot air-drying. These differences manifested themselves similarly in the water activity and in the moisture content of the samples, so they are associated with the effect of temperature on the desorption characteristics of the samples determined by air-drying kinetics and phase transitions along the process.

Table 3. Water activity (a_w), moisture content (x_w), (g water/g), soluble solids content (x_{ss}) (g soluble solids/g), fiber content (% of dry weight) and CIE L*a*b* coordinates of lulo bagasse powders. Mean ± standard deviation of three repetitions.

	AD60F	AD60C	AD70F	AD70C	LYO-F	LYO-C
a_w	0.119 ± 0.006 ^{a}	0.199 ± 0.006 ^{a}	0.258 ± 0.006 ^{b}	0.267 ± 0.007 ^{b}	0.166 ± 0.003 ^{b}	0.134 ± 0.003 ^{a}
x_w (g_w/g_{sample})	0.021 ± 0.004 ^{a}	0.035 ± 0.003 ^{a}	0.018 ± 0.002 ^{a,b}	0.015 ± 0.001 ^{b,c}	0.0194 ± 0.0006 ^{c}	0.022 ± 0.002 ^{d}
x_{ss} (g_{ss}/g_{total})	0.236 ± 0.005 ^{e}	0.149 ± 0.012 ^{a}	0.222 ± 0.006 ^{b}	0.146 ± 0.06 ^{c}	0.21 ± 0.11 ^{b}	0.26 ± 0.14 ^{d}
Fiber content						
Hemicellulose (%)	5.2 ± 0.2 ^{a}	11.3 ± 0.2 ^{e}	5.5 ± 0.2 ^{a}	11.30 ± 0.4 ^{c}	10.3 ± 0.2 ^{b}	10.7 ± 0.4 ^{b,c}
Cellulose (%)	18.6 ± 0.3 ^{a}	24.7 ± 0.4 ^{c}	21.3 ± 0.7 ^{b}	24.59 ± 0.05 ^{c}	22.1 ± 0.2 ^{b}	24.6 ± 0.3 ^{c}
Lignin (%)	10.1 ± 0.3 ^{a}	20.84 ± 3 ^{b}	17.6 ± 0.23 ^{a}	10.6 ± 0.2 ^{a}	8.5 ± 0.5 ^{b}	10.8 ± 0.4 ^{a}
Insoluble fiber (%)	35.2 ± 0.1 ^{b}	40.3 ± 0.4 ^{d}	33.3 ± 0.2 ^{a}	34.2 ± 0.2 ^{a,b}	38.4 ± 0.3 ^{c}	41.65 ± 1.01 ^{e}
Total fiber (%	42.6 ± 0.04 ^{b}	50.6 ± 0.6 ^{f}	41.4 ± 0.4 ^{a}	46.5 ± 0.3 ^{c}	46.0 ± 0.4 ^{d}	48.1 ± 0.3 ^{e}
Colour						
L*	58.4 ± 0.2 ^{c}	50.8 ± 0.2 ^{b}	53.3 ± 0.3 ^{c}	50.97 ± 0.06 ^{b}	63.134 ± 0.13 ^{d}	60.59 ± 0.05 ^{d}
a*	10.37 ± 0.02 ^{c}	10.22 ± 0.07 ^{c}	10.35 ± 0.11 ^{c}	10.76 ± 0.13 ^{d}	9.46 ± 0.07 ^{a}	9.98 ± 0.05 ^{b}
b*	38.22 ± 0.11 ^{d}	40.5 ± 0.3 ^{f}	39.5 ± 0.2 ^{e}	38.043 ± 0.10 ^{d}	36.74 ± 0.11 ^{c}	34.57 ± 0.06 ^{b}
C	39.60 ± 0.01 ^{d}	42.6 ± 0.3 ^{f}	40.8 ± 0.2 ^{e}	39.53 ± 0.12 ^{d}	37.94 ± 0.09 ^{c}	35.98 ± 0.06 ^{a}
h	74.84 ± 0.07 ^{c,d}	71.92 ± 0.04 ^{b}	75.32 ± 0.09 ^{d}	74.2 ± 0.2 ^{c}	75.56 ± 0.14 ^{d}	73.70 ± 0.02 ^{c}

AD60, AD70: hot air drying at 60 °C and 70 °C respectively; LYO: lyophilized; F: fine granulometry; C: coarse granulometry. ^{a,b,c} different letters on the same file indicate statistically significant differences at a 95% confidence level.

The different fiber fractions of lulo bagasse powders determined by the Van Soest method are summarized in Table 3. This method estimates the structural carbohydrates and indigestible substances linked to them that form the plant cell wall. Although the determination quantifies the fractions of hemicellulose, cellulose, and lignin, it does not allow for the quantification of components such as pectins and other polysaccharides that are also part of the dietary fiber. It is essentially the soluble fiber that is undervalued in the Van Soest method. It can be observed that the total fiber content varied from 34% to 47% being amounts lower than the dietary fiber obtained for other fruit waste powders such as pomegranate bagasses (45.6–50.3 g/100 g dry weight) [31], grape co-products (74.5 g/100 g dry weight) [32], apple pomace (51.1 g/100 g dry weight) [33], or banana peels (49.9 g/100 g dry weight) [34]. However, the total fiber content is similar to that for other fruits by-products such as grapefruit peel (44.2%) [35], mango (37.1%), and peach (37.6%) by-products [36]. Larrauri et al. [37] stated that products containing 50% of total fiber can be considered as a source rich in fiber. According to the Scientific Opinion on Dietary Reference Values for carbohydrates and dietary fiber [38], a daily dietary intake of at least 25 g of fiber is recommended. However, nutritionists recommend a daily fiber intake of 35 g per day [39]. It can be observed that in the three drying treatments, and practically for all fiber fractions, there are statistically significant differences between fine granulometry and coarse granulometry powders. This can be explained by the effect of milling and particle size reduction. Although health claims for fiber do not distinguish between insoluble and soluble fractions, it is accepted that the 'ideal dietary fiber' should have a balanced composition (insoluble and soluble fractions) [37]. Other authors accept that fiber sources suitable for use as food ingredients should have a ratio soluble to insoluble close to 1:2 to provide adequate physiological effects [40]. The results obtained for the lulo bagasse powders showed a much higher relationship between hemicellulose (part of the soluble fiber) and the insoluble fiber, although as previously mentioned, the Van Soest method used does not allow for the determination of other soluble fiber components.

Figure 3 shows the particle size distribution of lulo bagasse powders by the dry and wet methods. The particle size distribution of lulo bagasse powders by the dry method did not show remarkable differences between treatments.

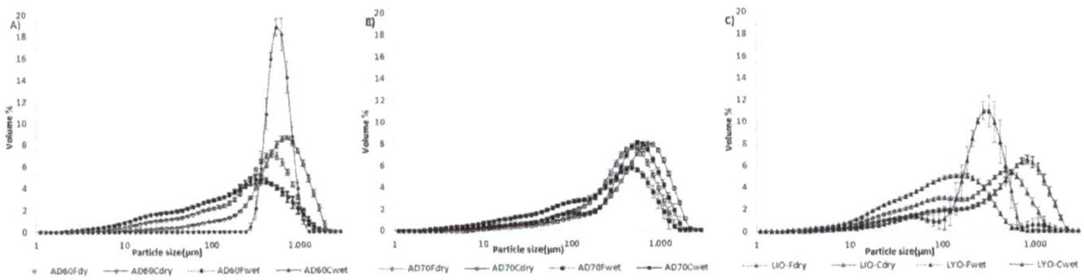

Figure 3. Particle size distribution of lulo bagasse powders by the dry method (empty markers) and wet method (full markers) for the three treatments considered: (**A**) air drying at 60 °C; (**B**) air drying at 70 °C; (**C**) lyophilization. AD60 and AD70: hot air-drying at 60 °C and 70 °C, respectively; LYO: lyophilized; F: fine granulometry; C: coarse granulometry.

A particle volume maximum appeared at 1000 μm for all treatments and a minor maximum at 100 μm only for the lyophilized powders. Conversely, the dehydration treatment applied largely determined the particle size distribution by the wet method. In the powders dehydrated by hot air at 60 °C (Figure 3B) and regardless of granulometry, no large changes in distribution were observed. Only a slight increase in the number of smaller particles (around 100 μm) was observed due to the solubilization of a part of the particles sized around 1000 μm. However, in the samples dried by hot air at 70 °C and lyophilized, the granulometry largely determined the changes in the distribution obtained by wet way (Figure 3A,C). In fine granulometry powders, the changes were very similar to those obtained for drying at 60 °C, with a shift in the distribution toward smaller particle sizes. This shift was more pronounced in the powders obtained by lyophilization, probably due to the higher solubility associated with increased porosity of the particles in this treatment. However, in coarse granulometry powders, the appearance of a much more pronounced peak at a particle size slightly smaller than 1000 μm was observed, reflecting an aggregation of smaller particles (Figure 3).

Table 3 includes L*a*b* coordinates and h and C parameters of the powders. For the luminosity values (L*), a significant decrease could be observed in the air-dried powders compared with the lyophilized ones. Regarding the parameters a* and b*, which represent (+red/green−) and (+yellow/blue−) respectively, all the values were found to be positive. Both parameters were significantly lower for lyophilized powders, while the differences among the air-dried ones were slight. Among the air-dried powders, differences between fine and coarse granulometry were higher than those caused by air temperature. Formation of brown compounds as a result of the Maillard reaction in air-dried powders is favored by the major exposure to oxygen in fine powders. Lyophilization greatly reduces the Maillard reactions [41]. As a result, air-dried powders tend to have a browner color and the lyophilized ones are a greenish color more similar to the fresh bagasse.

3.4. *Water Interaction and Emulsification Properties of Lulo Bagasse Powders*

Solubility is a physical property describing powder behavior in an aqueous solution. For fiber rich ingredients such as powders from fruit by-products, it mainly depends on fiber content, size, and physical characteristics of particles (porosity and physical stage of components). The solubility of lulo bagasse powders is included in Table 4.

Values were similar to that obtained for orange peel by Garau et al. [16], but higher to that obtained for tomato powders by Santos de Sousa et al. [42]. Statistical analysis showed granulometry (fine or coarse) as the most important factor determining significant differences ($p < 0.05$) among the samples. These results are consistent with those obtained for fiber content. The greater intensity of the milling treatment applied in fine powders produces a smaller particle size, a reduction in the content of the different insoluble fiber fractions, and therefore a greater solubility. Regarding the effect of the dehydration treatment, the lyophilized samples showed higher solubility levels, as a consequence of the more aggressive structural breaks suffered by the samples due to the sublimation of the water contained within the tissue. The decrease in solubility caused by temperature increase in hot air-dried samples could be due to the degradation of pectin substances and physical changes affecting other components during the drying process. Hygroscopicity is defined as the capability of a product to absorb water [43,44], and describe that a powder with low hygroscopicity, low humidity, low caking, and high solubility can be considered stable. The results obtained showed that hygroscopicity slightly decreased as the drying temperature increased. According to Ahmed et al. [45], powder moisture variations have a direct impact on hygroscopicity. Regarding swelling capacity, water holding, and retention capacities, they depend on microstructural properties of particles and the nature of the fiber content (soluble or insoluble).

Porous particles and soluble fibers have a high water holding capacity and swelling to form viscous solutions. Insoluble fibers can also adsorb and retain water within their fibrous matrix, but to a lesser extent. Lecumberri et al. [46] provides data on the swelling and water holding capacity of cellulose (insoluble fiber) and apple and orange pectin (soluble fiber), showing undetected or practically null values in the case of cellulose. The swelling value obtained for apple pectin (7.42 ± 1.15) was similar to that obtained for the lyophilized lulo bagasse powders and it was slightly lower in the hot air-dried ones. In all lulo bagasse powders, regardless of the dehydration treatment, water holding capacity was much lower than that for apple (16.51 ± 3.77) and orange (28.07 ± 5.37) pectin, although it was very similar to the value obtained for a fiber rich cocoa product. Additionally, for this property, the values were higher in lyophilized powders. According to Lecumberri et al. [47], wettability is strongly affected by size and porosity of particles. As a consequence of water sublimation, lyophilized powders showed a more porous particle surface, which makes water difficult to penetrate due to the greater surface tension that needs to be overcome. Wettability of lyophilized lulo bagasse powders showed much higher values than the air-dried ones.

Emulsifying properties concerning oil retention capacity and emulsifying activity and stability are required to use powders as an ingredient in fatty foods. These properties require the presence of long-chain chemical compounds with hydrophilic and hydrophobic groups. An effective emulsifying agent consists of water-soluble (hydrophilic) and water-insoluble (hydrophobic) domains. Pectin has an emulsifying capacity and has been traditionally used as a gelling and thickening reagent. Recently, reports of the remarkable emulsifying ability of pectin have attracted much attention, with sugar beet pectin [48] and pomegranate peel pectin [49] as examples. Non-conclusive results were obtained for emulsifying activity and stability of the powdered lulo bagasse. Oil retention capacity was very low and may be affected by the lower pectin content of the lulo bagasse. Similar powders from lemon, orange, peach, and apple bagasse have a greater oil retention capacity (contain 2.5 to 2.9 g of oil/g of sample), according to Martínez-Las Heras et al. [50].

Table 4. Hydration, water retention and emulsification properties of lulo bagasse powders. Mean ± standard deviation of three repetitions.

	AD60F	AD60C	AD70F	AD70C	LYO-F	LYO-C
Solubility (%)	35 ± 5 [b]	27 ± 6 [b]	32 ± 2 [d]	19 ± 4 [c]	45 ± 8 [a]	30 ± 4 [c]
Higroscopicity (g_{water}/100 g)	30.9 ± 0.4 [c]	23.0 ± 0.2 [a]	23.0 ± 0.2 [a]	22.7 ± 1.1 [a]	22.1 ± 0.20 [a]	25.32 ± 1.1 [b]
Wettability (s)	31.7 ± 0.6 [b]	8.7 ± 1.2 [a]	10.0 ± 1.0 [a]	11.0 ± 1.7 [a]	19.5 ± 2.6 [c]	17.0 ± 1.0 [b]
Swelling capacity (mL_{water}/g)	4.98 ± 0.02 [b]	4.46 ± 0.04 [a]	4.97 ± 0.02 [b]	4.98 ± 0.05 [b]	7.46 ± 0.05 [d]	5.48 ± 0.02 [c]
Water holding capacity ($g_{water}/g_{dry\ matter}$)	5.89 ± 0.10 [a]	5.7 ± 0.1 [a]	6.3 ± 0.2 [a]	7.6 ± 0.8 [b]	8.2 ± 0.7 [b]	6.4 ± 0.5 [a]
Water retention capacity ($g_{water}/g_{dry\ matter}$)	4.75 ± 0.02 [a]	4.5 ± 0.2 [a]	5.5 ± 0.2 [b]	5.83 ± 0.06 [c,d]	5.9 ± 0.1 [d]	5.9 ± 0.4 [b,c]
Emulsifying properties						
Oil retention capacity (g_{oil}/g_{sample})	0.142 ± 0.004 [a]	0.18 ± 0.03 [a,b]	0.20 ± 0.02 [b,c]	0.20 ± 0.02 [b]	0.24 ± 0.01 [c]	0.45 ± 0.04 [d]
Emulsification activity	N.D	N.D	N.D	N.D	N.D	N.D
Emulsification stability	N.D	N.D	N.D	N.D	N.D	N.D

AD60, AD70: hot air drying at 60 °C and 70 °C, respectively; LYO: lyophilized; F: fine granulometry; C: coarse granulometry. [a,b,c] different letters on the same column indicate statistically significant differences at a 95% confidence level. N.D: Not detected.

3.5. Antioxidant Properties

Total phenol and flavonoid content are represented in Figure 4.

Significant differences ($p < 0.05$) were observed between the hot air-dried powders and lyophilized ones. The low temperature and vacuum conditions in the lyophilization treatment limit oxidation reactions, thus preserving phenolic and flavonoid compounds [51] Results were similar to those reported by Crozier et al. [52] in pomegranate powder and red fruits. Some other authors have also evaluated the effect of different drying techniques (mainly hot air-drying and lyophilization) in phenol and flavonoid content of different fruit by-products such as orange by-products [16], bagasse of blackcurrant pomace [53], mango peel [54], and apple peel [55]. The results exhibited, in all cases, a decrease in antioxidant compounds as the temperature increased.

Results of the antiradical capacity by the DPPH and ABTS methods are included in Figure 4. The use of more than one single method is recommended to estimate the antioxidant activity of complex samples [56]. In this case, the ABTS free radical method, which has been reported to be more sensitive to hydrophilic antiradicals [57], was used in addition to the DPPH radical method. In coherence with total phenol and flavonoid content, both methods showed slightly higher values for lyophilized lulo bagasse powders. However, the differences were more marked for the results obtained by the ABTS radical method. Additionally, for the hydrophilic nature of antioxidants compounds in the lulo bagasse powders, differences between the DPPH and ABTS methods were probably due to their different sensitivities to the antiradical compounds that may be present in the sugarcane products. The time of reactions could also have produced some differences; nevertheless, the ABTS reaction is usually faster than the DPPH inhibition reaction [56].

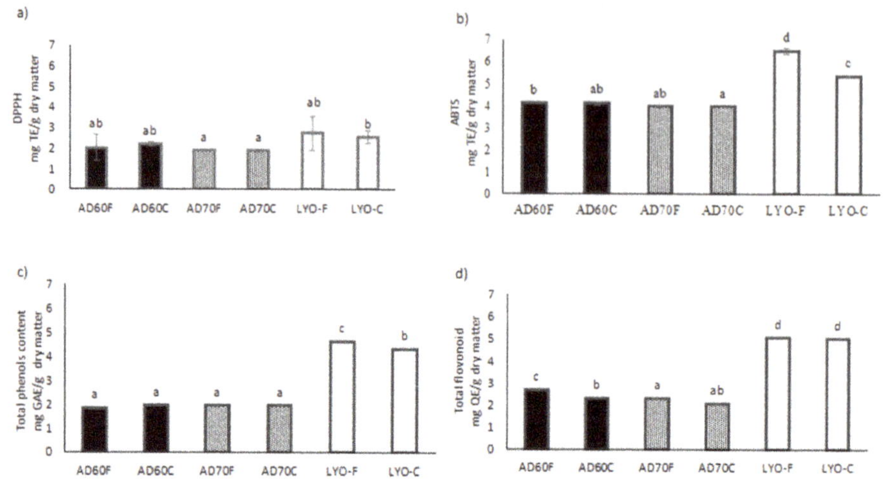

Figure 4. Antioxidant properties: results from DPPH method (**a**), results from ABTS method (**b**), total phenol content (**c**) and flavonoid content (**d**). AD60 and AD70: hot air-drying at 60 °C and 70 °C, respectively; LYO: lyophilized; F: fine granulometry; C: coarse granulometry.

3.6. Carotenoid Content

The β-cryptoxanthin, α-carotene and β-carotene content of lulo bagasse powders are summarized in Table 5.

Table 5. Carotenoids content (μg/g of dry sample) in the lulo bagasse powders. Mean ± standard deviation of three repetitions.

	AD60F	AD60C	AD70F	AD70C	LYO-F	LYO-C
β-cryptoxanthin	4.761 ± 0.014 ^{c}	4.06 ± 0.12 ^{b}	1.197 ± 0.008 ^{a}	1.193 ± 0.011 ^{a}	8.83 ± 0.04 ^{d}	8.72 ± 0.18 ^{d}
α-carotene	1.60 ± 0.07 ^{b}	1.577 ± 0.013 ^{b}	0.546 ± 0.017 ^{a}	0.581 ± 0.015 ^{a}	1.75 ± 0.01 ^{c}	1.73 ± 0.03 ^{c}
β-carotene	45.2 ± 0.4 ^{b}	45.1 ± 0.3 ^{b}	27.8 ± 0.3 ^{a}	27.61 ± 0.02 ^{a}	61.85 ± 0.25 ^{d}	61.15 ± 0.09 ^{c}
Total	51.5 ± 0.5 ^{b}	50.7 ± 0.2 ^{b}	29.51± 0.2 ^{a}	29.38 ± 0.2 ^{a}	72. 6 ± 0.4 ^{d}	71.6 ± 0.3 ^{c}

AD60 and AD70: Hot air-drying at 60 and 70 °C, respectively; LYO: Lyophilized; F: Fine granulometry; C: Coarse granulometry. ^{a,b,c} different letters on the same file indicate statistically significant differences at a 95% confidence level.

A significant effect ($p < 0.05$) of dehydration treatment and air-drying temperature was observed, while the granulometry did not affect the results significantly. Lyophilized powders had the highest content in all of the components analyzed. In the hot air-dried powders, the increase in air temperature had, in all cases, a negative effect. These results highlight, once again, the relevance of lyophilized process to preserve nutraceutical components due to the low temperature and the absence of oxygen in the drying chamber [41,58] quantified β-carotene in lyophilized by-products, showing lower values for guava by-product (26.67 μg/100 g dry matter), mango (58.26 μg/100 g dry matter), or passion fruit (53.93 μg/100 g dry matter). Regarding the effect of temperature in hot air-drying, the results were consistent with those reported by Albanese et al. [59] They demonstrated hot air-drying at 50 °C as a suitable method and alternative to freeze-drying to preserve carotenoid compounds and antioxidant activity in tomato peels. Heating of the extracts up to 100 °C caused a progressive reduction of total carotenoids up to about 30% after 250 min of treatment. In lulo bagasse powders, hot air-drying at 60 °C reduced the α-carotene content by less than 10% and the β-carotene content by less than 30% compared to the lyophilized powder. Increasing the drying temperature to 70 °C increased the losses of the two components to values greater than 50%. Although the hot air-drying treatment significantly reduced the content in the analyzed components, the beta-carotene content in the lyophilized powder was similar to the carrot juice provided by Bub et al. [60] in a human intervention study (330 mL of carrot juice providing 27.1 mg of β-carotene) that demonstrated the beneficial effects of this component in reducing the oxidation of low density lipoproteins.

4. Conclusions

Lulo bagasse is a suitable raw material for obtaining a powder rich in fiber and carotenoids useful as an ingredient in the food industry. The imbalance in the ratio between soluble and insoluble fiber should be taken into account in subsequent applications.

Drying kinetics showed that in the first stage of the air-drying process, water molecular transport from the innermost layers to the surface of the solid bed was not affected by air temperature. Structural characteristics and the composition of bagasse determine an internal control. In this sense, the air temperature could be reduced in this stage with energy savings as a consequence.

The slight differences observed in the monolayer water content provided by isotherms for the different processing conditions, showed the expected high stability of lulo bagasse powders. Although differences in the C parameter showed important structural and physical changes along drying, physico-chemical characteristics such as color were not highly affected.

Both granulometry and dehydration conditions determine the properties of the final powder. A finer granulometry leads, independent of the dehydration conditions, to better water interaction properties and, especially to greater solubility. In relation to dehydration methods, lyophilization is the method that provides better antioxidant properties and a higher carotenoid content, although carotenoid content is also acceptable in hot air-drying at 60 °C.

Author Contributions: Conceptualization, supervision, project administration, funding acquisition and review, N.B.; Data curation, S.D.S. and L.I.H.-C.; Writing—original draft preparation, L.I.H.-C.; Writing—review and editing, L.S. and C.B. All authors have read and agreed to the published version of the manuscript.

Funding: This research and APC were funded by Generalitat Valenciana, Project AICO/2017/'049.

Acknowledgments: The authors thank the research project "Fortalecimiento de los Encadenamientos Productivos de las Subregiones del Chocó" BPIN 2013000100284 Tecnológica del Chocó (in Spanish) by financial support to Leidy Indira Hinestroza-Córdoba.

Conflicts of Interest: The authors declare no conflict of interest.

References

1. Ministerio de Agricultura y Desarrollo Rural de Colombia Agronet Producción Nacional por Producto-Lulo. Available online: https://www.agronet.gov.co/Paginas/ProduccionNacionalProducto.aspx (accessed on 24 February 2020).
2. Forero, D.P.; Orrego, C.E.; Peterson, D.G.; Osorio, C. Chemical and sensory comparison of fresh and dried lulo (*Solanum quitoense* Lam.) fruit aroma. *Food Chem.* **2015**, *169*, 85–91. [CrossRef] [PubMed]
3. Gancel, A.L.; Alter, P.; Dhuique-Mayer, C.; Ruales, J.; Vaillant, F. Identifying carotenoids and phenolic compounds in naranjilla (*Solanum quitoense* Lam. var. Puyo hybrid), an Andean fruit. *J. Agric. Food Chem.* **2008**, *56*, 11890–11899. [CrossRef] [PubMed]
4. Forero, D.P.; Masatani, C.; Fujimoto, Y.; Coy-Barrera, E.; Peterson, D.G.; Osorio, C. Spermidine derivatives in lulo (*Solanum quitoense* Lam.) fruit: Sensory (taste) versus biofunctional (ACE-inhibition) properties. *J. Agric. Food Chem.* **2016**, *64*, 5375–5383. [CrossRef] [PubMed]
5. Departamento Nacional de Planeación de Colombia Colombianos Botan 9,76 Millones de Toneladas de Comida al año. Available online: https://www.dnp.gov.co/Paginas/Colombianos-botan-9,76-millones-de-toneladas-de-comida-al-a~no.aspx (accessed on 21 February 2020).
6. De Moraes Crizel, T.; Jablonski, A.; de Oliveira Rios, A.; Rech, R.; Flôres, S.H. Dietary fiber from orange byproducts as a potential fat replacer. *LWT Food Sci. Technol.* **2013**, *53*, 9–14. [CrossRef]
7. Karam, M.C.; Petit, J.; Zimmer, D.; Baudelaire Djantou, E.; Scher, J. Effects of drying and grinding in production of fruit and vegetable powders: A review. *J. Food Eng.* **2016**, *188*, 32–49. [CrossRef]

8. Majerska, J.; Michalska, A.; Figiel, A. A review of new directions in managing fruit and vegetable processing by-products. *Trends Food Sci. Technol.* **2019**, *88*, 207–219. [CrossRef]
9. AOAC. Official method 973.18, 2000. In *Official Methods of Analysis of the Association of Official Analytical Chemists*, 18th ed.; Association of Official Analytical Chemist: Arlington, VA, USA, 2000.
10. Mertens, D.; Allen, M.; Carmany, J.; Clegg, J.; Davidowicz, A.; Drouches, M.; Frank, K.; Gambin, D.; Garkie, M.; Gildemeister, B.; et al. Gravimetric Determination of Amylase-Treated Neutral Detergent Fiber in Feeds with Refluxing in Beakers or Crucibles: Collaborative Study. *J. AOAC Int.* **2002**, *85*, 1217–1240.
11. Mimouni, A.; Deeth, H.C.; Whittaker, A.K.; Gidley, M.J.; Bhandari, B.R. Rehydration process of milk protein concentrate powder monitored by static light scattering. *Food Hydrocoll.* **2009**, *23*, 1958–1965. [CrossRef]
12. Cai, Y.Z.; Corke, H. Production and properties of spray-dried Amaranthus betacyanin pigments. *J. Food Sci.* **2000**, *65*, 1248–1252. [CrossRef]
13. Freudig, B.; Hogekamp, S.; Schubert, H. Dispersion of powders in liquids in a stirred vessel. *Chem. Eng. Process. Process. Intensif.* **1999**, *38*, 525–532. [CrossRef]
14. Raghavendra, S.N.; Rastogi, N.K.; Raghavarao, K.S.M.S.; Tharanathan, R.N. Dietary fiber from coconut residue: Effects of different treatments and particle size on the hydration properties. *Eur. Food Res. Technol.* **2004**, *218*, 563–567. [CrossRef]
15. Robertson, J.A.; de Monredon, F.D.; Dysseler, P.; Guillon, F.; Amado, R.; Thibault, J.-F. Hydration Properties of Dietary Fibre and Resistant Starch: A European Collaborative Study. *LWT Food Sci. Technol.* **2000**, *33*, 72–79. [CrossRef]
16. Garau, M.C.; Simal, S.; Rosselló, C.; Femenia, A. Effect of air-drying temperature on physico-chemical properties of dietary fibre and antioxidant capacity of orange (*Citrus aurantium* v. Canoneta) by-products. *Food Chem.* **2007**, *104*, 1014–1024. [CrossRef]
17. Yasumatsu, K.; Sawada, K.; Moritaka, S.; Misaki, M.; Toda, J.; Wada, T.; Ishii, K. Whipping and Emulsifying Properties of Soybean Products. *Agric. Biol. Chem.* **1972**, *36*, 719–727. [CrossRef]
18. Singleton, V.L.; Orthofer, R.; Lamuela-Raventós, R.M. [14] Analysis of total phenols and other oxidation substrates and antioxidants by means of folin-ciocalteu reagent. *Methods Enzymol.* **1999**, *299*, 152–178. [CrossRef]
19. Wolfe, K.; Wu, X.; Liu, R.H. Antioxidant activity of apple peels. *J. Agric. Food Chem.* **2003**, *51*, 609–614. [CrossRef]
20. Luximon-Ramma, A.; Bahorun, T.; Crozier, A.; Zbarsky, V.; Datla, K.P.; Dexter, D.T.; Aruoma, O.I. Characterization of the antioxidant functions of flavonoids and proanthocyanidins in Mauritian black teas. *Food Res. Int.* **2005**, *38*, 357–367. [CrossRef]
21. Kuskoski, E.M.; Asuero, A.G.; Troncoso, A.M.; Mancini-Filho, J.; Fett, R. Aplicación de diversos métodos químicos para determinar actividad antioxidante en pulpa de frutos. *Food Sci. Technol.* **2005**, *25*, 726–732. [CrossRef]
22. Stratil, P.; Klejdus, B.; Kubáň, V. Determination of total content of phenolic compounds and their antioxidant activity in vegetables—Evaluation of spectrophotometric methods. *J. Agric. Food Chem.* **2006**, *54*, 607–616. [CrossRef]
23. Re, R.; Pellegrini, N.; Proteggente, A.; Pannala, A.; Yang, M.; Rice-Evans, C. Antioxidant activity applying an improved ABTS radical cation decolorization assay. *Free Radic. Biol. Med.* **1999**, *26*, 1231–1237. [CrossRef]
24. Rodrigues, E.; Mariutti, L.R.B.; Mercadante, A.Z. Carotenoids and phenolic compounds from Solanum sessiliflorum, an unexploited amazonian fruit, and their scavenging capacities against reactive oxygen and nitrogen species. *J. Agric. Food Chem.* **2013**, *61*, 3022–3029. [CrossRef] [PubMed]
25. Bunea, A.; Andjelkovic, M.; Socaciu, C.; Bobis, O.; Neacsu, M.; Verhé, R.; Camp, J. Van Total and individual carotenoids and phenolic acids content in fresh, refrigerated and processed spinach (*Spinacia oleracea* L.). *Food Chem.* **2008**, *108*, 649–656. [CrossRef] [PubMed]
26. Wolf, W.; Spiess, W.E.L.; Jung, G. Standardization of Isotherm Measurements (Cost Project 90 and 90 bis). In *Properties of Water in Foods*; En Simatos, D., Multon, J.L., Eds.; Springer: Dordrecht, The Netherlands, 1985; pp. 661–679.
27. Brunauer, S.; Deming, L.S.; Deming, W.E.; Teller, E. On a Theory of the van der Waals Adsorption of Gases. *J. Am. Chem. Soc.* **1940**, *62*, 1723–1732. [CrossRef]
28. Martínez-Las Heras, R.; Heredia, A.; Castelló, M.L.; Andrés, A. Moisture sorption isotherms and isosteric heat of sorption of dry persimmon leaves. *Food Biosci.* **2014**, *7*, 88–94. [CrossRef]

29. Brunauer, S.; Emmett, P.H.; Teller, E. Adsorption of Gases in Multimolecular Layers. *J. Am. Chem. Soc.* **1938**, *60*, 309–319. [CrossRef]
30. Vesterlund, S.; Salminen, K.; Salminen, S. Water activity in dry foods containing live probiotic bacteria should be carefully considered: A case study with Lactobacillus rhamnosus GG in flaxseed. *Int. J. Food Microbiol.* **2012**, *157*, 319–321. [CrossRef]
31. Viuda-Martos, M.; Ruiz-Navajas, Y.; Martin-Sánchez, A.; Sánchez-Zapata, E.; Fernández-López, J.; Sendra, E.; Sayas-Barberá, E.; Navarro, C.; Pérez-Álvarez, J.A. Chemical, physico-chemical and functional properties of pomegranate (*Punica granatum* L.) bagasses powder co-product. *J. Food Eng.* **2012**, *110*, 220–224. [CrossRef]
32. Llobera, A.; Cañellas, J. Dietary fibre content and antioxidant activity of Manto Negro red grape (*Vitis vinifera*): Pomace and stem. *Food Chem.* **2007**, *101*, 659–666. [CrossRef]
33. Sudha, M.L.; Baskaran, V.; Leelavathi, K. Apple pomace as a source of dietary fiber and polyphenols and its effect on the rheological characteristics and cake making. *Food Chem.* **2007**, *104*, 686–692. [CrossRef]
34. Happi Emaga, T.; Robert, C.; Ronkart, S.N.; Wathelet, B.; Paquot, M. Dietary fibre components and pectin chemical features of peels during ripening in banana and plantain varieties. *Bioresour. Technol.* **2008**, *99*, 4346–4354. [CrossRef]
35. Figuerola, F.; Hurtado, M.L.; Estévez, A.M.; Chiffelle, I.; Asenjo, F. Fibre concentrates from apple pomace and citrus peel as potential fibre sources for food enrichment. *Food Chem.* **2005**, *91*, 395–401. [CrossRef]
36. Amaya-Cruz, D.M.; Rodríguez-González, S.; Pérez-Ramírez, I.F.; Loarca-Piña, G.; Amaya-Llano, S.; Gallegos-Corona, M.A.; Reynoso-Camacho, R. Juice by—Products as a source of dietary fibre and antioxidants and their effect on hepatic steatosis. *J. Funct. Foods* **2015**, *17*, 93–102. [CrossRef]
37. Larrauri, J. New approaches in the preparation of high dietary fibre powders from fruit by-products. *Trends Food Sci. Technol.* **1999**, *10*, 3–8. [CrossRef]
38. Agostoni, C.; Bresson, J.-L.; Fairweather-Tait, S.; Flynn, A.; Golly, I.; Korhonen, H.; Lagiou, P.; Løvik, M.; Marchelli, R.; Martin, A.; et al. Scientific Opinion on Dietary Reference Values for carbohydrates and dietary fibre. *EFSA J.* **2016**, *8*, 1462. [CrossRef]
39. Pérez-Álvarez, J.A.; Viuda-Martos, M.; López-Marcos, M.C.; Fernández-López, J.; Sendra, E.; López-Vargas, J.H. Role of Fiber in Cardiovascular Diseases: A Reviewle. *Compr. Rev. Food Sci. Food Saf.* **2010**, *9*, 240–258.
40. Lucas-González, R.; Viuda-Martos, M.; Pérez-Álvarez, J.Á.; Fernández-López, J. Evaluation of Particle Size Influence on Proximate Composition, Physicochemical, Techno-Functional and Physio-Functional Properties of Flours Obtained from Persimmon (*Diospyros kaki* Trumb.) Coproducts. *Plant Foods Hum. Nutr.* **2017**, *72*, 67–73. [CrossRef]
41. Park, H.-J.; Lee, Y.; Eun, J.-B. Physicochemical characteristics of kimchi powder manufactured by hot air drying and freeze drying. *Biocatal. Agric. Biotechnol.* **2016**, *5*, 193–198. [CrossRef]
42. Santos de Sousa, A.; Vilela Borges, S.; Ferreira Magalhães, N.; Vaz Ricardo, H.; Damico Azevedo, A. Spray-Dried Tomato Powder: Reconstitution Properties and Colour. *Braz. Arch. Biol. Technol.* **2008**, *51*, 607–614. [CrossRef]
43. Bakar, J.; Ee, S.C.; Muhammad, K.; Hashim, D.M.; Adzahan, N. Spray-Drying Optimization for Red Pitaya Peel (*Hylocereus polyrhizus*). *Food Bioprocess Technol.* **2013**, *6*, 1332–1342. [CrossRef]
44. Bhusari, S.N.; Muzaffar, K.; Kumar, P. Effect of carrier agents on physical and microstructural properties of spray dried tamarind pulp powder. *Powder Technol.* **2014**, *266*, 354–364. [CrossRef]
45. Ahmed, A.M.; Ishida, Y.; Shimamoto, T. Molecular characterization of antimicrobial resistance in Salmonella isolated from animals in Japan. *J. Appl. Microbiol.* **2009**, *106*, 402–409. [CrossRef] [PubMed]
46. Lecumberri, E.; Mateos, R.; Izquierdo-Pulido, M.; Rupérez, P.; Goya, L.; Bravo, L. Dietary fibre composition, antioxidant capacity and physico-chemical properties of a fibre-rich product from cocoa (*Theobroma cacao* L.). *Food Chem.* **2007**, *104*, 948–954. [CrossRef]
47. Serna-Cock, L.; Torres-León, C.; Ayala-Aponte, A. Evaluación de polvos alimentarios obtenidos de cáscaras de mango (*Mangifera indica*) como fuente de ingredientes funcionales. *Inf. Tecnol.* **2015**, *26*, 41–50. [CrossRef]
48. Karnik, D.; Wicker, L. Emulsion stability of sugar beet pectin fractions obtained by isopropanol fractionation. *Food Hydrocoll.* **2018**, *74*, 249–254. [CrossRef]
49. Tao, Y.; Wu, Y.; Yang, J.; Jiang, N.; Wang, Q.; Chu, D.-T.; Han, Y.; Zhou, J. Thermodynamic sorption properties, water plasticizing effect and particle characteristics of blueberry powders produced from juices, fruits and pomaces. *Powder Technol.* **2018**, *323*, 208–218. [CrossRef]

50. Martínez-Las Heras, R.; Landines, E.F.; Heredia, A.; Castelló, M.L.; Andrés, A. Influence of drying process and particle size of persimmon fibre on its physicochemical, antioxidant, hydration and emulsifying properties. *J. Food Sci. Technol.* **2017**, *54*, 2902–2912. [CrossRef] [PubMed]
51. Mphahlele, R.R.; Fawole, O.A.; Makunga, N.P.; Opara, U.L. Effect of drying on the bioactive compounds, antioxidant, antibacterial and antityrosinase activities of pomegranate peel. *BMC Complement. Altern. Med.* **2016**, *16*, 143. [CrossRef]
52. Crozier, S.J.; Preston, A.G.; Hurst, J.W.; Payne, M.J.; Mann, J.; Hainly, L.; Miller, D.L. Cacao seeds are a "Super Fruit": A comparative analysis of various fruit powders and products. *Chem. Cent. J.* **2011**, *5*, 5. [CrossRef]
53. Michalska, A.; Wojdyło, A.; Lech, K.; Łysiak, G.P.; Figiel, A. Effect of different drying techniques on physical properties, total polyphenols and antioxidant capacity of blackcurrant pomace powders. *LWT Food Sci. Technol.* **2017**, *78*, 114–121. [CrossRef]
54. Dorta, E.; Lobo, M.G.; Gonzalez, M. Reutilization of mango byproducts: Study of the effect of extraction solvent and temperature on their antioxidant properties. *J. Food Sci.* **2012**, *77*, 80–88. [CrossRef]
55. Rana, S.; Gupta, S.; Rana, A.; Bhushan, S. Functional properties, phenolic constituents and antioxidant potential of industrial apple pomace for utilization as active food ingredient. *Food Sci. Hum. Wellness* **2015**, *4*, 180–187. [CrossRef]
56. Ozgen, M.; Reese, R.N.; Tulio, A.Z.; Scheerens, J.C.; Miller, A.R. Modified 2,2-azino-bis-3-ethylbenzothiazoline-6-sulfonic acid (ABTS) method to measure antioxidant capacity of selected small fruits and comparison to ferric reducing antioxidant power (FRAP) and 2,2′-diphenyl-1-picrylhydrazyl (DPPH) methods. *J. Agric. Food Chem.* **2006**, *54*, 1151–1157. [CrossRef] [PubMed]
57. Del Caro, A.; Piga, A.; Vacca, V.; Agabbio, M. Changes of flavonoids, vitamin C and antioxidant capacity in minimally processed citrus segments and juices during storage. *Food Chem.* **2004**, *84*, 99–105. [CrossRef]
58. Da Silva, L.M.R.; De Figueiredo, E.A.T.; Ricardo, N.M.P.S.; Vieira, I.G.P.; De Figueiredo, R.W.; Brasil, I.M.; Gomes, C.L. Quantification of bioactive compounds in pulps and by-products of tropical fruits from Brazil. *Food Chem.* **2014**, *143*, 398–404. [CrossRef] [PubMed]
59. Albanese, D.; Adiletta, G.; D'Acunto, M.; Cinquanta, L.; Di Matteo, M. Tomato peel drying and carotenoids stability of the extracts. *Int. J. Food Sci. Technol.* **2014**, *49*, 2458–2463. [CrossRef]
60. Bub, A.; Watzl, B.; Abrahamse, L.; Delincée, H.; Adam, S.; Wever, J.; Müller, H.; Rechkemmer, G. Moderate Intervention with Carotenoid-Rich Vegetable Products Reduces Lipid Peroxidation in Men. *J. Nutr.* **2000**, *130*, 2200–2206. [CrossRef]

© 2020 by the authors. Licensee MDPI, Basel, Switzerland. This article is an open access article distributed under the terms and conditions of the Creative Commons Attribution (CC BY) license (http://creativecommons.org/licenses/by/4.0/).

foods

MDPI

Article

Composition of Phenolic Acids and Antioxidant Properties of Selected Pulses Cooked with Different Heating Conditions

Yihan Liu [1,2], Sanaa Ragaee [1], Massimo F. Marcone [1] and El-Sayed M. Abdel-Aal [2,*]

[1] Department of Food Science, University of Guelph, Guelph, ON N1G 2W1, Canada; yliu55@uoguelph.ca (Y.L.); sragaee@uoguelph.ca (S.R.); mmarcone@uoguelph.ca (M.F.M.)
[2] Guelph Research and Development Centre, Agriculture and Agri-Food Canada, Guelph, ON N1G 5C9, Canada
* Correspondence: elsayed.abdelaal@canada.ca; Tel.: +1-226-217-8079; Fax: +1-226-217-8181

Received: 19 June 2020; Accepted: 7 July 2020; Published: 10 July 2020

Abstract: Pulses are recommended for healthy eating due to their high content of nutrients and bioactive compounds that can undergo changes during cooking. This study investigated the effects of four cooking methods (boiling, pressure, microwave, slow) and three heating solutions (water, salt, sugar) on the phenolic acids and antioxidant properties of three pulses (faba beans, lentils, peas). The composition of phenolic acids differed among the three pulses with *p*-coumaric and ferulic being the dominant acids. Cooking increased free phenolic acids and lessened bound phenolic acids in faba beans and peas, while decreased both free and bound phenolic acids in lentils. Cooking resulted in reductions in total phenol content (TPC) in faba bean methanol and bound extracts. Pressure and microwave cooking increased TPC in lentil methanol extracts, while pot boiling and slow cooking reduced TPC. Microwave cooking resulted in increases in TPC in bound phenolic extracts from lentils. For peas, cooking increased TPC in both methanol and bound phenolic extracts. Significant changes were also observed in the antioxidant capacity of cooked pulses based on the scavenging ability of DPPH, ABTS and peroxyl radicals subject to the type of pulse, polyphenol and antioxidant assay. Despite the significant reduction in antioxidants, high amounts of phenolics with potent antioxidant activities are still found in cooked pulses.

Keywords: traditional; slow; pressure and microwave cooking; polyphenols; antioxidant activity; faba bean; lentil; pea

1. Introduction

Consumption of pulses is recommended by health organizations due to their high-nutrient density and health benefits [1]. Pulses are good sources of protein, slowly digestible and resistant starch, dietary fiber, vitamins, minerals and bioactive compounds [2,3]. They are important dietary components worldwide as many countries rely on pulses as a source of inexpensive and plant protein. Seven pulses are important in human nutrition of which five (pea, bean, lentil, chickpea and faba beans) are significant in the global trade, while the other two (cowpea and pigeon pea) are usually consumed locally except for a few cases such as black-eyed pea that is traded in some parts of the Americas [4]. Pulses are commonly processed and/or cooked in a boiling water pot or pressure cooker prior to consumption which alters their nutrients and bioactive compounds and ultimately health benefits. Thus, it would be of interest to understand how various cooking methods can affect nutritional properties of pulses especially little or no information is currently available on the effect of microwave or slow cooking on the composition of phenolic compounds and antioxidant properties of commonly consumed pulses.

Pulses are good sources of various phenolic compounds including phenolic acids, flavonoids, isoflavones and tannins with ferulic being the most abundant phenolic acid in pulses, while flavonol glycosides, anthocyanins and tannins, which primarily exist in the seed coat of pulses, are present in high or low concentrations subject to pulse type and genotype [2,5]. The majority of phenolic compounds in lentils are flavonoids, including kaempeferol glycosides, catechin/epicatechin glucosides and procyanidins based on the analysis of 20 cultivars [6] and the composition of phenolic compounds vary with the seed coat color of lentils (e.g., green, gray, tan and brown) [7]. The total polyphenol content in Canadian pulses, including 14 peas, lentils, beans and chickpeas ranges from 1.2 to 7.5 mg/g and significantly associates with antioxidant capacity [8]. It has also been reported that phenolic compounds in pulses exhibit antioxidant activities against diverse free radical species [6,9–13]. As an antioxidanat agent, phenolic compounds are able to scavenge free radicals in the human body and combat oxidative damage; therefore, they could potentailly prevent various human diseases, such as cadiovascular diease, diabets and cancer [8,14–16].

Cooking can induce considerable changes in nutritional and structural properties of small and complex molecules in pulses, including phenolic compounds. The pot boiling and pressure cooking are the most common methods in processing and cooking of pulses with both methods have been found to trigger changes in the composition of phenolic compounds and antioxidant properties of pulses [17–19]. The boiling of lentils significantly reduces insoluble bound phenolic compounds and antioxidant capacity [17]. In the latter study, the sum of soluble and insoluble phenolic fractions has been reported to decrease due to the formation of irreversible covalent bond between liberated phenolic compounds and proteins. Industrial canning also causes changes in dry bean nutrients, bioactive compounds and antioxidant capacity [18]. For instance, canning increases protein and dietary fiber but diminishes minerals, phytates, trypsin inhibitor, lectin and antioxidant capacity. Similarly, the content of nutrients and bioactive compounds in faba bean, pea, lentil, chickpea, pinto bean, black-eyed bean, white bean and lupine which are regularly consumed in the Mediterranean countries has been found to alter after boiling but at different degrees subject to pulse type and the nature of component [19]. In general, the degree of changes in phenolic compounds and antioxidant capacities of pulses during processing or cooking depends on a number of factors, including cooking method or energy type, cooking time, heating solution and pre-treatments, such as soaking. Since little information has been reported on the effect of various cookers particularly microwave and slow cookers using different heating solutions on phenolic compounds and antioxidant capacity, the current study investigated effects of four cooking methods (traditional or boiling, pressure, microwave and slow) and three heating solutions (water, salt, sugar) on the composition of free and bound phenolic acids, the most abundant polyphenols in pulses, and antioxidant properties of three pulses (faba beans, peas and lentils). We propose that the addition of salt or sugar in water would speed up the cooking of pulses and reduce cooking time, resulting in improved cooking quality and nutritional properties.

2. Materials and Methods

2.1. Reagents

2,2-diphenyl-1-picrylhydrazyl (DPPH), 2,2-azino-bis (3-ethylbenzthiazoline-6sulfonic acid) (ABTS), fluorescein, 2,2′-Azobis (2-methylpropion-amidine) dihydrochloride (AAPH), Trolox, gallic, protocatechuic, *p*-hydroxybenzoic, gentisic, 3-hydroxybenzoic, catechuic, vanillic, caffeic, chlorogenic, syringic, *p*-coumaric, *o*-coumaric, trans-ferulic, *t*-iso-ferulic, sinapinic, and cinnamic and ferulic acids were purchased from Sigma (Sigma-Aldrich Canada Ltd., Oakville, ON, Canada). All the chemicals and reagents used in the study are analytical grade.

2.2. Pulses

Composite samples of faba beans (*Vicia faba*), lentils (*Lens culinaris*) and peas (*Pisum stativum*) (5 kg each) were obtained from harvested pulse crop breeding trials grown at Rosthern, Saskatchewan (peas and lentils) or Outlook, Saskatchewan (faba bean) in 2014 from the Crop Development Centre, University of Saskatchewan, Saskatoon, SK, Canada. The faba bean cultivar CDC Snowdrop (low tannin, small seed), lentil cultivar CDC Dazil (small red), and pea cultivar CDC Greenwater (green) were selected for this study based on their better cooking quality compared to other cultivars as measured in a previous study [20].

2.3. Cooking Methods

The four cooking methods (traditional or boiling, pressure, microwave and slow) were previously optimized in terms of cooking time and the concentration of salt and sugar heating solutions for each pulse in previous studies [20,21]. The cooking time of the traditional method was 22, 1 and 15 min for faba beans, lentils and peas [20]. Traditional cooking was carried out on a Salton Induction Cooktop. The concentrations of salt and sugar were 0.5 and 1.0%, respectively, and were chosen on the basis of the firmness of cooked pulses and cooking loss [21]. Since soaking is a common practice in pulse processing, the three pulses were pre-soaked for 24 h for the traditional, microwave and slow cooking methods. For the pressure cooking, the soaking time was 14 h for faba beans and peas, and no soaking was done for lentils since the pre-soaked lentils were mashed in the pressure cooker. The pressure cooking was performed using a Matfer 013203 Stainless Steel Pressure Cooker with Steamer Basket for 5, 3 and 9 min for pre-soaked faba beans and peas and non-soaked lentils, respectively. Microwave cooking was performed using a Panasonic NE-21521 Stainless Steel Commercial Microwave Oven at medium-high power level (power 7) for 7, 2 and 7 min for faba beans, lentils and peas, respectively. The slow cooking of faba beans, lentils and peas was carried out using a Hamilton Beach Programmable Slow Cooker at 80, 35 and 80 °C for 9, 1 and 9 h, respectively. The slow cooking of lentils at a high temperature (80 °C) produced mashed products because the lentil seeds are small and not as hard as faba beans or peas. Thus, the slow cooking of lentils was carried out at a lower temperature (35 °C) compared with faba beans and peas, but it was necessary to parboil or partially cook lentils for 1 min prior to slow cooking to improve cooking quality, i.e., having cooked lentils with the right firmness without being mashed. The cooked pulses were dried in an air oven at 50 °C overnight and the dried cooked pulses were milled to pass through a 500-µm mesh screen and stored dry in a desiccator until analysis.

2.4. Analysis of Phenolic Acids

Free and bound phenolic acids were extracted and analyzed according to the procedure described by Abdel-Aal and Rabalski [22]. Free phenolic acids were extracted twice from a 0.5-g sample in 80% methanol. The left-over pellet was immediately processed for the extraction of bound phenolic acids after defatting with hexane followed by alkaline treatment with 2M sodium hydroxide to liberate bound phenolic acids, then acidified to pH 2 with 2M hydrochloric acid solution to convert them to the acid form. The liberated phenolic acids were extracted three times with 10 mL of ethyl acetate and ethyl ether 1:1 ratio (*v*/*v*). The extracts were evaporated, and the residue was re-dissolved in 5 mL of nano-pure water, filtered through a 0.45-µm Acrodisc syringe filter, stored in a freezer prior to HPLC analysis. Following the completion of bound phenolic acid extracts, free phenolic acid extracts were concentrated and filtered prior to HPLC analysis. High-Performance Liquid Chromatograpy (HPLC) (Agilent Series 1100, Waldbronn, Germany) was used for the separation and quantification of phenolic acids in free (methanol) and bound extracts. A mixture of 12 authentic phenolic acid standards including gallic, protocatechuic, *p*-hydroxy-benzoic, gentisic, 3-hydroxy-benzoic, vanillic, caffeic, syringic, *p*-coumaric, ferulic, sinapinic and *o*-coumaric acids was used for calibration, identification and quantification. The detection of phenolic acids was performed at five wavelengths (260, 275, 300, 320, 330 nm) in which each phenolic acid was quantified at its maximum absorption wavelength. For example, *p*-hydroxybenzoic, protocatechuic and vanillic acids were quantified at 260 nm; syringic acid at 275 nm; and caffeic, *p*-coumaric, and ferulic acids at 320 nm to enhance the accuracy of the phenolic acid quantification.

2.5. Analysis of Total Phenol Content

Quantification of the total phenol content (TPC) in free and bound phenolic extracts was done by the Folin–Ciocalteu method using the procedure modified by Abdel-Aal and Rabalski [23]. The reaction mixture contained 50 μL of free or bound phenolic extract, 50 μL diluted Folin–Ciocalteu reagent and 100 μL of saturated sodium carbonate solution and the mixture was made up to 1.0 mL with distilled water. After a 30-min reaction in darkness, the absorbance at 725 nm was measured against a blank. A series of standard ferulic acid solutions were prepared with concentrations between 0–350 μg/mL and the absorbance was read at 725 nm against a reagent blank. The concentrations demonstrated a linear relationship with a determination coefficient (R^2) of 0.996 and regression equation $x = y - 0.0019/0.003$, where x is the concentration of ferulic acid (μg/mL) and y is the absorbance at 725 nm. The TPC is expressed as mg ferulic acid equivalent/g sample.

2.6. Antioxidant Assays

2.6.1. DPPH Scavenging Capacity

The scavenging capacity of free radicals was carried out using the stable 2,2-diphenyl-1-picrylhydrazyl radical (DPPH) [24]. The antioxidant reaction was initiated by transferring 25 or 50 μL of pulse extract into the 96 micro-plate well and adjusted to 50 μL with 50% ethanol, then 300 μL of freshly prepared DPPH solution (0.1 μmol/mL) was added. The reaction mixture was monitored by reading absorbance at 517 nm for 30 min at 1-min intervals. A blank reagent was used to study stability of DPPH over the test time. The absorbance measured at 30 min was used for the calculation of μmol DPPH scavenged by extracts. Trolox (6-hydroxy 2,5,7,8-tet-ramethylchroman-2-carboxylic acid) was used as an antioxidant reference. The scavenging capacity of pulses extract was calculated as the concentration of Trolox equivalent and expressed in μmol Trolox equivalents/g dry weight.

2.6.2. ABTS Scavenging Capacity

The scavenging capacity against radical cation (2,2′-azino-di-[3-ethyl benzthiazolinesulphonate] (ABTS) was measured using a Randox Laboratories assay kit (San Francisco, CA, USA) as outlined by Abdel-Aal and Rabalski [24]. A total of 30 μL of diluted extract or Trolox standard solution (0–140 μg/mL) was taken for reaction with 20 μL myoglobin working solution and 300 μL of ABTS solution. The reaction mixture was read at 405 nm every min for 10 min. The absorbance readings at 5 min were used for calculation. The Trolox provided in the kit was used as an antioxidant standard and used for the calculation of scavenging capacity of pulse extracts as a Trolox equivalent. The scavenging capacity was calculated as μmol Trolox equivalent/g sample dry weight.

2.6.3. Oxygen Radical Absorbance Capacity (ORAC)

The ORAC method was previously described by Abdel-Aal and Rabalski [24]. A total of 25 μL of sample extract, Trolox standard solution (0–140 μg/mL), or nano pure water (blank) were mixed with 150 μL fluorescein in each of the 96 micro-plate well. The mixture was conditioned at 37 °C for 15 min, then 25 μL of 2,2′-Azobis (2-methylpropion-amidine) dihydrochloride (AAPH) as a peroxyl radical generator was added to start the decaying of fluorescein. The degradation of fluorescein progressed for 60 min in the heated chamber of BioTech Synergy H4 with the following settings: fluorescence excitation at 485 nm, emission wavelength 528 nm, and reading was taken every min for 60 min. The microplate fluorescent reader was operated by Gen 5 software version 1.11.5 (BioTek). The data are presented as means of ORAC values in μmol Trolox equivalent/g sample dry weight.

2.7. Statistical Analysis

Analysis of variance (ANOVA) was performed to assess the effect of cooking methods on phenolic acids and antioxidant properties using IBM SPSS Statistics (Version 24.0, IBM Corp., Armonk, NY, USA). The least significant difference test was employed to determine significant differences between cooking methods ($p < 0.05$). The data are expressed as the means ± standard deviation of two (phenolic acids) or three (antioxidant assays) replicates. For phenolic acids analysis, each replicate is the average of two HPLC determinations, giving a total of 4 measurements.

3. Results and Discussion

3.1. Effects of Cooking on Phenolic Acids

Phenolic acids are the principal polyphenols found in grains and pulses, which primarily exist as bound derivatives. The main phenolic acids found in faba beans were the bound fraction, including *p*-coumaric and ferulic acid in addition to small concentrations of free phenolic acids (*p*-hydroxybenzoic, *p*-coumaric, ferulic and sinapinic acid) (Table 1). Earlier, Sosulski and Dabrowski [25] reported the presence of bound *p*-coumaric (16 µg/g), ferulic acid (15 µg/g), and *p*-hydroxybenzoic (trace amount) in raw faba bean flour in addition to a trace amount of syringic acid, which was not detected in the current study. The effect of cooking methods on composition of free and bound phenolic acids in faba beans showed significant differences among the four cooking methods subject to cooker type, heating solution and nature of phenolic compound (Table 1). Bound phenolic acids reduced after cooking at different extents depending on the type of cooking and heating solution. On the other hand, slight increases were observed in the free phenolic acids, particularly sinapic and ferulic acids. However, *p*-coumaric showed slight decreases. The pressure cooking or canning of Spanish common beans has been reported to reduce protocatechuic, *trans-p*-coumaric, *trans*-ferulic and sinapic acid [18]. In addition, boiling and autoclaving have shown to lower bioactive compounds in faba beans [26]. It seems that heating treatment in a pot, pressure, microwave or slow cooker liberates portion of bound phenolic compounds through thermal degradation, and the degree of dissociation may depend on the type of energy and heating solution as significant differences were observed among the cooking methods and heating solutions (Table 1). The slow cooker was more effective in liberating phenolic acids as higher reductions were found. Phenolic acids, particularly ferulic acid, exhibit beneficial health effects against oxidative stress, hypertension, type 2 diabetes and cardiovascular disease [27,28], and thus they are considered bioactive components. The current study shows changes in phenolic acids in faba beans with increased amounts of soluble phenolic acids. In a previous study [21], slow cooking has been found to hold a promise for improving nutritionally important starch fractions and the removal of flatus oligo-sugars, and its ability to release bound phenolic acids could strengthen its improving effects to support its implementation.

The phenolic acid composition of lentils was different from that of faba beans, but they were similar in that the majority of phenolic acids were present in the bound fraction (Table 2). Once again, *p*-coumaric and ferulic were the dominant phenolic acids in the bound fraction. In addition to those acids, protocatechuic and vanillic acids were also found at reasonable amounts in the bound fraction. Soluble or unbound phenolic acids in lentils were *p*-coumaric, *p*-hydroxybenzoic and ferulic acids. According to Alshikh and others [29], the phenolic compounds identified in five lentil cultivars, including gallic acid, protocatechuic acid, caffeic acid, *p*-coumaric acid, ferulic acid, sinapic acid, catechin and epicatechin. The hydroxybenzoic acids are dominant in the esterified and insoluble-bound forms compared with the free fraction. They also reported that the concentration of phenolic acids and other phenolic compounds are different among the lentil cultivars. Cooking methods and heating solutions significantly influenced the content of phenolic acids in lentils (Table 2). In lentil, the concentration of phenolic acids in both fractions decreased. This is in agreement with the results of Yeo and Shahidi [17] who suggested that the boiling of lentils reduces the sum of soluble and insoluble phenolic fractions due to the formation of irreversible covalent bond between liberated phenolic compounds and proteins. Despite the decrease in phenolic acids in lentils, there are still reasonable amounts of phenolic acids to boost the daily intake of these functional components.

Table 1. Effect of four cooking methods on the composition of phenolic acids in faba beans (μg/g, db).

Cooking Method	Solution	Free phenolic Acids				Bound Phenolic Acids	
		p-Hydroxybenzoic	*p*-Coumaric	Ferulic	Sinapic	*p*-Coumaric	Ferulic
Boiling	Water	1.24 ± 0.03	0.56 ± 0.02	2.98 ± 0.12	3.75 ± 0.13	13.81 ± 0.06	11.21 ± 0.12
	Salt	1.06 ± 0.04	0.35 ± 0.04	2.96 ± 0.15	3.00 ± 0.09	11.59 ± 0.09	9.79 ± 0.11
	Sugar	1.40 ± 0.02	0.30 ± 0.04	2.70 ± 0.10	3.55 ± 0.27	9.93 ± 0.02	8.39 ± 0.04
Pressure	Water	1.53 ± 0.04	0.48 ± 0.05	2.50 ± 0.14	3.03 ± 0.09	9.80 ± 0.05	8.43 ± 0.12
	Salt	1.16 ± 0.07	0.44 ± 0.02	3.47 ± 0.19	3.44 ± 0.07	9.96 ± 0.05	7.11 ± 0.03
	Sugar	1.12 ± 0.05	0.53 ± 0.03	2.97 ± 0.20	3.15 ± 0.18	10.41 ± 0.06	6.74 ± 0.03
Microwave	Water	1.34 ± 0.03	0.57 ± 0.11	3.51 ± 0.10	3.35 ± 0.08	11.52 ± 0.08	9.27 ± 0.04
	Salt	1.38 ± 0.10	0.42 ± 0.05	2.50 ± 0.17	3.47 ± 0.10	10.89 ± 0.01	9.25 ± 0.02
	Sugar	1.10 ± 0.06	0.40 ± 0.05	2.60 ± 0.05	3.66 ± 0.33	11.23 ± 0.05	8.79 ± 0.06
Slow	Water	1.27 ± 0.11	0.35 ± 0.02	2.29 ± 0.06	3.62 ± 0.13	9.91 ± 0.07	8.88 ± 0.04
	Salt	1.03 ± 0.03	0.31 ± 0.04	2.64 ± 0.13	2.93 ± 0.17	9.76 ± 0.05	8.39 ± 0.07
	Sugar	1.06 ± 0.05	0.49 ± 0.04	2.93 ± 0.42	3.46 ± 0.12	9.39 ± 0.06	8.61 ± 0.03
Raw		1.03 ± 0.04	0.63 ± 0.04	1.16 ± 0.07	1.67 ± 0.07	13.93 ± 0.04	11.33 ± 0.04
LSD at $p < 0.05$		0.25	0.16	0.55	0.39	1.75	1.66

Mean ± SD, $n = 4$.

As expected, the composition of phenolic acids in peas was also different from lentils and faba beans (Table 3). This indicates that the composition of phenolic acids differs among the three pulses. For the peas, the methanol extract or free fraction contained *p*-hydroxybenzoic, *p*-coumaric, caffeic and ferulic acids, while the bound fractions contained *p*-hydroxybenzoic, *p*-coumaric, ferulic and sinapic acids. The presence of vanillic, *p*-coumaric, ferulic and sinapic acids has been reported in raw pea extracts in addition to caffeic, quercetin, kaempherol, and procyanidins B2 and B3 [13]. Cooking with different cookers and heating solutions was found to change the concentration of phenolic acids in both fractions. Slight increases were observed in free *p*-hydroxybenzoic and ferulic acids, while bound ferulic, *p*-coumaric and sinapic acids decreased in cooked peas. The addition of salt or sugar in the cooking solution of the three pulses had different effects on phenolic acids as compared with the cooking in pure water. When salt or sugar is added to water, it increases its boiling point and the temperature of the solution will get hotter faster than pure water, which may induce a higher degree of thermal dissociation of bound molecules in pulses. This could speed up the cooking of pulses, resulting in improved cooking quality (improved firmness) and nutritional properties [21].

Table 2. Effect of four cooking methods on the composition of phenolic acids in lentils (μg/g, db).

Cooking Method	Solution	Free Phenolic Acids			Bound Phenolic Acids			
		p-Hydroxybenzoic	*p*-Coumaric	Ferulic	Protocatechuic	*p*-Coumaric	Ferulic	Vanillic
Boiling	Water	0.98 ± 0.09	1.65 ± 0.15	0.87 ± 0.04	5.88 ± 0.14	9.49 ± 0.15	9.10 ± 0.15	4.11 ± 0.03
	Salt	1.02 ± 0.10	1.47 ± 0.02	0.87 ± 0.05	5.92 ± 0.13	9.19 ± 0.11	9.23 ± 0.11	4.44 ± 0.04
	Sugar	1.16 ± 0.11	1.19 ± 0.03	0.64 ± 0.03	5.59 ± 0.16	9.21 ± 0.17	9.21 ± 0.11	4.01 ± 0.00
Pressure	Water	1.38 ± 0.12	1.68 ± 0.13	0.73 ± 0.05	5.67 ± 0.16	9.63 ± 0.10	8.81 ± 0.03	4.11 ± 0.03
	Salt	1.33 ± 0.08	1.32 ± 0.06	0.69 ± 0.01	5.61 ± 0.45	9.71 ± 0.05	8.93 ± 0.01	4.10 ± 0.04
	Sugar	1.38 ± 0.13	1.28 ± 0.15	0.68 ± 0.10	5.12 ± 0.29	9.59 ± 0.03	8.73 ± 0.01	4.13 ± 0.07
Microwave	Water	1.23 ± 0.15	1.02 ± 0.02	1.44 ± 0.08	5.55 ± 0.43	9.91 ± 0.05	8.69 ± 0.07	4.51 ± 0.03
	Salt	1.23 ± 0.11	1.42 ± 0.11	1.38 ± 0.02	5.51 ± 0.18	10.13 ± 0.12	8.81 ± 0.00	4.54 ± 0.04
	Sugar	1.21 ± 0.09	1.04 ± 0.05	1.25 ± 0.04	5.50 ± 0.84	9.75 ± 0.11	9.03 ± 0.07	4.33 ± 0.00
Slow	Water	1.33 ± 0.12	1.21 ± 0.09	1.01 ± 0.02	5.57 ± 0.26	10.22 ± 0.12	8.61 ± 0.11	4.42 ± 0.03
	Salt	1.23 ± 0.09	1.40 ± 0.09	1.03 ± 0.07	5.54 ± 0.18	9.81 ± 0.12	8.49 ± 0.07	4.41 ± 0.04
	Sugar	1.25 ± 0.13	1.15 ± 0.07	1.13 ± 0.09	5.32 ± 0.20	9.43 ± 0.06	8.47 ± 0.13	4.01 ± 0.05
Raw		1.55 ± 0.09	1.73 ± 0.07	0.85 ± 0.03	5.95 ± 0.35	13.61 ± 0.13	12.51 ± 0.14	6.16 ± 0.12
LSD at $p < 0.05$		0.23	0.35	0.39	0.36	1.69	1.53	0.81

Mean ± SD, $n = 4$.

Table 3. Effect of four cooking methods on the composition of phenolic acids in peas (μg/g, db).

Cooking Method	Solution	Free Phenolic Acids				Bound Phenolic Acids			
		p-Hydroxybenzoic	*p*-Coumaric	Caffeic	Ferulic	*p*-Hydroxybenzoic	*p*-Coumaric	Ferulic	Sinapic
Boiling	Water	1.14 ± 0.05	1.37 ± 0.06	1.01 ± 0.05	1.95 ± 0.09	1.88 ± 0.12	3.25 ± 0.02	11.16 ± 0.13	3.44 ± 0.14
	Salt	1.24 ± 0.06	1.36 ± 0.04	1.05 ± 0.03	1.96 ± 0.07	1.97 ± 0.11	3.33 ± 0.05	11.07 ± 0.14	3.34 ± 0.16
	Sugar	1.14 ± 0.07	1.08 ± 0.18	1.04 ± 0.04	1.88 ± 0.04	1.84 ± 0.09	3.17 ± 0.02	10.87 ± 0.26	3.26 ± 0.12
Pressure	Water	1.18 ± 0.07	1.40 ± 0.04	1.13 ± 0.04	1.86 ± 0.05	1.89 ± 0.11	3.29 ± 0.04	11.86 ± 0.40	3.25 ± 0.15
	Salt	1.35 ± 0.06	1.37 ± 0.11	1.11 ± 0.02	1.89 ± 0.10	1.92 ± 0.07	3.35 ± 0.05	11.79 ± 0.12	3.31 ± 0.15
	Sugar	1.24 ± 0.09	1.11 ± 0.07	1.09 ± 0.04	1.77 ± 0.06	1.79 ± 0.01	3.21 ± 0.03	11.77 ± 0.33	3.27 ± 0.17
Microwave	Water	1.11 ± 0.09	1.30 ± 0.05	1.24 ± 0.03	1.89 ± 0.10	1.95 ± 0.14	3.26 ± 0.03	11.89 ± 0.36	3.42 ± 0.09
	Salt	1.15 ± 0.09	1.30 ± 0.02	1.16 ± 0.07	1.91 ± 0.11	1.99 ± 0.13	3.33 ± 0.03	11.84 ± 0.48	3.37 ± 0.23
	Sugar	1.03 ± 0.07	1.13 ± 0.03	1.03 ± 0.01	1.79 ± 0.08	1.80 ± 0.18	3.19 ± 0.06	11.66 ± 0.11	3.28 ± 0.14
Slow	Water	1.16 ± 0.10	1.33 ± 0.04	1.04 ± 0.03	1.92 ± 0.05	1.92 ± 0.11	3.32 ± 0.05	11.69 ± 0.85	3.37 ± 0.15
	Salt	1.21 ± 0.08	1.31 ± 0.06	1.07 ± 0.05	1.94 ± 0.07	1.94 ± 0.10	3.33 ± 0.01	11.74 ± 0.02	3.29 ± 0.18
	Sugar	1.06 ± 0.07	1.04 ± 0.07	1.01 ± 0.08	1.84 ± 0.07	1.83 ± 0.09	3.21 ± 0.04	11.77 ± 0.40	3.25 ± 0.11
Raw		1.02 ± 0.05	1.51 ± 0.06	1.11 ± 0.02	1.36 ± 0.02	1.98 ± 0.11	4.29 ± 0.11	15.39 ± 0.29	5.63 ± 0.16
LSD at $p < 0.05$		0.15	0.23	0.13	0.25	0.12	0.44	1.71	0.93

Mean ± SD, $n = 4$.

3.2. Effect of Cooking on Total Phenolic Content

Pulses are a good source of phenolic compounds in addition to their high content of protein, resistant starch and dietary fiber which make them important dietary components. Total phenolic content (TPC) was determined in methanol and alkaline-liberated phenol extracts of raw and cooked pulses and the results are expressed as mg ferulic acid equivalent, the most common phenolic acid in pulses (Table 4). Unlike phenolic acids, TPC in methanol extracts was much higher than that of alkaline-liberated phenol extracts. This observation is expected since the method is not specific for phenolic compounds, but the method also measures other reducing substances in the extract such as flavonoids, proteins, sugars, etc. Despite the fact that the method is not specific for phenols, it is commonly used to measure TPC in grains and pulses and it is also considered an indication of antioxidant activity of plant extracts [30]. The TPC in methanol extract was the highest in faba beans compared with lentils and peas, while the total content of alkaline-liberated phenols was the highest in lentils followed by faba beans and lastly peas (Table 4). It appears that faba beans and lentils are better dietary sources of phenolic compounds compared with peas for those selected pulses and cultivars. Lentils have the highest total phenol content among a diverse array of pluses, including pinto beans, cowpeas, baby lima beans, lentils, chickpeas, small red kidney beans, black kidney beans, navy beans and mung beans [31]. A high value of TPC (>92.9 mg gallic acid equivalent per g) has been reported for 13 faba bean genotypes [32]. Nithiyanantham and others [33] reported similar TPC values for chickpeas and peas, but lower TPC values have been reported for cooked pulses that are usually consumed in the Mediterranean region [19]. It is not easy to compare results from different studies due to the absence of standardized method. In addition, the current study expressed TPC as a ferulic acid equivalent not gallic acid equivalent since ferulic acid is the most common polyphenol in most pulses [5].

The cooking of faba beans in various cookers, including a pot, pressure, microwave and slow cooker significantly affected its TPC content of methanol and alkaline-liberated phenol extracts (Table 4). The four cookers and different heating solutions resulted in reductions in the soluble reducing substances and alkaline-liberated phenolic acids. The reduction percentage ranged from 35–55% and 18–50% for soluble reducing substances and alkaline-liberated phenolic acids, respectively. The highest reduction of soluble reducing substances was in samples cooked in the slow cooker, while microwave cooking produced the highest reduction for alkaline-liberated phenolic acids. This is in agreement with the results reported by Siah et al. [26], who investigated the effect of boiling and pressure cooking on phenolic compounds in five pre-soaked faba bean genotypes. They reported that TPC dropped from 2.8–11.2 mg/g in unprocessed samples to 0.7–2.4 mg/g in boiled faba bean samples and 0.7–1.9 mg/g in pressure cooked faba bean samples. Soaking and heating caused leaching and/or the thermal degradation of phenolic compounds, which could explain the reduction in TPC.

The effect of cooking methods on phenolic compounds in lentils was different from that in faba beans, as the pressure and microwave cooking methods resulted in an increased amount of the total reducing substances measured in the methanol extracts from the cooked samples, while pot boiling and slow cooking reduced TPC in methanol extracts (Table 4). Microwave cooking caused higher increase percentages (59–69%) than that produced by the pressure cooker (3–20). These increases in the total reducing substance may be due to the thermal dissociation and/or degradation of macro-molecules, such as protein and starch, into smaller molecules having a higher reducing power. Microwave cooking also resulted in an increase in the alkaline-liberated phenolic acids at a percentage of 2–19%. On the other hand, pot boiling, pressure and slow cooking reduced the content of alkaline-liberated phenolic acids at reduction percentages of 35–45%, 12–49% and 5–37%, respectively. The addition of salt or sugar also affected the content of total phenols. Xu and Chang [34] reported a 56% reduction in TPC in lentil compared with unprocessed samples. The cooking of pulses breaks cell walls and releases phenolic compounds into the cooking solution.

For peas, cooking methods resulted in increases in both soluble reducing substances and alkaline-liberated phenolic acids (Table 4). This effect is different from that in faba beans but similar to the effect of microwave and pressure cooking in lentils. The increase ranges from 9–48% and 4–146% for soluble reducing substances and alkaline-liberated phenolic acids, respectively. This indicates significant variations among cookers and heating solutions. It has been reported that pressure cooking increases TPC in peas by about 30% [33]. This supports the current results. The increase in phenolic and reducing compounds may be attributed to the increase in reducing power of released compounds due to thermal effects.

Table 4. Effect of four cooking methods on the total phenol content of methanol and alkaline-liberated phenolic extracts from pulses (mg ferulic acid equivalents/g, db).

Cooking Method	Solution	Methanol Extract (ME)	Alkaline-Liberated Phenolic Extract (ALPE)	% Increase (+) or Decrease (−)	
				ME	ALPE
			Faba Bean		
Boiling	Water	23.88 ± 2.10	2.84 ± 0.18	−37.7	−39.2
	Salt	24.73 ± 0.53	3.06 ± 0.09	−35.5	−34.5
	Sugar	22.48 ± 0.37	2.90 ± 0.15	−41.4	−37.9
Pressure	Water	17.64 ± 0.99	3.81 ± 0.16	−54.0	−18.4
	Salt	19.45 ± 1.18	3.40 ± 0.14	−49.3	−27.2
	Sugar	19.49 ± 0.28	2.31 ± 0.14	−49.2	−50.5
Microwave	Water	22.45 ± 0.49	2.61 ± 0.05	−41.4	−44.1
	Salt	23.26 ± 0.27	2.55 ± 0.11	−39.3	−45.4
	Sugar	23.82 ± 0.26	2.51 ± 0.16	−37.9	−46.3
Slow	Water	19.15 ± 0.86	2.81 ± 0.09	−50.1	−39.8
	Salt	17.25 ± 0.46	2.95 ± 0.07	−55.0	−36.8
	Sugar	17.08 ± 0.56	2.62 ± 0.01	−55.4	−43.9
Raw		38.34 ± 3.23	4.67 ± 0.15	-	-
LSD at $p < 0.05$		7.71	0.90	-	-
			Lentil		
Boiling	Water	8.67 ± 0.14	4.14 ± 0.09	−38.6	−40.1
	Salt	9.10 ± 0.11	4.51 ± 0.49	−35.6	−34.7
	Sugar	9.56 ± 0.19	3.81 ± 0.19	−32.3	−44.9
Pressure	Water	14.48 ± 0.69	6.06 ± 0.54	2.5	−12.3
	Salt	16.49 ± 0.27	5.89 ± 0.70	16.7	−14.8
	Sugar	16.94 ± 2.47	3.56 ± 0.04	19.9	−48.5
Microwave	Water	22.45 ± 0.49	7.23 ± 0.80	58.9	4.6
	Salt	23.26 ± 0.07	8.22 ± 0.36	64.6	19.0
	Sugar	23.82 ± 0.26	7.05 ± 0.84	68.6	2.0
Slow	Water	7.89 ± 0.58	4.33 ± 0.81	−44.2	−37.3
	Salt	8.38 ± 0.01	4.78 ± 0.30	−40.7	−30.8
	Sugar	8.44 ± 0.01	6.60 ± 0.07	−40.3	−4.5
Raw		14.13 ± 0.90	6.91 ± 0.23	-	-
LSD at $p < 0.05$		8.45	2.13	-	-
			Pea		
Boiling	Water	13.24 ± 0.69	4.79 ± 0.22	28.5	146.9
	Salt	11.23 ± 0.17	3.97 ± 0.01	9.0	104.6
	Sugar	14.01 ± 0.03	2.69 ± 0.11	36.0	38.7
Pressure	Water	13.26 ± 0.50	3.83 ± 0.06	28.7	97.4
	Salt	11.27 ± 0.41	2.15 ± 0.09	9.4	10.8
	Sugar	11.16 ± 0.03	3.03 ± 0.15	8.3	56.2
Microwave	Water	14.23 ± 0.76	2.02 ± 0.01	38.2	4.1
	Salt	13.28 ± 0.53	2.77 ± 0.05	28.9	42.8
	Sugar	11.76 ± 0.19	2.81 ± 0.07	14.2	44.8
Slow	Water	11.84 ± 0.10	2.68 ± 0.25	15.0	38.1
	Salt	15.29 ± 1.65	4.09 ± 0.04	48.4	110.8
	Sugar	14.36 ± 3.76	3.81 ± 0.16	39.4	96.4
Raw		10.30 ± 0.06	1.94 ± 0.02	-	-
LSD at $p < 0.05$		2.17	1.25	-	-

Mean ± SD, $n = 4$.

3.3. *Effect of Cooking on Antioxidant Properties*

Due to the variety of antioxidants in food extracts and their different action mechanisms, three assays were used to assess antioxidant activity of aqueous methanol and alkaline-liberated phenolic extracts obtained from raw and cooked pulses. Figure 1 shows the radical scavenging capacities of free and bound phenolic extracts from faba beans cooked with various cookers and solutions, i.e., 12 different cooking conditions. The three assays used in the current study are based on the ability of extracts to scavenge DPPH (2,2-diphenyl-1-picrylhydrazyl), ABTS (2,2′-azino-di-[3-ethyl benzthiazolinesulphonate]) and peroxyl radicals. Significant differences were observed among the cooking conditions, indicating their impact on antioxidant properties of cooked pulses. In general, the cooking of pulses significantly reduced DPPH scavenging capacity compared with raw faba bean extracts that exhibited scavenging capacities of 22.6 and 3.5 μmol trolox equivalent/g for free and bound phenolic extracts, respectively. The reduction in antioxidant capacity occurred in both free and bound extracts. This reduction goes in parallel with the reduction in phenolic compounds and could be attributed to the formation of irreversible covalent bond between liberated phenolic compounds and proteins [17]. A similar trend was observed for ABTS scavenging capacity in the case of aqueous methanol or free extracts, as the ABTS scavenging capacity decreased for all the cooked samples in comparison with the raw faba bean sample (17.6 μmol trolox equivalent/g). For bound phenolic extracts, the ABTS scavenging capacity slightly increased after cooking. This could be due to the differences in the type of radicals and phenolic compounds in both extracts. Similar to DPPH assay, the ORAC of both free and bound extracts from cooked faba beans decreased compared with the raw samples, which had ORAC values of 188 and 34 μmol Trolox equivalent/g, respectively. It has been reported that boiling and autoclaving cooking methods significantly reduce the antioxidant capacities of faba beans based on DPPH radical scavenging capacity, Trolox equivalent antioxidant capacity (TEAC) and ORAC assays [26]. Despite the significant reduction, there were still high amounts of phenolic compounds with potent antioxidant activities found in the boiled and autoclaved faba beans, as well as their broths, similar to the current study. The study suggested that it may be desirable to consume cooked faba beans together with its broth to maximize the potential health benefits. Pedrosa and others [18] also reported that pressure cooking reduces ORAC values by about 38% in acidified methanol extracts of common bean. In general, the cooking of faba beans with a microwave seems to retain high antioxidant activity, followed by boiling and slow cooking.

The antioxidant properties of cooked lentils are shown in Figure 2. Once again, significant reductions were observed in the DPPH scavenging capacity of free and bound phenolic extracts in cooked lentils compared with raw lentils, which exhibited DPPH scavenging capacities of 5.7 and 4.6 μmol trolox equivalent/g for free and bound phenolic extracts, respectively. Xu and Chang [34] reported that 9.6% and 26.1% reductions in DPPH scavenging capacity in boiled and pressure-cooked lentils. Slight decreases in DPPH scavenging capacity in boiled lentils have also been found by Landi et al. [35]. Additionally, Yeo and Shahidi [17] reported that soluble and insoluble phenolic fractions in lentils decrease due to the formation of irreversible covalent bond between liberated phenolic compounds and proteins. Significant differences were found among the cooking methods and heating solutions with boiling retained the highest DPPH scavenging capacity in the case of methanol extracts. The presence of other phenolic compounds and non-phenolic antioxidants in the methanol extract could contribute to the antioxidant potential. In the case of bound phenolic extracts, microwave cooking retained the highest DPPH scavenging capacity when the lentils were cooked in water. For the ABTS assay, there were significant reductions in both free and bound phenolic extracts compared with the raw lentils (13.5 and 2.1 μmol trolox equivalent/g, respectively). Pressure cooking exhibited the best retention of ABTS scavenging capacity in both free and bound extracts among the cooking methods. The cooking methods also reduced the ORAC values of free and bound phenolic extracts from lentils except for pressure cooking and boiling in water by which ORAC increased for both free and bound phenolic extracts. In general, cooking reduces the antioxidant activity of pulses, but the cooked pulses still retain a considerable antioxidant capacity. In a few cases, there were increases in the antioxidant activity, particularly in the ABTS and ORAC of bound phenolic extracts, perhaps due to the contribution of released phenolic acids and other phenolic compounds such as flavonoids. It has been reported that bound phytochemicals contribute about 82–85% of the total antioxidant activity in lentils [36].

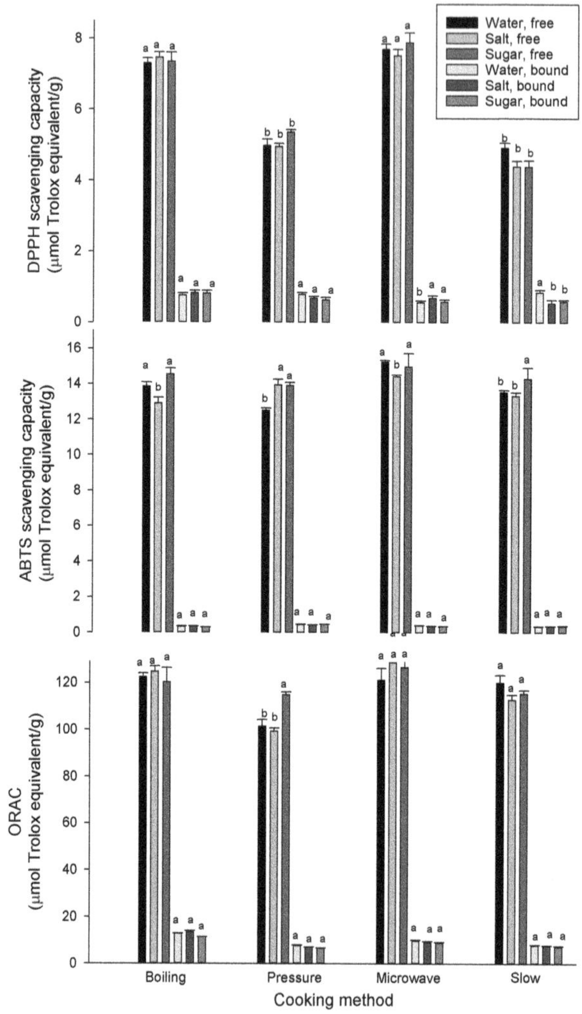

Figure 1. Antioxidant capacities of free and bound phenolic extracts of faba beans (μmol trolox equivalent/g, db) measured by DPPH (top), ABTS (middle) and Oxygen Radical Absorbance Capacity (ORAC) (bottom) assays. For free or bound phenolic extract, means followed by a different letter are significantly different at $p < 0.05$. Error bars represent standard deviation values ($n = 3$).

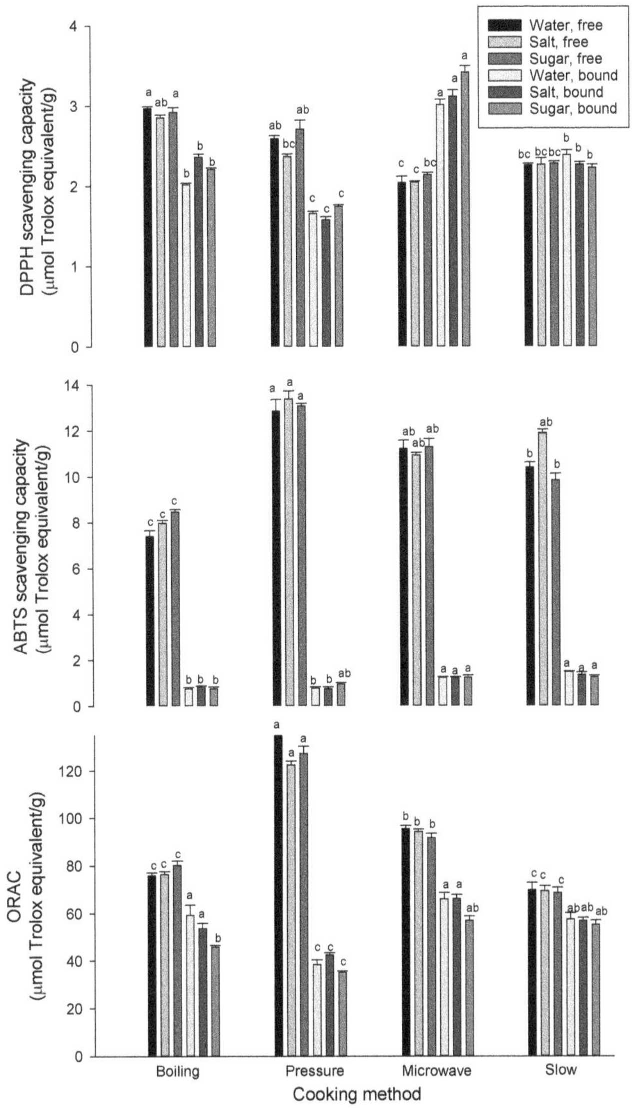

Figure 2. Antioxidant capacities of free and bound phenolic extracts of lentils (μmol trolox equivalent/g, db) measured by DPPH (top), ABTS (middle) and ORAC (bottom) assays. For free or bound phenolic extract, means followed by a different letter are significantly different at $p < 0.05$. Error bars represent standard deviation values ($n = 3$).

The antioxidant capacity of peas was also significantly impacted by cooking conditions showing decreases or increases subject to the antioxidant assay and type of extract, e.g., free versus bound phenolic extract (Figure 3). There were reductions in DPPH scavenging capacity in free extracts due to the cooking of peas with the four cookers, while increases were observed in the bound phenolic extracts as compared with raw peas free and bound phenolic extracts that had DPPH scavenging capacities of 2.3 and 0.5 μmol trolox equivalent/g, respectively. ABTS scavenging capacity of free and bound phenolic extracts increased in cooked pea. These results supported by the increase in both free and bound phenolic compounds (Table 4). ORAC values increased in bound phenolic extracts of cooked peas and decreased in free phenolic extracts. Raw peas had ABTS scavenging capacities of 9.2 and 0.4 μmol trolox equivalent/g and ORAC values of 69.5 and 15.8 μmol trolox equivalent/g in free and bound phenolic extracts, respectively. The antioxidant activity of free phytochemical extracts of green and yellow peas has been found to remain unchanged after cooking, while cooking decreases antioxidant activity in chickpeas by 30% and in soybeans by 38% and increases antioxidant activity in lentils by 10% [36]. They also reported insignificant changes in antioxidant activity of bound phytochemical extracts from peas, chickpeas and soybeans after cooking. The increase in antioxidant activity of free phytochemical extracts could be attributed to the release of bound phenolics, while the decrease could be due to the loss of soluble phytochemicals during cooking. The boiling and steaming of green and yellow peas, chickpeas and lentils decreases DPPH scavenging activity and ORAC values, while pressure cooking increases ORAC values [33]. It appears that changes (decreases or increases) in phenolic compounds and antioxidant activity are subject to the type of phenolic (e.g., free versus bound) and cooking conditions (e.g., the type of cooker and heating solution). Faba beans, lentils and peas also possessed different antioxidant activities due to their different phenolic and non-phenolic compositions, with faba beans and lentils having higher antioxidant activities than that of peas. It has been reported that processing can be an effective means to improve bioactive and anti-nutritional compound attributes in pulses [3,37].

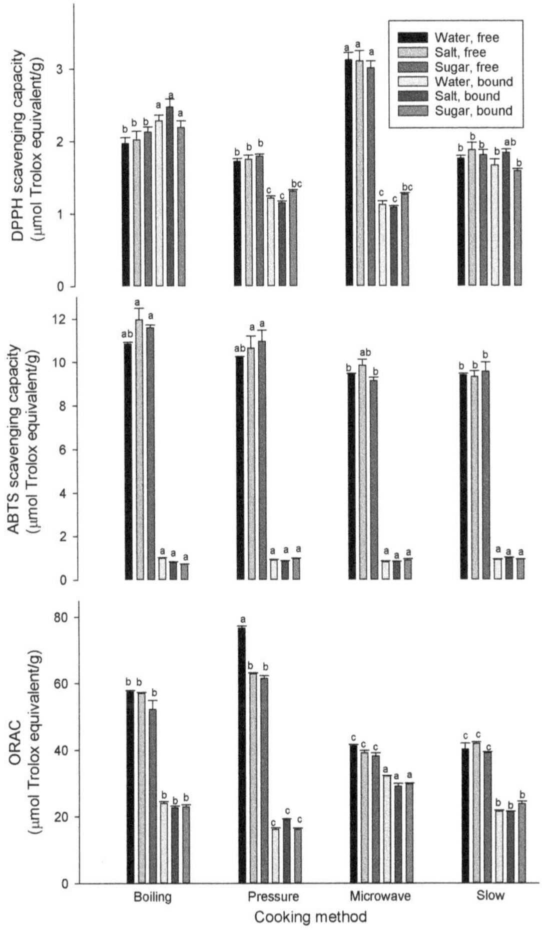

Figure 3. Antioxidant capacities of free and bound phenolic extracts of peas (µmol trolox equivalent/g, db) measured by DPPH (top), ABTS (middle) and ORAC (bottom) assays. For free or bound phenolic extract, means followed by a different letter are significantly different at $p < 0.05$. Error bars represent standard deviation values ($n = 3$).

4. Conclusions

Pulses are good sources of phenolic antioxidants in addition to other important phytochemicals and nutrients, such as flavonoids, dietary fiber, protein, resistant and slowly digestible starches, etc. Thus, it is crucial to understand how cooking could affect their nutrient composition and antioxidant properties. The current study demonstrates that changes in phenolic compounds and antioxidant activity are subject to the type of phenolic (e.g., free versus bound) and cooking conditions (e.g., type of cooker and heating solution). The four cooking methods showed significant differences among pulses based on their phenolic content and/or antioxidant activity. Differences in phenolic compounds and antioxidant activities indicate that their reactions to cooking conditions are unalike. Despite the significant reductions in phenolic antioxidants, there were still high amounts of phenolic compounds with potent antioxidant activities in cooked pulses. The results also indicate that faba beans and lentils are better dietary sources of phenolic antioxidants compared with peas. More research is needed to study the antioxidant activities of cooked pulses using simulated digestion models and in vivo studies (animal and human) to better understand their health effects.

Author Contributions: Conceptualization, E.-S.M.A.-A. and S.R.; Methodology, E.-S.M.A.-A.; Validation, E.-S.M.A.-A. and S.R.; Formal analysis, Y.L. and E.-S.M.A.-A.; Investigation, Y.L.; Resources, E.-S.M.A.-A., S.R. and M.F.M.; Writing—original draft preparation, Y.L.; Writing—review and editing, E.-S.M.A.-A., S.R. and M.F.M.; Visualization, E.-S.M.A.-A. and S.R.; Supervision, E.-S.M.A.-A., S.R. and M.F.M.; Project administration, E.-S.M.A.-A.; Funding acquisition, E.-S.M.A.-A. All authors have read and agreed to the published version of the manuscript.

Funding: This research was funded by Saskatchewan Pulse Growers, grant number PRO1533 and The APC was funded by Agriculture and Agri-Food Canada and University of Guelph. Check carefully that the details given are accurate and use the standard spelling of funding agency names at https://search.crossref.org/funding, any errors may affect your future funding.

Acknowledgments: This research was financially supported by Saskatchewan Pulse Growers. Thanks go to Iwona Rabalski from Guelph Research and Development Centre, Agriculture and Agri-Food Canada for her excellent technical support and T. Warkentin and A. Vandenberg for providing the pulse samples.

Conflicts of Interest: The authors declare no conflict of interest.

References

1. Leterme, P. Recommendations by health organizations for pulse consumption. *Br. J. Nutr.* **2002**, *88*, 239–242. [CrossRef] [PubMed]
2. Hall, C.; Hillen, C.; Garden Robinson, J. Composition, nutritional value, and health benefits of pulses. *Cereal Chem.* **2017**, *94*, 11–31. [CrossRef]
3. Abdel-Aal, E.-S.M. Processing Pulses to Enhance Bioactive and Anti-Nutritional Attributes. A Mini-Review. 2016. Available online: http://canadianfoodbusiness.com/2016/06/20/processing-pulses-to-enhance-bioactive-and-anti-nutritional-attributes/ (accessed on 9 July 2020).
4. Best, D. 10 things to know about pulses. *CFW* **2013**, *58*, 105–107. [CrossRef]
5. Singh, B.; Singh, J.P.; Shevkani, K.; Singh, N.; Kaur, A. Bioactive constituents in pulses and their health benefits. *J. Food Sci. Technol.* **2017**, *54*, 858–870. [CrossRef] [PubMed]
6. Zhang, B.; Deng, Z.; Ramdath, D.D.; Tang, Y.; Chen, P.X.; Liu, R.; Tsao, R. Phenolic profiles of 20 Canadian lentil cultivars and their contribution to antioxidant activity and inhibitory effects on α-glucosidase and pancreatic lipase. *Food Chem.* **2015**, *172*, 862–872. [CrossRef]
7. Mirali, M.; Purves, R.W.; Vandenberg, A. Profiling the phenolic compounds of the four major seed coat types and their relation to color genes in lentil. *J. Nat. Prod.* **2017**, *80*, 1310–1317. [CrossRef] [PubMed]
8. Padhi, E.; Liu, R.; Hernandez, M.; Tsao, R.; Ramdath, D. Total polyphenol content, carotenoid, tocopherol and fatty acid composition of commonly consumed Canadian pulses and their contribution to antioxidant activity. *J. Funct. Foods* **2017**, *38*, 602–611. [CrossRef]
9. Dueñas, M.; Sarmento, T.; Aguilera, Y.; Benitez, V.; Mollá, E.; Esteban, R.M.; Martín-Cabrejas, M.A. Impact of cooking and germination on phenolic composition and dietary fibre fractions in dark beans (*Phaseolus vulgaris* L.) and lentils (*Lens culinaris* L.). *LWT -Food Sci. Technol.* **2016**, *66*, 72–78. [CrossRef]
10. Ombra, M.; D'Acierno, A.; Nazzaro, F.; Riccardi, R.; Spigno, P.; Zaccardelli, M. Phenolic Composition and Antioxidant and Antiproliferative Activities of the Extracts of Twelve Common Bean (*Phaseolus vulgaris* L.) Endemic Ecotypes of Southern Italy before and after Cooking. *Oxid. Med. Cell Longev.* **2016**, *2016*. [CrossRef]
11. Zou, Y.; Chang, S.K.; Gu, Y.; Qian, S.Y. Antioxidant activity and phenolic compositions of lentil (*Lens culinaris* var. *Morton*) extract and its fractions. *J. Agric. Food Chem.* **2011**, *59*, 2268–2276. [CrossRef] [PubMed]
12. Beninger, C.W.; Gu, L.; Prior, R.L.; Junk, D.C.; Vandenberg, A.; Bett, K.E. Changes in polyphenols of the seed coat during the after-darkening process in pinto beans (*Phaseolus vulgaris* L.). *J. Agric. Food Chem.* **2005**, *53*, 7777–7782. [CrossRef] [PubMed]
13. Amarowicz, R.; Troszyńska, A. Antioxidant activity of extract of pea and its fractions of low molecular phenolics and tannins. *Pol. J. Food Nutr. Sci.* **2003**, *12*, 10–15.
14. Ramdath, D.; Renwick, S.; Duncan, A.M. The role of pulses in the dietary management of diabetes. *Can. J. Diabetes* **2016**, *40*, 355–363. [CrossRef] [PubMed]
15. Wang, Y.; Zhang, X.; Chen, G.; Yu, J.; Yang, L.; Gao, Y. Antioxidant property and their free, soluble conjugate and insoluble-bound phenolic contents in selected beans. *J. Funct. Foods* **2016**, *24*, 359–372. [CrossRef]

16. Aguilera, Y.; Estrella, I.; Benitez, V.; Esteban, R.; Martín-Cabrejas, M. Bioactive phenolic compounds and functional properties of dehydrated bean flours. *Food Res. Int.* **2011**, *44*, 774–780. [CrossRef]
17. Yeo, J.; Shahidi, F. Effect of hydrothermal processing on changes of insoluble-bound phenolics of lentils. *J. Funct. Foods* **2017**, *38*, 716–722. [CrossRef]
18. Pedrosa, M.; Cuadrado, C.; Burbano, C.; Muzquiz, M.; Cabellos, B.; Olmedilla-Alonso, B.; Asensio-Vegas, C. Effects of industrial canning on the proximate composition, bioactive compounds contents and nutritional profile of two Spanish common dry beans (*Phaseolus vulgaris* L.). *Food Chem.* **2015**, *166*, 68–75. [CrossRef]
19. Kalogeropoulos, N.; Chiou, A.; Ioannou, M.; Karathanos, V.; Hassapidou, M.; Andrikopoulos, N. Nutritional evaluation and bioactive microconstituents (phytosterols, tocopherols, polyphenols, triterpenic acids) in cooked dry legumes usually consumed in the Mediterranean countries. *Food Chem.* **2010**, *121*, 682–690. [CrossRef]
20. Abdel-Aal, E.; Ragaee, S.; Rabalski, I.; Warkentin, T.; Vandenberg, A. Nutrient content and viscosity of Saskatchewan-grown pulses in relation to their cooking quality. *Can. J. Plant Sci.* **2019**, *99*, 67–77. [CrossRef]
21. Liu, Y.; Ragaee, S.; Marcone, M.F.; Abdel-Aal, E. Effect of different cooking methods and heating solutions on nutritionally-important starch fractions and flatus oligosaccharides in selected pulses. *Cereal Chem.* **2020**. Under Review.
22. Abdel-Aal, E.; Rabalski, I. Effect of baking on free and bound phenolic acids in wholegrain bakery products. *J. Cereal Sci.* **2013**, *57*, 312–318. [CrossRef]
23. Abdel-Aal, E.S.M.; Rabalski, I. Bioactive compounds and their antioxidant capacity in selected primitive and modern wheat species. *Open Agric.* **2008**, *2*, 7–14. [CrossRef]
24. Abdel-Aal, E.S.M.; Rabalski, I. Antioxidant Properties of high-lutein grain-based functional foods in comparison with ferulic acid and lutein. *Am. J. Biomed. Sci.* **2013**, *5*, 109–125. [CrossRef]
25. Sosulski, F.W.; Dabrowski, K.J. Composition of free and hydrolyzable phenolic acids in the flours and hulls of ten legume species. *J. Agric. Food Chem.* **1984**, *32*, 131–133. [CrossRef]
26. Siah, S.; Wood, J.; Agboola, S.; Konczak, I.; Blanchard, C. Effects of soaking, boiling and autoclaving on the phenolic contents and antioxidant activities of faba beans (*Vicia faba* L.) differing in seed coat colours. *Food Chem.* **2014**, *142*, 461–468. [CrossRef]
27. Prabhakar, P.K.; Prasad, R.; Ali, S.; Doble, M. Synergistic interaction of ferulic acid with commercial hypoglycemic drugs in streptozotocin induced diabetic rats. *Phytomedicine* **2013**, *20*, 488–494. [CrossRef] [PubMed]
28. Srinivasan, M.; Sudheer, A.R.; Menon, V.P. Ferulic acid: Therapeutic potential through its antioxidant property. *J. Clin. Biochem. Nutr.* **2007**, *40*, 92–100. [CrossRef]
29. Alshikh, N.; de Camargo, A.; Shahidi, F. Phenolics of selected lentil cultivars: Antioxidant activities and inhibition of low-density lipoprotein and DNA damage. *J. Funct. Foods* **2015**, *18*, 1022–1038. [CrossRef]
30. Huang, D.; Ou, B.; Prior, R.L. The chemistry behind antioxidant capacity assays. *J. Agric. Food Chem.* **2005**, *53*, 1841–1856. [CrossRef]
31. Zhao, Y.; Du, S.; Wang, H.; Cai, M. In vitro antioxidant activity of extracts from common legumes. *Food Chem.* **2014**, *152*, 462–466. [CrossRef]
32. Chaieb, N.; González, J.L.; López-Mesas, M.; Bouslama, M.; Valiente, M. Polyphenols content and antioxidant capacity of thirteen faba bean (*Vicia faba* L.) genotypes cultivated in Tunisia. *Food Res. Int.* **2011**, *44*, 970–977. [CrossRef]
33. Nithiyanantham, S.; Selvakumar, S.; Siddhuraju, P. Total phenolic content and antioxidant activity of two different solvent extracts from raw and processed legumes, *Cicer arietinum* L. and *Pisum sativum* L. *J. Food Compos. Anal.* **2012**, *27*, 52–60. [CrossRef]
34. Xu, B.; Chang, S. Effect of soaking, boiling, and steaming on total phenolic contentand antioxidant activities of cool season food legumes. *Food Chem.* **2008**, *110*, 1–13. [CrossRef] [PubMed]
35. Landi, N.; Pacifico, S.; Piccolella, S.; Di Giuseppe, A.; Mezzacapo, M.; Ragucci, S.; Di Maro, A. Valle Agricola lentil, an unknown lentil (*Lens culinaris* Medik.) seed from Southern Italy as a novel antioxidant and prebiotic source. *Food Funct.* **2015**, *6*, 3155–3164. [CrossRef]

36. Han, H.; Baik, B. Antioxidant activity and phenolic content of lentils (*Lens culinaris*), chickpeas (*Cicer arietinum L.*), peas (*Pisum sativum L.*) and soybeans (*Glycine max*), and their quantitative changes during processing. *Int. J. Food Sci. Technol.* **2008**, *43*, 1971–1978. [CrossRef]
37. Malcolmson, L.; Han, J. Pulse processing and untilization of pulse ingredients in foods. In *Health Benefits of Pulses*; Dahl, W.J., Ed.; Springer Nature: Basel, Switzerland, 2019; Chapter 9; pp. 129–149. [CrossRef]

© 2020 by the authors. Licensee MDPI, Basel, Switzerland. This article is an open access article distributed under the terms and conditions of the Creative Commons Attribution (CC BY) license (http://creativecommons.org/licenses/by/4.0/).

foods

MDPI

Article

The Strength of the Nutrient Solution Modulates the Functional Profile of Hydroponically Grown Lettuce in a Genotype-Dependent Manner

Biancamaria Senizza [1,†], Leilei Zhang [1,†], Begoña Miras-Moreno [1], Laura Righetti [2], Gokhan Zengin [3], Gunes Ak [3], Renato Bruni [2], Luigi Lucini [1,4,*], Maria Isabella Sifola [5], Christophe El-Nakhel [5], Giandomenico Corrado [5] and Youssef Rouphael [5,*]

1 Department for Sustainable Food Process, Università Cattolica del Sacro Cuore, 29122 Piacenza, Italy; biancamaria.senizza@unicatt.it (B.S.); leilei.zhang@unicatt.it (L.Z.); mariabegona.mirasmoreno@unicatt.it (B.M.-M.)
2 Department of Food and Drug, University of Parma, 43124 Parma, Italy; laura.righetti@unipr.it (L.R.); renato.bruni@unipr.it (R.B.)
3 Department of Biology, Science Faculty, Selcuk University, 42130 Konya, Turkey; gokhanzengin@selcuk.edu.tr (G.Z.); akguneselcuk@gmail.com (G.A.)
4 Research Centre for Nutrigenomics and proteomics (PRONUTRIGEN), Università Cattolica del Sacro Cuore, 29122 Piacenza, Italy
5 Department of Agricultural Sciences, University of Naples Federico II, 80055 Portici, Italy; sifola@unina.it (M.I.S.); christophe.elnakhel@unina.it (C.E.-N.); giandomenico.corrado@unina.it (G.C.)
* Correspondence: luigi.lucini@unicatt.it (L.L.); youssef.rouphael@unina.it (Y.R.)
† These authors contributed equally.

Received: 22 July 2020; Accepted: 19 August 2020; Published: 21 August 2020

Abstract: Considering that functional components of plant foods are mainly secondary-metabolism products, we investigated the shaping of health-promoting compounds in hydroponically grown butterhead lettuce (*Lactuca sativa* L. var. *capitata*) as a function of the strength of the nutrient solution utilized. To this aim, untargeted metabolomics profiling, in vitro antioxidant capacity (total phenolics, 2,2-diphenyl-1-picrylhydrazyl (DPPH), 2,2′-azino-bis(3-ethylbenzothiazoline-6-sulphonic acid) (ABTS), cupric reducing antioxidant capacity (CUPRAC), and ferric reducing antioxidant power (FRAP) assays), and the inhibition of selected enzyme activities were investigated in two butterhead lettuce cultivars with different pigmentation, i.e., green and red Salanova. Full-strength nutrition, together with half- and quarter-strength solutions of macronutrients, was tested. Our results indicate that by reducing the nutrients strength, we could elicit a distinctive shaping of the phenolic profile of lettuce. It is noteworthy that only specific classes of phenolics (namely, lignans and phenolic acids, followed by flavones and anthocyanins) were modulated by the induction of nutritional eustress (fold-change values in the range between −5 and +11). This indicates that specific responses, rather than a generalized induction of phenolic compounds, could be observed. Nonetheless, a genotype-dependent response could be observed, with the red cultivar being much more responsive to nutritional deprivation than the green Salanova lettuce. Indeed, analysis of variance (ANOVA) confirmed a genotype x nutrition interaction in red Salanova ($p < 0.001$). As a consequence of the changes in phenolic composition, also the antioxidant capacity ($p < 0.001$) and amylase inhibition ($p < 0.001$) properties were affected by the growing conditions. However, the effect on cholinesterase and tyrosinase inhibition was poorly affected by the nutritional strength. Provided that yields are not compromised, the application of a controlled nutritional eustress in hydroponically cultivated lettuce may represent a valuable strategy to produce food with tailored functional features in a sustainable manner.

Keywords: *Lactuca sativa*; polyphenols; metabolomics; antioxidants; eustress

1. Introduction

As autotrophs, plants have evolved sophisticated mechanisms to take up inorganic molecules in solution from the soil [1,2]. Mineral absorption is therefore a selective and efficient process, of fundamental importance in plant physiology. In addition, knowledge in this scientific area is central to sustainable agriculture and environmental protection [3].

In recent years, the development of high-throughput analytic studies of small metabolites (i.e., metabolomics) has represented a major technical breakthrough in plant science [4]. Metabolomics has allowed scientific advances in understanding plants' response and adaptation to mineral availability [5–7], also because "omics" can provide a comprehensive view without any assumption about the levels and the effects of experimental factors [8,9]. Plant nutrition studies are, however, hindered by two factors. Soil is complex, variable, and dynamic in terms of its chemical, physical, hydraulic, and biological characteristics [10]. Moreover, plant mineral uptake and translocation, being strictly dependent on water movement, are highly dependent on environmental conditions, such as temperature and light (e.g., intensity and quality). Because of the plant-soil-atmosphere relationship, the molecular analysis of the effects of mineral nutrition in plants is typically performed in hydroponics. Although the term hydroponics may have different connotations, currently, it is mostly applied to indicate any plant production system using a water-based nutrient solution (NS) without soil [11]. For all these reasons, the response of various crops to a range of chemical stressors, such as mineral deficiency, toxicity, and osmotic potential, has been investigated using hydroponics [12–15].

Even if different formulations have been created, typically, the amount of mineral elements in the NS is set to the highest possible concentration without incurring toxicity and stress [16,17]. The NS in hydroponics is a manageable and flexible experimental factor useful to understand and modulate a plant's response to nutrient availability beyond mineral withholding, such as by varying the strength and the ion ratio of the NS [18]. Investigations aiming at elucidating the effects of global NS alterations have also a practical impact [19], considering that hydroponics is increasingly used in the commercial production of several crops (i.e., tomatoes, strawberries, cucumbers, peppers, eggplants, lettuce) [20].

There is a consolidated consensus that an adequate supply of mineral elements is central to crop yield. More recently, growing evidence has indicated that the reduction of nutrients to a level that does not result in deficiency symptoms has metabolic consequences of interest also for applied research [21,22]. Specifically, nutritional eustress activates physiological responses and molecular mechanisms that elicit the accumulation of health-promoting bioactive compounds (i.e., antioxidants) necessary for plant adaptation to suboptimal environments [23].

Evidently, these findings provide an opportunity to meet consumers' increasing demand of high-quality fresh vegetables associated with health benefits [24]. Indeed, epidemiological studies have shown a correlation between a plant-based diet and nutritional benefits associated with fruits and vegetables rich in phytochemical compounds. Lettuce (*Lactuca sativa* L.) is considered an important source of phytoconstituents such as carotenoids, vitamins, and polyphenols [25]. Although it is already widely consumed, biotic and abiotic elicitations have been effectively applied to increase its phytochemical content and thus its perception as a "healthier" food. Regarding the modulation of microconstituents in growth media, the total flavonoid content was significantly increased following Ca^{2+} supplementation and correlated with the antioxidant activity of lettuces [26]. On the other hand, macronutrients deprivation showed an increase in antioxidant compounds such as ascorbic acid, polyphenols, and carotenoids [27–29]. The biosynthesis of these antioxidant metabolites important for human diet, results from the activation of plant defense mechanisms to environmental stresses. Application of chemical elicitors, such as jasmonic acid, elicited the accumulation of phenolic compounds (i.e., flavonoids and phenolic acids) as well as of other phytochemicals such as vitamin C, chlorophylls, and carotenoids [28]. The content of these pigments was also modulated under different light wavelengths [29], with the highest increase induced by fluorescent light plus blue LED and monochromic red radiation. Furthermore, lettuce plants exposed to chilling stress (4 °C for 1 day) rapidly activated key genes involved in the biosynthesis of phenolic compounds, ascorbic acid, and α-tocopherol [30]. From these studies, it is evident that lettuce polyphenol content and, thus, polyphenol intake through the human diet, relevant for the prevention of several diseases, can be modulated by NS management.

Here, we aimed at investigating the changes in lettuce phytochemicals in response to a sub-optimal plant nutrient supply, with a focus on the shaping of phenolic compounds, antioxidant capacity, and inhibition potential towards health-related key enzymes. To test this hypothesis, we used lettuce as a model crop, considering two lettuce genotypes that underlie different leaf colorations. Lettuce is a major crop widely cultivated in hydroponics also because of its rapid growth, abundant edible leaves, and fast production cycle [26,31]. We focused on two similar butterhead lettuce varieties that in normal growing conditions develop red-pigmented and green-pigmented leaves. A factorial experimental design, with three progressively reduced concentrations of macronutrients and an invariant micronutrient supply, was chosen. Furthermore, high-resolution ultra-high performance liquid chromatography-quadrupole time-of-flight (UHPLC-QTOF) mass spectrometry-based untargeted metabolomics was used to comprehensively unravel the impact of the reduction of the NS strength on phytochemical profiles.

2. Materials and Methods

2.1. Plant Grow Conditions and Experimental Design

The work was carried out on two butterhead lettuce (*L. sativa* L. var. *capitata*) varieties, namely, green and red Salanova® (Rijk Zwaan Italia, Bologna, Italia), grown with a closed Nutrient Film Technique (NFT) in a 28 m^2 open gas-exchange growth chamber (Model Process-C5, Spagnol srl, Treviso, Italy). The experimental system and the environmental conditions were as previously described [28].

The experimental design was full factorial, with the factor "genotype" (G) having two levels (i.e., green and red Salanova), and the factor "strength of the nutrient solution" (S) having three levels, namely, full strength (FS), half strength (HS), and quarter strength (QS). The concentrations of macronutrients in the FS nutrient solution were: 9.0 mM nitrate, 1.0 mM phosphorous, 4.0 mM potassium, 2.0 mM sulfur, 4.0 mM calcium, 1.0 mM manganese, and 1.0 mM ammonium. The concentrations of the macronutrients in HS and QS nutrient solutions were half and a quarter, respectively, of those of the FS solution. In each nutrient solution, the concentrations of microelements were invariant (i.e., 15 μM iron, 9 μM manganese, 0.3 μM copper, 1.6 μM zinc, 20 μM boron, 0.3 μM molybdenum). The average values of the electrical conductivity of the FS, HS, and QS solutions measured in the NFT channels were 1.5 ± 0.1, 0.75 ± 0.1, and 0.5 ± 0.1 dS m^{-1}, respectively.

2.2. Sample Preparation

We used a randomized complete-block experimental design with the six treatments and three replicates for each cultivar, making a total of 18 experimental units, comprising 12 plants each ($n = 216$ plants). Determinations of total phenolic and total flavonoid content, antioxidant activity assays, and enzyme inhibitory activities were carried out using lyophilized biomass, whereas metabolomics analysis was done on liquid nitrogen-quenched fresh biomass. In both cases, the 12 plants within each replicate of each treatment were pooled.

2.3. Total Phenolic and Total Flavonoid Contents

The total phenolic and flavonoid contents were determined as previously reported [32], using the Folin-Ciocalteu and the $AlCl_3$ assays, respectively. Total phenolics were expressed as gallic acid equivalents (mg GAEs/g extract), while total flavonoids as rutin equivalents (mg REs/g extract).

2.4. UHPLC-QTOF Mass Spectrometry Profiling

Each lettuce sample (1.0 g) was extracted in triplicate in 20 mL of 80% methanol (LCMS grade, VWR, Milan, Italy), acidified with 1% v/v HCOOH using ultrasounds at the amplitude of 80% for 20 min (Fisher Scientific model FB120, Pittsburgh, PA, USA). The extracts were centrifuged at 8000 *g* for 15 min at 4 °C (Eppendorf 5810R, Hamburg, Germany), and the supernatants were directly filtered into HPLC glass vials through 0.22 μm cellulose syringe filters.

The phenolic profile was investigated in duplicate through ultra-high-pressure liquid chromatography electrospray ionization quadrupole-time-of-flight mass spectrometry (UHPLC-ESI/QTOF-MS) as previously reported [33], with the exception that an Agilent Pursuit 3 PFP column (100 × 2.0 mm, 3 μm) was used for chromatographic separation. Briefly, the positive scan mode was used to acquire monoisotopic accurate masses in the 100–1000 m/z range at a rate of 0.8 spectra/s. The binary gradient separation used a mixture of water and acetonitrile (both with 0.1% HCOOH, both LCMS grade, VWR, Milan, Italy) as mobile phase. Quality controls (QCs) were prepared by pooling aliquots of each sample and then analyzed in data-dependent tandem mass spectrometry. To this aim, 12 precursor ions per scan were selected for auto-MS/MS (absolute threshold = 1000; relative threshold = 0.001%; collision energies = 10, 20 and 40 V). The injection volume was 6 μL, nitrogen was used as both sheath gas (10 L/min and 350 °C) and drying gas (8 L/min and 330 °C), the nozzle voltage was 300 V, and the capillary voltage was 3.5 kV in both MS and MS/MS acquisitions.

Raw data were then processed in Profinder B.07 (from Agilent Technologies, Santa Clara, CA, United States), and annotation was done according to the whole isotope pattern (accurate mass, isotopic spacing, and isotopic ratio), with a mass accuracy of <5 ppm. Mass and retention time alignment, as well as filters by frequency (only those compounds identified within 100% of replications in at least one treatment were retained) were also adopted. The database exported from Phenol-Explorer 3.6 [34] was used as a reference for identification. Based on the Metabolomics Standards Initiative [35], an annotation of level 2 (i.e., putatively annotated compounds) was achieved. QCs were elaborated in MS-DIAL 3.98 (RIKEN Center for Sustainable Resource Science: Metabolome Informatics Research Team, Yokohama, Japan), using MS/MS spectra to confirm features obtained by Profinder B.07 [36]. For this purpose, publicly available MS/MS experimental spectra built in the software (e.g., MoNA, Mass bank of North America) were used.

The phenolic dataset was then semi-quantitatively elaborated [37]. Phenolic compounds were grouped into classes and subclasses, and cumulative intensities were quantified using calibration curves of pure phenolic standards (purity >98%; from Sigma-Aldrich, St. Louis, MO, USA), analyzed under the above-described analytical conditions. Ferulic acid (for hydroxycinnamic acids and other phenolic acids), quercetin (for flavonols), sesamin (furan and furofuran lignans), cyanidin (anthocyanins), catechin (flavanols), luteolin (flavones and other remaining flavonoids), resveratrol (stilbenes), and tyrosol (tyrosols and other polyphenols) were used as representatives of their corresponding classes. A linear regression curve (not weighed and not forced to origin) was built using a concentration range of 0.05–500 ppm, and results were expressed as mg equivalents/100 g dry weight (dw).

2.5. Determination of Antioxidant and Enzyme Inhibitory Effects

Antioxidant assays included metal chelating, phosphomolybdenum, ferric reducing antioxidant power (FRAP), cupric reducing antioxidant capacity (CUPRAC), 2,2′-azino-bis(3-ethylbenzothiazoline-6-sulphonic acid) (ABTS), and 2,2-diphenyl-1-picrylhydrazyl (DPPH) activities of the extracts as previously described [38]. Results were expressed as Trolox equivalents, and Ethylenediaminetetraacetic acid (EDTA) was used for metal chelating assays. The possible inhibitory effects of the extracts against acetylcholinesterase (AChE, from electric eel acetylcholinesterase (type-VI-S), EC 3.1.1.7), butyrylcholinesterase (BChE, from horse serum, EC 3.1.1.8)) (by Ellman's method), α-amylase (from porcine pancreas, EC. 3.2.1.1), α-glucosidase (from *Saccharomyces cerevisiae*, EC. 3.2.1.20), and tyrosinase (from mushroom, EC 1.14.18.1) were evaluated through in vitro bioassays.

The enzyme inhibitory effects were determined as equivalents of kojic acid (KAE) for tyrosinase, of galantamine (GALAE) for AChE and BChE, and of acarbose (ACAE) for α-amylase and α-glucosidase.

2.6. Statistical Analysis

PASW Statistics 25.0 (SPSS Inc., Segrate, Italy) was used for the analysis of variance (ANOVA; $p < 0.05$) in semiquantitative values of each representative phenolic class, for enzyme inhibitory assays, and for total flavonoids and total phenolics. Homogenous subclasses were investigated according to the Duncan's post-hoc test. Correlation coefficients between enzymatic assays and different phenolic subclasses were determined according to Pearson in PASW Statistics 25.0 ($p = 0.05$, two-tailed).

Metabolomics raw data were processed in Mass Profiler Professional B.12.06 (Agilent Technologies, Santa Clara, CA, United States) for mass and retention time alignment, for filtering by abundance and by frequency, for normalization and baselining, as previously reported [39]. Thereafter, a fold-change-based hierarchical cluster analysis (Euclidean distance, Ward's linkage) was carried out for unsupervised distribution. The dataset was next exported into SIMCA 16 (Umetrics, Malmo, Sweden) for orthogonal projection to latent structures discriminant analysis (OPLS-DA) supervised modelling. Outliers were investigated according to Hotelling's T2 (95% and 99% confidence limits for suspect and strong outliers, respectively; Supplementary Figure S1). CV-ANOVA ($p < 0.01$) cross-validation and permutation testing (for overfitting, N = 200) were also done from the OPLS-DA model. Thereafter, model fitness parameters (goodness-of-fit, R^2Y, and goodness-of-prediction, Q^2Y) were recorded, and the variable importance in projection (VIP) analysis was used to identify the most discriminant compounds (VIP score > 1.1) in lettuce grown under different conditions.

3. Results

3.1. Phenolic Profile of Salanova Lettuce

The polyphenolic composition of lettuce was explored using a metabolomics approach, resulting in the putative annotation of 327 phenolic compounds belonging to different classes. A wide diversity of phenolics could by annotated by our untargeted metabolomics analysis. In particular, 163 flavonoids (including flavanones, anthocyanins, flavonols, and flavanols), 69 phenolic acids, 67 low-molecular-weight phenolics, 19 lignans, and 9 stilbenes were found. The dataset, including composite mass spectrum and ontology classification (i.e., classes and subclasses) of annotated compounds, is provided as Supplementary Material (Supplementary Table S1 for MS annotations and Supplementary Table S2 for MS/MS-identified compounds, respectively).

A strong influence of genetic background could be observed. Indeed, the red cultivar featured a higher content of polyphenols, compared to the green one. The most abundant phenolic classes were lignans, tyrosols, and phenolic acids (Table 1). In particular, low-molecular-weight phenolics (expressed as tyrosol equivalents) reached the highest amounts under quarter-strength growth, with 829 and 1316 mg/ 100 g eq., followed by lignans compound with 765 and 1682 mg/100 g eq. and phenolics acids with 368 and 876 mg/100 g eq. in green and red Salanova lettuce, respectively.

Table 1. Semi-quantitative profile of phenolic compounds in green and red Salanova lettuce (S) in nutrient solutions with different macrocations concentration (C). Results are cumulative untargeted-metabolomics profiling data per each phenolic subclass.

Source of Variance	Anthocyanins Cyanidin eq. (mg/100 g)	Flavanols Catchin eq. (mg/100 g)	Flavones Luteolin eq. (mg/100 g)	Flavonols Quercetin eq. (mg/100 g)	Lignans Sesamin eq. (mg/100 g)	Tyrosols Tyrosols eq. (mg/100 g)	Phenolic Acids Ferulic eq. (mg/100 g)	Stilbenes Resveratrol eq. (mg/100 g)
Salanova (S)								
Green	26.18 ± 1.23 b	31.21 ± 1.83 b	143.5 ± 9.49 b	62.66 ± 2.85 b	724 ± 20.05 b	759 ± 22.66 b	328.2 ± 14.22 b	34.21 ± 2.37 b
Red	45.61 ± 2.37 a	49.61 ± 4.70 a	193.8 ± 13.19 a	106.5 ± 4.77 a	1116 ± 109.51 a	1227 ± 44.80 a	778.0 ± 29.06 a	58.22 ± 5.40 a
Concentration (C)								
Full strength	32.52 ± 1.63	35.18 ± 2.74 b	153.0 ± 14.18 b	72.00 ± 5.30 b	717 ± 37.71 b	897 ± 66.19 b	492.0 ± 51.78 b	48.30 ± 9.39
Half strength	37.16 ± 3.96	36.19 ± 6.57 b	194.0 ± 19.22 a	85.76 ± 9.55 a	820 ± 54.44 b	1022 ± 96.57 a	544.7 ± 81.41 b	43.02 ± 3.27
Quarter strength	38.01 ± 4.72	49.85 ± 4.38 a	159.0 ± 11.17 b	95.99 ± 7.55 a	1224 ± 147.49 a	1059 ± 76.23 a	622.5 ± 78.29 a	47.32 ± 4.45
C × S								
Green × Full strength	29.27 ± 2.15 bc	28.04 ± 2.53	121.0 ± 6.40 c	58.06 ± 5.48 c	729 ± 48.87 c	718 ± 47.40	329.2 ± 21.76 c	30.77 ± 1.54
Green × Half strength	25.97 ± 2.10 c	27.11 ± 2.55	141.4 ± 17.00 bc	56.68 ± 1.72 c	679 ± 29.54 c	728 ± 31.51	286.5 ± 6.08 c	36.12 ± 5.16
Green × Quarter strength	23.29 ± 1.67 c	38.47 ± 2.30	168.1 ± 19.08 bc	73.23 ± 4.05 b	765 ± 9.67 c	830 ± 21.59	368.8 ± 29.56 c	35.72 ± 4.97
Red × Full strength	35.76 ± 1.72 b	42.32 ± 2.52	185.1 ± 20.83 b	85.94 ± 4.01 b	704 ± 61.70 c	1077 ± 64.24	654.7 ± 27.10 b	65.83 ± 16.20
Red × Half strength	48.35 ± 3.80 a	45.27 ± 12.26	246.6 ± 15.14 a	114.8 ± 7.75 a	962 ± 64.65 b	1316 ± 73.80	803.0 ± 49.45 a	49.92 ± 1.12
Red × Quarter strength	52.73 ± 2.95 a	61.23 ± 5.24	149.8 ± 12.29 bc	118.8 ± 5.20 a	1682 ± 107.47 a	1288 ± 63.68	876.3 ± 18.27 a	58.92 ± 2.92
Significance								
Salanova (S)	***	***	**	***	***	***	***	***
Concentration (C)	ns	*	*	***	***	*	***	ns
S × C	***	ns	**	*	***	ns	**	ns

Data are mean ± standard error; $n = 6$. The symbols ns, *, **, and *** indicate a nonsignificant or a significant statistical difference at $p \leq 0.05$, 0.01, and 0.001, respectively. For each variable, letters indicate statisticallly homogenous groups according to Duncan's multiple-range test ($p = 0.05$). The effects of the factor Salanova were compared according to Student's t-test.

Salanova lettuce was found to be a rich source of flavonoids, including flavones (i.e., 121–222 and 150–264 mg/100 g eq., in green and red Salanova, respectively) and flavonols (i.e., 54–73 and 85–119 mg/100 g eq., in green and red Salanova, respectively). The less abundant phenolic classes were stilbenes, flavanols, and anthocyanins. Anthocyanins concentration ranged from 23 to 29 mg/100 g eq. and from 27 to 53 mg/100 g eq. in green and red Salanova lettuce, respectively. Cyanidin and pelargonidin derivatives were the most represented phenolics among anthocyanins.

Apart from the genotype, polyphenols content was influenced by the different growing treatments. The HS and QS reduced-nutrient solutions increased the accumulation of polyphenols compared to the full-strength solution. In particular, the highest increase was observed for lignans in red Salanova (+41%). A significant interaction between the red cultivar and the HS treatment was recorded for most of the phenolics classes (anthocyanins, flavones, flavonols, lignans, and phenolic acids). Cultivation in QS nutrient solution increased phenol concentration with respect to FS but not to HS.

Several lignan compounds could be annotated through our untargeted profiling approach, such as episesamin, conidendrin, pinoresinol, arctigenin, 7-oxomatairesinol, isohydroxymatairesinol, medioresinol, dimethylmatairesinol. Regarding phenolic acids, Salanova lettuce was mainly characterized by hydroxycinnamic acids such as chicoric acid, p-coumaric acid, caffeic acid, ferulic acid 4-*O*-glucoside. Low-molecular-weight phenolics were the most relevant phenolics in Salanova lettuce, enriched of subclasses such as alkylphenols, phenolic terpenes, hydroxybenzaldehydes, and hydroxyphenylpropenes. However, flavone compounds (like naringin 4′-*O*-glucoside, tetramethylscutellarein, apigenin 7-*O*-glucoside, luteolin 7-*O*-glucuronide) and flavonols (kaempferol 3-*O*-glucoside, quercetin 3-*O*-glucoside, myricetin 3-*O*-glucoside) were also represented.

From each supervised model (Figure 1), the most discriminating compounds were selected using the VIP approach with a threshold of >1.1 (Supplementary Table S3) followed by Veen analysis (Figure 2). The plant response to reduced nutrient concentrations appeared to be strongly influenced by the genotype, since only 28.9% of significant metabolites were common to both cultivars.

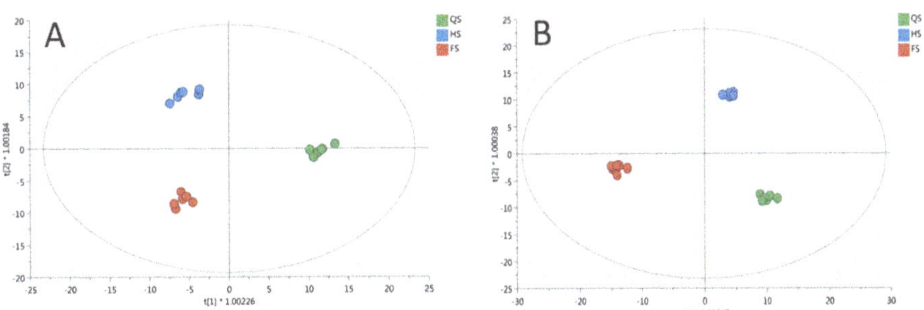

Figure 1. Orthogonal projection to latent structures discriminant analysis (OPLS-DA) built using (**A**) green and (**B**) red Salanova samples grown at reduced nutrient concentrations (full strength (FS), half strength (HS), and quarter strength (QS)).

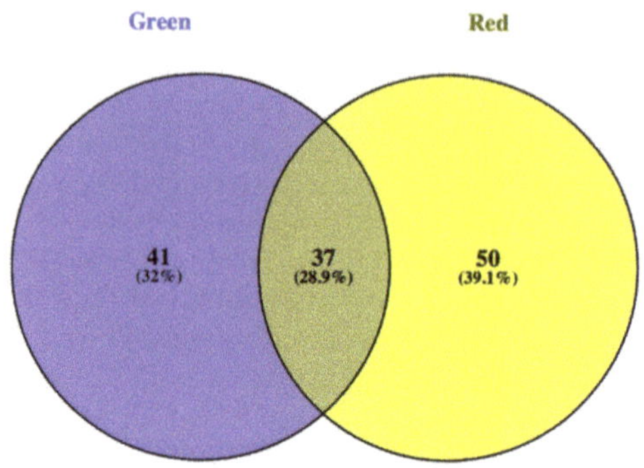

Figure 2. Venn analysis of variable importance in projection (VIP) compound (VIP > 1.1) common between red and green Salanova cultivars.

3.2. In Vitro Antioxidant Capacity

In this study, the in vitro antioxidant properties of both Salanova lettuce cultivars were assessed by different complementary methods, namely, the phosphomolybdenum assay, ABTS [2,2′-azinobis-(3-ethylbenzothiazoline-6-sulfonate)] and DPPH (2,2-diphenyl-1-picrylhydrazyl) scavenging capacity assays, CUPRAC and FRAP methods, and metal chelating activity assay (Table 2).

The half- and quarter-strength nutrient solutions increased the antioxidant activity compared to the full-strength solution. Notably, the red Salanova cultivar grown in half-strength nutrient solution showed two-fold higher values of scavenging ability, compared to the control. In particular, DPPH and ABTS assays recorded values of 32.46 mg TE/g and 41.73, respectively, whilst ferric reduction activity and cupric capacity showed values of 34.87 mg TE/g and 91.40 mg TE/g, respectively. Besides, the control sample was characterized by values of 9.53 mg TE/g and 21.14 mg TE/g for DPPH and ABTS, respectively, whilst ferric and cupric reduction activity values were 19.01 mg TE/g and 49.05 mg TE/g, respectively.

Turning to the green cultivar, different results were obtained. In fact, the quarter-strength solution reported the higher reducing power, especially when assessing cupric and ferric reduction activities (CUPRAC: 32.74 mg TE/g and FRAP: 14.45 mg TE/g), which were 1.2 times higher than those of the control FS solution (CUPRAC: 28.90 mg TE/g FRAP: 11.71 mg TE/g). Regarding the metal chelating activity, the HS solution recorded a 1.2-fold higher value (16.83 mg EDTAE/g) when compared to the FS solution (14.92 mg TE/g).

3.3. Enzyme Inhibitory Activity

In addition to the in vitro antioxidant capacity, cholinesterases (AChE and BChE), tyrosinase, α-amylase, and β-glucosidase inhibitory activities were evaluated for both red and green Salanova genotypes (Table 3). Our results revealed that the extracts showed moderate enzyme inhibitory activities, with differences as a function of the strength of the nutrient solution. In particular, the FS solution demonstrated activity against AChE (2.23 mg GALAE/g), while the HS solution was the most effective BChe inhibitor (6.16 mg GALAE/g) in red lettuce. Besides, the QS solution exhibited inhibition against glucosidase (0.81 mmol ACAE/g). Regarding green Salanova, a significant interaction between the genotype and the nutrient solution was recorded with respect to BChE. A significant increase was observed for cultivation in the HS solution compared to the FS solution, while the inhibitory activity did not increase by further decreasing the nutrients (QS solution).

Table 2. Total phenolic acids, total flavonoid content, and antioxidant activities in green and red Salanova lettuce (S) in nutrient solutions with different macrocations concentration (C). GAE, gallic acid equivalents, RE, rutin equivalents, TE, trolox equivalents, EDTAE, ethylenediaminetetraacetic acid equivalents.

Source of Variance	Total Phenolic Content (mg GAE)	Total Flavonoid Content (mg RE)	DPPH (mg TE)	ABTS (mg TE)	CUPRAC (mg TE)	FRAP (mg TE)	Phosphomolydenum (mmol TE)	Metal Chelating (mg EDTAE)
Salanova (S)								
Green	12.41 ± 0.28 b	22.17 ± 1.39 a	3.31 ± 0.86 b	12.28 ± 0.81 b	30.41 ± 1.47 b	12.73 ± 0.58 b	0.74 ± 0.02 b	15.61 ± 0.57 b
Red	22.31 ± 1.42 a	12.80 ± 1.80 b	22.64 ± 2.78 a	30.94 ± 2.34 a	69.33 ± 5.31 a	27.30 ± 2.00 a	1.11 ± 0.05 a	19.52 ± 0.54 a
Concentration (C)								
Full strength	14.87 ± 0.88 b	21.18 ± 1.57 a	5.71 ± 1.30 b	16.73 ± 1.70 c	38.98 ± 3.35 c	15.36 ± 1.17 b	0.85 ± 0.04 b	16.91 ± 0.88 b
Half strength	20.67 ± 2.75 a	14.16 ± 3.02 b	17.08 ± 5.04 a	26.45 ± 4.83 a	60.49 ± 10.23 a	23.46 ± 3.82 a	1.02 ± 0.10 a	19.36 ± 0.93 a
Quarter strength	16.55 ± 1.25 b	17.11 ± 2.04 ab	16.14 ± 3.26 a	21.64 ± 2.74 b	50.14 ± 6.05 b	21.23 ± 2.33 a	0.91 ± 0.07 b	16.42 ± 0.61 b
C × S								
Green × Full strength	12.12 ± 0.39 c	23.07 ± 0.59 a	1.89 ± 0.64 d	12.33 ± 2.10 d	28.90 ± 2.09 d	11.71 ± 0.55 d	0.79 ± 0.05 cd	14.92 ± 1.02
Green × Half strength	12.57 ± 0.57 c	23.90 ± 1.46 a	1.70 ± 0.23 d	11.18 ± 0.36 d	29.58 ± 1.95 d	12.04 ± 0.44 d	0.74 ± 0.02 cd	16.83 ± 1.01
Green × Quarter strength	12.54 ± 0.56 c	19.54 ± 3.89 ab	6.36 ± 2.07 cd	13.34 ± 1.30 d	32.74 ± 3.51 d	14.45 ± 1.42 cd	0.69 ± 0.05 d	15.08 ± 0.89
Red × Full strength	17.61 ± 0.49 b	19.30 ± 3.02 ab	9.53 ± 1.08 c	21.14 ± 0.74 c	49.05 ± 2.10 c	19.01 ± 0.58 c	0.90 ± 0.05 c	18.91 ± 0.88
Red × Half strength	28.77 ± 2.57 a	4.42 ± 0.10 c	32.46 ± 4.16 a	41.73 ± 3.00 a	91.41 ± 8.64 a	34.87 ± 3.43 a	1.30 ± 0.09 a	21.89 ± 0.47
Red × Quarter strength	20.55 ± 0.44 b	14.67 ± 0.84 b	25.92 ± 2.07 b	29.95 ± 1.95 b	67.53 ± 5.28 b	28.01 ± 1.84 b	1.12 ± 0.03 b	17.77 ± 0.36
Significance								
Salanova (S)	***	***	***	***	***	***	***	***
Concentration (C)	***	*	***	***	***	***	***	**
S × C	***	**	***	***	***	***	***	ns

Data are mean ± standard error; $n = 6$. The symbols ns, *, **, and *** indicate a nonsignificant or a significant statistical difference at $p \leq 0.05$, 0.01, and 0.001, respectively. For each variable, letters indicate statistical homogenous groups according to Duncan's multiple-range test ($p = 0.05$). The effects of factor Salanova were compared according to Student's t-test.

Table 3. Enzyme inhibitory activities in green and red Salanova lettuce (S) in nutrient solutions with different macrocations concentration (C). AChE, acetylcholinesterase, BChE, butyrylcholinesterase, GALAE, galantamine equivalents, KAE, kojic acid equivalents, ACAE, acarbose equivalents.

Source of Variance	AChE (mg GALAE)	BChE (mg GALAE)	Tyrosinase (mg KAE)	Amylase (mmol ACAE)	Glucosidase (mmol ACAE)
Salanova (S)					
Green	2.20 ± 0.14	5.58 ± 0.21 a	64.77 ± 0.90	0.35 ± 0.01	0.79 ± 0.01
Red	1.88 ± 0.21	4.55 ± 0.20 b	65.82 ± 1.37	0.35 ± 0.01	0.84 ± 0.04
Concentration (C)					
Full strength	1.98 ± 0.19	5.25 ± 0.12 a	66.62 ± 1.45	0.35 ± 0.01 ab	0.83 ± 0.02
Half strength	2.16 ± 0.25	5.36 ± 0.32 a	62.54 ± 1.47	0.37 ± 0.02 a	0.85 ± 0.04
Quarter strength	1.98 ± 0.20	4.59 ± 0.34 b	66.72 ± 1.00	0.33 ± 0.01 b	0.78 ± 0.04
C × S					
Green × Full strength	2.23 ± 0.12	5.30 ± 0.19 ab	63.08 ± 1.29	0.35 ± 0.02 b	0.79 ± 0.02
Green × Half strength	2.55 ± 0.05	6.16 ± 0.40 a	63.85 ± 0.65	0.34 ± 0.01 b	0.78 ± 0.02
Green × Quarter strength	1.79 ± 0.39	5.27 ± 0.36 ab	67.38 ± 2.00	0.35 ± 0.02 b	0.81 ± 0.02
Red × Full strength	1.23 ± 0.04	5.20 ± 0.15 ab	70.17 ± 1.59	0.35 ± 0.01 b	0.86 ± 0.02
Red × Half strength	1.90 ± 0.40	4.56 ± 0.18 bc	61.23 ± 2.90	0.40 ± 0.02 a	0.92 ± 0.07
Red × Quarter strength	2.18 ± 0.04	3.90 ± 0.44 c	66.06 ± 0.46	0.31 ± 0.01 b	0.74 ± 0.09
Significance					
Salanova (S)	ns	***	ns	ns	ns
Concentration (C)	ns	**	ns	***	ns
S × C	ns	*	ns	***	ns

All data are expressed as mean ± s.e; $n = 6$. The symbol ns, *, ** and *** indicate a nonsignificant or a significant statistical difference at $p \leq 0.05$, 0.01, and 0.001, respectively. For each variable, letters indicate statistical homogenous groups according to Duncan's multiple-range test ($p = 0.05$). The factor Salavanova was compared according to Student's *t*-test.

Regarding the red-pigmented lettuce, the HS solution was the most effective AChE inhibitor (1.90 mg GALAE/g), while the full-strength nutrient solution revealed higher potential against butyrylcholinesterase (5.20 mg GALAE/g) and tyrosinase, recording a value of 70.17 mg KAE/g, i.e., 10-fold higher when compared with the other activities. In addition, a significant interaction between the two experimental factors was recorded with respect to amylase for the HS treatment.

3.4. Pearson's Correlation Analysis

Aiming to inspect the contributions of each different class of phenolics to the biological activities we measured, Pearson's correlations coefficients were investigated (Supplementary Table S4). The most significant correlations ($p < 0.01$) were recorded between flavonols, flavonols, and phenolic acids and both DPPH values (0.74, 0.61, and 0.88 respectively) and ABTS values (0.68, 0.59, and 0.73). In addition, flavonols were correlated with the FRAP assay recorded value of 0.62. Regarding the in vitro enzymatic inhibition, we found a correlation between tyrosinase activity and flavonols (0.81) and phenolic acids (0.69) and a significant ($p < 0.01$) negative correlation between stilbenes and glucosidase activity (−0.62). A lack of correlation between polyphenols content and acethylcholinesterase inhibition and BChE activity was observed. Overall, more interesting results were found considering the red-pigmented lettuce cultivar. In fact, strong correlations ($p < 0.01$) were outlined between phenolic acids and tyrosols by the DPPH, ABTS, CUPRAC, and FRAP assays, and between flavones and metal chelating activity (0.65). Regarding flavonols, a correlation was outlined by DPPH (0.70) and ABTS (0.59) assays, thus confirming the role of polyphenols as the main contributors to the antioxidant properties of this plant food. Finally, no correlations between enzyme activity and polyphenols were detected, except for a negative correlation between butyrylcholinesterase and lignans (0.74).

4. Discussion

Lettuce is one of the most important salad vegetables, known as a rich source of vitamins, polyphenols, and antioxidant compounds. The most represented antioxidant compounds are polyphenols, whose concentration varies depending on environmental and genetic factors.

Because of the high content of bioactive compounds, including polyphenols, vegetable-rich diets have been associated to a low risk of chronic diseases, thus entailing strategies aimed at enriching the functional content of vegetables. Hydroponic cultivation has emerged as a promising tool to produce vegetables in a more sustainable and economically valuable manner [40–42]. It is noteworthy that hydroponic cultivation offers the possibility to precisely manage crop nutrients availability, hence opening the possibility to modulate the actual functional profile of the produce. In this study, two Salanova cultivars (i.e., green and red) and three levels of nutrients in solution were considered, aiming to assess the impact of nutrient concentration and composition on lettuce growth and bioactive compounds production, as a source of functional foods.

It is well known that most functional compounds, being plant secondary metabolites, result from the interaction between a genetic background and the environment [43–45]. Despite using two rather related cultivars, our results from the semi-quantitative analysis of polyphenols profile highlighted the strong influence of the genetic background on the phenolic profile. In particular, the red genotype showed a higher phenolic content compared to the green one. In general, the red genotype resulted more responsive to the applied treatments, in agreement with previous studies [27,32]. Nonetheless, each genotype exhibited a distinctive response to the decrease in nutrients strength, with the red cultivar exhibiting a strength-dependent increase in phenolic acids, lignans, and flavanols. Previously reported literature [46] suggested that the phenolic content is significantly influenced by cultivars, in addition to environmental stressor, which is in accordance with our findings. Polyphenols are abundant micronutrients in our diet, and evidence for their role in the prevention of degenerative diseases such as cancer and cardiovascular diseases is emerging [47]. However, their health-promoting effects depend on the amount consumed as well as on their bioavailability. Interestingly, we did not observe a generalized stress-induced induction, but metabolomic profiling indicated that the amounts of specific phenolic classes (lignans, flavones, anthocyanins, and phenolic acids, Table 1) increased. These low-molecular-weight phenolics have shown high bioaccessibility in simulated gastrointestinal digestion. In particular, plant lignans like sesamin are rapidily absorbed and are detected in the systemic circulation within a few hours from ingestion [48,49]. Furthermore, specific active metabolites are produced from lignans degradation by the colonic microflora, namely, the enterolignans enterolactone and enterodiol, and may have either agonistic or antagonistic effects on estrogens [50,51]. The content of this class of phytoestrogens was strongly increased under nutritional stress in red Salanova, in a stress-dependent manner. Therefore, this nutritional chemical eustress potentially represents a valuable tool to elicit the plant's metabolic responses leading to a higher accumulation of lignans. Apart from their role in ameliorating menopausal symptoms and their consequences [52], lignans guard against the accumulation of reactive oxygen species, whose overproduction can damage cellular constituents, and play a role in the pathogenesis of different disorders [53]. Indeed, the results we obtained from the radical scavenging assays indicated that the red Salanova HS extracts possessed a significant free-radical scavenging activity, with the highest significant correlations recorded between lignans and DPPH.

Cultivation of red Salanova in HS nutrient solutions also triggered the accumulation of flavonols, with kaempferol-glucoside, quercetin-glucoside, and myricetin-glucoside detected as the major metabolites. These metabolites are known to provide a strong antioxidant effect and can increase superoxide dismutase, catalase, and glutathione peroxidase activities. Accordingly, we outlined a positive correlation between flavonols and free-radical scavenging activity. According to the literature, the antioxidant activity of flavonoids is mainly related to their ability to act as hydrogen donors and efficient scavengers of free radicals from lipid peroxidation [54]. Furthermore [55,56], the absorption of these glycosides has been reported to occur in the small intestine, and its efficiency is higher than that for the aglycone itself.

The low concentration of anthocyanins we found in lettuce (including the red variety) might indicate a major contribution of lipophilic pigments, such as carotenoids, in leaf pigmentation. Indeed, we previously reported β-cryptoxanthin, violaxanthin, neoxanthin, lutein, and β-carotene as major carotenoids in red Salanova lettuce (ranging 614–1011 μg/g dw) compared to the non-pigmented genotype (ranging 289–444 μg/g dw) [27,32]. In our experimental conditions, the red genotype showed higher in vitro antioxidant activity, likely due to the elicitation of bioactive secondary metabolites like polyphenols, ascorbic acid, and caffeic acid derivatives, as previously investigated [27].

The half-strength nutrient solution was also found to increase α-amylase-inhibiting properties, compared to the full-strength solution, in red Salanova. This enzyme is involved in the digestion of carbohydrates, and its inhibition may represent a strategy to lower the levels of postprandial glycemia via the control of starch breakdown [57,58]. A previous study highlighted the role of the lettuce carotenoid lactucaxanthin as an α-amylase and α-glucosidase inhibitor [59], suggesting its nutritional relevance in the treatment of type 2 diabetes and obesity. On the other hand, both Salanova genotypes, irrespective of the nutritional treatment, showed moderate tyrosinase-inhibiting properties. This key enzyme catalyzes browning and melanin synthesis and has been proposed to manage hyperpigmentation [60].

Although the cultivars chosen were related to one another, their phenolic signatures were distinctive, with phenolic classes differing between the two cultivars ($p < 0.01$). Such genotype-related differences were maintained in response to nutritional stress, with elicitation of specific phenolic classes in each cultivar. The phenolic profile of the red genotype showed significant modulation for anthocyanins, lignans, flavonols, and phenolic acids ($p < 0.001$) in a stress-dependent manner. This is expected if we consider that, under the term phenolics, we include a rather broad diversity of compounds with only a partial sharing of their biosynthetic pathways. Consistently with our metabolomic profiling, colorimetric assays (total phenolics and in vitro antioxidant capacity) and amylase inhibition were significantly affected by the "genotype x nutrients strength" interaction ($p < 0.01$). Pearson's correlations evidenced that total phenolic content justified a large part of the antioxidant capacity, both in terms of radical scavenging and reducing power assays (correlations 0.79–0.96) and in terms of amylase inhibition (correlation 0.87). Nonetheless, individual phenolic classes showed a comparatively much lower degree of correlation to antioxidant capacity.

It is worthy to note that green Salanova presented some interesting features as well, having the highest content of total flavonoids and the highest inhibitory activity against BChE. Such features were cultivar-dependent and did not show a relationship with the strength of nutrient solution, except for an increase in BChE inhibition in half-strength nutrient solution. The inhibition of AChE and BChE is considered one of the possible therapeutic strategy against neurodegenerative disorders such as Alzheimer's disease [61], senile dementia, and ataxia. Recently, secondary metabolites from plants such as flavonoids and terpenoids have been proven to possess AChE and tyrosinase inhibitory activities [62–66]. Accordingly, green Salanova displayed higher total flavonoid content compared to the pigmented type, suggesting the putative role of these compounds in modulating cognitive functions.

Considering the mechanisms underlying the shaping of phenolic compounds in plants, it is known that vegetative growth is mainly supported under favorable conditions, while secondary metabolism receives metabolic allocation priority in resource-limited environments [63]. As an example, phenylalanine (a rate-limiting precursor of all phenylpropanoids) is also an essential amino acid for protein synthesis. Notwithstanding, the carbon-nitrogen balance hypothesis [64] suggests that carbon-based secondary metabolites (including phenolics) inversely correlate with nitrogen availability, whereas nitrogen-containing secondary metabolites (e.g., alkaloids) typically show a direct correlation. As reviewed by Heimler et al. [63], the increase in phenolics under low-nitrogen conditions is not generalized and involves specific classes of phenolics. Besides confirming this last point, here we show that nutrients availability modulates phenolics content (in turn modulating functional properties related to phenolics) in a genotype-dependent manner.

5. Conclusions

Our results indicate that by reducing the content of macroelements in the nutrient solution, we could significantly modulate the phenolic profile of our lettuce and, therefore, also its antioxidant capacity and amylase inhibition properties. It is noteworthy that the red cultivar was much more responsive to nutritional deprivation, compared to green Salanova lettuce. Furthermore, it is important to notice that only specific classes of phenolics (including lignans, flavones, anthocyanins, and phenolic acids) were shaped by the induction of nutritional eustress, indicating that specific responses should be considered, rather than focusing on a generalized modulation of phenolic compounds' production. The modulation of nutritional strength did not significantly affect cholinesterase and tyrosinase inhibition; on the other hand, such activities depend on several other phytochemicals in addition to phenolic compounds.

The application of a controlled nutritional eustress in hydroponically cultivated lettuce may represent a valuable strategy to produce food with tailored functional features. Interestingly, the approach proposed can also meet both the current need for an increased sustainability of agricultural production and the demand for healthy foods at consumer's level.

Supplementary Materials: The following are available online at http://www.mdpi.com/2304-8158/9/9/1156/s1, Table S1: Whole dataset produced from untargeted metabolomics analysis carried out in red and green Salanova subjected to different concentrations of nutrient solution. Compounds are presented with individual intensities and with composite mass spectra. Table S2: List of metabolites confirmed with their fragmentation pattern. Table S3: Statistically significant metabolites (VIP > 1.1) common to both green and red Salanova subjected to different concentrations of nutrient solution. Table S4: Correlation matrix (Pearson) showing correlation coefficients computed for the polyphenol classes, antioxidant capacity, and enzyme activity inhibition. Figure S1: Supervised model validation and outliers check. Permutation test outcome and analysis of outliers for A) green and B) red Salanova subjected to different concentrations of nutrient solution.

Author Contributions: Conceptualization, L.L., G.C., and Y.R.; methodology, B.M.-M., G.Z., L.L., M.I.S., C.E.-N., and Y.R.; formal analysis, B.S., G.A., and L.Z.; data curation, B.M.-M., L.L., L.R., M.I.S., C.E.-N., Y.R., and G.C.; writing—original draft preparation, B.S., L.R., and L.Z.; writing—review and editing, G.Z., R.B., L.L, G.C., and Y.R.; supervision, L.L. and Y.R. All authors have read and agreed to the published version of the manuscript.

Funding: This research received no external funding.

Acknowledgments: The authors acknowledge the "Fondazione Romeo and Enrica Invernizzi", Milan, for the generous support of PRONUTRIGEN Research Center equipment and activities.

Conflicts of Interest: The authors declare no conflict of interest.

References

1. Fageria, N.K.; Baligar, V.C.; Jones, C.A. *Growth and Mineral Nutrition of Field Crops*; CRC Press: Boca Raton, FL, USA, 2010.
2. Marschner, H. *Marschner's Mineral Nutrition of Higher Plants*; Academic Press: Waltham, MA, USA, 2011.
3. White, P.; Brown, P. Plant nutrition for sustainable development and global health. *Ann. Bot.* **2010**, *105*, 1073–1080. [CrossRef] [PubMed]
4. Hall, R.D. Plant metabolomics in a nutshell: Potential and future challenges. *Annu. Plant. Rev. Online* **2018**, *43*, 1–24. [CrossRef]
5. Nikiforova, V.J.; Kopka, J.; Tolstikov, V.; Fiehn, O.; Hopkins, L.; Hawkesford, M.J.; Hesse, H.; Hoefgen, R. Systems rebalancing of metabolism in response to sulfur deprivation, as revealed by metabolome analysis of arabidopsis plants. *Plant. Physiol.* **2005**, *138*, 304–318. [CrossRef]
6. Takahashi, H.; Imamura, T.; Miyagi, A.; Uchimiya, H. Comparative metabolomics of developmental alterations caused by mineral deficiency during in vitro culture of gentiana triflora. *Metabolomics* **2012**, *8*, 154–163. [CrossRef]
7. Ding, Z.; Jia, S.; Wang, Y.; Xiao, J.; Zhang, Y. Phosphate stresses affect ionome and metabolome in tea plants. *Plant. Physiol. Biochem.* **2017**, *120*, 30–39. [CrossRef]
8. Schauer, N.; Fernie, A.R. Plant metabolomics: Towards biological function and mechanism. *Trends Plant. Sci.* **2006**, *11*, 508–516. [CrossRef]
9. Rouached, H.; Rhee, S.Y. System-level understanding of plant mineral nutrition in the big data era. *Curr. Opin. Syst. Biol.* **2017**, *4*, 71–77. [CrossRef]
10. Maucieri, C.; Nicoletto, C.; Os, E.V.; Anseeuw, D.; Havermaet, R.V.; Junge, R. Hydroponic technologies. In *Aquaponics Food Production Systems: Combined Aquaculture and Hydroponic Production Technologies for the Future*; Goddek, S., Joyce, A., Kotzen, B., Burnell, G.M., Eds.; Springer International Publishing: Cham, Switzerland, 2019; pp. 77–110. [CrossRef]
11. Epstein, E. *Mineral Nutrition of Plants: Principles and Perspectives*; CABI Publishing: Wallingford Oxfordshire, UK, 1972.
12. Jones, J.B., Jr. *Hydroponics: A practical Guide for the Soilless Grower*; CRC Press: Boca Raton, FL, USA, 2016.
13. López-Millán, A.F.; Sagardoy, R.; Solanas, M.; Abadía, A.; Abadía, J. Cadmium toxicity in tomato (*Lycopersicon esculentum*) plants grown in hydroponics. *Environ. Exp. Bot.* **2009**, *65*, 376–385. [CrossRef]
14. Chatzigianni, M.; Ntatsi, G.; Theodorou, M.; Stamatakis, A.; Livieratos, I.; Rouphael, Y.; Savvas, D. Functional quality, mineral composition and biomass production in hydroponic spiny chicory (*Cichorium spinosum* L.) are modulated interactively by ecotype, salinity and nitrogen supply. *Front. Plant Sci.* **2019**, *10*, 1040. [CrossRef]
15. Sagardoy, R.; Morales, F.; López-Millán, A.F.; Abadía, A.; Abadía, J. Effects of zinc toxicity on sugar beet (*Beta vulgaris* l.) plants grown in hydroponics. *Plant. Biol.* **2009**, *11*, 339–350. [CrossRef] [PubMed]
16. Walters, K.J.; Behe, B.K.; Currey, C.J.; Lopez, R.G. Historical, current, and future perspectives for controlled environment hydroponic food crop production in the united states. *HortScience* **2020**, *55*, 758–767. [CrossRef]
17. Taiz, L.; Zeiger, E. *Plant Physiology*; Sinauer Associates: Sunderland, MA, USA, 2010.
18. Sanchez, D.H.; Siahpoosh, M.R.; Roessner, U.; Udvardi, M.; Kopka, J. Plant metabolomics reveals conserved and divergent metabolic responses to salinity. *Physiol. Plant.* **2008**, *132*, 209–219. [CrossRef] [PubMed]
19. Baron, D.; Amaro, A.C.; Campos, F.G.; Boaro, C.S.; Ferreira, G. Plant physiological responses to nutrient solution: An overview. In *Plant Metabolites and Regulation Under Environmental Stress*; Elsevier: Amsterdam, The Netherlands, 2018; pp. 415–425.
20. Kumar, P.; Saini, S. Nutrients for hydroponic systems in fruit crops. In *Hydrocultural and Hydroponics Systems*; IntechOpen: London, UK, 2020. [CrossRef]
21. Tyler, G.; Olsson, T. Concentrations of 60 elements in the soil solution as related to the soil acidity. *Eur. J. Soil Sci.* **2001**, *52*, 151–165. [CrossRef]

22. Rouphael, Y.; Kyriacou, M.C.; Carillo, P.; Pizzolongo, F.; Romano, R.; Sifola, M.I. Chemical eustress elicits tailored responses and enhances the functional quality of novel food perilla frutescens. *Molecules* **2019**, *24*, 185. [CrossRef]
23. Pannico, A.; El Nakhel, C.; Kyriacou, M.C.; Giordano, M.; Stazi, S.R.; De Pascale, S.; Rouphael, Y. Combating micronutrient deficiency and enhancing food functional quality through selenium fortification of select lettuce genotypes grown in a closed soilless system. *Front. Plant Sci.* **2019**, *10*, 1495. [CrossRef]
24. Rouphael, Y.; Kyriacou, M.C. Enhancing quality of fresh vegetables through salinity eustress and biofortification applications facilitated by soilless cultivation. *Front. Plant Sci.* **2018**, *9*, 1254. [CrossRef]
25. Mulabagal, V.; Ngouajio, M.; Nair, A.; Zhang, Y.; Gottumukkala, A.L.; Nair, M.G. In vitro evaluation of red and green lettuce (*Lactuca sativa*) for functional food properties. *Food Chem.* **2010**, *118*, 300–306. [CrossRef]
26. Kim, M.J.; Moon, Y.; Tou, J.C.; Mou, B.; Waterland, N.L. Nutritional value, bioactive compounds and health benefits of lettuce (*Lactuca sativa* L.). *J. Food Compos. Anal.* **2016**, *49*, 19–34. [CrossRef]
27. Jurić, S.; Sopko Stracenski, K.; Król-Kilińska, Ż.; Žutić, I.; Uher, S.F.; Đermić, E.; Topolovec-Pintarić, S.; Vinceković, M. The enhancement of plant secondary metabolites content in *Lactuca sativa* L. by encapsulated bioactive agents. *Sci. Rep.* **2020**, *10*, 1–12. [CrossRef]
28. El-Nakhel, C.; Pannico, A.; Kyriacou, M.C.; Giordano, M.; De Pascale, S.; Rouphael, Y. Macronutrient deprivation eustress elicits differential secondary metabolites in red and green-pigmented butterhead lettuce grown in a closed soilless system. *J. Sci. Food Agric.* **2019**, *99*, 6962–6972. [CrossRef]
29. Złotek, U.; Świeca, M.; Jakubczyk, A. Effect of abiotic elicitation on main health-promoting compounds, antioxidant activity and commercial quality of butter lettuce (*Lactuca sativa* L.). *Food Chem.* **2014**, *148*, 253–260. [CrossRef] [PubMed]
30. Chen, X.; Gou, W.; Xue, X.; Wang, L.; Qiao, X. Growth and quality responses of "Green Oak Leaf" lettuce as affected by monochromic or mixed radiation provided by fluorescent lamp (FL) and light-emitting diode (LED). *Sci. Hortic.* **2014**, *172*, 168–175. [CrossRef]
31. Oh, M.M.; Carey, E.E.; Rajashekar, C.B. Environmental stresses induce health-promoting phytochemicals in lettuce. *Plant Physiol. Biochem.* **2009**, *47*, 578–583. [CrossRef] [PubMed]
32. El-Nakhel, C.; Petropoulos, S.A.; Pannico, A.; Kyriacou, M.C.; Giordano, M.; Colla, G.; Troise, A.D.; Vitaglione, P.; De Pascale, S.; Rouphael, Y. The bioactive profile of lettuce produced in a closed soilless system as configured by combinatorial effects of genotype and macrocation supply composition. *Food Chem.* **2020**, *309*, 125713. [CrossRef] [PubMed]
33. Reyes-Chin-Wo, S.; Wang, Z.; Yang, X.; Kozik, A.; Arikit, S.; Song, C.; Xia, L.; Froenicke, L.; Lavelle, D.O.; Truco, M.-J. Genome assembly with in vitro proximity ligation data and whole-genome triplication in lettuce. *Nat. Commun.* **2017**, *8*, 1–11. [CrossRef]
34. Zengin, G.; Karanfil, A.; Uren, M.C.; Kocak, M.S.; Sarikurkcu, C.; Gungor, H.; Nancy Picot, C.M.; Mahomoodally, M.F. Phenolic content, antioxidant and enzyme inhibitory capacity of two: Trametes species. *RSC Adv.* **2016**, *6*, 73351–73357. [CrossRef]
35. Rocchetti, G.; Chiodelli, G.; Giuberti, G.; Lucini, L. Bioaccessibility of phenolic compounds following in vitro large intestine fermentation of nuts for human consumption. *Food Chem.* **2018**, *245*, 633–640. [CrossRef]
36. Rothwell, J.A.; Perez-Jimenez, J.; Neveu, V.; Medina-Remón, A.; M'Hiri, N.; García-Lobato, P.; Manach, C.; Knox, C.; Eisner, R.; Wishart, D.S.; et al. Phenol-Explorer 3.0: A major update of the Phenol-Explorer database to incorporate data on the effects of food processing on polyphenol content. *Database* **2013**, 1–8. [CrossRef]
37. Salek, R.M.; Steinbeck, C.; Viant, M.R.; Goodacre, R.; Dunn, W.B. The role of reporting standards for metabolite annotation and identification in metabolomic studies. *Gigascience* **2013**, *2*, 2–4. [CrossRef]
38. Tsugawa, H.; Cajka, T.; Kind, T.; Ma, Y.; Higgins, B.; Ikeda, K.; Kanazawa, M.; Vandergheynst, J.; Fiehn, O.; Arita, M. MS-DIAL: Data-independent MS/MS deconvolution for comprehensive metabolome analysis. *Nat. Methods* **2015**, *12*, 523–526. [CrossRef]
39. Rocchetti, G.; Lucini, L.; Rodriguez, J.M.L.; Barba, F.J.; Giuberti, G. Gluten-free flours from cereals, pseudocereals and legumes: Phenolic fingerprints and in vitro antioxidant properties. *Food Chem.* **2019**, *271*, 157–164. [CrossRef] [PubMed]
40. Sambo, P.; Nicoletto, C.; Giro, A.; Pii, Y.; Valentinuzzi, F.; Mimmo, T.; Lugli, P.; Orzes, G.; Mazzetto, F.; Astolfi, S.; et al. Hydroponic solutions for soilless production systems: Issues and opportunities in a smart agriculture perspective. *Front. Plant Sci.* **2019**, *10*, 923. [CrossRef]

41. Savvas, D. Hydroponics: A modern technology supporting the application of integrated crop management in greenhouse. *Food Agric. Environ.* **2003**, *1*, 80–86.
42. Gruda, N. Do soilless culture systems have an influence on product quality of vegetables. *J. Appl. Bot. Food Qual.* **2009**, *82*, 141–147.
43. Krauss, S.; Schnitzler, W.H.; Grassman, J.; Woitke, M. The influence of different electrical conductivity values in a simplified recirculating soilless system on inner and outer fruit quality characteristics of tomato. *J. Agric. Food Chem.* **2006**, *54*, 441–448. [CrossRef]
44. Rouphael, Y.; Kyriacou, M.C.; Petropoulos, S.A.; De Pascale, S.; Colla, G. Improving vegetable quality in controlled environments. *Sci. Hort.* **2018**, *234*, 275–289. [CrossRef]
45. Fallovo, C.; Rouphael, Y.; Rea, E.; Battistelli, A.; Colla, G. Nutrient solution concentration and growing season affect yield and quality of *Lactuca sativa* L. var. acephala in floating raft culture. *J. Sci. Food Agric.* **2009**, *89*, 1682–1689. [CrossRef]
46. Grochowski, D.M.; Uysal, S.; Aktumsek, A.; Granica, S.; Zengin, G.; Ceylan, R.; Locatelli, M.; Tomczyk, M. In vitro enzyme inhibitory properties, antioxidant activities, and phytochemical profile of Potentilla thuringiaca. *Phytochem. Lett.* **2017**, *20*, 365–372. [CrossRef]
47. Lucini, L.; Rocchetti, G.; Trevisan, M. Extending the concept of terroir from grapes to other agricultural commodities: An overview. *Curr. Opin. Food Sci.* **2020**, *31*, 88–95. [CrossRef]
48. Manach, C.; Scalbert, A.; Morand, C.; Rémésy, C.; Jiménez, L. Polyphenols: Food Sources Bioavailab. *Am. J. Clin. Nutr.* **2004**, *79*, 727–747. [CrossRef]
49. Bolvig, A.K.; Adlercreutz, H.; Theil, P.K.; Jorgensen, H.; Knudsen, K.E.B. Absorption of plant lignans from cereals in an experimental pig model. *Br. J. Nutr.* **2016**, *115*, 1711–1720. [CrossRef] [PubMed]
50. Kuijsten, A.; Arts, I.C.W.; Van't Veer, P.; Hollman, P.C.H. The relative bioavailability of enterolignans in humans is enhanced by milling and crushing of flaxseed. *J. Nutr.* **2005**, *135*, 2812–2816. [CrossRef] [PubMed]
51. Setchell, K.D.R.; Borriello, S.P.; Gordon, H.; Lawson, A.M.; Harkness, R.; Morgan, D.M.L.; Kirk, D.N.; Adlercreutz, H.; Anderson, L.C.; Axelson, M. Lignan formation in man-microbial involvement and possible roles in relation to cancer. *Lancet* **1981**, *318*, 4–7. [CrossRef]
52. Mousavi, Y.; Adlercreutz, H. Enterolactone and estradiol inhibit each other's proliferative effect on MCF-7 breast cancer cells in culture. *J. Steroid Biochem. Mol. Biol.* **1992**, *41*, 615–619. [CrossRef]
53. Sammartino, A.; Tommaselli, G.A.; Gargano, V.; Di Carlo, C.; Attianese, W.; Nappi, C. Short-term effects of a combination of isoflavones, lignans and Cimicifuga racemosa on climacteric-related symptoms in postmenopausal women: A double-blind, randomized, placebo-controlled trial. Gynecol. *Endocrinolgy* **2006**, *22*, 646–650. [CrossRef]
54. Jiang, K.; Song, Q.Y.; Peng, S.J.; Zhao, Q.Q.; Li, G.-D.; Li, Y.; Gao, K. New lignans from the roots of Schisandra sphenanthera. *Fitoterapia* **2015**, *103*, 63–70. [CrossRef] [PubMed]
55. Sharififar, F.; Dehghn-Nudeh, G.; Mirtajaldini, M. Major flavonoids with antioxidant activity from *Teucrium polium* L. *Food Chem.* **2009**, *112*, 885–888. [CrossRef]
56. Martijn, P. Absorption of dietary quercetin healthy ileostomy volunteers13 glycosides. *Am. J. Clin. Nutr.* **2018**, 1276–1282. [CrossRef]
57. Morand, C.; Manach, C.; Crespy, V.; Remesy, C. Quercetin 3-o-β-glucoside is better absorbed than other quercetin forms and is not present in rat plasma. *Free Radic. Res.* **2000**, *33*, 667–676. [CrossRef]
58. Villanueva-Suárez, M.J.; Mateos-Aparicio, I.; Pérez-Cózar, M.L.; Yokoyama, W.; Redondo-Cuenca, A. Hypolipidemic effects of dietary fibre from an artichoke by-product in Syrian hamsters. *J. Funct. Foods* **2019**, *56*, 156–162. [CrossRef]
59. Giuberti, G.; Rocchetti, G.; Lucini, L. Interactions between phenolic compounds, amylolytic enzymes and starch: An updated overview. *Curr. Opin. Food Sci.* **2020**. [CrossRef]
60. Gopal, S.S.; Lakshmi, M.J.; Sharavana, G.; Sathaiah, G.; Sreerama, Y.N.; Baskaran, V. Lactucaxanthin-a potential anti-diabetic carotenoid from lettuce (*Lactuca sativa*) inhibits α-amylase and α-glucosidase activity in vitro and in diabetic rats. *Food Funct.* **2017**, *8*, 1124–1131. [CrossRef] [PubMed]
61. Cui, H.X.; Duan, F.F.; Jia, S.S.; Cheng, F.R.; Yuan, K. Antioxidant and tyrosinase inhibitory activities of seed oils from torreya grandis Fort. ex Lindl. *Biomed. Res. Int.* **2018**. [CrossRef] [PubMed]
62. Willett, W.; Rockström, J.; Loken, B.; Springmann, M.; Lang, T.; Vermeulen, S.; Garnett, T.; Tilman, D.; DeClerck, F.; Wood, A. Food in the Anthropocene: The EAT–Lancet Commission on healthy diets from sustainable food systems. *Lancet* **2019**, *393*, 447–492. [CrossRef]

63. Heimler, D.; Romani, A.; Ieri, F. Plant polyphenol content, soil fertilization and agricultural management: A review. *Eur. Food Res. Technol.* **2017**, *243*, 1107–1115. [CrossRef]
64. Nguyen, P.M.; Niemeyer, E.D. Effects of nitrogen fertilization on the phenolic composition and antioxidant properties of basil (*Ocinum basilicum* L.). *J. Agric. Food Chem.* **2008**, *56*, 8685–8691. [CrossRef]
65. Erdogan Orhan, I.; Senol, F.S.; Ozturk, N.; Celik, S.A.; Pulur, A.; Kan, Y. Phytochemical contents and enzyme inhibitory and antioxidant properties of *Anethum graveolens* L. (dill) samples cultivated under organic and conventional agricultural conditions. *Food Chem. Toxicol.* **2013**, *59*, 96–103. [CrossRef]
66. Muñoz, E.; Avila, J.G.; Alarcón, J.; Kubo, I.; Werner, E.; Céspedes, C.L. Tyrosinase inhibitors from calceolaria integrifolia s.l.: Calceolaria talcana aerial parts. *J. Agric. Food Chem.* **2013**, *61*, 4336–4343. [CrossRef]

© 2020 by the authors. Licensee MDPI, Basel, Switzerland. This article is an open access article distributed under the terms and conditions of the Creative Commons Attribution (CC BY) license (http://creativecommons.org/licenses/by/4.0/).

foods

MDPI

Article

Accumulation of Phenolic Acids during Storage over Differently Handled Fresh Carrots

Jarkko Hellström [1], Daniel Granato [2] and Pirjo H. Mattila [3,*]

[1] Production Technologies, Natural Resources Institute Finland (Luke), Myllytie 1, FI-31600 Jokioinen, Finland; jarkko.hellstrom@luke.fi

[2] Production Technologies, Natural Resources Institute Finland (Luke), Tietotie 2, FI-02150 Espoo, Finland; daniel.granato@luke.fi

[3] Production Technologies, Natural Resources Institute Finland (Luke), Itäinen Rantakatu 4, FI-20520 Turku, Finland

* Correspondence: pirjo.mattila@luke.fi; Tel.: +358-40-7057708

Received: 30 September 2020; Accepted: 18 October 2020; Published: 21 October 2020

Abstract: Carrots contain a significant content of phenolic compounds, mainly phenolic acids. Technological processing of carrots inflicts wounding stress and induces accumulation of these compounds, especially caffeic acid derivatives, in the periderm tissue. In this study, the effect of minimal processing (polishing, washing, peeling, and grating) on the retention of soluble phenolic acids in carrots was monitored during cold storage. Storage for up to 4 weeks and 24 h was used for whole and grated carrot samples, respectively. Total phenolic acid levels found in differently processed carrots varied greatly at the beginning of the storage period and on dry weight basis they ranged from 228 ± 67.9 mg/kg (grated carrot) to 996 ± 177 mg/kg (machine washed). In each case, processing followed by storage induced phenolic acid accumulation in the carrots. At the end of the experiment (4 weeks at +8 °C), untreated and machine-washed carrots contained ca. 4-fold more phenolic acids than at day 0. Similarly, polished carrots contained 9-fold and peeled carrots 31-fold more phenolic acids than at day 0. The phenolic acid content in grated carrot doubled after 24 h storage at +4 °C. Individual phenolic acids were characterized by high resolution mass spectrometry. MS data strongly suggest the presence of daucic acid conjugates of phenolic acids in carrot. Storage time did not have statistically similar effect on all compounds and generally in a way that dicaffeoyldaucic acid had the highest increase. This research provides important information for primary production, packaging, catering, the fresh-cut industry and consumers regarding the selection of healthier minimally processed carrots.

Keywords: phenolic acids; food processing; minimally processed foods; UHLPC-MS/MS

1. Introduction

It is well known that fruits, berries and vegetables are important components of a healthy diet. Vegetables contain several phenolic compounds which are recognized as antioxidants and possess many beneficial health effects, for example, in reduction of the risk of cardiovascular diseases, cancers, neurodegenerative diseases, diabetes and osteoporosis [1–3]. In foods, phenolic compounds may contribute to bitterness, astringency, color, flavor and oxidative stability of products [4]. Carrots are among the richest vegetable sources of phenolic acids. The main phenolic acid aglycone in carrots is caffeic acid, of which the share is over 70% of all phenolic acid aglycones. Caffeic acid occurs in carrot in the soluble forms of caffeoylquinic acids. Other phenolic acid aglycones in carrots are *p*-OH-benzoic, ferulic, vanillic and *p*-coumaric acids [5].

Due to changes in consumer behavior, the demand for minimally processed fruit and vegetable products has increased dramatically in recent years. Fresh carrots are currently available as untreated, washed, polished, peeled, grated and fresh cut. Due to the widespread diversity of the fresh carrot products there is a need to understand the retention of phytochemicals during processing and storage [6].

Several research groups have stated that the content of phenolics and the antioxidative capacity of carrots subjected to minimal processing could increase during storage [7–16]. Most of these studies have followed the concentration of only a few or one phenolic acid or even the total phenolic content [17]. Becerra-Moreno et al. [9] studied the effect of glyphosate on the accumulation on the individual phenolic acids in wounded (shredded) carrots, while Klaiber et al. [13] studied influence of washing treatment on individual phenolic acid contents of carrot sticks. Simões et al. [16] studied the effect of storage of carrots under O_2 and CO_2 levels on the individual phenolic acid contents and Viacava et al. [18] studied the effect of wounding stress and extrusion on the free and bound phenolic profiles. However, it seems that there is still no consensus on the profile of phenolic acids in carrot, and more data on the effect of various commercial and catering processes on the content of individual phenolic acids are also needed.

In fact, the accumulation of phenolic compounds in carrot has been found to be linked with the wound-induced activation of phenylalanine ammonia-lyase (PAL) synthesis with the purpose to repair the wounding damage and to prevent invasion by pathogens [12]. These processes could occur between a few minutes to several hours after wounding [19]. This mechanism is dependent on several factors, including wounding intensity, initial levels, cut methods and temperature, for example [1,14,20]. Currently, polishing is one of the most used methods to process fresh carrots for the retail markets. The polishing system includes washing and mild brushing, and it is designed to remove the surface membrane (periderm) from carrots. The shelf life of the polished carrots is much longer than that of washed or peeled. Traditional methods of washing leave this membrane layer intact and as it dries out, it turns opaque leaving the appearance of carrots unpleasant.

Although several studies have been published about the effects of wounding over phenolic accumulation in carrot, to our knowledge, information about the effect of polishing and peeling is not available or scarce, respectively. In addition, inadequate information is available about the effect of wounding and storage on the intact phenolic acid compounds. The compartment of various commercial and catering processes such as polished, peeled, untreated, machine washed, peeled and shredded carrots would be valuable. Hence, our goal was to study the effect of various treatments on the accumulation of the intact soluble phenolic acids during storage of carrots. As the consumption of fresh minimal processed vegetable products increases, the potential to enhance the phenolic acid content and intake of these health-promoting compounds is tempting. This research provides important information for primary production, packaging, catering, fresh-cut industry and consumers towards the selection of healthier minimally processed carrot.

2. Materials and Methods

2.1. Samples

Harvest-fresh carrot samples (*Daucus carota*, cv. Panther) were provided from the local commercial farmer, packer and marketer, Karotia Ltd. Three kinds of samples were collected, namely untreated (harvest-fresh), machine washed and machine washed and polished carrots. All carrot samples were from the same batch, and they were approximately the same size. A portion of the washed carrots was further processed in a laboratory by peeling using a household peeling knife. The peels were also collected for phenolic acid analysis. A portion of the peeled carrots was grated using a household food processor (Philips, Andover, MA, USA) in the laboratory.

Untreated, washed, washed and polished and washed and peeled carrots were packed into retail plastic low-density polyethylene bags (LDPE; 0.5 kg, approximately 6–8 carrots/one plastic bag) and were stored at 8 °C in the dark cold storage. Samples for phenolic acid analysis were taken after 0, 2, 7, 14- and 28-days of storage (three replicate bags from all sample types, from all time points).

Peeled carrots were stored also under water at 8 °C and samples were taken after 0 and 2 days (three 0.5 kg replicates from every time point). At different time points, the whole carrots from every sample replicate were halved lengthwise, and the halves were chopped, mixed and freeze-dried prior to the phenolic acid analysis. Grated carrot was stored at 4 °C in three bowls covered with aluminum foil, all containing 1.5 kg, and 100 g-samples were taken from all bowls after 0, 1, 5- and 24-h storage to be freeze-dried before analyses. Prior to freeze-drying the samples were frozen at −21 °C. Freeze-drying was carried out at −40 °C under vacuum of 0.1–0.2 mbar. A schematic representation of the study protocol is shown in Figure 1.

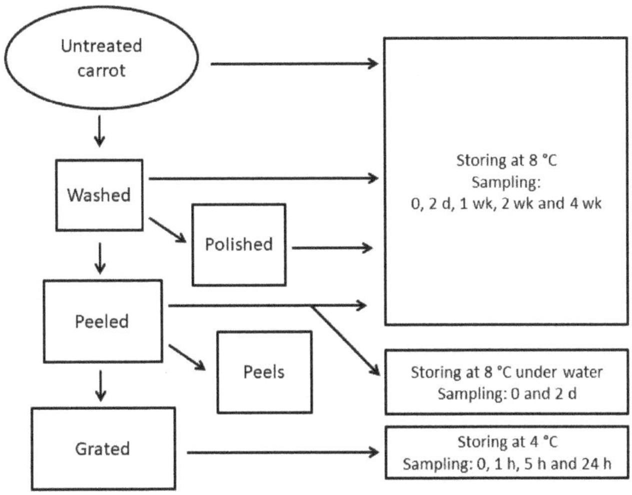

Figure 1. Scheme of carrot processing.

2.2. *Quantification of Soluble Phenolic Acids*

Freeze dried samples (0.5 g) were homogenized in 7 mL of a mixture of methanol, containing 2 g/L of butylated hydroxyanisole (BHA) obtained from Acros Organics (Geel, Belgium) and 10% acetic acid (85:15 *v*/*v*) with an IKA Ultra-Turrax T 25 homogenizer (IKA Werke GmbH & Co., Staufen, Germany). The homogenized sample extract was ultrasonicated for 30 min and increased to a volume of 10 mL with distilled water. After mixing, 1 mL of the solution was filtered (Acrodisc GHP, 0.2 μm, Port Washington, NY, USA). The analytical UHPLC system consisted of an Agilent 1290 Infinity Series ultra-high-performance liquid chromatograph (Agilent Technologies, Santa Clara, CA, USA) equipped with a diode array detector. Phenolic acid separation was done with a Zorbax Eclipse Plus C_{18} (2.1 × 50 mm, 1.8 μm) column (Agilent Technologies Inc., Santa Clara, CA, USA) with a C_{18} guard column. The temperature of the column oven was set at 35 °C. A gradient elution was employed with a mobile phase consisting of 50 mM H_3PO_4 at pH 2.5 (solution A) and acetonitrile (solution B) as follows: Isocratic elution 95% A, 0–1.2 min; linear gradient from 95% A to 85% A, 1.2–4.25 min; linear gradient from 85% A to 80% A, 4.25–10 min; linear gradient from 80% A to 50% A, 10–15 min; isocratic elution 50% A, 15–16.2 min; linear gradient from 50% A to 95% A, 16.2–17 min; post-time 2 min before the next injection. The flow rate of the mobile phase was 0.4 mL/min, and the injection volume was 2.0 μL. UV spectra of peaks were recorded between 190 and 400 nm. Chlorogenic acid (5-caffeoyl quinic acid) was used as an external standard for all caffeoyl quinic acids, caffeic acid for all other caffeic acid derivatives and ferulic acid for all ferulic acid derivatives with the assumption that the molar absorptivity at the detection wavelength (329 nm) depended solely on the structural cinnamic acid part in each compound. The results were then calculated according to the molecular masses of the actual compounds. All quantifications were based on peak area and the samples were analyzed in triplicate. In order to calculate the results in dry weight basis the residual moisture of freeze-dried carrots was determined by drying at 105 °C overnight (17 h).

2.3. Mass Spectrometric Identification

An Acquity UPLC—Xevo G2 QTOF high resolution mass spectrometer (Waters, Milford, MA, USA) operated by Waters MassLynx 4.1 software was used for the identification of phenolic acids using the analytical conditions as follows: Compounds were separated on Waters Acquity BEH C18 (1.7 µm, 2.1 mm × 150 mm) column using a gradient of 0.1% formic acid in H_2O (A) and of 0.1% formic acid in acetonitrile (B). The gradient program was as follows: 2%–60% of B in 24 min, 60%–100% of B in 24–31 min, held at 100% of B for 2 min, 100%–2% in 1 min and held at 2% of B for 4 min. The flow rate was 0.55 mL/min, temperature of the column oven was 45 °C and the injection volume was 2 µl. An electrospray interface (ESI) was used with capillary voltage of −1 kV in negative mode. The sampling cone was set to 35 V and extraction cone to 4 V. The cone and desolvation nitrogen gas flows were 15 and 990 l/h, respectively. The desolvation temperature was 550 °C. Source temperature was 150 °C. Argon was used as the collision gas. MS analyses were done by data independent acquisition (MSE) centroid data mode in a full scan *m*/*z* 50–1200 with 0.2 sec scan time. In the MSE function, the precursor ions from the low-collision energy MS-mode were fragmented using high collision energy ramped up from 15 to 40 eV.

2.4. Statistical Analyses

Results were expressed as means followed by the standard deviation (n = 3). For inferential analysis, one-way ANOVA was used and, when applicable, Tukey's test was applied to separate the means. The significance level was set to 0.05 to reject the null hypothesis (no difference between samples). TIBCO Statistica v. 13.3 (TIBCO Statistica, Palo Alto, CA, USA) was used in the analysis.

3. Results and Discussion

3.1. Identification and Quantification of Phenolic Acids

It is known that about 70% of carrot phenolics exist in soluble forms [5,18] and when carrots are treated with wounding stress, they produce high levels of caffeoylquinic acids, i.e., chlorogenic acid and its derivatives [13,14,20]. Hence, only soluble phenolic acids were studied in the present study. However, according to Viacava et al. [18] there is significant increase also in nonextractable phenolic acids during storage at 15 °C for 48 h.

Eleven different caffeic/ferulic acid derivatives were tentatively identified by UHPLC-DAD (Figure 2). They were further characterized by LC-QTOF-MS (Table 1) as 3-caffeoylquinic acid (3-CQA), 5-caffeoylquinic acid (5-CQA), feruloyl-rutinoside (FRut) caffeoyldaucic acid (CDA), feruloylquinic acid (FQA), dicaffeoyldaucic acid (diCDA), 3,5-dicaffeoylquinic acid(3,5-diCQA), 4,5-dicaffeoylquinic acid (4,5-diCQA), caffeoyl-feruloyldaucic acid (CFDA) and feruloyldaucic acid (FDA).

Accurate *m*/*z* values for the deprotonated caffeoylquinic acid isomers were 353.0867 and 353.0876 agreeing quite well with the theoretical monoisotopic mass (353.0873). The identification was further ensured by the MS^2 fragments of 191.06 and 179.03 corresponding to quinic acid and caffeic acid, respectively. The later eluting CQA had the same retention time with the reference standard confirming it was 5-CQA while the elution order suggested that the earlier eluting CQA was 3-CQA [21]. Two dicaffeoylquinic acids were detected at *m*/*z* values of 515.1169 and 515.1182 (predicted *m*/*z* 515.1190) and MS^2 fragments of 353.09, 191.05 and 179.03 corresponding to CQA, QA and CA, respectively. According to the retention order and previous studies [9,22] they were tentatively identified as 3,5-diCQA and 4,5-diCQA. Faisal et al. (2017) [23] characterized soluble phenolic acids in six carrot cultivars and all of them had 3-CQA, 5-CQA and two isomers of dicaffeoylquinic acid agreeing with our results. They also found two different feruloylquinic acid isomers in all cultivars while we could detect only one isomer (*m*/*z* 367.1042). Previous studies have also found 4-caffeoylquinic acid [24,25], diferuloylquinic acid [23] and caffeoylhexoside [26] in some carrot cultivars but we could not find them in our samples. However, a compound at *m/z* 501.1605 with a fragment at *m/z* 193.0491 (ferulic acid) was detected. These ions could be derived from feruloyl-rutinoside.

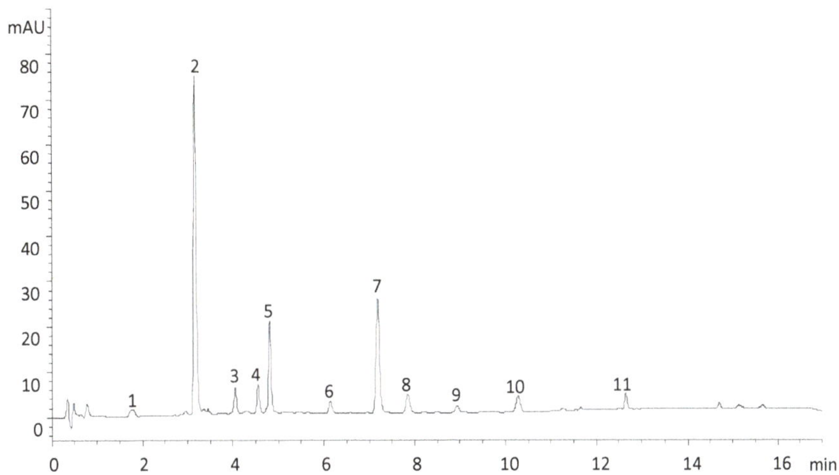

Figure 2. HPLC-DAD (λ = 329 nm) chromatogram of carrot extract. Peaks identified as 3-CQA (1), 5-CQA (2), FRut (3), FQA (4), CDA (5), 3,5-diCQA (6), diCDA (7), 4,5-diCQA (8), CFDA (9), FDA (10), diFDA (11).

Table 1. MS data of tentatively identified phenolic acids in carrot.

Phenolic Acid	Deprotonated Formula	Monoisotopic Mass	Detected Mass (*m/z*)	Mass Difference (ppm)	Characteristic Fragments (MS^2)
3-Caffeoylquinic acid, 3-CQA	$C_{16}H_{17}O_9$	353.0873	353.0876	0.85	191.0558, 179.0345
5-Caffeoylquinic acid, 5-CQA	$C_{16}H_{17}O_9$	353.0873	353.0867	−1.70	191.0551, 179.0348
Feruloyl-rutinoside	$C_{22}H_{29}O_{13}$	501.1608	501.1605	−0.60	193.0491
Caffeoyldaucic acid, CDA	$C_{16}H_{13}O_{10}$	365.0509	365.0499	−1.37	203.0183, 179.0337
Feruloylquinic acid, FQA	$C_{17}H_{19}O_9$	367.1029	367.1042	3.54	191.0556, 193,0514
Dicaffeoyldaucic acid, diCDA	$C_{25}H_{19}O_{13}$	527.0826	527.0816	−1.90	365.0507, 203.0190, 179.0345
3,5-Dicaffeoylquinic acid, 3,5-diCQA	$C_{25}H_{23}O_{12}$	515.1190	515.1169	−4.08	353.0897, 191.0536, 179.0354
4,5-Dicaffeoylquinic acid, 4,5-diCQA	$C_{25}H_{23}O_{12}$	515.1190	515.1182	−1.55	353.0888, 191.0540, 179.0305
Caffeoyl-feruloyldaucic acid, CFDA	$C_{26}H_{21}O_{13}$	541.0982	541.0969	−2.40	379.0662, 193.0504, 365.0496
Feruloyldaucic acid, FDA	$C_{17}H_{15}O_{10}$	379.0655	379.0687	0.44	193.0848, 203.0158
Diferuloyldaucic acid	$C_{27}H_{23}O_{13}$	555.1139	555.1155	2.88	ND

In the present study several daucic acid derivatives of hydroxycinnamic acids were tentatively identified in carrots, namely caffeoyldaucic acid, dicaffeoyldaucic acid, feruloyldaucic acid, diferuloyldaucic acid and caffeoyl-feruloyldaucic acid (Table 1). For diferolyldaucic acid the identification based solely on molecular ion since the signal was too weak for MS^2.

Kammerer et al. [26], and a few years later Kreutzmann et al. [24], reported caffeic acid derivatives at *m/z* values of 365 and 527 in carrots. However, they did not provide any suggestions for the molecular or structural formulas. Recently, Pace et al. [27] tentatively identified those compounds as N-tryptophan conjugates. This conclusion is not supported by the HRMS data of the current study. For instance, a molecular ion at *m/z* 527.0816 has a much better match with a formula $C_{25}H_{19}O_{13}$ (527.0826, deprotonated diDCA) than with $C_{26}H_{27}N_2O_{10}$ (527.1666, deprotonated caffeoyl-N-tryptophan-hexoside). Furthermore, the detected MS^2 fragments 365.05, 203.02 and 179.04 are in good accordance with the structures of CDA, daucic acid, and caffeic acid, respectively. Dicaffeoyldaucic acid was for the first time isolated and fully characterized in sweet potato (Dini et al., 2006) [28] and later Toffali et al. [29] showed that, also, carrot cells can produce diCDA. In the same study, also, methylated diCDA was reported as a carrot metabolite. The MS^2 fragment at *m/z* 193.05 detected in the present study indicates that the methylation appears most probably in caffeoyl moiety, i.e., the suggested compound at *m/z* 541.0969 is deprotonated caffeoyl-feruloyldaucic acid (Figure 3). Similarly, Kammerer et al. [26] reported a compound at *m/z* 541 with caffeoyl and feruloyl moieties. Albeit MS data strongly suggest that caffeic and ferulic acids in carrot appear to a considerable extent as daucic acid conjugates, further research, including NMR studies, is required to fully characterize these compounds.

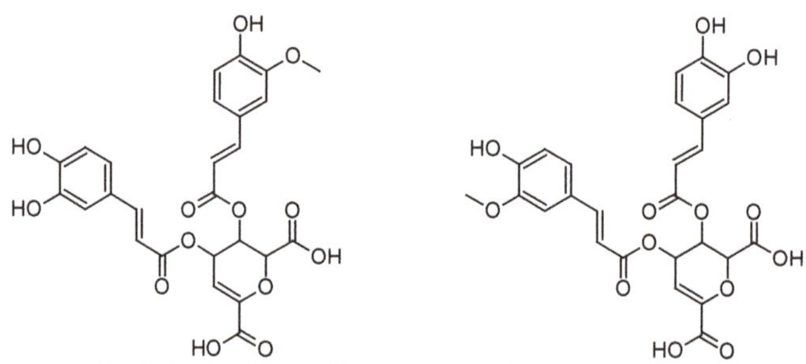

Figure 3. Two possible isomers of caffeoyl-feruloyldaucic acid.

3.2. *Effect of Carrot Processing on Total Phenolic Acids*

Total soluble phenolic acid levels (summed amount of phenolic acids with caffeoyl and feruloyl moieties) found in differently processed carrots varied much at the beginning of the storage on day 0 as calculated on a dry weight basis (Table 2). The dry weight of the whole and grated carrot samples varied from 11.4% to 12.5%, and the dry weight of carrot peels was 4.6%. At the beginning of the experiment untreated carrots contained total phenolic acids 893 ± 87 mg/kg DW (dry weight). Machine washing had no statistically significant effect ($p > 0.05$) on the contents (996 ± 177 mg/kg DW) and polishing of the carrots reduced the concentration moderately (666 ± 24 mg/kg DW, $p < 0.05$). Peeling had the most considerable effect, because total phenolic acid content of the peeled carrots was 246 ± 81 mg/kg DW at the beginning of the experiment, and this content did not appreciably diminish after grating (226 ± 68 mg/kg DW). The magnitude of these levels is in-line with earlier data [5,7]. The highest levels of phenolic acids were detected in peels (3270 ± 107 mg/kg DW). This was expected, because although phenolic acids are present in the root, they are mainly concentrated in the periderm [30]. According to Zhang and Hamauzu [31], although accounting for only 11% of the amount of the carrot fresh weight, peels could provide 54.1% of the amount of the phenolics in 100 g fresh weight of carrots.

Phenolic acids accumulated in all processed and the untreated carrots during cold storing at 8 °C. This was in accordance with several studies dealing with the induction of phenolic acids in wounded carrot products [7–16]. However, according to Alarcón-Flores et al. [7] commercial fresh-cut carrots contained lower levels of phenolic acids than the fresh counterpart. Figure 4 shows that the induction of the total phenolic acids was very strong during the first week, after which more differences were found between the treatments. In the case of untreated carrots, the concentrations of total phenolic acids reached a plateau after two weeks of cold storing at 8 °C. The phenolic acid concentration of polished carrots increased linearly during the two days and two weeks cold storage and prolonging the storing from two to four weeks had statistically no further effect ($p > 0.05$) on the content of phenolic acids. In peeled carrots the phenolic acid contents differed statistically from each other at all time points ($p < 0.05$). The phenolic acid contents increased during the whole four-week storing period, reaching 7547 ± 1090 mg/kg DW, which implied a 31-fold increase. If the peeled carrots were stored under the water at the same temperature for 2 days, the content of phenolic acids (403 ± 72 mg/kg) was much lower than in carrots stored under normal atmosphere (1283 ± 306 mg/kg). This is probably since PAL activity is inhibited by low oxygen environments [30]. At the end of the experiment (four weeks), untreated and machine-washed carrots contained roughly 4-fold (3336 ± 191 and 3990 ± 506 mg/kg DW, respectively), and the polished carrots 8-fold (5256 ± 252 mg/100 g DW) more phenolic acids than at day 0. The phenolic acid content in grated carrot was not significantly changed during the first 5 h, but after 24 h storage the concentration was doubled (495 ± 64 mg/kg DW). This was in line with Viacava et al. [18] who found that total free and bound phenolic acid contents increased 288.1% and 407.6%, respectively, in shredded carrot, when stored at 15 °C for 48 h.

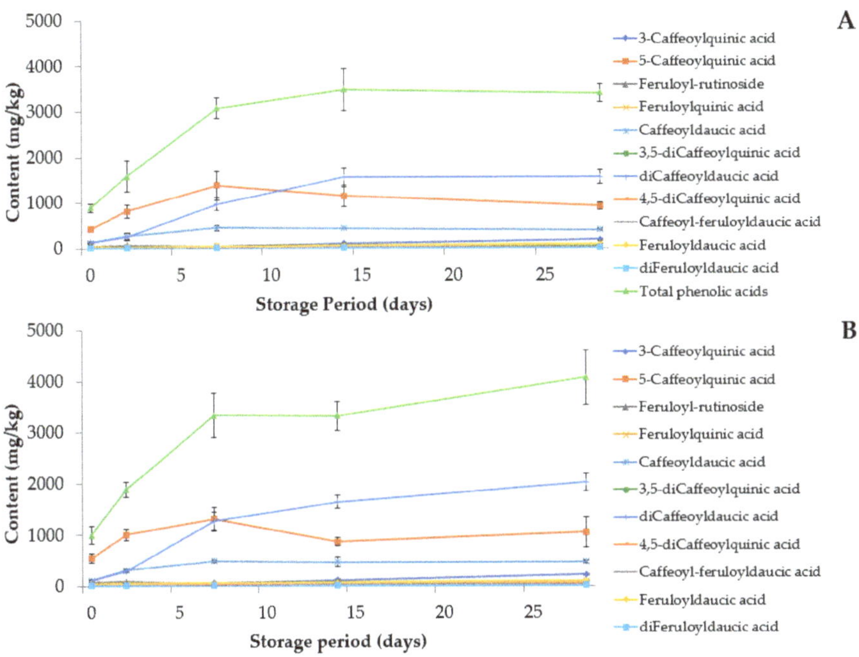

Figure 4. *Cont.*

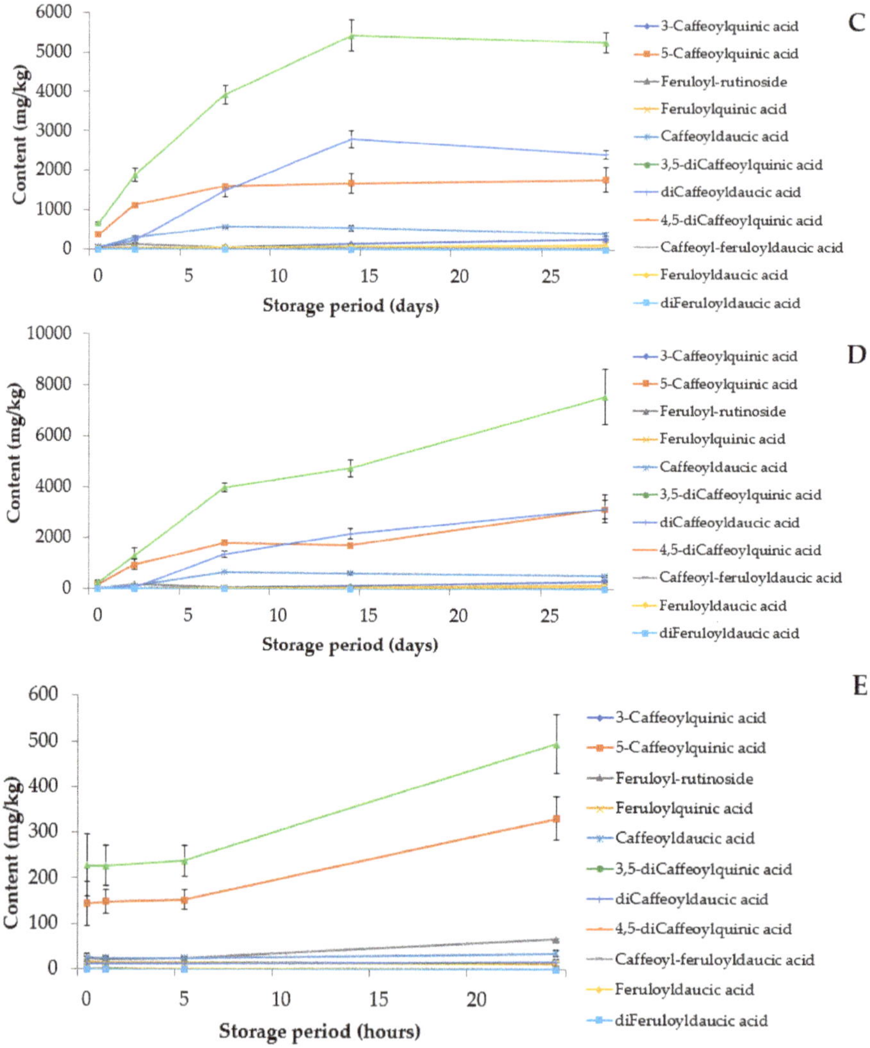

Figure 4. Phenolic acid composition (mg/kg DW) of untreated (**A**), washed (**B**), polished (**C**), peeled (**D**) and grated (**E**) carrots stored at 8 °C for either 28 days or 24 h.

Although the peeled carrot seems to be a rich source of phenolic acids, the shelf life of fresh-cut carrots is strongly dependent on sensory quality which was estimated for these kinds of products to be only 4 days [6]. Instead, the use of gentle processes and optimal packaging technologies will lead to good quality products high in terpene flavor [32].

Table 2. Initial contents of phenolic acids in untreated, washed, polished, peeled and grated carrots (mg/kg DW).

Phenolic Acid	Carrot Sample					
	Untreated	*Washed*	*Polished*	*Peeled*	*Grated*	*Peels*
3-Caffeoylquinic acid	19.1 ± 2.57	14.7 ± 1.3	10.2 ± 1.8	2.35 ± 0.7	2.07 ± 0.6	57.0 ± 1.8
5-Caffeoylquinic acid	429 ± 45.5	540 ± 90.7	374 ± 20.0	164 ± 59.5	144 ± 48.2	1760 ± 55
Feruloyl-rutinoside	45.4 ± 5.7	83.1 ± 24.9	69.6 ± 6.8	20.4 ± 6.3	23.3 ± 4.0	261 ± 7.2
Feroyl-quinic acid	41.9 ± 9.7	52.0 ± 13.5	43.5 ± 6.5	24.5 ± 6.7	16.6 ± 5.3	163 ± 19.0
Caffeoyldaucic acid	121 ± 6.2	112 ± 21.5	64.7 ± 6.7	20.5 ± 7.3	25.2 ± 9.7	396 ± 13.3
3,5-diCaffeoylquinic acid	12.9 ± 0.8	11.7 ± 2.8	7.54 ± 0.7	1.75 ± 0.8	1.40 ± 0.1	42.2 ± 2.4
Dicaffeoyldaucic acid	137 ± 21.7	113 ± 24.0	56.2 ± 3.7	8.10 ± 1.9	11.9 ± 7.3	356 ± 12.7
4,5-diCaffeoylquinic acid	26.5 ± 2.0	21.6 ± 1.8	12.4 ± 1.9	1.04 ± 0.1	1.04 ± 0.5	76.7 ± 5.0
Caffeoylferuloyldaucic acid	8.30 ± 1.6	7.48 ± 1.4	5.16 ± 1.0	0.64 ± 0.1	0.64 ± 0.4	25.9 ± 0.4
Feruloyldaucic acid	34.6 ± 3.5	28.7 ± 3.6	15.9 ± 0.7	1.76 ± 0.3	1.21 ± 0.4	85.9 ± 4.2
diFeruloyldaucic acid	16.9 ± 2.1	12.2 ± 0.6	6.42 ± 0.4	0.96 ± 0.1	0.66 ± 0.0	46.4 ± 2.8
Total	893 ± 87.2	996 ± 177.2	666 ± 23.7	246 ± 81.4	227 ± 67.9	3270 ± 106.6

3.3. *Effect of Carrot Processing on Individual Phenolic Acid Compounds*

Chlorogenic acid was the main phenolic acid in all samples on day 0, which is in-line with previous reports [33,34]. Similarly, 5-CQA has been reported as the major caffeoylquinic acid in carrots [23–26,35,36], while some sources have indicated that 3-CQA is the most abundant form [22,27,37,38]. In the samples of the present study 3-CQA was always a minor phenolic acid. It is obvious that some variation should occur in the phenolic acid profile between different carrot varieties, but it is also very easy to confuse 3-CQA with 5-CQA due to the historical ambiguity in the nomenclature of such compounds [39]. IUPAC specified the rules for nomenclature of chemical compounds in 1976 and at that moment 3-CQA was turned to 5-CQA and vice versa. According to IUPAC, compounds known with their trivial names as chlorogenic acid and neochlorogenic acid are 5-CQA and 3-CQA, respectively. However, in the studies performed by Formica-Oliveira et al. [22] and Pace et al. [27], the terms chlorogenic acid and 3-CQA are used as synonyms and thus, it is plausible to assume that they were using the previous nomenclature and the compound in question was actually 5-CQA according to current rules. This would be very understandable since most of the chemical suppliers have kept the old, yet incorrect, nomenclature in their product catalogues [39]. After 5-CQA the next highest amounts were determined for daucic acid derivatives of caffeic acid (CDA and diCDA) in peels and untreated and washed carrots but not in polished, peeled or grated carrots (Table 2, Figure 4). Polishing and peeling removed the skin efficiently, so it seems that CDA and diCDA are especially concentrated in the skin of carrot.

The content of most phenolic acids increased in all treatments during storage for 28 days. The concentrations of diCDA increased more considerably—being 11-fold, 18-fold, 43-fold, and 392-fold in untreated, washed, polished and peeled carrot samples, respectively, after 28 days of storage (Figure 4). At the end of the storage period, diCDA was the main phenolic acid in all samples other than in peeled carrot, wherein the contents of diCDA and 5-CQA acid were of the same level. The concentrations of 5-CQA and CDA also increased significantly, especially in peeled carrot. In most treatments the most intense increase in 5-CQA and DCA content appeared during the first week after which they reached the plateau, or their content started to decrease slightly (Figure 4). For diCDA the period of intense accumulation was usually longer, i.e., two weeks and then plateau was reached. An exception was the peeled carrot where the content of 5-CQA and diCDA tended to increase the whole four weeks. A sharp increase was observed for 3-caffeoylquinic acid, from 10-fold (untreated carrot) to 138-fold (peeled carrot) although the concentrations remained lower compared with the main three compounds. It has been shown that wounding stress increases the contents of different CQAs during storage of carrots [40,41]. Surjadinata et al. [20] monitored the fate of phenolics during the storage of shredded carrot, and after one-week, the content of 5-CQA was 7.3–11.3-fold more than the original content.

Accordingly, in the present study the content of 5-CQA in peeled carrot was ca. 11 times the initial content after one week storing.

Grated carrot is not expected to be stored for long periods due to its limited shelf life and hence, the accumulation of phenolic acids in grated carrot was followed only for 24 h in the present study. During the first 5 h, no changes in the phenolic composition were observed. After 24 h, the increase in the total phenolic acid content was almost entirely due to increase in the of 5-CQA. Accordingly, Becerra-Moreno et al. [41] and Viacava et al. [18] found that chlorogenic acid increased the most in shredded carrot. Interestingly, after 24 h next highest content after 5-CQA was found for the compound tentatively identified as feruloyl-rutinoside (FRut). In other carrot samples the FRut content seemed to be highest after two days storing after which it declined. It is possible that fast but temporary biosynthesis of FRut is among the first responses for wound-induced stress in carrot.

As soluble phenolic acids were converted to caffeic acid and ferulic acid equivalents it was evidential that caffeic acid was the most abundant phenolic acid aglycone in all samples (Table 3). This was in line with our previous study [5]. Additionally, the storage induced the biosynthesis of caffeic acid more remarkably than for ferulic acid. Increased biosynthesis of phenolic acids results from stress-induced increase in PAL activity as is well demonstrated in previous studies [12,42]. After four weeks of storage, untreated and machine-washed carrots contained roughly 5-fold, and the polished carrots 10-fold more caffeic acid compared to the content in the beginning of the storage (Table 3). In peeled carrots the caffeic acid content was over 40 times the initial content at the end of the experiment. The increase in ferulic acid content was much more modest being one and a half times the initial content in untreated and washed carrots, and two times and seven times the initial content in the polished and peeled carrots, respectively, at the end of the storing period. As a result, the proportion of caffeic acid increased from 82% or 85% to 94% or 97% of the combined caffeic and ferulic acid content during the experiment. In fact, an early study showed that the storage of carrots for 2 days increased the contents of caffeic acid derivatives in a more pronounced way compared to the ferulic acid content [41].

Table 3. Total caffeic and ferulic acid in carrot subjected to different technological procedures calculated as free acids.

Storage Period	Caffeic Acid (mg/kg)	Ferulic Acid (mg/kg)	Dry Matter (%)
Days		*Untreated Carrots*	
0	411 ± 42.8 [b]	71.8 ± 4.9 [b]	11.71 ± 0.49
2	777 ± 175.8 [b]	70.5 ± 11.5 [b]	12.15 ± 0.30
7	1657 ± 96.7 [a]	79.3 ± 17.3 [b]	11.81 ± 0.19
14	1970.2 ± 256.6 [a]	80.64 ± 16.3 [b]	11.68 ± 0.27
28	1898 ± 113.8 [a]	117.6 ± 11.9 [a]	11.46 ± 0.18
p-value [1]	<0.001	0.008	
Days		*Washed Carrots*	
0	440 ± 74.4 [b]	84.8 ± 17.3 [a]	11.99 ± 0.28
2	916 ± 57.1 [b]	90.2 ± 16.5 [a]	12.22 ± 0.37
7	1839 ± 244.5 [a]	77.3 ± 6.2 [a]	11.57 ± 0.35
14	1901 ± 169.0 [a]	86.6 ± 4.6 [a]	12.10 ± 0.61
28	2318 ± 278.3 [a]	117 ± 26.2 [a]	11.63 ± 0.37
p-value [1]	<0.001	0.100	
Days		*Polished Carrots*	
0	281 ± 5.3 [d]	63.8 ± 5.9 [b]	12.52 ± 0.23
2	891 ± 84.8 [c]	87.7 ± 16.1 [b]	11.99 ± 0.15
7	2170 ± 142 [b]	73.4 ± 8.7 [b]	11.85 ± 0.42

Table 3. *Cont.*

Storage Period	Caffeic Acid (mg/kg)	Ferulic Acid (mg/kg)	Dry Matter (%)
14	3150 ± 239 [a]	94.2 ± 9.4 [b]	11.97 ± 0.32
28	2963 ± 102 [a]	145 ± 20.0 [a]	11.89 ± 0.24
p-value [1]	<0.001	<0.001	
Days		*Peeled Carrots*	
0	103 ± 35.7 [c]	22.3 ± 6.0 [c]	12.25 ± 0.19
2	562 ± 127.5 [c]	75.7 ± 23.9 [c]	11.90 ± 0.64
7	2200 ± 104.7 [b]	45.3 ± 6.6 [b]	11.78 ± 0.87
14	2722 ± 191.1 [b]	53.4 ± 8.2 [b]	11.73 ± 0.96
28	4256 ± 620.7 [a]	146 ± 46.7 [a]	11.93 ± 0.35
p-value [1]	<0.001	<0.001	
Hours		*Grated Carrot*	
0	96.0 ± 35.0 [b]	18.7 ± 2.0 [b]	12.16 ± 0.32
1	97.5 ± 20.6 [b]	16.6 ± 2.4 [b]	12.28 ± 0.14
5	100 ± 14.5 [b]	17.1 ± 2.2 [b]	12.11 ± 0.26
24	200 ± 32.9 [a]	33.9 ± 1.7 [a]	12.08 ± 0.25
p-value [1]	0.004	<0.001	

Note: [1] Probability values obtained by one-way ANOVA. Different letters in the same column represent different means ($p < 0.05$).

4. Conclusions

In conclusion, minimally processed carrots can be an inexpensive, rich source of phenolic acids in the diet. By simply increasing wounding stress intensity it is possible to enhance the biosynthesis of phenolic acids. The phenolic acid content of harvest-fresh carrots multiplies during a few days of storage. Peeling greatly reduces the immediate phenolic content of carrots but a few days of storage will increase it efficiently. However, polishing seems to be an especially good practice to process carrots for retail markets. The appearance and shelf life of polished carrots are good, and their phenolic acid content increases greatly during storage. During storage, the greatest increases were found in the concentrations of dicaffeoyldaucic acid and 5-calffeoylquinic acid. The exception was grated carrot in which the largest increases were found in the concentration of 5-calffeoylquinic acid and feroyl-rutinoside.

Author Contributions: P.H.M.: Conceptualization, methodology, investigation, project administration, resources, writing—original draft preparation. J.H.: Conceptualization, methodology, investigation, formal analysis, writing—original draft preparation. D.G.: data curation, writing—original draft preparation. All authors have read and agreed to the published version of the manuscript.

Funding: This research received no external funding.

Conflicts of Interest: The authors declare no conflict of interest.

References

1. Han, X.; Shen, T.; Lou, H. Dietary polyphenols and their biological significance. *Int. J. Mol. Sci.* **2007**, *8*, 950–988. [CrossRef]
2. Scalbert, A.; Manach, C.; Morand, C.; Remesy, C. Dietary polyphenols and the prevention of diseases. *Crit. Rev. Food Sci. Nutr.* **2005**, *45*, 287–306. [CrossRef]
3. Granato, D.; Lorenzo, J.M.; Kovacevic, D.B.; Barba, F.J.; Cruz, A.G.; Putnik, P. Functional foods: Product development, technological trends, efficacy testing, and safety. *Annu. Rev. Food Sci. Technol.* **2020**, *11*, 93–118. [CrossRef]
4. Shahidi, F.; Naczk, M. *Food Phenolics*; Technomic Publishing Co. Inc.: Lancaster, PA, USA, 1995.
5. Mattila, P.; Hellström, J. Phenolic acids in potatoes, vegetables, and some of their products. *J. Food Compos. Anal.* **2007**, *20*, 152–160. [CrossRef]

6. Condurso, C.; Cincotte, F.; Tripodi, G.; Merlino, M.; Giarranta, F.; Verzera, A.; Dini, I.; Tenore, G.C.; Dini, A. A new approach for the shelf-life definition of minimally processed carrots. *Postharvest Biol Technol.* **2020**, *163*, 111138. [CrossRef]
7. Alarcón-Flores, M.I.; Romero-González, R.; Vidal, J.L.M.; González, F.J.E.; Frenich, A.G. Monitoring of phytochemicals in fresh and fresh-cut vegetables: A comparison. *Food Chem.* **2014**, *142*, 392–399. [CrossRef]
8. Babic, I.; Amiot, M.J.; Nguyen-The, C.; Aubert, S. Changes in phenolic content in fresh ready-to-use shredded carrots during storage. *J. Food Sci.* **1993**, *58*, 351–356. [CrossRef]
9. Becerra-Moreno, A.; Benavides, J.; Cisneros-Zevallos, L.; Jacobo-Velaázquez, D.A. Plants as Biofactories: Glyphosate-Induced Production of Shikimic Acid and Phenolic Antioxidants in Wounded Carrot Tissue. *J. Agric. Food Chem.* **2012**, *60*, 11378–11386. [CrossRef]
10. Cisneros-Zevallos, L. The use of controlled post-harvest abiotic stresses as a tool for enhancing the nutraceutical content and adding-value to fresh fruits and vegetables. *J. Food Sci.* **2003**, *68*, 1560–1565. [CrossRef]
11. Du, W.X.; Avena-Bustillos, R.J.; Breksa, A.P., III; McHugh, T.H. Effect of UV-B light and different cutting styles on antioxidant enhancement of commercial fresh-cut carrot products. *Food Chem.* **2012**, *134*, 1862–1869. [CrossRef]
12. Howard, L.R.; Griffin, L.E. Lignin formation and surface discolouration of minimally processed carrot sticks. *J. Food Sci.* **1993**, *58*, 1065–1067. [CrossRef]
13. Klaiber, R.G.; Baur, S.; Koblo, A.; Carle, R. Influence of washing treatment and storage atmosphere on phenylalanine ammonia-lyase activity and phenolic acid content of minimally processed carrot sticks. *J. Agric. Food Chem.* **2005**, *53*, 1065–1072. [CrossRef]
14. Reyes, L.F.; Villarreal, J.E.; Cisneros-Zevallos, L. The increase in antioxidant capacity after wounding depends on the type of fruit or vegetable tissue. *Food Chem.* **2007**, *101*, 1254–1262. [CrossRef]
15. Ruiz-Cruz, S.; Islas-Osuna, M.A.; Sotelo-Mundo, R.R.; Vázquez-Ortiz, F.; González-Aguilar, G.A. Sanitation procedure affects biochemical and nutritional changes of shredded carrots. *J. Food Sci.* **2007**, *72*, S146–S152. [CrossRef]
16. Simões, A.D.N.; Allende, A.; Tudela, J.A.; Puschmann, R.; Gil, M.I. Optimum controlled atmospheres minimise respiration rate and quality losses while increase phenolic compounds of baby carrots. *LWT-Food Sci. Technol.* **2011**, *44*, 277–283. [CrossRef]
17. Lòpez-Gámez, G.; Elez-Martínez, P.; Martín-Belloso, O.; Soliva-Fortuny, R. Enhancing phenolic content in carrots by pulsed electric fields during post-treatment time: Effects on cell viability and quality attributes. *Innov. Food Sci. Emerg. Technol.* **2020**, *59*, 102252. [CrossRef]
18. Viacava, F.; Santana-Gálvez, J.; Heredia-Olea, E.; Pérez-Carrillo, E.; Nair, V.; Cisneros-Zevallos, L.; Jacobo-Velázquez, D.A. Sequential application of postharvest wounding stress and extrusion as an innovative tool to increase the concentration of free and bound phenolics in carrots. *Food Chem.* **2020**, *307*, 125551. [CrossRef]
19. Leon, J.; Rojo, E.; Sanchez-Serrano, J.J. Wound signalling in plants. *J. Exp. Bot.* **2001**, *52*, 1–9. [CrossRef]
20. Surjadinata, B.B.; Cisneros-Zevallos, L. Biosynthesis of phenolic antioxidants in carrot tissue increases with wounding intensity. *Food Chem.* **2012**, *134*, 615–624. [CrossRef]
21. Clifford, M.N.; Johnston, K.L.; Knight, S.; Kuhnert, N. Hierarchical scheme for LC-MS^n identification for chlorogenic acids. *J. Agric. Food Chem.* **2003**, *51*, 2900–2911. [CrossRef]
22. Formica-Oliveira, A.C.; Martínez-Hernández, G.B.; Aguayo, E.; Gómez, P.A.; Artés, F.; Artés-Hernández, F. UV-C and hyperoxia abiotic stresses to improve healthiness of carrots: Study of combined effects. *J. Sci. Food Technol.* **2016**, *53*, 3465–3476. [CrossRef]
23. Faisal, N.A.; Chathta, S.A.S.; Hussain, A.I.; Ikram, M.; Bukhari, S.A. Liaison of phenolic acids and biological activity of escalating cultivars of *Daucus carota*. *Int. J. Food Prop.* **2017**, *20*, 2782–2792. [CrossRef]
24. Kreutzmann, S.; Christensen, L.P.; Edelenbos, M. Investigation of bitterness in carrots (*Daucus carota* L.) based on quantitative chemical and sensory analysis. *LWT-Food Sci. Technol.* **2008**, *41*, 193–205. [CrossRef]
25. Esatbeyoglu, T.; Rodríguez-Werner, M.; Schlösser, A.; Liehr, M.; Ipharraguerre, I.; Winterhalter, P.; Rimbach, G. Fractionation of plant bioactives from black carrots (*Daucus carota* subspecies *sativus* varietas *atrorubens* Alef.) by adsorptive membrane chromatography and analysis of their potential anti-diabetic activity. *J. Agric. Food Chem.* **2016**, *64*, 5901–5908. [CrossRef]

26. Kammerer, D.; Carle, R.; Scheiber, A. Characterization of phenolic acids in black carrots (*Daucus carota* ssp. *sativus* var. *atrorubens* Alef.) by high-performance liquid chromatography/electrospray ionization mass spectrometry. *Rapid Commun. Mass Spectrom.* **2004**, *18*, 1331–1340. [CrossRef]
27. Pace, B.; Capotorto, I.; Cefola, M.; Minasi, P.; Montemurro, N.; Carbone, V. Evaluation of quality, phenolic and carorotenoid composition of fresh-cut purple Polignano carrots stored in modified atmosphere. *J. Food Comp. Anal.* **2020**, *86*, 103363. [CrossRef]
28. Dini, I.; Tenore, G.C.; Dini, A. New polyphenol derivative in Ipomoea batatas tubers and its antioxidant activity. *J. Agric. Food Chem.* **2006**, *54*, 8733–8737. [CrossRef]
29. Toffali, K.; Ceoldo, S.; Stocchero, M.; Levi, M.; Guzzo, F. Carrot-specific features of the phenylpropanoid pathway identified by feeding cultured cells with defined intermediates. *Plant Sci.* **2013**, *209*, 81–92. [CrossRef]
30. Hager, T.J.; Howard, L.R. Processing effects on carrot phytonutrients. *Hortic. Sci.* **2006**, *41*, 74–79. [CrossRef]
31. Zhang, D.; Hamauzu, Y. Phenolics, ascorbic acid, carotenoids and antioxidant activity of broccoli and their changes during conventional and microwave cooking. *Food Chem.* **2004**, *88*, 503–509. [CrossRef]
32. Edelenbos, M.; Balasubramaniam, M.; Pedersen, H.T. Effects of minimal processing and packaking on volatile compounds and other sensory aspects in carrots. *Acta Hortic.* **2010**, *876*, 269–277. [CrossRef]
33. Ahmad, T.; Cawood, M.; Iqbal, Q.; Ariño, A.; Batool, A.; Tariq, R.M.S.; Azam, M.; Akhtar, S. Phytochemicals in *Daucus carota* and their health benefits—Review article. *Foods* **2019**, *8*, 424. [CrossRef]
34. Akhtar, S.; Rauf, A.; Imran, M.; Qamar, M.; Riaz, M.; Mubarak, M.S. Black carrot (*Daucus carota* L.), dietary and health promoting perspectives of its polyphenols: A review. *Trends Food Sci. Technol.* **2017**, *66*, 36–47. [CrossRef]
35. Alasalvar, C.; Grigor, J.M.; Zhang, D.; Quantick, P.C.; Shahidi, F. Comparison of volatiles, phenolics, sugars, antioxidant vitamins, and sensory quality of different colored carrot varieties. *J. Agric. Food Chem.* **2001**, *49*, 1410–1416. [CrossRef]
36. Frond, A.D.; Iuhas, C.I.; Stirbu, I.; Leopold, L.; Socaci, S.; Andreea, S.; Ayvaz, H.; Andreea, S.; Mihai, S.; Diaconeasa, Z.; et al. Phytochemical characterization of five edible purple-reddish vegetables: Anthocyanins, flavonoids, and phenolic acid derivatives. *Molecules* **2019**, *24*, 1536. [CrossRef]
37. del Rosario Cuellar-Villarreal, M.; Ortega-Hernández, E.; Becerra-Moreno, A.; Welti-Chanes, J.; Cisneros-Zevallos, L.; Jacobo-Velázquez, D.A. Effects of ultrasound treatment and storage time on the extractability and biosynthesis of nutraceuticals in carrot (*Daucus carota*). *Postharvest Biol. Technol.* **2016**, *119*, 18–26. [CrossRef]
38. Keser, D.; Gamze, G.; Kelebek, H.; Keksin, M.; Soysal, Y.; Sekerli, Y.E.; Arslan, A.; Selli, S. Characterization of aroma and phenolic composition of carrot (*Daucus carota* 'Nantes') powders obtained from intermittent microwave drying using GC-MS and LC-MS/MS. *Food Bioprod. Process.* **2020**, *119*, 350–359. [CrossRef]
39. Kremr, D.; Bajer, T.; Bajerová, P.; Surmová, S.; Ventura, K. Unremitting problems with chlorogenic acid nomenclature: A review. *Quim. Nova* **2016**, *39*, 530–533. [CrossRef]
40. Jacobo-Velázquez, D.A.; Martínez-Hernández, G.B.; del Rodríguez, S.; Cao, C.-M.; Cisneros-Zevallos, L. Plants as biofactories: Physiological role of reactive oxygen species on the accumulation of phenolic antioxidants in carrot tissue under wounding and hyperoxia stress. *J. Agric. Food Chem.* **2011**, *59*, 6583–6593. [CrossRef]
41. Becerra-Moreno, A.; Redondo-Gil, M.; Benavides, J.; Nair, V.; Cisneros-Zevallos, L.; Jaccobo-Velázquez, D.A. Combined effect of water loss and wounding stress on gene activation of metabolic pathways associated with phenolic biosynthesis in carrot. *Front. Plant Sci.* **2015**, *6*, 837. [CrossRef]
42. Alegria, C.; Gonçalves, E.M.; Moldão-Martins, M.; Cisneros-Zevallos, L.; Abreu, M. Peel removal improves quality without antioxidant loss, through wound-induced phenolic biosynthesis in shredded carrot. *Postharvest Biol. Technol.* **2016**, *120*, 232–239. [CrossRef]

Publisher's Note: MDPI stays neutral with regard to jurisdictional claims in published maps and institutional affiliations.

© 2020 by the authors. Licensee MDPI, Basel, Switzerland. This article is an open access article distributed under the terms and conditions of the Creative Commons Attribution (CC BY) license (http://creativecommons.org/licenses/by/4.0/).

foods

MDPI

Review

Innovative Hurdle Technologies for the Preservation of Functional Fruit Juices

Predrag Putnik [1,*], Branimir Pavlić [2], Branislav Šojić [2], Sandra Zavadlav [3], Irena Žuntar [4], Leona Kao [1], Dora Kitonić [1] and Danijela Bursać Kovačević [1,*]

1 Faculty of Food Technology and Biotechnology, University of Zagreb, Pierottijeva 6, 10000 Zagreb, Croatia; lkao@pbf.hr (L.K.); kitonicdora@gmail.com (D.K.)

2 Faculty of Technology Novi Sad, University of Novi Sad, Bulevar cara Lazara 1, 21000 Novi Sad, Serbia; bpavlic@uns.ac.rs (B.P.); sojic@tf.uns.ac.rs (B.Š.)

3 Department of Food Technology, Karlovac University of Applied Sciences, Trg J. J. Strossmayera 9, 47000 Karlovac, Croatia; sandra.zavadlav@vuka.hr

4 Faculty of Pharmacy and Biochemistry, University of Zagreb, Ante Kovačića 1, 10000 Zagreb, Croatia; izuntar@pharma.hr

* Correspondence: pputnik@alumni.uconn.edu (P.P.); dbursac@pbf.hr (D.B.K.)

Received: 27 April 2020; Accepted: 17 May 2020; Published: 1 June 2020

Abstract: Functional nutrition, which includes the consumption of fruit juices, has become the field of interest for those seeking a healthy lifestyle. Functional nutrition is also of great interest to the food industry, with the aims of improving human health and providing economic prosperity in a sustainable manner. The functional food sector is the most profitable part of the food industry, with a fast-growing market resulting from new sociodemographic trends (e.g., longer life expectancy, higher standard of living, better health care), which often includes sustainable concepts of food production. Therefore, the demand for hurdle technology in the food industry is growing, along with the consumption of minimally processed foods, not only because this approach inactivates microorganisms in food, but because it can also prolong the shelf life of food products. To preserve food products such as fruit juices, the hurdle technology approach often uses non-thermal methods as alternatives to pasteurization, which can cause a decrease in the nutritional value and quality of the food. Non-thermal technologies are often combined with different hurdles, such as antimicrobial additives, thermal treatment, and ultraviolet or pulsed light, to achieve synergistic effects and overall quality improvements in (functional) juices. Hence, hurdle technology could be a promising approach for the preservation of fruit juices due to its efficiency and low impact on juice quality and characteristics, although all processing parameters still require optimization.

Keywords: functional fruit juice; hurdle technology; non-thermal processing; preservation; quality; probiotic

1. Introduction

Functional foods are products with added active substances, which if consumed in adequate quantities not only provide basic nutrition, but also have a positive impact on general health or contribute to reduction of the risks for developing certain diseases [1]. Consumers are now increasingly looking for products that are safe and natural, have Generally Recognized as Safe (GRAS) status, and are produced using sustainable or ecological technology [2]. By 2020, it is estimated that the global market for functional foods will reach \$304.5 billion, with an average annual growth of 8.5%, despite the fact that the development of new (functional) foods is a fairly costly and complex business [3]. In light of these trends, consumers have a tendency to purchase fruit juices or beverages in order to satisfy their daily nutrient needs and improve their wellbeing in a quick and easy way. According to the National

Health Service (UK), the daily recommendation of 5 servings of fruits and vegetables ("five a day") can be adequately substituted with 150 mL of 100% fruit or vegetables juices [4]. It has been demonstrated that biological active compounds (BACs) in the organism are much more easily absorbed and digested if they are ingested from juices than those from whole plant tissue.

Hurdle technology is a set of methods used for inactivation of microorganisms in food preservation. Hurdles are food preservation factors, which are combined to achieve certain food quality and stability, in terms of the temperature, pH, redox potential, water activity, preservatives, and competitive microorganisms. Today, more than sixty hurdles are known to have an effect on food stability and quality, as well as microbial stability [5]. Not all hurdles are combined together, nor are they all used to preserve one food product. The effect that a hurdle has on the food depends on the intensity, which affects the microbial stability and means the intensity needs to increase, while if the hurdle affects the food quality the intensity needs to decrease. Classical heat treatment (pasteurization) of fruit juices disrupts the stability of the thermolabile BAC molecules, so the application of non-thermal technologies using the "hurdle concept" could be a promising solution in order to meet the consumer criteria of quality, nutritional value, and product safety.

To preserve food using hurdle technology, it is necessary to create a hostile environment for microorganisms. This environment can either cause the death of microorganisms or slow down their growth, largely depending on how microorganisms react to the treatment. These reactions are being researched as methods of food protection, together with the effects of hurdle technology on homeostasis, stress reactions, metabolic exhaustion of microorganisms, and multitarget preservation, all of which are essential for food safety.

The use of only a single technology may cause damage to microorganisms that they may recover from during food storage. However, if this recovery is effectively prevented by using a combination of additional hurdles, the microbial cells will not be able to grow and higher inactivation levels could be reached. In this respect, most of the current research studies are focused on high-power ultrasound (HPU), pulsed electric field (PEF), and high-pressure processing (HPP) [6]. These technologies have low greenhouse gas emissions and energy consumption and reduced environmental impacts; hence, they are considered as sustainable technologies that are in line with Agenda 2030, which is geared towards their use [7]. HPP represents a technology with a physical effect as the main preservation factor; thus, it is often used as a hurdle technology. When processing fruit juices, PEF has been applied as an electrotechnology within hurdle technology in combination with numerous other options, e.g., antimicrobial additives, thermal treatment, ultraviolet light (UV), and manothermosonication (MTS). Similarly to HPU, it is also commonly combined with numerous other technologies, e.g., UV-C light, pulsed light, and high pressure. [8]. However, interestingly the use of HPU and PEF as a combination hurdle technology has not been sufficiently explored in terms of the ecological and economical potential as a substitution for pasteurization in fruit juices. This is despite previous research results implicating these two non-thermal techniques as having synergistic potential for overall quality improvements of (functional) juices.

2. Functional Fruit Juices for Delivery of Bioactive Compounds, Probiotics, and Prebiotics

In today's world, there is a growing trend toward health and fitness. Accordingly, the food industry has directed their efforts towards producing healthier and more nutritious products that will satisfy consumer demands. One way to go about this is to design completely new products, while another way is to add functional ingredients to already existing products. Functional ingredients are substances that are beneficial for our health, e.g., carotenoids, phenolic acids, flavonoids, polyunsaturated fatty acids, probiotics, minerals, and vitamins. Because of the fast-paced Western lifestyle, people have less time to eat fresh fruit and prefer to consume fruit juice instead. Hence, juice represents an excellent vehicle for the delivery of bioactive ingredients.

Some fruit juices are already highly valued because of their unique nutritive composition. However, new products in the fruit sector are constantly being designed for several reasons. People are willing to purchase healthier products, such as those that are organically made, those they find appealing, and new and different fruit juices. The demand for functional fruit juices is increasing rapidly. Superfruits with high levels of bioactive compounds, such as goji berries, areola, noni, and acai, are already added to some fruit juices, as well as prebiotics, probiotics, omega-3 fatty acids, botanicals, and isoflavones from soy, in order to make products functional [9,10]. In 2019, one study tried to identify interactions between bioactive compounds and the production of functional fruit juice. Cranberry bush juice was added to pear juice in order to make it functional. Finally, juice made from 95% pear and 5% cranberry bush (PC2) and juice made from 97.5% pear and 2.5% cranberry bush (PC1) were the most popular products among consumers. These juices also had good total soluble solid/total titratable acidity (TSS/TTA) ratios, and for the PC2 juice the interactions between bioactive compounds were the highest [11].

The economic value of juice production was confirmed over the last 5 years with statistical data from the European Union (EU), where production of juices not made from concentrate increased by 5.4%, while production of freshly squeezed juices grew by 4.8%. Simultaneously, the production of nectars and other types of fruit beverages declined by 3.8% and 1.6%, respectively [12]. In this context, domestic autochthonous fruit varieties are attractive raw materials used for the production of functional fruit juices that provide profitable food products on the market, while preserving traditional agriculture, contributing to the strengthening of local economies, and consequently contributing to the welfare of society.

Fruits and their juices contain significant amounts of different BACs, which individually or mutually have a significant impact on improving human health. The most common types of BACs in the majority of fruit juices are polyphenols, which are antioxidants with anti-inflammatory activities. Polyphenols bind free radicals, participate in the inhibition of pro-oxidative processes, and significantly influence chronic inflammation or lessen the burden of various medical pathologies, such as cardiovascular diseases, forms of cancers [13–17], and lipid oxidation in plasma [16,18]. There is a large body of evidence showing that polyphenols are true aids to human health and wellbeing; however, there is controversy associated with their utilization in food production. In order to obtain true functional product with medicinal value, it has to be designed according to strict rules and assessed with randomized, double-blind, placebo-controlled clinical trials [19].

In addition to the usual nutritional value, in order to provide functionality and added value to fruit juices there is the possibility of adding prebiotics and probiotics, which have been examined recently. The term "probiotic" refers to mixtures made of one or more bacterial species that act positively on the health of the host to which they are introduced by improving the properties of the hosts intestinal microflora [20–22]. The most common beneficial effects of probiotics are the reduction of lactose intolerance; reduction of cholesterol levels; stimulation of the immune system; and increased absorption of minerals with antimicrobial, anticarcinogenic. and antihypertensive effects [23]. Probiotic bacteria are traditionally added to fermented dairy products; however, their applicability is limited by lactose intolerance or by diets that require cholesterol restriction [23]. One solution is fruit juice, which stands out as a new category of carriers for probiotic bacteria [24–27] and has the advantage of containing suitable nutrient contents for their growth, hence stabilizing the probiotic product [23]. The use of probiotics improves the nutritional properties of the product, while in a fruit juice the addition of probiotics enhances the antioxidant activity (limited to certain strains of lactic acid bacteria) [23]. The most commonly used probiotic bacteria in fruit preparations are the different strains of *Lactobacillus plantarum*, *Lactobacillus acidophilus*, *Lactobacillus helveticus*, *Lactobacillus casei*, *Lactobacillus paracasei*, *Lactobacillus rhamnosus*, *Lactobacillus reuteri*, *Bifidobacterium breve*, *Lactobacillus gasseri*, *Lactobacillus crispatus*, and others [28]. Selection of the right bacterial strains for the production of fruit preparations is crucial, as ensuring their stability, survival, and functionality is more challenging than in fermented dairy products [23,29]. Despite all the positive attributes that probiotics can bring to human wellbeing, there are number of challenges that are associated with their consumption and industrial application.

3. The Use of Non-Thermal Technologies within the Hurdle Concept for the Preservation of Functional Fruit Juice

Currently, consumers demand that fruit juices are minimally processed, contain no additives, and have an extended shelf life. Recent studies have confirmed that consumers are often ready to pay a higher price for premium products of high sensory quality; therefore, there is currently high motivation to develop and produce premium quality juices. Such fruit juices are produced directly by squeezing or cold-pressing and are not subjected to any further processing—they are simply stored at a temperature of 2–5 °C, meaning they last for only a few days (1–3 days). On the other hand, the results of a major study involving over 160 US and EU fruit juice producers stated that microbiological safety is a key factor in product protection within a market. As many as 92% of manufacturers have encountered subsequent contamination (yeast or mold) in their products [30]. Functional fruit juices are short-term products, so classical (thermal) pasteurization (70–121 °C/30–120 s) assures they have an adequate shelf life, even though this treatment reduces the nutritional value [31,32] and the quality of the juice [33]. To that end, a recent study aimed to discover how physical, chemical, and sensory properties of tropical fruits juices change after pasteurization (85 °C, 30 s) and during 90 and 180 days of storage [34]. In the aforementioned study, the authors used two formulations made from different ratios of acerola cherry (*Malpighia emarginata* D.C.), cashew apple (*Anacardium occidentale* L.), yellow mombin (*Spondias mombin*), pineapple (*Ananas comosus* L.), acai (*Euterpe oleracea*), and camu-camu (*Myrciaria dubia*) fruit. The results revealed that the pasteurization did not affect the total polyphenolic content or antioxidant activity of either formulation but caused a 7% reduction of ascorbic acid. A higher carotenoid content in the pasteurized samples as compared to the control sample was explained by the combination of homogenization and heat treatment, which caused disruptions of cell membranes and the protein–carotenoid complex, making carotenoids more accessible for extraction.

Non-thermal processing technologies are able to extend the shelf life while retaining the nutritional and sensory value. These technologies operate at lower temperatures and with shorter processing times, while assuring microbiological safety, inactivation of the enzymes, and higher stability of the BACs in the juice. Recently, for the purposes of extending the shelf life and retaining the quality of juices, combinations of thermal and non-thermal technologies were tested as "hurdle" technologies. Here, these technologies are applied in a certain sequence, so that each of the non-invasive ("milder") processing conditions gives the best results in terms of extending the shelf life and sensory quality of fruit juices [35,36]. In this respect, most of the research studies today are HPU, PEF, cold plasma (CP), high-voltage electrical discharge (HVED), and HPP technologies [6]. The specificity of these concepts is reflected in the synergistic effects of various inhibition mechanisms or by the inactivation of target microorganisms [37]. Therefore, the innovative "obstacle" technology is an advanced approach to fruit juice processing with the potential to meet the high demands of consumers and manufacturers [30].

In recent decades, there has been an extensive increase in the number of commercially available foods processed using non-thermal technologies. For instance, food products processed using non-thermal technologies such as HPP have been marketed in Japan since 1990 and in the United States and Europe since 1996 [38]. At the same time, consumers do not have clear knowledge and information about the new technological processes applied to food; therefore, their attitudes are often negative towards such processing. This is the reason why consumers often consider processed food unhealthy and are often skeptical about its quality [39]. A recent study revealed that Brazilian consumer perception of food technologies was influenced by the degree of food technology neophobia [40]. Fresh, cold-pressed, and non-pressurized juice were mostly associated with healthy and natural products, whereas pasteurized juice and pressurized juice were associated with processed products and unhealthiness.

3.1. Pulsed Electric Field Processing

The application of pulsed electric field processing technology as an alternative to pasteurization of fruit juices has shown good results, without significant deterioration of nutritional or sensory juice quality [41,42]. PEF processing involves the application of a high voltage (about 50 kV cm^{-1}) in a very short time (μs to ms) to food placed between two electrodes. Here, the main parameters are temperature and time, electric field strength, and energy input [30].

Microbial inactivation by PEF is based on the rupture of microbial cells due to differences in the potential outside vs. inside the cells [43]. When the difference in these potentials reaches a critical value characteristic for each microorganism, electroporation occurs, inducing irreversible structural changes and inactivation of microorganisms [44]. Electroporation can be reversible, meaning that the intensity of the treatment may not inflict sufficient damage to the cells or that the pores on the cellular membrane might be too small. This indicates that microbes can recover from partial inactivation, so PEF is frequently combined with different hurdles for full safety.

The efficiency of PEF depends on the experimental conditions, food composition, and electrical conductivity of the food. If the conductivity is higher than 6 μs, the food is harder to process with this technology [43]. Additionally, PEF depends on several factors, such as the electrical field strength; processing time; width, frequency, and shape of the pulse; polarity; energy; and applied temperature [43]. Studies have indicated that microbial characteristics (e.g., cell size and shape, type of cell envelope) alone seem not to have the expected impact on PEF resistance [45]. Furthermore, PEF could not stand alone as an efficient treatment for inactivation of spores under mild processing conditions, meaning it should be combined with other preservation techniques to achieve the required results. Moreover, the other technological limitations of PEF, such as the high initial cost of equipment, still do not allow the complete commercialization of food products processed using PEF.

Liquid foods without bubbles are the most suitable for PEF treatment due to the presence of charged molecules that can transfer electric charges formed during treatment [42]. This means that PEF can also be used for processing of dairy and alcoholic beverages. Several studies have confirmed the lower efficacy of PEF in treatment of juices containing higher amounts of macronutrients (e.g., fats and proteins) in comparison to simple microbial suspensions [46]. With respect to microbial inactivation, physicochemical juice parameters such as pH, water activity, and soluble solids are some of the main factors that can affect the PEF treatment. Table 1 gives an overview of the application of PEF for the preservation of functional fruit juices and beverages over the last 5 years.

Many studies have confirmed the superior polyphenolic, mineral, and free amino acid contents, as well as better preservation of native color in PEF-treated juices as compared to thermally treated juices. High-intensity PEF (HIPEF) was able to improve the bioaccessibility of bioactive antioxidants in several functional beverages [47,48]. Recent experiments showed that PEF preprocessing applied to whole raw tomatoes (*Lycopersicum esculentum* cv. Raf) increased concentrations of individual carotenoids (phytofluene, phytoene, lycopene, δ-carotene, lutein, γ-carotene, β-carotene) [49]. In particular, the concentrations of phytoene and phytofluene were increased by 178% and 131%, respectively, in tomato products treated with PEF and compared to untreated fruit. The authors explained this as being caused by activation of the secondary metabolism in the treated tomatoes as a strategy to overcome unfavorable stress conditions. Moreover, the authors revealed that PEF treatment applied to tomato fruit increased the bioaccessibility of δ-carotene (2%), β-carotene (53%), lutein (125%), lycopene (137%), and γ-carotene (527%). These results strongly confirmed that PEF could lead to an easier release of carotenoids from the tomato matrix [49]. On the other hand, PEF had a negative effect on the cellular membranes of tomatoes that led to softening of the vegetable [50]. In conclusion, PEF was implicated as a beneficial preparation technology for production of tomato-based products with high carotenoid concentration.

Table 1. The application of pulsed electric field (PEF) treatment for the preservation of functional fruit juices and beverages over the last 5 years.

Juice Type	PEF Conditions	Nutritive/Physicochemical Quality	Microbial Safety	Key Conclusions	Reference
Mango (*Mangifera indica* Linnaeus) and papaya (*Carica papaya* Linnaeus) juices with added stevia infusion	Square-wave bipolar pulses, with pulse width of 2.5 μs. 20–40 kV cm^{-1} 100–360 μs stevia leaf infusion: 0–2.5% (w/v)	-ascorbic acid -total anthocyanins (TA) -total carotenoids (TC) -steviol glycosides -total soluble solid content -CIELab -hydroxymethylfurfural content (HMF)	/	-higher electric field strengths revealed higher ascorbic acid reduction. -higher electric field strengths resulted in stevia beverages with higher TA and TC contents. -the ratio between rebaudioside A and stevioside increased after PEF treatments. -higher electric fields led to significantly higher HMF value. -HMF and color variations were greater in beverages without stevia. -optimum PEF conditions with respect to bioactive compounds: 21 kV cm^{-1} during 360 μs with 2.5% stevia	Carbonell-Capella et al. (2016) [51]
Sour cherry juice Apricot and peach nectars	Square-wave bipolar pulses with 3 μs duration and 20 μs delaying timeFlow rate: 50 mL min^{-1} 24 kV cm^{-1} 125 Hz (66 μs, 8.4 Js^{-1}) 250 Hz (131 μs, 16.8 Js^{-1}) 400 Hz (210 μs, 26.9 Js^{-1})	-titratable acidity (TA) -electrical conductivity (EC) -Commission Internationale de l'Éclairage LAB (CIELab) -non-enzymatic browning index (BI) -total ascorbic acid content (TAAC) -total β-carotene content (TBC) -total monomeric anthocyanin content (TMAC) -aroma compounds -sensory analysis	/	-PEF treatment did not change 94% of the sensory properties and 64% (sour cherry juice), 60% (apricot nectar), and 30% (peach nectar) of the physical properties. -aroma compounds were affected by. -in all investigated samples. PEF treatment significantly changed 57% of a total of 73 identified aroma compounds. -PEF could be applied with different treatment times for the pasteurization of all investigated samples.	Evrendilek (2016) [52]
Apple juice (unclarified)	Cyclic PEF treatment—each cycle consisted of 50 pulses (one pulse every 30 s). Design of experiments (DOE): −30 kV cm^{-1} -number of cycles: 4, 6, 8 (total of 200, 300, and 400 pulses, respectively) -storage: 24, 48, and 72 h under refrigeration. T < 35 °C	-total vitamin C content -total polyphenols -antioxidant activity (ABTS)	-Mesophilic and psychrophilic actinomycetes -Microscopic fungi -Yeasts -Enterococci -*Salmonella* -*Staphylococcus aureus*	-Regardless of the number of pulses, PEF did not affect the contents of vitamin C or total polyphenols during storage. -PEF treatment and the number of pulses influenced antioxidant activity, which decreased immediately after the treatment and after 24 h of storage. -PEF treatment successfully inactivated food spoilage microorganisms. -increased number of pulses positively affected the reduction in number of studied microorganisms.	Dziadek et al. (2019) [53]

Table 1. *Cont.*

Juice Type	PEF Conditions	Nutritive/Physicochemical Quality	Microbial Safety	Key Conclusions	Reference
Pinot noir juices (*Vitis vinifera* L.) obtained at different maceration times (0, 2, 4, 8, and 14 days) after PEF treatments	PEF operating variables: -constant pulse width 20 µs -50 Hz -1.5 kV cm^{-1} -243 pulses ("PEF Low") -1033 pulses ("PEF High") -estimated specific energy inputs were 14.48 ± 0.11 kJ/kg and 69.99 ± 0.52 kJ/kg for "PEF Low" and "PEF High", respectively T < 25 ± 2 °C	-vitamin C -total phenolic content -malvidin-3-*O* -glucoside content -DPPH scavenging activity -simulated in vitro human gastrointestinal digestion -cell culture experiments using Caco-2 cell lines -biomarkers for general cellular health and integrity	/	-PEF treatment increased juice yield and preserved intense juice color. -PEF pretreatment of grapes improved the release of malvidin-3-*O*-glucoside for 224%. -PEF treatment resulted in higher total phenolic content (+61%), vitamin C (+19%), DPPH scavenging activity (+31%), bioprotective capacity (+25% for cell viability and +30% for LDH leakage).	Ying Leong et al. (2016) [54]
Date juice (variety Bou-Hattem)	High-intensity pulsed electric field (HIPEF) operating variables: -bipolar square-wave pulses of 4 µs -35 kV cm^{-1} -100 Hz for 1000 µs T < 35 °C Thermal treatment at 90 °C for 60 s in a tubularheat exchanger. All samples were stored in darkness for 5 weeks at 4 °C.	-total phenolic compounds -CIELab color measurement -HMF determination -turbidity evaluation -pH -soluble solids determination	/	-HIPEF treatment preserved the nutritive and physicochemical quality of date juices during storage in comparison to thermally treated and control (untreated) samples. -after HIPEF treatment, juices revealed higher amounts of total phenols, which were better preserved during storage than that untreated and thermally processed samples. -HIPEF did not alter the color parameters. -HIPEF treatment reduced HMF content of date juice after processing and during storage in comparison to thermally treated samples. -all investigated physicochemical properties were better-preserved after HIPEF in comparison to thermally processed and control (untreated) samples.	Mtaoua et al. (2016) [55]
Orange juice Watermelon juice Coconut water	All PEF processing conditions werestudied in a continuous-flow system. Moderate-intensity PEF: bipolar square-wave pulses of E = 0.9 and 2.7 kV cm^{-1} pulse width: τ = 15, 100 or 1000 µs High-intensity PEF (used in industrial applications): monopolar square-wave pulses of E = 10 or 20 kV cm^{-1} and pulse width τ = 2 µs	/	*Escherichia coli* *Listeria monocytogenes* *Lactobacillus plantarum* *Salmonella* Senftenberg *Saccharomyces cerevisiae*	-moderate-intensity PEF was shown to be very effective and easy to scale, and thus could be an alternative for pasteurization of fruit juices. -Optimal PEF conditions, which could match those of pasteurization: E = 2.7 kV cm^{-1} τ = 1000 µs. -moderate-field PEF can be used for treatment of both high-acid and low-acid products, in contrast to high-intensity PEF, which is only suitable for high-acid products. -moderate-intensity PEF demonstrated slight differences in the degree of inactivation between the different microbial species tested, while high-intensity PEF indicated greater differences between the microbial species.	Timmermans et al. (2019) [56]

Table 1. *Cont.*

Juice Type	PEF Conditions	Nutritive/Physicochemical Quality	Microbial Safety	Key Conclusions	Reference
Cloudy apple juice	Low-intensity PEF: 12.5 kV cm^{-1} Flow 27.6 L/h Energy input 76.4 kJ/L Frequency 62 Hz T_{inlet} 37.6 °C T_{outlet} 59.5 °C Thermal pasteurization (TP): 72 °C/15 s High-intensity PEF: 12.3 kV cm^{-1} Flow 24.5 L/hEnergy input 132.5 kJ/L Frequency 94 Hz T_{inlet} 37.3 °C T_{outlet} 72.8–73.8 °C Thermal pasteurization (TP): 85 °C/30 s Storage: 3 weeks at 4 °C	-color measurement -turbidity and cloud stability -particle size distribution -polyphenol oxidase (PPO) activity -peroxidase (POD) activity -pectin methylesterase (PME) activity -total soluble solids (TSS) -sugar profile -pH, titratable acidity (TA) -organic acid profile -vitamin C -sensory analysis -volatile compounds		-PEF-treated juices differed from the untreated juice, showing higher lightness (L*) and redness (a*). -PPO, POD, and PME activities were greatly reduced by high-intensity PEF. -vitamin C and cloud stability decreased during storage. -significant changes in pH, titratable acidity, organic acid, and sugar contents were not observed. -esters noticeably increased in juices after PEF treatments in comparison to TP treatment, where ester degradation reactions occurred together with the formation of off flavors. -increased contents of fructose and glucose and decreased contents of sucrose were observed during storage in all juices.	Wibowo et al. (2019) [57]
Beverages formulated with a blend of fruit juices (orange, kiwi, pineapple, and mango) and water (WB), milk (MB), or soy milk (SB)	-High-intensity pulsed electric field (HIPEF): 35 kV cm^{-1} 4 μs bipolar pulses at 200 Hz for 1800 μs T < 35 °C Thermal treatment (TT): 90 °C for 1 min)	-in vitro gastrointestinal digestion -individual carotenoids -lipophilic antioxidant activity (LAA) -bioaccessibility	/	-after HIPEF treatment, the contents of several carotenoids increased by between 8% and 28%. -HIPEF was found to be more effective than TT in preserving the concentrations and bioaccessibility of carotenoids and other lipophilic compounds in terms of antioxidant activity of investigated beverages. -the beverage with the highest bioaccessibility of total carotenoids was MB, followed by SB and WB. -milk matrix (MB) in combination with HIPEF improved the bioaccessibility of carotenoids by 15% as compared with the untreated samples. -HIPEF and TT decreased the bioaccessibility of carotenoids in WB. -food matrixes and food processing are able to modify the bioaccessibility of carotenoids.	Rodríguez-Roque et al. (2015) [48]

Table 1. *Cont.*

Juice Type	PEF Conditions	Nutritive/Physicochemical Quality	Microbial Safety	Key Conclusions	Reference
Beverages formulated with a blend of fruit juices (orange, kiwi, pineapple, and mango) and water (WB), milk (MB), or soy milk (SB)	-High-intensity pulsed electric field (HIPEF): 35 kV cm^{-1} 4 µs bipolar pulses at 200 Hz for 1800 µs T < 35 °C Thermal treatment (TT): 90 °C for 1 min)	-vitamin C -individual phenolic compounds -total phenolic content (TPC) -hydrophilic antioxidant activity (HAA) -bioaccessibility	/	-HIPEF reduced the content of vitamin C (8%–15%) as compared with untreated samples. -TT negatively affected the stability of vitamin C (losses of 31%) in comparison to untreated samples. -HIPEF did not alter the bioaccessibility of vitamin C in comparison with untreated samples. -significant decrease in the vitamin C bioaccessibility was noticed in TT samples. -HIPEF treatment provoked increased content of several phenolic compounds in MB and SB. -food matrix and processing could modify the bioaccessibility of bioactive compounds	Rodríguez-Roque et al. (2015) [47]
Clarified pomegranate juice (Hicaz cultivar)	pulse duration: 3 µs pulse delay time: 20 µs frequency: 500 pps controlled flow rate: 60 mL min^{-1} DOE: 0, 17, 23, 30 kV cm^{-1} 5, 15, 25, 35 °C Total treatment time was estimated at 108.4 µs, with applied energies of 37.5, 50.3, and 65.3 J, respectively.	-pH -CIELab, browning index (BI) -total antioxidant capacity (TAC)–DPPH -total phenolic content (TPC) -total monomeric anthocyanins (TMAC) -total ascorbic acid (TAAC) -sensory evaluation	*E. coli* O157:H7 (EDL 931 04054) *S. aureus* (95047)	-electric field strength was the most significant factor in terms of bacterial inactivation. -the inactivation of *S. aureus* and *E. coli* O157:H7 in PEF-treated samples reached up to 4.47 and 5.43 log CFU/mL, respectively. -the decreases in the mean initial TAC, TMAC, and TAAC with increased temperature, electric field strength, and energy were not significant. -the sensory properties of flavor, taste, aftertaste, and overall acceptance were not affected by PEF alone or PEF mild heat treatment.	Evrendilek (2017) [45]

As mentioned earlier, PEF can inactivate microorganisms in juices while retaining quality (e.g., color, soluble solids, viscosity, content of bioactive compounds, etc.), which makes it an advantageous method for preserving foods over an extended shelf life [53,58]. The effects of PEF on the sensorial and nutritional quality of mango juice right after treatment and during storage (75 days) were recently tested [59]. Experiments showed that there were no significant changes in conductivity or in the pH of the samples after treatment. However, pH did decrease slightly during storage for 59 days. Furthermore, the Commission Internationale de l'Éclairage LAB (CIELab) lightness (L*) remained the same during storage, which indicated good preservation of color. The native yellow color of mango juice remained unchanged, as PEF inactivates polyphenoloxidase (PPO) and peroxidase (POD), which catalyze the browning reactions. Electrotechnologies may induce structural and conformational changes in food enzymes, which could limit their activity, thus strongly affecting product quality.

Another study aimed to evaluate and compare the effects of processing via high-intensity pulsed electric fields (HIPEF) vs. thermal pasteurization (90 °C/60 s) on the nutritive and physicochemical parameters of date juice (variety Bou-Hattem) before and after storage for 5 weeks at 4 °C. In comparison to pasteurization, HIPEF better preserved the physicochemical characteristics and even caused an increase in the concentrations of some bioactive compounds in fruit juices. Moreover, HIPEF-treated date juices, in comparison to untreated and thermally processed samples, retained quality better during storage. A possible reason could be the improved extraction of intracellular contents due to permeabilization of plant cells, which is enhanced by PEF [60]. As a result of food processing and storage, some biochemical reactions may occur, which could promote the formation of new compounds. Those new compounds are usually formed by constituents (e.g., polyphenols) that are stored in polymers, such as pectin or cellulose, and could be released during processing or storage [61].

Non-enzymatic browning reactions were slowed down by PEF, which resulted in a low concentration of hydroxymethylfurfural (HMF) in treated date juice. HMF, as an indicator of non-enzymic browning, is probably a result of Maillard reaction in the presence of a reducing sugars (e.g., glucose and fructose) and acids under high temperature [62]. The slow browning rates observed in HIPEF juices in comparison to pasteurized ones could be related to the high retention of acids in date juice [61]. The reason for this could be the preservation of ascorbic acid oxidation in HIPEF. Non-oxidized ascorbic acid cannot provide reactive carbonyl groups, which are precursors of non-enzymatic browning that lead to the formation of HMF.

Another study on Pinot noir grape juice was done to evaluate exposure to PEF and the effects this has on the quality and nutritive value (e.g., contents of anthocyanins, total phenolics, and vitamin C). Grapes have a number of anthocyanins, phenolic acids, flavonols, flavan-3-ols, and stilbenes. Prolonged maceration is commonly used to increase the concentrations of these compounds, which are released from grape skins, while PEF can be used to accelerate this process. To that end, after preparation the grape juice was treated with PEF (constant pulse width of 20 µs; pulse frequency of 50 Hz; electric field strength of 1.5 kV cm^{-1}; "PEF Low" pulse numbers of 243, and "PEF High" of 1033) and subsequently macerated. The grape juice treated with PEF had better yield during the first four days of maceration than untreated juice. Enhancement of juice yield and improvement of mass transport of juice from the grape pulp is credited to the PEF permeabilization of cell membranes. Juices that were processed with "PEF low" had increased vitamin C content during the first two days of maceration. However, when the maceration was longer than 8 days, PEF-treated juices exhibited higher degradation of vitamin C than non-treated juices. It seems that the higher release of vitamin C in early maceration increased the available total vitamin C for oxidation due to the higher susceptibility to oxidation and degradation with prolonged maceration. "PEF high" samples showed an increase of total phenols after two days of maceration. "PEF low" and "PEF high" juices had higher contents of malvidin-3-*O*-glucoside (anthocyanin) after eight days of maceration as compared to control samples, suggesting that PEF has the potential to maximize the release of phenolics much more effectively than conventional maceration alone. In addition, the maceration process was successfully shortened from 14 to 8 days, thus confirming the potential of PEF in grape juice processing [30].

3.2. High-Power Ultrasound Processing

The use of ultrasound (US) as an alternative to pasteurization in fruit juices has great potential, as the acoustic energy is immediately transmitted through the entire volume of juice, which significantly shortens the treatment time. In addition, its lower energy consumption and potentially lower costs makes this method appropriate for industrial usage [63]. Fruit juice processing commonly combines high-power ultrasound processing with high-intensity ultrasonic waves (10–1000 Wcm^{-2}) at low frequencies (20–100 kHz). Due to the high power, ultrasonic waves in juice cause physiochemical changes, accelerating biochemical reactions and causing enzymatic and microbial inactivation [64]. Expansion vortexes are formed due to the passage of the ultrasonic waves through the liquid, forming small bubbles and negative pressures, which is defined as the cavitation phenomenon [65]. The collapse of the cavitational bubbles inside and around microbial cells causes the pressure gradient, which induces mechanical damage to the cells. Additionally, free radicals that damage the cell walls are formed during HPU treatment, with spontaneous formation of bactericidal hydrogen peroxide further enhancing the microbial inactivation [64]. Studies have documented that ultrasound processing has the potential to achieve a 5-log reduction in foodborne pathogens in fruit juices. Although ultrasonic waves cannot guarantee bacterial spore inactivation, synergic effects with other technologies (e.g., in hurdle technology) could enhance the effectiveness of spore-forming bacteria inactivation [66,67]. The main disadvantages found during ultrasound processing are related to process scale-up and quality changes in the product. Parameters influencing the effectiveness of HPU treatment in fruit juice processing are the ultrasonic strength, frequency, ultrasonic probe diameter, amplitude, time of treatment, and temperature [64]. This technology makes it possible to obtain microbiologically stable fruit juice with preserved BACs and improved bioavailability [68,69]. Table 2 provides a summary of the application of HPU for the preservation of functional fruit juices and beverages over the last 5 years.

HPU can provoke microstructure changes of treated juice, particularly the rupturing of cellular walls. More precisely, the physical force of ultrasound causes the dispersion of the intracellular components into the juice, causing higher injury to the cellular structures, which occurred through elongating the HPU processing time [70]. These structural changes may greatly contribute to the improved physical properties of juice with respect to rheological properties, color attributes, cloudiness, and water-soluble pectin.

The addition of various prebiotic carbohydrates in orange juice, such as xylooligosaccharides (XOS), has been introduced through high-intensity ultrasound (HIUS) for the engineering of functional beverages with similar sensory characteristics to those of fresh products [71]. The quality parameters and XOS stability of functional beverages were investigated under the effects of nominal US power values (W) of 0 (control/untreated sample), 300, 600, 900, and 1200 W over 10 min. HIUS treatment over a temperature range of 51–88 °C did not affect the chemical stability of the XOS. Therefore, XOS-enriched beverages preserved the functional properties of added prebiotics. However, the negative effects of HIUS were noticed on the stability of ascorbic, malic, and citric acid contents, as well as on the total phenolic content and antioxidant activity.

Table 2. The application of ultrasound (US) treatment for the preservation of functional fruit juices and beverages over the last 5 years.

Juice Type	US Conditions	Nutritive/Physicochemical Quality	Microbial Safety	Key Conclusions	Reference
Strawberry juices (*Fragaria x ananassa* Duch. cv. Aromas) with prebiotic fiber (inulin and oligofructose)	Ultrasonic cleaning bath (TestLab, Argentina) −40 kHz −180 W Time: 0, 15, 30 min Preservation treatment included US + addition of geraniol. Storage: 14 days at 5 °C	-total phenolic content (TPC) -total flavonoids content (TFC) -total antioxidant capacity (TAC) -ascorbic acid -sensory evaluation	Mesophilic bacteria Yeasts Molds Inoculation with *E. coli* O157/H7 and *L. innocua*	-optimal preservation conditions: inulin: oligofructose = 5: 3, 0.225 µL/mL of geraniol and ultrasound time equal to 0. -the optimum preservation treatment was highly effective in reduction of native microflora and inhibition of inoculated pathogens in juices. -the optimum preservation treatment did not have a negative influence on bioactive compounds or antioxidant capacity. -the optimum preservation treatment promoted the stability of prebiotic fibers added into juices during storage.	Cassani et al. (2018) [72]
Cloudy apple juice (cv. Golden Delicious)	High-power ultrasound treatment (HPU) -ultrasound probe system (UP 100H, Hielscher Company, Teltow, Germany). 100 W, 30 kHz -probes with tip diameters of 10 and 20 mm -amplitudes: 40% and 80% -time: 3, 6, 9 min Storage: 7 days at 4 °C	-total phenols (TP) -total flavan-3-ols (TFL) -in vitro antioxidant capacity (DPPH, FRAP)	/	-in comparison to untreated samples, HPU treatment caused reductions in TP of 32.94% and TFL of 21.66%, while DPPH and FRAP values declined by 23.76% and 27.49%, respectively. -all HPU variables significantly affected phenolic stability and antioxidant capacity. -7 days of cold storage revealed the highest reduction of TP (89.21%), followed by TFL (82.80%), DPPH (79.51%), and FRAP (66.04%). -lower reductions of TFL (46.97%), DPPH (20.55%), and FRAP (24.16%) were observed during cold storage in untreated samples than in ultrasound treated.	Bursać Kovačević et al. (2019) [73]
Pear juice (*Pyrus bretschneideri* Read.)	-ultrasonic processor (VC 750, Sonics and Materials Inc., Newtown, CT, USA) 750 W, 20 kHz, 12.7 mm Amplitude: 70% US pasteurization at: 25, 45, and 65 °C (US25, US45, and US65) for 10 min. Conventional pasteurization: 65 °C/10 min (P1) 95 °C/2 min (P2)	-enzyme activity: POD, PPO, PME -ascorbic acid content -total phenols and flavonoids -soluble solids content, pH, acidity -total antioxidant capacity	-total plate count, yeast and mold	-significant reduction in residual activities of POD (43.2%), PPO (37.83%), and PME (40.22%) were observed in US45 juices. -the highest enzyme inactivation was found in US65 juices, which exhibited residual activities of POD, PME, and PPO of 4.3%, 3.25%, and 1.91%, respectively. -a complete inactivation of microbes was found in P95 and US65 treatments. -significant increase in the contents of ascorbic acid, total phenols, and flavonoids was detected in the US25 samples. -Ultrasound pasteurization at 65 °C for 10 min ensured the best retention of investigated bioactive compounds and enzymatic and microbial inactivation.	Saeeduddin et al. (2015) [74]

Table 2. *Cont.*

Juice Type	US Conditions	Nutritive/Physicochemical Quality	Microbial Safety	Key Conclusions	Reference
Clarified pomegranate juice	US generator (500 W, 20 kHz; Vibra-Cell 505, Sonics and Materials, Inc., Newtown, CT, USA) 500 W, 19 mm diameter probe, <35 °C Amplitude: 50%, 75%, 100% Time: 0, 6, 12, 18, 24, 30 min Pulse intervals of 5 s on and 5 s off.	-CIELab -total monomeric anthocyanins -total phenols -soluble solids content, pH	*E. coli* ATCC25922 *S. cerevisiae* ATCC 2366	-5-log reduction in *E. coli* ATCC 25,922 as a surrogate for *E.coli* O157:H7 and 1.36-log reduction of *S. cerevisiae* inoculated into pomegranate juice were achieved after US treatment (100%, 30 min). -the contents of anthocyanins were decreased upon US treatment (75% and 100%, t > 18 min) -total phenols did not change significantly during US treatment. -US technology showed the potential for enhancements in both the safety and quality of pomegranate juice.	Pala et al. (2015) [75]
Apple/grape juice blend (50:50)	Sonicator (JY96-IIN, Ningbo Scientz Biotechnology Co. Ltd. Ningbo, China), 25 kHz, 70% amplitude -Thermal treatment 100 °C/4 min (blanching) -high-temperature short-time (HTST) 72 °C/15 s -Ultra-sonication 5 and 10 min -Thermo-ultrasound 5 and 10 min/40 °C -Thermo-ultrasound 5 and 10 min/50 °C -Ultrasonic probe	-polyphenolic profile -total phenols -total flavonols -total flavonoids -DPPH-free radical scavenging activity -total antioxidant capacity -soluble solids content (SSC), viscosity, turbidity, pH, and acidity -CIELab	/	-ultrasonication (5 and 10 min) indicated significant increase in antioxidant activity and anthocyanin content of treated samples as compared to other treatments. -regarding individual bioactive compounds, significant differences were observed among all the treatments in this study. -insignificant influence of all the examined treatments on pH and titratable acidity (TA) was observed. -blanching, HTST, ultra-sonication, and thermo-ultra-sonication treatments had significant effects on viscosity, turbidity, SSC, and color parameters.	Aadil et al. (2020) [76]
Bayberry juice	Ultrasonic processor (600 W; BILON-600Y, Bilon Co., Ltd., Shanghai, China), 13 mm diameter probe tip, 20 kHz. -amplitudes of 20%, 40%, 60%, 80%, and 100%, with corresponding ultrasonic intensity levels of 90, 181, 271, 362, and 452 Wcm^{-2}, respectively. -treatment times of 2, 4, 6, 8, and 10 min, with pulse durations of 5 s on and 5 s off. Thermal processing (TP) 90 °C/1 min	-soluble solids content (SSC), pH, titratable acidity -ascorbic acid -monomeric anthocyanins-antioxidant activity (DPPH) -hydroxymethylfurfural (HMF) -browning degree (BD) -CIELab and polymeric color (PC)-superoxide dismutase (SOD) activity	/	-US did not affect pH, SSC, TA, or yellowness (b*) as compared to untreated samples. -HMF, PC, BD, and L* values increased with higher ultrasonic intensity and prolonged treatment time. -in contrast to anthocyanins, US at lower intensity for a short time did not affect the stability of ascorbic acid. -with increasing US intensity and time, bioactive compounds decreased. -SOD activity increased (21–28%) with short (2–6 min) treatment time and then decreased with extension of ultrasound processing. -US treatments (<450 Wcm^{-2}, 8 min) were optimal for preserving the quality of bayberry juice compared with TP treatment.	Cao et al. (2019) [77]

Table 2. *Cont.*

Juice Type	US Conditions	Nutritive/Physicochemical Quality	Microbial Safety	Key Conclusions	Reference
Grapefruit juice	Ultrasonic cleaner (SB-600 DTY, Ningbo Scientz Biotechnology Company Limited, Ningbo, China), 600 W 28 kHz, power radiation 70%, 20 °C, 30, 60, and 90 min	-total carotenoids -total lycopene -total anthocyanins -individual phenolic compounds -viscosity -sugars (sucrose, glucose, and fructose) -mineral elements	The microbiological analysis [78]	-significant increases in total carotenoids, lycopene, sugar contents, and phenolic compounds and decreases in viscosity and microbials were found in all the US samples as compared to control. -maximum improvement was observed in the US samples treated for 90 min. -complete microbial reduction was not achieved after US treatment.	Aadil et al. (2015) [79]
Strawberry juice	VC-750 US unit (Sonics and Materials, Inc., Newtown, CT, USA) 20 kHz, probe diameter 12.5 mm Sonication time: 5, 10, 15 min HTST: 72 °C/15 s Storage: 2 weeks at room temperature	-CIELab -ascorbic acid content (AAC) -antioxidant capacity (AOC) -total phenolic content (TPC)	*E. coli* O157:H7	-5-log reduction of *E. coli* O157:H7 was achieved by US treatment for 5 min (5.04 log CFU/mL), 10 min (5.36 log CFU/mL), and 15 min (6.08 log CFU/mL). -higher retentiona of color parameters were achieved during US as compared to HTST. -AAC decreased in all samples during storage, although US10 and US15 samples showed higher retention of AAC as compared to control samples. -US15 samples demonstrated the highest AOC and TPC during storage.	Yildiz et al. (2019) [80]
Blueberry juice (*Vaccinium corymbosum*)	Continuous ultra-sonication system (Model CPX 500, Cole Palmer Instruments, Vernon Hills, IL, USA) Probe diameter 10 mm, 20 kHz, 500 W Amplitude: 40%, 80%, 100% T < 25 °C	-total anthocyanin content (TAC) -total phenol content (TPC) -antioxidant activity (AA) -total soluble solids, pH, and titratable acidity -CIELab	-Aerobic plate count (APC) -Total coliforms (TC) -Yeasts and molds	-increased US intensity (amplitude) resulted in greater reductions of APC, TC, and yeast. -molds were not detected in the juice samples. -the highest log reduction in total aerobes (1.36 log CFU/mL) was achieved with high-intensity (100% amplitude) treatment. -US treatments did not affected TAC and color parameters of treated samples. -the TPC content of US treated samples significantly increased with flow rate and amplitude.	Mohideen et al. (2015) [81]
Peach juice	Ultrasonic tip (ECO-SONIC, QR1000 Model, Brazil) 1000 W, 20 kHz, 1.26 cm^2 titanium tip793.65 Wcm^{-2} Sonication time: 0, 3, 6, 10, 15 min Storage: 21 days at 25 °C	-microstructure, physical properties, and stability (optical microstructure, particle size distribution, pulp sedimentation, serum cloudiness) -CIELab -rheological properties	/	-pulp sedimentation was highly reduced by the US treatment. -juice consistency and serum cloudiness (turbidity) showed an increase upon US treatment, followed by a decrease and then a new increase with respect to processing time. -US could be used to improve the physical properties of peach juice without significant color changes during storage.	Rojas et al. (2016) [82]

HPU treatment may prevent the activity of browning enzymes such as PPO and POD, maintain the taste and nutritional value, and the improve stability of juice with respect to conventional heat treatment [83]. The efficiency of sonication depends on the intensity of treatment. Accordingly, it can cause inactivation of microorganisms, increase the concentration of bioactive compounds, or accelerate fermentation. The impact that ultrasound has on a juice depends on the juice itself. A recently conducted experiment on strawberry juice showed that ultrasound did not affect the color of the juice, nor did it form "off flavors". Moreover, it increased the acidity and sweetness of the juice and the scavenging activity (i.e., antioxidative capacity) of the samples by 75%, which continuously grew during storage. After the treatment, the concentration of phenolic compounds was higher than in fresh juice, likely caused by cavitation or addition of hydroxyl radicals to the phenolic rings during the treatment.

The positive aspects of ultrasound treatment were documented in a study on the effects of HPU on passionfruit juice. The juice was treated with a frequency of 20 kHz for 10 min, after which it was stored with untreated samples at 4 and 10 °C for 10 days. Samples were taken every other day and their color, pH, ascorbic acid content, and microbial count were measured. Results of this experiment showed that HPU did not cause sufficient cell damage to start immediate microbial inactivation. However, sonication treatment successfully reduced the aerobic mesophilic bacteria and yeast counts. Due to HPU treatment, the color parameters lightness (L*), a* (redness), and yellowness (b*) of passionfruit juice changed and the color was lighter and greener. The degradation of ascorbic acid was significant, but was lower than the decrease caused by pasteurization, hence making the HPU treatment a better method for preservation of fruit juices. Further studies showed that the juice was stable for 10 days after sonication because there were no significant changes in microbial quality of the product, as opposed to non-sonicated samples [84]. Aside from juices, the positive impacts of ultrasound extend to the quality of meat, fish, cheese, chocolate, vegetables, emulsions, and other products. Although ultrasound is commonly used as a replacement for pasteurization, it can also be applied as a pretreatment for foods that are processed by conventional methods, such as freezing, defrosting, drying, extrusion, slicing, and frying.

Non-clarified strawberry juice was exposed to HPU and thermal treatment (TT) in order to compare their influences of the quality and nutritional value during 10 days of storage. The evaluated properties were color, total soluble solids, total acidity, microbial growth, total phenolic content, and antioxidant activity (DPPH assay). The juice was sonicated at 40 kHz for 10 and 30 min at 90 °C for 60 s. All HPU samples showed a constant color right after the treatments, which started to degrade during storage (the largest drop was observed on the third day). TT samples had higher color changes after the treatment compared with untreated samples, but after the initial change the color remained stable over the storage period. All samples showed no "off odors", but HPU samples were sweeter than TT juice, probably due to the breakage of cellular structures that contained sugars. HPU and TT samples had similar total soluble solid contents, total acidity, and nutritive value, while TT inhibited microbial growth better than HPU [85].

Conversely, the negative side effects of ultrasound on juice quality can include changes of color, viscosity, formation of free radicals and off flavors, as well as degradation of lipids. In some juices, sonication reduced the amount of ascorbic acid, anthocyanins, and carotenoids. These factors, as well as the cost of ultrasound treatment, affect (and limit) the application of ultrasound in the food industry [86].

Due to the low lethality of ultrasound, a number of microorganisms, especially bacteria spores, can survive HPU treatment. To insure a higher lethality, HPU is often combined with pressure, heat, or both at the same time. These combinations are more energy-efficient than ultrasound alone, but cannot be used for preservation of thermosensitive products. For example, thermosonication can accelerate enzymatic inactivation and sterilization by shortening the treatment time and decreasing the intensity of sterilization and the corresponding negative side effects [87]. Moreover, the effectiveness of ultrasound can be improved when it is combined with chlorination, extremes of pH, and addition of preservatives [88].

3.3. High-Pressure Processing

High-pressure processing is considered the best non-thermal method used for food preservation. Food products that can be processed with HPP are low- to medium-moisture semisolids and solids (vacuum packed), high-moisture solids (vacuum packed), and high-moisture liquid food (bottled). Products are packed before the process, which decreases the chance of postcontamination. After this, products are placed in baskets and lowered into a vessel. The vessel is then filled with water and the products are treated with high hydrostatic pressure, which can be as high as 800 MPa. The pressure is evenly spread throughout the vessel, which assures that the product will be properly processed [89]. By applying HPP, non-covalent bonds in enzymes, ribosomes, and molecules of the cell membranes are ruptured, resulting in microbial inactivation. Since bacterial spores are quite resistant to HPP, they are often combined with high temperature to make the process more efficient [86]. The efficiency of HPP depends on the pressure, temperature, time, and pH [86].

Governmental authorities in the USA, Europe, and Canada have approved the use of HPP for reducing pathogens in juices and extending their shelf life. With an annual growth of at least 1%, fruit beverages show the fastest growth in the HPP food market, so such products may hold a higher market share than meat products that utilize HPP technology. However, the main technological limitations of HHP are the high investment costs for the equipment and the batch operation mode. The application of HPP for the preservation of functional fruit juices and beverages over the last 5 years is outlined in Table 3.

When HPP (190 MPa, 60 °C) was compared with pasteurization (90 °C, 5 min), reduced numbers of bacteria and yeast were shown in mango juice, and the shelf life of the juice was prolonged for up to 60 days while stored at 4 °C. Furthermore, losses of ascorbic acid, carotenoids, and total phenols were reduced as compared to control samples [86]. HPP has good selectivity for destruction of pathogenic bacteria while retaining good probiotic bacteria. For instance, after HPP treatment, *Lactobacillus rhamnoses* GG probiotic bacteria were preserved. Some studies even show that after 30 days of storage, the bacteria had a content of over 108 CFU/mL, with a well-preserved nutritive profile. Hence, it is not surprising that consumers showed tendencies toward purchasing HPP-treated probiotic products rather than thermally treated products [90].

The inactivation of quality degradation by endogenous enzymes in fruit juices is needed to obtain a high-quality product. The effects of HPP on the inactivation of various enzymes in fruit juices, e.g., PPO, POD, pectinmethylesterase (PME), β-glucosidase, polygalacturonase, lipoxygenase, amylase, and hydroperoxide lyase, have been extensively studied [91]. One recent study aimed to compare white grape juice samples treated with HPP and TT with respect to microbial levels (aerobic bacteria, coliform bacteria, and yeast/mold), physiochemical and antioxidant properties, enzymatic activities (PPO and POD), and sensory analysis during storage. After 20 days of storage, enzymatic activity decreased by 50% in HPP samples, with significant drops in microbial levels. Moreover, HPP juice received a similar sensory evaluation as the fresh juice, while TT the reduced sensory appeal in the samples.

Table 3. Application of high-pressure processing (HPP) for the preservation of functional fruit juices and beverages over the last 5 years.

Juice Type	HPP Conditions	Nutritive/Physicochemical Quality	Microbial Safety	Key Conclusions	Reference
Cloudy apple juice	400 MPa at room temperature for 3 min (HPP1) 600 MPa at room temperature for 3 min (HPP2) Storage: 3 weeks at 4 °C	-color measurement -turbidity and cloud stability -particle size distribution -polyphenol oxidase (PPO), peroxidase (POD), and pectin methylesterase (PME) activity -total soluble solid (TSS) -sugar profile -pH, titratable acidity (TA)–organic acid profile -vitamin C -sensory analysis -volatile compounds	/	-CIELab parameters did not significantly differ among HPP1- and HPP2-treated juices. -HPP did not completely inactivate PPO and POD. -due to high residual PME activity (>90%), cloud stability decreased during storage in HPP juices. -significant changes in pH, titratableacidity, organic acid, and sugar content between HPP1 and HPP2 were not observed. -increased content of fructose and glucose and decreased content of sucrose were observed during storage.	Wibowo et al. (2019) [57]
Beverages formulated with a blend of fruit juices (orange, kiwi, pineapple, and mango) and water (WB), milk (MB), or soy milk (SB)	HPP: 400 MPa at 40 °C for 5 min Thermal treatment (TT): 90 °C for 1 min)	-in vitro gastrointestinal digestion -individual carotenoids -lipophilic antioxidant activity (LAA) -bioaccessibility	/	-HPP improved the contents of *cis*-violaxanthin, anteraxanthin, lutein, and zeaxanthin in MB (between 12% and 37%) compared with the control (untreated) samples. -lower amounts of total carotenoids were observed in TT beverages in comparison to HIPEF ones. -HIPEF was found to be more effective than TT in preserving the concentrations and bioaccessibility of carotenoids and other lipophilic compounds with antioxidant activity in the investigated beverages. -the beverage with the highest bioaccessibility of total carotenoids was MB, followed by SB and WB. -HPP increased the bioaccessibility of carotenoids in SB beverages by 10%. -HPP and TT decreased the bioaccessibility of carotenoids in WB. -food matrix and food processing are able to modify the bioaccessibility of carotenoids.	Rodríguez-Roque et al. (2015) [48]

Table 3. *Cont.*

Juice Type	HPP Conditions	Nutritive/Physicochemical Quality	Microbial Safety	Key Conclusions	Reference
Beverages formulated with a blend of fruit juices (orange, kiwi, pineapple and mango) and water (WB), milk (MB), or soy milk (SB)	HPP: 400 MPa at 40 °C for 5 min Thermal treatment (TT): 90 °C for 1 min)	-vitamin C -individual phenolic compounds -total phenolic content (TPC) -hydrophilic antioxidant activity (HAA) -bioaccessibility	/	-HPP did not alter the content of vitamin C in comparison with untreated samples, with the exception of SB, where a decrease of 10.5% was found. -TT negatively affected the stability of vitamin C (losses of 31%) in comparison to untreated samples. -HPP did not change the bioaccessibility of vitamin C in comparison with control samples, except for MB, which increased by 8%. -significant decrease in the vitamin C bioaccessibility was noticed in TT samples. -HIPEF treatment provoked increased contents of several phenolic compounds in MB and SB. -food matrix and processing could modify the bioaccessibility of bioactive compounds	Rodríguez-Roque et al. (2015) [47]
Pomegranate juice (*P. granatum* L. cv. Hicaznar)	HHP: 200, 300, 400 MPa 5 °C, 15 °C, 25 °C 5 and 10 min Thermal treatment (TT): 85 °C/10 min	-pH, titrable acidity, °Brix -CIELab -total phenolic content (TPC) -total monomeric anthocyanin concentration (TMAC) -antioxidant (free radical scavenging) activity (AA) (DPPH) -ascorbic acid-mannitol	Mesophilic bacteria Yeasts Molds	-HHP juices indicated no significant decreases in AA, TPC, and TMAC as compared to TT samples. -HHP juices treated for 5 min exhibited had ascorbic acid but decreased content with HPP at 10 min. -lower mannitol content was detected in HHP juices as compared to control. -optimal HPP treatments regarding microbial inactivation: 400 MPa/15 °C/5 min and 400 MPa/5 °C/10 min.	Subasi and Alpas (2017) [92]
Açaí juice (*Euterpe oleracea*)	HPP: 400, 450, 500, 600 MPa 20 °C 5 min Thermal treatment (TT): 85 °C/1 min	-anthocyanins -non-anthocyanin phenolic compounds -tocopherols -antioxidant capacity toward oxygen and nitrogen reactive species (ORAC, HOCl-scavenging capacity, H_2O_2-scavenging capacity, effect on the formation of nitroso compounds (NOC)	/	-HPP technology could be successfully used to produce high-quality açaí juice, with better retention of anthocyanins and increases in the content of non-anthocyanin phenolic compounds and antioxidant capacity as compared to the untreated and TT juices. -α-tocopherol, γ-tocopherol, and vitamin E activity were not changed upon HPP treatment compared to the control. -in comparison to TT juices, only HPP-500 MPa revealed a significant decrease in α-tocopherol. -HPP optimal conditions: 500 MPa/5 min/20 °C with respect to nutritive quality.	da Silveira et al. (2019) [93]

Table 3. *Cont.*

Juice Type	HPP Conditions	Nutritive/Physicochemical Quality	Microbial Safety	Key Conclusions	Reference
Cloudy apple juice (*Malus domestica* Borkh. cv. Gloster)	HPP: 200, 300, 600 MPa 5, 25, 45 °C 1, 5, and 15 min Darkness storage at 4 ± 2 °C for 2, 4, 6, 8, and 12 weeks.	-total soluble solids and pH -sugar content -vitamin C -polyphenol oxidase (PPO) and peroxidase (POD) activity -total and individual polyphenols	/	-HPP decreased residual activity for polyphenol oxidases (<1%) and peroxidases (33%). -POD was found to be more pressure-, temperature-, and time-resistant compared to PPO. -HPP significantly decreased gallic acid, all flavanols, and dichydrochalcones. -storage time significantly affected the stability of individual polyphenols. -HPP enhanced the color stability in cloudy apple juices due to the inhibition of enzymatic reactions during storage time; thus, it can be considered for preservation of apple products.	Marszałek et al. (2019) [94]
Concord grape juice	HPP: 319, 350, 425, 500, 531 MPa 35, 60, 120, 180, 205 s 5 °C	/	Inoculation (10^7 CFU/mL): *E. coli* O157:H7 *S. enterica* *L. monocytogenes*	-the combined effects of high pressure (extrinsic factor), acid pH (3.39), and the phenolic compounds (intrinsic factors to Concord grape juice) may have had significant effects on achieving log reductions greater than 5-log for pathogens. -*E. coli* O157:H7 was found to be more resistant to HPP as compared to *S. enterica*, while *L. monocytogenes* did not show growth in any sample, either before or after HPP. -HPP at moderate pressure (400 MPa) and short time (2 min) was effective in reducing the pathogens tested in Concord grape juice.	Petrus (2020) [95]
Pitaya–pineapple (*Stenocereus* sp.–*Fragaria ananassa*) beverage	HPP: 400, 500, 600 MPa 2, 5, and 10 min 27, 29, and 31 °C Pressure come-up times (CUTs) were 2, 5, and 10 min.	A beverage composed of 55% pitaya pulp and 45% (w/w) pineapple pulp. -pH, total soluble solids (TSS), moisture content -vitamin C -total phenolic compounds (TPC) -betalains -pectin methyl esterase (PME)	/	-400 MPa/CUT treatment caused an increase of vitamin C (64%) compared with untreated beverages. -TPC decreased by 13%–48% at 400–600 MPa/CUT-6 min. -HPP did not alter the contents of betacyanin or betaxantin in beverages (near 100% retention). -the highest PME inactivation (23%) was achieved at 600 MPa/2–10 min. -the content of vitamin C increased from 5% (600 MPa-CUT) to 64% (400 MPa/CUT).	Sandate-Flores et al. (2017) [96]

Table 3. *Cont.*

Juice Type	HPP Conditions	Nutritive/Physicochemical Quality	Microbial Safety	Key Conclusions	Reference
Aloe vera–litchi mixed beverage	High-pressure thermal processing (HPTP) 400–600 MPa 30–60 °C 0–15 min Pressure come-up times (CUTs) were 87 s and 135 s Beverage formulations: aloe vera/litchi (v/v,%) Sample 1 (20:80) Sample 2 (25:75) Sample 3 (15:85) Sample 4 (0:100)	-sensory evaluation -pH, TSS, acidity -CIELab -ascorbic acid -total phenolic content -antioxidant capacity -enzyme activity (PME, PPO and POD)	-natural microbiota population	-the best beverage formulation was aloe vera/litchi (v/v,%) = 15:85. -HPTP minimally affected physicochemical properties of evaluated beverages. -the temperature applied during the HPTP had major impacts on ascorbic acid, phenolics, and antioxidants. -ascorbic acid in beverage samples was reduced by up to 40% after HPTP treatment. -PME was determined as the most resistant enzyme, with maximum inactivation achieved up to 53%. -the optimal HPTP conditions: 600 MPa/15 min/56 °C resulted in 49% inactivation of PME and 74% retention of ascorbic acid	Swami-Hulle et al. (2017) [97]
Keitt mango juice	Highly Hydrostatic Pressure (HHP) treatments: HHP1:200 MPa/15 min HHP2:400 MPa/15 min HHP3:600 MPa/15 min Thermal treatment (TT): 80 °C/30 min	-volatile compounds -sensory evaluation (QDA)	/	-total of 35 volatile compounds were detected in Keitt mango juice (hydrocarbons: 85.94%, alcohols: 5.07%, esters: 4.94%). -12 major aroma-active compounds characterize the typical flavor of Keitt mango juice. -all treatments (HHP and TT) could alter degrees of aroma loss compared to fresh juice. -ester contents were found to reduce after both TT and HHP. -results of QDA revealed that fresh mango juice was the most accepted, followed by HHP and TT juices.	Zhang et al. (2019) [98]

Bioaccessibility is another important nutritive parameter for evaluation of functional food products as fruit juices. A recent study confirmed that the bioaccessibility of phenolics in fruit juice may be enhanced by modulation of the process parameters and the composition of the food matrix. A comparison of the high-pressure homogenization (HPHP; 250 MPa) and T80 (80 °C/30 min) and T90 (90 °C/30 s) thermal treatments with the addition of milk was associated with phenolic bioaccessibility and activity in kiwi and pomelo juices. Both HPP and TT increased the contents of phenols in both juices, although the T80 sample was an exception because of the phenolic degradation in pomelo juice. Unfortunately, the bioaccessibility of free and total phenols did not increase as a result of the applied treatments. Moreover, in some samples the total phenolic content even decreased, while the addition of milk improved the bioaccessibility of free and total phenols. Here, the exception was soy milk, where the number of total phenols decreased. In kiwi juice samples, the best bioaccessibility was with the addition of semiskimmed milk, while in pomelo juice this resulted from the addition of the whole milk. The authors explained that the improvements were due to the fact that some phenols can create bonds with milk proteins, while others can create bonds with lipids. To conclude, under the tested circumstances the kiwi juice had better phenolic activity and bioaccessibility than pomelo juice [99].

Green asparagus juice was tested in terms of its microbial contamination, physiochemical properties, antioxidant activity, and changes in BAC content in HPP vs. TT treatments (121 °C, 3 min) and control (no treatment). The researchers used HPP at 200, 400, and 600 MPa for 10 and 20 min. HPP at 400 and 600 MPa reduced the number of microbes (total mesophilic bacteria) to the same values as the TT. Moreover, the physiochemical properties and pH did not change during the HPP treatment, which had more BACs (ascorbic acid, rutin, and total phenolic content) in comparison to the TT [100].

Apple juice samples (elephant apple) were treated with HPP (600 MPa/5 min) and TT (80 °C/60 s) for extended storage [101]. Samples were compared after treatment regarding quality parameters (pH, °Brix, total acidity, viscosity, color, and sensory analysis), nutritive value (antioxidant activity, total phenols, total flavonoids), and microbial load. Juice samples were analyzed over 60 days, during which pH, °Brix, and total acidity were similar among HPP and TT samples. However, color, antioxidant activity, total phenols, and total flavonoids were better preserved with HPP samples than TT samples. HPP samples had a lower microbial load than other samples. In conclusion, HPP was identified as a good technology for preservation of apple juice.

Cloudy apple juice was treated with HPP (600 MPa for 3 min) and TT (85 °C for 5 min) to examine changes in the quality and nutritive value. TT caused significant color changes, with an increase in L* and decreases in b* and a* CIELab values. Sugars and acids in the samples remained similar after both treatments. TT-treated juices showed higher amounts of aldehydes, ketones, alkanes, alcohols, and organosulfur compounds than HPP and untreated samples did. These likely resulted from Maillard and oxidative reactions, which are propagated with heating. The results showed that better retention of natural juice color was possible in HPP samples, while TT samples showed better inactivation of PPO and POD [102].

The advantages of using HPP in the food industry, aside from food conservation, include improved antioxidant properties in food components. For example, in strawberry juice, when both treatments HPP and TT are combined, this will decrease the total monomeric anthocyanin (TMA) and vitamin C contents. However, results showed that the TMA and vitamin C contents did not decrease with HPP but did during TT [103]. Another interesting nutritive benefit of HPP is the decrease of the glycemic index (GI). HHP processing of fresh mango puree may provide additional benefits for glycemic control, as it was shown that the mean GI for HHP-processed mango puree was significantly lower (32.7) in comparison with unprocessed puree (42.7) [104]. HPP can also increase the concentrations of amino acids, such as y-aminobutrimetric acid (GABA), which is created from glutaminic acid. To that end, brown rice was treated with HPP of 200 MPa. After 4 days of storage, the concentrations of all amino acids, including GABA, rose more significantly in HPP samples than in control samples [105]. HHP has been used as an alternative technology to improve the functional properties of various kinds of proteins. Bovine lactoferrin was treated by HPP at different pH values

(4.0 and 6.0), at 300–700 ± 10 MPa for 30 and 60 min at 25 °C. Here, higher pressure induced changes in the tertiary structure of lactoferrin at pH 4.0, indicating its partial denaturation and aggregation. After being treated by HHP at pH 6.0, insoluble protein aggregates modified secondary structure. Therefore, in milk-based products, replacing pasteurization with HPP can result in higher solubility, foaming, and emulsifying properties in lactoferrins.

The HPP, HPU, and PEF samples were tested for their ability to reduce the power of the microbial load of *Escherischia coli* in strawberry juice. Samples were treated with HPP at 200–400 MPa for 2 min; HPU with 120 µm, 24 kHz at T = 25–55 °C; PEF with electric field intensity at 25–35 kV cm^{-1} and treatment time = 5–27 µs; and TT at T = 72 °C. The focus was on the processing conditions needed for each technology to give a 5-log microbial reduction from a starting inoculum level of roughly 10^6 $CFUmL^{-1}$. HPP gave 5.75-log reduction with 300 MPa|t = 1 min; HPU (thermosonication) gave a 5.69-log reduction with 55 °C, t = 3 min; PEF gave a 5.53-log reduction with 35 kV cm^{-1}, t = 27 µs, 350 mL/min flow rate, 2 µs pulse width in monopolar mode; and TT contained no detectable *E. coli*. The authors concluded that the above non-thermal parameters had crucial roles in the microbial reduction ability, which was equivalent to thermal pasteurization.

The two technologies (HPP and HPU) were compared and tomato juice was treated with both to see which process had a better effect on the stability and in vitro bioaccessibility of carotenoids. Tomato juice was subjected to different HPP treatments at 200, 300, 400, and 500 MPa. HPU samples were treated with 200, 400, 600, and 800 W, with a constant frequency of 25 Hz. With the increase of pressure, the contents of all-*trans* lycopene decreased, while *cis*-lycopene isomers increased. The best HPP treatment was achieved at 200 MP, with a total lycopene content of 776.94 µg/100 g. The best HPU treatment was at 400 W, with a total lycopene content of 920.52 µg/100 g. The β-carotene and ζ-carotene contents were higher in HPP and HPU than in untreated samples, while HPP and HPU showed slight increases in the bioaccessibility of carotenoids. The highest lycopene bioaccessibility in HPP samples was obtained for all-*trans* (6.86%) and isomers (13.35%) at 500 MPa. In HPU samples, bioaccessibility peaks for all-*trans* (9.71%) and of isomers (15.82%) were achieved at 800 W. Afterwards, the HPP lycopene was fully liberated from the protein bonds and well dispersed in the tomato juice, making it fully available. On the other hand, ultrasound treatment caused less damage to the cell membranes, which resulted in lower pigment release. Therefore, it was concluded that HPP increases the antioxidant activity, while HPU slightly decreases it [106], which was attributed to their technological differences.

3.4. Hurdle Technologies Applied to Fruit Juices

In fruit juice manufacturing, important hurdle methods can be combined, which include: (i) application of low or high temperature; (ii) manipulation with water activity; (iii) acidity; (iv) redox potential; (v) application of chemical or natural preservatives; and (vi) application of competitive microorganisms (e.g., *Lactobacillus* species). These are valid methods for extending the shelf life and retaining the quality of a juice. The specificity of the hurdle concept is reflected in the synergistic effect of various mechanisms for inhibition or inactivation of target microorganisms [37]. Therefore, the innovative "obstacle" technology approach is an advanced approach to fruit juice processing with the potential to meet the high demands of consumers and manufacturers [107]. With the development of innovative process technologies, various novel hurdles have been investigated. Here, the technologies are applied in a certain sequence so that each of the non-invasive ("milder") processing conditions gives the best results in terms of extending the shelf life and sensory quality [35,36]. Table 4 presents selected examples of the application of the hurdle approach for the preservation of functional fruit juices and beverages over the last 5 years.

One of the main problems with the production of fruit juices is browning. For instance, to avoid negative changes of color, apple juice was treated with HPP with the addition of α- and β-cyclodextrins (CD). CDs are allowed as additives in the food industry as natural antibrowning agents in fruit and vegetables juices [108]. The α- and β-CDs are officially approved as food additives in the EU, USA, and Japan, and α-CD is classified as a novel food by the EU [109–111]. For β-CD (E459), an acceptable daily intake of 5 mg kg^{-1} body weight per day has been established.

The concentrations of α- and β-CDs that were added to the juice were 10, 30, and 40 mM; and 5, 10, and 15 mM, respectively. The juice samples were exposed to HPP and browning was measured with the CIELab method. The control sample was untreated apple juice, while the rest of the samples were treated with HPP at 300, 400, and 500 MPa for 5 min at T = 22 °C. The browning index was calculated from L*, a*, and b* parameters, and phenolic compounds were measured using UHPLC. The results showed that HPP treatment did not cause color changes. Phenolic levels were higher in HPP samples, as increased pressure fostered extraction process. The majority of samples that contained 30 mM of α-CD had the least browning, while samples with 10 mM of α-CD showed the same effect but at 400–500 MPa. It was concluded that browning enzymes degraded with exposure to HPP and with the addition of α-CD, with the best results reached at 500 MPa exposure and 30 mM α-CD. For β-CD, the best results were observed with 15 mM, as higher concentrations were not tested due to limited water solubility. It was recommended that the use of β-CD can be accompanied by processing at 500 MPa [30].

Some studies have shown limited inactivation of bacterial endospores under PEF treatment; however, when combined with thermal energy, successful inactivation could be achieved [121]. Inactivation of *Bacillus subtilis* spores were studied in Ringer solution at different pH values (4 and 7), sugar levels (5% and 10%), and conductivity levels (4 and 15 $mScm^{-1}$), as well as under various PEF operating conditions regarding the specific energy (up to 350 kJ/kg), electric field strength (6–11 kV/cm), and inlet temperature (56, 70, and 80 °C). The results revealed that the higher inlet temperature (80 °C) at 9 kV/cm and 10% of sugar content, which increased the electrical conductivity of the medium, was sufficient for successful microbial inactivation (3-log reduction) [121]. Additionally, it was found that *Bacillus subtilis* spores are more heat-sensitive in acidic environments [122].

The combined application of various technologies was further researched in blueberry juice [87]. The authors employed heat treatment (HT), sonication (SC), thermosonication (TS), manosonication (MS), manothermal (MT), and manothermosonication (MTS) to evaluate enzyme inactivation (PPO) and anthocyanin stability, as well as the reduction of *E. coli* O157:H7. Blueberry juice was treated with different temperatures (30, 40, 50, 60, 70, and 80 °C) and power settings (280, 420, 560, and 700 W) for 10 min. Afterwards, samples were treated with HT (80 °C), TS (40 °C/560 W), MT (350 MPa/40 °C), MS (560 W/5 min/350 MPa), or MTS (560 W/5 min/40 °C/350 MPa, 40 °C) for 5, 10, 15, and 20 min. Firstly, HT inactivated PPO (−2% after 5 min) and *E. coli* well, with 80% retention of anthocyanins. TS exposure slightly decreased the *E. coli* load by 0.17 log, with 23% PPO activity and 99% preservation of anthocyanins after the same length of time as for HT. MTS preserved 98% of anthocyanins and showed a 5.85-log reduction, with 11% PPO activity. Hence, combined treatment with sonication, high pressure, and mild thermal treatments has potential for the production of safe and nutritious blueberry juice [30].

The combined effects of ultrasound and PEF on quality of grapefruit juice were studied [123]. Grape juice was firstly treated with PEF (1 kHz, 20 kV cm^{-1}, 600 μs, T < 45 °C), and subsequently the juice sample was sonicated using an ultrasound bath cleaner (28 kHz, 420 W, 20 °C, 30 min). The obtained results revealed that the acidity, soluble solids content, pH, electrical conductivity, and color parameters after treatment were not significantly changed in comparison to the control (untreated sample). As this hurdle concept resulted in decreased viscosity and increased cloud values as compared to control samples, the authors concluded that combined treatment with PEF and ultrasound might be used for the processing of grapefruit juice.

Table 4. The application of the hurdle approach for the preservation of functional fruit juices and beverages over the last 5 years.

Juice Type	Hurdle Approach	Nutritive/Physicochemical/ Microbial Quality	Key Conclusions	Reference
Kiwifruit juice (*Actinidia deliciosa* cv. Hayward)	Ultrasound: 40 kHz, 180 W Sonication time: 10 min (US 10) 30 min (US 30) T < 20 °C Pomegranate extract (PE) 180 µg/mL US10 + PE180 US30 + PE180 Refrigerated storage: 0, 2, and 7 days	-individual phenolic compounds -sensory evaluation -antioxidant activity (AA) -CIELab -total aerobic mesophilic bacteria(MES) -yeast and molds (YM)	-at 7th day of refrigerated storage, US treatments for 10 and 30 min showed significant reductions on yeasts and molds counts as compared to control samples (0.96 and 1.40 log reductions, respectively). -the addition of a second hurdle technology to the US treatments increased the effectiveness in terms of microbial inactivation, which means that shorter US treatment time could be applied to juice when combined with PE. -addition of PE to US-treated juice can improve the retention of AA of sonicated juices. -this hurdle technology showed a potential for use in fruit juice industry.	Tomadoni et al. (2015)
Strawberry juice (*Fragaria x ananassa* Duch, cv. Camarosa) enriched with fiber (oligofructose)	Ultrasound: 40 kHz, 180 W Sonication time: 0, 15, 30 min Vanillin: 0–1.25 mg/mL Juice formulation: inulin/oligofructose proportion = 1:3, 1:1, 3:1 Refrigerated storage: 14 days	-*Saccharomyces* spp. strains -total aerobic mesophilic bacteria (MES) -*Enterobacteriaceae* and total coliforms (EB) -yeasts and molds (YM) -sensory evaluation (QDA)	-microbiological indices were strongly affected by addition of vanillin regardless of the US time and fibers proportion were assayed. -vanillin and US resulted in critical factors for allsensory attributes studied. -fibers proportion did not modify microbiological or sensory indices. -Optimal hurdle approach conditions: 1.25 mg/mL of vanillin, 7.5 min of ultrasound time, and 5:3 ratio of inulin/oligofructose	Cassani et al. (2017) [112]
Orange and cloudy apple juices	Ultrasound–ultraviolet irradiation treatment + additives Dynashock multifrequency ultrasound waves: 600 W 28, 45, and 100 kHz at 1 ms time Heat treatment: 45, 50, 52, 55, or 60 °C UV-C lamp: 15 W Additives: sodium benzoate, potassium sorbate, α- and β-pinene	-*E. coli* O157:H7	-at all tested heating temperatures, *E. coli* O157:H7 was inactivated exponentially as the organisms were heated in the suspending medium. -5-log reduction pf *E. coli* O157:H7 could be achieved at 45, 50, 52, 55, and 60 °C, equivalent to 481.5, 103.6, 45.0, 22.4, and 10.54 min, respectively. -during US treatment, the temperature increased faster in the orange juice, resulting in a faster inactivation rate as compared to apple juice. -heat liberated by US cavitation resulted in 85% of the reduction in E. coli population. -the greater efficiency of the combined treatments (US+UV-C) was found in apple rather than in orange juice. -for cloudy and pigmented samples, the efficacy of the UV-C treatment was limited in terms of reducing the microorganisms on the surfaces of the treated samples. -combined US+UV-C treatment resulted in significantly faster microbial inactivation than singular US or UV-C treatments, especially in apple juice.	Gabriel (2015) [113]

Table 4. *Cont.*

Juice Type	Hurdle Approach	Nutritive/Physicochemical/ Microbial Quality	Key Conclusions	Reference
Apple juice -commercial(CAJ) -freshly pressed (NAJ)	Continuous flow through pulsed light system (PLc, 0.73 Jcm^{-2}, 155 mL min^{-1}, EEO: 1.8×10^3–4.1×10^3 kW h/m^3/order), alone or combined with ultrasound (US, 30 min, EEO: 4.4×10^5–1.1×10^5 kW h/m^3/order) at ambient temperature. Cold storage (4 °C): 12 days	-*Escherichia coli* ATCC 35218 -*Salmonella* Enteritidis MA44 -*Saccharomyces cerevisiae* KE 162 -indigenous flora -color evolution-sensory shelf life and consumer sensory field studies	-single PL treatment did not cause differences between CAJ and NAJ, resulting in 1.8–4.2-log reductions. -obtained results revealed that combined treatment (US + PLc) caused 3.7–6.3-log reductions of tested microbes and positively affected the browning prevention during storage. -inhibitory synergistic effect between US and PLc was observed in postponed mold and yeast recovery for 7 days of cold storage. -processed NAJ was positively evaluated by a group of consumers, who emphasized its fresh natural apple taste.	Ferrairo et al. (2016) [114]
Prebiotic cranberry juice fortified with fructo-oligosaccharides (FOS)	The juice was subjected to HPP for 5 min (450 MPa) and to ultrasonic treatment for 5 min (18 kHz, 500 W, 600 and 1200 WL^{-1}), followed by HPP for 5 min (450 MPa).	-pH, total soluble solids, instrumental color parameters -organic acids -anthocyanins -antioxidant activity -fructo-oligosaccharides (FOS)	-combined treatment (US + HPP) is viable process for the treatment of prebiotic juices. -the combination of ultrasound (1200 WL^{-1}) followed by HPP increased the amounts of cyaniding, peonidin, and malvidine derivatives in the prebiotic juice, which corresponded to an increase in chroma and decrease in luminosity.	Gomes et al. (2017) [115]
Guava juice (*Psidium guajava* L.) Mango juice (*Mangifera indica* L.)	Nanoemulsions of *Mentha piperita* L. essential oil (n-MPEO) or suspensions of MPEO (s-MPEO) in combination with mild heat (MHT) (50, 52, 54 °C; 10 min), PEF (20, 25, 30 kV cm^{-1}; 150 μs), and HHP (150, 200, 300 MPa; 15 min) treatments.	-*Escherichia coli* O157:H7	-MPEO was found to be more effective in guava than in mango juice. -s-MPEO and n-MPEO exhibited a synergistic effect in combination with MHT, PEF, and HHP against *E. coli* O157:H7. -MHT at 54 °C for 10 min caused a 2.7-log and 2.3-log reduction of *E. coli* O157:H7 in guava and mango juices, respectively. -when s-MPEO or n-MPEO was combined with MHT, PEF, or HHP, an average 5-log reduction of *E.coli* O157:H7 was achieved, which varied with the tested concentrations, treatment intensity, and the food matrix. -n-MPEO in combination with MHT, PEF, and HHP can be considered as promising methods to ensure microbial safety in fruit juices.	De Carvahlo et al. (2018) [116]

Table 4. *Cont.*

Juice Type	Hurdle Approach	Nutritive/Physicochemical/ Microbial Quality	Key Conclusions	Reference
Apple juice	Ultrasound: 40 kHz, 700 W, 1, 2, 3, 4, and 5 min Fumaric acid (FA): 0%, 0.05%, 0.1%, and 0.15% (w/v)	Three strains each of: *E. coli* O157:H7 (ATCC 35150, ATCC 43889, and ATCC 43890) *S. Typhimurium* (ATCC, 19585, ATCC 43971, and DT 104) *L. monocytogenes* (ATCC 15313, ATCC, 19111, and ATCC, 19115) -color parameters, pH -non-enzymatic browning index -total phenolic content	-combined US + 0.15% FA treatment for 5 min achieved 5.67, 6.35, and 3.47 log reductions in *E. coli* O157:H7, *S. Typhimurium*, and *L. monocytogenes*, respectively, with the 1.55, 2.37, and 0.57 log CFU reductions attributed to the synergistic effect. -US + 0.15% FA treatment (5 min) did not affect the product quality. -obtained results suggest that simultaneous application of US and FA is a novel approach for ensuring the microbial safety of apple juice.	Park et al. (2019) [117]
Açai juice (*Euterpe oleracea*)	Combined ultrasound and ozone treatment. Ultrasound (US): 19 kHz 350 and 700 J mL^{-1} 5 min Processing temperature: 32 ± 1.2 °C Direct immersion of ozone gas for 5 or 10 min. Ozone concentration (O_3): 1.50 ppm Processing temperature: 25 °C	-pH and titratable acidity -cloud value and non-enzymatic browning -viscosity -antioxidant activity (DPPH and ABTS) -total phenolic contents (TPC) -total anthocyanin content (TAC) -peroxidase (POD) and polyphenoloxidase (PPO) activity -total mesophilic bacteria and mold and yeast counts	-as ozone single treatment decreased the TPC and single US increased the TPC in açai juice samples, the combined processes showed no significant difference in TPC when compared to control (untreated) samples. -two isolated processes (O_3, US) reduced TAC in the juice, therefore a high reduction of TAC in the açai juice was also observed after the combined treatments. -increase of the energy density (US) and the concentration of the ozone in combined treatment favored the reduction of the POD activity. -for the combined treatments, the PPO activity was higher than the treatments with single processes. -the combination of processes was shown to significantly reduce the contamination of açai juice.	Oliveira et al. (2018) [118]
Mixed Satsuma mandarin (*Citrus unshiu* Marc.) and Hallabong tangor (*Citrus kiyomi* x *Citrus ponkan*) juice (MH)	PEF combined with heat. Juice was heated to 55 °C and 70 °C prior to PEF treatment: PEF1-55 °C, 19 kV cm^{-1}, 170 kJL^{-1}, 24 µs, 166 kHZ PEF2-70 °C, 16 kV cm^{-1}, 100 kJL^{-1}, 30 µs, 115 kHZ PEF3-70 °C, 12 kV cm^{-1}, 100 kJL^{-1}, 30 µs, 320 kHZ	-total mesophilic aerobes, yeasts and molds, and coliforms counts -color -total soluble solid content and pH -ascorbic acid -antioxidant capacity	-H-PEF processing at 70 °C (16 kV cm^{-1}, 100 kJL^{-1}) preserved the physicochemical parameters and antioxidant capacity of MH juice -the same treatment influenced changes in juice color and browning degree, while demonstrating a strong inactivation effect on indigenous microorganisms (reduced the aerobe, yeast/mold, and coliform counts in MH juice by 3.9, 4.3, and 0.8 log CFUmL^{-1}, respectively). -high electric field strength enhanced microbial inactivation, even at a relatively low level of specific energy.	Lee et al. (2018) [119]

Table 4. *Cont.*

Juice Type	Hurdle Approach	Nutritive/Physicochemical/ Microbial Quality	Key Conclusions	Reference
Pineapple juice (*Ananas comosus*)	Combined pressure–thermal treatments: -600 MPa at 75, 85, and 95 °C for 0, 2, 5, 10, and 15 min -95 °C at 300, 450, and 600 MPa for 0, 2, 5, 10, and 15 min. Thermal treatment (TT): 75–95 °C for 0 to 60 min treatment times	-pH, acidity, and °Brix -ascorbic acid -kinetic models	-TT alone at 75 to 95 °C; the pineapple juice revealed loss in ascorbic acid content (2% to 5%), while significant loss was observed with increases in process temperature and treatment time, following First-order kinetics (loss up to 39%). -no significant difference in ascorbic acid content was observed between HPP (300–600 MPa at 30 °C) and control (untreated sample), irrespective of pressure holding times or pressure come-up time. -combined pressure–thermal treatment increased the ascorbic acid degradation rate and could be fitted by first-order fractional conversion.	Dhakal et al. (2018) [120]

A recent study combined the use of HHP and ultrasound at 75 °C to achieve inactivation of *Aliclyclobacillus acidoterrestris* spores in apple juice. The juice was pretreated at 600 MPa for 15 min to weaken the spores. After the pressure treatment, the partially damaged spores were exposed to power ultrasound at 24 kHz and 20.2 WmL^{-1} for 60 min to reach the final spore reduction of 4.2 log. Nevertheless, the HPP-75 °C treatment was found to be the most effective for inactivating spores, resulting in 3.3-log reductions after 10 min vs. no inactivation for thermosonication (TS) and thermal processing [124].

The effects of combined treatment with US and HPP on enzymes (polyphenolase, POD, and PME), physicochemical properties, bioactive compounds, antioxidant activity, and microorganisms in apple juice (*M. domestica* cv. Fuji) were evaluated [125]. The combined treatment started with US (25 kHz and 70% amplitude) at 20 °C for 60 min, with subsequent HPP treatment at 250, 350, and 450 MPa for 10 min at room temperature. The obtained results indicated that the combined US-HPP treatment at 450 MPa caused the highest inactivation of enzymes, with complete inactivation of total plate counts, yeasts, and molds. The synergistic impacts of sonication and HHP also significantly improved the contents of phenolic compounds, ascorbic acid, antioxidant capacity, and color properties.

A combination of different hurdles, such as mild heat (54 °C for 10 min) or PEF treatments (25 pulses; 25 kV cm^{-1}, 3.35 kJ cm^{-1} per pulse), and the addition of essential oil components (carvacrol, citral, and (+)-limonene) were employed to reduce spoilage in a clear apple juice [126]. For this purpose, PEF and the mild heat resistance of five strains of *Leuconostoc* spp. and five *Saccharomyces* spp. were tested, with different inactivation levels achieved for each treatment and strain. The authors concluded that combined processes could be useful alternatives for prolonging the shelf life of food products after moderate-intensity heat treatment, which would diminish the undesirable effects of high temperatures on food quality.

Alicyclobacillus acidoterrestris is a bacterium that is very resilient to high temperatures, as it is a thermoacidophilic, spore-forming, non-pathogenic microbe that grows at pH = 2.5–6.0 and temperatures ranging from 25 to 70 °C [67]. Acidic juices, such as apple juice, are a good medium for this microbe, which is able to cause spoilage. To that end, this study documented the exposure of apple juice to HPU and UV radiation. The juice was treated under UV-C radiation at 254 nm and 13.44 Wm^{-2}, while HPU was used at 35 kHz frequency and 120–480 W power levels. Different combinations of conditions were used, namely: (i) sole UV-C; (ii) sole HPU; (iii) exposure of 25 min of HPU and 5 min of UV-C radiation; (iv) exposure of 5 min of HPU and 25 min of UV-C; (v) exposure of 5 min of UV-C and 25 min of HPU; (vi) exposure to TT at 95 °C (control); and (vii) exposure of TS in a thermostatic water bath at 95 °C. Spore inactivation was higher when ultrasound and thermal treatment were applied in combination than when each procedure was used separately. The use of UV-C alone severely reduced the number of spores by 5 logs in a short period of time. For the above parameters, the combined treatment (i) showed around a 5-log reduction; (ii) showed a low 1-log reduction; (iii) showed a 2-log reduction; (iv) showed a 5-log reduction; (v) showed a 4-log reduction; (vi) showed a 4-log reduction; and (vii) showed a 5-log reduction. HPU treatment alone had the lowest ability to reduce the microbial count, while thermal treatment (control) was comparable to sole use of UV-C, sole use of thermosonication, and combined treatment with 5 min of HPU and 25 min of UV-C. It was concluded that the results of this study provided new options for assuring food safety and quality in food manufacturing.

In 2015, a group of Mexican and Spanish scientists used combined alterations in pH, preservatives, and PEF treatment to inactivate *Saccharomyces cerevisiae* and *E. coli* in prickly pear juices. They found that *S. cerevisiae* was more resilient to pH reduction and addition of preservatives than *E. coli*, so the PEF was mainly used to inactivate *S. cerevisiae*. Interestingly, here pH was not just used to inactivate microorganisms, but also to increase their sensitivity to PEF treatment. The highest reduction of activity was accomplished with treatment parameters of 15 ls, 50 Hz, 36 kV cm^{-1} [127].

To test how technology modifies bioaccessibility, researchers mixed 75% of blended fruit juice (composed of orange, kiwi, pineapple, and mango) with 17.5% of water, milk, or soy milk and 7.5%

of sugar, then the mixture was treated with PEF, HPP, or TT to evaluate the influence of carotenoids and antioxidant activity on bioaccessibility. PEF conditions were 35 kV cm^{-1} with 4 μs bipolar pulses at 200 Hz for 1800 μs, HPP conditions were 400 MPa at 40 °C for 5 min, while TT was performed at 90 °C t = 1 min. After the treatments, the bioaccessibility of all carotenoids was reduced in the samples, except in the case of *cis*-violaxanthin and neoxanthin, which increased by 79% in PEF and HPP samples. Losses of carotenoids were the highest during the TT, while samples with milk had the best results for bioaccessibility.

4. Conclusions

Juices are very important food commodities in the global market. Their preservation needs to account for microbial safety and thermally sensitive biologically active compounds. Pasteurization is a thermal treatment that has assured microbial safety in the past, but not without losses of the nutritive value of the juices. The hurdle technology approach, involving the combination of various preservation factors, has the potential to satisfy both aspects and to safely preserve the natural value of foods over extended periods of time.

Functional juices can be engineered from a number of biologically active ingredients, such as polyphenols, which have known antioxidative activity; or probiotic lactic acid bacteria, which need to be processed so that their positive aspects are retained in the final product.

Current markets demand juices with native nutritive value, sensory appeal, and an extended shelf life, but without additives. It is difficult to preserve products in such a way with the usual thermal treatments (e.g., pasteurization), as quite often higher temperatures have detrimental influences on the nutritive value of foods. Hence, there is an ongoing need to combine alternative non-thermal technologies for juice preservation. The pulsed electric field approach is an electrotechnology and a good option for the preservation of liquid foods that is able to ensure microbial safety and preserve the nutritive value of foods. However, it might need to be combined with other methods to achieve full microbial inactivation. Ultrasound-based technologies are valid options for microbial inactivation in juices, however as with PEF, sole application of such technology is not sufficient to achieve inactivation of microbes.

To reach the full potential of this technology, it is often combined with short thermal treatments to achieve complete product safety. Although it is generally good for preserving the nutritive and sensory attributes of foods, ultrasound still can generate "off flavors" in juices, undesired changes in viscosity, formation of free radicals, and lipid degradation [42,128]. High-pressure processing is one of the best options for the preservation of juices. It causes microbial inactivation with selective preservation of probiotic bacteria, improves the bioaccessibility of important nutrients, preserves and improves the native quality of products, and provides enzymatic inactivation. However, microbial spores are quite resilient to pressurization, so this treatment is often supplemented with thermal treatments to achieve the full effect. UV radiation shows good inactivation ability, hence it could be a good addition to hurdle technologies.

Although there have been numerous attempts to identify the ideal combination of technologies and their parameters in the processing of juices, there is still a gap in the research in terms of combining two of the most promising hurdle technology techniques, such as high-pressure processing and pulsed electric field techniques. This is somewhat unexpected, as both of these technologies are sustainable, selective, and thermally sensitive—processing factors that are currently cherished in food manufacturing. Some shortcomings that need to be overcome with these technologies include improvements in microbial inactivation, larger numbers of tests at the industrial scale, lowering implementation costs, and providing more funding to the scientific community for experimentation with innovative industry solutions.

Author Contributions: Conceptualization, P.P. and D.B.K.; writing—original draft preparation, P.P., D.B.K., L.K., D.K.; writing—review and editing, P.P., D.B.K., B.P., B.Š., S.Z., I.Ž.; project administration, P.P., D.B.K. All authors have read and agreed to the published version of the manuscript.

Funding: This research was funded by the Croatian Science Foundation through the funding of the "Hurdle Technology and 3D Printing for Sustainable Fruit Juice Processing and Preservation" project, number "IP-2019-04-2105".

Conflicts of Interest: The authors declare no conflict of interest.

References

1. Granato, D.; Nunes, D.S.; Barba, F.J. An integrated strategy between food chemistry, biology, nutrition, pharmacology, and statistics in the development of functional foods: A proposal. *Trends Food Sci. Technol.* **2017**, *62*, 13–22. [CrossRef]
2. Putnik, P.; Lorenzo, J.; Barba, F.; Roohinejad, S.; Režek Jambrak, A.; Granato, D.; Montesano, D.; Bursać Kovačević, D. Novel food processing and extraction technologies of high-added value compounds from plant materials. *Foods* **2018**, *7*, 106. [CrossRef] [PubMed]
3. Musina, O.; Putnik, P.; Koubaa, M.; Barba, F.J.; Greiner, R.; Granato, D.; Roohinejad, S. Application of modern computer algebra systems in food formulations and development: A case study. *Trends Food Sci. Technol.* **2017**, *64*, 48–59. [CrossRef]
4. NHS-UK. *5 A Day Portion Sizes*; NHS: Les, UK, 2015.
5. Leistner, L. Basic aspects of food preservation by hurdle technology. *Int. J. Food Microbiol.* **2000**, *55*, 181–186. [CrossRef]
6. Bevilacqua, A.; Petruzzi, L.; Perricone, M.; Speranza, B.; Campaniello, D.; Sinigaglia, M.; Corbo, M.R. Nonthermal technologies for fruit and vegetable juices and beverages: Overview and advances. *Compr. Rev. Food Sci. Food Saf.* **2018**, *17*, 2–62. [CrossRef]
7. Djekic, I.; Sanjuán, N.; Clemente, G.; Jambrak, A.R.; Djukić-Vuković, A.; Brodnjak, U.V.; Pop, E.; Thomopoulos, R.; Tonda, A. Review on environmental models in the food chain-Current status and future perspectives. *J. Clean. Prod.* **2018**, *176*, 1012–1025. [CrossRef]
8. Dixit, Y.; O'Sullivan, C.; Cullen, P.J.; Tiwari, B.K. Hurdle technologies for fruit juices. In *Fruit Juices Extraction, Composition, Quality and Analysis*, 1st ed.; Rajauria, G., Tiwari, B., Eds.; Academic Press: Cambridge, MA, USA, 2018; pp. 539–554.
9. Jayaprakasha, G.K.; Bhimanagouda, S.P. *Nutraceuticals and Functional Foods: Chemistry and Health Promoting Properties of Fruits and Beverages Involved in Prevention of Chronic Diseases*; Jayaprakasha, G.K., Patil, B.S., Eds.; EOLSS Publications: Abu Dhabi, UAE, 2015; p. 354.
10. Priyadarshini, A.; Priyadarshini, A. Market Dimensions of the Fruit Juice Industry. In *Fruit Juices-Extraction, Composition, Quality and Analysis*; Rajauria, G., Tiwari, B., Eds.; Academic Press: Cambridge, MA, USA, 2018; pp. 18–21.
11. Lachowicz, S.; Oszmiański, J. The influence of addition of cranberrybush juice to pear juice on chemical composition and antioxidant properties. *J. Food Sci. Technol.* **2018**, *55*, 3399–3407. [CrossRef]
12. European Fruit Juice Association. Liquid Fruit Market Report. Zagreb. 2017. Available online: http://www.aijn.org/publications/facts-and-figures/aijn-market-reports/ (accessed on 18 December 2017).
13. Fraga, C.G.; Croft, K.D.; Kennedy, D.O.; Tomás-Barberán, F.A. The effects of polyphenols and other bioactives on human health. *Food Funct.* **2019**, *10*, 514–528. [CrossRef]
14. Szymanowska, U.; Baraniak, B.; Bogucka-Kocka, A. Antioxidant, anti-inflammatory, and postulated cytotoxic activity of phenolic and anthocyanin-rich fractions from polana raspberry (*Rubus idaeus* L.) fruit and juice-in vitro study. *Molecules* **2018**, *23*, 1812. [CrossRef]
15. Quan, W.; Tao, Y.; Lu, M.; Yuan, B.; Chen, J.; Zeng, M.; Qin, F.; Guo, F.; He, Z. Stability of the phenolic compounds and antioxidant capacity of five fruit (apple, orange, grape, pomelo and kiwi) juices during in vitro -simulated gastrointestinal digestion. *Int. J. Food Sci. Technol.* **2018**, *53*, 1131–1139. [CrossRef]
16. Jakobek, L.; Šeruga, M.; Medvidović-Kosanović, M.; Novak, I. Anthocynin content and antioxidant activity of various red fruit juices. *Dtsch. Lebensmitt. Rundsch.* **2007**, *103*, 58–64.
17. Rasouli, H.; Farzaei, M.H.; Khodarahmi, R. Polyphenols and their benefits: A review. *Int. J. Food Prop.* **2017**. [CrossRef]
18. Yuan, L.; Liu, J.; Zhen, J.; Xu, Y.; Chen, S.; Halm-Lutterodt, N.V.; Xiao, R. Vegetable and fruit juice enhances antioxidant capacity and regulates antioxidant gene expression in rat liver, brain and colon. *Genet. Mol. Biol.* **2017**, *40*, 134–141. [CrossRef] [PubMed]

19. Granato, D.; Barba, F.J.; Bursać Kovačević, D.; Lorenzo, J.M.; Cruz, A.G.; Putnik, P. Functional foods: Product development, technological trends, efficacy testing, and safety. *Annu. Rev. Food Sci. Technol.* **2020**, *11*, 93–118. [CrossRef] [PubMed]
20. Šušković, J.; Brkić, B.; Matošić, S. Mehanizam probiotičkog djelovanja bakterija mliječne kiseline. *Mljekarstvo* **1997**, *47*, 57–73.
21. Frece, J.; Kos, B.; Beganović, J.; Vuković, S.; Šušković, J. In vivo testing of functional properties of three selected probiotic strains. *World J. Microbiol. Biotechnol.* **2005**, *21*, 1401–1408. [CrossRef]
22. Frece, J. *Sinbiotički Učinak Bakterija: Lactobacillus acidophilus M92, Lactobacillus plantarum L4 i Enterococcus faecium L3*; Prehrambeno-biotehnološki fakultet Sveučilišta u Zagrebu: Zagreb, Croatia, 2007.
23. Fernandes Pereira, A.L. Turning fruit juice into probiotic beverages. In *Fruit Juices Extraction, Composition, Quality and Analysis*, 1st ed.; Rajauria, G., Tiwari, B., Eds.; Academic Press: Cambridge, MA, USA, 2018; pp. 279–285.
24. Martins, E.M.F.; Ramos, A.M.; Vanzela, E.S.L.; Stringheta, P.C.; de Oliveira Pinto, C.L.; Martins, J.M. Products of vegetable origin: A new alternative for the consumption of probiotic bacteria. *Food Res. Int.* **2013**, *51*, 764–770. [CrossRef]
25. Nicolesco, C.L.; Buruleanu, L.C. Correlation of some substrate parameters in growing *Lactobacillus acidophilus* on vegetable and fruit cocktail juices. *Bull. Univ. Agric. Sci. Vet. Med. Cluj. Napoca.* **2010**, *67*, 352–359.
26. Nualkaekul, S.; Charalampopoulos, D. Survival of *Lactobacillus plantarum* in model solutions and fruit juices. *Int. J. Food Microbiol.* **2011**, *146*, 111–117. [CrossRef]
27. Sheela, T.; Suganya, R.S. Studies on anti-diarrhoeal activity of synbiotic plums juice. *Int. J. Sci. Res. Pub.* **2012**, *2*, 1–5.
28. Patel, A.R. Probiotic fruit and vegetable juices-recent advances and future perspective. *Int. Food Res. J.* **2017**, *24*, 1850–1857.
29. Shah, N.P. Functional cultures and health benefits. *Int. Dairy J.* **2007**, *17*, 1262–1277. [CrossRef]
30. Snyder, A.B.; Worobo, R.W. The incidence and impact of microbial spoilage in the production of fruit and vegetable juices as reported by juice manufacturers. *Food Control* **2018**, *85*, 144–150. [CrossRef]
31. Bursać Kovačevć, D.; Kljusurić, J.G.; Putnik, P.; Vukušić, T.; Herceg, Z.; Dragovic-Uzelac, V. Stability of polyphenols in chokeberry juice treated with gas phase plasma. *Food Chem.* **2016**, *212*, 323–331. [CrossRef] [PubMed]
32. Herceg, Z.; Kovačević, D.B.; Kljusurić, J.G.; Jambrak, A.R.; Zorić, Z.; Dragović-Uzelac, V. Gas phase plasma impact on phenolic compounds in pomegranate juice. *Food Chem.* **2016**, *190*, 665–672. [CrossRef]
33. Petruzzi, L.; Campaniello, D.; Speranza, B.; Corbo, M.R.; Sinigaglia, M.; Bevilacqua, A. Thermal treatments for fruit and vegetable juices and beverages: A literature overview. *Compr. Rev. Food Sci. Food Saf.* **2017**, *16*, 668–691. [CrossRef]
34. Wurlitzer, N.J.; Dionísio, A.P.; Lima, J.R.; Garruti, D.D.S.; Silva Araújo, I.M.D.; da Rocha, R.F.J.; Maia, J.L. Tropical fruit juice: Effect of thermal treatment and storage time on sensory and functional properties. *J. Food Sci. Technol.* **2019**, *56*, 5184–5193. [CrossRef]
35. Khan, I.; Tango, C.N.; Miskeen, S.; Lee, B.H.; Oh, D.-H. Hurdle technology: A novel approach for enhanced food quality and safety—A review. *Food Control* **2017**, *73*, 1426–1444. [CrossRef]
36. Singh, S.; Shalini, R. Effect of hurdle technology in food preservation: A review. *Crit. Rev. Food Sci. Nutr.* **2014**, *56*, 641–649. [CrossRef]
37. Rahman, M.S. Hurdle technology in food preservation. In *Minimally Processed Foods-Technologies for Safety, Quality, and Convenience*; Wasim, S.M., Rahman, S., Eds.; Springer: New York, NY, USA, 2015; pp. 17–33.
38. Galanakis, C.M.; Martinez-Saez, N.; del Castillo, M.D.; Barba, F.J.; Mitropoulou, V.S. Patented and commercialized applications. In *Food Waste Recovery-Processing Technologies and Industrial Techniques*; Galanakis, C.M., Ed.; Academic Press: Cambridge, MA, USA, 2015; pp. 321–361.
39. Deliza, R. Consumer perception of novel technologies. In *Fruit Preservation-Novel and Conventional Technologies*; Rosenthal, A., Deliza, R., Welti-Chanes, J., Barbosa-Cánovas, G.V., Eds.; Springer: New York, NY, USA, 2018; Volume 1, pp. 1–20.
40. Martins, I.B.A.; Oliveira, D.; Rosenthal, A.; Ares, G.; Deliza, R. Brazilian consumer's perception of food processing technologies: A case study with fruit juice. *Food Res. Int.* **2019**, *125*. [CrossRef]

41. Koubaa, M.; Barba, F.; Bursać Kovačević, D.; Putnik, P.; Santos, M.S.; Queirós, R.P.; Moreira, S.A.; Duarte, R.R.; Saraiva, J.A. Pulsed electric field processing of different fruit juices. In *Fruit Juices Extraction, Composition, Quality and Analysis*, 1st ed.; Rajauria, G., Tiwari, B., Eds.; Academic Press: Cambridge, MA, USA, 2018; pp. 437–445.
42. Gabrić, D.; Barba, F.; Roohinejad, S.; Gharibzahedi, S.M.T.; Radojčin, M.; Putnik, P.; Bursać Kovačević, D. Pulsed electric fields as an alternative to thermal processing for preservation of nutritive and physicochemical properties of beverages: A review. *J. Food Process Eng.* **2018**, *41*. [CrossRef]
43. Barba, F.J.; Parniakov, O.; Pereira, S.A.; Wiktor, A.; Grimi, N.; Boussetta, N.; Saraiva, J.A.; Raso, J.; Martin-Belloso, O.; Witrowa-Rajchert, D.; et al. Current applications and new opportunities for the use of pulsed electric fields in food science and industry. *Food Res. Int.* **2015**, *77*, 773–798. [CrossRef]
44. Barbosa-Cánovas, G.V. *Nonthermal Preservation of Foods*; Marcel Dekker: New York, NY, USA, 1998; p. 276.
45. Evrendilek, G.A. Impacts of pulsed electric field and heat treatment on quality and sensory properties and microbial inactivation of pomegranate juice. *Food Sci. Technol. Int.* **2017**, *23*, 668–680. [CrossRef]
46. Bendicho, S.; Espachs, A.; ArÁNtegui, J.; MartÍN, O. Effect of high intensity pulsed electric fields and heat treatments on vitamins of milk. *J. Dairy Res.* **2002**, *69*, 113–123. [CrossRef] [PubMed]
47. Rodríguez-Roque, M.J.; de Ancos, B.; Sánchez-Moreno, C.; Cano, M.P.; Elez-Martínez, P.; Martín-Belloso, O. Impact of food matrix and processing on the in vitro bioaccessibility of vitamin C, phenolic compounds, and hydrophilic antioxidant activity from fruit juice-based beverages. *J. Funct. Foods* **2015**, *14*, 33–43. [CrossRef]
48. Rodríguez-Roque, M.J.; de Ancos, B.; Sánchez-Vega, R.; Sánchez-Moreno, C.; Cano, M.P.; Elez-Martínez, P.; Martín-Belloso, O. Food matrix and processing influence on carotenoid bioaccessibility and lipophilic antioxidant activity of fruit juice-based beverages. *Food Funct.* **2015**, *7*, 380–389. [CrossRef]
49. González-Casado, S.; Martín-Belloso, O.; Elez-Martínez, P.; Soliva-Fortuny, R. Application of pulsed electric fields to tomato fruit for enhancing the bioaccessibility of carotenoids in derived products. *Food Funct.* **2018**, *9*, 2282–2289. [CrossRef]
50. González-Casado, S.; Martín-Belloso, O.; Elez-Martínez, P.; Soliva-Fortuny, R. Enhancing the carotenoid content of tomato fruit with pulsed electric field treatments: Effects on respiratory activity and quality attributes. *Postharvest Biol. Technol.* **2018**, *137*, 113–118. [CrossRef]
51. Carbonell-Capella, J.M.; Buniowska, M.; Cortes, C.; Zulueta, A.; Frigola, A.; Esteve, M.J. Influence of pulsed electric field processing on the quality of fruit juice beverages sweetened with *Stevia rebaudiana*. *Food Bioprod. Process.* **2017**, *101*, 214–222. [CrossRef]
52. Evrendilek, G.A. Change regime of aroma active compounds in response to pulsed electric field treatment time, sour cherry juice apricot and peach nectars, and physical and sensory properties. *Innov. Food Sci. Emerg. Technol.* **2016**, *33*, 195–205. [CrossRef]
53. Dziadek, K.; Kopec, A.; Drozdz, T.; Kielbasa, P.; Ostafin, M.; Bulski, K.; Oziemblowski, M. Effect of pulsed electric field treatment on shelf life and nutritional value of apple juice. *J. Food Sci. Technol. Mys.* **2019**, *56*, 1184–1191. [CrossRef] [PubMed]
54. Leong, S.Y.; Burritt, D.J.; Oey, I. Evaluation of the anthocyanin release and health-promoting properties of Pinot Noir grape juices after pulsed electric fields. *Food Chem.* **2016**, *196*, 833–841. [CrossRef] [PubMed]
55. Mtaoua, H.; Sánchez-Vega, R.; Ferchichi, A.; Martín-Belloso, O. Impact of high-intensity pulsed electric fields or thermal treatment on the quality attributes of date juice through storage. *J. Food Process Preserv.* **2017**, *41*. [CrossRef]
56. Timmermans, R.A.H.; Mastwijka, H.C.; Berendsen, L.B.J.M.; Nederhoff, A.L.; Matser, A.M.; Van Boekel, M.A.J.S.; Groot, M.N.N. Moderate intensity Pulsed Electric Fields (PEF) as alternative mild preservation technology for fruit juice. *Int. J. Food Microbiol.* **2019**, *298*, 63–73. [CrossRef]
57. Wibowo, S.; Essel, E.A.; De Man, S.; Bernaert, N.; Van Droogenbroeck, B.; Grauwet, T.; Van Loey, A.; Hendrickx, M. Comparing the impact of high pressure, pulsed electric field and thermal pasteurization on quality attributes of cloudy apple juice using targeted and untargeted analyses. *Innov. Food Sci. Emerg. Technol.* **2019**, *54*, 64–77. [CrossRef]
58. Buckow, R.; Ng, S.; Toepfl, S. Pulsed electric field processing of orange juice: A review on microbial, enzymatic, nutritional, and sensory quality and stability. *Compr. Rev. Food Sci. Food Saf.* **2013**, *12*, 455–467. [CrossRef]

59. Salinas-Roca, B.; Elez-Martínez, P.; Welti-Chanes, J.; Martín-Belloso, O. Quality changes in mango juice treated by high-intensity pulsed electric fields throughout the storage. *Food Bioproc. Technol.* **2017**, *10*, 1970–1983. [CrossRef]
60. Agcam, E.; Akyıldız, A.; Akdemir Evrendilek, G. Comparison of phenolic compounds of orange juice processed by pulsed electric fields (PEF) and conventional thermal pasteurisation. *Food Chem.* **2014**, *143*, 354–361. [CrossRef]
61. Patras, A.; Tiwari, B.; Brunton, N.; Butler, F. Modelling the effect of different sterilisation treatments on antioxidant activity and colour of carrot slices during storage. *Food Chem.* **2009**, *114*, 484–491. [CrossRef]
62. Önür, İ.; Misra, N.N.; Barba, F.J.; Putnik, P.; Lorenzo, J.M.; Gökmen, V.; Alpas, H. Effects of ultrasound and high pressure on physicochemical properties and HMF formation in Turkish honey types. *J. Food Eng.* **2018**, *219*, 129–136. [CrossRef]
63. Tiwari, B.K. Ultrasound: A clean, green extraction technology. *Trends Analyt. Chem.* **2015**, *71*, 100–109. [CrossRef]
64. Swamy, G.J.; Muthukumarappan, K.; Asokapandian, S. Ultrasound for fruit juice preservation. In *Fruit Juices Extraction, Composition, Quality and Analysis*, 1st ed.; Rajauria, G., Tiwari, B., Eds.; Academic Press: Cambridge, MA, USA, 2018; pp. 451–461.
65. Mason, T.J. Sonochemistry and sonoprocessing: The link, the trends and (probably) the future. *Ultrason. Sonochem.* **2003**, *10*, 175–179. [CrossRef]
66. Condón-Abanto, S.; Arroyo, C.; Álvarez, I.; Condón, S.; Lyng, J.G. Application of ultrasound in combination with heat and pressure for the inactivation of spore forming bacteria isolated from edible crab (*Cancer pagurus*). *Int. J. Food Microbiol.* **2016**, *223*, 9–16. [CrossRef] [PubMed]
67. Tremarin, A.; Brandão, T.R.S.; Silva, C.L.M. Application of ultraviolet radiation and ultrasound treatments for *Alicyclobacillus acidoterrestris* spores inactivation in apple juice. *LWT-Food Sci. Technol.* **2017**, *78*, 138–142. [CrossRef]
68. Ordóñez-Santos, L.E.; Martínez-Girón, J.; Arias-Jaramillo, M.E. Effect of ultrasound treatment on visual color, vitamin C, total phenols, and carotenoids content in Cape gooseberry juice. *Food Chem.* **2017**, *233*, 96–100. [CrossRef] [PubMed]
69. Margean, A.; Lupu, M.I.; Alexa, E.; Padureanu, V.; Canja, C.M.; Cocan, I.; Negrea, M.; Calefariu, G.; Poiana, M.-A. An overview of effects induced by pasteurization and high-power ultrasound treatment on the quality of red grape juice. *Molecules* **2020**, *25*, 1669. [CrossRef] [PubMed]
70. Wang, J.; Wang, J.; Vanga, S.K.; Raghavan, V. High-intensity ultrasound processing of kiwifruit juice: Effects on the microstructure, pectin, carbohydrates and rheological properties. *Food Chem.* **2020**, *313*. [CrossRef] [PubMed]
71. Silva, E.K.; Arruda, H.S.; Pastore, G.M.; Meireles, M.A.A.; Saldaña, M.D.A. Xylooligosaccharides chemical stability after high-intensity ultrasound processing of prebiotic orange juice. *Ultrason. Sonochem.* **2020**, *63*. [CrossRef]
72. Cassani, L.; Tomadoni, B.; Moreira, M.R.; Agüero, M.V. Improving quality parameters of functional strawberry juices: Optimization of prebiotic fiber enrichment and geraniol treatment. *Food Bioproc. Technol.* **2018**, *11*, 2110–2124. [CrossRef]
73. Bursać Kovačević, D.; Bilobrk, J.; Buntić, B.; Bosiljkov, T.; Karlović, S.; Rocchetti, G.; Lucini, L.; Barba, F.J.; Lorenzo, J.M.; Putnik, P. High-power ultrasound altered the polyphenolic content and antioxidant capacity in cloudy apple juice during storage. *J. Food Process Preserv.* **2019**, *43*. [CrossRef]
74. Saeeduddin, M.; Abid, M.; Jabbar, S.; Wu, T.; Hashim, M.M.; Awad, F.N.; Hu, B.; Lei, S.; Zeng, X. Quality assessment of pear juice under ultrasound and commercial pasteurization processing conditions. *LWT-Food Sci. Technol.* **2015**, *64*, 452–458. [CrossRef]
75. Pala, Ç.U.; Zorba, N.N.; Özcan, G. Microbial inactivation and physicochemical properties of ultrasound processed pomegranate juice. *J. Food Prot.* **2015**, *78*, 531–539. [CrossRef] [PubMed]
76. Aadil, R.M.; Khalil, A.A.; Rehman, A.; Khalid, A.; Inam-ur-Raheem, M.; Karim, A.; Gill, A.A.; Abid, M.; Afraz, M.T. Assessing the impact of ultra-sonication and thermo-ultrasound on antioxidant indices and polyphenolic profile of apple-grape juice blend. *J. Food Process Preserv.* **2020**. [CrossRef]
77. Cao, X.; Cai, C.; Wang, Y.; Zheng, X. Effects of ultrasound processing on physicochemical parameters, antioxidants, and color quality of bayberry juice. *J. Food Qual.* **2019**, *2019*. [CrossRef]

78. Tournas, V.; Stack, M.E.; Mislivec, P.B.; Koch, H.A.; Bandler, R. Yeasts, Molds, and Mycotoxins. In *Bacteriological Analytical Manual*, 8th ed.; Jinneman, K., Ed.; FDA Center for Food Safety and Applied Nutrition: Washington, DC, USA, 2001.
79. Aadil, R.M.; Zeng, X.-A.; Wang, M.-S.; Liu, Z.-W.; Han, Z.; Zhang, Z.-H.; Hong, J.; Jabbar, S. A potential of ultrasound on minerals, micro-organisms, phenolic compounds and colouring pigments of grapefruit juice. *Int. J. Food Sci. Technol.* **2015**, *50*, 1144–1150. [CrossRef]
80. Yildiz, G.; Aadil, R.M. Comparison of high temperature-short time and sonication on selected parameters of strawberry juice during room temperature storage. *J. Food Sci. Technol.* **2019**, *57*, 1462–1468. [CrossRef]
81. Mohideen, F.W.; Solval, K.M.; Li, J.; Zhang, J.; Chouljenko, A.; Chotiko, A.; Prudente, A.D.; Bankston, J.D.; Sathivel, S. Effect of continuous ultra-sonication on microbial counts and physico-chemical properties of blueberry (*Vaccinium corymbosum*) juice. *LWT-Food Sci. Technol.* **2015**, *60*, 563–570. [CrossRef]
82. Rojas, M.L.; Leite, T.S.; Cristianini, M.; Alvim, I.D.; Augusto, P.E.D. Peach juice processed by the ultrasound technology: Changes in its microstructure improve its physical properties and stability. *Food Res. Int.* **2016**, *82*, 22–33. [CrossRef]
83. Chen, L.; Bi, X.; Guo, D.; Xing, Y.; Che, Z. The effect of high-power ultrasound on the quality of carrot juice. *Food Sci. Technol. Int.* **2019**, *25*, 394–403. [CrossRef]
84. Gómez-López, V.M.; Buitrago, M.E.; Tapia, M.S.; Martínez-Yépez, A. Effect of ultrasonication on microbial quality, colour, and ascorbic acid content of passion-fruit juice during storage. *Acta Aliment.* **2017**, *46*, 470–480. [CrossRef]
85. Tomadoni, B.; Cassani, L.; Viacava, G.; Moreira, M.D.R.; Ponce, A. Effect of ultrasound and storage time on quality attributes of strawberry juice. *J. Food Process Eng.* **2017**, *40*. [CrossRef]
86. Roobab, U.; Aadil, R.M.; Madni, G.M.; Bekhit, A.E.-D. The impact of nonthermal technologies on the microbiological quality of juices: A review. *Compr. Rev. Food Sci. Food Saf.* **2018**, *17*, 437–457. [CrossRef]
87. Zhu, J.; Wang, Y.; Li, X.; Li, B.; Liu, S.; Chang, N.; Jie, D.; Ning, C.; Gao, H.; Meng, X. Combined effect of ultrasound, heat, and pressure on Escherichia coli O157:H7, polyphenol oxidase activity, and anthocyanins in blueberry (*Vaccinium corymbosum*) juice. *Ultrason. Sonochem.* **2017**, *37*, 251–259. [CrossRef] [PubMed]
88. Sánchez-Rubio, M.; Taboada-Rodríguez, A.; Cava-Roda, R.; López-Molina, D.; Marín-Iniesta, F. Combined use of thermo-ultrasound and cinnamon leaf essential oil to inactivate *Saccharomyces cerevisiae* in culture broth and natural orange juice. *J. Food Sci. Technol.* **2018**, *55*, 4623–4633. [CrossRef] [PubMed]
89. Abera, G. Review on high-pressure processing of foods. *Cogent. Food Agric.* **2019**, *5*, 1568725. [CrossRef]
90. Moreira, R.M.; Martins, M.L.; Leite Júnior, B.R.d.C.; Martins, E.M.F.; Ramos, A.M.; Cristianini, M.; Campos, A.N.D.R.; Stringheta, P.C.; Silva, V.R.O.; Canuto, J.W.; et al. Development of a juçara and Ubá mango juice mixture with added Lactobacillus rhamnosus GG processed by high pressure. *LWT* **2017**, *77*, 259–268. [CrossRef]
91. Chakraborty, S.; Kaushik, N.; Rao, P.S.; Mishra, H.N. High-pressure inactivation of enzymes: A review on its recent applications on fruit purees and juices. *Compr. Rev. Food Sci. Food Saf.* **2014**, *13*, 578–596. [CrossRef]
92. Subasi, B.G.; Alpas, H. Effect of high hydrostatic pressure processing and squeezing pressure on some quality properties of pomegranate juice against thermal treatment. *High Press. Res.* **2017**, *37*, 78–92. [CrossRef]
93. da Silveira, T.F.F.; Cristianini, M.; Kuhnle, G.G.; Ribeiro, A.B.; Filho, J.T.; Godoy, H.T. Anthocyanins, non-anthocyanin phenolics, tocopherols and antioxidant capacity of açaí juice (*Euterpe oleracea*) as affected by high pressure processing and thermal pasteurization. *Innov. Food Sci. Emerg. Technol.* **2019**, *55*, 88–96. [CrossRef]
94. Marszałek, K.; Szczepańska, J.; Starzonek, S.; Woźniak, Ł.; Trych, U.; Skąpska, S.; Rzoska, S.; Saraiva, J.A.; Lorenzo, J.M.; Barba, F.J. Enzyme inactivation and evaluation of physicochemical properties, sugar and phenolic profile changes in cloudy apple juices after high pressure processing, and subsequent refrigerated storage. *J. Food Process Eng.* **2019**, *42*. [CrossRef]
95. Petrus, R.; Churey, J.; Worobo, R. Searching for high pressure processing parameters for *Escherichia coli* O157:H7, *Salmonella enterica* and *Listeria monocytogenes* reduction in Concord grape juice. *Br. Food J.* **2020**, *122*, 170–180. [CrossRef]
96. Sandate-Flores, L.; Rostro-Alanis, M.d.J.; Mancera-Andrade, E.I.; Esquivel-Hernandez, D.A.; Brambila-Paz, C.; Parra-Saldívar, R.; Welti-Chanes, J.; Escobedo-Avellaneda, Z.; Rodríguez-Rodríguez, J. Using high hydrostatic pressures to retain the antioxidant compounds and to reduce the enzymatic activity of a pitaya–pineapple (*Stenocereus* sp.–*Fragaria ananassa*) beverage. *J. Food Sci. Technol.* **2017**, *54*, 611–619. [CrossRef]

97. Swami Hulle, N.R.; Chakraborty, S.; Rao, P.S. Effect of high pressure thermal processing on the quality attributes of *Aloe vera*-litchi mixed beverage. *Innov. Food Sci. Emerg. Technol.* **2017**, *40*, 68–77. [CrossRef]
98. Zhang, W.; Dong, P.; Lao, F.; Liu, J.; Liao, X.; Wu, J. Characterization of the major aroma-active compounds in Keitt mango juice: Comparison among fresh, pasteurization and high hydrostatic pressure processing juices. *Food Chem.* **2019**, *289*, 215–222. [CrossRef]
99. Quan, W.; Tao, Y.; Qie, X.; Zeng, M.; Qin, F.; Chen, J.; He, Z. Effects of high-pressure homogenization, thermal processing, and milk matrix on the in vitro bioaccessibility of phenolic compounds in pomelo and kiwi juices. *J. Funct. Foods* **2020**, *64*. [CrossRef]
100. Chen, X.H.; Qin, W.D.; Ma, L.H.; Xu, F.; Jin, P.; Zheng, Y.H. Effect of high pressure processing and thermal treatment on physicochemical parameters, antioxidant activity and volatile compounds of green asparagus juice. *LWT-Food Sci. Technol.* **2015**, *62*, 927–933. [CrossRef]
101. Nayak, P.K.; Rayaguru, K.; Radha Krishnan, K. Quality comparison of elephant apple juices after high-pressure processing and thermal treatment. *J. Sci. Food Agric.* **2017**, *97*, 1404–1411. [CrossRef]
102. Yi, J.; Kebede, B.T.; Hai Dang, D.N.; Buvé, C.; Grauwet, T.; Van Loey, A.; Hu, X.; Hendrickx, M. Quality change during high pressure processing and thermal processing of cloudy apple juice. *LWT-Food Sci. Technol.* **2017**, *75*, 85–92. [CrossRef]
103. Aaby, K.; Grimsbo, I.H.; Hovda, M.B.; Rode, T.M. Effect of high pressure and thermal processing on shelf life and quality of strawberry purée and juice. *Food Chem.* **2018**, *260*, 115–123. [CrossRef]
104. Elizondo-Montemayor, L.; Hernández-Brenes, C.; Ramos-Parra, P.A.; Moreno-Sánchez, D.; Nieblas, B.; Rosas-Pérez, A.M.; Lamadrid-Zertuche, A.C. High hydrostatic pressure processing reduces the glycemic index of fresh mango puree in healthy subjects. *Food Funct.* **2015**, *6*, 1352–1360. [CrossRef]
105. Shigematsu, T.; Murakami, M.; Nakajima, K.; Uno, Y.; Sakano, A.; Narahara, Y.; Hayashi, M.; Ueno, S.; Fujii, T. Bioconversion of glutamic acid to γ-aminobutyric acid (GABA) in brown rice grains induced by high pressure treatment. *Jpn. J. Food Eng.* **2010**, *11*, 189–199. [CrossRef]
106. Zhang, W.; Yu, Y.; Xie, F.; Gu, X.; Wu, J.; Wang, Z. High pressure homogenization versus ultrasound treatment of tomato juice: Effects on stability and in vitro bioaccessibility of carotenoids. *LWT-Food Sci. Technol.* **2019**, *116*. [CrossRef]
107. Rahman, S.M.E.; Jin, Y.-G.; Oh, D.-H. Combined effects of alkaline electrolyzed water and citric acid with mild heat to control microorganisms on cabbage. *J. Food Sci.* **2010**, *75*, M111–M115. [CrossRef]
108. Astray, G.; Gonzalez-Barreiro, C.; Mejuto, J.C.; Rial-Otero, R.; Simal-Gándara, J. A review on the use of cyclodextrins in foods. *Food Hydrocoll.* **2009**, *23*, 1631–1640. [CrossRef]
109. EC. Commission decision of 26 May 2008 authorising the placing on the market of alpha-cyclodextrin as a novel food ingredient under Regulation (EC) no 258/97 of the European Parliament and of the Council. *Off. J. Eur. Union* **2008**, *146*, 12–15.
110. EC. Regulation (EC) no 1333/2008 of the European Parliament and of the council of 16 December 2008 on food additives. *Off. J. Eur. Union* **2008**, *354*, 16–33.
111. Mortensen, A.; Aguilar, F.; Crebelli, R.; Di Domenico, A.; Dusemund, B.; Frutos, M.J.; Galtier, P.; Gott, D.; Gundert-Remy, U.; Leblanc, J.C.; et al. Re-evaluation of β-cyclodextrin (E 459) as a food additive. *EFSA J.* **2016**, *14*. [CrossRef]
112. Cassani, L.; Tomadoni, B.; Moreira, M.R.; Ponce, A.; Agüero, M.V. Optimization of inulin: Oligofructose proportion and non-thermal processing to enhance microbiological and sensory properties of fiber-enriched strawberry juice. *LWT* **2017**, *80*, 446–455. [CrossRef]
113. Gabriel, A.A. Combinations of selected physical and chemical hurdles to inactivate *Escherichia coli* O157:H7 in apple and orange juices. *Food Control* **2015**, *50*, 722–728. [CrossRef]
114. Ferrario, M.; Guerrero, S. Effect of a continuous flow-through pulsed light system combined with ultrasound on microbial survivability, color and sensory shelf life of apple juice. *Innov. Food Sci. Emerg. Technol.* **2016**, *34*, 214–224. [CrossRef]
115. Gomes, W.F.; Tiwari, B.K.; Rodriguez, Ó.; de Brito, E.S.; Fernandes, F.A.N.; Rodrigues, S. Effect of ultrasound followed by high pressure processing on prebiotic cranberry juice. *Food Chem.* **2017**, *218*, 261–268. [CrossRef] [PubMed]

116. de Carvalho, R.J.; de Souza, G.T.; Pagán, E.; García-Gonzalo, D.; Magnani, M.; Pagán, R. Nanoemulsions of *Mentha piperita* L. essential oil in combination with mild heat, pulsed electric fields (PEF) and high hydrostatic pressure (HHP) as an alternative to inactivate *Escherichia coli* O157: H7 in fruit juices. *Innov. Food Sci. Emerg. Technol.* **2018**, *48*, 219–227. [CrossRef]
117. Park, J.-S.; Ha, J.-W. Ultrasound treatment combined with fumaric acid for inactivating food-borne pathogens in apple juice and its mechanisms. *Food Microbiol.* **2019**, *84*. [CrossRef] [PubMed]
118. Oliveira, A.F.A.; Mar, J.M.; Santos, S.F.; da Silva Júnior, J.L.; Kluczkovski, A.M.; Bakry, A.M.; Bezerra, J.d.A.; Nunomura, R.d.C.S.; Sanches, E.A.; Campelo, P.H. Non-thermal combined treatments in the processing of açai (*Euterpe oleracea*) juice. *Food Chem.* **2018**, *265*, 57–63. [CrossRef]
119. Lee, S.J.; Bang, I.H.; Choi, H.-J.; Min, S.C. Pasteurization of mixed mandarin and Hallabong tangor juice using pulsed electric field processing combined with heat. *Food Sci. Biotechnol.* **2018**, *27*, 669–675. [CrossRef]
120. Dhakal, S.; Balasubramaniam, V.M.; Ayvaz, H.; Rodriguez-Saona, L.E. Kinetic modeling of ascorbic acid degradation of pineapple juice subjected to combined pressure-thermal treatment. *J. Food Eng.* **2018**, *224*, 62–70. [CrossRef]
121. Siemer, C.; Toepfl, S.; Heinz, V. Inactivation of *Bacillus subtilis* spores by pulsed electric fields (PEF) in combination with thermal energy–I. Influence of process- and product parameters. *Food Control* **2014**, *39*, 163–171. [CrossRef]
122. Siemer, C.; Toepfl, S.; Heinz, V. Inactivation of *Bacillus subtilis* spores by pulsed electric fields (PEF) in combination with thermal energy II. Modeling thermal inactivation of *B. subtilis* spores during PEF processing in combination with thermal energy. *Food Control* **2014**, *39*, 244–250. [CrossRef]
123. Aadil, R.M.; Zeng, X.-A.; Sun, D.-W.; Wang, M.-S.; Liu, Z.-W.; Zhang, Z.-H. Combined effects of sonication and pulsed electric field on selected quality parameters of grapefruit juice. *LWT-Food Sci. Technol.* **2015**, *62*, 890–893. [CrossRef]
124. Evelyn; Kim, H.J.; Silva, F.V.M. Modeling the inactivation of *Neosartorya fischeri* ascospores in apple juice by high pressure, power ultrasound and thermal processing. *Food Control* **2016**, *59*, 530–537. [CrossRef]
125. Abid, M.; Jabbar, S.; Hu, B.; Hashim, M.M.; Wu, T.; Wu, Z.; Khan, M.A.; Zeng, X. Synergistic impact of sonication and high hydrostatic pressure on microbial and enzymatic inactivation of apple juice. *LWT-Food Sci. Technol.* **2014**, *59*, 70–76. [CrossRef]
126. Chueca, B.; Ramírez, N.; Arvizu-Medrano, S.M.; García-Gonzalo, D.; Pagán, R. Inactivation of spoiling microorganisms in apple juice by a combination of essential oils' constituents and physical treatments. *Food Sci. Technol. Int.* **2015**, *22*, 389–398. [CrossRef] [PubMed]
127. García-García, R.; Escobedo-Avellaneda, Z.; Tejada-Ortigoza, V.; Martín-Belloso, O.; Valdez-Fragoso, A.; Welti-Chanes, J. Hurdle technology applied to prickly pear beverages for inhibiting *Saccharomyces cerevisiae* and *Escherichia coli*. *Lett. Appl. Microbiol.* **2015**, *60*, 558–564. [CrossRef] [PubMed]
128. Pingret, D.; Fabiano-Tixier, A.-S.; Chemat, F. Degradation during application of ultrasound in food processing: A review. *Food Control* **2013**, *31*, 593–606. [CrossRef]

© 2020 by the authors. Licensee MDPI, Basel, Switzerland. This article is an open access article distributed under the terms and conditions of the Creative Commons Attribution (CC BY) license (http://creativecommons.org/licenses/by/4.0/).

foods

MDPI

Article

Does High Voltage Electrical Discharge Treatment Induce Changes in Tannin and Fiber Properties of Cocoa Shell?

Veronika Barišić [1], Ivana Flanjak [1,*], Mirela Kopjar [1], Mirta Benšić [2], Antun Jozinović [1], Jurislav Babić [1], Drago Šubarić [1], Borislav Miličević [1,3], Kristina Doko [4], Midhat Jašić [5] and Đurđica Ačkar [1]

1 Faculty of Food Technology Osijek, Josip Juraj Strossmayer University of Osijek, Franje Kuhača 18, 31000 Osijek, Croatia; veronika.barisic@ptfos.hr (V.B.); mirela.kopjar@ptfos.hr (M.K.); antun.jozinovic@ptfos.hr (A.J.); jurislav.babic@ptfos.hr (J.B.); drago.subaric@ptfos.hr (D.Š.); bmilicevic@ptfos.hr (B.M.); dackar@ptfos.hr (Đ.A.)
2 Department of Mathematics, Josip Juraj Strossmayer University of Osijek, Trg Ljudevita Gaja 6, 31000 Osijek, Croatia; mirta@mathos.hr
3 Polytechnic in Požega, Vukovarska 17, 34000 Požega, Croatia
4 Federal Agro Mediterranean Institute, Biskupa Čule 10, 88000 Mostar, Bosnia and Herzegovina; kristina.skender@gmail.com
5 Faculty of Technology Tuzla, Univerzitetska 8, 75000 Tuzla, Bosnia and Herzegovina; jasic_midhat@yahoo.com
* Correspondence: ivana.flanjak@ptfos.hr

Received: 4 June 2020; Accepted: 18 June 2020; Published: 19 June 2020

Abstract: Cocoa shell is a by-product of the chocolate industry that is rich in dietary fiber and bioactive components. In this research, the influence of high voltage electric discharge (HVED) treatment on chemical and physical characteristics of the cocoa shell, i.e., the effects of applied time and frequencies on grinding ability, water binding capacity (WBC), dietary fibers and tannin content was investigated. HVED had a significant influence on the chemical and physical properties of cocoa shell, all of which could be linked to changes in fiber properties. Along with the fiber content, grinding ability and water binding capacity were increased. These properties have already been linked to fiber content and soluble/insoluble fiber ratio. However, this research implies that change in fiber properties could be linked to tannin formation via complexation of other polyphenolic components. Additional research is needed to verify this effect and to establish mechanisms of tannin formation induced by HVED and its influence on fiber quantification.

Keywords: cocoa shell; high voltage electrical discharge; tannin; dietary fiber; water binding capacity; grindability

1. Introduction

Cocoa shell is the major by-product of the cocoa processing industry. It is a part of the cocoa bean that is separated from cotyledon during pre-roasting or after the roasting of beans [1]. It has been reported that several tons of cocoa shell need to be disposed annually, which poses a large problem [2,3]. Cocoa beans are rich in bioactive compounds, which are stored in the cotyledon. During fermentation, these components diffuse into cocoa shell, which becomes rich in bioactive compounds [4]. In addition, cocoa shell is rich in dietary fiber, mainly consisting of cellulose, carbohydrates and pectic polysaccharides [2] and presents great material for use in food industry and enrichment of food poor in dietary fibers. In the last few years, cocoa shell has been used as a raw biomass material, feedstuff, adsorbent, soil conditioner, garden mulch or burnt for fuel [5,6].

High voltage electric discharge (HVED) is a non-thermal process that has been used for the treatment of waste products from the food industry in the last few years [7]. It is also used as an extraction method, because it can disrupt the cellular walls and increase the overall mass transfer of the cellular content [8]. HVED is an innovative technique that interjects energy directly in aqueous solution between electrodes that are submerged. Electric discharge in water consists of two phases: corona streamer discharge process and arc discharge process. For the streamer discharge process weak shock waves are characteristic, as well as small number of bubbles and active radicals. When transiting to arc discharge process, number of bubbles is increased, shock waves become stronger, turbulence and concentration of free radicals are increased [9]. These shock waves and explosions of cavitation bubbles can affect particle size by fragmentation of cell membranes [10]. Electric discharge directly in water leads to production of molecular oxygen and hydrogen, hydrogen peroxide, hydroxyl radicals and oxygen radical ion, all of which are very reactive species.

Since HVED can disrupt cellular walls, which are in cocoa shell predominantly composed of cellulose with lesser amounts of hemicellulose and pectin [3], the aim of this study was to evaluate HVED influence on cocoa shell dietary fiber content and properties related to it. For use of cocoa shell in food industry, dietary fiber content, grindability, water and oil binding capacity and content of bioactive components are very important.

2. Materials and Methods

2.1. Preparation of Cocoa Shell

Cocoa shell samples were obtained after roasting fermented cocoa beans (West Africa mix supplied by Huyser, Möller B.V., Edam, Holland) at 135 °C for 55 min in custom made roaster (Metal workshop "ILMA", Požega, Croatia). After that, the cocoa shell was easily separated by hand from the cotyledon.

Untreated cocoa shell (UCS) sample was obtained by grinding cocoa shell attained after separation from the cotyledon. Control samples were obtained by mixing the unmilled cocoa shell in water for 15, 30 and 45 min at concentrations of 1.5% and 3.0%. After mixing, the shell was separated from water and dried in the laboratory oven (Memmert, UFE 500, Schwabach, Germany) at 40 °C. Dry samples were ground in the laboratory mill (IKA, M20, Staufen, Germany) (25 g for 2 min with cooling) to obtain a fine powder (composite sample obtained by repeated grinding) and as such were frozen and stored for analyses. The grinded untreated cocoa shell was also frozen and stored for analysis in the same way as a cocoa shell mixed in water.

2.2. HVED Treatment

High-voltage electrical discharge equipment which was described by Barišić et al. [11] includes a chamber connected to a high-voltage pulse generator of 30 kV (the device was custom made by Inganiare CPTS1, Osijek, Croatia for the Faculty of Food Technology Osijek). Treatment chamber contains a stainless steel cylindrical needle (diameter 2.5 mm), and the ground electrode in the form of a plate (diameter 45 mm). Mixing of samples is achieved by magnetic stirrer. The distance between the electrodes during all treatments was 2 cm. Electric field density was 15 kV/cm during all treatments. HVED energy input ranged between 13.11–79.80 kJ/kg.

Unmilled cocoa shell (same as control samples prepared in water) was treated in HVED device at concentrations of 1.5% (6 g in 400 mL of distilled water) and 3.0% (12 g in 400 mL of distilled water). The treatment time was 15, 30 and 45 min, and the used frequencies were 40 and 80 Hz. Each sample (HVED, control or untreated) was treated until 200 g of sample was gained which gave us uniform sample for all analyses. The cocoa shell treated with HVED was dried, grind and stored until analyses in the same way as the control samples. Control samples (mixed in water) and HVED treated samples were dried to a dry matter content of 86.00 ± 0.85%.

2.3. Tannin Content

2.3.1. Extraction

Each sample was weighed (2 g) and extracted three times with 10 mL of *n*-hexane (Carlo Erba Reagents, Val de Reuil, France) to remove lipids. Samples were dried at air over night and extracted with 5 mL 70% methanol (J. T. Baker, Deventer, Netherland) in ultrasound bath. After that, samples were centrifuged for 10 min at 3000 rpm (Sigma 2-16, Sigma, Osterode, Germany). Supernatant was decanted in 10 mL volumetric flask. That procedure was repeated twice after which flask with supernatant was filled up with 70% methanol.

2.3.2. Spectrophotometric Analysis

Tannin content was determined by method described by Amorim et al. [12]. Method is based on binding of tannins with casein. Calibration curve was created with the standard solutions of tannic acid (Sigma-Aldrich, St. Louis, USA) in the range of concentrations from 0.5 to 3 mg/mL ($y = 0.9011x + 0.0095$; $R^2 = 0.9993$). Total phenol content and residual phenol content (obtained after complexation of tannin and casein) were determined spectrophotometrically at 760 nm according to the method of Singleton et al. [13]. Tannin content in prepared extracts was calculated Equation (1) as the difference between total phenol content and residual phenol content. Results are presented as mg of tannic acid per g of defatted sample (mg TA/g) and as a percentage of tannin in total phenol content (%).

$$\text{Tannin}\left(\frac{\text{mg TA}}{\text{g}}\right) = \text{total phenol content} - \text{residual phenol content} \tag{1}$$

2.4. Determination of Dietary Fibers

Dietary fibers were determined according to gravimetric AOAC method 991.43 [14]. Samples were treated with thermostable α-amylase, protease and amyloglucosidase (Megazyme Total Dietary Fiber Assay Kit, Megazyme Ltd., Bray, Ireland). The share of insoluble dietary fibers (IDF, %) was determined gravimetrically after filtration, and soluble dietary fibers (SDF, %) were determined by precipitation from the obtained filtrate. After correction for undigested protein (Kjeldahl method) and ash (mineralization at 525 °C), total dietary fibers were calculated Equations (2) and (3) as a sum of IDF and SDF. The values were calculated on the dry matter of the sample.

$$\text{Total Dietary Fibre } (\%) = \frac{\frac{R_1+R_2}{2} - p - A - B}{\frac{m_1+m_2}{2}} \times 100 \tag{2}$$

$$B = \frac{BR_1 + BR_2}{2} - BP - BA \tag{3}$$

where: R_1 = residue weight 1 from m_1; R_2 = residue weight 2 from m_2; m_1 = sample weight 1; m_2 = sample weight 2; A = ash weight from R_1; P = protein weight from R_2; B = blank; BR = blank residue; BP = blank protein from BR_1; BA = blank ash from BR_2.

2.5. Grindability of Cocoa Shell

The grindability of cocoa shell was determined by sieving the powdered cocoa shell samples on analytical sieve shaker (Retsch GmbH, AS200, Haan, Germany) and measurement of mass of obtained fractions. A total of 50 g of the sample was sieved through six sieves (50, 71, 100, 125, 200 and 315 μm) during 15 min. After weighing each fraction, results were expressed as percentages of cocoa shell mass that was weighted on each sieve (%).

2.6. Water Binding Capacity (WBC) and Oil Binding Capacity (OBC)

For determination of WBC standard AACC Method 88-04 [15], was used. To 2.5 g of cocoa shell sample 30 mL of water was added. These solutions were left to stand at room temperature with periodic mixing. After that, the samples were centrifuged at 3000 rpm for 15 min (Centra-MP4R, IEC, Mumbai, India). The supernatant was decanted, and the remaining residue was weighted. The analysis was performed in two repetitions. The results were calculated Equation (4) and were expressed as grams of H_2O absorbed per gram of cocoa shell (g/g).

$$\mathrm{WBC}\left(\frac{\mathrm{g}}{\mathrm{g}}\right) = \frac{\text{gel mass}}{\text{dry matter mass in the initial sample}} \quad (4)$$

For determination of OBC same procedure was used. The only difference was that for OBC instead of water cold pressed rapeseed oil was used. Results were expressed as grams of oil absorbed per gram of cocoa shell (%) obtained by formula Equation (5):

$$\mathrm{OBC}\left(\frac{\mathrm{g}}{\mathrm{g}}\right) = \frac{\text{gel mass}}{\text{dry matter mass in the initial sample}} \quad (5)$$

2.7. Fourier Transform Infrared Spectroscopy with Attenuated Total Reflection (FTIR-ATR) Analysis

FTIR–ATR spectra were recorded with a Cary 630 spectrometer (Agilent, Santa Clara, CA, USA) in wavenumber range from 4000 to 650 cm^{-1}. For each sample, 32 scans were recorded and averaged with a spectral resolution of 16 cm^{-1}.

2.8. Statistical Analysis

Statistical analysis was conducted using Statistica®, Version 13.4.0.14 (1984–2018 TIBCO Software Inc, Palo Alto, CA, USA). To determine the statistically significant difference of treatment effects, main effects and factorial analysis of variance (ANOVA) were used. P-value that was considered significant was 0.05. In addition, Pearson's correlation coefficients was determined ($p < 0.05$).

3. Results and Discussion

3.1. Tannin Content

Tannin content of untreated cocoa shell and treated samples are shown in Figure 1 where results for tannin content (mg TA/g of defatted sample) and percentages of tannins in total phenols (%) are presented. It can be seen that the untreated shell had the lowest content of tannins, and the tannin content increased with all treatments. In samples treated with HVED, share of tannins in total phenols ranged from 45.03 to 65.09%. In our previous research [16], we have measured the decrease of content of all major polyphenolic compounds in cocoa shell treated with HVED (catechin, epicatechin, epicatechin gallate, gallic acid, caffeic acid and *p*-coumaric acid). These components are extractable by water, and the decrease in treated shell may have been the consequence of extraction, as reported by Jokić et al. [17], however, they are also prone to reactions of condensation in suitable conditions (Figure 2). Since the aim of this study was not to establish the effect of HVED treatment on extraction of bioactive compounds, cocoa shell was not milled before treatment, unlike in research of Jokić et al. [17]. Extraction yield is also dependent on electric field intensity, contact surface between material and solvent, liquid to solid ratio, etc. Considering above mentioned, HVED conditions applied in this research are not favorable for extraction [9]. Thus, extraction of polyphenolic compounds was aggravated. HVED generates different reactive species, which may have induced polymerization. Hence, HVED is a source of radicals that can easily oxidize tannins, which leads to an increase in their rigidity. Contrary to our results, Delsart et al. [18] and Lukić et al. [19] reported decrease of total tannin content in red wine treated by HVED and cold plasma, respectively, ascribing it to oxidation of tannins during treatment.

However, one has to bear in mind that cold plasma and HVED treatment differ in that cold plasma includes gas introduction into liquid, and that there are major differences in chemical composition, mainly polyphenolic profile, of the treated samples. In addition, since treatment time in this research was significantly longer, oxidized tannins and other phenols might have been involved in mutual reactions, mainly because the oxidized phenols are hydrophobic. It has been reported that hydrophobic reactions can occur among polyphenols and induce their aggregation [20].

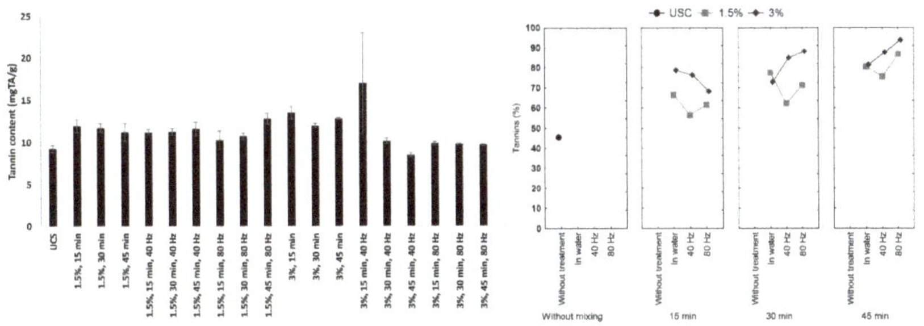

Figure 1. Tannin content (expressed on defatted sample weight) in cocoa shell before and after the high voltage electric discharge (HVED) treatment and percent of tannin in total phenols.

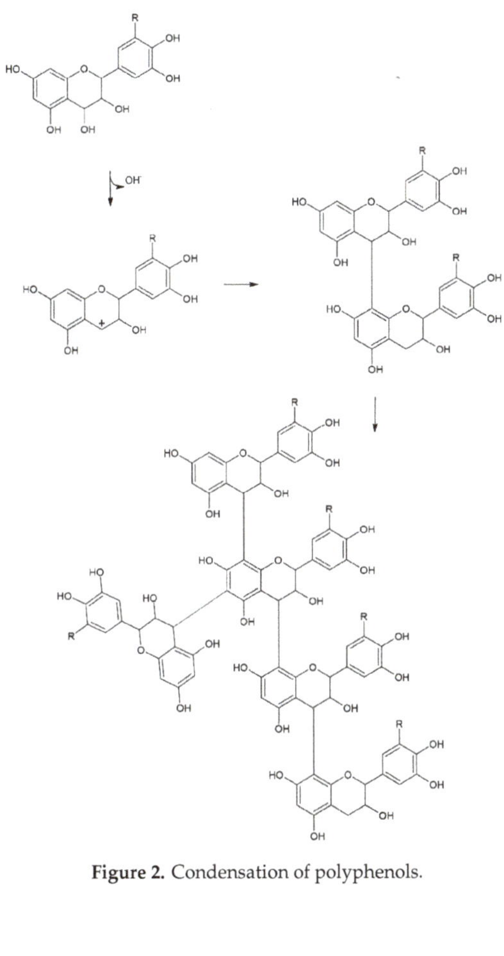

Figure 2. Condensation of polyphenols.

Results of tannin content, both in sample and in total phenols show that tannins are very resistant to HVED treatment. With the exception of 3.0% sample treated at 40 Hz, where significant reduction of tannin content occurred with the increase of treatment time, a slight increase of their content after treatment was observed (Figure 1), possibly due to oxidation and aggregation of tannins, but also due to the loss of a portion of soluble substances during treatment, which led to a change in the ratio of components in the samples.

Statistical analysis confirmed that tannins are very resistant to HVED treatment, since statistical significance was not established. Furthermore, factorial analysis of variance showed that influence of concentration and mixing time on percent of tannin in total phenols is statistically significant (Table 1). Coefficient of correlation is showing that tannin (%) is in relation with the smallest and largest particles, insoluble and total fibers. This may indicate that tannins have an impact on the proportion of fibers in cocoa shells, since the content of insoluble and total fibers increase as the proportion of tannin increases.

Table 1. Factorial analysis of variance.

		Sum of Squares	DF	Mean Square	F-Value	p-Value
OBC (g/g)	Intercept	392.4578	1	392.4578	533,103.9	<0.001 *
	Concentration (C)	0.2885	1	0.2885	391.8	<0.001 *
	Mixing (M)	0.1274	2	0.0637	86.5	<0.001 *
	Treatment (T)	0.1614	2	0.0807	109.6	<0.001 *
	C*M	0.0200	2	0.0100	13.6	<0.001 *
	C*T	0.0474	2	0.0237	32.2	<0.001 *
	M*T	0.0516	4	0.0129	17.5	<0.001 *
	C*M*T	0.0821	4	0.0205	27.9	<0.001 *
	Error	0.0133	18	0.0007		
WBC (g/g)	Intercept	2384.278	1	2384.278	313,752.4	<0.001 *
	Concentration (C)	15.149	1	15.149	1993.6	<0.001 *
	Mixing (M)	3.762	2	1.881	247.6	<0.001 *
	Treatment (T)	0.040	2	0.020	2.6	0.101638
	C*M	0.130	2	0.065	8.5	0.002470 *
	C*T	0.177	2	0.088	11.6	<0.001 *
	M*T	0.266	4	0.066	8.8	<0.001 *
	C*M*T	0.516	4	0.129	17.0	<0.001 *
	Error	0.137	18	0.008		
Tannin (mg TA/g of defatted sample)	Intercept	4714.410	1	4714.410	1021.198	<0.001 *
	Concentration (C)	0.030	1	0.030	0.006	0.936726
	Mixing (M)	13.343	2	6.671	1.445	0.261805
	Treatment (T)	17.066	2	8.533	1.848	0.186168
	C*M	24.240	2	12.120	2.625	0.099895
	C*T	10.946	2	5.473	1.186	0.328333
	M*T	29.441	4	7.360	1.594	0.218875
	C*M*T	24.668	4	6.167	1.336	0.294923
	Error	83.098	18	4.617		
Tannin (% of total polyphenols)	Intercept	114,239.9	1	114,239.9	3614.463	<0.001 *
	Concentration (C)	202.5	1	202.5	6.406	0.020914 *
	Mixing (M)	453.1	2	226.5	7.167	0.005134 *
	Treatment (T)	84.5	2	42.3	1.337	0.287391
	C*M	75.8	2	37.9	1.200	0.324306
	C*T	49.7	2	24.9	0.786	0.470508
	M*T	146.0	4	36.5	1.155	0.363192
	C*M*T	102.3	4	25.6	0.809	0.535434
	Error	568.9	18	31.6		

OBC: oil binding capacity; WBC: water binding capacity; DF: degree of freedom; * $p < 0.05$ statistically significant.

3.2. Dietary Fibers

Proportions of soluble, insoluble and total fibers of cocoa shell samples are shown in Figure 3. It can be seen that content of insoluble and total fibers in treated samples is higher than in untreated cocoa shell. The effect of HVED on soluble dietary fibers was not statistically significant (Table 2). An increase of insoluble fiber share after treatment had statistical significance for mixing time and there is a visible trend. A greater effect on increase of insoluble fiber content was in HVED treated samples at 1.5% concentration than at 3.0% due to greater energy input at lower sample concentration.

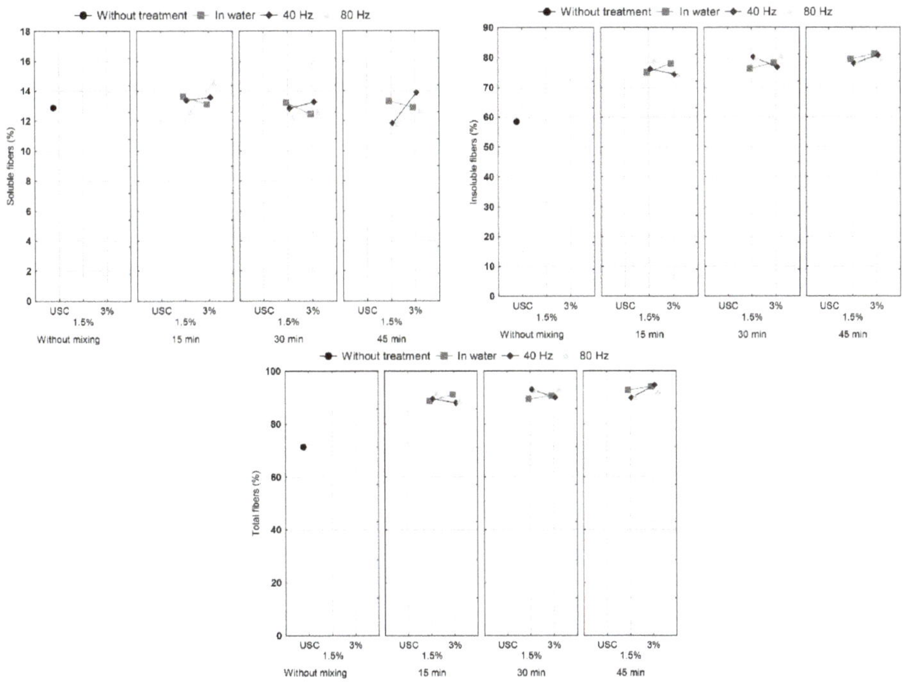

Figure 3. Content of insoluble, soluble and total fibers in cocoa shell before and after the HVED treatment.

Increasing the fiber content in treated cocoa shells can be explained by the fact that during various treatments, fiber probably bonded with other components of the cocoa shell. However, some researches have shown that results obtained by gravimetric determination of fiber may be increased due to presence of insoluble proteins and condensed tannins [21–23]. The method used in this research has a step to exclude undigested proteins from the results for fiber content, however, condensed tannins may have an effect on the observed increase.

Condensed tannins are, along with resistant protein and Maillard reaction products, part of so-called Klason lignin [21], which is not always considered as a fiber. As shown by Perez et al. [24], roasted husk contains large amounts of free amino-acids and sugars, and our previous research [16] showed significant amounts of catechins. HVED generates free radicals and charged particles and highly reactive species (H^+, OH^-, H_2O_2), which may have induced advanced Maillard reactions and reactions of condensation of catechins to condensed tannins (Figure 2).

Our previous research showed that most likely 5-HMF and acrylamide are reacting with free radicals created by HVED and generating new compounds, which could be a part of Klason lignin [25]. This could contribute to increase of insoluble dietary fibers especially because condensed tannins and products of advanced Maillard reactions are insoluble in water.

In addition, we noticed that HVED treated samples had more undigested proteins than non-treated samples (results not shown). Decreased digestibility of proteins can be result of complex formation with tannins especially because HVED treatment generates change in pH and surface charge, which could be favorable conditions for complexation. Reduced in vitro and in vivo digestibility of proteins due to formation of complexes with tannins has already been reported for sorghum and several Acacia species. In addition, protein-protein complexation induced by tannins, and enzyme inhibition by tannins were also reported [26]. Although corrections for proteins were made, the other components that were bonded to them were not included here.

Table 2. Main effects analysis of variance.

	Effect	Sum of Squares	DF	Mean Square	F-Value	*p*-Value
0–50 µm	Intercept	169.1845	1	169.1845	59.27547	0.000006 *
	Concentration	20.7446	1	20.7446	7.26806	0.019455 *
	Mixing	2.4918	2	1.2459	0.43652	0.656146
	Treatment	2.8862	2	1.4431	0.50561	0.615433
	Error	34.2505	12	2.8542		
51–71 µm	Intercept	2025.733	1	2025.733	451.2232	<0.001 *
	Concentration	54.266	1	54.266	12.0875	0.004574 *
	Mixing	12.171	2	6.085	1.3555	0.294609
	Treatment	1.952	2	0.976	0.2174	0.807671
	Error	53.873	12	4.489		
72–100 µm	Intercept	1371.127	1	1371.127	705.1271	<0.001 *
	Concentration	18.601	1	18.601	9.5659	0.009312 *
	Mixing	0.303	2	0.152	0.0780	0.925440
	Treatment	1.729	2	0.864	0.4446	0.651247
	Error	23.334	12	1.945		
101–125 µm	Intercept	525.4466	1	525.4466	970.2258	<0.001 *
	Concentration	0.9016	1	0.9016	1.6648	0.221262
	Mixing	0.5083	2	0.2542	0.4693	0.636443
	Treatment	0.8918	2	0.4459	0.8234	0.462285
	Error	6.4989	12	0.5416		
126–200 µm	Intercept	3189.472	1	3189.472	1644.420	<0.001 *
	Concentration	1.189	1	1.189	0.613	0.448736
	Mixing	2.527	2	1.264	0.651	0.538775
	Treatment	2.636	2	1.318	0.679	0.525389
	Error	23.275	12	1.940		
201–315 µm	Intercept	5128.784	1	5128.784	3745.356	<0.001 *
	Concentration	0.147	1	0.147	0.107	0.748864
	Mixing	3.778	2	1.889	1.379	0.288933
	Treatment	4.432	2	2.216	1.618	0.238628
	Error	16.432	12	1.369		
>315 µm	Intercept	31,757.57	1	31,757.57	1505.873	<0.001 *
	Concentration	26.88	1	26.88	1.275	0.280959
	Mixing	45.68	2	22.84	1.083	0.369462
	Treatment	52.88	2	26.44	1.254	0.320301
	Error	253.07	12	21.09		
Insoluble fibers	Intercept	109,774.9	1	109,774.9	30,704.36	<0.001
	Concentration	0.1	1	0.1	0.02	0.883234
	Mixing	36.8	2	18.4	5.15	0.024326 *
	Treatment	2.7	2	1.4	0.38	0.691002
	Error	42.9	12	3.6		
Soluble fibers	Intercept	3038.191	1	3038.191	7378.913	<0.001 *
	Concentration	0.962	1	0.962	2.336	0.152303
	Mixing	2.116	2	1.058	2.570	0.117785
	Treatment	0.501	2	0.250	0.608	0.560488
	Error	4.941	12	0.412		
Total fibers	Intercept	149,338.0	1	149,338.0	43,452.47	<0.001 *
	Concentration	1.6	1	1.6	0.47	0.508137
	Mixing	21.7	2	10.9	3.16	0.079032
	Treatment	1.0	2	0.5	0.14	0.867489
	Error	41.2	12	3.4		

DF: degree of freedom; * $p < 0.05$ statistically significant.

There are already some researches investigating the effect of electrical discharge on fibers. Yuan et al. [27] concluded that plasma improves the tensile strength and surface roughness, which leads to higher interfacial contact. In addition, during the treatment, it came to oxidation of fibers. Sinha and Panigrahi [28] observed increased hydrophobicity of jute fibers after plasma treatment, probably because of oxidation or decrease of phenolic and secondary alcoholic groups. Improved flexural strength of fibers occurred because of better adhesion between fibers and matrix. Bozaci et al. [29] and Karahan and Özdoğan [30] came to the conclusion that fibers have increased hydrophilicity, rougher surface and higher proportion of damaged fibers after plasma treatment.

Additional research is needed to reveal whether proposed mechanisms may be applicable to influence of HVED on fibers in cocoa shell.

3.3. *Grindability of Cocoa Shell*

The largest change in share of particles after HVED treatment was in the particle size ranges 0–50 μm and >315 μm (Table 3). Untreated cocoa shell had the largest percentage of particles between 0 and 50 μm and the smallest percentage of particles larger than 315 μm compared to treated and control samples. Any treatment, either only in water or with HVED, has led to an increase in the share of particles with size greater than 315 μm and a reduction in the share of particles with size less than 50 μm which was proven by coefficient of correlation (Table 4). There is a relation between the smallest and the largest particles. Main effect analysis of variance showed that there was a statistically significant difference between different sample concentrations during treatment for particle sizes of 0–50 μm, 51–71 μm and 72–100 μm (Table 2). In all treated samples decrease in the percentage of smaller particles and an increase in the percentage of larger particles was observed. The minimum change occurred in the sample 1.5%, 30 min, 40 Hz. Statistical analysis shows that there was a correlation between particle sizes and dietary fibers implying that difficulty to grind HVED treated cocoa shell can be caused by increased content of fibers.

Table 3. Grindability of cocoa shell samples before and after the HVED treatment.

Sample	0–50 μm (%)	51–71 μm (%)	72–100 μm (%)	101–125 μm (%)	126–200 μm (%)	201–315 μm (%)	>315 μm (%)
UCS	15.19	21.89	11.83	7.94	18.24	14.10	10.81
1.5%, 15 min	3.63	14.70	8.27	5.29	14.66	17.87	35.58
1.5%, 30 min	3.12	12.64	7.83	5.07	13.12	16.76	41.47
1.5%, 45 min	1.71	10.64	7.33	4.74	12.11	15.44	48.03
1.5%, 15 min, 40 Hz	5.64	13.78	7.42	5.52	13.62	17.42	36.61
1.5%, 30 min, 40 Hz	8.39	13.47	7.33	5.16	13.35	17.51	34.80
1.5%, 45 min, 40 Hz	3.67	13.67	7.42	5.15	12.65	16.94	40.50
1.5%, 15 min, 80 Hz	5.48	10.55	6.28	4.66	11.59	15.75	45.69
1.5%, 30 min, 80 Hz	2.64	11.71	8.59	5.50	13.33	16.75	41.48
1.5%, 45 min, 80 Hz	2.98	9.95	8.93	5.54	13.07	16.67	42.87
3.0%, 15 min	3.52	9.98	7.22	4.64	11.72	15.36	47.56
3.0%, 30 min	2.28	9.19	9.33	5.77	13.10	16.40	43.93
3.0%, 45 min	2.16	9.26	10.98	6.40	15.05	17.31	38.84
3.0%, 15 min, 40 Hz	1.36	8.89	11.45	5.79	15.12	18.76	38.64
3.0%, 30 min, 40 Hz	0.64	4.04	11.23	7.40	15.23	18.69	42.77
3.0%, 45 min, 40 Hz	2.08	8.84	10.14	5.24	12.78	16.18	44.75
3.0%, 15 min, 80 Hz	1.50	11.73	10.83	6.07	15.48	19.18	35.20
3.0%, 30 min, 80 Hz	1.32	6.51	9.05	4.92	12.47	15.72	50.01
3.0%, 45 min, 80 Hz	3.07	11.41	7.46	4.41	11.17	15.14	47.33

UCS: untreated cocoa shell.

Table 4. Pearson's coefficients of correlation.

Variable	0–50 μm	51–71 μm	72–100 μm	101–125 μm	126–200 μm	201–315 μm	>315 μm	WBC (g/g)	OBC (g/g)	Insoluble Fibers (%)	Soluble Fibers (%)	Total Fibers (%)	Tannin (mg TA/g)	Tannin (%)
0–50 μm	1.000													
51–71 μm	0.839	1.000												
72–100 μm	0.006	−0.083	1.000											
101–125 μm	0.364	0.199	0.819	1.000										
126–200 μm	0.430	0.402	0.803	0.908	1.000									
201–315 μm	−0.456	−0.329	0.312	0.210	0.303	1.000								
>315 μm	−0.809	−0.795	−0.467	−0.717	−0.851	0.018	1.000							
WBC (g/g)	−0.252	−0.106	−0.473	−0.422	−0.391	0.096	0.349	1.000						
OBC (g/g)	−0.827	−0.838	−0.103	−0.479	−0.600	0.114	−0.862	0.233	1.000					
Insoluble fibers (%)	−0.765	−0.714	−0.455	−0.674	−0.751	0.264	0.883	0.450	0.776	1.000				
Soluble fibers (%)	−0.519	−0.351	0.098	−0.179	−0.023	0.582	0.268	0.306	0.334	0.334	1.000			
Total fibers (%)	−0.791	−0.723	−0.425	−0.666	−0.722	0.318	0.875	0.464	0.780	0.994	0.435	1.000		
Tannin (mg TA/g)	−0.244	−0.154	0.058	−0.089	0.035	0.357	0.097	−0.118	−0.100	0.123	0.065	0.127	1.000	
Tannin (%)	−0.768	−0.747	0.004	−0.365	−0.511	0.018	0.762	−0.014	0.844	0.635	0.167	0.627	0.067	1.000

Bold values were considered significant at $p < 0.05$.

3.4. Water and Oil Binding Capacity

WBC and OBC are important parameters for processing of food and any change in these properties influences production process. Water binding capacity (WBC) and oil binding capacity (OBC) of cocoa shell samples are shown in Figure 4. It is visible that the sample of untreated cocoa shell had the lowest WBC and OBC. Samples treated at a concentration of 1.5% had the higher WBC compared to samples treated at a concentration of 3.0%. OBC showed the opposite trend, where samples treated at 3.0% had higher capacity for binding oil than samples treated at 1.5%. The largest increases can be observed in samples treated for 45 min in water and with HVED. Statistical analysis showed that there was a statistically significant difference between tested concentrations and shearing time but treatment (with or without HVED) did not show statistical significance. All combinations of these effects have proven to be significant (Table 1).

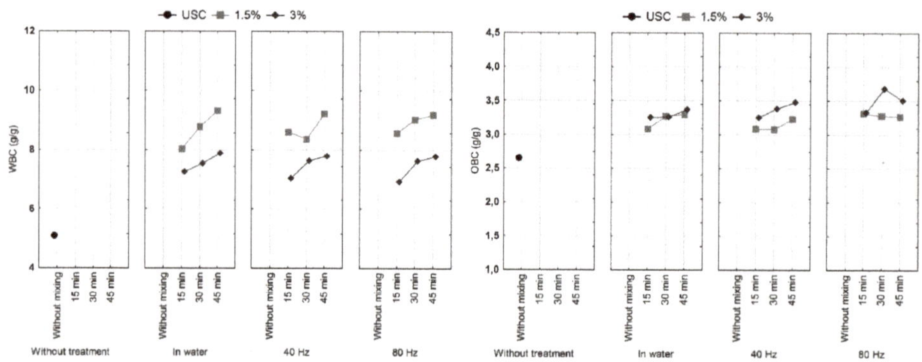

Figure 4. Oil and water binding capacity of cocoa shell before and after the HVED treatment.

According to Sangnark and Noomhorm [31], water and oil binding capacity are correlated to particle size. This research also revealed correlation of OBC with particle sizes (Table 4). Additionally, porosity, overall charge density and hydrophobic properties of fibers, all of which may be changed by HVED treatment, can greatly affect WBC and OBC [31,32]. This may also be substantiated by correlation of OBC with total fiber, insoluble fiber and tannin content in this research.

3.5. FTIR-ATR

The changes in chemical composition by HVED treatment were supported by FTIR-ATR analysis. All the treatments had similar trend so only representative spectra are shown in Figure 5.

In untreated cocoa shell C=O stretching at 1737 cm^{-1} is presented only with a shoulder, and there is a peak at 1602.8 cm^{-1}. After the treatment, a small peak appears at 1737 cm^{-1}. Karahan and Özdoğan [30] ascribed this peak to ester groups of pectin. This is implying that increased content of soluble fibers may be linked to the appearance of this peak after the treatment. However, according to Günzler and Gremlich [33] and Grillo et al. [34], this is also C=O stretch in unconjugated esters, carboxylic acids, aldehydes and ketones.

C-H asymmetric deformation vibrations in untreated shell are presented through a shoulder at 1410 cm^{-1}, whereas after the treatments peak appears at 1431.3 cm^{-1}.

In the untreated cocoa shell there is a peak at 1028.7 cm^{-1} with two shoulders at 1096 cm^{-1} and 1148 cm^{-1}. Treatments did not change shoulder at 1096 cm^{-1}, unlike the other one that has shifted to 1155 cm^{-1} (C-H deformation) and a small peak appears there. This is also close to peak (1152 cm^{-1}) of C-O-C asymmetric vibration in carbohydrates and glucosides according to Grillo et al. [34].

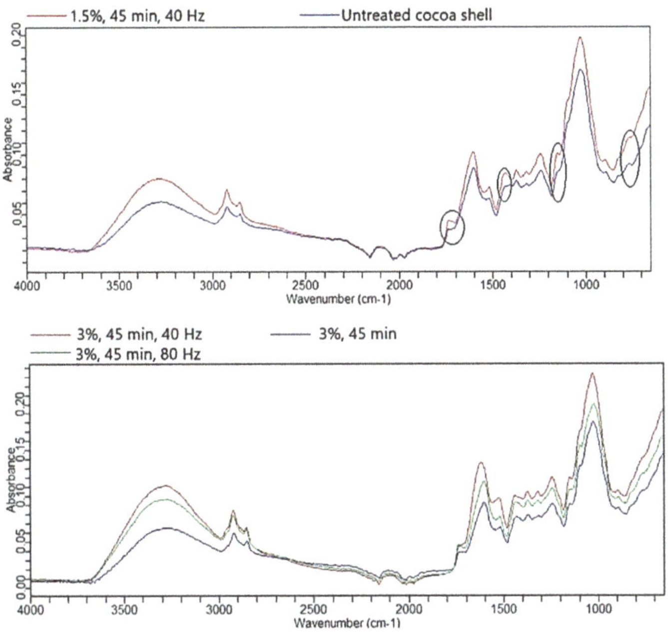

Figure 5. Representative Fourier transform infrared spectroscopy with attenuated total reflection (FTIR-ATR) spectra of cocoa shell before and after the HVED treatment.

Untreated cocoa shell had small peak at 760 cm^{-1} (ring deformation vibrations). Treatments transfers this to shoulder.

These changes in spectra are the result of combined effect of changes in fiber composition (insoluble:soluble ratio) and phenol changes. Bozaci et al. [29] also observed shift of the bands after cold plasma treatment of jute fibers. They assigned this to reaction of fibers with active species from the plasma.

4. Conclusions

In essence, our study showed influence of HVED on fiber properties (soluble, insoluble and total fiber content) and related physical properties—occurrence of larger particle size and increase of water and oil binding capacity. In addition, it has been established that changes in fiber properties correlate to changes in tannin content. It is evident that HVED has a significant influence on the physical and chemical characteristics of cocoa shells due to formation of large number of reactive species, including free radicals and ions, and the reactions occurring during treatment need to be further examined in order to see why such changes are taking place and to reveal actual mechanisms that are involved. Other chemical characteristics of the modified cocoa shell in future studies should be considered as well. In addition, the effect of change in physical and chemical properties of shell on its applicability in different foods needs to be examined because its properties such as grinding, taste, color etc. are the main reasons for its non-use in food production. This research is valuable for future applications of untreated and cocoa shells treated with HVED in the food industry.

Author Contributions: Conceptualization, V.B. and Đ.A.; methodology, I.F., M.K.; investigation, V.B., I.F., M.B., A.J.; writing—original draft preparation, V.B., Đ.A.; writing—review and editing, A.J., J.B., D.Š., B.M., Đ.A.; visualization, A.J., K.D., M.J.; funding acquisition, Đ.A. All authors have read and agreed to the published version of the manuscript.

Funding: This work has been supported in part by Croatian Science Foundation under the project UIP 2017-05-8709.

Conflicts of Interest: The authors declare no conflicts of interest.

References

1. Arlorio, M.; Coisson, J.D.; Travaglia, F.; Varsaldi, F.; Miglio, G.; Lombardi, G.; Martelli, A. Antioxidant and biological activity of phenolic pigments from *Theobroma cacao* hulls extracted with supercritical CO_2. *Food Res. Int.* **2005**, *38*, 1009–1014. [CrossRef]
2. Martin-Cabrejas, M.A.; Valiente, C.; Esteban, R.M.; Molla, E. Cocoa hull: A potential source of dietary fibre. *J. Sci. Food Agric.* **1994**, *66*, 307–311. [CrossRef]
3. Redgwell, R.; Trovatoa, V.; Merinat, S.; Curti, D.; Hedigera, S.; Manez, A. Dietary fibre in cocoa shell: Characterisation of component polysaccharides. *Food Chem.* **2003**, *81*, 103–112. [CrossRef]
4. Hernández-Hernández, C.; Viera-Alcaide, I.; Morales-Sillero, A.M.; Fernández-Bolaños, J.; Rodríguez-Gutiérrez, G. Bioactive compounds in Mexican genotypes of cocoa cotyledon and husk. *Food Chem.* **2017**, *240*, 831–839. [CrossRef] [PubMed]
5. Chun, K.S.; Husseinsyah, S.; Osman, H. Utilization of cocoa pod husk as filler in polypropylene biocomposites: Effect of maleated polypropylene. *J. Compos. Mater.* **2015**, *28*, 1507–1521. [CrossRef]
6. Panak Balentić, J.; Ačkar, Đ.; Jokić, S.; Jozinović, A.; Babić, J.; Miličević, B.; Šubarić, D.; Pavlović, N. Cocoa shell: A by-product with great potential for wide application. *Molecules* **2018**, *23*, 1404. [CrossRef] [PubMed]
7. Boussetta, N.; Turk, M.; De Taeye, C.; Larondelle, Y.; Lanoiselle, J.L.; Vorobiev, E. Effect of high voltage electrical discharges, heating and ethanol concentration on the extraction of total polyphenols and lignans from flaxseed cake. *Ind. Crop. Prod.* **2013**, *49*, 690–696. [CrossRef]
8. Yan, L.-G.; Deng, Y.; Ju, T.; Wu, K.; Xi, J. Continuous high voltage electrical discharge extraction of flavonoids from peanut shells based on "annular gap type" treatment chamber. *Food Chem.* **2018**, *256*, 350–357. [CrossRef]
9. Li, Z.; Fan, Y.; Xi, J. Recent advances in high voltage electric discharge extraction of bioactive ingredients from plant materials. *Food Chem.* **2019**, *277*, 246–260. [CrossRef]
10. Boussetta, N.; Vorobiev, E. Extraction of valuable biocompounds assisted by high voltage electrical discharges: A review. *C. R. Chim.* **2014**, *17*, 197–203. [CrossRef]
11. Barišić, V.; Jozinović, A.; Flanjak, I.; Šubarić, D.; Babić, J.; Miličević, B.; Doko, K.; Ačkar, Đ. Difficulties with Use of Cocoa Bean Shell in Food Production and High Voltage Electrical Discharge as a Possible Solution. *Sustainability* **2020**, *12*, 3981. [CrossRef]
12. Amorim, E.L.C.; Nascimento, J.E.; Monteiro, J.M.; Sobrinho, T.J.S.P.; Araújo, T.A.S.; Albuquerque, U.P. A simple and accurate procedure for the determination of tannin and flavonoid levels and some applications in ethnobotany and ethnopharmacology. *Funct. Ecosyst. Communities* **2008**, *21*, 88–94.
13. Singleton, V.L.; Orthofer, R.; Lamuela-Raventós, R.M. Analysis of total phenols and other oxidation substrates and antioxidants by means of Folin-Ciocalteu reagent. *Methods Enzymol.* **1999**, *299*, 152–178. [CrossRef]
14. AOAC 991.43. Total, soluble, and insoluble dietary fiber in foods: Enzymatic-gravimetric method, MES-TRIS buffer. In *AOAC Official Methods, Suppl*; Association of Official Analytical Chemists: Rockville, MD, USA, March 1995; Chapter 32; pp. 7–9.
15. AACC Method 88-04. In *AACC Methods Manual, Revised Ed.*; American Association of Cereal Chemists: St. Paul, MN, USA, 1983.
16. Barišić, V.; Flanjak, I.; Križić, I.; Jozinović, A.; Šubarić, D.; Babić, J.; Miličević, B.; Ačkar, Đ. Impact of high-voltage electric discharge treatment on cocoa shell phenolic components and methylxanthines. *J. Food Process Eng.* **2020**, *43*, e13057. [CrossRef]
17. Jokić, S.; Pavlović, N.; Jozinović, A.; Ačkar, Đ.; Babić, J.; Šubarić, D. High-voltage electric discharge extraction of bioactive compounds from the cocoa bean shell. *Chem. Biochem. Eng. Q.* **2019**, *33*, 271–280. [CrossRef]
18. Delsart, C.; Grimi, N.; Boussetta, N.; Miot Sertier, C.; Ghidossi, R.; Vorobiev, E.; Mietton Peuchot, M. Impact of pulsed-electric field and high-voltage electrical discharges on red wine microbial stabilization and quality characteristics. *J. Appl. Microbiol.* **2015**, *120*, 152–164. [CrossRef]
19. Lukić, K.; Vukušić, T.; Tomašević, M.; Ćurko, N.; Gracin, L.; Kovačević Ganić, K. The impact of high voltage electrical discharge plasma on the chromatic characteristics and phenolic composition of red and white wines. *Innov. Food Sci. Emerg. Technol.* **2019**, *53*, 70–77. [CrossRef]

20. Guyot, S.; Bernillon, S.; Poupard, P.; Renard, M.G.C.C. Multiplicity of phenolic oxidation products in apple juices and ciders, from synthetic medium to commercial products. In *Recent Advances in Polyphenol Research*; Daayf, F., Lattanzio, V., Eds.; Wiley-Blackwell: Hoboken, NJ, USA, 2008; pp. 278–293.
21. Lecumberri, E.; Mateos, R.; Izquierdo-Pulido, M.; Rupérez, P.; Goya, L.; Bravo, L. Dietary fibre composition, antioxidant capacity and physico-chemical properties of a fibre-rich product from cocoa (*Theobroma cacao* L.). *Food Chem.* **2007**, *104*, 948–954. [CrossRef]
22. Bravo, L.; Grades, N.; Saura-Calixto, F. Composition and potential uses of mesquite pods (Prosopis pallida L): Comparison with carob pods (Ceratonia siliqua L). *J. Sci. Food Agric.* **1994**, *65*, 303–306. [CrossRef]
23. Saura-Calixto, F. Dietary fibre complex in a sample rich in condensed tannins and uronic acids. *Food Chem.* **1987**, *23*, 95–103. [CrossRef]
24. Perez, E.; Mendez, A.; Leon, M.; Hernandez, G.; Sivoli, L. Proximal composition and the nutritional and functional properties of cococa by-products (pods and husks) for their use in the food industry. In *Chocolate-Cocoa Byproducts Technology, Rheology, Styling, and Nutrition*; Perez Sira, E., Ed.; Nova: New York, NY, USA, 2015; pp. 219–230.
25. Barišić, V.; Flanjak, I.; Tot, A.; Budeč, M.; Benšić, M.; Jozinović, A.; Babić, J.; Šubarić, D.; Miličević, B.; Ačkar, Đ. 5-hydroxymethylfurfural and acrylamide content of cocoa shell treated with high voltage electrical discharge. *Food Cont.* **2020**, 107043. [CrossRef]
26. Ozdal, T.; Capanoglu, E.; Altay, F. A review on protein–phenolic interactions and associated changes. *Food Res. Int.* **2013**, *51*, 954–970. [CrossRef]
27. Yuan, X.; Jayaraman, K.; Bhattacharyya, D. Effects of plasma treatment in enhancing the performance of woodfibre-polypropylene composites. *Compos. Part A Appl. Sci. Manuf.* **2004**, *35*, 1363–1374. [CrossRef]
28. Sinha, E.; Panigrahi, S. Effect of plasma treatment on structure, wettability of jute fiber and flexural strength of its composite. *J. Compos. Mater.* **2009**, *43*, 1791–1802. [CrossRef]
29. Bozaci, E.; Sever, K.; Demir, A.; Seki, Y.; Sarikanat, M.; Ozdogan, E. Effect of the atmospheric plasma treatment parameters on surface and mechanical properties of jute fabric. *Fibers Polym.* **2009**, *10*, 781–786. [CrossRef]
30. Karahan, H.A.; Özdoğan, E. Improvements of surface functionality of cotton fibers by atmospheric plasma treatment. *Fibers Polym.* **2008**, *9*, 21–26. [CrossRef]
31. Sangnark, A.; Noomhorm, A. Effect of dietary fibre from sugarcane bagasse and sucrose esteron doughand bread properties. *LWT* **2004**, *37*, 697–704. [CrossRef]
32. Ulbrich, M.; Flöter, E. Impact of high pressure homogenization modification of a cellulose based fiber product on water binding properties. *Food Hydrocoll.* **2014**, *41*, 281–289. [CrossRef]
33. Günzler, H.; Gremlich, H. *IR-Spektroskopie: Eine Einführung*, 4th ed.; WILEY-VCH GmbH & Co. KGaA: Weinheim, Germany, 2003.
34. Grillo, G.; Boffa, L.; Binello, A.; Mantegna, S.; Cravotto, G.; Chemat, F.; Dizhbite, T.; Lauberte, L.; Telysheva, G. Analytical dataset of Ecuadorian cocoa shells and beans. *Data Brief.* **2019**, *22*, 56–64. [CrossRef] [PubMed]

© 2020 by the authors. Licensee MDPI, Basel, Switzerland. This article is an open access article distributed under the terms and conditions of the Creative Commons Attribution (CC BY) license (http://creativecommons.org/licenses/by/4.0/).

foods

MDPI

Article

Effect of Carrier Agents on the Physicochemical and Technofunctional Properties and Antioxidant Capacity of Freeze-Dried Pomegranate Juice (*Punica granatum*) Powder

Adegoke Olusesan Adetoro [1], Umezuruike Linus Opara [1] and Olaniyi Amos Fawole [2,*]

1 Postharvest Technology Research Laboratory, South African Research Chair in Postharvest Technology, Department of Horticultural Science, Stellenbosch University, Private Bag X1, Stellenbosch 7602, South Africa; 21412944@sun.ac.za (A.O.A.); Opara@sun.ac.za (U.L.O.)

2 Postharvest Research Laboratory, Department of Botany and Plant Biotechnology, University of Johannesburg, P.O. Box 524, Auckland Park, Johannesburg 2006, South Africa

* Correspondence: olaniyif@uj.ac.za; Tel.: +27-11-559-7237; Fax: +27-11-559-2411

Received: 3 September 2020; Accepted: 28 September 2020; Published: 1 October 2020

Abstract: The physicochemical and technofunctional properties and antioxidant capacity of freeze-dried "Wonderful" pomegranate juice powder (PJP), produced with different carrier agents, were investigated. Powders were produced using maltodextrin, gum Arabic, and waxy starch as carrier agents and characterised by scanning electron microscopy (SEM) and particle size distribution. Results showed that PJP produced with maltodextrin had the highest yield (46.6%), followed by gum arabic (40.6%), while waxy starch had the least yield (35.4%). Powders produced with maltodextrin (96.5%) and gum arabic (96.1%) were highly soluble, which indicates better reconstitution properties. Waxy starch-added PJP had the lowest hygroscopicity (4.7%), which offers good stability during storage and a lower degree of caking compared to maltodextrin (10.2%) and gum arabic (12.6%) powders. Powders obtained from maltodextrin and gum arabic exhibited larger particle diameters ranging between 12 to 120 μm while the lowest particle diameter range was with powders formed from waxy starch (8–40 μm). Freeze-dried pomegranate powder produced with maltodextrin retained more redness (a^*) by approximately 44%, compared to gum arabic. Similarly, PJP with maltodextrin and gum arabic had higher total soluble solids (10.3 and 10.4 °Brix), respectively. Total anthocyanin content was 54% more in PJP with maltodextrin than waxy starch PJP. Similarly, the powder produced with maltodextrin had higher radical scavenging activity (33.19 mM TE/g dry matter; DM) compared to gum arabic (28.45 mM TE/g DM) and waxy starch (26.96 mM TE/g DM). Overall, maltodextrin reflected the most suitable carrier agent to produce PJP.

Keywords: total soluble solids; particle size distribution; total anthocyanin content; antioxidant capacity

1. Introduction

Pomegranate (*Punica granatum* L.) belongs to the Punicaceae family and is widely grown in many parts of the world, such as Europe, Asia, North Africa, the Mediterranean basin and, in recent times, South Africa [1,2]. The increased commercial production of pomegranate from 828 ha in 2016 to 1024 ha in 2019 is highly related to its rich phytochemical compounds in the edible part of the fruit [3,4]. Polyphenols, such as flavonoids, condensed tannins and hydrolysable tannins, are major components found in pomegranate arils or juice [5]. They are the major source of protective compounds that work against the damaging effects of free radicals [6]. Pomegranate fruit is mostly consumed as fresh juice, flavourings, colourings, concentrates and jellies for recipes [7]. Epidemiological studies have associated the consumption of pomegranate fruit to a reduced risk of coronary heart disease, diseases that are not transmissible, such as cancer, and diabetes as a result of its high antioxidant capacity [8,9]. Pomegranate fruit was noted to be actively used in folk medicine as a result of its high anthocyanin compositions of cyanidin, delphinidin and pelargonidin, which are attributed to the red colour of fruit and aril [10]. It is essentially useful in the cure of many parasitic diseases such as ulcers, diarrhoea, acidosis, dysentery, and haemorrhage [11].

Due to its health and nutritional benefits, pomegranate fruit is in demand throughout the year. Thus, the food industry desires a novel process aimed at increasing the shelf-life and improving the pigment stability of pomegranate products. The drying of fruit juice into powder form is a novel way to extend the shelf-life [12]. As opposed to dried pomegranate arils, pomegranate juice powders have the advantages of easier storage and distribution. Furthermore, the powders can be used as an ingredient to formulate foods. Spray drying is a commonly used method in many food industries for producing food powders and agglomerates [13,14]. Along with being an attractive feature of this technological process, the scarce heat damage to the product is essential [15]. Another challenging factor during spray drying is the clogging of nozzles, especially when drying sugar and acid-rich foods such as honey and natural fruit juices [16]. The low operating conditions involved in freeze-drying could be an appropriate drying method to produce niche fruit powders from sugar and acid-rich fruit like pomegranate. Freeze-drying is one of the techniques used to produce high-value powder products [12]. It is a method that results in high-quality dehydrated products due to the low operating temperatures required in the process and the absence of liquid water [17]. This method reduces thermal damage of nutrients and preserves flavour and colour components of the product [18].

Studies have reported some factors to be considered during the production of fruit powders: the stickiness of powder particles and safe handling and storage [19]. Stickiness during drying is mainly due to the high content of sugars such as fructose, glucose, sucrose and acid materials; for example, organic acids such as citric, malic and tartaric acids, which are attributed low molecular weight, contribute more than 90% of solids in fruit juices [20,21]. In order to overcome the sticky behaviour of fruit juice powder, high molecular weight carriers or drying aids such as maltodextrin, gum arabic, waxy starch, pectin, vegetable fibres, and starches as encapsulation agents are added [16,21–23]. Studies have shown that carrier agents further preserve some sensitive properties of the food material, such as carotenoids and flavours, and minimise volatile and reactive properties. For instance, mango juice powder obtained through maltodextrin, gum arabic, and waxy starch resulted in characteristic amorphous particles [24–27].

Yousefi et al. [28] reported that gum arabic showed a high colour change and increased glass transition temperature (Tg) of pomegranate powder. Similarly, Seerangurayar et al. [19] reported that carrier-agent-added date powders had lower hygroscopicity, which offers good storage stability. Fazaeli et al. [18] reported that additives enhanced the properties of the final product as a result of an increase in Tg and contributed to the high stability of quality attributes of black mulberry juice powder during storage. However, there are limited scientific studies specifically on the processing of pomegranate juice with the use of a freeze-dryer. To further examine the field of application for pomegranate products, this work investigates the freeze-drying of pomegranate juice to evaluate the influence of different carrier agents (maltodextrin, gum arabic and waxy starch) on the physicochemical and technofunctional properties and antioxidant activities of the powders.

2. Materials and Methods

2.1. Raw Material and Sample Preparation

Pomegranate fruit (cv. Wonderful) were harvested at commercial maturity from Blydeverwacht orchard, Wellington, South Africa. The fruit were sorted for uniformity of size, shape, and colour and transported in an air-conditioned vehicle to the Postharvest Technology Laboratory at Stellenbosch University. Fruits were washed, and the juice was extracted using a hand-operated domestic press and frozen at −20 °C for about 24 h.

The fresh juice was thawed and clarified using a centrifuge system (5810 R Eppendorf AG, Hamburg, Germany) at 10,000 rpm for 20 min. The cold, sterile single strength clarified juice with 16.2 °Brix (total soluble solids) was diluted and standardised with distilled water to 12 °Brix and rapidly frozen at −80 °C until experiments were carried out.

In order to obtain a flowable powder from pomegranate juice, a preliminary study was conducted to investigate the amount of carrier that would be added to the pomegranate juice. Each of the carriers (maltodextrin—Sigma Aldrich Co., St. Louis, MO, USA; gum arabic—Sigma Aldrich Co., France; waxy starch—Sigma Aldrich Co., USA) was incorporated in 100 mL pomegranate juice at a range between 10 to 40 g to select a suitable concentration of carrier agent. A 30 g concentration of (maltodextrin, gum arabic or waxy starch)/100 mL pomegranate juice was observed to produce a flowable powder, which was added after standardisation. The mixture was homogenised using a laboratory homogeniser for 5 min [28].

2.2. Freeze-Drying Procedure

The pomegranate juice was placed in a 90-mL specimen jar and frozen in a static-air freezer at −80 °C. Freeze-drying of frozen samples was carried out in triplicates. A specimen jar containing the samples was carefully taken to a laboratory-scale freeze-dryer (VirTis Co., Gardiner, NY, USA) operating at condenser temperature −85 °C and pressure 18 mTorr and drying continued for 72 h. Dried samples were removed from the freeze-dryer and ground by electrical blender into free-flowing powder (Figure 1). The pomegranate juice powders (PJPs) were transferred and sealed in plastic bags in a desiccator that contained phosphorus pentoxide to prevent moisture absorption from the surrounding air until further analysis.

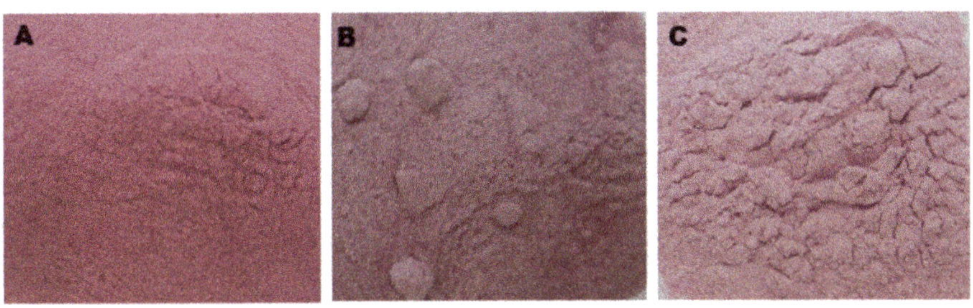

Figure 1. Freeze-dried pomegranate powder produced with (**A**) maltodextrin, (**B**) gum arabic and (**C**) waxy starch.

2.3. Yield, Water Activity and Physicochemical Attributes of PJP

2.3.1. Powder Yield Determination

The percentage yield of powder was calculated based on the fresh weight [29]

$$\text{Yield}(\%) = \frac{\text{Weight of powder (g)}}{\text{Fresh weight (g)}} \times 100 \quad (1)$$

2.3.2. Determination of Water Activity and Moisture Content

The water activity (a_w) of PJP was determined with an electronic dew point water activity meter (CH 8853 Novasina AG, Lachen, Switzerland). The final moisture content of the PJP was measured using a moisture analyser (KERN DBS 60-3 Balingen, Germany) at 120 °C.

2.3.3. Colour Measurement

Colour of PJP was determined by direct reading using a chromometer (Minolta model CR-200, Osaka, Japan) to obtain the colour values: L^* (brightness/darkness), a^* (redness/greenness), and b^* (yellowness/blueness). The measurements were taken at three different times from a colourless petri dish and averaged. The maximum for 'L^*' value is 100 (white), and the minimum is zero (black). The colour attributes chroma C^*, hue angle $h°$ and total colour difference (TCD) were calculated [1,30].

$$h^{o} = \tan^{-1}\frac{b^*}{a^*} \tag{2}$$

$$C^* = \sqrt{a^{*2} + b^{*2}} \tag{3}$$

$$\text{TCD} = \left((\Delta L^*)^2 + (\Delta a^*)^2 + (\Delta b^*)^2\right)^{\frac{1}{2}} \tag{4}$$

L^*, a^* and b^* represent the value after drying at each treatment level and results were expressed as means ± SE of the determinations obtained.

2.3.4. Determination of Total Soluble Solids (TSSs), Titratable Acidity (TA) and pH

Five grams of PJP were extracted in 50 mL distilled water. For 5 min, the mixture was vortexed with the use of a vortex and sonicated for 15 min in an ultrasonic bath (Separation Scientific, Cape Town, South Africa). This was followed by centrifugation at 10,000 rpm for 25 min and recovery of the supernatant for TSS, TA and pH measurements. TSS measurement was determined with the use of a digital hand refractometer (model PT-32; ATAGO, Tokyo, Japan) blanked with distilled water. For TA, 2 mL of supernatant was diluted in 70 mL of distilled water and titrated against 0.2 N of sodium hydroxide (NaOH) to a pH of 8.2 with the use of a Metrohm 862 compact titrosampler (Herisau, Switzerland).

2.4. Technofunctional Characterisation of PJP

2.4.1. Solubility

Solubility (%) was determined using the Eastman and Moore method [24] and modified slightly. One gram of the sample was uniformly dispersed in H_2O of 50 mL and distilled in a vortex for 30 s. At 3000 rpm for 5 min under 25 °C, the solution was carefully placed in a tube and centrifuged. A 25 mL aliquot of the supernatant was transferred to preweighed Petri dishes and the sample was immediately dried at 105 °C for 5 h. Solubility (%) was determined by subtracting the initial weight from the final weight divided by the initial weight.

2.4.2. Hygroscopicity

Hygroscopicity was calculated according to [31], with slight modifications. Two grams of the sample were placed inside a hermetic bottle that was controlled with NaCl-saturated solution in a constant relative humidity chamber (MLR-352 H Versatile Environmental Test Chamber, Kyoto, Japan) set at 68.9% RH and 25 °C [32]. The weight of the sample was calculated to validate the condition for equilibrium between the samples and the environment. The hygroscopicity was expressed as % moisture on wet basis (w.b.).

2.4.3. Bulk Density

In a 100 mL graduated cylinder, PJP (20 g) was weighed and carefully dropped 10 times from a height of 15 cm using a rubber mat. The bulk density was determined by the division of the mass of powder and the volume estimated from the cylinder [33].

2.4.4. Water- and Oil-Holding Capacity Determination

According to Jalal et al. [34], the water-holding capacity (WHC) and oil-holding capacity (OHC) of PJP were calculated. A mixture of 25 mL distilled water or sunflower oil and 250 mg of dry sample were slightly vortexed and left at room temperature for 1 h. The solution was placed in a tube and centrifuged at 4000 rpm for 10 min at 25 °C, after which the residue was weighed. The water/oil holding capacity was expressed as g of water/oil held per g of sample. The formula to calculate WHC/OHC is as follows:

$$\text{WHC/OHC (g/g)} = \frac{\text{residue fresh weight} - \text{residue dry weight}}{\text{residue dry weight}} \quad (5)$$

2.4.5. Particle Size Distribution

The particle size of the powder was determined with the use of a laser light diffraction instrument (Mastersizer S, model MAM 5005; Malvern Instruments, Malvern, UK). Under magnetic agitation, a small amount of PJP was homogenised in 99% isopropanol, following careful monitoring of the distributed particle size, which was taken in three successive measurements. De Brouckere's mean diameter was used to express the particle size and the mean diameter over the volume distribution. This is mostly used to characterise a particle [35].

2.4.6. Microstructure

The microstructure of PJP was examined with the use of a scanning electron microscope (X-Max 51, Oxford Instruments, Concord, MA, USA). SEM images of powder were obtained from uniformly mixed powder samples. Under a high vacuum condition, the samples were coated with a very thin layer of gold. This is often used to provide a reflective surface for the electron beam. The gold coating was carried out in a sputter coater (ACE200 LEICA Mikrosysteme GmbH, Vienna, Austria) under a low vacuum condition while inert argon gas was present. Subsequent viewing of the gold-coated samples was carried out under the microscope.

2.5. Phenolic Contents and Antioxidant Capacity

2.5.1. Determination of Total Phenolic Content (TPC)

TPC was determined by the Folin-Ciocalteu method using a methanolic extract of PJP [1]. In a test tube, the supernatant (0.05 mL) was mixed with 0.45 mL (50% methanol), followed by adding 0.5 mL Folin-Ciocalteu after 2 min. The mixture was then vortexed and kept in the dark for 10 min before adding 2% Na_2CO_3 and further incubation for 40 min in the dark. The absorbance of each sample was read at 520 nm in a UV-vis spectrophotometer (Thermo Scientific Technologies, Madison, WI, USA) against a blank containing 50% methanol. Absorbance was compared with a standard curve (Gallic acid, 0–10 mg), and results were expressed as mg gallic acid equivalent per gram dry matter (mg GAE/g DM).

2.5.2. Total Anthocyanin Content

Total anthocyanin content (TAC) was quantified differentially by using the pH method [36]. In triplicates, 1 mL extract was mixed with 9 mL of pH 1.0 and pH 4.5 buffers in separate conditions. In pH 1.0 and 4.5 buffers, absorbance was measured at 520 and 700 nm, expressing the result (cyanidin 3-glucoside) using Equation (6).

$$A = (A_{510} - A_{700})_{\text{pH } 1.0} - (A_{510} - A_{700})_{\text{pH } 4} \quad (6)$$

$$\text{Total monomeric anthocyanin (mg/mL)} = \frac{A \times \text{MW} \times \text{DF}}{\varepsilon \times L} \quad (7)$$

where A = absorbance, ε = cyd-3-glucoside molar absorbance (26,900), MW = anthocyanin molecular weight (449.2), DF = dilution factor, and L = cell path length (1 cm). Final results are expressed as equivalent per gram dry matter (mg C_3gE/g DM).

2.5.3. Radical-Scavenging Activity (RSA)

In triplicate, the RSA assay was carried out according to Fawole and Opara [1]. Briefly, in test tubes, an aqueous methanolic extract of PJP (0.015 mL) was diluted with methanol (0.735 mL) and methanolic DPPH solution (0.75 mL, 0.1 mM) was immediately added. The mixtures were incubated in the dark and at room temperature for 30 min. The absorbance was measured at 517 nm using a UV-vis spectrophotometer (Helios Omega, Thermo Scientific, Waltham, MA, USA) and compared with the standard curve (Trolox equivalent, 0–2.0 mM). The free-radical activity of PJP was expressed as Trolox equivalent (mM) per gram dry matter (mM TE/g DM).

2.5.4. Ferric-Ion Reducing Antioxidant Power (FRAP)

The antioxidant power of PJP was measured using the calorimetric method, according to [1,37]. The FRAP working solution was freshly prepared in mixtures of 300 mM acetate buffer (50 mL), 10 mM 2,4,6-tripyridyl-s-triazine (TPTZ; 5 mL) and 20 mM ferric chloride (5 mL) at 37 °C. Diluted aqueous methanolic PJP extracts (0.15 mL) were added to 2.85 mL of the FRAP working solution in triplicates, followed by incubating the mixture in the dark for 30 min. Measurement of the absorbance at 593 nm was carried out to monitor the reduction of the Fe^{3+}-TPTZ complex to a coloured Fe^{2+}-TPTZ complex at low pH by PJP extracts. Trolox (0–10 mM) was used for the calibration curve, and the results were expressed as Trolox (mM) equivalents per gram dry matter (mM TE/g DM).

2.6. Statistical Analysis

Data were analysed using STATISTICA (Statistica 13.0, StatSoft Inc., Tulsa, OK, USA) and presented as means ± standard error. All analyses were done in triplicates. Data were subjected to analysis of variance (ANOVA), and means were separated according to Fisher's LSD test at a level of significance of 95%. The graphical presentations were processed by using GraphPad Prism software 4.03 (GraphPad Software, Inc., San Diego, CA, USA). Principal component analysis (PCA) was carried out using XLSTAT software version 2012.04.1 (Addinsoft, Bordeaux, France).

3. Results and Discussion

3.1. Quality Attributes of Pomegranate Juice

Table 1 shows the quality attributes of the pomegranate juice used for freeze-drying. It was observed that the pomegranate juice had strong acidity (pH value 3.10), which implied that it is less susceptible to microbial growth. The high level of TSSs (15.3 °Brix) is typical of the pomegranate cv. Wonderful as it is considered as sweet-sour cultivar [38]. Predominant sugars in pomegranate juice are glucose, fructose and sucrose [39] and are reported to contribute to powder stickiness during drying [21,40]. The phenolic contents (TPC and TAC) have additive effects on the pharmacological properties of pomegranates and gives the characteristic red colouration to the juice [39,41]. Values of TPC and TAC correspond with that of [38]. The contents of antioxidants serve as indicators for nutrient retention during the freeze-drying process. Depending on the cultivar, the colour of pomegranate juice could range from very deep to slightly red. Based on the moderately low values for the colour parameters (L^*, a^*, C^* and $h°$), the investigated pomegranate juice could be considered as slightly red in colour. Properties of colour are quality indicators that are necessary to reflect sensory attractiveness; thus, the production of phenolic-rich coloured PJP was of high priority in this study.

Table 1. Quality attributes of pomegranate juice (cv Wonderful) processed into powders.

Quality Attributes	Values (Means ± SE)
TSS (°Brix)	15.3 ± 0.07
TA (% citric acid)	1.35 ± 0.01
pH	3.10 ± 0.01
TPC (mg GAE/100 mL)	22.6 ± 2.82
TAC (mg C3gE/100 mL)	7.4 ± 1.95
L^*	23.0 ± 2.21
a^*	15.4 ± 1.37
C^*	18.6 ± 1.44
$h°$	34.5 ± 0.94

L^*, lightness; a^* redness; C^* chroma; $h°$, hue angle; TSSs, total soluble solids; TA, titratable acidity; TPC, total phenolic content; TAC, total anthocyanin content. SE: standard error.

3.2. Yield, Moisture Content and Water Activity

The effect of carrier agents on the yield of freeze-dried PJP is presented in Table 2. The most amount of powder was obtained with maltodextrin (46.6%), while waxy starch yielded the least (35.4%). This disagrees with Yousefi et al. [28], who reported that gum arabic had the highest yield in their study of the effect of carrier type and spray drying of pomegranate juice. This could be due to differences in the drying methods of the production of PJP. Additionally, the differences in the yield of PJP could be due to the configuration of the carrier agents. The least yield found in waxy starch was also noted by Yousefi et al. [28] due to its crystalline nature.

Table 2. Yield, moisture content, water activity and colour attributes of freeze-dried pomegranate juice powder (PJP).

Carrier	Yield %	MC %	a_w	L^*	a^*	C^*	$h°$	TCD
Maltodextrin	46.6 ± 0.04 a	0.7 ± 0.02 b	0.31 ± 0.00 b	69.0 ± 1.42 b	29.3 ± 0.49 a	29.3 ± 0.48 a	0.6 ± 0.19 c	49.2 ± 1.39 b
Gum arabic	40.6 ± 0.12 b	1.8 ± 0.02 a	0.49 ± 0.01 a	64.6 ± 0.39 c	16.3 ± 0.24 c	16.7 ± 0.23 c	11.9 ± 0.37 a	42.2 ± 0.40 c
Waxy starch	35.4 ± 0.30 c	0.2 ± 0.02 c	0.05 ± 0.00 c	78.7 ± 1.20 a	18.6 ± 0.29 b	18.6 ± 0.29 b	4.0 ± 0.31 b	56.5 ± 1.19 a

MC, moisture content; L^*, lightness; a^* redness; C^* chroma; $h°$, hue angle; a_w, water activity; TCD, total colour difference. Presented as means ± SE in each column, data followed by different letters are significantly different ($p < 0.05$) according to Fisher's LSD.

The moisture content (MC) and water activity (a_w) of freeze-dried PJP were significantly ($p < 0.05$) different among carriers (Table 2). Gum-arabic-added PJP had the highest moisture (1.8%) after drying, while waxy-starch-added PJP had the least value of moisture (0.2%). Similarly, the highest values of water activity were observed in the powder produced with gum arabic (0.49), followed by maltodextrin-added PJP, which had 0.31, while waxy starch-added PJP had the least water activity (0.20).

In dried food materials, moisture content is one of the factors related to drying efficiency [31]. From this study, the powder produced with waxy starch had low moisture ($p < 0.05$) and water activity and was strongly and positively correlated ($r = 0.947$). This was expected as the decrease in moisture content could directly be linked to reduced water activity and, therefore, the powder produced with waxy starch could perform better in storage stability than powders obtained with other drying agents. Decreased water activity prevents the growth of most bacteria, yeasts, and moulds, which are not capable of growing below water activity values of 0.87, 0.88, and 0.80, respectively [42]. Furthermore, Laroche et al. [43] noted that water activity values ranging between 0.20–0.50 prevented microbial infestation in food powders. Daza et al. [44] also reported that freeze-dried samples with values of water activity lower than 0.3 were less susceptible to microbial attack. Results from this study also support the findings by Mosquera et al. [45], who reported that the lower the critical water activity in freeze-dried strawberry powder, the better the stability during storage. Lower moisture prevents the agglomeration of particles which hinders the caking of powder, thereby reducing the retention of active components and other powder properties such as flowability and dispersion [46]. Lower moisture content in dried fruit is related to its low water activity [47]. The high moisture content and water activity observed in powder produced with gum arabic may be explained due to the difficulty for water to diffuse through the carrier agent, where crusts are formed around the surface particle [48].

3.3. Colour Attributes

Fresh pomegranate juice has a favourable red colour due to the rich content of anthocyanins [1]. Lightness (L^*) of freeze-dried PJP was significantly ($p < 0.05$) different among carrier agents (Table 2). Waxy starch appeared lighter (78.7) than maltodextrin and gum arabic, with lightness values of 69.0 and 64.6, respectively. According to Comunian et al. [49], the increased lightness in powder obtained from waxy starch was as a result of the dilution effect, which was the pure white colour of the carrier, and this could be responsible for the lighter colour of the PJP. In contrast, the darker colour of gum arabic could be responsible for the darker red colour of the powder. The effect of carrier type on the quality properties of powdered and reconstituted pomegranate juice (cv. Malas) was previously assessed by Yousefi et al. [28] and an increase in the values of L^* with the use of waxy starch and maltodextrin was observed in comparison to gum arabic.

The characteristic red colouration of pomegranate powder measured as a^* was also significantly ($p < 0.05$) different among carrier agents (Table 2). Juice powder produced with maltodextrin had the highest redness (29.3), while gum arabic had the least value (16.3). Furthermore, a significant ($p < 0.05$) difference was observed between carrier agents for chroma (C^*) of freeze-dried pomegranate powder (Table 2). PJP produced with maltodextrin had the highest C^*, followed by those produced by waxy starch and gum arabic. Hue angle (h^o) of PJP was significantly ($p < 0.05$) different among carrier agents. PJP produced with maltodextrin had the least colour purity ($h°$; 0.6), closer to 0°, which suggests a higher degree of redness compared to PJP produced with gum Arabic, with highest hue angle (11.9). This suggests that changes observed in the red colour of PJP depend on the type of carrier agent [14,50]. The powder produced with waxy starch had the highest TCD (56.5), followed by maltodextrin (49.2) while powders produced with waxy starch showed the lowest TCD (42.2). Overall, the changes in colour attributes could be attributed to the addition of carrier and alteration in polyphenols during drying [51].

3.4. Total Soluble Solids (TSSs) and Titratable Acidity (TA)

Table 3 shows the effect of carrier agents on the total soluble solids (TSSs), titratable acidity (TA) and pH of PJP.

Table 3. Physicochemical attributes of freeze-dried PJP.

Carrier	TSSs (°Brix)	TA (% Citric Acid)
Maltodextrin	10.3 ± 0.17 [a]	0.24 ± 0.04 [a]
Gum arabic	10.4 ± 0.21 [a]	0.24 ± 0.02 [a]
Waxy starch	8.6 ± 0.20 [b]	0.18 ± 0.01 [b]

TSSs, total soluble solids; TA, titratable acidity. Presented as means ± SE in each column, data followed by different letters are significantly different ($p < 0.05$), according to Fisher's LSD.

PJP produced with maltodextrin and gum arabic had higher TSSs (10.3 and 10.4 °Brix, respectively) than those produced with waxy starch (8.6 °Brix; Table 3). Increased soluble solids composition observed for PJP produced with maltodextrin and gum arabic were approximately 17% higher than waxy starch. Similarly, there was a significant ($p < 0.05$) difference among carrier agents in the titratable acidity (TA) of PJP (Table 3). PJP produced with maltodextrin had the highest TA (0.24%), while powder produced with waxy starch had the least (0.18%). Rahman and Lamb [52] stated that soluble solids, organic acids, amino acids, soluble pectin and mineral salts, among several others, are the major chemical constituents found in fruit. Higher values of total soluble solids and titratable acidity observed in the powders produced with maltodextrin and gum arabic can be attributed to the crystalline nature of carrier agents. Carrier agents or additives are different in molecular weight compounds and crystalline configuration [28,53–55].

3.5. Technofunctional Properties

3.5.1. Solubility

The solubility index is an important feature to characterise the wettability and dispersibility of powders in solutions. Solubility differed significantly ($p < 0.05$) among the juice powders (Table 4).

Table 4. Technofunctional properties of freeze-dried PJP.

Carrier	Solubility %	Hygroscopicity %	Bulk Density (g cm^{-3})	WHC (g/g)	OHC (g/g)
Maltodextrin	96.5 ± 0.09 ^{a}	10.2 ± 0.04 ^{b}	0.77 ± 0.01 ^{a}	0.67 ± 0.03 ^{b}	1.64 ± 0.01 ^{b}
Gum arabic	96.1 ± 0.46 ^{a}	12.6 ± 0.01 ^{a}	0.74 ± 0.02 ^{a}	0.25 ± 0.01 ^{c}	1.96 ± 0.01 ^{a}
Waxy starch	35.4 ± 0.09 ^{b}	4.7 ± 0.07 ^{c}	0.64 ± 0.02 ^{b}	1.84 ± 0.01 ^{a}	1.45 ± 0.01 ^{c}

Presented as means ± SE in each column, data followed by different letters are significantly different ($p < 0.05$), according to Fisher's LSD.

Powders produced with maltodextrin and gum arabic showed similar results with higher solubility (96.5 and 96.1%, respectively), while waxy starch had the least solubility (35.4%). Higher values observed in the powders produced with maltodextrin and gum arabic could be related to the crystalline nature of the powder. Cano-Chauca et al. [24], in their study on spray drying of mango juice powder, also recorded a higher value of up to 95% for maltodextrin- and gum-arabic-produced powders, which is similar to the values generated in this study. Similarly, the solubility values of pineapple and cashew juice powders were also higher, with average values of 81.56 and 95.1%, respectively [56,57]. The low solubility value observed in waxy starch, reported in this study, was also supported by Mishra and Rai [58], who reported less solubility in the powder produced with waxy starch.

3.5.2. Hygroscopicity

Table 4 shows significant ($p < 0.05$) differences among carrier agents in the hygroscopic nature of freeze-dried PJP. The powder produced with gum arabic had the highest hygroscopicity (12.6%), followed by maltodextrin (10.2%), while samples produced with waxy starch showed the lowest hygroscopicity (4.7%). Differences in hygroscopic values could be due to the nature of the powders and the rate at which the powders produced hold molecules of water from the surrounding air. The result from this study was similar to the study by Tonon et al. [59], who reported that gum arabic showed the highest percentage of hygroscopicity in comparison with maltodextrin 10DE, maltodextrin 20DE and tapioca starch in acai powder. The authors noted that hygroscopicity of powder could be used to explain the mechanisms of water adsorption in powder materials as being attributed to the number of hydrophilic groups appearing in the structure of each carrier.

Furthermore, a higher number of hydrophilic groups are present in maltodextrin and gum arabic, which relates to the easy absorption of moisture from the atmosphere [60]. The authors also explained that the dynamics of moisture adsorption by carbohydrate material is duly associated with the links between the hydrogen available in the molecules of water and the hydroxyl groups present in the amorphous region of the substrate and the crystalline region. Similarly, a high hygroscopic nature of powder as a result of water absorbed from the surrounding air could also be used to explain the high moisture content of the powder [14]. However, the moisture–hygroscopicity relationship cannot be made general for all powder samples. For instance, a study by Ahmed et al. [61] noted that the hygroscopic nature of spray-dried sweet potato was highly influenced by drying agents and could not be directly related to varying moisture content.

3.5.3. Bulk Density

The results of the bulk density of freeze-dried PJP produced with different carrier agents are also shown in Table 4. Powders produced with maltodextrin exhibited the highest value of bulk density (0.77 g cm^{-3}), followed by gum arabic (0.74 g cm^{-3}), while waxy starch had the least bulk density (0.64 g cm^{-3}). Bulk density is the "addition of the mass of solid particles and moisture per total volume occupied by the particles, surface moisture and all pores, closed or open, in the surrounding atmosphere and is generally used to characterise the final product obtained by milling or drying" [28,62]. Furthermore, Chegini and Ghobadian [63] reported that powder with higher moisture content is usually associated with higher bulking weight because of the minute volume of water attracted from the atmosphere, which is considerably denser than the dry solid material. This report is in line with the results obtained in this study. PJP produced with maltodextrin and gum arabic showed higher moisture content and higher bulk density. This also supported the findings by Ferrari et al. [35], who related a higher moisture content with bulk density in powders produced with gum arabic as well as the mixture of both maltodextrin and gum arabic.

3.5.4. Water- and Oil-Holding Capacity

There was a significant ($p < 0.05$) difference among carrier agents in the water-holding capacity (WHC) of freeze-dried PJP (Table 4). Results showed that waxy starch had the highest WHC (1.84 g/g), while gum arabic had the least (0.25 g/g). Furthermore, it was observed that waxy starch held water for up to 63.6% and 86.4% more than maltodextrin and gum arabic, respectively. The higher performance of waxy starch to hold more water than other carrier agents could be attributed to the particle structure of the carrier agents. An increase in the particle density of the powder is associated with a reduction in water-holding capacity [64]. Hong and Zhang [65] also reported similar results in their study on the effect of ultra-fine pulverisation on the particle structure of soybean dietary fibre.

There was a significant ($p < 0.05$) difference among carrier agents in the oil-holding capacity (OHC) of freeze-dried PJP (Table 4). Gum arabic had the highest values for oil-holding capacity (1.96 mL/g), followed by maltodextrin with 1.64 mL/g, while waxy starch had the least (1.45 mL/g). This indicates that the higher the WHC of PJP, the lower the OHC for the studied carrier agents. The nature of the carrier agents could also be related to the OHC of PJP. A study by Chau et al. [66] noted that particle size and processing technique, such as the addition of additives or carriers, could be traceable to the effective increase in the oil-holding capacity of powder.

3.5.5. Particle Size Distribution

Figure 2 shows the particle size distribution of freeze-dried pomegranate powders obtained using maltodextrin, gum arabic and waxy starch as carrier agents. A normal distribution curve was observed for all the carriers. The particles produced with waxy starch showed the highest volume (59.8%) and the lowest particle diameter within the range of 8 to 40 μm. The powder produced with maltodextrin and gum arabic presented volumes of 59.3% and 52.5%, respectively, while their particle diameters ranged between 12 and 120 μm.

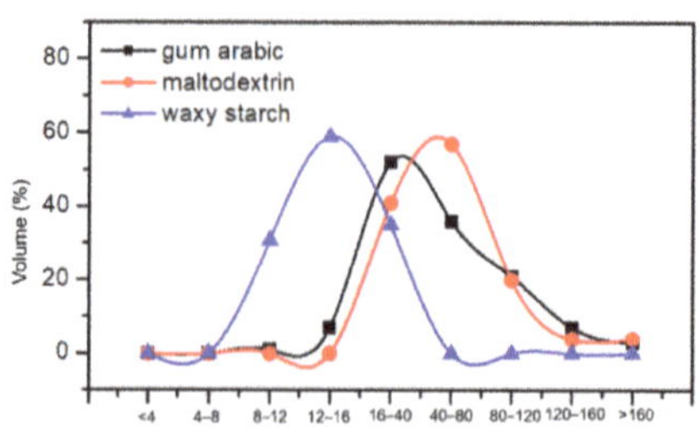

Figure 2. Particle size distribution of freeze-dried pomegranate powders produced with different carriers.

Particles obtained from maltodextrin and gum arabic exhibited larger size ranges that are normal in the case of powder analysis since a higher proportion of smaller particles occupies or fills up the spaces in between the larger ones. The formation of larger particles is not only attributed to agglomeration but also the molecular size of the carriers [59]. The process of agglomeration breaks down the powders' exposure to oxygen, and, thereby, anthocyanin pigments are protected. The higher retention of anthocyanin in powders formed with maltodextrin and gum arabic could be explained by the characteristic feature of powder agglomeration [31]. The mean diameter of pomegranate powders obtained with maltodextrin and gum arabic were different to that of other fruit powders such as blackberry (13.0–34.2 µm), and raspberry (14.6–18.3 µm) [67,68]. This could be due to the low operating temperature of the freeze-dryer, which makes the initial phase of agglomeration easier as a result of irreversibly bound particles during drying, resulting in larger particle sizes [67]. Additionally, according to Kurozawa et al. [69], the solubility and flowability of spray-dried powder reduced with a decrease in the size of the particle. This is similar to the observed solubility result presented in this study (Table 4). A direct particle size-solubility relationship was observed in the present study.

3.5.6. Microstructure

Figure 3 presents the scanning electron microscopy (SEM) microstructure of the powders produced with different carrier agents. The powders produced had particles of different sizes for the carrier agents. Powder production using maltodextrin and gum arabic resulted in the smaller particles adhering strongly to the surface of the larger ones (agglomeration), which is in agreement with the results obtained for particle size distribution. There were similarities in the powders produced with maltodextrin and gum Arabic, both showing predominantly angular shapes, while particles prepared with waxy starch have a higher degree of uniformity with a spherical shape, as shown in Figure 3c. This result also supported the findings by Leonel [70], who evaluated tapioca starch morphology and observed a rounded shape and smooth surface. Loksuwan [71] also reported a similar structure when investigating the morphology of tapioca starch during β-carotene encapsulation.

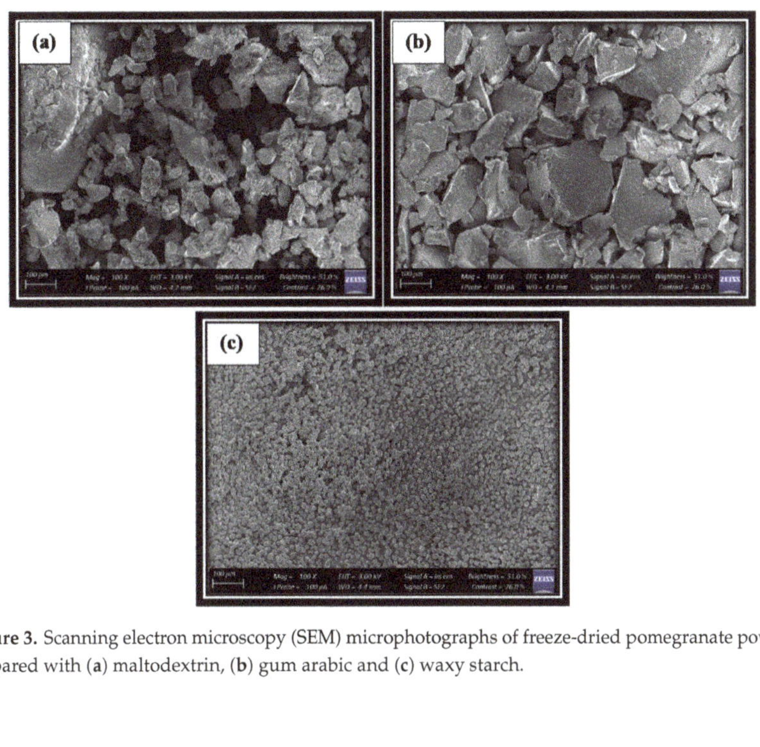

Figure 3. Scanning electron microscopy (SEM) microphotographs of freeze-dried pomegranate powder prepared with (**a**) maltodextrin, (**b**) gum arabic and (**c**) waxy starch.

3.6. Total Phenolic and Anthocyanin Contents

Graphical representation of total phenolic content and total anthocyanin content observed is shown in Figure 4a,b.

(**a**)

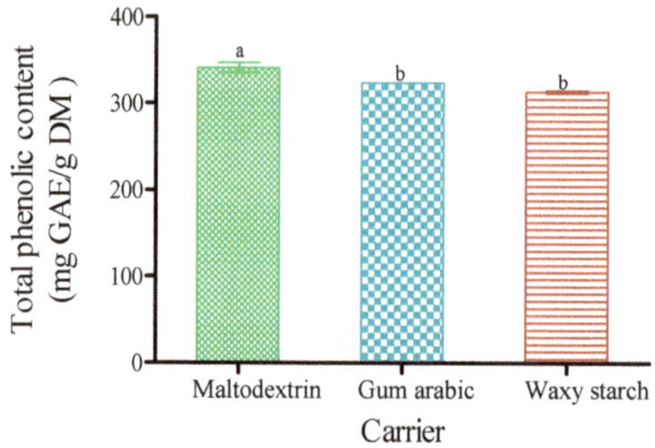

(**b**)

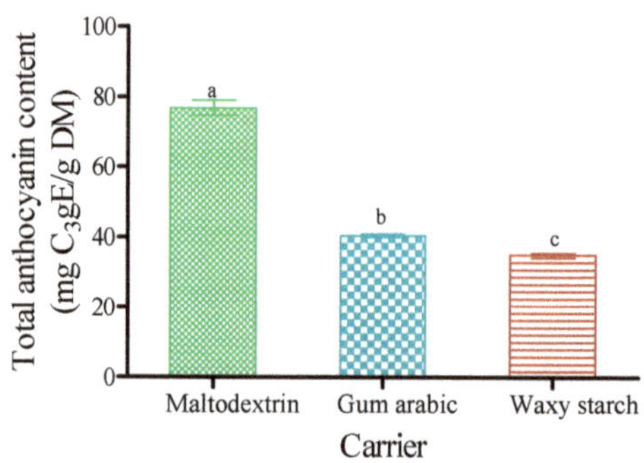

Figure 4. Total phenolic content (**a**) and total anthocyanin content (**b**) of freeze-dried pomegranate powder using different carriers. Bars with different letters are significantly different ($p < 0.05$) according to Fisher's LSD. Vertical bars indicate the standard error of the mean.

Results indicated that total phenolic content (TPC) of freeze-dried PJP was significantly ($p < 0.05$) different among carrier agents (Figure 4a). PJP produced with maltodextrin had the highest TPC (341.8 mg GAE/g DM); however, PJP produced with gum arabic resulted in a lower total phenolic content (323.8 mg GAE/g DM) that was not statistically different from the powder produced with waxy starch, which had the least TPC (313.3 mg GAE/g DM). Furthermore, total phenolic content was approximately 8.3% more in the powder produced with maltodextrin than waxy starch.

As observed in this study, the carrier agents related differently to the production of PJP, which could be a result of their structural complexities due to their soluble or insoluble nature. For instance, the nature of powder produced with maltodextrin and gum arabic appeared coarse with larger particle sizes, where the specific surface area could be lower than powder produced with waxy starch, with a finer appearance. Results showed that the interactions might interfere with polyphenol extraction and determination in the powder samples. Du et al. [14] noted the different interactions in three carbohydrate carriers (maltodextrin, gum arabic and starch sodium octenyl succinate) in the production of persimmon pulp powders. The authors noted that carriers showing the least polyphenol retention had the smallest particle size, with more surfaces exposed to oxygen, thus resulting in lower polyphenol retention. However, Tonon et al. [59] reported that gum arabic showed greater potential compared to maltodextrin and tapioca starch with regard to polyphenol retention in spray-dried acai pulp powders. This markedly distinct character may be due to the different samples, structure of carriers, and drying condition used.

Similarly, for the total anthocyanin content (TAC), there was a significant ($p < 0.05$) difference among carrier agents (Figure 4b). The powder produced with maltodextrin had the highest TAC (76.91 mg C3gE/g DM), whereas waxy starch had the lowest TAC (35.01 mg C3gE/g DM). Furthermore, an approximate 54% higher in total anthocyanin content was observed for maltodextrin compared to waxy starch. Yousefi et al. [28] investigated the use of maltodextrin, gum arabic, and waxy starch as carrier agents in the spray drying of pomegranate juice. The authors found that maltodextrin was more effective than gum arabic and waxy starch with regard to the preservation of anthocyanins, which supports the result of this study. Similarly, Tonon et al. [72] reported the lowest anthocyanin retention in the powder produced with starch.

3.7. DPPH Radical Scavenging Activity and FRAP Activity

The radical scavenging activity (RSA) of PJP was significantly ($p < 0.05$) different among carrier agents (Figure 5a). The powder produced with maltodextrin had higher antioxidant capacity RSA (33.19 mM TE/g DM), compared to gum arabic (28.45 mM TE/g DM) and waxy starch (26.96 mM TE/g DM). The lower value in powder produced with waxy starch could be traceable to the insoluble nature of the carrier.

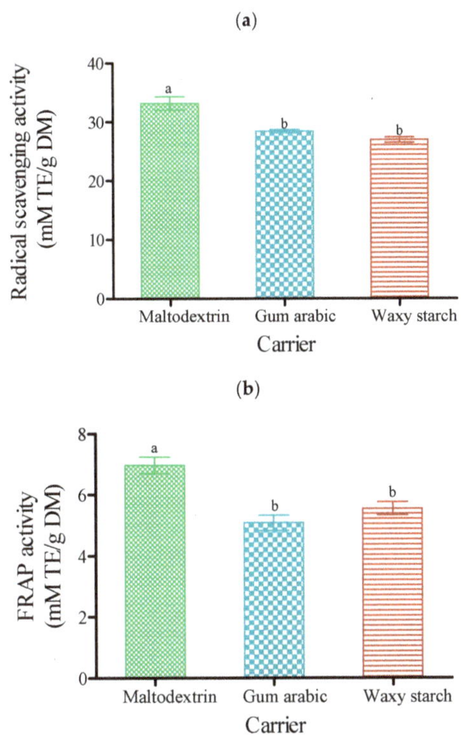

Figure 5. Antioxidant capacity (**a**) radical scavenging activity (RSA) and (**b**) ferric-reducing antioxidant power (FRAP) activity of freeze-dried pomegranate powder using different carriers. Bars with different letters are significantly different ($p < 0.05$) according to Fisher's LSD. Vertical bars indicate the standard error of the mean.

Similarly, the ferric-reducing power (FRAP) of PJP was significantly ($p < 0.05$) different among carrier agents (Figure 5b). PJP produced with maltodextrin had the highest FRAP (6.97 mM TE/g DM), while powder produced with gum arabic had the least (5.09 mM TE/g DM). The higher value of the powder produced with maltodextrin could be traceable to the high soluble nature of the carrier agent. Tonon et al. [72] reported that powders produced with maltodextrin showed higher antioxidant capacity in the spray-dried acai powder due to its high soluble nature, which supports the results of this study. Lim et al. [73] also noted that powder produced with maltodextrin increased the antioxidant capacity of spray-dried blueberry by-products. Furthermore, maltodextrin as a drying agent significantly increased the antioxidant activity of spray-dried amla juice powder [74].

3.8. Principal Component Analysis

The results show the average of phenolic contents, antioxidant capacity, technofunctional properties and colour coordinates of pomegranate of freeze-dried powder. The two principal components (F1 and F2) explain 100.0% of the total data variance (Figure 6).

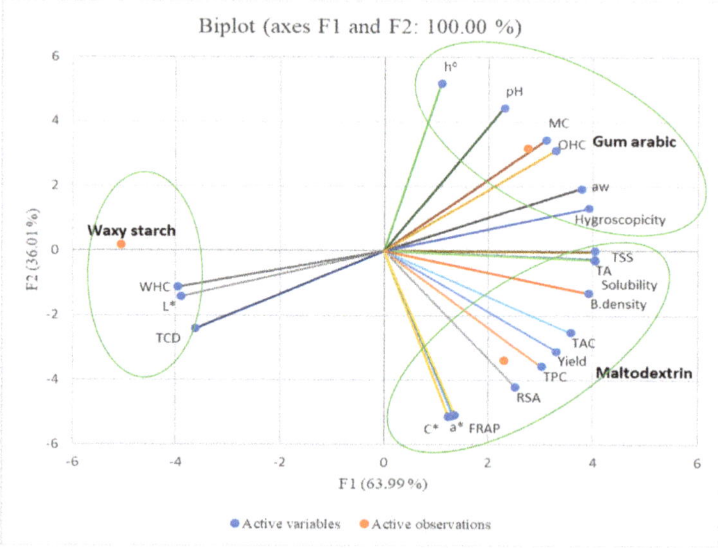

Figure 6. Principal component analysis of the first two factors (F1 and F2) based on physicochemical properties, phenolic contents, antioxidant capacity and technofunctional properties of pomegranate powder cv. Wonderful obtained from maltodextrin, gum arabic and waxy starch. MC, moisture content; L^*, lightness; a^*, redness; C^*, chroma; $h°$, hue angle; TCD, total colour difference; RSA, radical scavenging activity; FRAP, ferric reducing antioxidant power; TPC, total phenolic content; TAC, total anthocyanin content; TSSs, total soluble solids; TA, titratable acidity; a_w, water activity; WHC, water holding capacity; OHC, oil holding capacity.

As observed, F1 explained 63.99% of the total variance, while F2 explained only 36.01% of the total variability, which showed that the disparity among freeze-dried pomegranate powder was described by F1 (Figure 6). The observations indicated that the powders produced with maltodextrin and gum arabic had higher positive scores along the F1 plane and could be associated with moisture content (MC), water activity (a_w), total soluble solids (TSSs), titratable acidity (TA), pH, yield, total phenolic content (TPC), radical scavenging activity (RSA), total anthocyanin content (TAC), solubility, hygroscopicity, bulk density and oil-holding capacity (OHC) (Table 5).

Table 5. Factor loadings, eigenvalue, cumulative variance (%) and score for the first two principal (F1–F2) components based on different carrier agents.

Loadings	F1	F2
L^*	−0.965	−0.263
a^*	0.306	−0.952
C^*	0.324	−0.946
$h°$	0.270	0.963
TCD	−0.895	−0.446
MC	0.769	0.640
a_w	0.933	0.359
TSS	1.000	0.000
TA	0.999	−0.050
pH	0.570	0.822
Yield	0.818	−0.576
TPC	0.748	−0.664
RSA	0.624	−0.781
FRAP	0.336	−0.942
TAC	0.884	−0.468
Solubility	0.999	−0.054
Hygroscopicity	0.969	0.247
Bulk density	0.970	−0.243
WHC	-0.978	−0.208
OHC	0.815	0.580
Scores		
Maltodextrin	2.309	−3.377
Gum arabic	2.744	3.188
Waxy starch	−5.053	0.189

MC, moisture content; L^*, lightness; a^*, redness; C^*, chroma; $h°$, hue angle; TCD, total colour difference; RSA, radical scavenging activity; FRAP, ferric reducing antioxidant power; TPC, total phenolic content; TAC, total anthocyanin content; TSSs, total soluble solids; TA, titratable acidity; a_w, water activity; WHC, water-holding capacity; OHC, oil-holding capacity.

In addition, higher negative scores along F1 (Table 5; Figure 6) correspond to TCD, water-holding capacity and the lightness of the powder produced with waxy starch. Along F1 (Figure 6), lower positive scores correspond to redness (a^*), chroma (C^*), hue ($h°$) and ferric-reducing antioxidant power (FRAP) of freeze-dried powder produced with maltodextrin and gum arabic. Likewise, high positive scores along F2 (Table 5) is associated with hue ($h°$), moisture content (MC), pH and oil-holding capacity (OHC) of the freeze-dried powder produced with gum arabic (Figure 6). Additionally, along the F2, high negative scores (as shown in Figure 6 and Table 5) for maltodextrin could characterise the powder for having high yield, redness (a^*), chroma (C^*), total phenolic content (TPC), radical scavenging activity (RSA) and ferric-reducing antioxidant power (FRAP). However, lower positive scores along F2 were from freeze-dried powder from gum arabic (associated with water activity (a_w), total soluble solids (TSSs) and hygroscopicity). The lower negative scores (Figure 6) along F2 (Table 5) were from maltodextrin and waxy starch (associated with lightness (L^*), TCD, titratable acidity (TA), solubility, total anthocyanin content (TAC), bulk density and water-holding capacity (WHC). The results from the PCA showed that powders produced with carrier agents (maltodextrin, gum arabic and waxy starch) resulted in significantly different properties.

4. Conclusions

The use of three carrier agents (maltodextrin, gum arabic and waxy starch) in the production of freeze-dried PJP was investigated. The results indicated that maltodextrin was more effective in enhancing the yield as well as the physicochemical properties of the PJP, such as colour, TSSs and TA. Similarly, maltodextrin and gum arabic performed better as carriers agents in enhancing the solubility of freeze-dried PJP compared to waxy starch. Maltodextrin was better in the preservation of phenolic content and antioxidant capacity of PJP. Therefore, it could be inferred that maltodextrin resulted in the best carrier agent that retained biochemical activities and maintained the technofunctional properties in the production of freeze-dried pomegranate powder. This study has shown that maltodextrin is the most suitable carrier agent for the formulation or fortification of pomegranate-based food products for baking, candies and ice-cream. This study reports the results of powder produced in a laboratory-scale freeze-dryer. However, a scale-up can be investigated in order to produce, on an industrial scale, powders with similar characteristics. Moreover, further research is required to investigate the storability and optimisation of PJP.

Author Contributions: The following authors contributed to the work: conceptualization, O.A.F. and U.L.O.; methodology, A.O.A. and O.A.F.; software, A.O.A.; formal analysis, A.O.A.; investigation, A.O.A.; resources, O.A.F. and U.L.O.; writing—original draft preparation, A.O.A.; writing—review and editing, O.A.F. and U.L.O.; supervision, O.A.F. and U.L.O.; project administration, O.A.F.; funding acquisition, O.A.F. and U.L.O. All authors have read and agreed to the published version of the manuscript.

Funding: The financial support of the World Academy of Science and the National Research Foundation of South Africa through the award of a doctoral scholarship to Adegoke Olusesan Adetoro is gratefully acknowledged (IUD: 105483).

Acknowledgments: This work is based on research supported by the National Research Foundation (NRF) of South Africa (IUD: 64813 and IUD: 105722).

Conflicts of Interest: There is no conflict of interest between the authors. The opinions, findings and conclusions or recommendations expressed are those of the author(s) alone, and the NRF accepts no liability whatsoever in this regard.

Abbreviations

PJP	pomegranate juice powder
MC	moisture content
L*	lightness
a*	redness
C*	chroma
h°	hue angle
TCD	total colour difference
RSA	radical scavenging activity
FRAP	ferric reducing antioxidant power
TPC	total phenolic content
TAC	total anthocyanin content
TSS	total soluble solids
TA	titratable acidity
a_w	water activity
WHC	water holding capacity
OHC	oil holding capacity
SEM	scanning electron microscopy
w.b.	wet basis
RH	relative humidity

References

1. Fawole, O.A.; Opara, U.L. Effects of maturity status on biochemical concentration, polyphenol composition and antioxidant capacity of pomegranate fruit arils (cv. Bhagwa). *S. Afr. J. Bot.* **2013**, *85*, 23–31. [CrossRef]
2. Mphahlele, R.R.; Caleb, O.J.; Fawole, O.A.; Opara, U.L. Effects of different maturity stages and growing locations on changes in chemical, biochemical and aroma volatile composition of 'Wonderful' pomegranate juice. *J. Sci. Food Agric.* **2016**, *96*, 1002–1009. [CrossRef] [PubMed]
3. Gil, M.I.; Tomas-Barberan, F.A.; Hess-Pierse, B.; Holcroft, D.M.; Kader, A.A. Antioxidant activity of pomegranate juice and its relationship with phenolic composition and processing. *J. Agric. Food Chem.* **2000**, *48*, 4581–4589. [CrossRef] [PubMed]

4. Pomegranate Association of South Africa (Pomasa) Pomegranate Industry Statistics. Paarl, South Africa, 2019. Available online: https://www.sapomegranate.co.za/statistics-and-information/pomegranate-industry-overview/ (accessed on 18 September 2020).
5. Seeram, N.P.; Aviram, M.; Zhang, Y.; Henning, S.M.; Feng, L.; Dreher, M.; Heber, D. Comparison of antioxidant potency of commonly consumed polyphenol-rich beverages in the United States. *J. Agric. Food Chem.* **2008**, *56*, 1415–1422. [CrossRef] [PubMed]
6. Cao, G.; Sofic, E.; Prior, R. Antioxidant capacity of tea and common vegetables. *J. Agric. Food Chem.* **1996**, *44*, 3426–3431. [CrossRef]
7. Opara, U.L.; Al-Ani, M.R.; Al-Shuaibi, Y.S. Physico-chemical properties, vitamin C content and antimicrobial properties of pomegranate fruit (*Punica granatum* L.). *Food Bioprocess Technol.* **2009**, *2*, 315–321. [CrossRef]
8. Malik, A.; Afaq, F.; Sarfaraz, S.; Adhami, V.M.; Syed, D.N.; Mukhtar, H. Pomegranate fruit juice for chemoprevention and chemotherapy of prostate cancer. *Proc. Natl. Acad. Sci. USA* **2005**, *102*, 14813–14818. [CrossRef] [PubMed]
9. Caleb, O.J.; Opara, U.L.; Witthuhn, C.R. Modified atmosphere packaging of pomegranate fruit and arils: A review. *Food Bioprocess Technol.* **2012**, *5*, 15–30. [CrossRef]
10. Kulkami, A.P.; Aradhya, S.M. Chemical changes and antioxidant activity in pomegranate arils during fruit development. *Food Chem.* **2005**, *93*, 319–324.
11. Longtin, R. The pomegranate: Nature's power fruit. *J. Natl. Cancer Inst.* **2003**, *95*, 346–348. [CrossRef]
12. Różyło, R. Recent trends in methods used to obtain natural food colorants by freeze-drying. *Trends Food Sci. Technol.* **2020**, *102*, 39–50. [CrossRef]
13. Sagar, V.R.; Kumar, P.S. Recent advances in drying and dehydration of fruits and vegetables: A review. *J. Food Sci. Technol.* **2010**, *47*, 15–26. [CrossRef] [PubMed]
14. Du, J.; Ge, Z.Z.; Xu, Z.; Zou, B.; Zhang, Y.; Li, C.M. Comparison of the efficiency of five different drying carriers on the spray drying of persimmon pulp powders. *Dry. Technol.* **2014**, *32*, 1157–1166. [CrossRef]
15. Patel, K.C.; Chen, X.D. Drying of aqueous lactose solutions in a single stream dryer. *Food Bioprod. Process.* **2008**, *86*, 185–197. [CrossRef]
16. Sablani, S.S.; Shrestha, A.K.; Bhandari, B.R. A new method of producing date powder granules: Physicochemical characteristics of powder. *J. Food Eng.* **2008**, *87*, 416–421. [CrossRef]
17. Jayaraman, K.S.; Das Gupta, D.K. Dehydration of fruits and vegetables-recent developments in principles and techniques. *Dry. Technol.* **1992**, *10*, 1–50. [CrossRef]
18. Fazaeli, M.; Emam-Djomeh, Z.; Ashtari, A.K.; Omid, M. Effect of spray drying conditions and feed composition on the physical properties of black mulberry juice powder. *Food Bioprod. Process.* **2012**, *90*, 667–675. [CrossRef]
19. Seeranguravar, T.; Manickavasagan, A.; Al-Ismaili, A.M.; Al-Mulla, Y.A. Effect of carrier agents on physicochemical properties of foam-mat freeze-dried date powder. *Dry. Technol.* **2018**, *36*, 1292–1303. [CrossRef]
20. Dolinsky, A.; Maletskaya, K.; Snezhkin, Y. Fruit and vegetable powders production technology on the bases of spray and convective drying methods. *Dry. Technol.* **2000**, *18*, 747–758. [CrossRef]
21. Adhikari, B.; Howes, T.; Bhandari, B.R.; Troung, V. Effect of addition of maltodextrin on drying kinetics and stickiness of sugar and acid-rich foods during convective drying: Experiments and modelling. *J. Food Eng.* **2004**, *62*, 53–68. [CrossRef]
22. Osorio, C.; Forero, D.P.; Carriazo, J.G. Characterisation and performance assessment of guava (*Psidium guajava* L.) microencapsulates obtained by spray-drying. *Food Res. Int.* **2011**, *44*, 1174–1181. [CrossRef]
23. Wang, W.; Zhou, W. Characterization of spray-dried soy sauce powders using maltodextrin as carrier. *J. Food Eng.* **2012**, *109*, 399–405. [CrossRef]
24. Cano-Chauca, M.; Stringheta, P.C.; Ramos, A.M.; Cal-Vidal, J. Effect of the carriers on the microstructure of mango powder obtained by spray drying and its functional characterization. *Innov. Food Sci. Emerg. Technol.* **2005**, *5*, 420–428. [CrossRef]
25. Silva, M.A.; Sobral, P.J.A.; Kieckbusch, T.G. State diagrams of freeze-dried camu-camu (*Myrciaria dubia* (HBK) Mc Vaugh) pulp with and without maltodextrin addition. *J. Food Eng.* **2006**, *77*, 426–432. [CrossRef]
26. Phisut, N. Spray drying technique of fruit juice powder: Some factors influencing the properties of product. *Int. Food Res. J.* **2012**, *4*, 1297–1306.

27. Etzbach, L.; Meinert, M.; Faber, T.; Klein, C.; Schieber, A.; Weber, F. Effects of carrier agents on powder properties, stability of carotenoids, and encapsulation efficiency of goldenberry (*Physalis peruviana* L.) powder produced by co-current spray drying. *Curr. Res. Food Sci.* **2020**, *3*, 73–81. [CrossRef]
28. Yousefi, S.; Emam-Djomeh, Z.; Mousavi, S.M. Effect of carrier type and spray drying on the physicochemical properties of powdered and reconstituted pomegranate juice (*Punica granatum* L.). *J. Food Sci. Technol.* **2011**, *48*, 677–684. [CrossRef]
29. Muzaffar, K.; Kumar, P. Parameter optimization for spray drying of tamarind pulp using response surface methodology. *Powder Technol.* **2015**, *279*, 179–184. [CrossRef]
30. Pathare, P.B.; Opara, U.L.; Al-Said, F.A.J. Colour measurement and analysis in fresh and processed foods: A review. *Food Bioproc. Technol.* **2013**, *6*, 36–60. [CrossRef]
31. Ferrari, C.C.; Germer, S.P.M.; Aguirre, J.M. Effects of spray drying conditions on the physicochemical properties of blackberry powder. *Dry. Technol.* **2011**, *30*, 154–163. [CrossRef]
32. Largo Avila, E.; Cortes Rodríguez, M.; Velásquez, C.; José, H. Influence of maltodextrin and spray drying process conditions on sugarcane juice powder quality. *Rev. Fac. Nac. Agron. Medellin* **2015**, *68*, 7509–7520. [CrossRef]
33. Goula, A.M.; Adamopoulos, K.G. Effect of maltodextrin addition during spray drying of tomato pulp in dehumidified air: II. Powder properties. *Dry. Technol.* **2008**, *26*, 726–737. [CrossRef]
34. Jalal, H.; Pal, M.A.; Ahmad, S.R.; Rather, M.; Andrabi, M.; Hamdani, S. Physico-chemical and functional properties of pomegranate peel and seed powder. *J. Pharm. Innov.* **2018**, *7*, 1127–1131.
35. Ferrari, C.C.; Germer, S.P.M.; Alvim, I.D.; Vissotto, F.Z.; de Aguirre, J.M. Influence of carrier agents on the physicochemical properties of blackberry powder produced by spray drying. *J. Food Sci. Technol.* **2012**, *47*, 1237–1245. [CrossRef]
36. Wrolstad, R.E. *Colour and Pigment Analyses in Fruit Products*; Agricultural Experiment Station, Oregon State University Station: Corvallis, OR, USA, 1993; Volume 624.
37. Benzie, I.F.F.; Strain, J.J. The ferric reducing ability of plasma (FRAP) as a measure of "antioxidant power". The FRAP assay. *Anal. Biochem.* **1996**, *239*, 70–76. [CrossRef]
38. Fawole, O.A.; Opara, U.L. Physico-mechanical, phyto-chemical, and free radical scavenging properties and volatile compounds in eight pomegranate cultivars and classification by principal component and cluster analyses. *Br. Food J.* **2013**, *116*, 544. [CrossRef]
39. Fawole, O.A.; Opara, U.L. Changes in physical properties, chemical and elemental composition and antioxidant capacity of pomegranate (cv. Ruby) fruit at five maturity stages. *Sci. Hortic.* **2013**, *150*, 37–46. [CrossRef]
40. Quek, S.Y.; Chok, N.K.; Swedlund, P. The physicochemical properties of spray-dried watermelon powder. *Chem. Eng. Process.* **2007**, *46*, 386–392. [CrossRef]
41. Seeram, N.P.; Adams, L.S.; Henning, S.M.; Niu, Y.; Zhang, Y.; Nair, M.G.; Heber, D. In vitro anti-proliferative, apoptotic and antioxidant activities of punicalagin, ellagic acid and a total pomegranate tannin extract are enhanced in combination with other polyphenols as found in pomegranate juice. *J. Nutr. Biochem.* **2005**, *16*, 360–367. [CrossRef]
42. Bourdoux, S.; Li, D.; Rajkovic, A.; Devlieghere, F.; Uyttendaele, M. Performance of drying technologies to ensure microbial safety of dried fruits and vegetables. *Compr. Rev. Food Sci.* **2016**, *15*, 1056–1066. [CrossRef]
43. Laroche, C.; Fine, F.; Gervais, P. Water activity affects heat resistance of microorganisms in food powders. *Int. J. Food Microbiol.* **2005**, *97*, 307–315. [CrossRef] [PubMed]
44. Daza, L.D.; Fujita, A.; Fávaro-Trindade, C.S.; Rodrigues-Ract, J.N.; Granato, D.; Genovese, M.I. Effect of spray drying conditions on the physical properties of Cagaita (Eugenia dysenterica DC.) fruit extracts. *Food Bioprod. Process.* **2016**, *97*, 20–29. [CrossRef]
45. Mosquera, L.H.; Moraga, G.; Martınez-Navarrete, N. Critical water activity and critical water content of freeze-dried strawberry powder as affected by maltodextrin and arabic gum. *Food Res. Int.* **2011**, *47*, 201–206. [CrossRef]
46. da Silva Delabona, P.; Pirota, R.D.P.B.; Codima, C.A.; Tremacoldi, C.R.; Rodrigues, A.; Farinas, C.S. Effect of initial moisture content on two Amazon rainforest Aspergillus strains cultivated on agro-industrial residues: Biomass-degrading enzymes production and characterization. *Ind. Crop. Prod.* **2013**, *42*, 236–242. [CrossRef]
47. Hammami, C.; René, F. Determination of freeze-drying process variables for strawberries. *J. Food Eng.* **1997**, *32*, 133–154. [CrossRef]

48. Goula, A.M.; Adamopoulos, K.G. A new technique for spray drying orange juice concentrate. *Innov. Food Sci. Emerg. Technol.* **2010**, *11*, 342–351. [CrossRef]
49. Comunian, T.A.; Monterrey-Quintero, E.S.; Thomazini, M.; Balieiro, J.C.D.C.; Piccone, P.; Pittia, P.; Favaro-Trindade, C.S. Assessment of production efficiency, physicochemical properties and storage stability of spray-dried chlorophyllide, a natural food colourant, using gum Arabic, maltodextrin and soy protein isolate-based carrier systems. *Int. J. Food Sci. Technol.* **2011**, *46*, 1259–1265. [CrossRef]
50. Kha, T.C.; Nguyen, M.H.; Roach, P.D. Effects of spray drying conditions on the physicochemical and antioxidant properties of the Gac (*Momordica cochinchinensis*) fruit aril powder. *J. Food Eng.* **2010**, *98*, 385–392. [CrossRef]
51. Horuz, E.; Altan, A.; Maskan, M. Spray drying and process optimization of unclarified pomegranate (*Punica granatum*) juice. *Dry. Technol.* **2012**, *30*, 787–798. [CrossRef]
52. Rahman, S.; Lamb, J. Air drying behaviour of fresh and osmotically dehydrated pineapple. *J. Food Process. Eng.* **1991**, *14*, 163–171. [CrossRef]
53. Karatas, S.; Esin, A.A. A laboratory scraped surface drying chamber for spray drying of tomato paste. *Lebensm. Wiss. Technol.* **1990**, *23*, 354–357.
54. Bhandari, B.R.; Datta, N.; Crooks, R.; Howes, T.; Rigby, S. A Semi-empirical Approach to optimize the quantity of drying aids required to spray dry sugar rich foods. *Dry. Technol.* **1997**, *15*, 2509–2525. [CrossRef]
55. Shrestha, A.K.; Ua-Arak, T.; Adhikari, B.P.; Howes, T.; Bhandari, B.R. Glass transition behavior of spray dried orange juice powder measured by differential scanning calorimetry (DSC) and thermal mechanical compression test (TMCT). *Int. J. Food Prop.* **2007**, *10*, 661–673. [CrossRef]
56. Abadio, F.D.B.; Domingues, A.M.; Borges, S.V.; Oliveira, V.M. Physical properties of powdered pineapple (*Ananas comosus*) juice- Effect of malt dextrin concentration and atomization speed. *J. Food Eng.* **2004**, *64*, 285–287. [CrossRef]
57. De Oliveira, M.A.; Maia, G.A.; De Figueiredo, R.W.; De Souza, A.C.R.; De Brito, E.S.; De Azeredo, H.M.C. Addition of cashew tree gum to maltodextrin-based carriers for spray drying of cashew apple juice. *J. Food Sci. Technol.* **2009**, *44*, 641–645. [CrossRef]
58. Mishra, S.; Rai, T. Morphology and functional properties of corn, potato and tapioca starches. *Food Hydrocoll.* **2006**, *20*, 557–566. [CrossRef]
59. Tonon, R.V.; Baroni, A.F.; Brabet, C.; Gibert, O.; Pallet, D.; Hubinger, M.D. Water sorption and glass transition temperature of spray dried açai (*Euterpe oleracea* Mart.) juice. *J. Food Eng.* **2009**, *94*, 215–221. [CrossRef]
60. Gabas, A.L.; Telis, V.R.N.; Sobral, P.J.A.; Telis-Romero, J. Effect of maltodextrin and arabic gum in water vapour sorption thermodynamic properties of vacuum dried pineapple pulp powder. *J. Food Eng.* **2007**, *82*, 246–252. [CrossRef]
61. Ahmed, M.; Akter, M.S.; Eun, J.B. Impact of alpha-amylase and maltodextrin on physicochemical, functional and antioxidant capacity of spray dried purple sweet potato flour. *J. Sci. Food Agric.* **2010**, *90*, 494–502. [CrossRef]
62. Barbosa-Canovas, G.V.; Juliano, P. Physical and chemical properties of food powders. In *Encapsulated and Powdered Foods*; Onwulata, C., Ed.; Taylor & Francis: Boca Raton, FL, USA, 2005; pp. 39–71.
63. Chegini, G.R.; Ghobadian, B. Effect of spray-drying conditions on physical properties of orange juice powder. *Dry. Technol.* **2005**, *23*, 656–668. [CrossRef]
64. Zhang, L.H.; Li, S.F. Effects of micronization on properties of Chaenomeles sinensis (Thouin) Koehne fruit powder. *Innov. Food Sci. Emerg. Technol.* **2009**, *10*, 633–637. [CrossRef]
65. Hong, J.; Zhang, S.Y. Effect of ultra-fine pulverization by wet processing on particle structure and physical properties of soybean dietary fiber. *J. China Agric. Univ.* **2005**, *10*, 90–94. (In Chinese)
66. Chau, C.F.; Wang, Y.T.; Wen, Y.L. Different micronization methods significantly improve the functionality of carrot insoluble fibre. *Food Chem.* **2007**, *100*, 1402–1408. [CrossRef]
67. Tonon, R.V.; Freitas, S.S.; Hubinger, M.D. Spray drying of açai (*Euterpe oleraceae* Mart.) juice: Effect of inlet air temperature and type of carrier agent. *J. Food Process. Preserv.* **2011**, *35*, 691–700. [CrossRef]
68. Syamaladevi, R.M.; Insan, S.K.; Dhawan, S.; Andrews, P.; Sablani, S.S. Physicochemical properties of encapsulated red raspberry (*Rubus idaeus*) powder: Influence of high-pressure homogenization. *Dry. Technol.* **2012**, *30*, 484–493. [CrossRef]
69. Kurozawa, L.E.; Morassi, A.G.; Vanzo, A.A.; Park, K.J.; Hubinger, M.D. Influence of spray drying conditions on physicochemical properties of chicken meat powder. *Dry. Technol.* **2009**, *27*, 1248–1257. [CrossRef]

70. Leonel, M. Ana´lise da forma e tamanho de gra^nulos de amidos de diferentes fontes botânicas. *Ciênc. Tecnol.* **2007**, *23*, 579–588. [CrossRef]
71. Loksuwan, J. Characteristics of microencapsulated b-carotene formed by spray drying with modified tapioca starch, native tapioca starch and maltodextrin. *Food Hydrocoll.* **2007**, *21*, 928–935. [CrossRef]
72. Tonon, R.V.; Brabet, C.; Hubinger, M.D. Anthocyanin stability and antioxidant activity of spray dried ac¸ai (*Euterpe oleracea* Mart.) juice powder produced with different carrier agents. *Int. Food Res. J.* **2010**, *43*, 907–914. [CrossRef]
73. Lim, K.; Ma, M.; Dolan, K.D. Effects of spray drying on antioxidant capacity and anthocyanidin content of blueberry by-products. *J. Food Sci.* **2011**, *76*, 156–164. [CrossRef]
74. Mishra, P.; Mishra, S.; Mahanta, C.L. Effect of maltodextrin concentration and inlet temperature during spray drying on physicochemical and antioxidant properties of amla (*Emblica officinalis*) juice powder. *Food Bioprod. Process.* **2014**, *92*, 252–258. [CrossRef]

© 2020 by the authors. Licensee MDPI, Basel, Switzerland. This article is an open access article distributed under the terms and conditions of the Creative Commons Attribution (CC BY) license (http://creativecommons.org/licenses/by/4.0/).

foods

MDPI

Article

Drying Habanero Pepper (*Capsicum chinense*) by Modified Freeze Drying Process

Cicerón González-Toxqui [1], Álvaro González-Ángeles [1,*], Roberto López-Avitia [1] and David González-Balvaneda [2]

[1] Facultad de Ingeniería, Universidad Autónoma de Baja California, Blvd. Benito Juárez s/n, Mexicali 21280, B.C., Mexico; ciceron.gonzalez@uabc.edu.mx (C.G.-T.); ravitia@uabc.edu.mx (R.L.-A.)

[2] Instituto Tecnológico de Mexicali, Ave. Tecnológico s/n Colonia Elías Calles Cp., Mexicali 21376, B.C., Mexico; gonzadav@itmexicali.edu.mx

* Correspondence: gangelesa@gmail.com; Tel.: +52-1-6862110294

Received: 7 February 2020; Accepted: 1 April 2020; Published: 5 April 2020

Abstract: Freeze drying process was applied to habanero pepper and modified, in order to reduce energy expenditure on frozen and dehydration techniques. Six alkaline solutions, olive oil, avocado oil, coconut oil, grape oil, sesame oil and safflower oil, were used to reduce time on vacuum chamber. Also, frozen step was modified by using dry ice (CO_2) obtaining 43% of energy saving. The final product had high quality, moisture within 3% to 7% range, low microorganisms number, without organoleptic attributes damage and having all the characteristics of a fresh product by rehydrating. Dried sample was rehydrated by immersion in water at 40 °C for 5 min, obtaining 75% of initial humidity.Markedchanges on rehydrated final product was not perceived. The most effective oil to reduce the moisture was safflower followed by coconut and sesame, whilst the least effective were olive, followed by avocado and grape oils.

Keywords: dehydration; conserving vegetables; improving shelf-life; rehydrated pepper; histological preparation; green practices

1. Introduction

Mexico stands out in the generation of chili varieties in the world. Around 90% of chili consumed worldwide is of Mexican origin. Other producing countries are China, Indonesia, Turkey, Spain, the United States, and Nigeria. Additionally, 80% of the production of habanero pepper is marketed as fresh and the remaining 20% goes to the preparation of sauces and pasta, oris dehydrated. It exports mainly to the United States, Japan, South Korea, Italy, and Germany [1,2].

Habanero pepper is an herbaceous plant or shrub, branched, reaching a size up to 2.5 meters high. The immature specimens of habanero peppers are green, but their color varies at maturity. The most common colors are orange (semi-ripe) and red when ripe. According to scientific research, the origins of habanero pepper are found in the zone that extends from southern Brazil to northern Argentina, through eastern Bolivia and western Paraguay [3].

Likewise, due to its different properties, habanero pepper is used in different areas as gastronomy, medicine within which its components are used to make ointments that relieve the severe pain caused by arthritis; within the chemical industry, it is used to make the base of some paintings, as well as to make tear gas [4].

Due to the fact that fresh chili is perishable, it is necessary to extend its shelf life by choosing suitable preserving methods, minimizing loss or damage of active ingredient, and consuming the least energy in the process.

There are a lot of drying methods (sun drying, hot air drying, spray drying, vacuum drying, freeze-drying, hybrid methods among others) used to decrease microbial activity, to storage for a long time, and transport the product around the world [5–9].

All methods have advantages and disadvantages, some are cheap andeco friendly and others are expensive like freeze-drying; however, freeze-dryinghas a peculiarity. The final dehydrated product could be rehydrated again, being almost as good as the fresh product.This peculiarity is because the dried product keeps its original shape, something that does not happen in other methods, where the final dried product shrinks.

In literature, there isa lot ofresearch where the freeze-dried method was modified for different aims, for instance, to investigate the effect of various cryoprotectants on cell viability during freeze-drying [10], to assess physico-chemical properties and the antioxidant profile of oyster mushrooms, where the transparent lid of the drying chamber was covered with aluminium foil to prevent the degradation of antioxidants by light oxidation [11]. Another investigation was realized to evaluate the application of CO_2 laser microperforations to blueberry skin. Under the same set of freeze-drying conditions, blueberries with and without perforations were processed. The results showed that the primary drying time was significantly reduced from 17 ± 0.9 h for nontreated berries to 13 ± 2.0 h when nine microperforations per berry fruit were made. Concomitantly, the quality was also significantly improved, as the percentage of nonbusted blueberries at the end of the process increased from an average of 47% to 86%. It was demonstrated that CO_2-laser microperforation has high potential as a skin pretreatment for the freeze-drying of blueberries [12].

The aim of this studyis to modify the freeze-drying methodto obtain dried productsin a faster and cheaper way, contributing to green practices.

2. Materials and Methods

Fresh habanero pepper was obtained in the local market. The specimens were carefully selected by good condition, no punched or damaged surface, showing good quality evidence. The fresh product was cleaned with water and dried at room temperature.

To determine percent solids and weight (in grams) of the sample as well its moisture percent, a thermo balance was used (Ohaus MB-35 model with halogen source heating). The specimen was heated at 105 °C for 15 min (fresh samples) and 3.5 min (freeze-dried samples) [13].

Six alkaline emulsions were prepared (olive oil, avocado oil, coconut oil, grape oil, sesame oil, and safflower oil) by mixing 1.0 liter of alkaline water (pH 10), 50 g of K_2CO_3 (with 98% of purity, ASC reagent, Aldrich Co., Toluca, Mexico) and 5 g of olive oil, avocado oil, coconut oil, grape oil, sesame oil, and safflower oil. The final emulsion obtained pH 12. All samples were immersed in pH 12 solution (above described) for 1 minute at room temperature [14–17].

The freeze-drying process equipment for the test was the Sanshon equipment model FDG-O.5, with 31 × 54 cm process area, 0.5 m^2 of drying area, and 5 liters of water capacity every 8 h in the drying process. It has a vacuum pump of 6×10^{-1} mbar model TRIVAC D16B, 3-ph, 240-265/415-480 V and 60 Hz. It also contains a vacuum chamber, equipped with three heating plates and three shelves where the specimens are placed, a freezing chamber (−40 °C), and a water-heating system on plates (105 °C).

The dehydration process on industrial equipment was performed by 3 stages: first, the product was frozen for 3 h at −35 °C. Once the sample is frozen, a vacuum pressure of 6×10^{-1} mbar is applied. Later, two temperature ramps were applied automatically: one for the sublimation of frozen water at 80 °C for 7 h and another for drying the remnant water at 60 °C for 13 h (total drying time 20 h for untreated samples) [18].

In order to reduce energy expenditure on dehydration processes, six pretreatments solutions were used, as well as the frozen stage changed by using dry ice (CO_2) [6].

The samples were submerged on alkaline solutions for one minute after they passed to the frozen stage for 10 min at −79 °C using dry ice (CO_2). Once the samplewas frozen, a vacuum pressure of $6–10^{-1}$ mbar was applied. Then, two temperature ramps were applied automatically, one for sublimation at 80 °C for 7 h and another at 60 °C for 6 h.

The dried sample was rehydrated by immersion in water at 40 °C for 5 min.

The histological cutting preparation was made following the next steps:

Fixation: treat the tissue with chemical substances to keep the properties of the organic tissue intact, inactivating enzymes that degenerate the tissue.Washing: The excess fixative is removed. Clarification: After dehydrating the tissue, a liquid paraffin solution is passed as an inclusion medium and xylene or xylol as a miscible medium [19].Inclusion: The tissue sample is placed in a container and molten paraffin is added at 60 °C, placing the sample in an oven 30 min at 60 °C for 6 h.Cut: The cube obtained can be cut into sections thin enough to allow the passage of light, having a thickness of 5–10 micrometers.Mounting: Using a mounting solution, 1% gelatin at 38 °C, using it as an adhesive for the strip obtained from cutting on a slide.Coloring: To be observed under a microscope, it must be colored or contrasted [20].

Observation of alkaline emulsion penetration on sample.

Alkaline emulsion of ethyl oleate with pH of (12) changes of color when phenolphthalein 1% is added. This approach could help to visually assess the penetration of the solution (alkaline emulsion of ethyl oleate) in grapes.

Habanero samples were immersed in alkaline emulsion of ethyl oleatefor 1 min. Then, 500 µl of phenolphthalein 1% was added to samples. Images were taken under a dissecting microscope (Stemi DV4, Carl Zeiss, Germany)

For bacterial analysis, serial dilutions of 1:10 of samples in distillated water were performed and 0.1 mL of each dilution was extended in a plate; the plates were incubated until the colonies were appreciable for counting.

The bacteria were isolated in pure culture (standard), in a Petri box and incubated for a period of 24 h at 45 °C.The board was read and the total colony forming units were counted. The specimens were analyzed in an optic microscope (Carl Zeiss, Germany, under visible light).

To observe the morphology of the studied sample, the scanning electronic microscope JEOL JSM- IT 100 (JEOL, Peabody, MA, USA was used. Microchemical analysis of the studied sample was performed by energy-dispersive X-ray spectroscopy (JEOL, Peabody, MA, USA).

3. Results and Discussion

After completing the dehydration process for the six different samples, the final moisture percentage was assessed and compared in Table 1.

Table 1. Final freeze time process, drying time process, and final moisture product for treated samples at different alkaline emulsions.

Habanero Pepper Samples	Freezing Process (Minutes)	Drying Process (Hours)	Final Moisture(%)	Energy Saving (%)	Effectiveness of Emulsions (Adimensional)
Samples without pretreatment	180	20	4.6	0	
Olive oil	10	14	6.55	39	0.083
Coconut oil	10	13	4.08	43	0.470
Avocado oil	10	14	6.04	39	0.154
Grape oil	10	14	5.10	39	0.286
Sesame oil	10	14	4.97	39	0.304
Safflower oil	10	14.5	3.01	36	0.564

The initial moisture of the pepper was between 84%–87%. The dehydrated samples obtained moisture 3%–7% range, which is the most cited in the literature [21].It is important to mention that to reach the final time where the sample obtains the desired humidity (3%–7%), that is, the drying process was stopped every hour less than the original time, after which the moisture of the sample was measured, and if the moisture was too great, we kept drying to obtain values inside the range.

Since most of the emulsions presented similar drying times, their effectiveness was sought in the final moisture of the sample, whichis calculated under the next formula: Effectiveness= 1 − (drying process × final moisture/100). It can be seen in Table 1 that themost effective oil to reduce moisture is safflower, followed by coconut and sesame, whilst the least effective are olive, followed by avocado and grape oils.

The energy saving was calculated in relation to the total time of dried sample with treatment. Comparing the total time on the freeze-drying process of sampleswithout pretreatment against olive oil, avocado oil, coconut oil, grape oil, sesame oil, and safflower oil, it observes an energy saving of 36% to 43%. The best alkaline pretreatment for saving energy was coconut oil with a 43%, safflower oil beingable to contribute the least to save energy.

The reduction in drying time protruded with the alkaline solution, probably due to the alkaline solution breaking the wax on the skin surface that acts as a control of moisture diffusion in leaves and fruits, allowing quicker extraction of the water [14].

On the other hand, it is important to mention that the frozen stage must be on a quick mode of freezing; smaller ice crystals will form inside and, therefore, there will be less damage in the wall cells. Due to the above mentioned facts, the freeze-drying process was modified in the first stage, using dry ice (CO_2) for frozen biological samples [22,23].

Figure 1 shows habanero pepper sections obtained after being immersed in alkaline solution and phenolphthalein in order to determine the penetration of alkaline solution in biological samples. Magenta color reveals the presence of alkaline solution in Figure 1; it can be appreciated that there is no alkaline solution penetration, and the solution remains on skin surface(exocarp), with no impact on habanero pepper endorcarp and mesocarp, just in the cover skin [19].

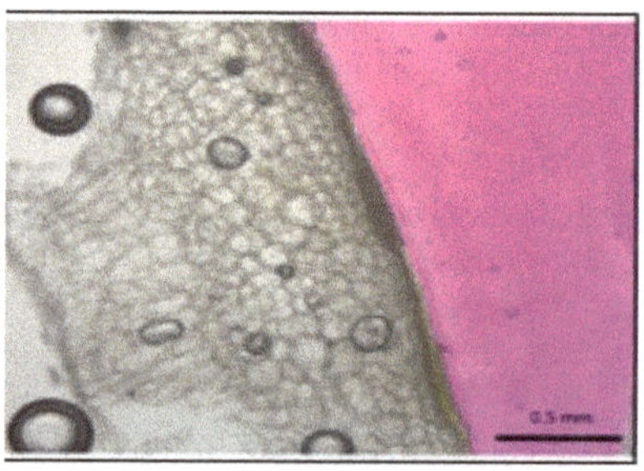

Figure 1. Dried sample section obtained after being immersed in the alkaline solution and phenolphthalein in order to determine the penetration of the alkaline solution. The phenolphthalein turned bright magenta with alkaline solutioncontact; structure viewed at 40× magnification microscopy. The figure shows no alkaline solution penetration in the sample.

Due to the hygroscopic property on freeze-dried products, the product was carefully packaged and sealed after the freeze-drying process.

After the freeze-drying process, the organoleptic properties (taste, odor, color, and appearance) show optimal conditions for human consumption. Samples have a typical strong flavor and extremely concentrated odor, more than the fresh samples. It has a slightly foamy appearance, crispy consistency, and intense color, as shown in Figure 2.

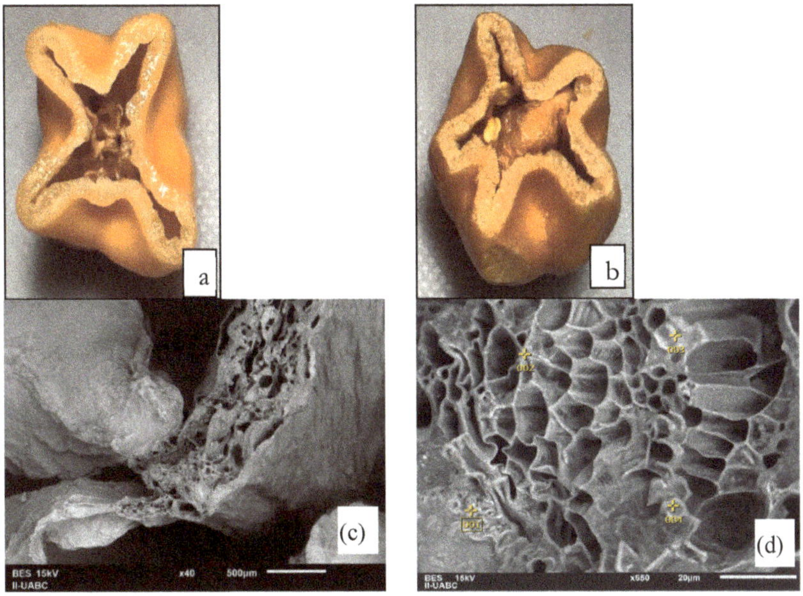

Figure 2. Images of fresh and freeze-dried habanero pepper: (**a**) before freeze-drying, (**b**) after freeze-drying, (**c**) SEM image of dried sample at 40×, (**d**) the place where energy-dispersive X-ray analyses (EDX) were performed (650×) for a dried sample.

Figure 3 shows the energy-dispersive X-ray analysis. It can be seen that the spectra are composed mainly of carbon and oxygen, follow bypotassium and chlorine, and little quantities of phosphorus and sulfur (in this hierarchy, according to peak intense) [24,25].

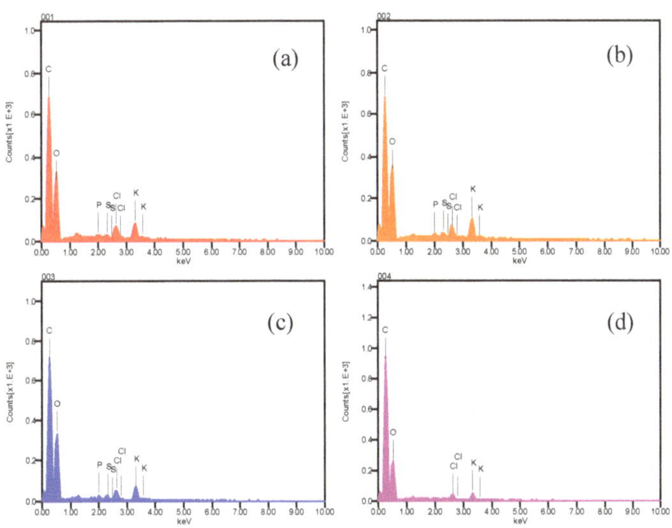

Figure 3. EDX spectrum of atomic elements present in chili, obtained on the dried sample (Figure 2d) with a scanning electronic microscope.(**a**) EDX spectra of atomic elements presents in zone 001 of Figure 2d. (**b**) EDX spectra obtained in zone 002. (**c**) and (**d**) energy-dispersive X-ray spectra obtained in points 003 and 004 of dried sample respectively.

The taste on final freeze-dried product, using pretreatments solutions, can be evaluated as well; comparing flavors between products is highly important, which is vital for a good-quality product.

Once dehydrated, habanero pepper should be stored properly for quality properties. During storage, a variation of the carotenoid pigments of the skin occurs due to an oxidation process. This is increased by the action of external agents of a physical nature such as temperature, humidity, and light, or chemical nature such as metal ions, enzymes, peroxides and free oxygen. These variations affect essentially the color, odor, composition and visual appearance. On the other hand, these changes may also affect the microbiological properties of the product; the isolation on these factors can be useful to ensure product quality, storage-selecting conditions can predict the useful product life evolution, and avoiding moisture in storage must be considered [18,26].

It is important to remember that any fresh product subjected to freezing and thawing does not have the same fresh look, as there is a broken cell membrane by the crystal-freezing process. Figure 4 shows how the freeze-drying process helps to decrease the microbial load number. Low microbial number of microorganisms is a particular interest for the food industry, due to the fact that lyophilized products could be used in hospitals and proportioned to patients with low defenses in childhood, and seniors [18].

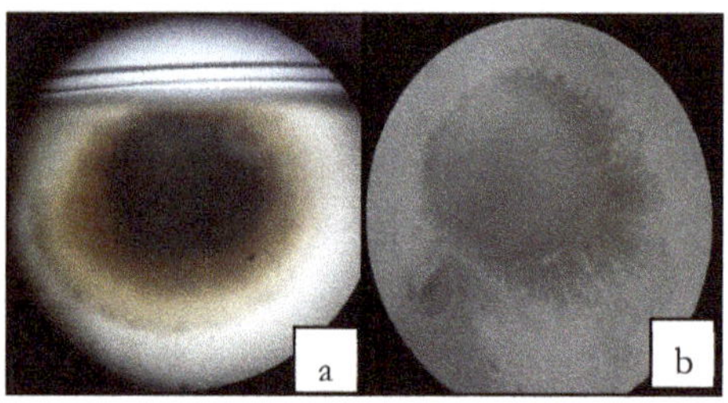

Figure 4. Microbial load habanero pepper sample without alkaline emulsion (**a**), before the freeze-drying process, (**b**)after the freeze-drying process.Structure viewed at 16× magnification microscopy. It can be seen how the microbial load decreases after the sample is dried.

Finally, dried samples were rehydrated by immersion in water at 40 °C for 5 min, obtaining a humidity of 75%,very similar tothe initial humidity of fresh habanero samples(84%–87%). Changes on rehydrated final product were not perceived [27].The idea of rehydration is due to customers consuming sliced pepper in salads or cooking it in pieces and not only in powder form as a dried condiment.

4. Conclusions

This investigation shows six pretreatment alkaline solution developments and a CO_2 frozen method on the habanero pepper freeze-drying process as an alternative, adding improvement generating 36% to 43% in energy savings. The best alkaline emulsion to reducedrying-time was coconut oil. On the other hand, the most effective oil to reduce the moisture was safflower, followed by coconut and sesame, whilst the least effective were olive, followed by avocado and grape oils.

Final moisture for dried samples obtained 3% to 7% after the freeze-drying process. The humidity achieved for rehydrated samples was 75%,similarto the initial humidity of fresh habanero samples.

After the freeze-drying process, products have hygroscopic properties and high considerations on storage conditions must be applied, such as storage in a cool, dry place away from sunlight, sealed vacuum packing; thus, properties will be preserved to eitherrehydrationorconsumption. Finally, if all enterprises involved in green manufacturing take into account these findings and constantly improve their processes, they will stop emitting several kg of pollutioninto the atmosphere.

Author Contributions: C.G.-T. accomplished the study. Á.G.-Á. drafted the manuscript. R.L.-A. contributed substantially to the interpretation of results. D.G.-B. facilitated the freeze-drying equipment. All authors have critically revised and contributed to the final version of this manuscript.

Funding: This research was funded by CONAHCYT, grant number 692364/ PhD. scholarship

Acknowledgments: The author would like to thank the National Humanities Science and Technology Council of Mexico (CONAHCYT) for the support given under grant number692364/ PhD. scholarship.

Conflicts of Interest: The authors declare no conflict of interest.

References

1. Melorose, J.; Perroy, R.; Careas, S. México, líder mundial en exportación de chile: SAGARPA. *Statew. Agric. L. Use Baseline* **2015**, *1*, 2015.
2. SAGARPA Un panorama del cultivo del chile. *Serv. Inf. Agroaliment. Pesq.* **2010**, *1*, 20.
3. SAGARPA Agrícola Nacional. *Planeación Agrícola Nac. 2017-2030* **2017**, *1*, 1–14.
4. Restrepo Gallego, M.; Llanos Ríos, N.; Fonseca Echeverri, C.E. Composición de las oleorresinas de dos variedades de ají picante (habanero y tabasco) obtenidas mediante lixiviación con solventes orgánicos* Composition of oleoresins from two kinds of chili pepper (habanero and tabasco) obtained by lixiviation with orga. *Rev. Lasallista Investig.* **2007**, *4*, 14–19.
5. Fernandes, F.A.N.; Rodrigues, S.; Law, C.L.; Mujumdar, A.S. Drying of Exotic Tropical Fruits: A Comprehensive Review. *Food Bioprocess Technol.* **2011**, *4*, 163–185. [CrossRef]
6. Sagar, V.R.; Suresh Kumar, P. Recent advances in drying and dehydration of fruits and vegetables: A review. *J. Food Sci. Technol.* **2010**, *47*, 15–26. [CrossRef]
7. Chandrasekaran, S.; Ramanathan, S.; Basak, T. Microwave food processing—A review. *FRIN* **2013**, *52*, 243–261. [CrossRef]
8. Belessiotis, V.; Delyannis, E. Solar drying. *Sol. Energy* **2011**, *85*, 1665–1691. [CrossRef]
9. De la Fuente-Blanco, S.; De Sarabia, E.R.F.; Acosta-Aparicio, V.M.; Blanco-Blanco, A.; Gallego-Juárez, J.A. Food drying process by power ultrasound. *Ultrasonics* **2006**, *44*, 523–527. [CrossRef]
10. Nikolaou, A.; Sgouros, G.; Mitropoulou, G.; Santarmaki, V. Freeze-Dried Immobilized Kefir Culture in Low. *Foods* **2020**, *9*, 115. [CrossRef]
11. Piskov, S.; Timchenko, L.; Grimm, W.; Rzhepakovsky, I. Effects of Various Drying Methods on Some Physico-Chemical Properties and the Antioxidant Profile and ACE Inhibition Activity of Oyster Mushrooms (Pleurotus Ostreatus). *Foods* **2020**, *9*, 160. [CrossRef] [PubMed]
12. Munzenmayer, P.; Ulloa, J.; Pinto, M.; Ramirez, C.; Valencia, P.; Simpson, R.; Almonacid, S. Freeze-Drying of Blueberries: Effects of Carbon Dioxide (CO2) Laser Perforation as Skin Pretreatment to Improve Mass Transfer, Primary Drying Time, and Quality. *Foods* **2020**, *9*, 1–18. [CrossRef] [PubMed]
13. OHAUS, "Combining Speed and Precision in One Moisture Analyzer" 2020. Available online: https://us.ohaus.com/en-US/Products/Balances-Scales/Moisture-Analyzers (accessed on 3 April 2020).
14. Doymaz, İ.; Pala, M. The effects of dipping pretreatments on air-drying rates of the seedless grapes. *J. Food Eng.* **2002**, *52*, 413–417. [CrossRef]
15. Arancibia, C.; Riquelme, N.; Zúñiga, R.; Matiacevich, S. Comparing the effectiveness of natural and synthetic emulsifiers on oxidative and physical stability of avocado oil-based nanoemulsions. *Innov. Food Sci. Emerg. Technol.* **2017**, *44*, 159–166. [CrossRef]
16. Berasategi, I.; Barriuso, B.; Ansorena, D.; Astiasarán, I. Stability of avocado oil during heating: Comparative study to olive oil. *Food Chem.* **2012**, *132*, 439–446. [CrossRef]
17. Beaulieu, J.C.; Park, H.S.; Mims, A.G.B.; Kuk, M.S. Extension of green bell pepper shelf life using oilseed-derived lipid films from soapstock. *Ind. Crops Prod.* **2009**, *30*, 271–275. [CrossRef]
18. Faustino, J.M.F.; Barroca, M.J.; Guiné, R.P.F. Study of the drying kinetics of green bell pepper and chemical characterization. *Food Bioprod. Process.* **2007**, *85*, 163–170. [CrossRef]
19. Serafin-Higuera, N.; Hernandez-Sanchez, J.; Ocadiz-Delgado, R.; Vazquez-Hernandez, J.; Albino-Sanchez, M.E.; Hernandez-Pando, R.; Gariglio, P. Retinoic acid receptor β deficiency reduces splenic dendritic cell population in a conditional mouse line. *Immunol. Lett.* **2012**, *146*, 15–24. [CrossRef]
20. Trentham, W.R.; Sams, C.E.; Conway, W.S. Histological effects of calcium chloride in stored apples. *J. Am. Soc. Hortic. Sci.* **2008**, *133*, 487–491. [CrossRef]
21. De Torres, C.; Schumacher, R.; Alañón, M.E.; Pérez-Coello, M.S.; Díaz-Maroto, M.C. Freeze-dried grape skins by-products to enhance the quality of white wines from neutral grape varieties. *Food Res. Int.* **2015**. [CrossRef]

22. Di Matteo, P.; Donsì, G.; Ferrari, G. The role of heat and mass transfer phenomena in atmospheric freeze-drying of foods in a fluidised bed. *J. Food Eng.* **2003**, *59*, 267–275. [CrossRef]
23. Ghio, S.; Barresi, A.A.; Rovero, G. A Comparison of Evaporative and Conventional Freezing Prior to Freeze-Drying of Fruits and Vegetables. *Food Bioprod. Process.* **2000**, *78*, 187–192. [CrossRef]
24. González-Zamora, A.; Sierra-Campos, E.; Pérez-Morales, R.; Vázquez-Vázquez, C.; Gallegos-Robles, M.A.; López-Martínez, J.D.; García-Hernández, J.L. Measurement of capsaicinoids in chiltepin hot pepper: A comparison study between spectrophotometric method and high performance liquid chromatography analysis. *J. Chem.* **2015**, *2015*. [CrossRef]
25. Igual, M.; García-Martínez, E.; Camacho, M.M.; Martínez-Navarrete, N. Effect of thermal treatment and storage on the stability of organic acids and the functional value of grapefruit juice. *Food Chem.* **2010**, *118*, 291–299. [CrossRef]
26. Samoticha, J.; Wojdyło, A.; Lech, K. The influence of different the drying methods on chemical composition and antioxidant activity in chokeberries. *LWT Food Sci. Technol.* **2016**, *66*, 484–489. [CrossRef]
27. Park, J.N.; Sung, N.Y.; Byun, E.H.; Byun, E.B.; Song, B.S.; Kim, J.H.; Lee, K.A.; Son, E.J.; Lyu, E.S. Microbial analysis and survey test of gamma-irradiated freeze-dried fruits for patient's food. *Radiat. Phys. Chem.* **2015**, *111*, 57–61. [CrossRef]

© 2020 by the authors. Licensee MDPI, Basel, Switzerland. This article is an open access article distributed under the terms and conditions of the Creative Commons Attribution (CC BY) license (http://creativecommons.org/licenses/by/4.0/).

foods

MDPI

Article

Vacuum and Infrared-Assisted Hot Air Impingement Drying for Improving the Processing Performance and Quality of *Poria cocos* (Schw.) Wolf Cubes

Weipeng Zhang [1], Chang Chen [2], Zhongli Pan [2] and Zhian Zheng [3,*]

[1] School of Artificial Intelligence, Beijing Technology and Business University, 11 Fucheng Road, Beijing 100048, China; zhangwp@btbu.edu.cn

[2] Department of Biological and Agricultural Engineering, University of California, Davis, One Shields Avenue, Davis, CA 95616, USA; cgchen@ucdavis.edu (C.C.); zlpan@ucdavis.edu (Z.P.)

[3] College of Engineering, China Agricultural University, 17 Qinghua Donglu, Beijing 100083, China

* Correspondence: zhengza@cau.edu.cn

Abstract: The objective of this study was to develop an efficient drying technology for poria cubes in order to improve product quality. Poria cubes were dried using different methods, including air impingement drying, infrared-assisted air impingement drying, vacuum drying, two-stage vacuum drying, and infrared-assisted air impingement drying. The results were compared with those from hot air drying. For the two-stage drying, the tested conditions were the first stage of vacuum drying with temperatures between 65–85 °C and a switching moisture ratio of 70–90%. The second stage infrared-assisted air impingement drying also had temperatures 65–85 °C. The drying kinetics (effective moisture diffusivity (D_{eff}), Biot number (Bi), and mass transfer coefficient (k) were studied via the product qualities (broken ratio, firmness, microstructure, and water-soluble polysaccharide content) and specific energy consumption (SEC) of the drying processes. The results showed that two-stage drying led to the lowest drying time and energy consumption, and also obtained the best qualities. Box–Behnken experimental design with response surface methodology (RSM) was used to optimize the two-stage operating conditions as 82 °C under vacuum drying until a moisture content of 81% and a temperature of 69 °C with infrared-assisted air impingement drying was achieved. These findings suggested that two-stage vacuum and infrared-assisted air impingement drying is a promising method for producing high quality and energy efficient dried poria cubes.

Keywords: vacuum; physicochemical properties; poria cubes; optimization; stage drying

Citation: Zhang, W.; Chen, C.; Pan, Z.; Zheng, Z. Vacuum and Infrared-Assisted Hot Air Impingement Drying for Improving the Processing Performance and Quality of *Poria cocos* (Schw.) Wolf Cubes. *Foods* **2021**, *10*, 992. https://doi.org/10.3390/foods10050992

Academic Editor: Danijela Bursać Kovačević and Predrag Putnik

Received: 25 March 2021
Accepted: 28 April 2021
Published: 1 May 2021

Publisher's Note: MDPI stays neutral with regard to jurisdictional claims in published maps and institutional affiliations.

Copyright: © 2021 by the authors. Licensee MDPI, Basel, Switzerland. This article is an open access article distributed under the terms and conditions of the Creative Commons Attribution (CC BY) license (https://creativecommons.org/licenses/by/4.0/).

1. Introduction

Poria cocos (Schw.) Wolf is a fungus that is usually found on the roots of pine trees that grow in mountainous and hilly regions in the southern China. It is a traditional herb and is commonly used for treating insomnia, urinary dysfunction, and cancer [1]. Pharmacological studies have identified that the functionality and biomedical activities of *poria cocos* should be attributed to water-soluble polysaccharides, including (1,3)-β-D-glucan, (1,3)-α-D-glucan (1,3)-β-D-glucose, and more [2]. Nowadays, poria cubes or slices are consumed as an important daily functional food for its health benefit.

Due to the high moisture content and perishability, freshly harvested *poria cocos* is vulnerable to quality deterioration and spoilage. Drying is an essential operation in the postharvest processing of *poria cocos* products because it prevents microorganisms to grow and extends shelf life. The quality attributes of poria cubes are usually affected during the drying process. Particularly, dried poria cubes that are broken or cracked usually have particles diameter of less than 5 mm and much lower market prices than intact ones [3]. In addition, dried products with low contents of water-soluble polysaccharides have fewer nutritional values, which are less acceptable to consumers.

Hot air drying (HAD) is one of the most commonly used drying methods for poria cubes due to its low equipment cost and simplicity. However, it causes the non-uniform distribution of heat and moisture in materials, thus resulting in the breaking and cracking of poria cubes. The HAD process is also time-consuming and energy-intensive, which leads to low processing throughputs and high production costs [4,5]. The demands for higher drying efficiency, lower energy consumption, and better product quality have led to increasing interests in developing alternative drying methods for poria cubes.

Compared to the HAD, the air impingement drying (AID) process is worth examining. Therein, hot air impinges on the surface of materials at a high velocity, enhancing heat and mass transfer rates and rapidly removing the water from the material's surface [6]. Deng et al. [7] observed that hot air impingement drying (AID) significantly reduced the thickness of the thermal boundary layer on the surface of pepper and shortened its drying time. Xiao et al. [8] considered AID as one of the fastest drying methods for materials with small dimensions. Moreover, the reduction in drying time was beneficial for saving energy during the drying process.

For infrared (IR) drying, thermal energy is transferred from the heating element to a product's surface via radiation. Consequently, IR drying has the advantages of uniform heating, a low processing time, and a high heat transfer rate [9]. However, because of the limited penetration depth, IR may not be used as a single heating source during drying processes [10]. IR-assisted hot air impingement drying (IR-AID) adopts the advantages of both IR and AID, resulting as a uniform, rapid, and energy-efficient drying process [11]. Zhang et al. [12] reported that IR-AID was a promising drying technology for high-quality sponge gourd slices, as it significantly reduced its energy consumption. Supmoon [13] showed that IR-AID led to a higher drying rate, less shrinkage, lower hardness, and less color deterioration for the production of healthy potato chips. However, the study of IR-AID for poria cubes is still limited [3].

The presence of oxygen in the atmosphere during air drying usually has negative effects on the quality attributes of poria cubes, particularly causing the oxidation of water-soluble polysaccharides [14]. Therefore, vacuum drying (VD) may preserve the quality of poria cubes due to its oxygen-deficient drying conditions. However, the VD process is slow and consumes large amounts of energy, which may not be economically feasible for large-scale productions. Therefore, applying a two-stage drying method (i.e., VIR-AID) by using the VD and IR-AID may fuse the advantages of both drying methods, creating a synergistic effect for efficiently drying poria cubes with improved quality. The VIR-AID drying strategy method was proposed as a promising efficient drying method for poria cubes. Response surface methodology (RSM) is a widely used approach for optimizing [15] the drying processes of different types of foods, such as mushrooms [16], bee pollen [17], and potato slices [18]. Currently, there is no available information in the literature about the optimization of VIR-AID drying conditions for poria cubes.

The knowledge of drying kinetics is important for determining suitable operating conditions for improved drying efficiency [19]. Dincer and Hussain [20] developed a semi-analytical drying model with new drying parameters, namely drying coefficients and lag factors, which considered both external and internal resistances to moisture transfer. Mass transfer parameters—such as Biot (Bi) number, effective moisture diffusivity (D_{eff}), and moisture transfer coefficient (k)—during the drying process could be determined in a simple and accurate manner. The drying model has been successfully applied to the drying kinetics of yam slices [21], apple slices [22], and passionfruit peel [23].

Therefore, the firmness, integrity, and water-soluble polysaccharide contents of dried poria cubes are used as key quality indicators in the poria processing industry. The objectives of this study were to: (1) study the drying characteristics, drying kinetics, product qualities, and energy consumption of poria cubes under HAD, AID, IR-AID, VD, and VIR-AID; and to (2) optimize the operating conditions of the VIR-AID method in order to minimize the total drying time, broken ratio, and SEC, as well as to maximize the water-soluble polysaccharide content via RSM.

2. Materials and Methods

2.1. Raw Material

Freshly harvested *poria cocos* produced in Jinzai (Anhui, China) were provided by Qiaokang Technology Co., Ltd. Ripe *poria cocos* were manually dug out of the soil. The *poria cocos* with a uniform shape and weight of 2.9 ± 0.3 kg were selected for performing the experiments and stored in a refrigerator at 4 °C to maintain freshness and ensure consistency. Poria cubes were produced with a procedure shown in Figure 1. Specifically, whole fresh *poria cocos* were stored in a stainless-steel tub at room temperature for 2 h to reach a uniform initial temperature of 24 ± 2 °C. Then, the brown peel of *poria cocos* at the outer layer was manually removed with a knife during the first and second peeling processes. The white *poria cocos* chunks were obtained and covered with a polyethylene plastic film to avoid moisture loss. Poria cubes with the side length of 14 mm were produced using a cookie-cutter (Yuexi Mechanical Technology Co. Ltd., Jinzai, Anhui, China). Finally, the poria cubes without damaged edges and corners were selected for the drying experiments. The initial moisture content (MC_0) of the poria cubes was 0.51 ± 0.01 kg·kg^{-1} on a wet basis (1.04 kg·kg^{-1} on dry basis), which was determined with a vacuum oven drying at 75 °C for 24 h according to the Official Analytical Chemists method no.934.06. Triplicate measurements were conducted.

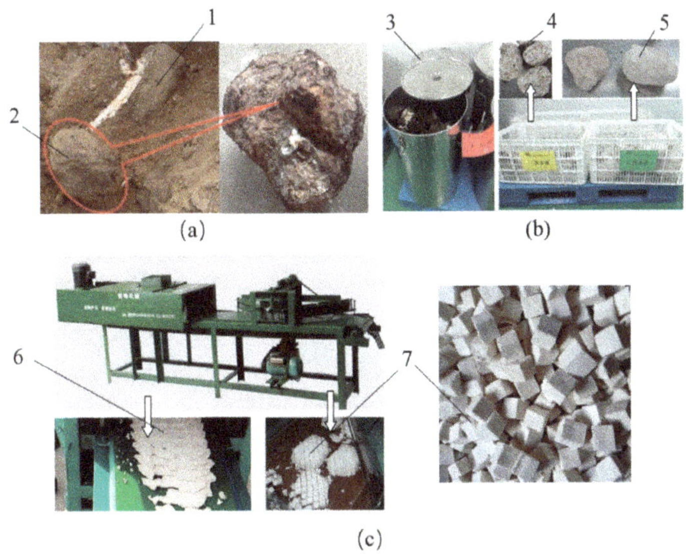

Figure 1. Production procedures of fresh poria cubes: (**a**) Raw material; (**b**) peeling process; (**c**) slicing and cubing processes. (1) Pine tree; (2) *poria cocos*; (3) stainless-steel container; (4) *poria cocos* after first peeling; (5) *poria cocos* chunks after second peeling; (6) poria slices; (7) poria cubes.

2.2. Drying Equipment

Drying tests were carried out using four high-precision computer-controlled drying systems at the processing workshop of Qiaokang Technology Co., Ltd. processing workshop (Anhui, China). As shown in Figure 2, digital load cell systems with a precision of 0.01 g (HYLF-010, Meikong, Hangzhou, China) were installed in the four dryers to track the weight change and moisture loss of samples during the drying process. The energy consumptions of different drying processes were measured by a watt-hour meter (DTSU1717-4P, HangLong, Shanghai, China). The change of weight and energy consumption were monitored and recorded by the logging system automatically and continuously.

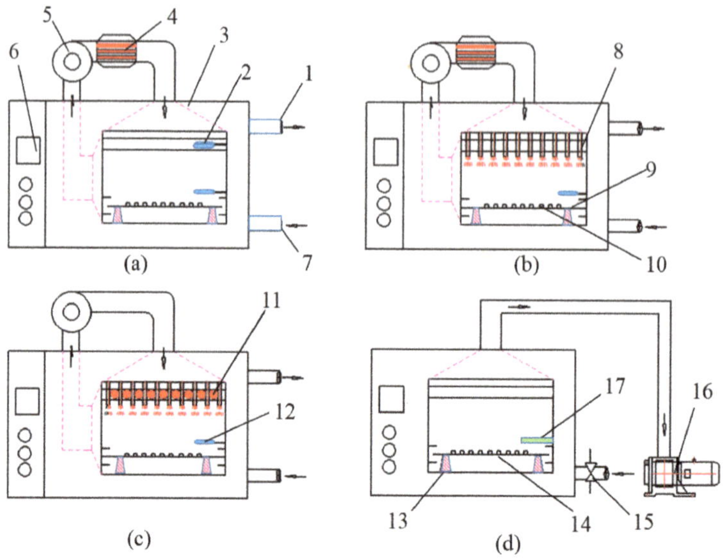

Figure 2. Schematic diagrams of lab-scale drying system: (**a**) HA dryer; (**b**) AID dryer; (**c**) IR-AID; and (**d**) VD dryer. (1) Air outlet; (2) temperature and humidity sensor; (3) air distribution chamber; (4) electric assistance; (5) centrifugal fan; (6) control system; (7) air inlet; (8) nozzles; (9) trays; (10) poria cubes; (11) IR emitters; (12) drying temperature sensor; (13) load cell; (14) electric heating board; (15) fast solenoid valve; (16) vacuum pump; (17) pressure sensor.

The HA dryer shown in Figure 2a was mainly comprised of an electric heater and an air conditioning unit. An electric heater was installed between the fan and drying chamber to heat the air to the target temperature. A centrifugal fan was installed above the dryer to draw air from the air inlet and blow hot air into the drying chamber via an air distribution chamber. The airflow was measured using an anemometer (SW-6086, Suwei, Hangzhou, China) and adjusted by an inverter (EV4300, Taida, Shanghai, China). A thermocouple with an accuracy of ±0.1 °C was installed on the sample tray to measure and control the air temperature. As shown in Figure 2b,c, the air impingement system in the AID dryer and IR-AID dryer had similar configurations. Compared with the HA dryer, a series of round nozzles were installed in the air distribution chamber of the AID and IR-AID dryer. The centrifugal fan delivered the vertical airflow with 0.6 $m \cdot s^{-1}$ velocity, which was speed up to 3.0 $m \cdot s^{-1}$ as it passed through the nozzle and impinged the poria cubes. The distance between the nozzles and drying samples was fixed at 120 mm. Since the vertical airflow affected the load cell system's measuring precision, the centrifugal fan stopped for 20 s when the sample was weighed every 15 min. For the IR-AID dryer, a row of infrared emitters (IR-0H03, Hongyao, Jiangshu, China) was installed on top of the tray to ensure uniform heating.

As shown in Figure 2d, the VD system mainly consisted of a drying chamber and a vacuum pump (2BV2070, Bosan, Shandong, China). The drying temperature of the electric heating board was controlled by a PID controller (model E5CN, Omron, Tokyo, Japan) with a sensitivity of ±0.5 °C. Chamber pressure was maintained at 5 ± 3 kPa and measured with a pressure sensor (MIK-P300, Meikong, Hangzhou, China). It had an accuracy of ±2 kPa. The boiling point of water under 5 kPa was nearly 33 °C [16].

2.3. Drying Experiments

The pre-drying temperature in all four drying scenarios was 65 °C. Before beginning the drying experiments, four dryers ran for 30 min to achieve a stable temperature. In each drying test, 5000 ± 5 g poria cubes were spread in a single layer on a stainless-steel tray with a loading capacity of 5 kg·m^{-2}. The samples were dried to a final moisture content of 0.17 ± 0.005 kg·kg^{-1} on a dry basis for safe storage [24].

All experiments were repeated three times. After drying was finished, the dried poria cubes were cooled in a desiccator for 30 min. Then, they were vacuum-sealed in polyethylene bags to prevent moisture absorption. Further, they were stored in a refrigerator (4 °C) for no longer than 3 days.

2.4. Drying Kinetics

The moisture content (MCs) of poria cubes on a dry basis was determined in a vacuum oven at 75 °C for 24 h according to the official analytical chemists method No. 934.06. The moisture ratio (MR) and drying rate (DR) were calculated using the following equations [7]:

$$\mathrm{MR} = \frac{MC_t - MC_e}{MC_0 - MC_e} \tag{1}$$

$$\mathrm{DR} = \frac{MC_{t_1} - MC_{t_2}}{t_1 - t_2} \tag{2}$$

where MC_t is the moisture content on a dry basis at a particular drying time t, kg·kg^{-1} dry mass. MC_0 is the initial MC on dry basis, kg·kg^{-1} dry mass. MC_e is the equilibrium MC under different drying methods, around 0.08 kg·kg^{-1} dry mass. MC_{t1} and MC_{t2} are the MCs of *poria cocos* cubes on a dry basis at drying times t_1 and t_2, respectively, kg·kg^{-1} dry mass. t_1 and t_2 are the minimum drying times.

Fick's second law is widely used to describe the effective moisture diffusivity (D_{eff}) with which one can make assumptions about neglected shrinkage, constant temperature, and diffusion coefficients, as well as uniform initial moisture distribution. D_{eff} can be calculated using the following equations [8]:

$$\mathrm{MR} = \frac{8}{\pi^2}\sum_{n=0}^{\infty}\frac{1}{(2n+1)^2}\exp\left[-\frac{(2n+1)^2\pi^2 D_{eff}}{4L^2}t\right] \approx \frac{8}{\pi^2}\exp(-\frac{\pi^2 D_{eff}}{4L^2}t) \tag{3}$$

$$K_0 = -\frac{\pi^2 D_{eff}}{4L^2} \tag{4}$$

where D_{eff} is the effective moisture diffusivity, m^2·s^{-1}. n is an integer. L is the sample thickness, m. t is the drying time, s. K_0 is calculated by plotting ln (MR) versus time t.

2.5. Dincer Drying Model

The drying kinetics of poria cubes under different drying methods were studied using the Dincer model. Specifically, the experimental data of *MR*s were inserted into Equation (3).

According to the Dincer model, *MR* could be expressed in the exponential equation as follows [20]:

$$MR = G\cdot\exp(-St) \tag{5}$$

where G is the lag factor, indicating the resistance of moisture transfer during drying, and S is the drying coefficient, 1·s^{-1}, indicating the drying capability of wet materials.

The effective moisture diffusivity (D_{eff}, m^2·s^{-1}) was calculated based on the developed relationship [21]:

$$D_{eff} = S(L/\mu_1)^2 \tag{6}$$

The coefficient μ_1 was determined by the following equation [25]:

$$\mu_1 = -419.24G^4 + 2013.8G^3 - 3615.8G^2 + 2880.3G - 858.94 \tag{7}$$

The mass transfer Biot number was determined using the *Bi-G* correlation [26]:

$$Bi = 0.0576G^{26.7} \tag{8}$$

The moisture transfer coefficient (k, m·s^{-1}) was determined as follows [27]:

$$k = D \cdot Bi/L \quad (9)$$

The goodness of fir for the Dincer model was evaluated based on the adjusted coefficient of determination (*adj*-R^2), reduced chi-square (χ^2), and the root mean square error (RMSE) [28]. The qualified fit should have the highest *adj*-R^2 and lowest χ^2. The nonlinear curve that fit to the experimental data was performed using the MATLAB software (Math Works Inc., Model-R2013a, Natick, MA, USA).

$$adj\text{-}R^2 = 1 - \frac{\left(1 - \frac{\sum_{i=1}^{N}(MR_{pre,i} - MR_{exp,i})^2}{\sum_{i=1}^{N}(MR_{pre,i} - \overline{MR})^2}\right)(N-1)}{N-k-1} \quad (10)$$

$$\chi^2 = \frac{\sum_{i=1}^{N}(MR_{exp,i} - MR_{pre,i})^2}{N-n} \quad (11)$$

$$RMSE = \left[\frac{\sum_{i=1}^{N}(MR_{exp,i} - MR_{pre,i})^2}{N}\right]^{1/2} \quad (12)$$

where $M_{pre,i}$ is the ith predicted MR, $M_{exp,i}$ is the ith experimental MR, $\overline{MR}$ is the mean of the experimental MR, and N is number of observations. N and k are the number of constants and independent variables in the drying model, respectively.

2.6. Response Surface Methodology

Based on the drying experiments, a two-stage drying strategy using VD followed by IR-AID was proposed as a promising and efficient drying method for poria cubes. In the first stage, VD was applied to dry the poria cubes to a critical moisture ratio (MR) before the cubic shape started to significantly change. Then, the IR-AID method was used to complete the drying in the second stage.

A three-level Box–Behnken experimental design with three factors was applied to study the two-stage VIR-AID process of poria cubes. The heating temperature during the VD stage, the sample moisture ratio (MR) at the time of switching the drying method, and the heating temperature during the IR-AID stage were selected as our independent experimental variables. The factor levels were selected as follows: VD temperature (T_{VD}, 65–85 °C), MR at the switch point (MR_{switch}, 70–90%), and IR-AID temperature ($T_{\text{IR-AID}}$, 65–85 °C). The response variables were the overall drying time (VD time + IR-AID time), broken ratio, water-soluble polysaccharide content, and SEC of poria cubes. The experimental design consisted of 17 experiments with 5 replicates at the central point. As shown in Table 1, the independent variables were coded with values of −1, 0, and 1.

Table 1. Coded and actual values of independent variables.

Drying Parameters		Coded Level of Variables		
		−1	0	+1
T_{VD} (°C)	X_1	65	75	85
MR_{switch} (%)	X_2	70	80	90
$T_{\text{IR-AID}}$ (°C)	X_3	65	75	85

The second order polynomial coefficients were calculated and analyzed using the Design Expert software v.10 Trial (Stat-Ease, Minneapolis, MN, USA). The general form of the second-degree polynomial was as follows [5]:

$$Y = \beta_0 + \sum_{i=1}^{n}\beta_i x_i + \sum_{i=1}^{n}\beta_{ii} x_i^2 + \sum_{i=1}^{n-1}\sum_{j=i+1}^{n}\beta_{ij} x_i x_j \quad (13)$$

where Y is the predicted response, x_i and x_j are input variables input variables that influence the response variable Y; β_0 is the offset term; β_i, β_{ii}, and β_{ij} are the linear coefficient, quadratic coefficient, and interaction coefficient, respectively.

The numerical optimization of process variables based on multiple responses was performed using the Design Expert software. Desired goals (minimization of drying time, broken ratio, SEC, and the maximization of water-soluble polysaccharide content) were used to perform optimization of factors and the response.

2.7. Broken Ratio

The dried poria cubes were sorted by using a vibrating sieve (MGSXJ-12W, Guanghe, China), as shown in Figure 3. Based on the dimensions of samples, the dried poria cubes were classified into three grades: I, II, and III. The shape of grade I products was nearly cubic with slight side and corner damage. For grade II, obvious breakage and cracking were observed. Broken particles with diameters less than 5 mm were defined as grade III products.

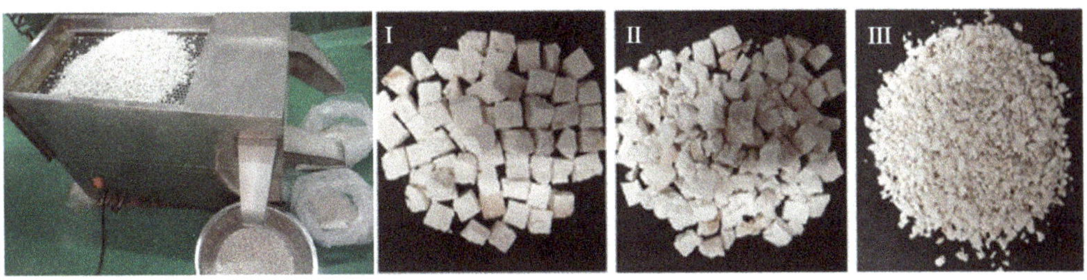

Figure 3. Vibrating sieve and typical images of I, II, and III grade products.

The broken ratio (μ,%) was described using the following equation:

$$\mu = \frac{m_{\mathrm{II}} + m_{\mathrm{III}}}{m_{\mathrm{I}} + m_{\mathrm{II}} + m_{\mathrm{III}}} \tag{14}$$

where m_{I}, m_{II}, and m_{III} were the sample masses (kg) of I, II, and III grade poria cubes, respectively.

2.8. Firmness

The firmness of poria cubes was determined via the compression test using a texture analyzer (TAPlus, Godalming, Surrey, UK). After sorting, the grade I poria cubes were randomly selected for the compression test. The diameter of two cylindrical probes was 35 mm and the compression parameters were as follows: maximum load 500 N, compression distance 8 mm, trigger force 0.1 g, pre-test speed 5 mm·s^{-1}, test speed 1 mm·s^{-1}, and post-test speed 5 mm·s^{-1}. The force-distance curve was recorded and the load force of the first peak was defined as the firmness. The compression tests for each drying condition were repeated 30 times and the average value was calculated and reported.

2.9. Water-Soluble Polysaccharide Content

The water-soluble polysaccharide content was determined via the phenol-sulfuric acid method by using D-glucose as the standard [29]. Specifically, dried poria cubes were ground up and passed through a 60-mesh screen. The powder (0.5 ± 0.001 g) was mixed with 25 mL distilled water in a 50 mL volumetric flask. The suspension was then heated in a 60 °C water bath with ultrasonic for 30 min to accelerate the extraction rate of polysaccharides, and then cooled to room temperature. The solution was vacuum filtrated to remove the solid residual. The solid residual was then thrice washed with distilled water. The washed water was then combined with the filtrated solution and diluted to

100 mL (v_1). Next, 0.5 mL(v_2) of filtrated solution was pipetted into a 10 mL tube and 1 mL of 5% phenol was added. Then, the mixture was shaken for 2 min. Further, 5 mL of sulfuric acid (98% $v \cdot v^{-1}$) was added into the mixture and shaken for 5 min. The absorbance at 490 nm was then determined using a UV-Vis spectrophotometer (Shimadzu UV-2600, Hangzhou, Zhejiang, China). Distilled water was used as a blank. The total water-soluble polysaccharides content (φ, mg·g^{-1}) of the dried poria cubes was calculated with the following equation:

$$\phi = \frac{200 \cdot f}{m_1} \cdot (\frac{y - 0.0882}{7.5486}) \tag{15}$$

where y is the absorbance of the filtrated solution at 490 nm; m_1 is the mass of the poria powder sample, 5 g; and f is the conversion factor, 1.28 [14].

2.10. Specific Energy Consumption

The specific energy consumption (SEC, MJ·kg^{-1}), which was the energy needed to remove 1 kg of water from poria cubes was calculated using Equation (16) [28]:

$$SEC = \frac{E}{m_{water}} \tag{16}$$

where E is the energy consumed during drying, MJ, and m_{water} is the mass of moisture removal during the drying, kg.

2.11. Statistical Analysis

The experimental data were calculated as means ± standard deviation (SD). Analysis of variance (ANOVA) was performed using Duncan's multiple range test with a significance level of 0.05. The F test was carried out to assess the homoscedasticity of the residuals by using the Design Expert software v.10 trial (Stat-Ease).

3. Results and Discussion

3.1. Drying Characteristics of Poria Cubes under Different Drying Methods

As shown in Figure 4a, the *MR* of poria cubes exponentially decreased with drying time, which was commonly observed during the food drying process [30]. The times required to dry the poria cubes from their initial *MCs* (1.04 kg·kg^{-1} on dry basis) to the desired final *MCs* (0.17 kg·kg^{-1} on dry basis) for HAD, AID, IR-AID, and VD were 400 min, 240 min, 185 min, and 340 min, respectively. Compared with the HAD, the AID and IR-AID reduced the drying time by 32.5% and 46.3%, respectively. Such results should mainly be attributed to air impingement, which sped up the airflow and significantly decreased the thicknesses of the heat and mass transfer boundary layer between the material surface and air, thus improving the drying rate. The results showed similar trends as reported in previous studies on the drying of Monukka seedless grapes [31]. In addition, the drying time when using IR-AID was 55 min shorter than when using AID, which could be attributed to the improved heating intensity and rate of moisture evaporation due to IR heating.

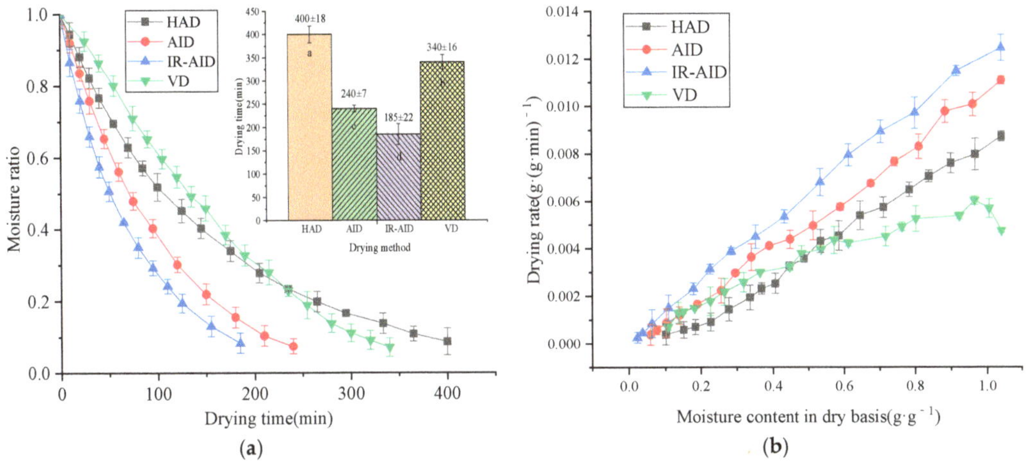

Figure 4. *MR* curves and drying rate curves of poria cubes under different drying methods: (**a**) *MR* curves; (**b**) drying rate curves.

The drying rate of VD was lower than HA drying during the early stage. It then became higher later. The VD drying time was shorter than the HA drying time. Such results were mainly due to the shape of the poria cubes, which remained intact during VD due to the lower temperature gradient and less thermal strain. While under HA, AID, and IR-AID, the poria cubes showed apparent cracking and braking, which contributed to more exposed surfaces for moisture evaporation when drying began. Similar findings were observed by Seremet et al. [32], who found that pumpkin slice structure changed under different drying methods, significantly affecting drying time. As shown in Figure 4b, drying occurred mainly in the falling-rate stage, except for the VD, where a short period of acceleration was observed when drying began. The moisture inside the poria cubes could be classified into intercellular water (loosely bonded water) and intracellular water (strongly bound water) [33]. The higher drying rates during the initial drying stages could be due to the abundant intercellular water in the poria cubes. During later drying stages, the drying rates were mainly controlled by the diffusion of intracellular water, which had a higher resistance to be removed when dried. Similar trends were observed in the HAD of apple and potato cubes, according to Khan et al. [34].

3.2. Drying Kinetics

The Dincer model (Equation (5)) was used to fit the drying curves under different drying methods. The lag factors (*G*) and drying coefficients (*S*) were determined and summarized (Table 2). The high values of $adj\text{-}R^2$ and low values of χ^2 and RMSE indicated good fits of the model. The lag factor (*G*) was directly related to the mass transfer Biot number (*Bi*), as expressed in Equation (8). *Bi* is one of the most important dimensionless numbers in drying that represents the magnitude of resistance to moisture diffusion inside the material. $Bi < 0.1$ indicates negligible internal resistance, whereas $Bi > 100$ indicates negligible external resistance. In general, *Bi* numbers were ranged between $0.1 < Bi < 100$ [20]. In this study, the calculated *Bi* values ranged between 0.1030 to 0.5626, indicating that both internal and external resistance to the moisture transfer existed during the poria cubes drying processes [22]. The drying coefficients (*S*) varied from 1.036×10^{-4} to 2.197×10^{-4} ($1{\cdot}s^{-1}$), which represented the drying capability of poria cubes. The infrared heating and hot air impingement drying enhanced drying capability, as indicated by the higher *S* values in AID and IR-AID.

Table 2. Drying kinetics parameters of the Dincer model and Fick's second law.

Drying Methods	Drying Constant		D_{eff} ($m^2 \cdot s^{-1}$)	D^*_{eff} ($m^2 \cdot s^{-1}$)	Bi	k ($m \cdot s^{-1}$)	$Adj\text{-}R^2$	χ^2	$RMSE$
	G	$S \times 10^{-4}$ (s^{-1})							
HAD	1.089	1.036	8.148×10^{-9}	7.531×10^{-9}	0.5626	6.548×10^{-7}	0.996	1.92×10^{-4}	0.014
AID	1.023	1.688	5.768×10^{-8}	1.310×10^{-8}	0.1057	8.701×10^{-7}	0.998	1.32×10^{-4}	0.011
IR-AID	1.022	2.197	7.901×10^{-8}	3.767×10^{-8}	0.1030	1.162×10^{-6}	0.997	0.44×10^{-4}	0.006
VD	1.087	1.044	8.401×10^{-9}	8.335×10^{-9}	0.5320	6.385×10^{-7}	0.995	2.83×10^{-4}	0.017

Note: D^*_{eff} is the effective moisture diffusivity calculated by Fick's second law.

D_{eff} characterizes the rate of moisture diffusion within the food material [35]. As shown in Table 1, the magnitudes of D_{eff} during the poria cube drying processes under HAD, AID, IR-AID, and VD were 8.148×10^{-9}, 5.768×10^{-8}, 7.901×10^{-9}, and 8.401×10^{-9} $m^2 \cdot s^{-1}$, respectively. Meanwhile, the mass transfer coefficient (k) ranged from 6.385×10^{-7} to 1.162×10^{-6} $m \cdot s^{-1}$. Among the four drying methods, the highest effective moisture diffusivity was obtained from the IR-AID process, which showed the best potential to improve the drying efficiency of poria cubes. The D_{eff} calculated by Fick's second law ranged from 8.335×10^{-9}~3.767×10^{-8} $m^2 \cdot s^{-1}$, which had a similar trend compare with the results of the Dincer drying model. The D_{eff} was in the order of IR-AID > AID > VD > HAD. The difference of D_{eff} determined from the Dincer and Fickian models may have been caused by the different mathematical models. This phenomenon was consistent with the drying of kiwifruit slices. Moreover, the specific values of D_{eff} calculated via the Dincer drying model and Fick's second law are not the same [25].

3.3. Broken Ratio and Firmness

The broken ratio was used to characterize the degree of structural destruction of dried poria cubes. Usually, the firmness of a material had a major influence on the broken ratio, and dried products with higher firmness usually had a lower broken ratio [36]. Figure 5a,b shows the broken ratio and firmness of poria cubes obtained from the four drying methods. The average broken ratios of dried poria cubes were 50.3%, 64.7%, 44.6%, and 3.4% under HAD, AID, IR-AID, and VD, respectively. The average firmness values were 9.94, 6.91, 14.65, and 43.48 kg for HA, AID, IR-AID, and VD, respectively. Meanwhile, Figure 5c shows the appearance of the dried poria cubes as a result of different drying methods. The results showed that poria cubes dried via the AID method had the highest broken ratio and lowest firmness. That was due to rapid airflow, which caused the material's surface to quickly dry and stress contraction on the sample surface to intensify in the early stage of drying. The non-uniform distribution of temperature and moisture in poria cubes led to severe surface cracking. The broken ratio of IR-AID was nearly 20% lower than that of the AID samples. Such results should be attributed to the improvement of heating uniformity within poria cubes via IR heating [37]. The gradients of temperature and moisture within poria cubes were reduced and thus cracking was mitigated. The broken ratio of HAD dried samples was nearly 10% lower than that from the AID, which was mainly due to the lower drying rates and less moisture gradient within the poria cubes. The lowest broken ratio and highest firmness were obtained in VD samples. As shown in Figure 5c, the VD samples had intact cube shapes without obvious damaged edges or corners. The results should mainly be attributed to the fact that the material temperature was typically low, as vacuums tend to reduce the boiling point of water or solvents, thus reducing thermal stresses and over-drying caused by high material temperature [38]. This improved the firmness and texture of dried poria cubes. Compared with the other three drying methods, the lower drying rate during VD also decreased the moisture gradient and the internal stress concentration in poria cubes. Due to the significant reduction of the boiling point of water, VD has been widely applied to eliminate internal fissures and bubbles in drying samples [39].

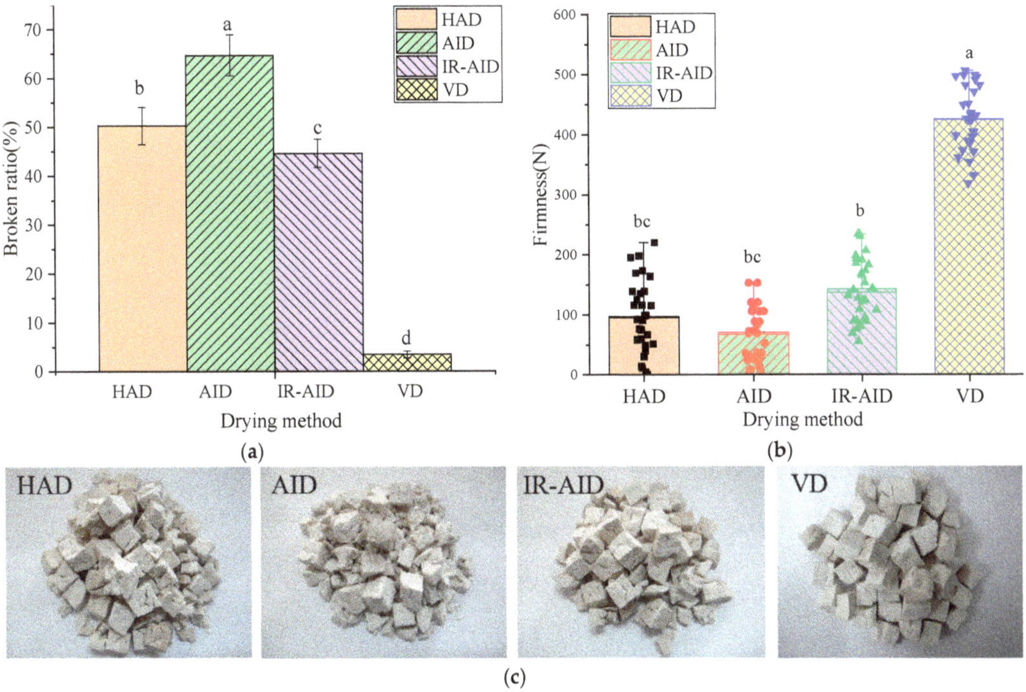

Figure 5. Broken ration, firmness, and images of poria cubes under different drying methods: (**a**) Broken ratio; (**b**) firmness; (**c**) images of poria cubes under different drying method. Different lowercase superscripts denote significant differences ($p < 0.05$).

The better integrity and higher firmness of VD-dried poria cubes should also be attributed to the uniform volume shrinkage during the drying process [40]. During the food drying process, volumetric shrinkages mainly occurred in two stages: major shrinkage occurred during the preheating and constant-rate drying period, and minor shrinkage subsequently occurred during the falling-rate drying period [41]. During the initial drying stage of VD, the moisture rapidly evaporated and the volume shrinkage of the poria cubes was proportional to the volume of moisture evaporation [42]. The end of the first stage was marked by a critical *MR* in the poria cubes. Below the critical *MR*, the volumetric change of poria cubes was minor and no longer proportional to the volume of moisture evaporation, and the cracking and breaking were also minor. At this stage, the VD should be switched to other rapid drying methods to reduce the overall drying time [43].

3.4. Retention of Water-Soluble Polysaccharide Contents

Figure 6 shows the contents of water-soluble polysaccharide in the dried poria cubes under different drying methods. HAD resulted in the lowest water-soluble polysaccharide content of 2.62 mg·g^{-1}. The water-soluble polysaccharide contents in the AID and IR-AID dried samples were 13.26% and 52% higher than that of HAD. Such results could be due to the prolonged exposure to oxygen during the air drying (HAD, AID, and IR-AID), causing the oxidative degradation of water-soluble polysaccharides [44]. During VD, the exposure of poria cubes to oxygen was limited, which was helpful for the retention of water-soluble polysaccharides and preservation of bioactive activities. Such results showed similar trends as reported by Yan et al. [45], who found that vacuum drying benefited the retention of water-soluble polysaccharides in bitter gourd (*Momordica charantia* L.) slices. Liu et al. [46] also proved that the physicochemical and biological activities of polysaccharides in Lentinula were strongly dependent on the type of drying method deployed.

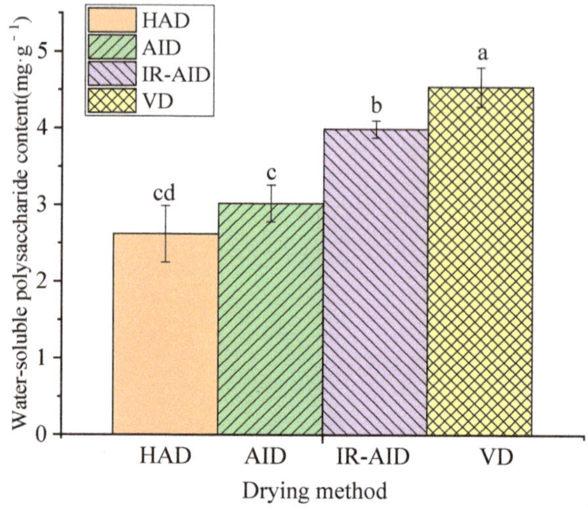

Figure 6. Water-soluble polysaccharide content of different drying methods. Different lowercase superscripts denote significant differences ($p < 0.05$)

3.5. *Specific Energy Consumption*

As shown in Figure 7, the SEC of the HAD, AID, IR-AID, and VD processes were 2.82, 2.33, 1.38, and 6.21 MJ·kg^{-1}, respectively. The VD process resulted in the highest SEC, which was likely due to the longer drying time and high energy consumption of the vacuum pump. It is worth noting that IR-AID consumed the lowest energy, which was directly associated with its shortest drying time. Besides, IR radiation could be directly absorbed by samples with minimum ambient, which usually leads to higher energy efficiency [47]. Chen et al. [48] reported that infrared heating could be used in conjunction with HA drying to improve drying efficiency and reduce energy consumption of carrot snacks. VD resulted in the highest retention of water-soluble polysaccharides and the best integrity in poria cubes. IR-AID led to the lowest energy consumption and drying time. Therefore, a two-stage drying strategy using VD and followed by IR-AID was proposed for poria cubes.

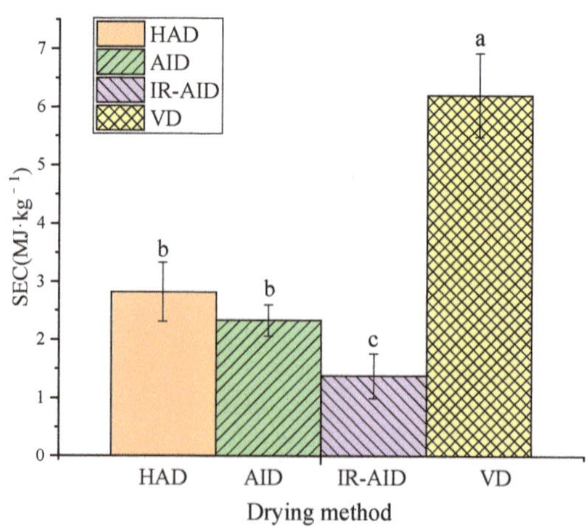

Figure 7. Specific energy consumption of different drying methods. Different lowercase superscripts denote significant differences ($p < 0.05$).

3.6. Two-Stage Vacuum and Infrared-Assisted Hot Air Impingement Drying

Based on the results of the single factor experiments, a three-level Box–Behnken experimental design with three factors was applied to study the two-stage VD and IR-AID processes of poria cubes. As shown in Table 3, the total drying time, broken ratio, water-soluble polysaccharide content, and SEC ranged from 172–343 min, 3.79–44.20%, 1.25–4.45 mg·g^{-1}, and 1.24–4.04 MJ·kg^{-1}, respectively.

Table 3. Box–Behnken experimental design with natural and coded drying conditions and values of response variables.

No.	Drying Conditions			Response			
	T_{VD} (°C)	MR_{switch} (%)	$T_{IR\text{-}AID}$ (°C)	Total Drying Time (min)	Broken Ratio (%)	Water-Soluble Polysaccharide Content (mg·g^{-1})	SEC (MJ·kg^{-1})
1	85	90	75	187	43.50	6.77	1.88
2	75	70	85	287	5.23	10.69	2.97
3	75	80	75	245	22.10	8.10	2.25
4	75	90	85	172	44.20	4.46	1.24
5	65	70	75	325	4.35	14.54	4.04
6	85	70	75	288	3.79	10.40	2.89
7	75	90	65	228	43.79	5.83	1.62
8	75	80	75	245	21.10	8.17	2.27
9	75	80	75	248	21.70	8.10	2.25
10	75	70	65	343	4.17	12.24	3.40
11	85	80	85	212	20.90	5.36	1.49
12	65	80	85	249	21.21	8.32	2.31
13	85	80	65	256	20.52	6.70	1.86
14	75	80	75	242	21.06	8.93	2.48
15	65	90	75	213	44.10	7.31	2.03
16	65	80	65	297	20.95	9.50	2.64
17	75	80	75	251	21.37	7.70	2.14

The response variables were fitted with second-order polynomial models using the least square method (MLS). The estimated regression coefficients, adjusted coefficient of determination (*Adj-R*^{2}), and coefficient of variation (*C.V.*) values are summarized in Table 4. The *Adj-R*^{2} values on total drying time, broken ratio, water-soluble polysaccharide content, and SEC were all higher than 0.97, indicating that the polynomial models fit well with the experimental results. Moreover, the lack of fit test results showed that there was no lack of fit for the models. Relatively low values (<10%) of C.V. for all response variables implied a high degree of precision and good reliability of the experimental values.

The regression equations between response variables and coded factors were also established. The total drying time (Y_1, min), broken ratio (Y_2, %), water-soluble polysaccharide content (Y_3, mg·g^{-1}), and SEC (Y_4, MJ·kg^{-1}) are shown in Equations (17)–(20).

$$Y_1 = 246.2 - 17.63x_1 - 55.37x_2 - 25.50x_3 + 2.75x_1x_2 + 1.00x_1x_3 + 1.52x_1^2 + 5.52x_2^2 + 5.78x_3^2 \quad (17)$$

$$Y_2 = 21.47 - 0.24x_1 + 19.76x_2 + 0.26x_3 - 0.01x_1x_2 + 0.03x_1x_3 - 0.16x_2x_3 - 0.49x_1^2 + 2.96x_2^2 - 0.079x_3^2 \quad (18)$$

$$Y_3 = 3.01 + 0.017x_1 - 0.052x_2 - 0.956x_3 + 0.17x_1x_2 + 0.41x_1x_3 - 0.21x_2x_3 - 0.27x_1^2 + 0.48x_2^2 - 0.2x_3^2 \quad (19)$$

$$Y_4 = 8.20 - 1.31x_1 - 2.94x_2 - 0.68x_3 + 0.9x_1x_2 - 0.036x_1x_3 + 0.045x_2x_3 + 0.36x_1^2 + 1.20x_2^2 - 1.09x_3^2 \quad (20)$$

where x_1 represents T_{VD}, °C, x_2 represents MR_{switch}, %, and x_3 represented $T_{IR\text{-}AID}$, °C.

Table 4. Analysis of variance (ANOVA) of the polynomial regression models for total drying time, broken ratio, water-soluble polysaccharide content, SEC.

Source	Total Drying Time (min)			Broken Ratio (%)			Water-Soluble Polysaccharide Content (mg·g^{-1})			SEC (MJ·kg^{-1})		
	DF *	Sum of Squares	*p*-Value	DF *	Sum of Squares	*p*-Value	DF *	Sum of Squares	*p*-Value	DF *	Sum of Squares	*p*-Value
Model	9	32,556.01	<0.0001	9	3160.99	<0.0001	9	9.71	<0.0001	9	100.69	<0.001
x_1	1	2485.13	<0.0001	1	0.45	0.1137	1	2.450×10^{-3}	0.6237	1	13.62	<0.0001
x_2	1	24,531.12	<0.0001	1	3122.48	<0.0001	1	0.022	0.1676	1	69.08	<0.0001
x_3	1	5202.00	<0.0001	1	0.56	0.0848	1	7.33	<0.0001	1	3.69	<0.0001
$x_1 \times x_2$	1	30.25	0.2892	1	4.000×10^{-4}	0.9586	1	0.12	0.0097	1	3.24	0.0033
$x_1 \times x_3$	1	4.00	0.6892	1	3.600×10^{-3}	0.8763	1	0.67	<0.0001	1	5.184×10^{-3}	0.0046
$x_2 \times x_3$	1	0.000	1.0000	1	0.11	0.4110	1	0.18	0.0029	1	8.100×10^{-3}	0.8745
x_1^2	1	9.79	0.5350	1	1.02	0.0300	1	0.30	0.0008	1	0.54	0.8435
x_2^2	1	128.53	0.0501	1	36.91	<0.0001	1	0.99	<0.0001	1	6.02	0.1374
x_3^2	1	140.42	0.0428	1	0.026	0.6750	1	0.17	0.0038	1	5.00	0.0008
Residual	7	161.05	-	7	0.97		7	0.065	-	7	1.35	-
Lack of Fit	3	114.25	0.1420	3	0.20	0.7903	3	0.034	0.3576	3	0.55	0.5047
Pure Error	4	46.80	-	4	0.76	-	4	0.031	-	4	0.80	-
Total	16	32,717.06	-	16	3161.96	-	16	9.78	-	16	102.04	-
Adj- R²	-	0.9887	-	-	0.9993	-	-	0.9848	-	-	0.9697	-
Pre- R²	-	0.9419	-	-	0.9986	-	-	0.9398	-	-	0.9009	-
Adeq Precision	-	43.968	-	-	141.174	-	-	43.991	-	-	30.168	-
C.V.%	-	1.90	-	-	1.65	-	-	3.20	-	-	5.22	-
PRESS	-	1901.12	-	-	4.44	-	-	0.59	-	-	10.11	-

Note: * DF represented degree of freedom; x_1 represented T_{VD}, °C; x_2 represents MR_{switch}, %; and x_3 represented $T_{IR\text{-}AID}$, °C.

The regression analysis results of Equation (17) showed that drying time was significantly affected by linear terms of T_{VD}, MR_{switch}, and $T_{IR\text{-}AID}$. An increase of VD temperature and IR-AID drying temperature, as well as a decrease of MR_{switch}, reduced the total drying time. The higher VD temperature and IR power could accelerate the water molecules present in the poria cubes, thus leading to faster evaporation and drying [12,16]. ANOVA results of Equation (18) suggested that MR_{switch} had a significant effect on the broken ratio of poria cubes during the VIR-AID drying process. The broken ratio of poria cubes decreased when MR_{switch} decreased. Table 3 also shows that when the MR_{switch} was higher than 80%, the broken ratio of products was higher than 20%. However, when MR_{switch} was 70%, the broken ratio of poria cubes was less than 5%. Such results should be due to the glass transition of poria cubes, from a rubbery state to a glassy state during the VD stage at an MR ranging between 70–80% [3]. Equation (19) showed that a lower VD temperature and MR_{switch} led to higher water-soluble polysaccharide content. This was consistent with the VD drying characteristics of poria cubes. Equation (20) indicated that SEC increased when $T_{IR\text{-}AID}$ increased and decreased when T_{VD} and MR_{switch}. In addition, the interaction terms between T_{VD} and MR_{switch} had a negative effect on SEC. Since the vacuum pump had a high energy consumption, the shorter VD drying time could significantly reduce energy consumption.

3.7. Numerical Optimization and Verification

Optimization targets minimized the total drying time, broken ratio, and SEC, as well as maximized the water-soluble polysaccharide content. The optimum operating conditions were obtained using Design Expert software (Stat-Ease). All responses were simultaneously optimized according to the goals and weights as presented in Table 5.

Table 5. Simultaneously optimized combined drying conditions with target and weight of investigated responses.

Response	Optimized Direction	Lower Limit	Upper Limit	Weight
Total Drying time (min)	Minimize	100	400	0.25
Broken ratio (%)	Minimize	2.0	50.0	0.25
Water-soluble polysaccharide content (mg·g^{-1})	Maximize	1.0	5.0	0.25
SEC (MJ·kg^{-1})	Minimize	1.0	5.0	0.25

As shown in Table 6, the optimal conditions obtained for the given criteria were VD at 82.17 °C until the MR of fresh *proia cocos* cubes were reduced to 81.11%. Then, the drying was switched to IR-AID at 69.04 °C. Validation experiments were performed for t oaf 82 °C, MR_{switch} at 81%, and $T_{IR\text{-}AID}$ at 69 °C. The total drying time, broken ratio, water-soluble polysaccharides content, and SEC were thus determined. Results showed the validation results were consistent with the predicted values, which proved the validity of optimization. Compared with IR-AID drying at 69 °C, the broken ratio of combined drying poria cubes decreased nearly 20%, and the drying time and SEC only increased by 58 min (0.80 MJ·kg^{-1}). Compared with VD drying at 82 °C, the SEC by the two-stage drying was only one-third of the VD process. The results indicated that the two-stage VD and IR-AID process was more effective in drying poria cubes with lower energy consumption and better product qualities.

Table 6. Prediction and validation results of response variables at the optimum condition.

Results	Operating Conditions			Response Variables			
	T_{VD} (°C)	MR_{switch} (%)	$T_{IR\text{-}AID}$ (°C)	Total Drying Time (min)	Broken Ratio (%)	Water-Soluble Polysaccharide Content (mg·g^{-1})	SEC (MJ·kg^{-1})
Prediction	82.17	81.11	69.04	245.29	23.07	3.24	2.01
Validation	82	81	69	255	24.88	3.32	2.04
IR-AID	—	100	69	197	53.34	3.34	1.24
VD	82	9	—	322	3.75	4.39	6.07

4. Conclusions

Fresh poria cubes were dried using the HAD, AID, IR-AID, and VD methods. To improve the drying efficiency and product qualities of poria cubes, as well as to reduce the energy consumption during the drying process, a two-stage drying was developed. It adopted advantages of VD and IR-AID. The operating conditions of the two-stage drying process were optimized as T_{VD} at 82 °C, MR_{switch} at 81%, and $T_{IR\text{-}AID}$ at 69 °C for the minimized drying time (255 min), broken ratio (24.88%), and SEC (2.04 MJ·kg^{-1}), as well as the maximized water-soluble polysaccharide content (3.32 mg·g^{-1}). The findings from this study indicate that the new two-stage VD and IR-AID is a promising technology for improving the drying efficiency and product quality, as well as for reducing the energy consumption of poria cubes during drying. The heating source and drying method usually affect physicochemical reactions and, ultimately, the nutritional value and sensory attributes of agricultural products. The new drying technologies, which used more than one drying method, was capable of producing products with the desired moisture content and drying quality in its final products.

Author Contributions: Conceptualization, W.Z., C.C., Z.P. and Z.Z.; methodology, C.C. and Z.P.; software, C.C.; validation, W.Z. and Z.Z.; formal analysis, W.Z. and C.C.; investigation, Z.Z. and Z.P.; resources, Z.Z; data curation, W.Z.; writing—original draft preparation, W.Z.; writing—review and editing, C.C. and Z.P.; visualization, Z.Z. and Z.P.; supervision, W.Z. and C.C.; project administration, W.Z. and Z.Z.; funding acquisition, W.Z., Z.Z., and Z.P. All authors have read and agreed to the published version of the manuscript.

Funding: This research was funded by the special fund support for the construction of modern agricultural industrial technology system (CARS-21); Natural Science Foundation of Beijing (No. 6204035); and the Excellent Youth Program of the Beijing Organization Department (2018000020124G034).

Institutional Review Board Statement: Not applicable.

Informed Consent Statement: Not applicable.

Data Availability Statement: All the results showed in the manuscript could be requested to the corresponding author who would provide them.

Acknowledgments: The authors would like to acknowledge the technical help from Nanjing Tengfei Drying Equipment Co. LTD, Nanjing, China, in helping to run the experiments.

Conflicts of Interest: The authors declare no conflict of interest.

References

1. Wu, K.; Fan, J.; Huang, X.; Wu, X.; Guo, C. Hepatoprotective effects exerted by Poria Cocos polysaccharides against acetaminophen-induced liver injury in mice. *Int. J. Biol. Macromol.* **2018**, *114*, 137–142. [CrossRef]
2. Zou, Y.T.; Zhou, J.; Wu, C.Y.; Zhang, W.; Shen, H.; Xu, J.D.; Zhang, Y.Q.; Long, F.; Li, S.L. Protective effects of Poria cocos and its components against cisplatin-induced intestinal injury. *J. Ethnopharmacol.* **2021**, *269*, 113722. [CrossRef]
3. Zhang, W.; Pan, Z.; Xiao, H.; Zheng, Z.; Chen, C.; Gao, Z. Pulsed vacuum drying (PVD) technology improves drying efficiency and quality of Poria cubes. *Dry. Technol.* **2017**, *36*, 908–921. [CrossRef]
4. Chen, C.; Venkitasamy, C.; Zhang, W.; Deng, L.; Meng, X.; Pan, Z. Effect of step-down temperature drying on energy consumption and product quality of walnuts. *J. Food Eng.* **2020**, *285*, 110105. [CrossRef]
5. Darvishi, H.; Farhudi, Z.; Behroozi-Khazaei, N. Multi-objective optimization of savory leaves drying in continuous infrared-hot air dryer by response surface methodology and desirability function. *Comput. Electron. Agric.* **2020**, *168*, 105112. [CrossRef]
6. Acar, C.; Dincer, I.; Mujumdar, A. A comprehensive review of recent advances in renewable-based drying technologies for a sustainable future. *Dry. Technol.* **2020**, 1–27. [CrossRef]
7. Deng, L.Z.; Mujumdar, A.S.; Zhang, Q.; Yang, X.H.; Wang, J.; Zheng, Z.A.; Gao, Z.J.; Xiao, H.W. Chemical and physical pretreatments of fruits and vegetables: Effects on drying characteristics and quality attributes—A comprehensive review. *Crit. Rev. Food Sci. Nutr.* **2019**, *59*, 1408–1432. [CrossRef] [PubMed]
8. Xiao, H.-W.; Bai, J.-W.; Xie, L.; Sun, D.-W.; Gao, Z.-J. Thin-layer air impingement drying enhances drying rate of American ginseng (Panax quinquefolium L.) slices with quality attributes considered. *Food Bioprod. Process.* **2015**, *94*, 581–591. [CrossRef]
9. Wang, H.-C.; Zhang, M.; Adhikari, B. Drying of shiitake mushroom by combining freeze-drying and mid-infrared radiation. *Food Bioprod. Process.* **2015**, *94*, 507–517. [CrossRef]
10. Salehi, F. Recent Applications and Potential of Infrared Dryer Systems for Drying Various Agricultural Products: A Review. *Int. J. Fruit Sci.* **2019**, *20*, 586–602. [CrossRef]
11. Lao, Y.; Zhang, M.; Chitrakar, B.; Bhandari, B.; Fan, D. Efficient Plant Foods Processing Based on Infrared Heating. *Food Rev. Int.* **2019**, *35*, 640–663. [CrossRef]
12. Zhang, Y.; Zhu, G.; Li, X.; Zhao, Y.; Lei, D.; Ding, G.; Ambrose, K.; Liu, Y. Combined medium- and short-wave infrared and hot air impingement drying of sponge gourd (Luffa cylindrical) slices. *J. Food Eng.* **2020**, *284*. [CrossRef]
13. Supmoon, N.; Noomhorm, A. Influence of Combined Hot Air Impingement and Infrared Drying on Drying Kinetics and Physical Properties of Potato Chips. *Dry. Technol.* **2013**, *31*, 24–31. [CrossRef]
14. Wan, J.Y.; Fan, Y.; Yu, Q.T.; Ge, Y.Z.; Yan, C.P.; Alolga, R.N.; Li, P.; Ma, Z.H.; Qi, L.W. Integrated evaluation of malonyl ginsenosides, amino acids and polysaccharides in fresh and processed ginseng. *J. Pharm. Biomed. Anal.* **2015**, *107*, 89–97. [CrossRef] [PubMed]
15. Ouakhssase, A.; Ait Addi, E. Mycotoxins in food: A review on liquid chromatographic methods coupled to mass spectrometry and their experimental designs. *Crit. Rev. Food Sci. Nutr.* **2020**, 1–21. [CrossRef]
16. Šumić, Z.; Tepić, A.; Vidović, S.; Vladić, J.; Pavlić, B. Drying of shiitake mushrooms in a vacuum dryer and optimization of the process by response surface methodology (RSM). *J. Food Meas. Charact.* **2016**, *10*, 425–433. [CrossRef]
17. Thakur, M.; Pant, K.; Naik, R.R.; Nanda, V. Optimization of spray drying operating conditions for production of functional milk powder encapsulating bee pollen. *Dry. Technol.* **2020**, *39*, 777–790. [CrossRef]
18. Onu, C.E.; Igbokwe, P.K.; Nwabanne, J.T.; Nwajinka, C.O.; Ohale, P.E. Evaluation of optimization techniques in predicting optimum moisture content reduction in drying potato slices. *Artif. Intell. Agric.* **2020**, *4*, 39–47.

19. Castro, A.M.; Mayorga, E.Y.; Moreno, F.L. Mathematical modelling of convective drying of fruits: A review. *J. Food Eng.* **2018**, *223*, 152–167. [CrossRef]
20. Dincer, I.; Hussain, M.M. Development of a new Biot number and lag factor correlation for drying applications. *Int. J. Heat Mass Transf.* **2004**, *47*, 653–658. [CrossRef]
21. Ju, H.-Y.; El-Mashad, H.M.; Fang, X.-M.; Pan, Z.; Xiao, H.-W.; Liu, Y.-H.; Gao, Z.-J. Drying characteristics and modeling of yam slices under different relative humidity conditions. *Dry. Technol.* **2015**, *34*, 296–306. [CrossRef]
22. Rajoriya, D.; Shewale, S.R.; Hebbar, H.U. Refractance Window Drying of Apple Slices: Mass Transfer Phenomena and Quality Parameters. *Food Bioprocess Technol.* **2019**, *12*, 1646–1658. [CrossRef]
23. Bezerra, C.V.; Meller da Silva, L.H.; Corrêa, D.F.; Rodrigues, A.M.C. A modeling study for moisture diffusivities and moisture transfer coefficients in drying of passion fruit peel. *Int. J. Heat Mass Transf.* **2015**, *85*, 750–755. [CrossRef]
24. Shi, H.; Zhang, M.; Devahastin, S. New Development of Efficient Processing Techniques on Typical Medicinal Fungi: A Review. *Food Rev. Int.* **2020**, *36*, 39–57. [CrossRef]
25. Mohammadi, I.; Tabatabaekoloor, R.; Motevali, A. Effect of air recirculation and heat pump on mass transfer and energy parameters in drying of kiwifruit slices. *Energy* **2019**, *170*, 149–158. [CrossRef]
26. Onwude, D.I.; Hashim, N.; Abdan, K.; Janius, R.; Chen, G. Modelling the mid-infrared drying of sweet potato: Kinetics, mass and heat transfer parameters, and energy consumption. *Heat Mass Transf.* **2018**, *54*, 2917–2933. [CrossRef]
27. Rahman, M.S.; Al-Shamsi, Q.H.; Bengtsson, G.B.; Sablani, S.S.; Al-Alawi, A. Drying Kinetics and Allicin Potential in Garlic Slices during Different Methods of Drying. *Dry. Technol.* **2009**, *27*, 467–477. [CrossRef]
28. Wang, H.; Liu, Z.L.; Vidyarthi, S.K.; Wang, Q.H.; Gao, L.; Li, B.R.; Wei, Q.; Liu, Y.H.; Xiao, H.W. Effects of different drying methods on drying kinetics, physicochemical properties, microstructure, and energy consumption of potato (Solanum tuberosum L.) cubes. *Dry. Technol.* **2020**, *39*, 418–431. [CrossRef]
29. McMinn, W.A.M. Prediction of moisture transfer parameters for microwave drying of lactose powder using Bi–G drying correlation. *Food Res. Int.* **2004**, *37*, 1041–1047. [CrossRef]
30. Zhang, W.; Chen, C.; Pan, Z.; Xiao, H.-W.; Xie, L.; Gao, Z.-J.; Zheng, Z.-A. Design and performance evaluation of a pilot-vacuum infrared drying (PVID) system for drying of berries. *Dry. Technol.* **2020**, *38*, 1340–1355. [CrossRef]
31. Xiao, H.-W.; Pang, C.-L.; Wang, L.-H.; Bai, J.-W.; Yang, W.-X.; Gao, Z.-J. Drying kinetics and quality of Monukka seedless grapes dried in an air-impingement jet dryer. *Biosyst. Eng.* **2010**, *105*, 233–240. [CrossRef]
32. Seremet Ceclu, L.; Botez, E.; Nistor, O.V.; Andronoiu, D.G.; Mocanu, G.D. Effect of different drying methods on moisture ratio and rehydration of pumpkin slices. *Food Chem.* **2016**, *195*, 104–109. [CrossRef] [PubMed]
33. Joardder, M.U.; Kumar, C.; Karim, M.A. Food structure: Its formation and relationships with other properties. *Crit. Rev. Food Sci. Nutr.* **2017**, *57*, 1190–1205. [CrossRef] [PubMed]
34. Khan, M.I.H.; Nagy, S.A.; Karim, M.A. Transport of cellular water during drying: An understanding of cell rupturing mechanism in apple tissue. *Food Res. Int.* **2018**, *105*, 772–781. [CrossRef]
35. Chen, C.; Upadhyaya, S.; Khir, R.; Pan, Z. Simulation of walnut drying under hot air heating using a nonequilibrium multiphase transfer model. *Dry. Technol.* **2020**, 1–15. [CrossRef]
36. Chen, L.; Opara, U.L. Texture measurement approaches in fresh and processed foods—A review. *Food Res. Int.* **2013**, *51*, 823–835. [CrossRef]
37. Puente-Diaz, L.; Spolmann, O.; Nocetti, D.; Zura-Bravo, L.; Lemus-Mondaca, R. Effects of Infrared-Assisted Refractance Window Drying on the Drying Kinetics, Microstructure, and Color of Physalis Fruit Puree. *Foods* **2020**, *9*, 343. [CrossRef] [PubMed]
38. Beaudry, C.; Raghavan, G.S.V.; Ratti, C.; Rennie, T.J. Effect of Four Drying Methods on the Quality of Osmotically Dehydrated Cranberries. *Dry. Technol.* **2004**, *22*, 521–539. [CrossRef]
39. Hao, Q.B.; Li, C.S.; Liu, G.Q.; Xu, X.Y.; Jiao, G.F.; Zheng, H.L.; Zhang, S.N.; Bai, L.F. Effect of high-temperature vacuum degassing on microstructure and current-carrying capacity of the Bi-2212 wire. *J. Alloys Compd.* **2018**, *742*, 897–902. [CrossRef]
40. Joardder, M.U.H.; Brown, R.J.; Kumar, C.; Karim, M.A. Effect of Cell Wall Properties on Porosity and Shrinkage of Dried Apple. *Int. J. Food Prop.* **2015**, *18*, 2327–2337. [CrossRef]
41. Souraki, B.A.; Ghavami, M.; Tondro, H. Correction of moisture and sucrose effective diffusivities for shrinkage during osmotic dehydration of apple in sucrose solution. *Food Bioprod. Process.* **2014**, *92*, 1–8. [CrossRef]
42. Goula, A.M.; Chasekioglou, A.N.; Lazarides, H.N. Drying and Shrinkage Kinetics of Solid Waste of Olive Oil Processing. *Dry. Technol.* **2015**, *33*, 1728–1738. [CrossRef]
43. Mancuhan, E.; Özen, S.; Sayan, P.; Sargut, S.T. Experimental investigation of green brick shrinkage behavior with Bigot's curves. *Dry. Technol.* **2016**, *34*, 1535–1545. [CrossRef]
44. Hentati, F.; Delattre, C.; Ursu, A.V.; Desbrieres, J.; Le Cerf, D.; Gardarin, C.; Abdelkafi, S.; Michaud, P.; Pierre, G. Structural characterization and antioxidant activity of water-soluble polysaccharides from the Tunisian brown seaweed Cystoseira compressa. *Carbohydr Polym.* **2018**, *198*, 589–600. [CrossRef] [PubMed]
45. Yan, J.K.; Wu, L.X.; Qiao, Z.R.; Cai, W.D.; Ma, H. Effect of different drying methods on the product quality and bioactive polysaccharides of bitter gourd (Momordica charantia L.) slices. *Food Chem.* **2019**, *271*, 588–596. [CrossRef] [PubMed]
46. Liu, Y.; Luo, M.; Liu, F.; Feng, X.; Ibrahim, S.A.; Cheng, L.; Huang, W. Effects of freeze drying and hot-air drying on the physicochemical properties and bioactivities of polysaccharides from Lentinula edodes. *Int. J. Biol. Macromol.* **2020**, *145*, 476–483. [CrossRef] [PubMed]

47. Wang, B.; Venkitasamy, C.; Zhang, F.; Zhao, L.; Khir, R.; Pan, Z. Feasibility of jujube peeling using novel infrared radiation heating technology. *LWT Food Sci. Technol.* **2016**, *69*, 458–467. [CrossRef]
48. Chen, J.; Venkitasamy, C.; Shen, Q.; McHugh, T.H.; Zhang, R.; Pan, Z. Development of healthy crispy carrot snacks using sequential infrared blanching and hot air drying method. *LWT Food Sci. Technol.* **2018**, *97*, 469–475. [CrossRef]

foods

MDPI

Article

Natural Drying of Astringent and Non-Astringent Persimmon "Rojo Brillante". Drying Kinetics and Physico-Chemical Properties

Cristina M. González [1], Rebeca Gil [2], Gemma Moraga [1,*] and Alejandra Salvador [2]

1 Departamento de Tecnología de Alimentos, Universitat Politècnica de Valencia, Camí de Vera s/n, 46021 Valencia, Spain; crima13c@upvnet.upv.es
2 Centro de Tecnología Postcosecha, Instituto Valenciano de Investigaciones Agrarias, CRTA, Moncada-Náquera km 4.5, Moncada, 46113 Valencia, Spain; gil_reb@gva.es (R.G.); salvador_ale@gva.es (A.S.)
* Correspondence: gemmoba1@tal.upv.es

Abstract: Persimmon (*Diospyros kaki* L.f.) crop has markedly increased in Spain, and "Rojo Brillante" persimmon is the main cultivated variety. This astringent cultivar requires de-astringency treatment before commercialization, which may involve an extra cost. Its short commercial season implies handling large volumes of fruits with consequent postharvest losses. Therefore, the development of derived added-value products is of much interest. In this study, astringent and non-astringent "Rojo Brillante" persimmons were dehydrated by following a natural drying method used in Asia. The drying kinetics and physico-chemical properties were analyzed for 81 days. The results indicated subsequent reductions in weight, water content, and water activity throughout the drying process, and the equatorial diameter decreased. All the employed thin-layer mathematical models were suitable for representing the drying characteristics of both products with similar behavior. The effective water diffusivity values were 5.07×10^{-11} m^2 s^{-1} and 6.07×10^{-11} m^2 s^{-1} for astringent and non-astringent persimmon samples, respectively. The drying treatment significantly decreased the soluble tannins content, and the astringent samples obtained similar values to those obtained for the non-astringent samples in 20 days. The external and internal flesh of the astringent fruit remained orange through the drying period, while brown coloration in the non-astringent fruit was observed after 57 drying days. Therefore, prior de-astringency treatment would not be necessary.

Keywords: *Diospyros kaki*; post-harvest losses; dehydrated persimmon; thin-layer modeling; drying rate

Citation: González, C.M.; Gil, R.; Moraga, G.; Salvador, A. Natural Drying of Astringent and Non-Astringent Persimmon "Rojo Brillante". Drying Kinetics and Physico-Chemical Properties. *Foods* **2021**, *10*, 647. https://doi.org/10.3390/foods10030647

Academic Editor: Danijela Bursać Kovačević and Predrag Putnik

Received: 26 February 2021
Accepted: 15 March 2021
Published: 18 March 2021

Publisher's Note: MDPI stays neutral with regard to jurisdictional claims in published maps and institutional affiliations.

Copyright: © 2021 by the authors. Licensee MDPI, Basel, Switzerland. This article is an open access article distributed under the terms and conditions of the Creative Commons Attribution (CC BY) license (https://creativecommons.org/licenses/by/4.0/).

1. Introduction

In Spain, persimmon (*Diospyros kaki* L.f.) production has markedly increased over the last 20 years, and the cultivation area has expanded almost 8-fold, from about 2253 ha in 2002 to over 18,000 ha in 2019. As a result, with close to 500,000 tons, Spain is now the second most important persimmon producing country worldwide after China (FAOSTAT, 2018). At present, cultivation is mainly based in "Rojo Brillante" cultivar, with a production of around 429,000 tons in Valencia Community (E Spain) and around 50 tons in Andalusia (S Spain) [1].

"Rojo Brillante" is an astringent persimmon cultivar, which involves its presenting high soluble tannins at harvest [2] and, therefore, the postharvest de-astringency treatment is required before commercialization. The introduction of postharvest techniques based on exposing the fruit to high CO_2 concentrations to eliminate astringency has been one of the main causes of the expansion of persimmon production in Spain in recent years. With this de-astringency method, it is possible to obtain a fruit without astringency while preserving a firm texture [3,4]. Currently, the Spanish production is mainly destined for exportation markets where there is a demand for persimmon as fresh fruit with a firm texture according to the current quality standards (UNECE 2016) [5]. It is noteworthy that the short commercial season of this cultivar (between mid-October and the end of December) implies the postharvest handling of large fruit volumes with consequent product loss and without achieving the quality required by the fresh fruit market. Therefore, one of the current challenges for the persimmon industry is the search for strategies that increase the value of the discarded fruit.

Drying is a reliable preservation method for fruits in technical feasibility and nutritional quality terms. Unlike expensive energy-intensive artificial drying, natural drying can provide an alternative with adequate drying capacity [6]. Even though natural drying generates a significant loss of bioactive compounds, dried fruit can still be a valuable source of dietary fiber, minerals, and antioxidants. Based on scientific evidence, persimmon can be considered a functional food due to its high contents of bioactive compounds that help reduce the risk of cardiovascular diseases, as well as kidney, colon, and rectal cancer, etc. Hence, dried fruit might be a potential snack that is healthier than most regular snacks [7,8]. In some Asian countries such as China, South Korea, and Japan, dehydrated persimmon is often consumed and commercially produced [9–11]. The general procedure followed to make this dried product comprises removing the sepals of the calix and skin, and then hanging the fruit on strings. In China and Japan near the end of the drying period, the dried fruit is kneaded to distribute moisture uniformly in the fruit, and to produce the shape of the final product. However, in South Korea, they are left to hang in a well-ventilated place [12].

Presently, although this drying technology is not applied to "Rojo Brillante", it would be a good strategy to enhance the surplus fruit and to increase the value of the discarded fruit.

It is necessary to study drying kinetics to know the drying time required to attain a product of adequate quality. Semitheoretical models, based on a serial development of Fick's second law of diffusion, are the most widely used for food products [13–15]. Several studies have focused on drying kinetics in different persimmon formats and varieties, along with different treatments other than drying. García-Pérez, Cárcel, Benedito, and Mulet (2007) [16] and Bozkir et al. (2019) [17] studied the influence of ultrasound or osmo-convective pretreatments on drying cubes of and cylindrical-shaped persimmon. Sampaio et al. (2017) [15] obtained the mathematical model of drying kinetics for the Fuyu persimmon variety in an osmo-convective drying procedure. Çelen et al. (2019) [18] focused on the microwave effects on the drying characteristics of persimmon slices, whereas Doymaz (2012) [19] assessed the drying kinetics and activation energy of persimmon slices using hot air drying. Nevertheless, scarce information on drying kinetics applied in the whole persimmon fruit is reported. Demiray and Tulek (2017) [20] studied the effect of different pretreatments and hot air-drying temperatures on the drying kinetics of whole persimmon. For the specific case of "Rojo Brillante", no studies have addressed the drying kinetics during the natural drying process of whole persimmon.

In this context, the aim of this research was to study the drying kinetics of persimmon "Rojo Brillante" when the natural drying method (hanging in a well-ventilated place) is applied to astringent and non-astringent fruits (submitted previously to the deastringency treatment). Moreover, the physico-chemical changes that occur during drying was also studied.

2. Materials and Methods

Persimmon fruits "Rojo Brillante" were harvested from commercial orchards in Valencia (E Spain) on 20 December 2018 at a commercial maturity stage (Color index (1000 a/Lb) between 15 and 17; firmness values between 33 and 35 N; initial water content between 77% and 78%). After harvest, the fruits were transported to the Instituto Valenciano de Investigaciones Agrarias (IVIA), where they were carefully selected for uniformity and separated into two groups. The first group was submitted to the astringency removal treatment in closed containers under standard conditions (95–98% CO_2 for 24 h at 20 °C). The second group was not subjected to the de-astringency treatment. One hundred fruits from both the astringent and non-astringent groups (submitted to the de-astringency treatment) were manually peeled and immersed for 10 min in a 4.5% sodium metabisulfite ($Na_2S_2O_5$) solution used as a disinfectant. The fruits were individually hung by the pedicel for natural drying in the IVIA pilot plant. In order to study the drying kinetics and the evolution of the physico-chemical properties, one sample of 10 fruits was taken every 17–20 days up to a period of 57 days and at the end of the drying process, two more samples were taken at 67 and 81 days. The sampling dates were as follows: Day 0 (20 December 2018), Day 20 (9 January 2019), Day 40 (29 January 2019), Day 57 (15 February 2019), Day 67 (25 February 2019), Day 81 (11 March 2019).

The average temperature and relative humidity during the drying period were taken from the IVIA weather station, and ranged from 10.3 to 11.3 °C and from 68.6% to 75%, respectively.

2.1. Weight Loss and Equatorial, Longitudinal Diameters

The fruits were individually weighed with an Absolute Digimatic caliper (PB3002-S/FACT, Mettler Toledo, Switzerland). The equatorial and longitudinal diameters were measured with a pachymeter (Mitutoyo 500-171-20, Coventry, UK). Ten replicates were performed.

2.2. External Color

The external color was evaluated by a Minolta Colorimeter (model CR-300, Ramsey, NY, USA) on 10 fruits. "L", "a", "b" Hunter parameters were measured, and the results were expressed as a skin color index: (1000a)/(Lb) [3].

2.3. Total Soluble Solids (TSS) and Soluble Tannins (ST)

Three samples of three individual fruits were used to determine TSS and ST. The fruits were cut into four longitudinal parts with the two opposite ends sliced and frozen at -20 °C to determine the ST content. The other opposite fruit parts were placed in an electric juice extractor (model 753, Moulinex, Barcelona, Spain) and filtered through a cheese cloth. The obtained juice was then used to determine the TSS content. ST were evaluated until Day 40 by the Folin–Denis method described by Arnal and Del Río (2004) [21], and the results were expressed as a percent of fresh weight. The TSS juice was measured in triplicate with a digital refractometer (model PR-1, Atago, Japan) and expressed as Brix.

2.4. Water Content and Water Activity

Three fruits were individually ground in a crushing machine. The water content and water activity (a_w) were measured using a vacuum oven (Vaciotem-T, J.P Selecta, Abrera, Barcelona, Spain) (60 ± 1 °C and pressure <100 mm Hg) and an Aqualab CX-2 (Decagon Devices Inc., Pullman, WA, USA), respectively. Three replicates were measured per sample.

2.5. Mathematical Modeling of Drying Curves

To investigate the drying characteristics of persimmon "Rojo Brillante", six commonly thin-layer drying semitheoretical models (Table 1) were used to fit the experimental drying data [22–27].

Table 1. Mathematical models given by several authors for drying curves.

Model	Mathematical Equation	References
Newton	$MR = \exp(-kt)$	[22]
Page	$MR = \exp(-kt^n)$	[23]
Midilli et al.	$MR = a\exp(-kt^n) + bt$	[24]
Logarithmic	$MR = a\ exp\ (-kt) + c$	[25]
Henderson and Pabis	$MR = a\exp(-kt)$	[26]
Verma model	$MR = a\ exp\ (-kt) + (1-a)\exp(-gt)$	[27]

k, n, a, g, c, b: Constants of each model applied; t: Time in days.

The non-linear least squares regression analysis was determined by the statistical software Solver (Excel 2016). In these models, *MR* is the dimensionless moisture ratio in Equation (1):

$$MR = \left(\frac{M_i - M_e}{M_0 - M_e}\right) \tag{1}$$

where M_i and M_0 are the moisture content (on a dry basis) at any drying time and at the initial time, respectively. M_e is the equilibrium moisture content and is relatively low (about 3%, wb) [28], so it can be neglected. Therefore, *MR* can be expressed as M_i/M_0.

The determination coefficient (R^2) is one of the primary criteria for selecting the best model to define the drying curves. Reduced chi-square (X^2), mean bias error (*MBE*), and root-mean-square error (*RMSE*) are used to determine the quality of fit. These parameters can be calculated using Equations (2)–(4):

$$X^2 = \frac{\sum_{i=1}^{N}\left(MR_{exp,i} - MR_{pre,\,i}\right)^2}{N - z} \tag{2}$$

$$MBE = \frac{1}{n}\sum_{i=1}^{N}\left(MR_{pre,i} - MR_{exp,\,i}\right) \tag{3}$$

$$RMSE = \left[\frac{1}{n}\sum_{i=1}^{N}\left(MR_{pre,i} - MR_{exp,\,i}\right)^2\right]^{1/2} \tag{4}$$

The higher the R^2, and the lower X^2, *MBE*, and *RMSE*, the better the mathematical model fits the experimental data [28]. $MR_{exp,I}$ is the experimental moisture ratio, $MR_{pre,I}$ is the predicted moisture ratio, N is the number of observations, and z is the number of constants.

Another criterion, the relative percent error (PE), is used to evaluate the predictive precision of models [14]. Lower relative PE values give better fitting models (Equation (5)):

$$PE\ (\%) = \frac{100}{N}\sum_{i=1}^{N}\frac{\left|MR_{exp,i} - MR_{pre,i}\right|}{MR_{exp,i}} \tag{5}$$

The drying rate is represented as $\Delta M/\Delta t$ (the water content to time ratio to the product's average water content between two consecutive weight control times) vs. *MR* (dimensionless moisture ratio) [29].

The experimental drying data for determining the effective water diffusivity were interpreted by Fick's second law of diffusion. To model the total amount of diffusing water entering the astringent and non-astringent persimmon samples, the equation in spheres for "long times" was applied [30]. The effective water diffusivity coefficient (D_e) was obtained by fitting the corresponding linear equation (Equation (6)), where r is the radius obtained from the longitudinal diameter (m), t is the time in days, and Y is the reduced driving force defined by Equation (7), in the dry basis moisture content terms.

$$Ln\ Y = Ln\left(\frac{6}{\pi^2}\right) - \left(\frac{\pi^2 D_e t}{r^2}\right) \tag{6}$$

$$Y = \frac{M_i - M_0}{M_e - M_0} \tag{7}$$

M (g water/g dry solids) at each dehydration (M_i) time in the initial product (M_0) and at the equilibrium time (M_e).

2.6. Statistical Analysis

Data were subjected to an analysis of variance (ANOVA) using the least significant difference (LSD) test with a 95% confidence interval to compare the test averages (Statgraphics Centurion XVII Manugistics, Inc., Rockville, MA, USA).

3. Results and Discussion

3.1. Physico-Chemical Determinations

The drying process brought about a marked gradual weight loss for the first 57 days. Thereafter, the fruit weight decreased only slightly until the end of the assay (Figure 1a). Weight loss paralleled the reduced water content (Figure 1b). The water content dropped from 78% at harvest to 25% after 67 days, before lowering to 15% at 81 days. No significant differences in water loss were found between the astringent and non-astringent fruits during the whole study period.

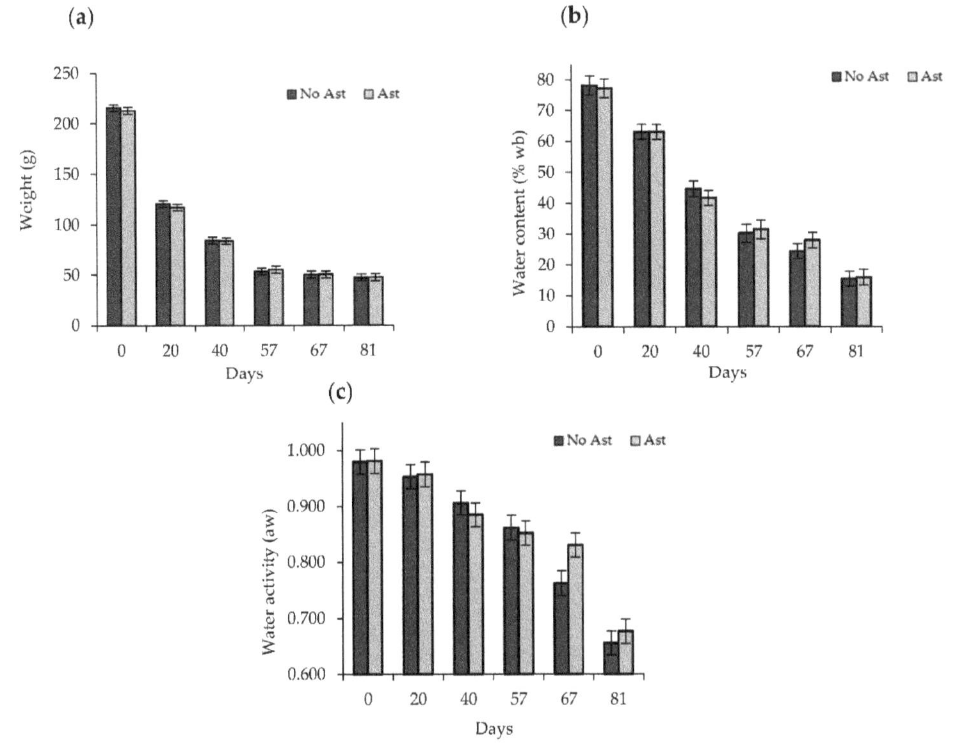

Figure 1. Weight loss (**a**), water content (g/100 g product on a wet basis) (**b**) and water activity (a_w) (**c**) of the non-astringent (no Ast) and astringent (Ast) persimmon samples during the drying treatment. Bars represent the least significant difference (LSD) intervals ($p \leq 0.05$).

The water activity gradually decreased (Figure 1c) from values of 0.980 on Day 0 to values of 0.860 after 57 drying days, with no differences between the astringent and non-astringent fruits. Unlike the water content, the most marked drop in water activity was detected after 57 days. On Day 67, the a_w of the astringent fruit (0.830) was higher than that of the non-astringent fruit (0.760).

According to previous authors, dried persimmon products are classified as semidried or dried depending on the water content [31]. The final water content of the S Korean semidried and dried persimmons are approximately 50% and 30%, respectively, with drying periods usually lasting 25 days to achieve 50% and approximately 60 days to accomplish 30% [10]. Similarly in our study, on Day 40, fruit samples showed 45% water content, which was 30% on Day 57. After 81 days, a drier product was obtained with 15% water content in both the astringent and non-astringent samples. The drying kinetics of this process could be the key to adjust the drying treatment.

Figure 2 illustrates the images of the astringent and non-astringent whole persimmon samples, which are cut longitudinally during the drying treatment from Day 0 to 81.

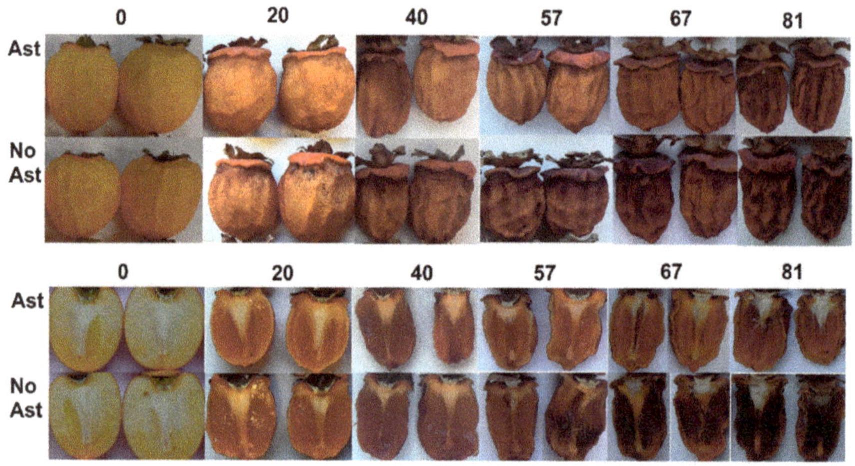

Figure 2. Images of the astringent and non-astringent persimmon cv. "Rojo Brillante" during the drying process.

The water content loss brought about a major reduction in the equatorial diameter up to 57 days (Figure 3a). The longitudinal diameter slightly lowered after 20 days to remain stable during the subsequent drying periods (Figure 3b). The minor changes in the longitudinal diameter that took place during the drying process were due to the position in which the fruits were hung. No significant differences were observed in the shape changes between the astringent and non-astringent fruits. These changes were accompanied by fruit shrinkage, warping, and wrinkling, which became more evident with the drying time.

During the drying process, the external fruit color darkened (Figure 2). At the beginning of the process, the fruit color index (CI) came close to 18 and reached values near 30 after 40 days with no differences between the astringent and non-astringent fruits. Nevertheless, after 57 days the CI values were significantly higher in the astringent fruit (CI = 46) than in the non-astringent fruit (CI = 40). These differences were still found after 81 days (Figure 3c). The color changes that occurred during the drying period were the result of several biochemical reactions, such as degradation of carotenoids, decomposition of other color pigments, and enzymatic and non-enzymatic reactions [32]. Yamada et al. (2009) [33] suggested that the oxidative and non-oxidative degradation of ascorbic acid would contribute markedly to the browning of this product type, while enzymatic browning, by polyphenol oxidase, and the Maillard reaction, between amino acids and reducing sugars would not play a key role. From Day 67 to 81 of drying, no significant changes ($p > 0.05$) in the color index of the astringent samples were detected, while the non-astringent samples continued to change.

A marked change in the internal flesh structure was also observed after 20 days by showing gelling symptoms, which became much more evident while the drying process prolonged. Mamet, Yao, Li, and Li (2017) [34] reported that persimmon tannins enhance the gel properties of pectin, even though mechanisms remain unclear. It is noteworthy that while the internal flesh color remained orange throughout the drying period in the astringent fruit, the flesh acquired a brown coloration from 57 drying days in the non-astringent fruit. This is consistent with the darker external coloration of non-astringent fruit and the significant difference in water activity at 67 days, which may be related to changes in both structure and water retention capacity.

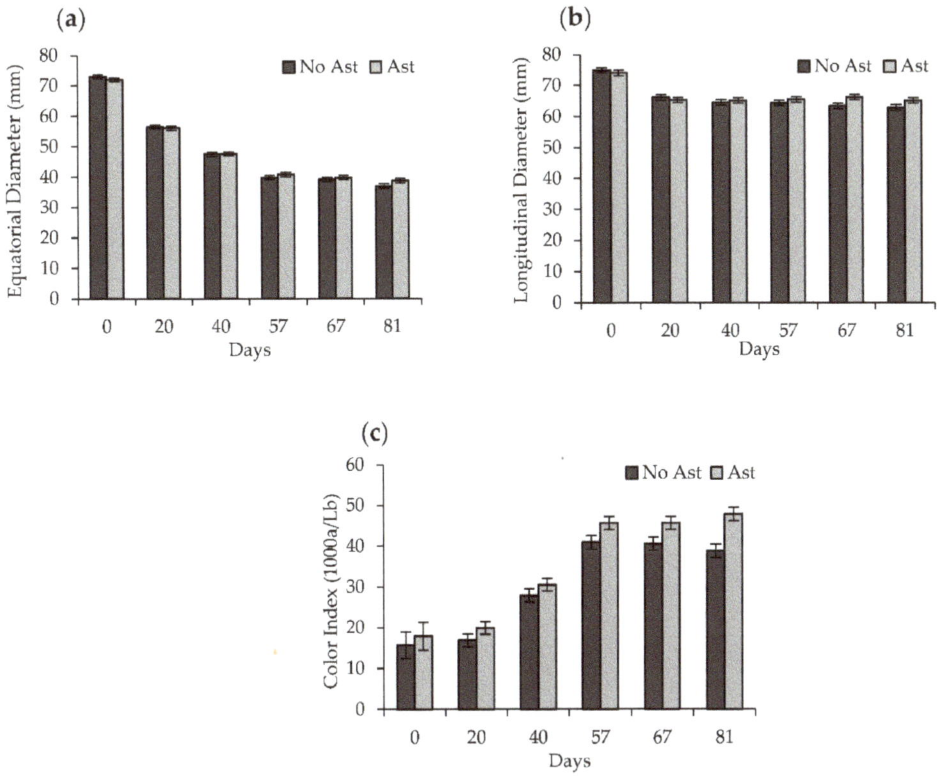

Figure 3. Equatorial diameter (**a**), longitudinal diameter (**b**), and external color index (**c**) of the non-astringent (no Ast) and astringent (Ast) persimmon samples during the drying treatment. Bars represent the least significant difference (LSD) intervals ($p \leq 0.05$).

For TSS (Figure 4a), a gradual increase was observed as the drying process advanced, with values going from close to 18 °Brix on Day 0 to close to 55 °Brix after 57 days, with no significant differences between the astringent and non-astringent fruits. After 81 days, the astringent fruit had higher TSS values (63 °Brix) than the non-astringent fruit (57 °Brix). As drying progressed, soluble solids became concentrated due to the fact that the water loss and new solids were also generated [6]. Similar TSS content have been reported in semidried and dried persimmon from South Korea [10,35].

Initially on Day 0, the astringent persimmons, not previously submitted to the de-astringency treatment, had an ST content of 0.6% (Figure 4b). These values fall within the range found by most previous studies conducted on "Rojo Brillante", which have been related to high astringency levels in fruits [3,36,37]. In contrast, the non-astringency fruit, submitted to the de-astringency treatment with a high CO_2 concentration, gave ST content values of 0.02%, which are sensory non astringency values for "Rojo Brillante" [3,36]. After 20 drying days, the ST content values in the astringency fruit notably dropped to 0.03%. The ST values were similar to those of the non-astringent fruit. Tannin insolubilization during the drying process could be associated with structural flesh changes (Figure 2), which happens during natural persimmon fruit ripening [2,32,38]. In astringent cultivars, the ripening process is accompanied by gradual tannin insolubilization, which leads to a progressive decline in ST with subsequent astringency reduction [3]. The softening that occurs during fruit ripening leads to pectin solubilization, which forms a complex with tannins and brings about their insolubilization [37]. Asgar and Yamauchi, Kato (2004) [38] have also related the flesh structural changes found during the sun-drying of Japanese persimmon to the solubilization and depolymerization of pectin polysaccharides.

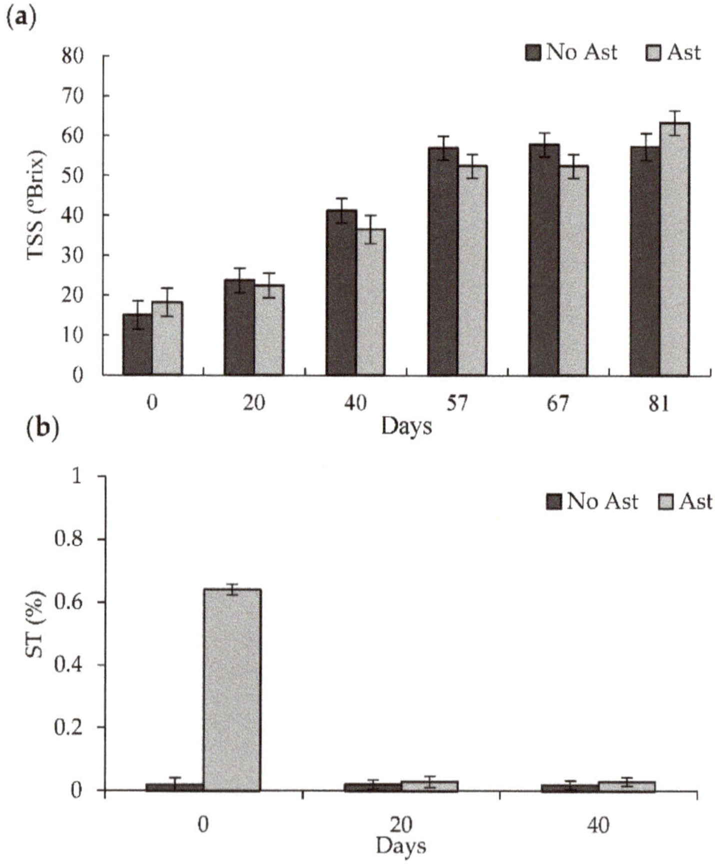

Figure 4. Total soluble solids (TSS) of the non-astringent (no Ast) and astringent (Ast) persimmon samples during the drying treatment (**a**). Total soluble tannin content (ST) of the non-astringent (no Ast) and astringent (Ast) persimmon samples up to the drying treatment at Day 40 (**b**). Bars represent the least significant difference (LSD) intervals ($p \leq 0.05$).

3.2. Fitting of Drying Curves and Drying Rate Determinations

The water content data obtained at the different drying times were converted into a dimensionless moisture ratio (Equation (1)) and then fitted to six thin-layer drying models (Table 1). These models have been used for agricultural products [13,15]. To estimate the parameters from those six models, a non-linear regression analysis was used with both the astringent and non-astringent persimmon samples. The statistical results of the models are summarized in Table 2 (astringent samples) and Table 3 (non-astringent samples). The best models describing the thin-layer drying characteristics of the persimmon samples were chosen with the highest R^2 values and the lowest X^2, *MSE*, *RMSE*, and *PE* values.

In both cases, R^2 values were higher than 0.995, while X^2, MBE, and $RMSE$ were ≤ 0.001, ≤ 0.018, and 0.000, respectively. PE values were between 2–10% for all the models assessed in both samples. These results were in agreement with those found in fruits such as cape gooseberry [39], pomegranate [40], apple or pumpkin [13].

Table 2. Values of the parameters of the models for astringent persimmon "Rojo Brillante".

	Models Parameters						Statistical Parameters				
Models	k	n	a	g	c	b	R^2	X^2	MBE	$RMSE$	$PE\%$
Newton	0.035	-	-	-	-	-	0.997	0.001	0.017	0.000	7.071
Page	0.030	1.043	-	-	-	-	0.997	0.001	0.017	0.000	9.021
Midilli et al.	0.027	1.101	1.028	-	-	0.000	0.998	0.000	0.011	0.000	10.243
Logarithmic	0.039	-	1.022	-	0.019	-	0.998	0.000	0.012	0.000	7.965
Henderson and Pabis	0.037	-	1.037	-	-	-	0.998	0.000	0.012	0.000	7.586
Verma model	0.036	-	0.007	0.036	-	-	0.998	0.001	0.018	0.000	7.663

k, n, a, g, c, b: Constants of each model applied. R^2: determination coefficient; X^2: reduced chi-square; MBE: mean bias error; $RMSE$: root-mean-square error; $PE\%$: relative percent error.

Table 3. Values of the parameters of the models for non-astringent persimmon "Rojo Brillante".

	Models Parameters						Statistical Parameters				
Models	k	n	a	g	c	b	R^2	X^2	MBE	$RMSE$	$PE\%$
Newton	0.037	-	-	-	-	-	0.999	0.000	0.008	0.000	2.305
Page	0.032	1.040	-	-	-	-	0.999	0.000	0.010	0.000	4.632
Midilli et al.	0.041	0.983	1.042	-	-	0.000	0.999	0.000	0.002	0.000	1.943
Logarithmic	0.039	-	1.028	-	0.011	-	0.999	0.000	0.002	0.000	2.032
Henderson & Pabis	0.038	-	1.037	-	-	-	0.999	0.000	0.004	0.000	3.095
Verma model	0.037	-	0.017	0.037	-	-	0.999	0.000	0.008	0.000	2.305

k, n, a, g, c, b: Constants of each model applied. R^2: determination coefficient; X^2: reduced chi-square; MBE: mean bias error; $RMSE$: root-mean-square error; $PE\%$: relative percent error.

Figure 5a,b shows the Midilli, Vermal, and Logarithmic models selected to represent the drying characteristics of the whole astringent and non-astringent persimmon samples. After fitting the experimental data, it took 34 days to reach a water content of 50% and 57 days to obtain 30%. Under these conditions, the desirable semidried and dried persimmon would be obtained. Figure 5c shows the effect of $\Delta M/\Delta t$ vs. MR on the drying rate of the astringent and non-astringent samples, where a good correlation was obtained ($R^2 = 0.9855$). In both the sample types, the drying rate was rapid during the initial period, but then slowed down in the later stages, and no constant rate of the drying period was observed. The entire drying process occurred during the falling-rate period. When the decrease in the drying rate was linear with the water content, water evaporation in the material continued as during the constant rate period. This indicates that mass transfer is governed by intrinsic product properties and internal resistance to water diffusion to the surface [41]. This result was similar to those reported for the thin-layer drying of other biomaterials [29,42,43].

In Figure 5d, $Ln\ Y$ vs. t/r^2 is plotted to determine, from the slope ($\pi^2 D_e$), the effective water diffusivity of both the astringent and non-astringent persimmon samples. Fick's second diffusion law has been widely used to describe the drying process during the falling-rate period for biological material [44]. The effective water diffusivity results were similar in both the astringent and non-astringent persimmon samples, with values of 5.07×10^{-11} and 6.08×10^{-11} m^2 s^{-1}, respectively. The R^2 was 0.996. The D_e values fell within the general range of 10^{-12}–10^{-8} m^2 s^{-1} in food materials [45]. Similar results were reported by Doymaz I (2012) [19] in persimmon slices (between 7.05×10^{-11} and 2.34×10^{-10} m^2 s^{-1}).

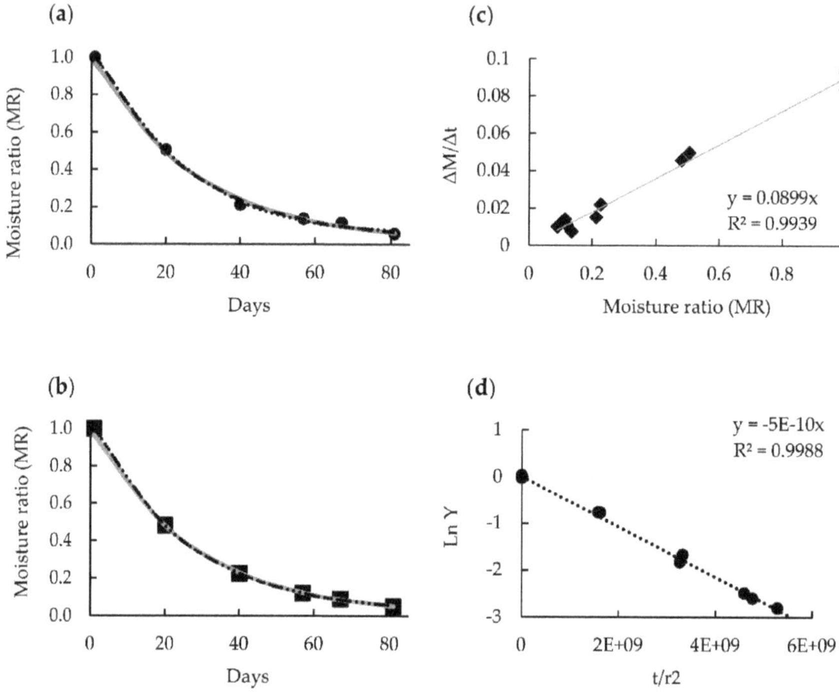

Figure 5. Modeling the drying curves of the astringent (**a**) and non-astringent (**b**) persimmon fruits with the Midilli (dotted line), Logarithmic (dashed line), and Verma models (gray line). Drying rate of the astringent and non-astringent persimmon samples (**c**). Effective water diffusivity (D_e) determination by equation $Ln\ Y = Ln\left(\frac{6}{\pi^2}\right) - \left(\frac{\pi^2 D_e t}{r^2}\right)$ in the astringent and non-astringent persimmon samples (**d**).

4. Conclusions

For the first time, the kinetics during the natural drying method followed in Asian countries (hanging the whole fruit in a well-ventilated place) was studied in "Rojo Brillante" persimmon. Since "Rojo Brillante" is an astringent cultivar, the behavior of the astringent and non-astringent fruits (submitted to the CO_2 treatment) was compared. The used thin-layer mathematical models were suitable for fitting the drying kinetics. No significant differences between the astringent and non-astringent fruits were found. Around 34 days were needed to reach a final water content of 50% and 57 days to reach one of 30%. This drying treatment was able to produce a natural decrease in ST contents. The astringent and non-astringent fruits obtained similar values in just 20 days. Different behaviors between the astringent and non-astringent samples were observed in a_w, external and internal color at 57 drying days. The astringent fruit remained orange, while brown coloration developed on the non-astringent fruit. Hence, the de-astringency treatment is not recommended. This natural drying technology, not yet applied to the "Rojo Brillante" persimmon industry, could be a good strategy to enhance the surplus of this seasonal fruit.

Author Contributions: Investigation, validation, methodology, formal analysis, writing—original draft preparation, C.M.G.; investigation, validation, methodology, formal analysis, R.G.; supervision, resources, conceptualization, funding acquisition, writing—review and editing, G.M.; supervision, resources, conceptualization, funding acquisition, writing—review and editing, A.S. All authors have read and agreed to the published version of the manuscript.

Funding: This work was supported by the Ministerio de Ciencia, Innovación y Universidades and Fondo Europeo de Desarrollo Regional (FEDER) through project RTA2017-00045-C02.

Institutional Review Board Statement: Not applicable.

Informed Consent Statement: Not applicable.

Data Availability Statement: Not applicable.

Acknowledgments: The authors thank Phillip Bentley for assistance in correcting the English manuscript.

Conflicts of Interest: The authors declare no conflict of interest.

References

1. Perucho, R. Evolution of production of the 'Rojo Brillante' cultivar in Spain and its impact on markets. *Acta Hortic.* **2018**, *1195*, 1–8. [CrossRef]
2. Tessmer, M.A.; Besada, C.; Hernando, I.; Appezzato-da-Glória, B.; Quiles, A.; Salvador, A. Microstructural changes while persimmon fruits mature and ripen. Comparison between astringent and non-astringent cultivars. *Postharvest Biol. Technol.* **2016**, *120*, 52–60. [CrossRef]
3. Salvador, A.; Arnal, L.; Besada, C.; Larrea, V.; Quiles, A.; Pérez-Munuera, I. Physiological and structural changes during ripening and deastringency treatment of persimmon fruit cv. "Rojo Brillante". *Postharvest Biol. Technol.* **2007**, *46*, 181–188. [CrossRef]
4. Novillo, P.; Salvador, A.; Magalhaes, T.; Besada, C. Deastringency treatment with CO2 induces oxidative stress in persimmon fruit. *Postharvest Biol. Technol.* **2014**, *92*, 16–22. [CrossRef]
5. UNECE Standard FFV-63 Concerning the Marketing and Commercial Quality Control of Persimmons. Available online: https://www.unece.org/fileadmin/DAM/trade/agr/standard/standard/fresh/FFV-Std/English/50Apples_2014.pdf (accessed on 10 March 2021).
6. Mat Desa, W.N.; Mohammad, M.; Fudholi, A. Review of drying technology of Figure. *Trends. Food Sci. Technol.* **2019**, *88*, 93–103. [CrossRef]
7. Sijtsema, S.J.; Jesionkowska, K.; Symoneaux, R.; Konopacka, D.; Snoek, H. Perceptions of the health and convenience characteristics of fresh and dried fruits. *LWT Food Sci. Technol.* **2012**, *49*, 275–281. [CrossRef]
8. Yaqub, S.; Farooq, U.; Shafi, A.; Akram, K.; Murtaza, M.A.; Kausar, T.; Siddique, F. Chemistry and functionality of bioactive compounds present in persimmon. *J. Chem.* **2016**, *16*, 1–13. [CrossRef]
9. Matsumoto, T.; Matsuzaki, H.; Takata, K.; Tsurunaga, Y.; Takahashi, H.; Kurahashi, T.; Maki, S.; Fujiwara, K. Inhibition of astringency removal in semidried japanese persimmon fruit by 1-methylcyclopropene treatment. *HortScience* **2007**, *42*, 1493–1495. [CrossRef]
10. Sun-wun, L.; Hye-Kyung, M.; Won-Young, L.; Jong-Kuk, K. Physicochemical characteristics of cold-air dried persimmons and traditional dried persimmons. *Korean Soc. Food Preserv.* **2011**, *18*, 481–487.
11. Oh, H.J.; Park, M.S.; Cho, K.J.; Choi, S.I.; Kang, H.M.; Kim, H. Geographical indication and development plans in South Korea: A study on dried persimmons. *Forest Sci. Technol.* **2018**, *14*, 41–46. [CrossRef]
12. Hur, S.; Kang, B.; Lee, D.; Lee, S.; Lee, J. Quality characteristics of domestic dried persimmon and imported dried persimmon. *Korean Soc. Food Preserv.* **2014**, *21*, 140–145. [CrossRef]
13. Akpinar, E.K. Determination of suitable thin layer drying curve model for some vegetables and fruits. *J. Food Eng.* **2006**, *73*, 75–84. [CrossRef]
14. Roberts, J.S.; Kidd, D.R.; Padilla-Zakour, O. Drying kinetics of grape seeds. *J. Food Eng.* **2008**, *89*, 460–465. [CrossRef]
15. Sampaio, R.M.; Neto, J.P.M.; Perez, V.H.; Marcos, S.K.; Boizan, M.A.; Da Silva, L.R. Mathematical modeling of drying kinetics of persimmon fruits (*Diospyros kaki* cv. Fuyu). *J. Food Process. Preserv.* **2017**, *41*, 1–7. [CrossRef]
16. García-Pérez, J.V.; Cárcel, J.A.; Benedito, J.; Mulet, A. Power ultrasound mass transfer enhancement on food drying. *Food Bioprod. Process.* **2007**, *85*, 247–254. [CrossRef]
17. Bozkir, H.; Rayman Ergün, A.; Serdar, E.; Metin, G.; Baysal, T. Influence of ultrasound and osmotic dehydration pretreatments on drying and quality properties of persimmon fruit. *Ultrason. Sonochem.* **2019**, *54*, 135–141. [CrossRef] [PubMed]
18. Çelen, S. Effect of microwave drying on the drying characteristics, color, microstructure, and thermal properties of Trabzon persimmon. *Foods* **2019**, *8*, 84. [CrossRef]
19. Doymaz, I. Evaluation of some thin-layer drying models of persimmon slices (*Diospyros kaki* L.). *Energy Convers. Manag.* **2012**, *56*, 199–205. [CrossRef]
20. Demiray, E.; Tulek, Y. The effect of pretreatments on Air drying characteristics of persimmons. *Heat Mass Transf.* **2017**, *53*, 99–106. [CrossRef]
21. Arnal, L.; Del Río, M.A. Effect of cold storage and removal astringency on quality of persimmon fruit (*Diospyros kaki*, L.) cv. Rojo brillante. *Food Sci. Technol. Int.* **2004**, *10*, 179–185. [CrossRef]
22. Liu, Q.; Bakker-Arkema, F.W. Stochastic modelling of grain drying: Part 2. Model development. *J. Agric. Eng. Res.* **1997**, *66*, 275–280. [CrossRef]
23. Page, G.E. Factors Influencing the Maximum rates of Air Drying Shelled Corn in Thin Layers. Master's Thesis, Purdue University, West Lafayette, IN, USA, 1949; pp. 355–367.
24. Midilli, A.; Kucuk, H.; Yapar, Z. A new model for single-layer drying. *Dry. Technol.* **2002**, *20*, 37–41. [CrossRef]

25. Togrul, I.T.; Pehlivan, D. Mathematical modelling of solar drying of apricots in thin layers. *J. Food Eng.* **2002**, *55*, 209–216. [CrossRef]
26. Wang, Z.; Sun, J.; Liao, X.; Chen, F.; Zhao, G.; Wu, J.; Hu, X. Mathematical modeling on Hot Air drying of thin layer apple pomace. *Food Res. Int.* **2007**, *40*, 39–46. [CrossRef]
27. Verma, L.R.; Bucklin, R.A.; Endan, J.B.; Wratten, F.T. Effects of drying air parameters on rice drying models. *Trans. Am. Soc. Agric. Eng.* **1985**, *28*, 296–301. [CrossRef]
28. Fang, S.; Wang, Z.; Hu, X. Hot Air drying of whole fruit chinese jujube (Zizyphus jujuba Miller): Thin-layer mathematical modelling. *Int. J. Food Sci. Technol.* **2009**, *44*, 1818–1824. [CrossRef]
29. Contreras, C.; Martín-Esparza, M.E.; Chiralt, A.; Martínez-Navarrete, N. Influence of microwave application on convective drying: Effects on drying kinetics, and optical and mechanical properties of apple and strawberry. *J. Food Eng.* **2008**, *88*, 55–64. [CrossRef]
30. Martinez-Navarrete, N.; Chiralt, A. Water diffusivity and mechanical changes during hazelnut hydration. *Food Res. Int.* **1999**, *32*, 447–452. [CrossRef]
31. Choi, J.Y.; Lee, H.J.; Cho, J.S.; Lee, Y.M.; Woo, J.H.; Moon, K.D. Prediction of shelf-life and changes in the quality characteristics of semidried persimmons stored at different temperatures. *Food Sci. Biotechnol.* **2017**, *26*, 1255–1262. [CrossRef] [PubMed]
32. Jia, X.; Katsuno, N.; Nishizu, T. Changes in the physico-chemical properties of persimmon (Diospyros kaki thunb.) during drying and quality deterioration during storage. *Rev. Agric. Sci.* **2019**, *8*, 1–14. [CrossRef]
33. Yamada, H.; Ando, T.; Tsutani, K.; Amano, S.; Yamamoto, Y. Mechanism of browning occurring during the processing of semi-dried persimmons. *J. Jpn. Soc. Hortic. Sci.* **2009**, *78*, 124–130. [CrossRef]
34. Mamet, T.; Yao, F.; Li, K.; Li, C. mei Persimmon tannins enhance the gel properties of high and low methoxyl pectin. *LWT Food Sci. Technol.* **2017**, *86*, 594–602. [CrossRef]
35. Im, J.; Lee, M. Physicochemical Compositions of Raw and Dried Wolha Persimmons. *Korean Soc. Food Preserv.* **2007**, *14*, 611–616.
36. Arnal, L.; Del Río, M.A. Removing astringency by carbon dioxide and nitrogen-enriched atmospheres in persimmon fruit cv. "Rojo brillante". *J. Food Sci.* **2003**, *68*, 1516–1518. [CrossRef]
37. Taira, S.; Ono, M.; Matsumoto, N. Reduction of persimmon astringency by complex formation between pectin and tannins. *Postharvest Biol. Technol.* **1997**, *12*, 265–271. [CrossRef]
38. Asgar, M.A.; Yamauchi, R.; Kato, K. Structural features of pectins from fresh and sun-dried Japanese persimmon fruit. *Food Chem.* **2004**, *87*, 247–251. [CrossRef]
39. Vega-Gálvez, A.; Puente-Díaz, L.; Lemus-Mondaca, R.; Miranda, M.; Torres, M.J. Mathematical Modeling of Thin-Layer Drying Kinetics of Cape Gooseberry (*Physalis peruviana* L.). *J. Food Process. Preserv.* **2014**, *38*, 728–736. [CrossRef]
40. Mundada, M.; Hathan, B.S.; Maske, S. Convective dehydration kinetics of osmotically pretreated pomegranate arils. *Biosyst. Eng.* **2010**, *107*, 307–316. [CrossRef]
41. Mulet, A. Drying modelling and water diffusivity in carrots and potatoes. *J. Food Eng.* **1994**, *22*, 329–348. [CrossRef]
42. Doymaz, I. Pretreatment effect on sun drying of mulberry fruits (*Morus alba* L.). *J. Food Eng.* **2004**, *65*, 205–209. [CrossRef]
43. Doymaz, I. Sun drying of figs: An experimental study. *J. Food Eng.* **2005**, *71*, 403–407. [CrossRef]
44. Doymaz, I. Convective drying kinetics of strawberry. *Chem. Eng. Process. Process Intensif.* **2008**, *47*, 914–919. [CrossRef]
45. Zogzas, N.; Maroulis, Z.B. Moisture diffusivity data compilation in foodstuff. *Dry. Technol.* **1996**, *14*, 2225–2253. [CrossRef]

foods

MDPI

Article

Tailoring the Functional Potential of Red Beet Purées by Inoculation with Lactic Acid Bacteria and Drying

Gabriel-Dănuț Mocanu, Ana Cosmina Chirilă, Aida Mihaela Vasile, Doina Georgeta Andronoiu, Oana-Viorela Nistor, Vasilica Barbu and Nicoleta Stănciuc *

Department of Food Science, Food Engineering, Biotechnology and Aquaculture, Faculty of Food Science and Engineering, "Dunarea de Jos" University of Galați, 800201 Galați, Romania; Danut.Mocanu@ugal.ro (G.-D.M.); cosminachana@gmail.com (A.C.C.); Aida.Vasile@ugal.ro (A.M.V.); Georgeta.Andronoiu@ugal.ro (D.G.A.); Oana.Nistor@ugal.ro (O.-V.N.); Vasilica.Barbu@ugal.ro (V.B.)

* Correspondence: Nicoleta.Stanciuc@ugal.ro; Tel.: +4-0336-130-183

Received: 16 October 2020; Accepted: 5 November 2020; Published: 6 November 2020

Abstract: This study was focused on a comparative analysis of two drying methods, such as convective and infrared drying, on the red beetroot purées with lactic acid bacteria, as a strategy for tailoring the health benefits of the selected plant. For both varieties, the total betalain contents varied from 13.95 ± 0.14 mg/g dry weight in *Beta vulgaris* var. *cylindra* when compared with 11.09 ± 0.03 mg/g dry weight in *Beta vulgaris* var. *vulgaris*, whereas significant differences were found in total phenolic and flavonoid contents. Significant drying induced changes were found in selected bioactives, in terms of total betalains, flavonoids, and polyphenols, which influenced the antioxidant activities of the purées, structure, and color parameters. In general, infrared technology was more protective, leading to an increase of 20% in flavonoids content. One logarithmic decrease in cell viability was observed in all powders samples. After the in vitro digestion, the betalains decreased, in both gastric and intestinal simulated juices, with a more pronounced profile in infrared processed purées. Textural and rheological analysis of the dried purées highlighted that the infrared drying is milder compared to the conventional one, allowing us to obtain powders with enhanced functional properties, in terms of bioactives content, cell viability, color, and structural and rheological behavior.

Keywords: beetroot; convective drying; infrared drying; purée; antioxidant activity; Fourier-transform infrared spectroscopy; confocal scanning microscopy; texture

1. Introduction

Considered in recent years as key components providing substantial benefits to human health, biologically active compounds, such as polyphenols and betalains, have recently gained particular interest from both academia and relevant industry from the perspective of developing foods and ingredients with positive effects for health [1].

Red beet (*Beta vulgaris*) are among the most popular varieties of vegetables, not only because of its significant content in biologically active compounds, such as polyphenols and betalains, but also their intense aromatic flavor [2]. Betalains are known as water-soluble nitrogen-containing pigments, consisting of two sub-classes: betacyanins (red-violet pigments) and betaxanthins (yellow-orange pigments) [3]. The health benefits of betalains are described as antimicrobial [4], inhibition of cell proliferation of human tumor cells [5], prevention of diseases like cancer, cardiovascular diseases [6], anti-inflammatory effects, antiradical and antioxidant activity [7], and enriching human low-density lipoproteins, which increase resistance to oxidation [8].

It is well known that color is an important quality indicator that determines the consumers' acceptance of foods. A more extensive application of natural colorants over synthetic ones has been observed recently on the market, especially because natural coloring is commonly associated with positive effects on human health and can also be used as functional additives. The use of pigments from a natural origin is usually limited due to their low stability, solubility, bioavailability, interaction with other components of the food matrix, etc. [9]. From the perspective of developing food products with enhanced properties, from both nutritional and functional point of view, it is necessary to choose appropriate methods for obtaining and stabilizing these valuable compounds.

Several colorants may be used for coloring foods, such as anthocyanins and carotenoids. The use of betalains for coloring food products is approved by the European Union and betalains are labeled as E-162; however, the use of betalains for specific uses provide several advantages, since these pigments, having the ability to cover a wide color range, from red to purple, are considered to be more stable to pH and temperature-induced changes. Due to the higher stability in a wider pH range, the use of betalains as a dye allows the replacement of anthocyanins for low-acid foods [3]. When comparing to carotenoids, which exhibit a yellow-orange color range, their applications are limited due to poor solubility in water; however, betaxanthins could be successfully used in application as yellow-orange food colorants [10].

Drying is a classical method of food preservation, providing several advantages such as smaller space for storage, lighter weight for transportation, and longer shelf-life [11]. Additionally, dried vegetables may be used as ingredients in ready-to-eat meals in order to add value by improving their functional quality due to health benefit compounds (vitamins, phytochemicals, and dietary fibers), and therefore, drying methods have acquired particular importance from the perspective of selecting methods that preserve the untainted qualities of the products [12]. One of the most used methods for drying vegetables and fruits is convective drying, which uses hot air to heat and remove water from the product [13].

The infrared assisted drying technology is one of the growingly popular approaches to provide heat for the drying of moist materials [14], allowing infrared radiation energy to be transmitted from the heating agent to the product. The technology brings some advantages, such as quick and homogeneous heating of the material without heating the surrounding air [15] where the generated heat in a layer below the surface is transferred to the material's center and surface. Additionally, due to the moisture transfer from the material's center to the surface, the heat and mass transfers are concurrent and countercurrent in layers, close to the material's surface and its other parts, respectively [16].

The main objective of this study was the comparative analysis of two techniques for drying red beetroot purées, impregnated with lactic acid bacteria to obtain powders with functional potential. Two drying methods were used: convective and infrared assisted drying methods. The red beetroot purées were analyzed for betalains content, antioxidant activity, cell viability, color, and in vitro release of total betalains. The structural and morphological particularities of the powders after drying were analyzed by using confocal laser microscopy. The drying induced changes in powders were analyzed by Fourier transform infrared spectroscopy (FT-IR). Textural properties and non-linear viscoelastic behavior were also analyzed for the rheological characterization of the powders.

2. Materials and Methods

2.1. Plant Materials

Two types of fresh red beet (taproots of *Beta vulgaris* L. var. *vulgaris* and *Beta vulgaris* L. var. *cylindra*) were obtained from the local vegetable market in Galați, Romania respectively Cahul, Republic of Moldova and stored at 4 °C before the analysis (not exceeding 48 h from procurement). The initial moisture content of the samples was 88.05 ± 0.20% and 88.17 ± 0.40% for *Beta vulgaris* L. var. *vulgaris* and *Beta vulgaris* L. var. *cylindra*, respectively, while the final moisture content was 8.03 ± 0.30% and 8.12 ± 0.20%.

2.2. *Lactic Bacteria*

The commercial culture *Lactobacillus casei* ssp. *paracasei* (*L. casei* 431®) was provided by Chr. Hansen (Hoersholm, Denmark).

2.3. *Sample Preparation*

2.3.1. Inoculum Preparation

The freeze-dried strain was reactivated in MRS (de Man, Rogosa, and Sharpe agar) broth (Merk, Darmstadt, Germany), at 37 °C for 24 h and grown under these conditions until achieving high cell concentration (>10^9 CFU/mL). The method described by Begot et al. [17] was used to estimate the bacterial cell concentration by the turbidimetric method. The logarithmic value of bacterial cell concentration was obtained by dilution and pour plate counting after incubation in MRS agar at 37 °C for 48 h under aerobic conditions [18].

2.3.2. Red Beet Pureés Preparation

Fresh beetroots were washed, peeled, and cut into small pieces. The purée was prepared by blending 1 kg from each variety of fresh red beet with 87.5 g of double distilled water for 5 min using a kitchen blender Philips HR2100/40. An inoculum of 1% of *L. casei* 431® was added into the purée samples. Six variants of red beet purée were obtained, coded as follows: BV_0—fresh sample of *Beta vulgaris* L. var. *vulgaris*, BC_0—fresh sample of *Beta vulgaris* L. var. *cylindra*, BV_C, BC_C—red beet purée dried by convection method, BV_{IR}, BC_{IR}—red beet purée dried by infra-red (IR) method.

2.4. *Drying of the Samples*

Drying experiments were carried out in an oven dryer with 5 perforated trays (mesh trays) (Concept SO4000 Infra 500W, Chocen, Czech Republic). The red beet purée samples were placed on a stainless steel tray with a surface area of 0.072 m^2. To ensure the optimal parameters of the drying process, the dryer was preheated for 1 h before use. The red beet purée (approximately 350 g for each variant) was spread on the tray surface obtaining a 1.5 mm thick layer using a digital micrometer. The layer thickness was measured in five points of the baking paper (in the four corners and the middle). The drying experiment was performed at a constant air velocity of 1.1 $m{\cdot}s^{-1}$, a relative humidity of 11.2%, and a drying temperature of 45 °C, while the ambient air temperature was 20 °C. The air velocity was measured with a VT 115 hotwire thermo-anemometer (Kimo Instruments, Millgrove, Ontario, Canada). The relative humidity was measured with a thermo hygrometer EE33 Series, fitted with a sensing probe (E + E Electronik Ges.m.b.H. Engerwitzdorf, Austria). The drying of red beet purée samples was carried out until the equilibrium humidity (3 h and 30 min in the case of convective drying and 2 h and 30 min for infrared drying). After each drying experiments, all the samples were cooled under laboratory conditions and stored in airtight containers. All drying experiments were performed in triplicate.

2.5. *Extraction of Betalains*

The extraction step of betalains was performed as described by Ravichandran et al. [19]. Briefly, 0.1 g of dried samples were mixed with 10 mL of 50% ethanol, agitated for 10 s, followed by centrifugation at 6000× *g* for 10 min. To increase the yield of betalains extraction, the supernatant was collected, and the extraction was repeated 3 more times. The supernatant was further used for the determination of betalains.

2.6. *Determination of Betalain Compounds by Spectrophotometric Methods*

The content of betaxanthins and betacyanins in the extracts was determined spectrophotometrically at 538 and 480 nm with a UV–Vis spectrometer.

The total betalain content (*BC*) was calculated using Equation (1):

$$BC\left(\frac{mg}{g}\right) = \frac{A \cdot DF \cdot M_w \cdot 1000}{ex1} \quad (1)$$

where *A* is the absorption, *DF* the dilution factor, and l the pathlength (1 cm) of the cuvette. For quantification of betacyanins and betaxanthins, the molecular weights (M_w) and molar extinction coefficients (*e*) (M_w of 550 g/mol; *e* of 60,000 L/mol·cm in H_2O) and (M_w of 308 g/mol; *e* of 48,000 L/mol·cm in H_2O) were applied.

2.7. Antioxidant Activity

The antioxidant activity of the ultrasound-assisted extracts dissolved in ethyl acetate was measured by using the modified ABTS (2,2′-Azino-bis(3-ethylbenzothiazoline-6-sulfonic acid) radical decolorization assay according to the method described by Miller and Rice—Evans [20]. The experiments were performed in triplicate.

2.8. Cells Viability

For viable cell counting of *L. casei* 431®, 10-fold serial dilutions of the samples were performed using sterile physiological serum (0.9 g NaCL%, *w/v*). The pour plate technique was employed. The viable cell number was determined by estimating the number of colony-forming units on the MRS-agar plates after incubation at 37 °C for 48 h. The counts were expressed as colony-forming units per gram. The viability of *L. casei* 431® was determined immediately after manufacturing and during storage at 7, 14, and 21 days.

2.9. In Vitro Release of the Betalains

The protocol described by Oancea et al. [21] was used to perform red beetroot dried purées to a simulated in vitro gastrointestinal digestion process, mimicking gastric and intestinal phases. The post-hydrolysis fractions were collected at every 30 min of digestion, centrifuged (10,000× *g* for 10 min), and analyzed for betalains content.

2.10. Color Parameters

The color parameters of fresh and dried samples were performed using a MINOLTA Chroma Meter CR-410 (Konica Minolta, Osaka, Japan). For color analysis, it is necessary to have a homogeneous powder the dried samples were ground for 20 s with a grinder (Gorenje SMK150B, Velenje, Republic of Slovenia). To obtain fresh red beetroot purée, the red beet was blended at 1900 rpm for 3 min with a laboratory blender (Philips HR2100/40, EC) to achieve a uniform color. The color parameters determined for fresh and dried red beetroot were L^* (lightness/darkness), a^* (red/green), and b^* (yellow/blue). The total color difference (ΔE) between samples was calculated according to Equation (2) [22]:

$$\Delta E = \sqrt{\left(L_0^* - L^*\right)^2 + \left(a_0^* - a^*\right)^2 + \left(b_0^* - b^*\right)^2} \quad (2)$$

Subscript 0 refers to the color of the fresh sample. The color intensity (C^*) and visual color appearance (h^*) were calculated according to Equations (3) and (4) [22]:

$$C^* = \sqrt{a^{*2} + b^{*2}} \quad (3)$$

$$h^* = tan^{-1}\left(\frac{b^*}{a^*}\right) \quad (4)$$

L^*, a^*, b^* values were used to establish the whiteness index according to the following Equation (5) [23]:

$$WI = 100 - \left[(100 - L^*)^2 + a^{*2} + b^{*2} \right]^{1/2} \quad (5)$$

According to Maskan [24] the browning index (BI) and the yellowness index (YI) were calculated using Equations (6) and (7):

$$BI = 100 \times \left(\frac{X - 0.31}{0.17} \right) \quad (6)$$

where

$$X = \frac{(a^* + 1.75 \cdot L^*)}{(5.645 \cdot L^* + a^* - 3.012 \cdot \hat{b})} \text{ and } YI = \frac{142.86 \cdot b^*}{L^*} \quad (7)$$

The color parameters are dimensionless. The color analysis was performed in triplicate.

2.11. Structural and Morphological Properties of the Dried Powders

To highlight the functional potential of red beet purées inoculated with *L. casei* 431 strain, the confocal laser scanning microscopy (CLSM) analysis was performed using an LSM 710 system (Carl Zeiss MicroImagining, Göttingen, Germany) equipped with a diode laser (405 nm), Ar-laser (458 nm, 488 nm, 514 nm), diode-pumped solid-state laser (DPSS; 561 nm) and HeNe-laser (633 nm). The 3D images were acquired with an AxioObserver Z1 inverted microscope (20x apochromatic objective, numerical aperture 1.4) and analyzed by ZEN 2012 SP1 software (black edition; Carl Zeiss MicroImagining, Göttingen, Germany). The autofluorescence of the samples was captured in the range of 500–660 nm wavelengths (emission spectrum of betalains) [25]. The acquisition parameters of the images were: mean method, line scan mode, speed 6, and 12-bit depth. In order to increase the signal-to-noise ratio, a frame average of eight scans was used.

2.12. FT-IR

The infrared spectra were collected using a Nicolet iS50 FT-IR spectrometer (Thermo Scientific, Waltham, MA, USA) equipped with a built-in ATR (Attenuated Total Reflectance) accessory, DTGS (Deuterated Triglycine Sulfate) detector, and KBr beam splitter. 32 scans were co-added over the range of 4000–400 cm^{-1} with a resolution of 4 cm^{-1}. Air was taken as the reference for the background spectrum before each sample. After each spectrum, the ATR plate was cleaned with ethanol solution. In order to verify that no residue from the previous sample remained, a background spectrum was collected each time and compared to the previous background spectrum. The FT-IR spectrometer was sited in a room that was air-conditioned with controlled temperature (21 °C).

2.13. Textural and Oscillatory Measurements

2.13.1. Texture Analysis

Fresh and reconstituted beetroot purée samples were poured into cylindrical plastic containers with 40 mm diameter and 50 mm length so that the height of the samples was 40 mm. A double penetration test was applied, using a 25.4 mm acrylic cylinder of a Brookfield CT3 texture analyzer (AMETEK Brookfield, Middleboro, MA, USA). The testing parameters were set as follows: target distance 10 mm, trigger load 0.067 N, pretest speed 2 mm/s, test speed 1 mm/s, return speed 1 mm/s, and load cell 1000 g. The textural parameters (firmness, adhesiveness, cohesiveness, and springiness) were determined with TexturePro CT V1.5 software, provided by Brookfield Engineering Labs. Inc., (Middleborough, MA, USA).

2.13.2. Rheological Analysis

The rheological analysis was performed with an AR2000ex controlled rheometer (TA Instruments, New Castle, DE, USA). Firstly, to identify the linear viscoelastic region, a strain sweep test was applied, maintaining a frequency of 1 Hz and varying the strain between 0.01 and 100%. Then, a dynamic frequency sweep test was applied, between 0.1 and 100 Hz, at a constant strain of 0.2 Hz. In all cases, storage modulus (G′) and loss modulus (G″) were registered. All the measurements were made at 25 °C, with a 40 mm diameter geometry and o closure gap of 2 mm.

2.14. Statistical Analysis

All analyses were performed in triplicate and data reported as mean ± standard deviation (SD). To identify significant differences, experimental data were subjected to one-way analysis of variance (ANOVA) after running the normality and homoscedasticity tests. The Tukey method with a 95% confidence interval was employed for post-hoc analysis; $p < 0.05$ was considered to be statistically significant. The statistical analysis was carried out using Minitab 18 software.

3. Results and Discussion

3.1. The Content of Selected Phytochemicals in Fresh Red Beet Purées

Initially, the betalains content of the fresh red beet pureés was measured by using spectrophotometric methods. A significantly ($p < 0.05$) higher content of total betalains was found in BC_0 pureés of 13.95 ± 0.14 mg/g DW when compared with 11.09 ± 0.03 mg/g DW in BV_0. Both varieties showed (Table 1) a similar content in betaxanthins, of 4.71 ± 0.06 mg/g DW and 4.06 ± 0.01 mg/g DW, whereas significant differences ($p < 0.05$) were found in betacyanins, with 9.23 ± 0.07 mg/g DW and 7.03 ± 0.02 mg/g DW, respectively. The total polyphenolic content (TPC) showed significant differences ($p < 0.05$), with a more than two-twice higher concentration of total polyphenols in BC_0 (71.94 ± 2.21 mg GAE/g DW) when compared with BV_0 (32.88 ± 0.34 mg GAE/g DW). Total flavonoids were found in a higher concentration in BV_0 of 36.00 ± 1.78 mg CE/g DW, when compared with 23.31 ± 1.34 mg CE/g DW in BC_0. The antioxidant profile showed significant differences ($p < 0.05$), with values of 93.46 ± 2.51 mMol Trolox/g DW in BC_0 and a lower value in BV_0 of 53.94 ± 2.87 mMol Trolox/g DW, suggesting that the betacyanins and polyphenols were the primary compounds responsible for antioxidant activity. The global phytochemical profile of the two varieties studied may be a result of varietal diversity, the influence of vegetation season, as well as climatic and cultivation conditions [26].

Table 1. Phytochemical profile of fresh and dried red beetroot purées.

Samples Code	Betalains		Total Polyphenols, mg GAE/g DW	Total Flavonoids, mg EC/g DW	ABTS Radical Scavenging Activity, mMol Trolox/g DW
	β-cyanin, mg/g DW	β-xanthin, mg/g DW			
BV_0	7.03 ± 0.02 ^{C}	4.06 ± 0.01 ^{C}	32.88 ± 0.34 ^{B}	23.31 ± 1.34 ^{B,C}	53.94 ± 2.87 ^{A}
BC_0	9.23 ± 0.07 ^{C}	4.71 ± 0.06 ^{C}	71.94 ± 2.21 ^{B}	36.00 ± 1.78 ^{B,C}	93.46 ± 2.51 ^{A}
BV_C	4.25 ± 0.02 ^{C}	2.53 ± 0.02 ^{C}	29.01 ± 0.97 ^{B}	16.22 ± 0.81 ^{B,C}	26.19 ± 0.32 ^{A}
BV_{IR}	4.49 ± 0.03 ^{C}	2.49 ± 0.02 ^{C}	27.76 ± 1.23 ^{B}	26.19 ± 1.50 ^{B,C}	92.21 ± 0.68 ^{A}
BC_C	4.19 ± 0.02 ^{C}	2.21 ± 0.01 ^{C}	26.05 ± 0.24 ^{B}	14.02 ± 0.49 ^{B,C}	55.95 ± 0.57 ^{A}
BC_{IR}	4.29 ± 0.02 ^{C}	2.60 ± 0,08 ^{C}	26.47 ± 1.18 ^{B}	27.06 ± 1.57 ^{B,C}	94.21 ± 0.38 ^{A}

BV_0—fresh sample of *Beta vulgaris* L. var. *vulgaris*; BC_0—fresh sample of *Beta vulgaris* L. var. *cylindra*; BV_C, BC_C—red beet purée dried by convection method; BV_{IR}, BC_{IR}—red beet purée dried by IR method. Means that do not share a letter as a superscript (A, B, C) are significantly different.

Regarding the individual betalains, Sawicki et al. [27] studied thirteen varieties and root parts of red beet and identified betanin and isobetanin from betacyanin group and vulgaxanthin I from betaxanthins as the major compounds. These authors reported betaxanthins levels ranging from 2.71 to 4.25 mg/g DW, whereas betacyanins content varied between 8.30 and 13.50 mg/g DW. The scavenging capacity of red beetroot varieties determined by the ABTS assay was found by these authors within the range from 37.68 to 49.71 mol Trolox/g DW, being positively correlated with the total betalain and betacyanin contents. In some other studies, it has been reported that the concentrations of betacyanins and betaxanthins in the roots of red beet varied from 400 and 2100 mg/kg fresh weight and between

200 and 1400 mg/kg fresh weight, respectively [28]. Moreover, the betalain content differs depending on the red beetroot variety [27,29].

3.2. The Content of Selected Phytochemicals in Dried Red Beet Purées

It is well known that food drying has as the main purpose the removal of free water from products to a level considered critical for chemical and microbiological reactions while reducing weight and volume, intended to reduce transportation and storage costs [13]. In our study, two drying methods were used for red beetroot purées, namely convective and infrared drying. Following the drying, the phytochemicals profile (Table 1) of the resulting powders were analyzed, to evaluate the losses of the main red beet pigments. A significant decrease in total betalains of approximatively 50% and 43% were observed in BV_C and BC_C, respectively. The infrared technology caused a less intense degradation effect in the total betalains content, with approximatively 50% and 38% in BV_{IR} and BC_{IR}, respectively.

A comparative analysis of the two drying methods on the betalains showed a more protective effect of the infrared technology, with a decrease of approximately 39% in BC_{IR} and 36% in BV_{IR} for betaxanthins, whereas convection drying leads to a significantly higher decrease of approximately 46% for both varieties. The convection drying caused a significant decrease in the betacyanin, with 40% in both varieties, whereas infrared drying leads to a significantly different decrease patterns in the selected purées, with a decrease of 51% in BC_{IR} and approximately 39% in BV_{IR}.

Drying by convection of BC_C purées caused a significant decrease in polyphenols and flavonoids, of approximately 60% and 52%, respectively. The selected phytochemicals were more stable in BV_C, with a reduction of approximately 21% and 30%, respectively. However, when drying by infrared technology, polyphenols and flavonoids were found to decrease in BC_{IR} by 62% and 23%, respectively, whereas in BV_{IR} a slight decrease in polyphenols (of 19%) and an increase of 20% in flavonoids content was observed.

A significant decrease of approximatively 72% was found in antioxidant activity of BC_C purées caused by convective drying, whereas heating by infrared of BV_C leads to an increase of 70% ABTS radical scavenging activity, due probably to the increase in flavonoid content. The hydrolysis of C-glycosides in the flavonoid composition leads to the formation of monomers that increase the total amount of flavonoids, as well as the antioxidant capacity [30].

Guldiken et al. [31] pointed out that during processing, including drying, the phytochemicals in the red beet products undergo both increases and decrease. For example, Sawicki et al. [32] analyzed three types of processing parameters of red beetroot (boiling, fermentation, and microwave-vacuum treatment) and suggested that the main compound among the betacyanins group in the analyzed products was betanin, while the predominant compound from the betaxanthins group was vulgaxanthin I. This phenomenon occurs as a result of red beet exposure to light and temperature, causing the transformation of betacyanins into decarboxylated forms. A significant reduction of total betalains with approximatively 54% was also observed by these authors when boiling whole roots for 45 min, while the total betacyanin and betaxanthin contents decreased by approximately 43% and 87%, respectively. Similar results were reported by Sawicki and Wiczkowski [33], suggesting a significant 54% reduction in the total betalains content after boiling of whole roots for 60 min.

3.3. Cells Viability

During the storage time (28 days at 4 °C), a decrease in the number of *L. casei 431*® viable cells of all samples was observed. Thus, the number of *L. casei* 431® cells decreased after 21 days from 8.79 log CFU/g to 7.17 log CFU/g for samples BV_C and BV_{IR}. For samples BC_C and BC_{IR} in the same conditions, the number of *L. casei* 431® cells decreased from 8.58 log CFU/g to 7.89 log CFU/g. A similar tendency was reported by Paraschiv et al. [34] for a probiotic *L. casei* 431® strain, after 21 days, during the storage in refrigeration conditions. The drying method did not significantly affect the cell viability. However, higher viability by 1 log CFU/g was observed in BC purée samples compared with BV purée samples.

For a food to be considered probiotic, it must contain viable probiotic cultures in populations over 6 log CFU/g during the shelf life of the product [35]. *L. casei* 431® remained viable in populations of more than 7 log CFU/g during 21 days of storage for all samples, which makes the red beet purée suitable to be defined as a functional product.

3.4. In Vitro Release of Betalains

The in vitro digestion of purées was determined by simulating the gastric and intestinal phases. The content of the betalains in the digestion phases resulting from in vitro gastrointestinal digestion of red beet products was analyzed. As suggested by Sawicki et al. [32], the obtained results may offer the opportunity to evaluate the changes in the total content of betalains after in vitro digestion and the potential for in vitro bioaccessibility of the main phytochemicals from beetroot dried powders. As shown in Figure 1a, the total betalains content decrease from convection dried red beet powders in BC_C after in vitro gastric digestion up to 20% during 120 min of reaction.

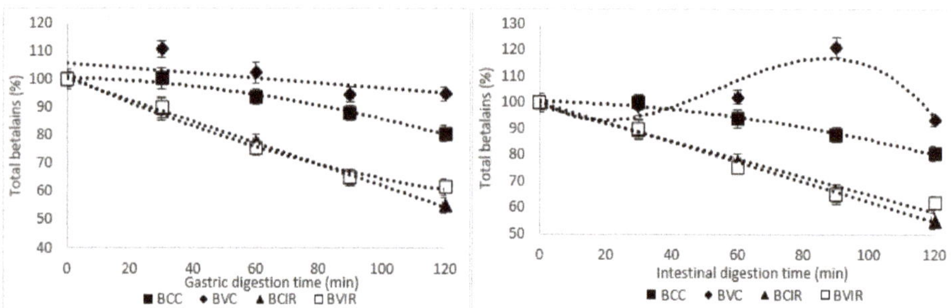

(a) (b)

Figure 1. The in vitro digestion of total betalains in powders samples in gastric simulated juice (**a**) and intestinal simulated juice (**b**): BV_C, BC_C—red beet purée dried by convection method; BV_{IR}, BC_{IR}—red beet purée dried by infra-red (IR) method.

The total concentration of betalains from BV_C in the gastric phases were increased by up to 10% after 30 min of digestion and decreased up to a maximum of 5% after 120 min (Figure 1a). Significantly, the decrease in total betalains content continued in intestinal simulated juice for BC_C, ranging from 6% at the beginning of digestion to a maximum of 19% after 120 min of reaction. In BV_C, a release in the total betalains was observed in the first 90 min of intestinal digestion, up to 21%, followed by a decrease in total betalains content (Figure 1b).

The infrared dried pattern of the total betalains in in vitro digestion is given in Figure 1a. In gastric digestion, BC_{IR} and BV_{IR} showed a significant decrease in total betalains, of approximately 45% and 38%, respectively. In simulated intestinal juice, the decrease reached a maximum value of 23% and 18% BC_{IR} and BV_{IR}, respectively, after 120 min of reaction (Figure 1b).

It has been suggested that betacyanins and betaxanthins have broad pH stability in the range pH 3.0–7.0 [36]. However, in a very acidic or basic environments, the molecules are involved in molecular cleavage, decarboxylation, and formation of other products [36]. Therefore, the significant decrease of total betalains in the gastric environment is probably due to the high instability at acidic conditions. Based on our results, a protective effect of the matrices may be considered in intestinal simulated juice. Our results are in good agreement with Sawicki et al. [32] and Tesoriere et al. [37], suggesting the importance of the food matrix in affecting the stability of betalains to the acidic pH.

3.5. Structural and Morphological Properties of the Dried Powders

The confocal analysis revealed the presence of many vegetal tissue fragments, coming from beetroot. These fragments present parenchymal cells, with dimensions of 104.91–110.84 μm (for BV sample) and 44.54–116.28 μm (for BC sample). The cellular content is rich in betalains which have an absorption peak at 536–540 nm and a wide emission spectrum, between 506–660 nm [25], and are biologically active compounds responsible for the remarkable antioxidant properties for which red beet is in the top 10 the most valuable vegetables [38].

Debris of vascular tissue (fragments of the tracheal elements of xylem), spirally ornamented, are often found, which is an important source of fiber for consumers (Figure 2, BC_{IR}). In comparison with the fresh purees, the reconstituted samples coming from IR drying possess a higher number of lysed cells, with fragmented cell walls, as a consequence of rapid dehydration, which also affected the cellular constituents (Figure 2, BV_{IR} and BC_{IR}).

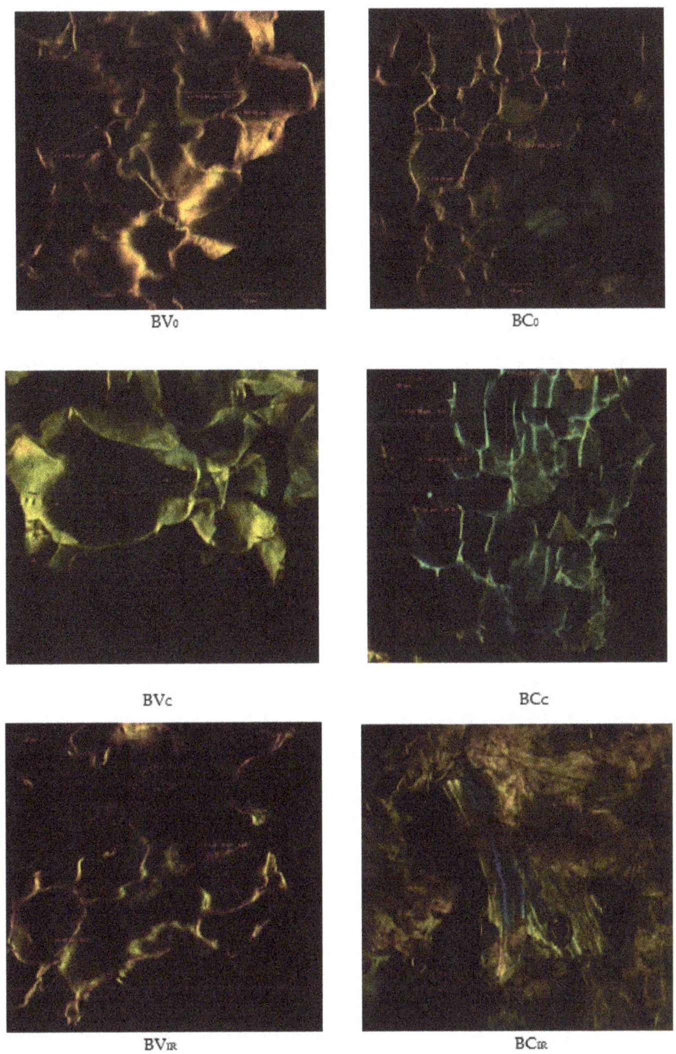

Figure 2. Confocal laser microscopy images of fresh and reconstituted red beetroot purée samples obtained with objective lenses of 20x: BV_0—fresh sample of *Beta vulgaris* L. var. *vulgaris*; BC_0—fresh sample of *Beta vulgaris* L. var. *cylindra*; BV_C, BC_C—red beet purée dried by convection method; BV_{IR}, BC_{IR}—red beet purée dried by infra-red (IR) method; *scale bar* 50 μm.

The convection drying technology is milder, a fact proved by the confocal images captured from the reconstituted BV_C and BC_C samples in which large cells can be visualized after rehydration (187.12 μm, respectively 137.33 μm), with rich content in valuable vegetable pigments (in green-yellow). In the BV_C samples, acquired with ZEN 2012 SP1 Black software, zoom 1.2, it can be seen even the microcolonies/biofilms resulting from the aggregation of probiotic lactic acid bacteria (*L. casei* 431 strain) from the functional product. Carbohydrates and biologically active compounds from plant tissue are an excellent support for maintaining the viability of lactic acid bacteria, so they will act synergistically and symbiotically. This result is also supported by previous studies by Barbu et al. [39] who used a strain of *Lb. plantarum* BL3 to obtain three variants of functional products based on beetroot.

3.6. FTIR

The FTIR spectra of all the red beetroot tested samples were recorded in the spectral range of 400–4000 cm^{-1} are shown in Figure 3. As can be seen, the main differences between the FTIR spectra of convection and IR dried beetroot samples were observed in the spectral ranges of 1028–1039. The FTIR spectra recorded weak to wide bands at 3550 and 1623 cm^{-1}, which correspond to stretching vibrations of OH groups and H-O-H bending arising from the moisture content of the red beetroot test samples, respectively. The weak bands were due to the loss of moisture content generated by the drying process. Similar findings were reported by Nesakumar et al. [40] in a study over the spoilage of red beetroot and by Singh et al. [41] for the extraction of betalain pigments from beta Vulgaris peels by microwave pretreatment.

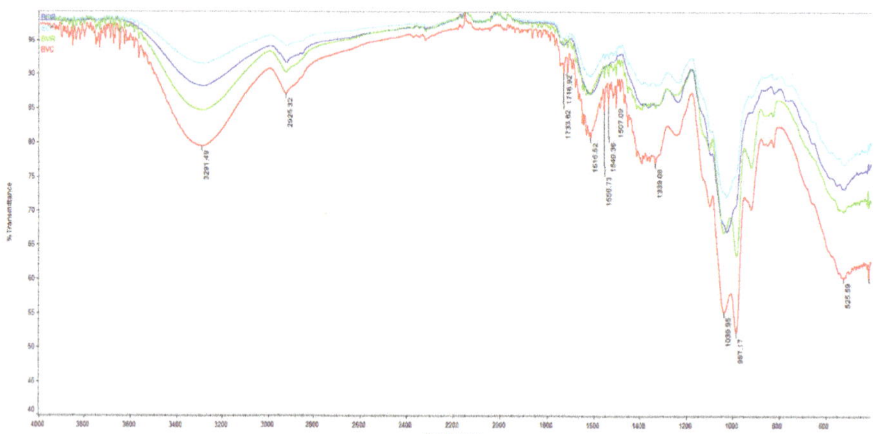

Figure 3. Fourier-transform infrared spectroscopy spectra of red beetroot dried samples: BV_C, BC_C—red beet purée dried by convection method; BV_{IR}, BC_{IR}—red beet purée dried by infra-red (IR) method.

3.7. Color Parameters

According to Henao-Ardila et al. [42] for acceptability of a powdered product, color represents an important parameter considered essential by the consumers, so reconstituted powder should have the same color as the fresh product. A significant effect on the color changes of the red beet is to be found in the processing temperatures and drying methods. The results of the color measurement for fresh and dried red beet purées are shown in Table 2.

Table 2. Effects of drying treatment on color parameters of fresh and dried red beetroot purées.

Color Parameters	Fresh Samples		Dried Samples			
	BV_0	BC_0	BV_C	BV_{IR}	BC_C	BC_{IR}
L^*	21.16 ± 0.3 ^{B}	21.19 ± 0.15 ^{B}	25.78 ± 0.15 ^{B}	30.67 ± 0.02 ^{B}	29.68 ± 0.12 ^{B}	32.15 ± 0.12 ^{B}
a^*	5.00 ± 0.2 ^{B,C}	8.32 ± 0.03 ^{B,C}	10.29 ± 0.22 ^{B,C}	14.19 ± 0.02 ^{B,C}	25.66 ± 0.26 ^{B,C}	29.14 ± 0.04 ^{B,C}
b^*	1.45 ± 0.04 ^{C}	1.41 ± 0.02 ^{C}	5.07 ± 0.21 ^{C}	6.56 ± 0.12 ^{C}	6.02 ± 0.03 ^{C}	7.78 ± 0.04 ^{C}
ΔE	-	-	8.02 ± 0.27 ^{B,C}	14.84 ± 0.28 ^{B,C}	19.84 ± 0.28 ^{B,C}	24.37 ± 0.21 ^{B,C}
C^*	5.21 ± 0.20 ^{B,C}	8.44 ± 0.03 ^{B,C}	11.47 ± 0.24 ^{B,C}	15.64 ± 0.03 ^{B,C}	26.35 ± 0.04 ^{B,C}	30.16 ± 0.25 ^{B,C}
h^*	16.17 ± 0.31 ^{B,C}	9.62 ± 0.11 ^{B,C}	26.23 ± 0.18 ^{B,C}	24.81 ± 0.47 ^{B,C}	13.20 ± 0.06 ^{B,C}	14.95 ± 0.10 ^{B,C}
BI	28.57 ± 0.00 ^{A}	35.29 ± 0.00 ^{A}	47.05 ± 0.00 ^{A}	52.94 ± 0.00 ^{A}	76.47 ± 0.00 ^{A}	88.23 ± 0.00 ^{A}
WI	20.98 ± 0.29 ^{B,C}	20.74 ± 0.15 ^{B,C}	24.89 ± 0.11 ^{B,C}	28.93 ± 0.03 ^{B,C}	24.90 ± 0.11 ^{B,C}	25.75 ± 0.04 ^{B,C}
YI	9.78 ± 0.28 ^{B,C}	9.51 ± 0.09 ^{B,C}	28.09 ± 0.28 ^{B,C}	30.56 ± 0.60 ^{B,C}	28.97 ± 0.13 ^{B,C}	34.57 ± 0.22 ^{B,C}

BV_0—fresh sample of *Beta vulgaris* L. var. *vulgaris*; BC_0—fresh sample of *Beta vulgaris* L. var. *cylindra*; BV_C, BC_C—red beet purée dried by convection method; BV_{IR}, BC_{IR}—red beet purée dried by IR method. Means that do not share a letter as a superscript (A, B, C) are significantly different. L^*—clarity; a^*—red/green colour component; b^*—blue/yellow colour component; ΔE—total colour difference; C^*—chroma; h^*—hue angle; BI—browning index; WI—whiteness index; YI—yellowness index.

The color values (for each type of drying method and plant material) were registered and indicated that the L^*, a^*, and b^* parameters increased for all powder samples as compared to the fresh samples. L^* values, which show the whiteness of the product, ranged between 21.16 ± 0.21 and 30.67 ± 0.02 for samples obtained from BV and between 21.19 ± 0.15 and 32.15 ± 0.12 for samples obtained from BC. After drying, the color of powder samples became lighter and reddish probably due to the betacyanins. Ng and Sulaiman [43] observed similar results for beetroot powder while Ochoa-Martinez et al. [44] for beetroot-orange juice. They also reported that thermal processing causes some modifications in red-orange betanins, by increasing the values of the red/green color parameter (a^*). The color coordinate (a^*) varied between 5.00 ± 0.21 (BV_0) and 29.14 ± 0.04 (BC_{IR}). In the case of dried samples, a maximum change of color was about 51.72% and this percent was established for the sample BC_{IR} compared with the value of the fresh product. The IR drying process may induce rapid heat penetration directly into the product. Therefore, a higher color change in IR dried samples compared with convective dried samples was noticed. b^* values of dried red beetroot samples increased compared with the values of the fresh samples. The increase of the b^* value was reflected by the decomposition of betanin with the occurrence of yellow compounds as betalamic acid, or condensation of betalamic acid with amino compounds (betaxanthins) or cyclo-DOPA (neobetacianins) [45].

The C^* values of the fresh red beetroot purée increase after the convection and IR drying process. In general, the C^* value ranged from 5.21 ± 0.20 to 30.16 ± 0.25. When the value of the hue angle (h^*) presented in Table 2 was checked, it can be concluded that this value tended to grow for all dried samples in the same mode as for the chroma (C^*) values. The visual color appearance value (h^*) in beetroot powders was the highest in the samples processed by convection method (26.23 ± 0.18 for BV_C sample), followed by the fresh sample and the other IR treated samples. The appearance of the product, i.e., the color, is very important for the consumers; for that reason, ΔE represents an important color parameter of the dried product, which reveals the human eye's capacity to differentiate the colors of the samples. The total color change (ΔE) of the dried red beetroot samples varied from 8.02 ± 0.27 to 24.37 ± 0.21 units, influenced by the applied drying method and temperature. The L^*, a^*, and b^* parameters were used to appreciate browning index (BI), which indicates the purity of the brown color, whiteness index (WI), which is an indicator of enzymatic discoloration and yellowness index (YI). YI is associated with product degradation by light, chemical exposure, and processing. Regarding the BI, the fresh red beetroot purées showed the lowest values (28.57 ± 0.00 for BV_0 and 35.29 ± 0.00 for BC_0). In our study, the values of BI increased, due to the enzymatic and non-enzymatic reactions that took place at a low temperature—short period of time. The increased YI values for all dried samples compared to the fresh samples can be assigned to the pigment decomposition and the occurrence of Maillard reactions.

3.8. Textural and Oscillatory Measurements

3.8.1. Texture Analysis

The texture analysis of the purée samples (Table 3) revealed that IR drying induced fewer damages in the vegetal tissue, comparing with convection drying. This could be concluded from the values of firmness, which are the lowest in the case of fresh pureés (0.76 ± 0.03 N for BV_0 and 0.72 ± 0.05 N for BC_0). The rehydrated samples did not manage to absorb the same amount of water which was removed during the drying process. Therefore, the vegetal particles are firmer and tend to gather at the bottom of the container, leading to increased firmness. At the same time, at rehydration, the purée particles did not manage to recover the bonds between them, due to cell war disruptions and fissures as reported by several microstructure studies [12,30]. This fact leads to lower adhesiveness, springiness, and cohesiveness values in the rehydrated samples.

Table 3. The values of textural parameters.

Parameter	BV_0	BV_C	BV_{IR}	BC_0	BC_C	BC_{IR}
Firmness, N	0.76 ± 0.03 ^{B}	1.31 ± 0.15 ^{B}	0.97 ± 0.04 ^{B}	0.72 ± 0.05 ^{B}	1.26 ± 0.01 ^{B}	1.02 ± 0.04 ^{B}
Adhesiveness, mJ	3.38 ± 0.29 ^{B}	0.69 ± 0.02 ^{B}	0.35 ± 0.01 ^{B}	4.58 ± 0.31 ^{B}	0.44 ± 0.02 ^{B}	0.25 ± 0.02 ^{B}
Cohesiveness, -	0.72 ± 0.10 ^{B}	0.47 ± 0.01 ^{B}	0.46 ± 0.09 ^{B}	0.67 ± 0.02 ^{B}	0.45 ± 0.01 ^{B}	0.42 ± 0.01 ^{B}
Springiness, mm	9.18 ± 0.19 ^{A}	5.77 ± 0.07 ^{A}	6.23 ± 0.10 ^{A}	8.72 ± 0.10 ^{A}	5.29 ± 0.08 ^{A}	6.06 ± 0.07 ^{A}

BV_0—fresh sample of *Beta vulgaris* L. var. *vulgaris*; BC_0—fresh sample of *Beta vulgaris* L. var. *cylindra*; BV_C, BC_C—red beet purée dried by convection method; BV_{IR}, BC_{IR}—red beet purée dried by infra-red (IR) method. Means that do not share a letter as a superscript (A, B, C) are significantly different.

When comparing the values of the textural parameters obtained for convectional drying with those obtained for IR drying, it can be noticed that the latter are closer to fresh samples. It can be concluded that IR drying is milder compared to the convectional one. As for beetroot varieties, from a textural point of view, no major differences between samples were observed.

3.8.2. Rheological Analysis Results

The rheological behavior of the fresh and reconstituted beetroot purée samples is presented in Figures 4 and 5. For both of the beetroot varieties purées, the fresh samples registered lower resistance to the applied stress when compared to the reconstituted samples. This behavior could be explained by the insoluble particles presented in the rehydrated samples, which lead to a more rigid network than in the fresh purée, as observed by Leverrier et al. [46] for apple purée.

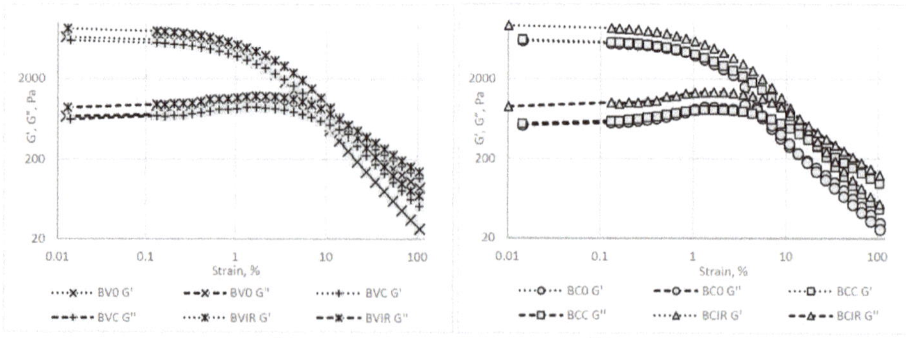

(a) (b)

Figure 4. Rheological behavior as revealed by the strain sweep test: (**a**) *Beta vulgaris L. var. vulgaris*, (**b**) *Beta vulgaris L. var. cylindra.*

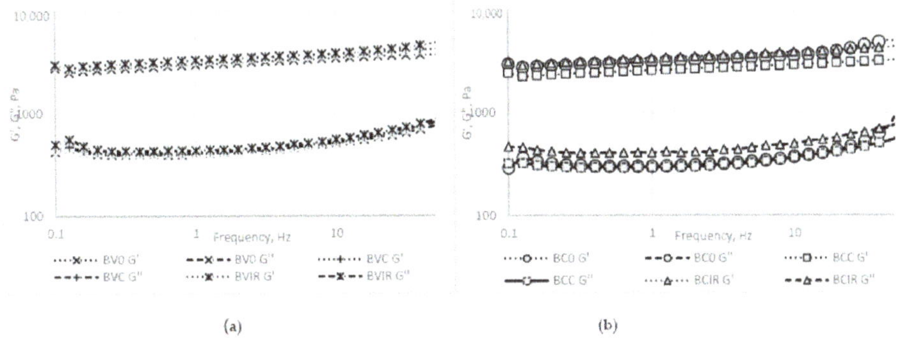

Figure 5. Rheological behavior as revealed by the frequency sweep test: (**a**) *Beta vulgaris L. var. vulgaris*, (**b**) *Beta vulgaris L. var. cylindra*.

The frequency sweep test revealed a gel-like behavior for all purée samples, as the elastic modulus (G′) shows higher values than the loss modulus (G″).

The IR drying determined a reconstituted purée sample that presented the highest consistency, stiffness, and resistance to flow marked by the G′/G″ intersection point, usually associated with the structure breakdown and beginning of the flow.

4. Conclusions

Two varieties of red beetroot were studied as potential sources of betalains, polyphenols, and flavonoids in fresh and dried forms of purées, from the perspectives of developing foods or food-grade ingredients with functional values, in terms of phytochemical and lactic acid bacteria. The drying of red beetroot influenced the phytochemical profile of both varieties, with a more protective effect registered for infrared drying processing. However, when drying by infrared technology, polyphenols and flavonoids were found to decrease in *Beta vulgaris* L. var. *cylindra*, whereas in *Beta vulgaris* L. var. *vulgaris*, a slight decrease in polyphenols and an increase in flavonoids content was observed. The decrease in phytochemicals significantly affected the antioxidant activity of the purées, regardless of the drying technology. In all the dried powders, a one logarithmic decrease in lactic acid bacteria viability was registered. After drying, the color of beetroot powder samples became lighter and reddish.

Our study brings some valuable results involving the optimization of drying processes to validate the quality of phytochemicals retention. The high antioxidant activity of red beet powders combined with a 10^7 CFU/g for lactic acid bacteria represents a good base for the development of a biosynthetic process to obtain additives with a potential for application in the food and pharmaceutical industry.

Author Contributions: Conceptualization, N.S. and G.-D.M.; formal analysis, A.C.C., A.M.V., D.G.A., O.-V.N., and V.B.; writing—original draft preparation, N.S., G.-D.M., A.M.V., D.G.A., O.-V.N., and V.B.; writing—review and editing, N.S. All authors have read and agreed to the published version of the manuscript.

Funding: This research received no external funding.

Acknowledgments: This work was supported by the project "Excellence, performance and competitiveness in the Research, Development and Innovation activities at "Dunarea de Jos" University of Galati", acronym "EXPERT", financed by the Romanian Ministry of Research and Innovation in the framework of Programme 1-Development of the national research and development system, Sub-programme 1.2-Institutional Performance-Projects for financing excellence in Research, Development and Innovation, Contract no. 14PFE/17.10.2018. The authors are grateful for the technical support offered by the Grant POSCCE ID 1815, cod SMIS 48745 (www.moras.ugal.ro). The *Integrated Center for Research, Expertise and Technological Transfer in Food Industry* is acknowledged for providing technical support.

Conflicts of Interest: The authors declare no conflict of interest.

References

1. Angelino, D.; Godos, J.; Ghelfi, F.; Tieri, M.; Titta, L.; Lafranconi, A.; Marventano, S.; Alonzo, E.; Gambera, A.; Sciacca, S.; et al. Fruit and vegetable consumption and health outcomes: An umbrella review of observational studies. *Int. J. Food Sci. Nutr.* **2019**, *70*, 652–667. [CrossRef]
2. Rocchetti, G.; Tomas, M.; Zhang, L.; Zengin, G.; Lucini, L.; Capanoglu, E. Red beet (*Beta vulgaris*) and amaranth (*Amaranthus* sp.) microgreens: Effect of storage and in vitro gastrointestinal digestion on the untargeted metabolomic profile. *Food Chem.* **2020**, *332*, 127415. [CrossRef] [PubMed]
3. Stintzing, F.C.; Carle, R. Functional properties of anthocyanins and betalains in plants, food and in human nutrition. *Trends Food Sci. Technol.* **2004**, *15*, 19–38. [CrossRef]
4. Strack, D.; Vogt, T.; Schliemann, W. Recent advances in betalain research. *Phytochemistry* **2003**, *62*, 247–269. [CrossRef]
5. Reddy, K.M.; Ruby, L.; Lindo, A.; Nair, G.M. Relative inhibition of lipid peroxidation, cyclooxygenase enzymes and human tumor cells prolifieration by natural food color. *J. Agric. Food Chem.* **2005**, *53*, 9268–9273. [CrossRef]
6. Delgado-Vargas, F.; Jiménez, A.R.; Paredes-López, O. Natural pigments: Carotenoids, anthocyanins, and betalains-characteristics, biosynthesis, processing, and stability. *Crit. Rev. Food Sci. Nutr.* **2000**, *40*, 173–289. [CrossRef] [PubMed]
7. Gentile, C.; Tesoriere, L.; Allegra, M.; Livrea, M.A.; Alessio, P.D. Antioxidant betalains from cactus pear (*Opuntia ficus-indica*) inhibit endothelial ICAM-1expression. *Ann. N. Y. Acad. Sci.* **2004**, *1028*, 481–486. [CrossRef]
8. Tesoriere, L.; Allegra, M.; Butera, D.; Livrea, M.A. Absorption, excretion, and distribution of dietary antioxidant betalains in LDLs Potential health effects of betalains in humans. *Am. J. Clin. Nutr.* **2004**, *80*, 941–945. [CrossRef]
9. Różyło, R. Recenttrends in methods used to obtain natural food colorants by freeze-dryin. *Trends Food Sci. Technol.* **2020**, *102*, 39–50. [CrossRef]
10. Azeredo, H.M.C. Betalains: Properties, sources, applications, and stability: A review. *Int. J. Food Sci. Tech.* **2009**, *44*, 2365–2376. [CrossRef]
11. Dandamrongrak, R.; Young, G.; Mason, R. Evaluation of various pre-treatments for the dehydration of banana and selection of suitable drying models. *J. Food Eng.* **2002**, *55*, 136–146. [CrossRef]
12. Vallespir, F.; Rodríguez, O.; Eim, V.S.; Rosselló, C.; Simal, S. Effects of freezing treatments before convective drying on quality parameters: Vegetables with different microstructures. *J. Food Eng.* **2019**, *249*, 15–24. [CrossRef]
13. Kaleta, A.; Górnicki, K. Some remarks on evaluation of drying models of red beet particles. *Energy Convers. Manag.* **2010**, *51*, 2967–2978. [CrossRef]
14. Torki-Harchegani, M.; Ghanbarian, D.; Maghsoodi, V.; Moheb, A. Infrared thin layer drying of saffron (*Crocus sativus* L.) stigmas: Mass transfer parameters and quality assessment. *Chin. J. Chem. Eng.* **2017**, *25*, 426–432. [CrossRef]
15. Kipcak, A.S.; Doymaz, İ.; Derun, E.M. Infrared drying kinetics of blue mussels and physical properties. *Chem. Ind. Chem. Eng. Q.* **2019**, *25*, 1–10. [CrossRef]
16. Jafari, F.; Movagharnejad, K.; Sadeghi, E. Infrared drying effects on the quality of eggplant slices and process optimization using response surface methodology. *Food Chem.* **2020**, *333*, 127423. [CrossRef]
17. Begot, C.; Desnier, I.; Daudin, J.D.; Labadie, J.C.; Lebert, A. Recommendations for calculating growth parameters by optical density measurements. *J. Microbiol. Methods* **1996**, *25*, 225–232. [CrossRef]
18. Wehr, H.M.; Frank, J.F. *Standard Methods for the Examination of Dairy Products*, 17th ed.; APHA Press-American Public Health Association: Washington, DC, USA, 2004.
19. Ravichandran, K.; Saw, N.M.M.T.; Mohdaly, A.A.A.; Gabr, A.M.M.; Kastell, A.; Riedel, H.; Cai, Z.; Knorr, D.; Smetanska, I. Impact of processing of red beet on betalain content and antioxidant activity. *Food Res. Int.* **2013**, *50*, 670–675. [CrossRef]
20. Miller, N.J.; Rice-Evans, C.A. Factors influencing the antioxidant activity determined by the $ABTS^{\cdot+}$ radical cation assay. *Free Radic. Res.* **1997**, *26*, 195–199. [CrossRef]

21. Oancea, A.-M.; Mahadi, H.; Vasile, A.M.; Barbu, V.; Enachi, E.; Bahrim, G.E.; Râpeanu, G.; Silvi, S.; Stănciuc, N. Functional evaluation of microencapsulated anthocyanins from sour cherries skins extract in whey proteins isolate. *LWT* **2018**, *95*, 129–135. [CrossRef]
22. *[CIE LEAD, EN ISO/CIE 11664-4:2019] Colorimetry—Part 4: CIE 1976 L * a * b * Colour Space*; ISO/CIE 11664-4:2019; ISO Copyright Office: Geneva, Switzerland, 2019.
23. Bolin, H.R.; Huxsoll, C.C. Control of minimally processed carrot (*Daucus carota*) surface discoloration caused by abrasion peeling. *J. Food Sci.* **1991**, *46*, 416–418. [CrossRef]
24. Maskan, M. Kinetics of colour change of kiwifruits during hot air and microwave drying. *J. Food Eng.* **2001**, *48*, 169–175. [CrossRef]
25. Miguel, M.C. Betalains in Some Species of the Amaranthaceae Family: A Review. *Antioxidants* **2018**, *7*, 53. [CrossRef]
26. Wiczkowski, W.; Topolska, J.; Honke, J. Anthocyanins profile and antioxidant capacity of red cabbages are influenced by genotype and vegetation period. *J. Funct. Foods* **2014**, *7*, 201–211. [CrossRef]
27. Sawicki, T.; Baczek, N.; Wiczkowski, W. Betalain profile, content and antioxidant capacity of red beetroot dependent on the genotype and root part. *J. Funct. Foods* **2016**, *27*, 249–261. [CrossRef]
28. Ninfali, P.; Bacchiocca, M.; Antinelli, A.; Biagiotti, E.; Di Gioacchino, A.M.; Piccoli, G.; Stocchi, V.; Brandi, G. Characterization and biological activity of the main flavonoids from Swiss chard (*Beta vulgaris* subspecies *cycla*). *Phytomed. Int. J. Phytother. Phytopharm.* **2007**, *14*, 216–221. [CrossRef] [PubMed]
29. Slatnar, A.; Stampar, F.; Vebaric, R.; Jakopic, J. HPLC-MSn identification of betalains profile of different beetroot (*Beta vulgaris* L. *ssp. vulgaris*) parts and cultivars. *J. Food Sci.* **2015**, *80*, 1952–1958. [CrossRef]
30. Seremet, L.; Nistor, O.V.; Andronoiu, D.G.; Mocanu, G.D.; Barbu, V.V.; Maidan, A.; Rudi, L.; Botez, E. Development of several hybrid drying methods used to obtain red beetroot powder. *Food Chem.* **2020**, *310*, 125637. [CrossRef] [PubMed]
31. Guldiken, B.; Toydemir, T.; Memis, K.N.; Okur, S.; Boyacioglu, D.; Capanoglu, E. Home-processed red beetroot (*Beta vulgaris* L.) products: Changes in antioxidant properties and bioaccessibility. *Int. J. Mol. Sci.* **2016**, *17*, 858. [CrossRef]
32. Sawicki, T.; Martinez-Villaluenga, C.; Frias, J.; Wiczkowski, W.; Peñas, E.; Baczek, N.; Zieliñski, H. The effect of processing and in vitro digestion on the betalain profile and ACE inhibition activity of red beetroot products. *J. Funct. Foods* **2019**, *55*, 229–237. [CrossRef]
33. Sawicki, T.; Wiczkowski, W. The effects of boiling and fermentation on betalain profiles and antioxidant capacities of red beetroot products. *Food Chem.* **2018**, *259*, 292–303. [CrossRef]
34. Paraschiv, D.; Vasile, A.; Constantin, M.; Ciobanu, A.; Bahrim, G. Study of physiological properties of some probiotics in multiple cultures with mesophilic lactic acid bacteria by Flora Danica Ch. Hansen commercial starter. *Ann. Univ. Dunarea Jos Galati Fasc. VI Food Technol.* **2011**, *35*, 56–65.
35. Trisnawita, Y.; Silalahi, J.; Sinaga, S.M. The effect of storage condition on viability of lactic acid bacteria in probiotic product. *Asian J. Pharm. Clin. Res.* **2018**, *11*, 84–86. [CrossRef]
36. Herbach, K.M.; Stintzing, F.C.; Carle, R. Betalain stability and degradation. Structural and chromatic aspects. *J. Food Sci.* **2006**, *71*, R41–R50. [CrossRef]
37. Tesoriere, L.; Fazzari, M.; Angileri, F.; Gentile, C.; Livrea, M.A. *In Vitro* Digestion of Betalainic Foods. Stability and Bioaccessibility of Betaxanthins and Betacyanins and Antioxidative Potential of Food Digesta. *J. Agric. Food Chem.* **2008**, *56*, 10487–10492. [CrossRef] [PubMed]
38. Vinson, J.A.; Hao, Y.; Su, X.; Zubik, L. Phenol antioxidant quantity and quality in foods: Vegetables. *J. Agric. Food Chem.* **1998**, *46*, 3630–3634. [CrossRef]
39. Barbu, V.; Cotârleț, M.; Bolea, C.A.; Cantaragiu, A.; Andronoiu, D.G.; Bahrim, G.E.; Enachi, E. Three Types of Beetroot Products Enriched with Lactic Acid Bacteria. *Foods* **2020**, *9*, 786. [CrossRef]
40. Nesakumar, N.; Baskar, C.; Kesavan, S.; Rayappan, J.B.B.; Alwarappan, S. Analysis of Moisture Content in Beetroot using Fourier Transform Infrared Spectroscopy and by Principal Component Analysis. *Sci. Rep.* **2018**, *8*, 1–10. [CrossRef] [PubMed]
41. Singh, A.; Ganesapillai, M.; Gnanasundaram, N. Optimizaton of extraction of betalain pigments from beta vulgaris peels by microwave pretreatment. *Mater. Sci. Eng.* **2017**, *263*, 1–10. [CrossRef]
42. Henao-Ardila, A.; Quintanilla-Carvajal, M.X.; Moreno, F.L. Combination of freeze concentration and spray drying for the production of feijoa (*Acca sellowiana* b.) pulp powder. *Powder Technol.* **2019**, *344*, 190–198. [CrossRef]

43. Ng, M.L.; Sulaiman, R. Development of beetroot (*Beta vulgaris*) powder using foam mat drying. *LWT* **2018**, *88*, 80–86. [CrossRef]
44. Ochoa-Martinez, L.A.; Garza-Juarez, S.E.; Rocha-Guzman, N.E.; Morales-Castro, J.; Gonzalez-Herrera, S.M. Functional Properties, Color and Betalain Content in Beetroot-Orange Juice Powder Obtained by Spray Drying. *Res. Rev. J. Food Dairy Technol.* **2015**, *3*, 30–36.
45. Antigo, J.L.D.; Bergamasco, R.; Scaramal Madrona, G. Effect of ph on the stability of red beet extract (*Beta vulgaris* L.) microcapsules produced by spray drying or freeze drying. *Food Sci. Technol. Camp.* **2018**, *38*, 72–77. [CrossRef]
46. Leverrier, C.; Almeida, G.; Espinosa-Munoz, L.; Cuvelier, G. Influence of Particle Size and Concentration on Rheological Behaviour of Reconstituted Apple Purees. *Food Biophys.* **2016**, *11*, 235–247. [CrossRef]

Publisher's Note: MDPI stays neutral with regard to jurisdictional claims in published maps and institutional affiliations.

© 2020 by the authors. Licensee MDPI, Basel, Switzerland. This article is an open access article distributed under the terms and conditions of the Creative Commons Attribution (CC BY) license (http://creativecommons.org/licenses/by/4.0/).

foods

MDPI

Review

Sous-Vide as a Technique for Preparing Healthy and High-Quality Vegetable and Seafood Products

Sandra Zavadlav [1,*], Marijana Blažić [1], Franco Van de Velde [2,3], Charito Vignatti [2,3], Cecilia Fenoglio [2], Andrea M. Piagentini [2], María Elida Pirovani [2], Cristina M. Perotti [3,4], Danijela Bursać Kovačević [5] and Predrag Putnik [6,7,*]

1 Department of Food Technology, Karlovac University of Applied Sciences, Trg J. J. Strossmayera 9, 47 000 Karlovac, Croatia; marijana.blazic@vuka.hr
2 Facultad de Ingeniería Química, Instituto de Tecnología de Alimentos, Universidad Nacional del Litoral (FIQ, UNL), 1° de Mayo 3250, Santa Fe 3000, Argentina; francovandevelde@hotmail.com (F.V.d.V.); cvignatti@fiq.unl.edu.ar (C.V.); cfenoglio@gmail.com (C.F.); ampiagen@fiq.unl.edu.ar (A.M.P.); mpirovan@fiq.unl.edu.ar (M.E.P.)
3 Consejo Nacional de Investigaciones Científicas y Técnicas (CONICET), Santiago del Estero 2829, Santa Fe 3000, Argentina; cperotti@fiq.unl.edu.ar
4 Facultad de Ingeniería Química, Instituto de Lactología Industrial (INLAIN), Universidad Nacional del Litoral (FIQ, UNL/CONICET), Santiago del Estero 2829, Santa Fe 3000, Argentina
5 Faculty of Food Technology and Biotechnology, University of Zagreb, Pierottijeva 6, 10000 Zagreb, Croatia; dbursac@pbf.hr
6 Greenbird Medical Inc., Trg dr. Žarka Dolinara 18, 48 000 Koprivnica, Croatia
7 Department of Food Technology, University North, Trg dr. Žarka Dolinara 1, 48000 Koprivnica, Croatia
* Correspondence: sandra.zavadlav@vuka.hr (S.Z.); pputnik@alumni.uconn.edu (P.P.)

Received: 20 September 2020; Accepted: 20 October 2020; Published: 25 October 2020

Abstract: *Sous-vide* is a technique of cooking foods in vacuum bags under strictly controlled temperature, offering improved taste, texture and nutritional values along with extended shelf life as compared to the traditional cooking methods. In addition to other constituents, vegetables and seafood represent important sources of phytochemicals. Thus, by applying *sous-vide* technology, preservation of such foods can be prolonged with almost full retention of native quality. In this way, *sous-vide* processing meets customers' growing demand for the production of safer and healthier foods. Considering the industrial points of view, *sous-vide* technology has proven to be an adequate substitute for traditional cooking methods. Therefore, its application in various aspects of food production has been increasingly researched. Although *sous-vide* cooking of meats and vegetables is well explored, the challenges remain with seafoods due to the large differences in structure and quality of marine organisms. Cephalopods (e.g., squid, octopus, etc.) are of particular interest, as the changes of their muscular physical structure during processing have to be carefully considered. Based on all the above, this study summarizes the literature review on the recent *sous-vide* application on vegetable and seafood products in view of production of high-quality and safe foodstuffs.

Keywords: *sous-vide* cooking; vegetables; seafood; cephalopods; safety; nutritive quality

1. The Perspective of the *Sous-Vide* Technique

Sous-vide is a professional cooking technology that finds its application in home, catering, molecular gastronomy and food industry, also known as lapping, vacuum cooking, vacuum-packed cooking or baking-cooling with vacuum packages [1–3]. On the one side, *sous-vide* has been demonstrated as the secret of great chefs worldwide for decades and, on the other side, this technique was the result of increased customer demand for "fresh-like" and good-quality processed foods [4,5].

Sous-vide means cooking under vacuum and includes a process where raw foods or half-cooked foods are placed in a plastic pouch or bag, hermetically sealed, and cooked slowly in a water bath at 65–95 °C over prolonged time (usually from 1 to 7 h). However, this can take up to 48 h or even more for some foods. With this method, native food juiciness can be retained while avoiding overcooking [6]. Among many advantages, it is important to point out that *sous-vide* cooking requires low-cost operations and equipment to provide consumers with high-quality ready-to-eat food products [7].

Sous-vide foods can be classified into groups based on the magnitude of applied heat treatment [2,8]. For instance, low temperature and longtime cooking method (LTLT) results in foods with favorable characteristics such as enhanced flavor and aroma, increased tenderness and desirable texture, reduced lipid oxidation which leads to extended shelf life, reduced losses of (volatile) flavor (due to vacuum packaging) and improved color and visual appeal [9–12]. Although the nutritional content in *sous-vide* cooked food is maintained and no additives or preservatives are required, this cooking technique modifies nutrients (e.g., proteins) in order to improve functionality [13].

At the point of reaching the optimal internal temperature and the desired textural properties, the food is quickly cooled and then stored refrigerated or frozen until the time of serving, regardless of whether it refers to restaurants or domestic use [6,14,15].

The assessment of the microbial safety is of great importance for this culinary method; therefore, it is essential to know the effects of such treatments on the microorganisms for evaluating products' safety [14]. Several studies have shown that the presence of pathogens in *sous-vide* foods at the time of consumption originates from the raw materials as they survive cooking [16]. The optimal temperature for the growth of the most pathogenic bacteria is between 30 and 50 °C, where inhibition of bacterial reproduction and growth starts. Thus, the core temperature of food during processing should not fall below 54.4 °C and cooking must be held for up to 6 h to ensure inactivation of the food pathogens such as *Salmonella* species, *Listeria monocytogenes*, and the pathogenic strains of *Escherichia coli*. This is a critical point only in case the food is not previously pasteurized.

For the case when the *sous-vide* product needs to be frozen for further use, freezing must be conducted rapidly and immediately after the treatment to prevent or limit possible health hazards. Besides, storage of *sous-vide* food in vacuum pouches is appropriate for preventing recontamination [6,8]. Considering safety, special attention is given to the toxicity derived from the spore-forming pathogenic bacteria that are capable to withstand the mild heat treatments, and hence able to germinate during refrigeration.

For foods with a shelf life of less than 10 days, it is recommended to cook at 70 °C for 2 min or equivalent to reach a 6-log reduction of the most heat-resistant vegetative pathogen (*Listeria monocytogenes*). In case of foods with a shelf life of more than 10 days, cooking at 90 °C for 10 min is recommended (or equivalent) to reach a 6-log reduction of spores *Clostridium botulinum*. However, nowadays, there is a trend of applying low temperatures in food processing (e.g., from 40 to 70 °C), which are considered as a "danger zone" for microbial growth, thus being difficult to assure product safety [14,15].

Stringer et al. [5] thoroughly reported available information about bacterial contamination in the *sous-vide* procedure and compared it to the traditional cooking methods. Heat resistance of pathogens has been investigated extensively at traditional cooking temperatures and powerful mathematical models were used to predict the growth of microorganisms, without the need to test every product and sets of parameters. Nevertheless, there is a lack of data for thermal inactivation of pathogens in low-temperature conditions under vacuum, similar to those applied in *sous-vide* [8,17]. Accordingly, the aim of the current study was to review modern and useful *sous-vide* techniques for preparing healthy and high-quality vegetables and seafood with respect to nutrition and safety.

2. *Sous-Vide* Application for Vegetable Processing

Sous-vide is different from minimally processed produce, as according to the International Fresh-cut Produce Association (IFPA), minimally processed vegetables are defined as "any fresh fruit or vegetable or any combination thereof that has been physically altered from its original form, but remains in a fresh state" [18]. Minimally processed vegetables are consumed raw and their processing involves several preparation steps that do not include cooking, nor do these types of products need additional preparations prior to consumption.

Oppositely, vegetables that have been processed with *sous-vide*, even if they have been treated with a low temperature, are not "fresh" anymore and therefore not minimally processed. The raw vegetables are processed with various operations, such as sorting, peeling, washing, chopping, cutting, blanching or pre-cooking. Then, raw or pre-cooked vegetables are packaged in plastic pouches or bags. Immediately before sealing under vacuum, air is mechanically removed from the package and then thermally processed. Sometimes *sous-vide* is a pasteurization step that reduces the microbial load, nonetheless this is not sufficient to prevent spoilage throughout the shelf life if stored under ambient conditions. When heating temperature is chosen for convectively heated foods, usually at the set temperature, the focus is only given to a length of holding. However, equally important is the time needed for both, to heat and to cool foods at desired temperatures [19]. The challenge is to define cooking conditions suitable to preserve the high quality of vegetables. After heating, vegetables are rapidly cooled down (within 2 h or less) below 3.3 °C, at this point, product temperature should be kept constant during entire storage and distribution.

The application of *sous-vide* technology has been studied for various vegetables [20–24], but not for fruits. Fruits are generally eaten raw, however chefs sometimes cook apples and pears until they are tender [6]. In the case of vegetables, the low amount of oxygen inside the pack will tend to preserve nutritional value and sensory quality, as compared to the other conventional methods such as boiling, steaming or microwaving. Vegetables that are treated by these methods will tend to lose their nutrients, as cellular walls are damaged by heat, which allows water and nutrients to leak out. The *sous-vide* procedure leaves vegetable cell walls mostly intact and makes food tender by dissolving the cementing material (pectin) that holds the cells together [25]. As pectin begins to dissolve between 82 and 85 °C, its depolymerization is bound to lead to texture degradation that might not be a desired consequence of *sous-vide* treatment [26]. These values, therefore, constitute the lowest viable temperature for the *sous-vide* cooked vegetables [6]. Thus, unlike other products, such as meat, for which the adequate temperature is 65–70 °C, *sous-vide* cooking of vegetables, due to various thermal diffusivity, must be performed at temperatures that are closer to 100 °C in order to inactivate two major foodborne pathogens, *E. coli* and *Salmonella* [21,25].

2.1. Changes in Physicochemical Properties and Sensory Quality of Sous-Vide Vegetable Products

Interestingly enough, *sous-vide*, in some cases with nutrient retention, even intensifies characteristic flavors, as it happens with rutabaga and turnip [27]. Additionally, other physical properties of vegetables, mainly texture and color, are greatly influenced by these treatments [28]. Regarding physicochemical properties, after the processing, pH and water activity remain almost the same as in raw vegetables, probably because they are prepared without additives. Even salt addition, as a taste enhancer, does not significantly modify water activity of the product [17]. However, dos Reis et al. [29] reported that pH of broccoli and cauliflower inflorescences diminished half a point after *sous-vide* processing (90 °C, 20 min) when compared to the fresh samples. This is probably due to some cellular ruptures in the walls that released inner acids.

Sous-vide treatment may reduce the weight loss of cooked vegetable products. Gonnella et al. [23] reported that asparagus spears' weight loss was 2.1% after *sous-vide* processing (80 °C, 15 min). Further, after microwaving (900 W, 2450 Hz, 1.5 min), weight reduction was 11.9% as a result of a more efficient moisture removal from the vegetable tissues [30].

Besides, it is well known that *sous-vide* treatment causes changes in the color of vegetables. For instance, L* value of asparagus spears decreased from 54.1 to 42.6, making them darker after the processing [23]. Furthermore, this value was even lower than with boiled asparagus (44.4). A reduction in hues parameter from 115.8 (raw) to 112.8 (*sous-vide* samples) was noticed due to conversion from green to olive-green h°. This was a consequence of chlorophyll transforming to pheophytin [31], which was also lower than the value obtained for asparagus after boiling (113.9). In the case of broccoli florets, a 60% increase in a* value was observed after *sous-vide* treatment (90 °C, 15 min), while boiling (100 °C, 3.5 min) reduced the value of this parameter by 36% [28]. Moreover, boiling did not modify h° parameter when compared to the raw florets and stems but *sous-vide* did reduce this value by 13% and 19%, respectively. dos Reis et al. [29] analyzed color parameters in organic cauliflower inflorescences after boiling (100 °C, 5 min), steaming (final temperature 95 °C), microwaving (800 W, 4 min) and *sous-vide* processing (90 °C, 20 min). In all cases, an L* parameter reduction was observed. As regards parameter a*, *sous-vide* cauliflower registered lower values compared to the other cooking methods.

Carrot color is attributed to the presence of carotenoid pigments and a* value correlates well with sensory acceptability. Patras et al. [32] reported that when compared to a fresh sample, carrot slices prepared via *sous-vide* (90 ° C, 10 min) decreased 12% in this parameter, while for samples boiled in water (until core temperature was 70 °C), the reduction was 30%. Moreover, the authors also observed that the losses of color red continued during storage. They reported a decrease of 27% for *sous-vide* carrot disks and a reduction of 32% for boiled samples after 20 and 5 days of storage, respectively. Consequently, it was concluded that *sous-vide* carrots were superior in color as compared to water-immersed cooked samples.

Trejo Araya et al. [33] compared the appearance of *sous-vide* carrot sticks (90 °C, 5 min in contact with water) to raw samples. Intensity of orange, visual firmness, brightness, surface moisture and flexibility (judged using finger touch) were used as indices. Panelists assigned higher scores to *sous-vide* carrots in all categories except for visual firmness. Concerning perception of texture, the processed vegetables were classified as more fibrous than raw samples but not significantly different. Furthermore, release of higher levels of moisture were perceived in the mouth. For instance, when compared to raw carrots the crunchiness and chewing time of *sous-vide* samples decreased because of water release during mastication.

Relating to textural aspects of broccoli, *sous-vide* samples (90 °C, 15 min) registered less stem softening than boiled ones (100 °C, 3.5 min). Shear force values decreased by 49.0% for the former and 65.7% for the latter, as compared to the initial stem firmness (36.5 N). It was speculated that when lower temperature and vacuum packaging were applied, cell wall disruption was minimized, and stem firmness was less affected [28]. Similarly, dos Reis et al. [29] measured values of 72.6 N for fresh organic broccoli, 16.8 N for broccoli submitted to *sous-vide* processing (90 °C, 20 min) and 12.7 N for broccoli subjected to conventional boiling (100 °C, 5 min). In contrast, for cauliflower, the same authors reported that firmness was less affected by boiling (33.5 N) than by *sous-vide* (15.6 N). The lowest shear force exhibited by *sous-vide* inflorescences could be associated with 3.2% water losses measured for these samples, against an increment of moisture by 1.0% for boiled samples.

Regarding aroma, studies using broccoli florets, green beans and carrots cooked via *sous-vide*, authors stated that these samples retained more aromatic volatile components than boiled samples [20,34,35]. Moreover, Rinaldi et al. [36] reported that *sous-vide* Brussel sprouts and carrots had different volatile profiles as compared to the steamed samples. This could be attributed to reduced degradation due to the lower presence of oxygen because of the vacuuming.

In conclusion, findings generally agree that *sous-vide* vegetables retain more aroma and taste than conventional cooked samples. Certainly, preparation via *sous-vide* avoids leaking of hydrophilic components into boiling water and these substances are related to the perceived flavor by consumers. Finally, to the best of our knowledge, no data are available regarding the effects of *sous-vide* on fruits' quality. Even though fruits are usually eaten raw, some varieties can be cooked. For instance, apples and pears could be submitted to *sous-vide* treatment to potentiate flavor and promote consumer acceptance, hence in this sense more studies are required.

2.2. Changes in Nutrients and Phytochemicals of Sous-Vide Vegetables

There are very few available reports about the impact of *sous-vide* on vegetables on the bioactive compounds such as carotenoids, phenolic compounds, vitamin C and/or their antioxidant capacity. Chiavaro et al. [37] studied the changes in the phytochemical contents and antioxidant capacity of carrot slices and Brussel sprouts after *sous-vide* processing (100 °C, 20 min) and refrigerated storage at 4 °C for 1, 5 and 10 days, with a reheated final step for 20 min in a water bath at 60 °C. Authors compared the results of *sous-vide* processing/refrigerated/reheated products with the corresponding raw and oven-steamed products. In this regard, *sous-vide* carrots and Brussel sprouts at day 1 of storage at 4 °C showed higher contents of carotenoids in comparison to raw and steamed products. For instance, *sous-vide* carrots exhibited carotenoids content 1.8 and 1.1 times higher than raw carrots and those prepared by steaming. This increase was particularly evident for α- and ß- carotenes, and it was probably due to the reheating of the *sous-vide* processing that may more efficiently release carotenes that normally reside in cellular crystals and are bound by the complexes with protein and/or residual membranes [37,38]. Moreover, carotenoids were released from both, carrots and Brussel sprouts, during storage beyond day 1. Here, a significant increase was observed in *sous-vide* samples as compared to the raw and steamed carrots (total carotenoid content in *sous-vide* carrots at day 10 of storage was 2.3 and 1.2 times higher than the total carotenoids in a raw and steamed carrots, respectively). Similarly, *sous-vide* carrot slices (90 °C, 10 min) preserved the total carotenoid content better than conventional cooking (samples were boiled in water until core temperature reached 70 °C) [32]. Carotenoids are lipophilic compounds are less prone to leakage during *sous-vide* processing, however they are still sensitive to oxidation. Thus, *sous-vide* vegetables are more protected against oxidative carotenoid degradation, not only during the cooking, but also throughout the refrigerated storage, as their contact with oxygen is limited due to vacuum-packaging.

Phenolic compounds seem to be better preserved in the vegetable products during *sous-vide* as compared to conventional cooking methods, e.g., boiling or steaming. Similar to carotenoids, this was probably due to limiting oxidation of these compounds under vacuum conditions [37]. In that sense, Martínez-Hernández et al. found a slight increase (lower than 1.4 times) in the phenolic compounds of kailan-hybrid broccoli cooked by *sous-vide* (90 °C, 15 min) as compared to the raw produce [28]. Moreover, Štěrbová et al. [39] reported a gain of 5% in the total phenolic content of Sacha inchi kernels cooked by *sous-vide* at 100 °C for 135 min. Additionally, Chiavaro et al. stated that *sous-vide* better preserved phenolics in carrots slices with slight increase in their content (of less than 5% as compared to the raw samples). This was particularly emphasized for some flavonoids (quercetin, kaempferol and luteolin) and hydroxycinnamic acid derivates, such as caffeic and ferulic acids [37]. On the other hand, Baardseth et al. [40] reported that *sous-vide* (100 °C, 15 min) of blanched/frozen green beans had no effect on the concentration of total phenolics. It was concluded that *sous-vide* or other types of cooking do not increase the content of phenolic compounds in vegetables, rather they facilitate the extraction by increasing the yield in the extracts (not in the vegetables).

However, the high water solubility of phenolic compounds and their thermal instability (of some of them) with the higher temperatures and the negative pressures of vacuum may cause, during *sous-vide*, forced lixiviation to the extra-cellular media and/or their possible thermal degradation. In this way, the phenolic compounds of inflorescences and co-products of several Brassica vegetables were significantly reduced with *sous-vide* (T = 80 °C, t = 15 min for inflorescences and leaves, and T = 80 °C, t = 90 min for stems) in relation to raw products [41,42]. Furthermore, Patras et al. [32] reported a decrease of 29.2% of the total phenolic content in carrot slices that were cooked by *sous-vide*. This was in addition to Chiavaro et al. [37] who found that *sous-vide* in Brussel sprouts decreased phenolics by 14% in comparison to the raw samples.

Vitamin C includes ascorbic acid and its oxidation product, dehydroascorbic acid [43]. It was believed that both of these compounds have several biological activities that include cancer-protective capacities in the body. The content of vitamin C in fruits and vegetables can be significantly reduced during processing and storage due to its solubility in water, thermal sensitivity and proneness to oxidation. Thus, it is expected that vegetables cooked by *sous-vide* may have decreased vitamin C content, but still at lower levels than conventional cooking. In a way, this effect is similar to other bioactive compounds, i.e., as the absence or the very low presence of oxygen in the pouches may mitigate ascorbic acid oxidation. So it is somewhat expected that thermal processing by *sous-vide* significantly reduced the vitamin C content of Brassica vegetables and co-products [41]. For instance, *sous-vide* of inflorescence and stem of Broccoli *cv.* 'Parthenon' experienced vitamin C degradation near to 84% and 67%, respectively. However, authors observed that the reduction was significantly higher after steaming when compared to *sous-vide* for some Brassica vegetables. This was probably due to the reduced amount of oxygen present when cooking by *sous-vide* [41,42]. On the other hand, good retention of vitamin C was observed for all *sous-vide* carrot slices with a significant slight reduction at the end of procedure (at day 10 of storage at 4 °C) of about 6%. Moreover, *sous-vide* of blanched/frozen green beans slightly decreased the levels of ascorbic acid, while conventional cooking by boiled water induced a loss of vitamin C close to 50% [40]. Therefore, *sous-vide* processing has potential to preserve vitamin C in vegetable foods. Probably, the percentage of retention depends on quality of vacuuming in the package and harshness of thermal treatment.

Lastly, some authors studied the effects of *sous-vide* on the antioxidant capacity content of several vegetable products. *Sous-vide* resulted in a significant reduction in the antioxidant potential trough the FRAP (Ferric Reducing Antioxidant Power) assay of all studied inflorescence and leaves of Brassica vegetables. Moreover, for some parts of some varieties, including the inflorescences of broccoli cv. 'Marathon' and broccoli cv. 'Parthenon', *sous-vide* resulted in losses that ranged from 40% to 50% of antioxidant potential as compared to the raw products. However, the antioxidant activity of the samples measured by the DPPH assay (2,2,1-diphenyl-1-picrylhydrazyl radical scavenging) increased after *sous-vide*. For instance, this increase in antioxidant activity was higher after the *sous-vide* processing of the inflorescences of broccoli cv. 'Marathon' (4 times higher than the raw product) and the stems and inflorescences of broccoli cv. 'Pastoret' (5 and 1.3 times higher than raw products). The increase in the antioxidant activity could be caused by the release of insoluble antioxidants or by the formation of new ones from temperature-dependent reactions, and/or due to water loss during processing [41,44]. It is possible that the differences between the antioxidant activities obtained using FRAP and DPPH might be due to the different principles on which these methods are based: acceptance of hydrogen atoms and electrons from antioxidants for DPPH and FRAP assays, respectively [45]. In agreement, Martínez-Hernández et al. [28] reported that cooking kailan-hybrid broccoli by *sous-vide* increased between 4.7- to 5.4-folds of the initial total antioxidant capacity as compared to uncooked samples. Moreover, *sous-vide* induced a great increase of the total antioxidant activity in carrot slices and Brussel sprouts by the great enhancement of carotenoids and flavonoids (only for carrots), as well as the marked retention of ascorbic acid shown by these samples [37].

In conclusion, *sous-vide* vegetables had higher amounts of bioactive compounds (e.g., carotenoids, phenolics and ascorbic acid) than conventionally cooked alternatives (e.g., boiled or steamed). In addition to this, *sous-vide* vegetables retained higher levels of antioxidant activity. Moreover, high temperature also induced modifications in the vegetable matrix than can be positive for the release of bioactive compounds. The *sous-vide* packaging provides protection from oxidation, and hence provides consumers a vegetable with richer phytochemical content. Lastly, further investigations are needed to evaluate if changes in the vegetable matrix are due to *sous-vide* processing, in other words, to clarify if this type of processing provides phytochemicals that are more easily released from the matrix, absorbed in the gastrointestinal tract (bioaccessibility), and available for physiological processes (bioavailability).

2.3. Microbiological Concerns of Sous-Vide Vegetable Products

The safety concerns of *sous-vide* products, particularly concerning spore-pathogen bacteria, needs to be carefully evaluated on a product-by-product basis. The maximum growth temperature for many pathogenic microorganisms growing on food products is between 42 and 49 °C, and some of them have been observed to grow slowly at temperatures between 50 and 55 °C. Therefore, the temperatures being used for *sous-vide* cooking might be close to, or overlap, the growth temperature ranges of foodborne pathogens [46].

The most often contaminants found in vegetables include *E. coli*, *Enterobacter* spp., *Klebsiella* spp., *Salmonella typhi*, *Serratia* spp., *Providencia* spp., *Staphylococcus aureus* and *Pseudomonas aeruginosa*, and other potentially pathogenic microbes [47]. In addition to these, some other types of vegetables are more susceptible to spoilage by other types of microorganisms, like *Bacillus cereus*, *Campylobacter jejuni*, *Clostridium botulinum*, *E. coli O157: H7*, *L. monocytogenes*, *Salmonella* spp., *Shigella*, *Staphylococcus* and *Vibrio cholera* [48]. Most of them are facultative anaerobes, which means that the cells are able to survive and grow in an environment with or without oxygen. Table 1 summarizes selected foodborne bacteria with possible growth and toxin production in temperature ranges below 34 °C; thus, the temperature range is shown to indicate the upper limits of temperature where growth has been observed to occur.

Slow heating of the product to the cook temperature may provoke a heat shock response by the microbes, making them more heat-tolerant to the cooking temperature. Therefore, good operating practice is oriented toward pre-heating the water bath to the appropriate cook temperature. This is especially important when the cook temperature is close to the upper growth temperature of a given microb as this may result in a decrease in the rate of its inactivation. At temperatures below 55 °C, spore-forming bacteria may survive and promote germination, resulting with an increase in bacterial cell number during the cooking, and consequently, increase the incidence of foodborne illness. To date, there is little to no scientific evidence to support any prediction model for foodborne pathogens' inactivation in vegetables and/or seafood at cooking temperatures at or below 55 °C, therefore, the safety of *sous-vide* cooked at temperatures below 55 °C cannot be assured. An alternative could be to add bio-preservatives (nisin and organic acids) additionally to apply non-thermal hurdles by using some innovative non-thermal technologies or to use time-temperature indicators in the packages for recording the storage history of a product [49,50].

A recent line of investigation [51] introduced the use of rosemary essential oil (REO) as a natural antibacterial and antifungal to process fresh-cut potatoes. Firstly, this essential oil would allow avoiding the use of synthetic preservatives due to the antimicrobial activity related to the presence of components such as 1,8-cineole, α-pinene, borneol, ver-benone, and camphor [52]. Secondly, the combination of REO with the characteristic aroma of vegetables could potentiate the final product flavor. The results showed that the synergistic use of REO and vacuum packaging, combined with refrigerated storage, controlled the growth of mesophilic bacteria and *Enterobacteriaceae* in minimally processed potatoes destined for cooking with the *sous-vide* method after 11 days of storage.

Sebastiá et al. [53] evaluated the microbiological quality of broccoli, courgette, potatoes and carrots processed via the *sous-vide* method (100 °C, 15–20 min, except for courgette, which was heated for 5 min), chilled below 3 °C and stored (0, 15 and 30 days). Broccoli had the highest aerobic plate counts in all the storage periods. Authors attributed these results to the inherent morphology of broccoli. Broccoli inflourescences have hydrophobic pockets, which develop isolated areas that are not reached in the washing process. Therefore, it is necessary to reduce the organic material present in the samples with chlorine and improve the disinfection stage [54–56].

Table 1. Temperature range (above 33 °C) of possible growth and toxin production for selected foodborne bacteria.

Bacteria		34	35	36	37	38	39	40	41	42	43	44	45	46	47	48	49	51	52	53	54	55	56
Bacillus cereus	G																						
	T																						
Campylobacter	G																						
Clostridium botulinum	G																						
	T																						
Clostridium perfringens	G																						
Listeria monocytogenes	G																						
Salmonella spp.	G																						
Shigella	G																						
Staphylococcus aureus	G																						
	T																						
STEC	G																						
Yersinia	G																						
Vibrio parahaemolyticus	G																						

* STEC-shiga toxin-producing *E. coli.* G—growth ; Optimun for growth ; T—toxin production in food .

Neither enterotoxigenic staphylococci nor staphylococcal enterotoxins were detected in any of the four *sous-vide* cook–chill preserved vegetables. Detection of both staphylococci and staphylococcal enterotoxins were done because of the several coagulase-negative staphylococci also detected as enterotoxins [57,58]. Moreover, the presence of coagulase-negative staphylococci as *S. epidermidis*, a normal skin commensal, in foods reflects poor hygiene. Both enterotoxigenic staphylococci and their toxins can be used as indicators to assess the risks of vegetable contamination by staphylococci [58,59]. Additionally, there are bacteria that grow at low temperatures and survive mild heat treatments, e.g., *L. monocytogenes* and *E. coli* [60,61]. However, neither of these species were detected in the studied vegetables. Authors asserted that the absence of these microorganisms in the samples is mainly due to the microbial quality of raw vegetables combined with strict temperature control during the process. This is in agreement with the Sous-Vide Advisory Committee (SVAC, 1991) [62] considerations that indicated that microbiological safety depends on thermal process intensity, cooling speed, the final temperature, temperature monitoring and time of refrigerated storage. Moreover, it is important to check the packaging integrity throughout the entire process to assure storage and safety [53].

Rinaldi et al. [63] carried out microbiological analyses (aerobic and anaerobic total plate counts, mesophilic lactic acid bacteria, yeasts and molds) of steamed and *sous-vide* carrots and Brussel sprouts (20 min under steam at 100 °C) after 1, 5 and 10 days of refrigerated storage. Both group of carrots showed microbiological counts that were always lower than 1-log colony-forming units per gram (CFU/g), and even after 10 days of storage at 4 °C. Therefore, both thermal treatments appeared efficiently to diminish the initial counts. Similarly, Sebastiá et al. [53] reported aerobic total plate counts of <1 log CFU/g for carrots submitted to the *sous-vide* process for up to 30 days under refrigerated storage. In the case of Brussel sprouts, both steamed and *sous-vide* samples showed a decrease of all microbiological counts. The values of aerobic total plate counts were 3.46 log CFU/g for steamed samples and 2.34–3.15 log CFU/g for *sous-vide* sprouts. Moreover, *sous-vide* Brussel sprouts exhibited all other microbiological counts (<1 log CFU/g) lower than the values observed for steamed samples for up to 10 days of refrigerated storage. Authors also associated these results with an appropriate time–temperature combination, particularly, controlled heating/chilling steps that reduced initial flora more efficiently than steaming [53].

The table below illustrates the recent literature review on the influence of sous-vide cooking on the quality parameters of vegetables (Table 2).

Table 2. Literature review on quality indices of *sous-vide* cooked vegetables over last five years.

Vegetable Sample	Cooking Treatment	Quality Parameters	Conclusion Remarks	Reference
Potato slices added with rosemary essential oil (REO) Six potato cultivars: Arinda, Elodie, Erika, Fontane, Marabel, Ranomi	Dipping pre-treatments: (i) peanut seed oil with 0.5% (v/v) rosemary essential oil (REO) (ii) peanut seed oil—control *Sous-vide* cooking: 105 °C for 30 min	Ascorbic Acid Total phenols Antioxidant activity (DPPH)	-The addition of REO had no influence on the nutritional content of cooked potato slices. -Ascorbic acid, total phenols and antioxidant activity were noticeably reduced during *sous-vide* cooking. -Although total phenols were well retained after cooking, the antioxidant activity indicated reduction of a mean value of 48%.	Amoroso at al. (2019) [24]
-Cauliflower (*Brassica oleracea* var. *botrytis*)—white rose -Romanesco-type cauliflower (green rose) -Brussel sprouts (*Brassica oleracea* var. *gemmifera*) -Broccoli (*Brassica oleracea* var. *botrytis italica*)	-*Sous-vide* cooking: 90 °C for 45 min (cauliflowers and broccoli) 90 °C for 50 min (Brussel sprouts) -Traditional cooking: unsalted water for 10 min (cauliflowers and broccoli) or 15 min (Brussel sprouts) -Steam cooking: 100 °C for 7 min -Storage: at 2 ± 1 °C for 5 days	The content of dry matter Total ash content Mineral compounds contents (K, Na, Ca, Mg, Mn, Fe, Zn, Cu) Organoleptic properties	-Losses of dry matter were minor in *sous-vide* cooked vegetables, whereas the traditional cooking led to a significant decrease in the dry matter content as compared to the raw material for all vegetable samples. -Treatment type demonstrated a distinct effect on the retention of micro and macro elements in all *Brassica* samples. -In comparison to steam cooking, *sous-vide* allowed higher preservation of the minerals contained in vegetable samples. -*Sous-vide* was the only cooking treatment that resulted in positive organoleptic properties. -The only benefit of boiling in water was the improved process yield.	Florkiewicz and Berski (2017) [64]

Table 2. *Cont.*

Vegetable Sample	Cooking Treatment	Quality Parameters	Conclusion Remarks	Reference
-Cauliflower (*Brassica oleracea* var. *botrytis*)—white rose -Romanesco-type cauliflower (green rose) -Brussel sprouts (*Brassica oleracea* var. *gemmifera*) -Broccoli (*Brassica oleracea* var. *botrytis italica*)	-*Sous-vide* cooking: 90 °C for 45 min (cauliflowers and broccoli) 90 °C for 50 min (Brussel sprouts) -Traditional cooking: unsalted water for 10 min (cauliflowers and broccoli) or 15 min (Brussel sprouts) -Steam cooking: 100 °C for 7 min	Microbiological analysis Vitamin C (L-ascorbic acid) content HPLC analysis of glucosinolates (GLS): glucoiberin, progoitrin, sinirgin, glucoraphainin, gluconapin, gluconasturtin, glucobrassicin, 4-metoxyglukobrassicin, neoglucobrassicin	-The use of a lower temperature during *sous-vide* cooking did not affect the quality and microbiological safety of the vegetables. -In comparison to raw vegetable samples, higher concentrations of GLS were determined in steamed vegetables. -Six glucosinolates from 9 identified (glucoraphanin, glucoiberin, progoitrin, gluconapin, glucobrassicin, 4-metoxyglucobrassicin) were found in higher amounts in broccoli prepared by the *sous-vide* method, compared to the samples traditionally cooked. - *Sous-vide* cooking of Brussel sprouts and Romanesco-type cauliflower resulted in greater losses of GLS, in comparison with the traditional cooking. -*Sous-vide* cooking can be an advanced processing method of broccoli intended for direct consumption.	Florkiewicz et al. (2017) [65]
-Cauliflower (*Brassica oleracea* var. *botrytis*)—white rose -Romanesco-type cauliflower (green rose) -Brussel sprouts (*Brassica oleracea* var. *gemmifera*) -Broccoli (*Brassica oleracea* var. *botrytis italica*)	-*Sous-vide* cooking: 90 °C for 45 min (cauliflowers and broccoli) 90 °C for 50 min (Brussel sprouts) -Traditional cooking: unsalted water for 10 min (cauliflowers and broccoli) or 15 min (Brussel sprouts) -Steam cooking: 100 °C for 7 min -Storage: at 2 ± 1 °C for 0, 48 and 120 h	Total phenolic content HPLC analysis of phenolic compounds: sinapic acid, caffeic acid, *p*-coumaric acid, gallic acid, protocatechuic acid Antioxidant activity (ABTS)	-*Sous-vide* appeared to be the most advantageous with regard to caffeic, *p*-coumaric and gallic acids' stability. -A significant positive correlation was found between antioxidant activity and the total phenolic compounds in raw and thermally treated vegetables. -*Sous-vide* method is considered as the optimal thermal technique for *Brassica* vegetables' processing with regard to phenolic compounds' preservation.	Florkiewicz et al. (2018) [66]

Table 2. *Cont.*

Vegetable Sample	Cooking Treatment	Quality Parameters	Conclusion Remarks	Reference
Pumpkin (*Cucurbita moschata* cv. Leite)	Cooking with the addition of 0.2% of salt (sodium chloride): Boiling—in water, 8 min Steaming—95 °C, 12 min Microwaving—2450 MHz, 10 min *Sous-vide*—90 °C, 30 min	Ascorbic acid content Total phenols Total flavonoids Total anthocyanins Total carotenoids Color analysis Sensory evaluation	-All cooking methods revealed losses of about 50% for ascorbic acid when compared to raw samples. -*Sous-vide* method affected the reduction of total flavonoids the most (30.27%). -Microwaved samples exhibited the highest level of anthocyanins and carotenoids, whereas the *sous-vide* samples had the most reduced values for both types of pigments with losses of 54.37% and 50.0%, respectively. -For all cooking methods, the total polyphenols content was significantly reduced from 49.68% to 64.94%. -The microwaved pumpkin showed the highest sensory applicability, followed by boiling, steaming and *sous-vide* cooking.	Da Silva et al. (2019) [67]
Asparagus spears (*Asparagus officinalis* L., cv Grande)	Boiling (B): 99.0 ± 1.0 °C for 5 min Steaming (S): 99.0 ± 1.0 °C for 6 min Conventional microwaving (MW): 900 W, 2450 MHz, 1.5 min *Sous-vide* boiling (SV-B): 99.0 ± 1.0 °C for 5 min *Sous-vide* (SV): 80 °C, 15 min *Sous-vide* microwaving (SV-MW): 900 W, 2450 MHz, 1.5 min	Color parameters Inorganic ion content Soluble sugars Ascorbic acid content Carotenoid content Chlorophyll content Rutin content Sensory evaluation	-MW resulted with the highest weight change, dry weight increase and the greatest total color difference as compared to raw samples. -Although all cooking methods were rated as sensory acceptable, SV-MW showed the best preferences. -SV-MW better preserved nutritive quality and color characteristics in comparison to other cooking methods. -In comparison to raw samples, SV-MW samples displayed increased violaxanthin content by 42%. - Rutin level was not statistically influenced by the cooking methods. -SV-MW was found to be the most suitable method for preservation of asparagus spears.	Gonnella et al. (2018) [23]

Table 2. *Cont.*

Vegetable Sample	Cooking Treatment	Quality Parameters	Conclusion Remarks	Reference
22 vegetables: sweet potato, broccoli, beetroot, white onion, red onion, garlic, kale, cauliflower, kohlrabi, red cabbage, carrot, red bell pepper, green bell pepper, yellow bell pepper, parsley root, tomato, leek, celeriac, celery, shallot—onion of Ascalan, spinach, potato	Conventional cooking: 100 °C, 2–20 min (in dependence of vegetable type) *Sous-vide* cooking: 84 °C, 30 or 60 min (in dependence of vegetable type)	Antioxidative activity (DPPH, FRAP)	-With no effect of the type of processing (conventional vs. *sous-vide* cooking) and determination method (FRAP vs. DPPH), the antioxidative potential of two vegetables (kohlrabi and red pepper) increased when compared to raw vegetable samples. -*Sous-vide* method resulted in higher antioxidative potential after processing for the case of kale, beetroot, red bell pepper, sweet potato, carrot, cauliflower and kohlrabi as compared to their raw samples. -In comparison to the conventional cooking method, improved antioxidative potential after cooking using the *sous-vide* method was detected for red onion, shallot, broccoli, tomato, parsley root and cauliflower. -When comparing the two types of cooking, the obtained results suggest that an increase in the antioxidant potential was higher for the *sous-vide* technique.	Kosewski et al. (2018) [22]
Tomato powder	*Sous-vide* cooking: 60 °C, 4 h	Amadori compounds (LC-MS/MS) L-ascorbic content Total phenolic content Lycopene content Antioxidant activities (DPPH, ORAC, FRAP, ABTS)	-After the *sous-vide* treatment of tomato powder, losses for the content of L-ascorbic acid (20.35%), total phenolic content (15.98%) and lycopene (10.93%) were determined. -The contents of Amadori compounds in the tomato powder subsequently after *sous-vide* treatment was 2.2 times of that before treatment. -*Sous-vide*-treated tomato powder indicated higher antioxidant activity than that from untreated samples measured by all four assays.	Yang et al. (2020) [44]

* DPPH—2,2,1-diphenyl-1-picrylhydrazyl radical scavenging; FRAP—The ferric reducing/antioxidant power assay; ABTS—2,2′-azino-bis-3-ethylbenzthiazoline-6-sulphonic acid; ORAC—Oxygen radical absorbance capacity.

3. *Sous-Vide* Applications in Seafood Processing

The demand for natural and lightly processed convenient seafood products is constantly growing, therefore the efforts in the processing of seafood and fish that will ensure safe and high-quality products resulted in the development of alternative process technologies such minimal cooking techniques [68]. Marine organisms have been recognized as a valuable dietary source of high quality bioactive components such as long-chain omega-3 fatty acids (PUFAs), easily digestible proteins, non-protein nitrogen compounds, fiber, taurine, sterol, and pigments. They also contain unique components that are not present in terrestrial organisms [69]. However, because marine organisms muscle lipids are highly prone to oxidation due to their high content of polyunsaturated fatty acids, their sensory and nutritional quality can be rapidly destroyed without suitable handling and processing, leading to rancidity and development of off-flavors. The most common methods that are used for the vegetable processing (e.g. stewing, microwaving, roasting, boiling and steaming) [70] could be used in seafood processing too [71–73]. Nevertheless, cooking facilitates several undesirable physicochemical reactions, among which, lipid oxidation is with the most detrimental consequences on cell membranes and denaturation of heme-proteins [74]. During cooking of vegetables, meat or seafood, water-soluble nutrients e.g., vitamins, minerals are typically lost at higher temperatures through evaporation and as exudates leave the food. This includes bioactive compounds and antioxidants, which are essential for maintaining a healthy immune system [75]. Some water-soluble proteins also may be lost with the water during cooking process. Higher cooking temperatures lead to myofibrillar protein shrinkage while decreased binding force between proteins and water that results with the decreased water holding capacity (WHC) in the myofibrils [76]. Sous-vide cooking could better preserve the stability of a secondary structure of proteins as compared to classical cooking procedures. Accordingly, Wan et al. [76] concluded that sous-vide cooking could be used as healthy alternative as it was proved to be helpful in maintaining the quality of largemouth bass fillets. However, in order to preserve sensory and nutritional quality during thermal processing of marine organisms, careful observation of operating technological parameters is required.

It was established that both temperature and cooking time have an effect on lipid oxidation in seafood products [7]. Furthermore, higher temperatures induce various biochemical reactions, protein aggregations and conformations, which change the tissue gaps in fish muscles [76]. Głuchowski et al. [77] by using higher process parameters in the sous-vide cooking of Atlantic Salmon (Salmo salar) achieved a similar intensity of cooked fish odor and flavor without significant degradation in the texture. However, vacuum pretreatment can be utilized to isolate oxygen thus avoid biochemical reactions that require oxygen, but also to minimize the reduction of lipids damage during heating process [76].

Dominguez-Hernandez et al. [10] confirmed that low-temperature over long-time cooking (LTLT) of meat offers multiple advantages over traditional high temperature cooking [6], such as reduced heat degradation of proteins and lipids and lower loss of liquid nutrients [78]. In a line with this, sous-vide might be useful even for processing of marine organisms like cephalopod or fish and seafood maintaining their fresh-like characteristics as with the produce [79]. However, there is a scarcity of literature concerning the sous-vide applications in seafood processing.

Mouritsen and Styrbæk [80] recently reviewed the future perspective of novel trends in cephalopod processing with respect to gastronomy. In this case increased demand for slightly processed seafood with a prolonged shelf life led to the application of sous-vide in order to obtain high quality seafood products [81]. Nevertheless, sous-vide has shown some limitations in seafood processing. When considering vacuuming, its lower degree implies that the higher pressure is applied, and high pressure is not recommended for sous-vide processing of fish fillets as the texture of fish is very delicate and gentle, and high pressure might initiate unwanted damages of the tissue. Consequently, such type of food cannot be completely vacuumed and the residual pressure inside the package is typically around 100–120 mbar, that reflects the difficulties for sous-vide processing [16]. Non-desirable reactions during sous-vide cooking like degradation of vaccum-seal bag, might also take place. Seafood contamination can easily occur by migration of plastic derived compounds into product during sous-vide cooking [82].

Cooking and preparing squid and squid or octopus by sous-vide technology are one of the biggest challenges today, as both of them are not always of uniform quality [83]. Like the abalone and clam, squid and octopus must be cooked very slowly to prevent the muscle fibers from toughening. Moderate temperature during sous-vide cooking of cephalopods retains its structure. Cuttlefish is quite difficult to cook, because it gets tough or rubbery, therefore must be tenderized before cooking. Cooking with sous-vide technology is beneficial as the low temperatures prevent the meat from contracting and gives a soft, tender food while plastic foil and vacuum prevents the loss of aromatic compounds.

In order to preserve the quality and safety of seafood sous-vide products, combinations with other processing steps such as different packaging technologies are often required. There is a possibility for sous-vide cooked marine organisms to be stored at modified atmosphere packaging (MAP) to achieve prolonged shelf life combined with low storage temperatures [84]. Lightly processed salmon (Salmo salar L.) by sous-vide cooking (45 °C for 15 min, 55 °C for 18 min, 65 °C for 21 min) was stored under modified atmosphere (MA) (60% CO_2 balanced with N_2) and soluble gas stabilization (SGS) (100% CO_2) at 4 °C for up to 24 days [85]. Authors found that SGS significantly improved shelf life of processed salmon by prolonging the lag-phase and slowing the growth rate of naturally occurring and inoculated bacteria. This study confirms that sous-vide processing at lower temperatures could provide inhibition of the bacterial growth, while improving chemical quality compared to traditional processing.

3.1. Changes in Physicochemical Properties and Sensory Quality of Sous-Vide Seafood Products

Recently, sous-vide has been considered as a potential technique that could have an impact on the improvement of taste. In this regard, the intensification of umami taste by producing more glutamate as a result of tenderizing meat upon the sous-vide cooking at 54 °C or 64 °C during 0.5, 1, 1.5, 2, 3, 4, 6, and 8 h has been explored [86]. However, authors concluded that aside from texture analysis and amino acid composition, full chemical analysis with determination of free nucleotides combined with a sensory evaluation could be useful approach to support this hypothesis. Another research suggested that the preparation of fermented fish products, such as fish sauces, could be used to provide free glutamate and inosinate. Fresh fish such as mackerel, anchovies and sardines are a particularly good source of adenosine triphosphate (ATP) that can be enzymatically turned into inosinate, soon after the fish is captured, and then fermented to create large quantities of glutamate [87].

In contrast to fish, cephalopods can be completely vacuumed inside the package, prior to sous-vide thermal treatment, because, this organisms have strong muscular tissue and firmness. The firmness is probably associated with changes on connective tissue structure, which lies immediately below the skin and is reported to be resistant to processes of the autolysis [88,89]. Muscle firmness is closely related to collagen content in seafood thus the squids are especially difficult to cook. Most methods require either quick frying or poor and slow roasting which can still result in hardening of the hard chewing food. Mouritsen and Styrbæk [80] reported that the mantle from cephalopods, like Sepia officinalis and Loligo forbesii, need very short heating at low temperatures (50–60 °C) to become more tender and ensure succulent structure. Low temperatures in sous-vide cooking of squids may ensures a soft and melt-in-the-mouth texture. On the other side, increased temperatures during sous-vide cooking enhance protein denaturation and coagulation of sarcoplasmic proteins on the surface and significantly alter the color. Moreover, protein denaturation plays a major role in the toughening of the texture of muscle products [85].

Cooking methods were also shown to affect the pH values. For instance, the pH of the largemouth bass increased during *sous-vide* cooking [76] due to the formation of disulfide bonds during the cooking process [90]. Croptova et al. [7] found increased pH values for *sous-vide* cooked mackerel samples at 60 °C for 10 min, while lower pH values were recorded in *sous-vide* cooked samples at 75 °C and 90 °C for 10 min on the seventh day of chilled storage. Authors supported previous explanation of this trend by the generation of trimethylamine and total volatile base nitrogen from either microbial or endogenous enzymatic degradation [91]. In addition, cooking losses were found to be significantly affected by *sous-vide* operating conditions and chilled storage. Especially, higher temperatures of *sous-vide* cooking enhanced cooking losses in mackerel samples, consequently negatively influencing the product quality. On contrary, the length of chilled storage had a positive contribution to the retention of water in the fish, probably due to structural changes occurring in protein matrix of the fish muscles and connective tissues after cooking [92], including reabsorption of water released by unfolded myofibrillar proteins and its distribution between intracellular and extracellular compartments [93].

Color changes that may occur during cooking are mainly attributed to protein denaturation that is reflected in meat by a more white color during the cooking process [94]. Both instrumental color parameters, lightness (L*) and yellowness (b*), revealed a significant increase throughout chilled storage of *sous-vide* as compared to the color parameters of the raw mackerel fillets [7]. Increased temperature of *sous-vide* significantly affected increased L* value. This is possible due to higher denaturation and aggregation of sarcoplasmic and myofibrillar proteins that reflects in increased light scattering [95]. Wan et al. [76] observed significant increase of whiteness in *sous-vide* cooked largemouth bass samples as compared with the fresh samples. Moreover, in the same study, authors concluded that the texture attribute of elasticity was greater pronounced in the cooked samples than in the raw ones, but in other indicators (hardness, chewiness and resilience) results were opposite. The decreasing of hardness and chewiness in cooked fish may be attributed to the lipid oxidation, which can produce numerous products (mainly aldehydes) that might cause the myofibrillar protein cross-linkages, thus leading to structural changes in proteins.

Regarding textural changes occurred during *sous-vide* cooking, recent study demonstrated a significantly lower values of breaking strength during chilled storage in comparison to the fresh samples of mackerel fillets. Interestingly, the fillets' firmness increased in proportion with raised heating temperature (60 °C *vs.* 90 °C), that may be explained by the heat-induced hardening of the fish muscles after denaturation of myofibrillar and sarcoplasmic proteins. Water losses from the muscle tissue were more intensive at higher temperature (90°C), so contributing to mackerel samples toughening [7].

Humaid et al. [81] investigated the use of high pressure processing (HPP) at 150 and 350 MPa for 5 and 10 min for *sous-vide* cooked lobster tails (*Homarus americanus*) at 65 °C/10 min. Results showed that the use of moderate pressures significantly influenced the texture and color of lobster tails, whereas processing time had a milder effect. HPP at 350 MPa significantly increased L* values of *sous-vide* cooked lobster tails, but without considerable impacts on the overall acceptability by a consumer panel. Another study of Humaid et al. [96] established the effectiveness of HPP in extending refrigerated shelf life of vacuum-packaged raw lobster tails while HPP pretreatment did not positively contribute to additional shelf life extension for *sous-vide* samples. Finally, the authors concluded that the use of HPP to vacuum-packaged lobsters for subsequent *sous-vide* cooking has a potential in the development of novel ready-to-eat functional seafood products able to be stable during refrigerated storage.

3.2. Changes in Nutrients and Phytochemicals of Sous-Vide Seafood Products

Next to the temperature/time regimes with sous-vide, the occurrence and the extent of the reactions related to changes of nutritive quality will depend on the heat transfer medium (liquid water or steam) and from exposure to oxygen (meat vacuum packed or not). Numerous reactions could occur during the cooking, affecting lipids and the volatile profile of seafood such as lipid hydrolysis, lipid oxidation and degradation of *nitrogenous* compounds (proteins, amino acids and trimethylamine N-oxide, TMAO) via Maillard reactions or other types of deteriorative reactions [82].

Protein denaturation is a major event in the cooking of meat or seafood, which is less pronounced in sous-vide. Hence food structure is somewhat better retained along with color and taste which can be very attractive for consumers [97]. Plastic foil prevents the loss of aromatic volatile compounds and water during the sous-vide, which contributes to juiciness and tenderness of meat and enhanced sensory attributes [16]. Roldán et al. [98] reported the formation of volatile compounds in the amino acid-involved reactions during sous-vide processing (60 °C for 6 and 24 h, 80 °C for 6 h) of lamb meat.

As suggested, the aromatic volatiles were associated with a specific stronger meaty flavor and roast notes, due to which fewer spices and less salt was required.

A healthy diet (lower salt diet) may help prevent certain long-term (chronic) diseases thus, it is of great interest to find the right balance between these different nutrients to achieve maximum health benefits. Djordjević et al. [99] revealed that fish as a food, in addition to the valuable content of proteins, minerals and vitamins, is particularly attractive to the consumers as it represents a very rich source of essential fatty acids, which play a role in the prevention of many human diseases [100].

Marine organisms have noticeable content of valued fatty acids and high temperature treatments may negatively affect the stability of these essential nutrients. Except for octopus, cephalopod is considered as a rich source of unsaturated fatty acids, in particular unsaturated omega-3 docosahexaenoic acid (DHA) and eicosapentaenoic acid (EPA) [101]. Therefore, cephalopods comprise great source of lean low-fat protein food and knowledge about the physical structure of the cephalopod's muscles (squid, octopus and cuttlefish) can be a useful guide for gastronomy and nutrition.

A recent study evaluated the impact of mild culinary treatments such as boiling (100 °C, 10 min), steaming (100 °C, 10 min), and *sous-vide* cooking (85 °C, 20 min), on the lipid composition and volatile components of farmed and wild European sea bass (*Dicentrarchus labrax*) [82]. Obtained results showed that the mild oxidative conditions of the three performed cooking methods did not provoke the hydrolysis of sea bass triglycerides, phospholipids or retinyl esters. In addition, no significant lipid oxidation took place during cooking. However, a slight oxidation of unsaturated acyl groups during cooking has been noticed, yielding several volatile secondary oxidation compounds of low molecular weight. Steaming and *sous-vide* cooking, in contrast to boiling, resulted in a slight oxidation of unsaturated acyl groups, leading to the formation of alcohols, aldehydes, ketones, alkyl furans and acids, that consequently positively affected aromatic profile. Volatile profiles of cooked samples were also enriched by degradation of nitrogenated fish components (includes Maillard-type reactions) that was considered as acceptable from a sensory point of view.

3.3. Microbiological Concerns of Sous-Vide Seafood Product

Microbiological deterioration in perishable products such as seafood occurs rapidly due to neutral pH, high water activity, and nutritional composition. When observing seafood safety of sous-vide seafood products, pathogenic bacteria that must be considered are classified into three groups. Namely: (i) bacteria naturally present in the habitat of the consumed species, such as Vibrio spp., non-proteolytic Clostridium botulinum type B, E and T, Plesiomonas shigelloides and Aeromonas spp; (ii) bacteria present in the environment in general (Listeria monocytogenes, proteolytic Clostridium botulinum type A and B, Clostridium perfringens and Bacillus spp; and iii) bacteria which have their usual habitat in man or animals (Salmonella spp., Shigella spp., Escherichia coli, Campylobacter jejuni and Staphylococcus aureus) [102]. The presence of bacteria group (i) and (ii) in live or fresh raw fish is not a common safety concern because tissue concentrations are too low to produce disease. Heat is lethal to microorganisms, but each species has its own particular heat tolerance, and there are many factors affecting their thermal resistance. The process is dependent both on the exposure time and on temperature required to achieve the desired death rate. Therefore, it is essential to determine the thermal death kinetics (D and z-values) of target bacteria in different food substrates and to characterize the time durations to be applied at certain temperatures according to this data. The decimal reduction time (D value) is the time in minutes at a given temperature required to destroy 1 log cycle (90 %) of the target microorganism. The z-value reflects the degrees of temperature change necessary to change the D-value by a factor of 10 [103]. For example, time (minutes) sufficient to achieve a pathogenic load reduction of six orders of magnitude, or a 6D reduction (six logarithms, e.g., from 10^3 to 10^{-3}) of Listeria monocytogenes for meat, fish, or poultry in water baths from 60 °C to 66 °C based on 2 min of cooking at 70 °C with z = 7.5 °C is given in Table 3.

Table 3. Hold time for sufficient thermal 6-log reduction of Listeria monocytogenes for meat, fish or poultry in water baths from 60 to 66 °C [6,104].

Thickness (mm)	Temperature						
	60 °C	61 °C	62 °C	63 °C	64 °C	65 °C	66 °C
5	0:51	0:40	0:31	0:25	0:20	0:17	0:14
15	1:13	1:02	0:53	0:47	0:42	0:38	0:35
25	1:41	1:30	1:21	1:16	1:08	1:03	0:59
35	2:09	1:56	1:46	1:38	1:31	1:26	1:21
45	2:42	2:29	2:17	2:08	2:00	1:53	1:48
55	3:26	3:11	2:58	2:47	2:38	2:30	2:23
65	4:15	3:58	3:43	3:31	3:20	3:11	3:02

However, the accumulation of large numbers of pathogens (Vibrio spp.) in filter-feeding shellfish represents a risk, especially since shellfish are commonly eaten raw. Pre-harvest contamination with pathogens from the group (iii) may present a risk since in some cases a very low infective dosage is sufficient to cause serious disease [102]. Predictive models for thermal inactivation and growth of microbes under sous-vide conditions were recently reviewed. Here some of the limitations of current modelling approaches were also observed, particularly for a longer processing at lower temperatures [8]. Though regular cooking procedures will eliminate the risks of contamination, advanced knowledge is required regarding their origin, biology, physiology, ecology, survival, growth and prevalence in seafood and related products, along with the epidemiology and symptomatology of the diseases with which they are associated [105].

Fish or seafood sous-vide processing requires lower cooking temperatures (50–75°C) with cooking time for several hours or even days. Sous-vide cooking is considered relatively safe for fish and seafood due to the fact that the food is consumed immediately after preparation, usually with a delay no longer than two hours, during which it is stored at temperatures above 54.4°C to prevent or slow down the reproduction of pathogenic bacteria [16]. The European Union guidelines recommend that the minimum heat treatment for sous-vide pasteurization should be equivalent to heating at 70 °C which should be reached throughout 2 minutes [15]. Picouet et al. [106] for sous-vide cooked salmon loins established 4.5 log CFU/g for TVC (Total viable aerobic count) and 3.0 log CFU/g for Enterobacteriaceae under at 40.7 ± 0.1 °C for 19 min of cooking conditions. Such a successful sous-vide cooking at a lower temperature was attributed to the influence of additional high pressure processing (210, 310 and 400 MPa for 5 min at 10 °C).

However, insufficient heat treatment is the major problem in sous-vide processes applied at low temperatures. Therefore, sous-vide cooking has been lately combined with the use of natural antioxidants to improve the efficiency of cooking process in terms of food safety during storage [107,108]. This is in compliance with the latest trends in the food industry where different processing procedures could be combined to promote the process efficiency [50]. Also, there is an increased awareness towards replacing the use of synthetic antioxidants in food processing by the addition of natural ones [109], accordingly sous-vide cooking supplemented by the addition of natural antioxidants could be perspective tool for foods preservation. Several studies have been shown this concept as a promising alternative in preserving quality of fish products. Alves et al. [110] among oregano, basil and rosemary extract confirmed oregano as the most effective for preservation of sous-vide cooked tilapia fillets. Though the addition of oregano essential oil (EO) lowered the pH of sous-vide cooked salmon, which may favor the microbial inactivation, Dogruyol et al. [108] found that L. monocytogenes was more rapidly inactivated due to antibacterial effect of different compounds present in EO, like carvacrol and thymol [108].

Limiting storage times is another way to control the growth of pathogens in sous-vide seafood. Table 4 provides the key findings regarding the impact of shelf life on the quality parameters of sous-vide seafood products. Here, storage for extended periods of time is not recommended unless the product is frozen [111]. Cooled sous-vide seafood products should be stored in cold holding units and maintained at an internal temperature of 3 °C or lower to prevent growth of anaerobic spore-forming pathogens such as non proteolytic Clostridium botulinum, Clostridium perfringens and Bacillus cereus. If the food has a pH > 4.6 or available water ≥0.92, cold holding times should be limited, whereas Baldwin [6] recommends that such sous-vide products can be kept for maximum of 30 days.

Table 5 summarizes the recent literature review on the influence of *sous-vide* cooking on the quality parameters of seafood.

Table 4. The impact of shelf life on the quality parameters of thermally treated seafood products.

Sample	Heat treatment	Shelf life	Key Findings	Reference
Salmon (*Salmo Salard*)	Par-roasting: 300 °C for 3 min *Sous-vide*: 80 °C for 43 min	Anaerobic conditions: 2 °C for 0, 4, 8, 12, 15, 18, 22 and 25 days	-The presence of *Enterobacteriaceae* was only detected on days 18, 22 and 25, but they were always below the minimal detection limit (<10 CFU/g). - *Sous-vide* cooking was found to be efficient in the growth inhibition of *Enterobacteriaceae* in salmon stored at 2 °C for up to 25 days. -The shelf life of the *sous-vide* salmon based on sensory analysis was established at 18 days.	Díaz et al. (2009) [112]
Bonito (*Sarda sarda*, Bloch, 1793	*Sous-vide*: 70 °C for 10 min	4 and 12 °C, 42 days	-The *sous-vide* cooking at 70 °C for 10 min reduced the mesophilic (3.46-log CFU/g) and psychrophilic (2.72-log CFU/g) bacterial counts of the raw material to an undetectable level (<1.00-log CFU/g). -*Sous-vide* bonitos were considered highly acceptable in quality until the 15th day of storage at 12 °C. -The shelf life of cold-stored (4 °C) *sous-vide* bonitos is 28 days.	Mol et al. (2011) [113]
Pirarucu (*Arapaima gigas*)	*Sous-vide*: 60 °C for 9.48 min	2 °C, 49 days	-The dorsal cut of raw pirarucu was the most appropriate cut for developing the *sous-vide* product in comparison to other cuts from raw pirarucu. -On the day 0, the *sous-vide* product reached sensory scores for acceptance ≥ 7 considering the hedonic scale, while on the 49th day, the attributes were scored with 5 on average. -Mesophilic and psychrotrophic anaerobes remained during storage within the acceptable limits.	Pino-Hernández et al. (2020) [114]
Lobster (*Homarus americanus*)	High-Pressure Processing (HPP): 150 MPa or 350 MPa for 10 min at 4 °C *Sous-vide*: 65 °C for 10 min	28 days storage at 2 °C	-Raw lobster pressurized at 350 MPa or *sous-vide* cooked maintained significantly lower microbial counts during storage. -HPP pretreatment did not affect additional shelf life extension for *sous-vide* cooked products. -*Sous-vide* can promote the commercial availability of refrigerated lobster tails in terms of the development of diverse lobster products that are more convenient than live lobsters and have better quality than frozen products.	Humaid et al. (2020) [96]
Atlantic mackerel (*Scomber scombrus*)	*Sous-vide*: 60, 75 and 90 °C for 10, 15 and 20 min	1, 3 and 7 days at 4 ± 1°C	-The strongest effect on the generation of primary and secondary products of lipid oxidation was found to be the duration of chilled storage. -Prolonged chilled storage of *sous-vide* cooked samples had a negative impact on its physicochemical parameters. -*Sous-vide* cooking decreased the firmness of the fish muscle during storage.	Cropotova et al. (2019) [7]

Table 5. Literature review on quality indices of *sous-vide* cooked seafood over the last five years.

Seafood Sample	Cooking Treatment	Quality Parameters	Conclusion Remarks	Reference
European sea bass (*Dicentrarchus labrax*)	*Sous-vide* cooking: 90 °C for 10 min The ratio of fish/ingredients was 1:0.002. Addition of: laurel (*Laurus nobilis*) and curcuma (*Curcuma longa*) Storage: 3 ± 1 °C for 60 days	pH Total volatile basic nitrogen (TVB-N) Trimethil amine–nigrogen (TMA-N) Microbial analysis: Total mesophilic aerobic (TMAB) Total psychrophilic aerobic bacteria (TPAB) Members of *Enterobacteriaceae* family Sensory evaluation	-The quality of the *sous-vide* seafood products strongly depends on initial quality parameters (microbiological, chemical and sensory) of the raw material. -All products were microbiologically safe during the storage period (<7.00-log cfu/g). -Aside from *sous-vide* processing, addition of laurel and curcuma could prolong shelf life by approximately 4 and 10 days, respectively. -Higher concentrations of laurel and curcuma could promote extended shelf life, but it might have a negative effect on the sensory perception.	Bolat et al. (2019) [107]
Largemouth bass (*Micropterus salmoides*)	Boiling (BT): 85 °C, 4 min Steaming: 100 °C, 4 min Vacuum boiling (VB): 85 °C, 5 min Vacuum steaming (VS): 100 °C, 5 min *Sous-vide* cooking (SV): 85 °C, 5 min	Color measurements Texture analysis The thiobarbituric acid (TBA) Water migration and distribution Microstructural changes	-VS and SV samples reached desirable quality, displaying more stable protein secondary structure and lower lipid oxidation in comparison to other cooking methods. -Protein structure was less damaged in VB, VS and SV samples compared with other cooked samples. -The VS and SV treatments both showed more immobilized water in comparison to other cooked samples.	Wan et al. (2019) [76]
Atlantic mackerel (*Scomber scombrus*)	*Sous-vide cooking*: 70 and 80 °C for 10 and 20 min with and without use of commercial antioxidants (TR25—rosemary extract and mix of tocopherols and RPT40—rosemary extract, α-tocopherol and ascorbyl palmitate). Storage: 4 °C, 9 days	Primary and secondary products of lipid oxidation Color parameters	-*Sous-vide* cooking and chilled storage negatively influence oxidative lipid stability in mackerel fillets with respect to primary and secondary lipid oxidation products. -Natural antioxidants positively affect the slower rate of lipid oxidation in cooked samples during chilled storage. -The b* value (yellowness) of the fish flesh significantly correlated with conjugated trienes generated from thermal polymerization of lipids during chilled storage of *sous-vide* cooked fish. -Irrespective of antioxidants used, higher temperature and prolonged cooking times enhanced lipid oxidation in mackerel samples.	Cropotova et al. (2019) [115]
Atlantic mackerel (*Scomber scombrus*)	*Sous-vide* cooking: 60, 75 and 90 °C for 10, 15 and 20 min Storage: 1, 3 and 7 days at 4 ± 1 °C	pH Water content and cook loss Water- and salt-soluble proteins Texture analysis Color parameters Lipid oxidation products	-*Sous-vide* cooking time and temperature showed the minimal influence on the formation of primary and secondary products of lipid oxidation and increase in b* value (yellowness) of the fish samples. -Length of chilled storage led to a significant intensification in oxidation and b* value (yellowness). -Length of chilled storage also had an impact on the structural and textural properties of the fish muscle, leading to a decreased cook loss.	Cropotova et al. (2019) [7]

Table 5. *Cont.*

Seafood Sample	Cooking Treatment	Quality Parameters	Conclusion Remarks	Reference
Atlantic salmon (*Salmo salar* Linnaeus, 1758)	*Sous-vide* cooking: 55, 57.5, 60 and 62.5 °C for 0.08 to 250 min Addition of antioxidants: -non-treated control (C) -0.5% (w/w) citric acid (S) -1% (v/w) oregano essential oil (O) -0.5% (w/w) citric acid + 1% (v/w) oregano essential oil added (OS)	*Listeria monocytogenes* ATCC 7644 inoculation pH value	- The inactivation times of *L. monocytogenes* in control group (C) were Table 4 -The inactivation times of *L. monocytogenes* in control samples (C) were significantly higher than all other treated samples (S, O, OS). -Addition of oregano oil (O), citric acid (S) and their combination (OS) significantly reduced the time required to inactivate *L. monocytogenes*. -Combined treatment (OS) was proven to improve the microbial inactivation at 57.5 and 60 °C better than each of the treatments alone did.	Dogruyol et al. (2020) [108]
Tilapia fillets (*Oreochromis niloticus*)	*Sous-vide* cooking: 60.5 °C for 41 min Addition of antioxidants: T1: Control—no herbs added, T2: added extract of oregano. T3: added extract of rosemary and T4: added extract of basil	Centesimal composition Microbiological analysis Lipid oxidation through Thiobarbituric acid reactive substances (TBARS)	-All samples with added extracts showed significantly higher moisture content, while in control samples, higher protein content was observed. -All samples were in accordance to microbiological standards recommended by legislation. -Control samples exhibited high values from Malondialdehyde (MDA)/kg, demonstrating oxidative rancidity characteristics. -Addition of plant extracts as natural antioxidants prolonged shelf life of *sous-vide* treated tilapia fillets.	Alves et al. (2020) [110]
Cephalic part of tuna (*Thunnus maccoyii*)	*Sous-vide* cooking: (1) 59.5 °C for 13 min (2) 59 °C for 39 min (3) 50 °C for 31 min (4) 50 °C for 62 min	Cooking loss, moisture and crude fat content Thermal protein denaturation (TPD) Color analysis Texture analysis Analysis of ATP-related compounds	-The analysis of TPD showed two peaks at approximately 71 and 48 °C (for actin and myosin, respectively). -Based on obtained results from kinetics analysis, the estimation of TPD under different processing conditions for each protein can be evaluated. -Texture changes were more induced by actin denaturation than myosin denaturation, while myosin denaturation was mostly responsible for changes in color and appearance.	Llave et al. (2018) [116]

4. Conclusions

Sous-vide cooking is becoming increasingly popular as a convenient and reliable method to produce healthy and high-quality vegetable and seafood products in the home, food service environment or food industry. Lately, due to the lack of time available for the consumption and preparation of meals and foods, *sous-vide* is gaining in popularity among consumers as an advantageous approach over conventional thermal treatments. *Sous-vide* cooking employs much lower temperatures than traditional cooking, and therefore it is possible to obtain more nutritious food products with well-retained bioactive compounds, which has significance from the health perspective. Moreover, studies have shown that *sous-vide* cooking could provide foods with higher nutritive value and more pronounced color characteristics, texture properties and sensory attributes than corresponding raw untreated food. Currently, marine organisms are much more in demand than vegetables for *sous-vide* processing due to higher risk of spoilage during shelf life. Therefore, major concerns related to *sous-vide* processing involve the microbiological safety of the products. In order to improve food safety, recent research combines *sous-vide* technology with natural antioxidants or innovative (non)thermal technologies that resulted in beneficial effects for both quality and safety issues. In conclusion, *sous-vide* has a great potential for future applications and to achieve safe foods with improved sensory and nutritive characteristics.

Author Contributions: Conceptualization, S.Z. and M.B.; writing—original draft preparation, S.Z., M.B, F.v.d.V., C.V., C.F., A.M.P., M.E.P., and C.M.P.; writing—review and editing, D.B.K. and P.P.; project administration, S.Z. and M.B. All authors have read and agreed to the published version of the manuscript.

Funding: This paper was produced as part of the "Modification of cheese ripening process and development of whey-based products—SIRENA" project, co-financed by the European Union from the European Structural and Investment Funds in the financial period 2014–2020, and the Operational Programme Competitiveness and Cohesion. Contract No: KK.01.1.1.04.0096.

Conflicts of Interest: The authors declare no conflict of interest.

References

1. Nyati, H. An evaluation of the effect of storage and processing temperatures on the microbiological status of sous vide extended shelf-life products. *Food Control* **2000**, *11*, 471–476. [CrossRef]
2. Todd, E. Foodborne Diseases. Overview of Emerging Food Technologies. In *Encyclopedia of Food Safety*; Motarjemi, Y., Moy, G., Todd, E., Eds.; Academic Press: London, UK, 2014; Volume 1, pp. 253–261.
3. Yıkmış, S.; Aksu, H.; Gökçe Çöl, B.; Demirçakmak, L. Evaluation of *sous-vide* technology in gastronomy. *Int. J. Agric. Life Sci.* **2018**, *4*, 226–231. [CrossRef]
4. García-Linares, M.C.; Gonzalez-Fandos, E.; García-Fernàndez, M.C.; García-Arias, M.T. Microbiological and nutritional quality of *sous vide* or traditionally processed fish: Influence of fat content. *J. Food Qual.* **2004**, *27*, 371–387. [CrossRef]
5. Stringer, S.C.; Fernandes, M.A.; Metris, A. *Safety of Sous-Vide Foods: Feasibility of Extending Combase to Describe the Growth/Survival/Death Response of Bacterial Foodborne Pathogens between 40 °C and 60 °C*; Institute of Food Research Report for Foods Standards Agency; Foods Standards Agency: London, UK, 2012.
6. Baldwin, D.E. Sous vide cooking: A review. *Int. J. Gastron. Food Sci.* **2012**, *1*, 15–30. [CrossRef]
7. Cropotova, J.; Mozuraityte, R.; Standal, I.B.; Rustad, T. The influence of cooking parameters and chilled storage time on quality of sous-vide Atlantic mackerel (*Scomber scombrus*). *J. Aquat. Food Prod. Technol.* **2019**, *28*, 505–518. [CrossRef]
8. Stringer, S.C.; Metris, A. Predicting bacterial behaviour in sous vide food. *Int. J. Gastron. Food Sci.* **2018**, *13*, 117–128. [CrossRef]
9. Vaudagna, S.R.; Sanchez, G.; Neira, M.S.; Insani, E.M.; Picallo, A.B.; Gallinger, M.M.; Lasta, J.A. Sous vide cooked beef muscles: Effects of low temperature-long time (LT-LT) treatments on their quality characteristics and storage stability. *Int. J. Food Sci. Tech.* **2002**, *37*, 425–441. [CrossRef]
10. Dominguez-Hernandez, E.; Salaseviciene, A.; Ertbjerg, P. Low-temperature long-time cooking of meat: Eating quality and underlying mechanisms. *Meat Sci.* **2018**, *143*, 104–113. [CrossRef]

11. Becker, A.; Boulaaba, A.; Pingen, S.; Krischek, C.; Klein, G. Low temperature cooking of pork meat—Physicochemical and sensory aspects. *Meat Sci.* **2016**, *118*, 82–88. [CrossRef]
12. Christensen, L.; Gunvig, A.; Tørngren, M.A.; Aaslyng, M.D.; Knøchel, S.; Christensen, M. Sensory characteristics of meat cooked for prolonged times at low temperature. *Meat Sci.* **2012**, *90*, 485–489. [CrossRef]
13. Kehlet, U.; Mitra, B.; Ruiz Carrascal, J.; Raben, A.; Aaslyng, M. The satiating properties of pork are not affected by cooking methods, sousvide holding time or mincing in healthy men—A randomized cross-over meal test study. *Nutrients* **2017**, *9*, 941. [CrossRef] [PubMed]
14. Ruiz, J.; Calvarro, J.; Sánchez del Pulgar, J.; Roldán, M. Science and technology for new culinary techniques. *J. Culin. Sci. Technol.* **2013**, *11*, 66–79. [CrossRef]
15. Stringer, S.C.; Metris, A. Spore forming pathogens and sous vide food safety. In *Sous Vide and Cook-Chill Processing for the Food Industry*, 2nd ed.; Ghazala, S., Ed.; Springer: New York, NY, USA, 2018.
16. Kilibarda, N.; Brdar, I.; Baltic, B.; Markovic, V.; Mahmutovic, H.; Karabasil, N.; Stanisic, S. The safety and quality of sous vide food. *Meat Technol.* **2018**, *59*, 38–45. [CrossRef]
17. Carlin, F. Microbiology of Sous-vide Products. In *Encyclopedia Food Microbiology*; Robinson, R., Batt, C., Batt, C.A., Eds.; Academic Press: London, UK, 2014; Volume 2, pp. 621–626.
18. Garrett, E.H. Fresh-cut produce. In *Principles and Applications of Modified Atmosphere Packaging of Foods*; Blakistone, B.A., Ed.; Aspen Publishers: Gaithersburg, Maryland, 1999; pp. 125–134. [CrossRef]
19. Gazzala, S. Developments in cook-chill and sous-vide processing. In *Improving the Thermal Processing of Foods*; Richardson, P., Ed.; Woodhead Publishing: Cambridge, UK, 2004; pp. 152–173.
20. Iborra-Bernad, C.; Tárrega, A.; García-Segovia, P.; Martínez-Monzó, J. Advantages of sous-vide cooked red cabbage: Structural, nutritional and sensory aspects. *LWT-Food Sci. Technol.* **2014**, *56*, 451–460. [CrossRef]
21. Guillén, S.; Mir-Bel, J.; Oria, R.; Salvador, M.L. Influence of cooking conditions on organoleptic and health-related properties of artichokes, green beans, broccoli and carrots. *Food Chem.* **2017**, *217*, 209–216. [CrossRef] [PubMed]
22. Kosewski, G.; Górna, I.; Bolesławska, I.; Kowalówka, M.; Więckowska, B.; Główka, A.K.; Morawska, A.; Jakubowski, K.; Dobrzyńska, M.; Miszczuk, P.; et al. Comparison of antioxidative properties of raw vegetables and thermally processed ones using the conventional and sous-vide methods. *Food Chem.* **2018**, *240*, 1092–1096. [CrossRef]
23. Gonnella, M.; Durante, M.; Caretto, S.; D'Imperio, M.; Renna, M. Quality assessment of ready-to-eat asparagus spears as affected by conventional and sous-vide cooking methods. *LWT* **2018**, *92*, 161–168. [CrossRef]
24. Amoroso, L.; Rizzo, V.; Muratore, G. Nutritional values of potato slices added with rosemary essential oil cooked in sous vide bags. *Int. J. Gastron. Food Sci.* **2019**, *15*, 1–5. [CrossRef]
25. Sila, D.N.; Smout, C.; Elliot, F.; Loey, A.V.; Hendrickx, M. Non-enzymatic de-polymerization of carrot pectin: Toward a better understanding of carrot texture during thermal processing. *J. Food Sci.* **2006**, *71*, 1–9. [CrossRef]
26. Fraeye, I.; Deroeck, A.; Duvetter, T.; Verlent, I.; Hendrickx, M.; Vanloey, A. Influence of pectin properties and processing conditions on thermal pectin degradation. *Food Chem.* **2007**, *105*, 555–563. [CrossRef]
27. Baldwin, D.E. *Sous Vide for the Home Cook*; Paradox Press LLC: New York, NY, USA, 2010.
28. Martínez-Hernández, G.B.; Artés-Hernández, F.; Gómez, P.A.; Artés, F. Quality changes after vacuum-based and conventional industrial cooking of kailan-hybrid broccoli throughout retail cold storage. *LWT-Food Sci. Technol.* **2013**, *50*, 707–714. [CrossRef]
29. dos Reis, L.C.R.; de Oliveira, V.R.; Hagen, M.E.K.; Jablonski, A.; Flôres, S.H.; de Oliveira Rios, A. Effect of cooking on the concentration of bioactive compounds in broccoli (*Brassica oleracea* var. Avenger) and cauliflower (*Brassica oleracea* var. Alphina F1) grown in an organic system. *Food Chem.* **2015**, *172*, 770–777. [CrossRef] [PubMed]
30. Renna, M.; Gonnella, M.; Giannino, D.; Santamaria, P. Quality evaluation of cook-chilled chicory stems (*Cichorium intybus* L., Catalogna group) by conventional and *sous vide* cooking methods. *J. Sci. Food Agr.* **2014**, *94*, 656–665. [CrossRef] [PubMed]
31. Lau, M.H.; Tang, J.; Swanson, B.G. Kinetics of textural and color changes in green asparagus during thermal treatments. *J. Food Eng.* **2000**, *45*, 231–236. [CrossRef]
32. Patras, A.; Brunton, N.P.; Butler, F. Effect of water immersion and sous-vide processing on antioxidant activity, phenolic, carotenoid content and color of carrot disks. *J. Food Process Preserv.* **2010**, *34*, 1009–1023. [CrossRef]

33. Trejo Araya, X.I.; Smale, N.; Zabaras, D.; Winley, E.; Forde, C.; Stewart, C.M.; Mawson, A.J. Sensory perception and quality attributes of high pressure processed carrots in comparison to raw, sous-vide and cooked carrots. *Innov. Food Sci. Emerg.* **2009**, *10*, 420–433. [CrossRef]
34. Iborra-Bernad, C.; Philippon, D.; García-Segovia, P.; Martínez-Monzó, J. Optimizing the texture and color of sous-vide and cook-vide green bean pods. *LWT-Food Sci. Technol.* **2013**, *51*, 507–513. [CrossRef]
35. Petersen, M.A. Influence of sous vide processing, steaming and boiling on vitamin retention and sensory quality in broccoli florets. *Z. Für Lebensm. Unters. Forsch.* **1993**, *197*, 375–380. [CrossRef]
36. Rinaldi, M.; Dall'Asta, C.; Meli, F.; Morini, E.; Pellegrini, N.; Gatti, M.; Chiavaro, E. Physicochemical and microbiological quality of sous-vide-processed carrots and brussels sprouts. *Food Bioprocess Tech.* **2012**, *6*, 3076–3087. [CrossRef]
37. Chiavaro, E.; Mazzeo, T.; Visconti, A.; Manzi, C.; Fogliano, V.; Pellegrini, N. Nutritional quality of sous vide cooked carrots and brussels sprouts. *J. Agr. Food Chem.* **2012**, *60*, 6019–6025. [CrossRef]
38. Kalt, W. Effects of production and processing factors on major fruit and vegetable antioxidants. *J. Food Sci.* **2005**, *70*, R11–R19. [CrossRef]
39. Štěrbová, L.; Hlásná Čepková, P.; Viehmannová, I.; Huansi, D.C. Effect of thermal processing on phenolic content, tocopherols and antioxidant activity of Sacha inchi kernels. *J. Food Process Preserv.* **2017**, *41*, 1–8. [CrossRef]
40. Baardseth, P.; Bjerke, F.; Martinsen, B.K.; Skrede, G. Vitamin C, total phenolics and antioxidative activity in tip-cut green beans (*Phaseolus vulgaris*) and swede rods (*Brassica napus* var.napobrassica) processed by methods used in catering. *J. Sci. Food Agr.* **2010**, *90*, 1245–1255. [CrossRef] [PubMed]
41. Lafarga, T.; Bobo, G.; Viñas, I.; Zudaire, L.; Simó, J.; Aguiló-Aguayo, I. Steaming and sous-vide: Effects on antioxidant activity, vitamin C, and total phenolic content of *Brassica* vegetables. *Int. J. Gastron. Food Sci.* **2018**, *13*, 134–139. [CrossRef]
42. Lafarga, T.; Viñas, I.; Bobo, G.; Simó, J.; Aguiló-Aguayo, I. Effect of steaming and sous vide processing on the total phenolic content, vitamin C and antioxidant potential of the genus *Brassica*. *Innov. Food Sci. Emerg.* **2018**, *47*, 412–420. [CrossRef]
43. Davey, M.W.; Montagu, M.V.; Inzé, D.; Sanmartin, M.; Kanellis, A.; Smirnoff, N.; Benzie, I.J.J.; Strain, J.J.; Favell, D.; Fletcher, J. Plant L-ascorbic acid: Chemistry, function, metabolism, bioavailability and effects of processing. *J. Sci. Food Agr.* **2000**, *80*, 825–860. [CrossRef]
44. Yang, C.; Zhang, S.; Shi, R.; Yu, J.; Li, S.; Tao, G.; Tsao, R.; Zhang, J.; Zhang, L. LC-MS/MS for simultaneous detection and quantification of Amadori compounds in tomato products and dry foods and factors affecting the formation and antioxidant activities. *J. Food Sci.* **2020**, *85*, 1007–1017. [CrossRef]
45. Moon, J.-K.; Shibamoto, T. Antioxidant assays for plant and food components. *J. Agr. Food Chem.* **2009**, *57*, 1655–1666. [CrossRef]
46. Hudson, J.A. *Maximum Growth Temperatures of Foodborne Pathogens and Appropriate Temperatures for Hot Holding*; Institute of Environmental Science and Research: Christchurch, New Zealand, 2011.
47. Azi, F.; Odo, M.O.; Okorie, P.A.; Njoku, H.A.; Nwobasi, V.N.; David, E.; Onu, T.C. Heavy metal and microbial safety assessment of raw and cooked pumpkin and *Amaranthus viridis* leaves grown in Abakaliki, Nigeria. *Food Sci. Nutr.* **2018**, *6*, 1537–1544. [CrossRef]
48. Balali, G.I.; Yar, D.D.; Afua Dela, V.G.; Adjei-Kusi, P. Microbial contamination, an increasing threat to the consumption of fresh fruits and vegetables in today's world. *Int. J. Microbiol.* **2020**, *2020*, 1–13. [CrossRef]
49. Hyytiä-Trees, E.; Skyttä, E.; Mokkila, M.; Kinnunen, A.; Lindström, M.; Lähteenmäki, L.; Ahvenainen, R.; Korkeala, H. Safety evaluation of sous vide-processed products with respect to nonproteolytic *Clostridium botulinum* by use of challenge studies and predictive microbiological models. *Appl. Environ. Microb.* **2000**, *66*, 223–229. [CrossRef] [PubMed]
50. Putnik, P.; Pavlić, B.; Šojić, B.; Zavadlav, S.; Žuntar, I.; Kao, L.; Kitonić, D.; Kovačević, D.B. Innovative hurdle technologies for the preservation of functional fruit juices. *Foods* **2020**, *9*, 699. [CrossRef]
51. Rizzo, V.; Amoroso, L.; Licciardello, F.; Mazzaglia, A.; Muratore, G.; Restuccia, C.; Lombardo, S.; Pandino, G.; Strano, M.G.; Mauromicale, G. The effect of sous vide packaging with rosemary essential oil on storage quality of fresh-cut potato. *LWT* **2018**, *94*, 111–118. [CrossRef]
52. Jiang, Y.; Wu, N.; Fu, Y.-J.; Wang, W.; Luo, M.; Zhao, C.-J.; Zu, Y.-G.; Liu, X.-L. Chemical composition and antimicrobial activity of the essential oil of Rosemary. *Environ. Toxicol. Pharmacol.* **2011**, *32*, 63–68. [CrossRef]

53. Sebastiá, C.; Soriano, J.M.; Iranzo, M.; Rico, H. Microbiological quality of sous vide cook-chill preserved food at different shelf life. *J. Food Process Preserv.* **2010**, *34*, 964–974. [CrossRef]
54. Guerzoni, M.E.; Gianotti, A.; Corbo, M.R.; Sinigaglia, M. Shelf-life modelling for fresh-cut vegetables. *Postharvest Biol. Technol.* **1996**, *9*, 195–207. [CrossRef]
55. Beuchat, L.R.; Nail, B.V.; Adler, B.B.; Clavero, M.R.S. Efficacy of spray application of chlorinated water in killing pathogenic bacteria on raw apples, tomatoes, and lettuce. *J. Food Prot.* **1998**, *61*, 1305–1311. [CrossRef] [PubMed]
56. Soriano, J.M.; Rico, H.; Moltó, J.C.; Mañes, J. Assessment of the microbiological quality and wash treatments of lettuce served in University restaurants. *Int. J. Food Microbiol.* **2000**, *58*, 123–128. [CrossRef]
57. Marin, M.E.; De La Rosa, M.C.; Cornejo, I. Enterotoxigenicity of *Staphylococcus* strains isolated from Spanish dry-cured hams. *Appl. Environ. Microb.* **1992**, *58*, 1067–1069. [CrossRef]
58. Soriano, J.M.; Font, G.; Rico, H.; MoltÓ, J.C.; MaÑEs, J. Incidence of enterotoxigenic staphylococci and their toxins in foods. *J. Food Prot.* **2002**, *65*, 857–860. [CrossRef]
59. Bennett, R.W. Atypical toxigenic staphylococcus and non-staphylococcus aureus species on the horizon? An update. *J. Food Prot.* **1996**, *59*, 1123–1126. [CrossRef]
60. Church, I.J.; Parsons, A.L. Review: Sous vide cook-chill technology. *Int. J. Food Sci. Technol.* **1993**, *28*, 563–574. [CrossRef]
61. Bolton, D.J.; McMahon, C.M.; Doherty, A.M.; Sheridan, J.J.; McDowell, D.A.; Blair, I.S.; Harrington, D. Thermal inactivation of *Listeria monocytogenes* and *Yersinia enterocolitica* in minced beef under laboratory conditions and in sous-vide prepared minced and solid beef cooked in a commercial retort. *J. Appl. Microbiol.* **2000**, *88*, 626–632. [CrossRef]
62. *Code of Practice for Sous Vide Catering System*; SVAC (Ed.) Sous Vide Advisory Committee: Tetbury, Gloucestershire, UK, 1991.
63. Rinaldi, M.; Caligiani, A.; Borgese, R.; Palla, G.; Barbanti, D.; Massini, R. The effect of fruit processing and enzymatic treatments on pomegranate juice composition, antioxidant activity and polyphenols content. *LWT-Food Sci. Technol.* **2013**, *53*, 355–359. [CrossRef]
64. Florkiewicz, A.; Berski, W. Application of sous vide method as an alternative to traditional vegetable cooking to maximize the retention of minerals. *J. Food Process Preserv.* **2017**, 42. [CrossRef]
65. Florkiewicz, A.; Ciska, E.; Filipiak-Florkiewicz, A.; Topolska, K. Comparison of Sous-vide methods and traditional hydrothermal treatment on GLS content in Brassica vegetables. *Eur. Food Res. Technol.* **2017**, *243*, 1507–1517. [CrossRef]
66. Florkiewicz, A.; Socha, R.; Filipiak-Florkiewicz, A.; Topolska, K. Sous-vide technique as an alternative to traditional cooking methods in the context of antioxidant properties of *Brassica* vegetables. *J. Sci. Food Agr.* **2018**, *99*, 173–182. [CrossRef]
67. da Silva, M.D.F.G.; Sousa, P.H.M.D.; Figueiredo, R.W.; Gouveia, S.T.; Lima, J.S.S. Cooking effects on bioactive compounds and sensory acceptability in pumpkin (*Cucurbita moschata* cv. Leite). *Rev. Ciência Agronômica* **2019**, *50*, 394–401. [CrossRef]
68. Banerjee, R.; Verma, A.K. Minimally Processed Meat and Fish Products. In *Minimally Processed Foods: Technologies for Safety, Quality, and Convenience*; Siddiqui, M.W., Rahman, M.S., Eds.; Springer: New York, NY, USA, 2014; pp. 193–250.
69. Hosomi, R.; Yoshida, M.; Fukunaga, K. Seafood consumption and components for health. *Glob. J. Health Sci.* **2012**, 4. [CrossRef]
70. Fabbri, A.D.T.; Crosby, G.A. A review of the impact of preparation and cooking on the nutritional quality of vegetables and legumes. *Int. J. Gastron. Food Sci.* **2016**, *3*, 2–11. [CrossRef]
71. Shie, J.S.; Park, J.W. Physical characteristics of surimi seafood as affected by thermal processing conditions. *J. Food Sci.* **1999**, *64*, 287–290. [CrossRef]
72. Barbosa, V.; Maulvault, A.L.; Alves, R.N.; Kwadijk, C.; Kotterman, M.; Tediosi, A.; Fernández-Tejedor, M.; Sloth, J.J.; Granby, K.; Rasmussen, R.R.; et al. Effects of steaming on contaminants of emerging concern levels in seafood. *Food Chem. Toxicol.* **2018**, *118*, 490–504. [CrossRef]
73. Pilavtepe-Celik, M.; Ozer, N.P.; Ozkoc, S.O.; Seyhun, N.; Dede, N. Infrared assisted microwave cooking of Atlantic salmon (*Salmo salar*). *Turk. J. Fish. Aquat. Sci.* **2014**, *14*. [CrossRef]

74. Kristinova, V.; Mozuraityte, R.; Aaneby, J.; Storrø, I.; Rustad, T. Iron-mediated peroxidation in marine emulsions and liposomes studied by dissolved oxygen consumption. *Eur. J. Lipid Sci. Technol.* **2014**, *116*, 207–225. [CrossRef]
75. Puertollano, M.A.; Puertollano, E.; Alvarez de Cienfuegos, G.; Pablo, M.A. Dietary antioxidants: Immunity and host defense. *Curr. Top. Med. Chem.* **2011**, *11*, 1752–1766. [CrossRef]
76. Wan, J.; Cao, A.; Cai, L. Effects of vacuum or sous-vide cooking methods on the quality of largemouth bass (*Micropterus salmoides*). *Int. J. Gastron. Food Sci.* **2019**, 18. [CrossRef]
77. Głuchowski, A.; Czarniecka, S.; Wasiak, Z.; Nowak, D. Effect of various cooking methods on technological and sensory quality of Atlantic salmon (*Salmo salar*). *Foods* **2019**, *8*, 323. [CrossRef]
78. Singh, C.B.; Kumari, N.; Senapati, S.R.; Lekshmi, M.; Nagalakshmi, K.; Balange, A.K.; Chouksey, M.K.; Venkateshwarlu, G.; Xavier, K.A.M. Sous vide processed ready-to-cook seerfish steaks: Process optimization by response surface methodology and its quality evaluation. *LWT* **2016**, *74*, 62–69. [CrossRef]
79. Gupta, V.; Kohli, V. Sous-vide: An emerging technology for fish preservation. *Fish. Technol.* **2019**, *56*, 93–100.
80. Mouritsen, O.G.; Styrbæk, K. Cephalopod Gastronomy—A Promise for the Future. *Front. Commun.* **2018**, 3. [CrossRef]
81. Humaid, S.; Nayyar, D.; Bolton, J.; Skonberg, D.I. Physicochemical properties and consumer acceptance of high-pressure processed, sous vide-cooked lobster tails. *J. Food Sci.* **2019**, *84*, 3454–3462. [CrossRef] [PubMed]
82. Nieva-Echevarría, B.; Manzanos, M.J.; Goicoechea, E.; Guillén, M.D. Changes provoked by boiling, steaming and sous-vide cooking in the lipid and volatile profile of European sea bass. *Food Res. Int.* **2017**, *99*, 630–640. [CrossRef] [PubMed]
83. Mo, H.; Nikitchina, T.; Manoli, T.; Cui, Z. Trends of processed products of squid. *Food Sci. Technol.* **2020**, 14. [CrossRef]
84. DeWitt, C.; Oliveira, A. Modified atmosphere systems and shelf life extension of fish and fishery products. *Foods* **2016**, *5*, 48. [CrossRef] [PubMed]
85. Abel, N.; Rotabakk, B.T.; Rustad, T.; Ahlsen, V.B.; Lerfall, J. Physiochemical and microbiological quality of lightly processed salmon (*Salmo salar* L.) stored under modified atmosphere. *J. Food Sci.* **2019**, *84*, 3364–3372. [CrossRef]
86. Clausen, M.P.; Christensen, M.; Djurhuus, T.H.; Duelund, L.; Mouritsen, O.G. The quest for umami: Can sous vide contribute? *Int. J. Gastron. Food Sci.* **2018**, *13*, 129–133. [CrossRef]
87. Mouritsen, O.G.; Duelund, L.; Calleja, G.; Frøst, M.B. Flavour of fermented fish, insect, game, and pea sauces: Garum revisited. *Int. J. Gastron. Food Sci.* **2017**, *9*, 16–28. [CrossRef]
88. Zavadlav, S.; Janči, T.; Lacković, I.; Karlović, S.; Rogulj, I.; Vidaček, S. Assessment of storage shelf life of European squid (cephalopod: *Loliginidae, Loligo vulgaris*) by bioelectrical impedance measurements. *J. Food Eng.* **2016**. [CrossRef]
89. Zavadlav, S.; Lacković, I.; Bursać Kovačević, D.; Greiner, R.; Putnik, P.; Vidaček Filipec, S. Utilizing impedance for quality assessment of European squid (*Loligo Vulgaris*) during chilled storage. *Foods* **2019**, *8*, 624. [CrossRef] [PubMed]
90. Sánchez del Pulgar, J.; Gázquez, A.; Ruiz-Carrascal, J. Physico-chemical, textural and structural characteristics of sous-vide cooked pork cheeks as affected by vacuum, cooking temperature, and cooking time. *Meat Sci.* **2012**, *90*, 828–835. [CrossRef] [PubMed]
91. Özyurt, G.; Kuley, E.; Balikçi, E.; Kaçar, Ç.; Gökdogan, S.; Etyemez, M.; Özogul, F. Effect of the icing with rosemary extract on the oxidative stability and biogenic amine formation in sardine (*Sardinella aurita*) during chilled storage. *Food Bioprocess Technol.* **2011**, *5*, 2777–2786. [CrossRef]
92. Ofstad, R.; Kidman, S.; Myklebust, R.; Olsen, R.L.; Hermansson, A.-M. Liquid-holding capacity and structural changes in comminuted salmon (*Salmo salar*) muscle as influenced by pH, salt and temperature. *LWT-Food Sci. Technol.* **1995**, *28*, 329–339. [CrossRef]
93. Lorenzo, I.; Serra, P.; Yébenes, J.C. The role of water homeostasis in muscle function and frailty: A review. *Nutrients* **2019**, *11*, 1857. [CrossRef] [PubMed]
94. Hughes, J.M.; Oiseth, S.K.; Purslow, P.P.; Warner, R.D. A structural approach to understanding the interactions between colour, water-holding capacity and tenderness. *Meat Sci.* **2014**, *98*, 520–532. [CrossRef]
95. Christensen, L.; Ertbjerg, P.; Aaslyng, M.D.; Christensen, M. Effect of prolonged heat treatment from 48°C to 63°C on toughness, cooking loss and color of pork. *Meat Sci.* **2011**, *88*, 280–285. [CrossRef] [PubMed]

96. Humaid, S.; Nayyar, D.; Bolton, J.; Perkins, B.; Skonberg, D.I. Refrigerated shelf-life evaluation of high pressure processed, raw and sous vide cooked lobster. *High Press. Res.* **2020**, 1–20. [CrossRef]
97. Ayub, H.; Ahmad, A. Physiochemical changes in sous-vide and conventionally cooked meat. *Int. J. Gastron. Food Sci.* **2019**, 17. [CrossRef]
98. Roldán, M.; Ruiz, J.; del Pulgar, J.S.; Pérez-Palacios, T.; Antequera, T. Volatile compound profile of sous-vide cooked lamb loins at different temperature–time combinations. *Meat Sci.* **2015**, *100*, 52–57. [CrossRef]
99. Djordjevic, J.; Pavlicevic, N.; Boskovic, M.; Baltic, T.; Ivanovic, J.; Glamoclija, N.; Baltic, M.Z. Sensory evaluation of cold-smoked trout packaged in vacuum and modifi ed atmosphere. *Meat Technol.* **2016**, *57*, 141–149.
100. Alabdulkarim, B.; Bakeet, Z.A.N.; Arzoo, S. Role of some functional lipids in preventing diseases and promoting health. *J. King Saud Univ. Sci.* **2012**, *24*, 319–329. [CrossRef]
101. Ozogul, Y.; Duysak, O.; Ozogul, F.; Özkütük, A.S.; Türeli, C. Seasonal effects in the nutritional quality of the body structural tissue of cephalopods. *Food Chem.* **2008**, *108*, 847–852. [CrossRef] [PubMed]
102. Huss, H.H.; Ababouch, L.; Gram, L. *Assessment and Management of Seafood Safety and Quality*; FAO Fisheries Technical Paper 444; FAO: Rome, Italy, 2003; pp. 26–84.
103. Silva, F.V.M.; Gibbs, P.A. Thermal pasteurization requirements for the inactivation of Salmonella in foods. *Food Res. Int.* **2012**, *45*, 695–699. [CrossRef]
104. Leonard, B. Pathogenic bacteria survival through cooking or pasteurization. In *Fish and Fishery Products Hazards and Controls Guidance*; Diane Publshing: Darby, PA, USA, 2011; p. 315.
105. dos Santos, C.A.M.L.V.; Regine, H.S. Fernandes. Bacteriological hazards and risks associated with seafood consumption in Brazil. *Rev. Inst. Med. Trop. São Paulo* **2013**, *55*, 219–228. [CrossRef]
106. Picouet, P.A.; Cofan-Carbo, S.; Vilaseca, H.; Ballbè, L.C.; Castells, P. Stability of sous-vide cooked salmon loins processed by high pressure. *Innov. Food Sci. Emerg.* **2011**, *12*, 26–31. [CrossRef]
107. Bolat, Y.; GenÇ, İ.Y.; Tunca, Y.; Demirayak, M. Effect of laurel (*Laurus nobilis*) and curcuma (*Curcuma longa*) on microbiological, chemical and sensory changes in vacuum packed sous-vide european sea bass (*Dicentrarchus labrax*) under chilled conditions. *Food Sci. Technol.* **2019**, *39*, 159–165. [CrossRef]
108. Dogruyol, H.; Mol, S.; Cosansu, S. Increased thermal sensitivity of *Listeria monocytogenes* in sous-vide salmon by oregano essential oil and citric acid. *Food Microbiol.* **2020**, 90. [CrossRef]
109. Bursać Kovačević, D.; Barba, F.J.; Lorenzo, J.M.; Rocchetti, G.; Lucini, L.; Putnik, P. Innovative technologies for fruit extracts: Value-added opportunities in the meat industry. *Iop Conf. Ser. Earth Environ. Sci.* **2019**, *333*. [CrossRef]
110. Alves, L.F.S.; Corrêa, S.S.; Rocha, J.D.M.; Amado, D.A.V.; Cottica, S.M.; Souza, M.L.R. Use of natural antioxidants in sous vide tilapia fillet. *Bol. De Indústria Anim.* **2020**, *77*. [CrossRef]
111. Johnston, W.A. *Freezing and Refrigerated Storage in Fisheries*; Food and Agriculture Organization of the United Nations: Rome, Italy, 1994.
112. Díaz, P.; Nieto, G.; Bañón, S.; Garrido, M.D. Determination of shelf life of sous vide salmon (*Salmo Salard*) based on sensory attributes. *J. Food Sci.* **2009**, *74*, S371–S376. [CrossRef]
113. Mol, S.; Ozturan, S.; Cosansu, S. Determination of the quality and shelf life of sous vide packaged bonito (*Sarda Sarda*, Bloch, 1793) stored at 4 and 12 C. *J. Food Qual.* **2012**, *35*, 137–143. [CrossRef]
114. Pino-Hernández, E.; de Carvalho Júnior, R.N.; Barata Alves, R.C.; Peixoto Joele, M.R.S.; da Silva e Silva, N.; Costa da Silva, E.V.; Henriques Lourenço, L.d.F. Evaluation of muscle cuts of pirarucu (*Arapaima gigas*) and sous vide product characterization and quality parameters. *Int. J. Gastron. Food Sci.* **2020**, 20. [CrossRef]
115. Cropotova, J.; Mozuraityte, R.; Standal, I.B.; Rustad, T. Assessment of lipid oxidation in Atlantic mackerel (*Scomber scombrus*) subjected to different antioxidant and sous-vide cooking treatments by conventional and fluorescence microscopy methods. *Food Control.* **2019**, *104*, 1–8. [CrossRef]
116. Llave, Y.; Shibata-Ishiwatari, N.; Watanabe, M.; Fukuoka, M.; Hamada-Sato, N.; Sakai, N. Analysis of the effects of thermal protein denaturation on the quality attributes of sous-vide cooked tuna. *J. Food Process Preserv.* **2018**, 42. [CrossRef]

Publisher's Note: MDPI stays neutral with regard to jurisdictional claims in published maps and institutional affiliations.

© 2020 by the authors. Licensee MDPI, Basel, Switzerland. This article is an open access article distributed under the terms and conditions of the Creative Commons Attribution (CC BY) license (http://creativecommons.org/licenses/by/4.0/).

foods

MDPI

Article

Evaluation of the Physicochemical and Structural Properties and the Sensory Characteristics of Meat Analogues Prepared with Various Non-Animal Based Liquid Additives

Gihyun Wi [1], Junhwan Bae [1], Honggyun Kim [2], Youngjae Cho [1] and Mi-Jung Choi [1,*]

[1] Department of Food Science and Biotechnology of Animal Resources, Konkuk University, Seoul 05029, Korea; dnlrlgus@hanmail.net (G.W.); royal__style@naver.com (J.B.); moonjae@konkuk.ac.kr (Y.C.)

[2] Department of Food Science and Biotechnology, Sejong University, Seoul 05006, Korea; vollry@nate.com

* Correspondence: choimj@konkuk.ac.kr; Tel.: +82-2-450-3048

Received: 10 February 2020; Accepted: 28 March 2020; Published: 8 April 2020

Abstract: This study investigates the effects of various non-animal-based liquid additives on the physicochemical, structural, and sensory properties of meat analogue. Meat analogue was prepared by blending together textured vegetable protein (TVP), soy protein isolate (SPI), and other liquid additives. Physicochemical (rheological properties, cooking loss (CL), water holding capacity (WHC), texture and color), structural (visible appearance and microstructure), and sensory properties were evaluated. Higher free water content of meat analogue due to water treatment resulted in a decrease in viscoelasticity, the highest CL value, the lowest WHC and hardness value, and a porous structure. Reversely, meat analogue with oil treatment had an increase in viscoelasticity, the lowest CL value, the highest WHC and hardness value, and a dense structure due to hydrophobic interactions. SPI had a positive effect on the gel network formation of TVP matrix, but lecithin had a negative effect resulting in a decrease in viscoelasticity, WHC, hardness value and an increase in CL value and pore size at microstructure. The results of sensory evaluation revealed that juiciness was more affected by water than oil. Oil treatment showed high intensity for texture parameters. On the other hand, emulsion treatment showed high preference scores for texture parameters and overall acceptance.

Keywords: meat analogue; liquid additives; soy protein isolate; lecithin; emulsion

1. Introduction

Increasing concerns about wellbeing in ethical, social, health, and ecological aspects have resulted in an increase in the number of vegetarians. The replacement of meat protein with plant protein has therefore become an important research topic [1,2]. However, meat analogues are still known to be different from meat in terms of mouthfeel, texture, taste, and flavor [3,4].

Many researchers have reported that fat plays a major role in meat or meat analogue quality such as juiciness, tenderness, mouthfeel, and flavor release of the product [5–9]. However, vegetable oils differ considerably in their physicochemical properties from animal fats, and this can therefore negatively affect juiciness and texture parameters [10,11].

Many studies have reported that an emulsion can be used as fat replacement in meat products owing to its positive effect on texture, water holding ability, and reduced fat content. Pietrasik et al. [12] reported that beef steak injected with 20% oil-in-water emulsion shows a decreased shear force and a high score of juiciness and tenderness in sensory evaluation. Serdaroğlu et al. [13] reported that using double emulsion as a beef fat replacement results in high water holding capacity (WHC), but decreased hardness. These results are due to not only the water content in the emulsion, but also depend on the type of oil and emulsifier [14].

An emulsion needs to amphiphilic emulsifier because it is a mixture of immiscible liquids such as oil and water. Emulsifiers have both a hydrophilic head group and lipophilic tail group, which can adsorb at the water–oil interface and act to lower interfacial tension. In addition, it can interact with the protein surface and cause structural changes in the protein, which affects the quality of meat protein [15,16]. Especially concerning soy protein isolate (SPI) and lecithin which are used as natural emulsifiers, Wang et al. [15] stated that heat-induced SPI affects the myofibrillar protein (MP), increasing the WHC, hardness, and springiness by forming a stable mixed gel by enhanced hydrophobic interaction and hydrogen bonds. On the contrary, Xia et al. [16] reported that the addition of lecithin to MP results in decreased hardness and springiness while the WHC increases as it increased hydrophobic interactions and hydrogen bonds, but decreased disulfide bonds. However, there are very few studies using emulsions and emulsifiers as a fat substitute for vegetable meat.

This study, therefore, aimed to analyze the effect of various non-animal-based liquid additives and their emulsifiers such as water, canola oil, O/W emulsion, SPI, and lecithin in meat analogues so as to compensate for the shortcomings of meat analogues.

2. Materials and Methods

2.1. Materials

Textured vegetable protein (TVP, SUPRO® MAX 5050 and SUPRO® MAX 5010, DuPont Korea, Seoul, Korea) and binder (GRINDSTED® Meatline 2714, DuPont Korea, Seoul, Korea) were selected as the meat analogue base. Textured vegetable protein (TVP) contains 55%–60% of SPI, 40%–45% of wheat gluten and wheat starch. Furthermore, the binder is a mixture of egg white powder, glucose, soy protein, locust bean gum, carrageenan and guar gum (supplied information by DuPont Korea). SPI (90% of protein on a dry matter basis, Avention, Incheon, Korea) and lecithin (from soybean, Samchun, Seoul, Korea) were used as emulsifiers. Canola oil was purchased from Samchun (Seoul, Korea).

2.2. Sample Preparation

The flow diagram for manufacturing the meat analogue is described in Figure 1. The base of the TVP matrix was prepared using TVPs, SPI, and binder, following which the various non-animal-based liquid additives were added. A total of 100 g of TVP was immersed into water (10 times in volume) for hydration for two hours. Thereafter, the hydrated TVP was dehydrated in a centrifugal dehydrator (ws-6600, Hanil Electric, Seoul, Korea) for 5 min at 1200 rpm. Further, 100 g of swollen and dehydrated TVP was mixed with different quantities of liquid additives. The mixing ratio and ingredients are described in Table 1. Each of the liquid additives was individually prepared before blending with the processed TVP. SPI and lecithin (1% by weight) were dissolved in water and canola oil, respectively and stirred for 12 h. O/W emulsion and OW additives were both prepared by mixing together the same concentrations of water, oil, SPI, and lecithin. The OW additive was not pre-emulsified like the O/W emulsion preparation. The emulsion preparation required 1% (*w*/*w*) SPI to be dissolved in distilled water as the aqueous phase and 1% (*w*/*w*) lecithin to be dissolved in canola oil as the oil phase. The emulsification process entailed the two phases to be first mixed at 4:6 (oil phase: aqueous phase) of weight ratio and homogenized at 14,000 rpm for 5 min using a high-speed homogenizer (T25 digital ULTRA-TURRAX®, IKA, Staufen, Germany). This emulsion was then emulsified further using a high-intensity ultrasonicator with 75% power at 150 W capacity for 1 min (SONOPLUS HD2200, Bandelin®, Berlin, Germany).

All prepared liquids were mixed with 100 g of processed TVP, and then blended for 60 s using a hand blender (550 W, Multiquick 3 Vario, Braun, Kronberg im Taunus, Germany). The TVP matrix was molded in a cylindrical mold (30 mm of diameter and 25 mm of height). The matrix was then cooked in an oven (M4207, Simfer, Istanbul, Turkey) at 180 °C for 14 min and cooled to room temperature for 30 min before being used for further analysis.

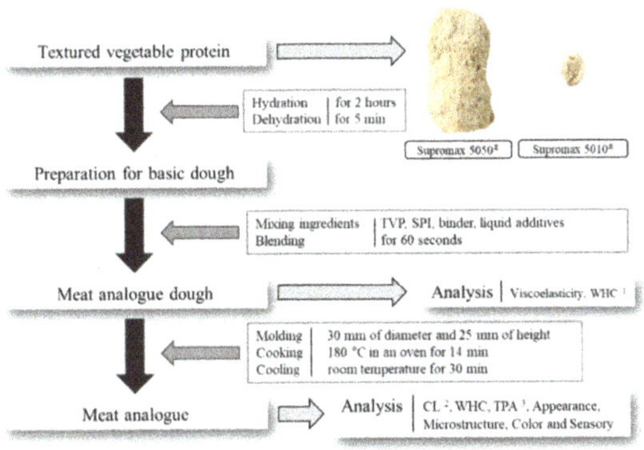

Figure 1. Flow diagram for manufacturing the meat analogue. [1] WHC: water holding capacity. [2] CL: cooking loss. [3] TPA: texture profile analysis.

Table 1. Mixing ratios of additives in meat analogues.

Types of Liquid Additives [3]	Mixing Ratios			
	Liquid Additive Concentration (g)	TVP [1] (g)	SPI [2] (g)	Binder (g)
W	15	100	4.5	3.0
	20	100	4.5	3.0
	25	100	4.5	3.0
	30	100	4.5	3.0
	35	100	4.5	3.0
WS [4]	15	100	4.5	3.0
	20	100	4.5	3.0
	25	100	4.5	3.0
	30	100	4.5	3.0
	35	100	4.5	3.0
O	15	100	4.5	3.0
	20	100	4.5	3.0
	25	100	4.5	3.0
	30	100	4.5	3.0
	35	100	4.5	3.0
OL [4]	15	100	4.5	3.0
	20	100	4.5	3.0
	25	100	4.5	3.0
	30	100	4.5	3.0
	35	100	4.5	3.0
E [5]	15	100	4.5	3.0
	20	100	4.5	3.0
	25	100	4.5	3.0
	30	100	4.5	3.0
	35	100	4.5	3.0
OW [6]	15	100	4.5	3.0
	20	100	4.5	3.0
	25	100	4.5	3.0
	30	100	4.5	3.0
	35	100	4.5	3.0

[1] TVP: textured vegetable protein. [2] SPI: soy protein isolate. [3] W: water, WS: water + SPI, O: canola oil, OL: canola oil + lecithin, E: O/W emulsion, OW: water + canola oil + SPI + lecithin. [4] One weight percent of emulsifier per added water or oil.[5] Mixing ratio of pre-emulsified liquid additives (water with 1% SPI and oil with 1% lecithin is 6:4). [6] Same amount and ingredients as emulsion, but each of the compounds was not pre-emulsified.

2.3. Dynamic Rheological Properties

The rheological properties of the TVP matrix were measured with a rheometer, (MCR 302, Anton Paar, Graz, Austria) comprising of parallel-plate geometries (diameter 25 mm, gap 2.5 mm) and 2.8 g of the dough was molded in a cylindrical mold and trimmed with a spatula. The strain sweeps test was performed from 0.01% to 100% strain at 10 rad/s to obtain linear viscoelastic regions while an angular frequency sweep test was carried out from 1 to 100 rad/s at 0.1% strain at 25 °C. The storage modulus (G′), loss modulus (G″) and tan δ were recorded continuously to assess the rheological properties of the TVP matrix.

2.4. Cooking Loss (CL)

Cooking method and conditions were determined based on the study done by Pathare and Roskully [17]. The internal temperature of the meat analogue was measured by inserting thermocouples and memorizing the value in connected data loggers (Data Acquisition-MX 100, Yokogawa, Japan). The cooking conditions were set at a temperature of 180 °C for 14 min depending on the temperature of the meat analogue in its center reaching 80 °C. After cooking, the samples were cooled at room temperature for 30 min. CL was calculated as the percentage weight difference between the dough before cooking and after cooking, using the following formula:

$$\text{CL } (^\circ\text{C}) = (W_1 - W_2)/W_1 \times 100$$

W_1: weight of meat analogue dough (g).
W_2: weight of cooked meat analogue (g).

2.5. Water Holding Capacity

WHC was measured by modifying a method described previously [18]. Briefly, 2 g of the dough and cooked samples of the meat analogue were put in a 15 mL conical tube with gauze underneath. The tube was then centrifuged at 3000 rpm for 10 min at 35 °C (centrifuge-1736R, LaboGene, Daejeon, Korea). WHC was calculated by comparing the weight of the samples before and after centrifugation, using the following formula:

$$\text{WHC } (\%) = (W_2/W_1) \times 100$$

W_1: weight of meat analogue before centrifugation (g).
W_2: weight of meat analogue after centrifugation (g).

2.6. Texture Profile Analysis (TPA)

TPA was measured by modifying a method described previously by Lin et al. [19] using a texture analyzer (CT3, Brookfield Engineering Labs Inc., Stoughton, MA, USA). The samples were shaped like a cylindrical column (30 mm in diameter and 25 mm in height). They were compressed to 50% deformation of its original thickness and were measured under the conditions of test speed (2.5 mm/s) and trigger load (100 g). A cylindrical probe (38.1 mm in diameter) was used for the test and the hardness, cohesiveness, springiness, and chewiness data were recorded.

2.7. Visible Appearance

The appearance of the meat analogue was assessed by adding 25 g of various liquid additives. The external and internal appearance was filmed with a digital camera (α350, Sony, Tokyo, Japan) and the characteristics were observed.

2.8. Color Analysis

The color of the meat analogue was analyzed by adding 25 g of the various liquid additives. The color of the inner cross-section from the cooked meat analogue was determined using a colorimeter

(Chroma Meter CR-400, Konica Minolta, Tokyo, Japan). The color analysis results were expressed according to the Commission International de l'Eclairage (CIE) system and reported as Hunter L* (lightness), a* (redness), and b* (yellowness).

2.9. Microstructure

The microstructure of the meat analogue was obtained by a field emission scanning electron microscope (TM4000Plus, Hitachi, Seoul, Korea) with an accelerating voltage of 15 kV. The micrograph of the samples was taken in 100× and 300× magnification. Sample preparation was performed according to the method suggested by Samard and Ryu [4]. Briefly, the cooked samples were cut into thin slices (approximately 10 × 10 × 1 mm^3) and were frozen at −100 °C for 24 h in a deep freezer (CLN, NIHON-FREEZER, Tokyo, Japan). Frozen samples were dried in a freeze-dryer (FDCF-12012, Operon, Gyeonggi-do, Korea) at a pressure of 5 Pa and a temperature of −80 °C for 48 h.

2.10. Sensory Evaluation

The meat analogue samples added with water, canola oil, O/W emulsion, and OW treatments were evaluated by ten experienced estimators who were graduate students and staff from the Department of Food Science and Biotechnology of Konkuk University. Sensory evaluation was conducted in individual booths on the basis of firmness, elasticity, stickiness, compactness, roughness, soy taste, oil taste, juiciness, and overall acceptance. A seven-point scoring test was used for evaluating the parameter intensity and sensory preference (7: very strong and good, 1: very weak and unacceptable). The Institutional Review Board (IRB) approved the consent procedure for sensory evaluation (nos. 700355-201901-HR-294). The written consents from all the participants were acquired before conducting the sensory evaluation.

2.11. Statistical Analysis

All experiments were analyzed using SPSS statistics (ver. 24.0, SPSS Inc., Chicago, IL, USA) except for sensory evaluation. The significance of the results was analyzed using one-way ANOVA and Duncan's multiple range test which were conducted at the $p < 0.05$ level to verify the statistical significance of each sample.

3. Results and Discussion

3.1. Dynamic Rheological Properties

The measurement of the dynamic rheological properties is important to control the quality of processed meat product or meat analogues. Although there are many studies reporting on TVP protein properties regarding the extrusion method and modified vegetable proteins [2,4,19], there are few analyses considering the physical properties of emulsifier effects, or considering preformulated emulsions to improve their properties. Therefore, in order to investigate the effect of emulsifier or different liquid additives on TVP matrix, firstly, we tried to analyze the rheological properties of non-heated meat analogues using a Rheometer®.

As shown in Figure 2, the values of G' and G'' were seen to decrease with increasing concentration of liquid additives in all the treatments. The value of tan δ tended to increase, which meant that the degree of decrease of the G' value was larger than the one for the G'' value, except for oil treatment. Therefore, with an increasing concentration of liquid additives, the non-heated meat analogue acquires the viscosity properties (liquid-like) of TVP except for oil treatment. A review report by Song and Zheng [20] explained that water has an important role in controlling the viscoelastic properties of dough, and as water content increases, the G' and G'' values decrease due to a lubricant effect. Likewise, Roccia et al. [21] reported that free water in the food system is known to reduce the elastic and viscous properties of dough due to its lubricating action. It is therefore supposed that the increase in liquid, especially free water, causes a lubricating action in the TVP dough.

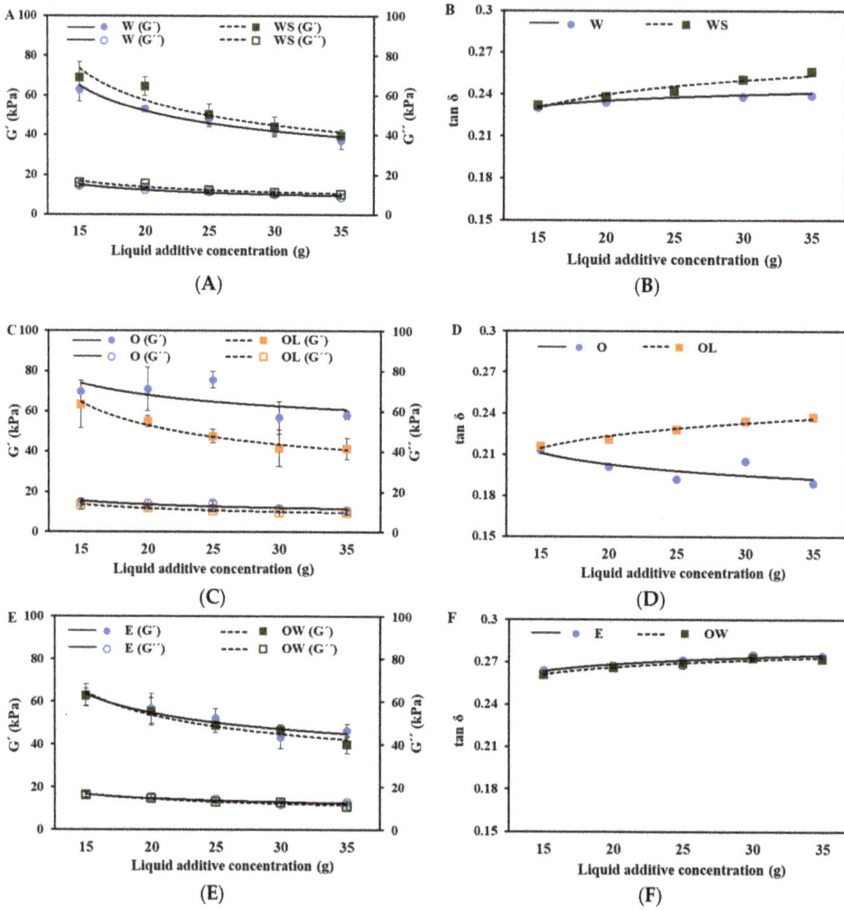

Figure 2. The differences in the rheological behavior (G′: the storage modulus; G′′: the loss modulus (A, C and E); tan δ: loss tangent (B, D and F)) of the meat analogue dough based on the liquid additives and emulsifier type.

The tan δ value of oil treatment without lecithin was seen to decrease with increasing oil concentration, although the G′ and G′′ values decreased with an increasing oil concentration similar to the other treatments. As the oil concentration was increased in the dough, the oil binding to the hydrophobic amino acid of proteins by hydrophobic interactions resulted in the formation of a uniform gel network [11], and it is supposed that the elastic behavior of dough increased significantly. Song and Zheng [20] also reported that an addition of oil promotes the aggregation of gluten and plays a role in increasing the elastic behavior of dough. Therefore, the values of G′ are relatively higher in oil treatment than in water treatment.

The values of G′, G′′, and tan δ (G′′/G′) of TVP were slightly increased after the addition of SPI in the water as a liquid additive. Studies done by Roccia et al. [21] have indicated that SPI can greatly uptake water owing to the hydrogen bond of SPI. Therefore, its high absorption of water contributes to its elastic properties, resulting in a solid-like substance [21]. Wang et al. [15] has reported that SPI, when mixed with wheat gluten, reduces the water availability caused by the disulfide bonds of wheat gluten. The rheological changes, therefore, describe the relative distribution of "elasticity" as compared to "viscosity" during the TVP gel matrix formation when mixed with water and SPI. This is presumed to be due to a reduction in lubricating action and water availability by having a higher content of bound water in the SPI.

The G′ and G″ values of oil-lecithin-added meat analogues were significantly decreased when compared to only oil-added samples resulting in an increased tan δ value ($p < 0.05$). The elasticity and viscosity properties of oil-lecithin-treated meat analogues become weak owing to the larger tan δ value. Lecithin is an amphiphilic surfactant, therefore, the hydrophilic region (choline and phosphate group) might be conjugated to water molecules or the hydrogen group in protein, while the hydrophobic region (monosaturated or saturated fatty acid) might be conjugated to oil molecules by hydrophobic interactions. When all the components are mixed, lecithin is adsorbed at the interface between water and oil, causing the mixture to become a more emulsified system, which is homogenously dispersed resulting in high lubrication properties [16], and it is the reason for an increase in the tan δ value. Azizi et al. [22] also reported that the extensibility of the flat bread dough decreased with the addition of lecithin.

The tan δ of emulsion or OW treatments were the highest among all treatments. This indicates that the addition of water and oil affect the rheological changes ranging from "elasticity" to "viscosity" markedly. However, there was no significant difference in the tan δ value between the emulsion-added TVP and the OW-added TVP.

3.2. Cooking Loss

CL presents the degree of meat shrinkage during cooking, which is an important indicator to evaluate the meat quality related to the juiciness and yield of the final product. In general, the CL of processed meat products is affected by preparation parameters like constituents of composite materials. Therefore, CL of meat analogues was determined depending on the various liquid additives, and has been shown in Figure 3.

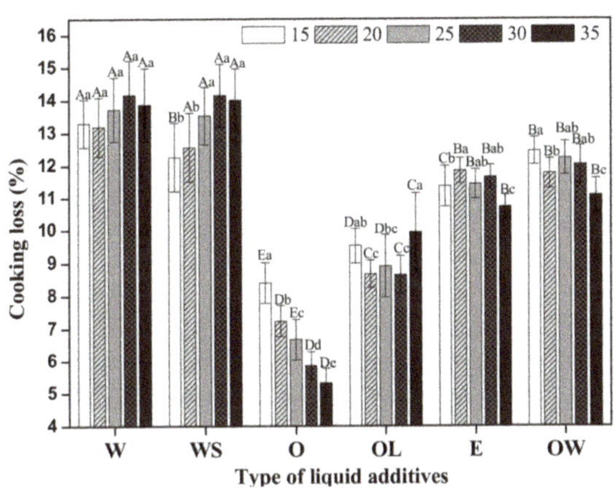

Figure 3. The CL value of meat analogues based on the type and concentration of liquid additives. [a–e] indicates significant differences in treatment concentrations ($p < 0.05$); [A–E] indicates significant differences in treatment types ($p < 0.05$).

The CL of water-added meat analogue with or without SPI was the highest among all treatments. As the amount of water increased, CL of meat analogues showed a typical increase from 12.5% to 14.5% regardless of SPI addition. The results are in accordance with those indicated by the researchers who cooked meat analogues at 70 °C for 150 min under sous-vide conditions [23]. However, there was no influence of SPI on CL of meat analogues, except for a loss of 15 g content. Although elasticity properties of meat analogues were enhanced when compared to the previous results by water uptake of SPI, SPI contributed to no significant improvement in CL. Wang et al. [15] reported that when SPI was heated at a high temperature, the hydrophobic amino acid residues get exposed by reducing the β-sheet content. Therefore, the heated-SPI cannot play a role in water uptake resulting in high CL.

On the other hand, CL of oil-added meat analogues (without lecithin) was seen to decrease with increasing oil concentration until the lowest value was attained. Interestingly, the lecithin influence on the CL of meat analogue was reflected by the significant increase when compared to oil treatment without lecithin ($p < 0.05$). Without lecithin, the oil induced hydrophobic interactions between protein and oil, or between protein and protein, enhancing the elastic behavior and gel strength. This means that the retained water in meat analogue mixture could be entrapped into gel structures bounded by hydrophobic interactions. This may explain why the CL of oil containing meat analogue was the lowest among all the samples. Lecithin is amphiphilic and attracts both water and fatty substances. Therefore, when lecithin was mixed with the oil and base of TVP, the hydrophobic region in lecithin may have bound to the canola oil or protein by hydrophobic interactions [16,24,25]. Meanwhile, the choline or phosphate groups of the hydrophilic part of lecithin binds to retained water by hydrogen bonds in the hydrated TVP. Huang et al. [26] reported that hydrogen bonds had a significantly negative correlation with gel strength. This study showed that hydrogen bonds were weakened as the temperature increased. Therefore, it is a possibility that once the water-oil-TVP protein network was set by lecithin, a high temperature caused an increase in hydrophobic interactions as decreasing hydrogen bonds caused an intense "shrinkage" of the network, resulting in a leakage of water.

The value of CL in emulsion-added samples presented was between the WS-added sample and the OL-added sample ($p < 0.05$). The CL of emulsion treatment was slightly higher than that in the OL treatment. Moreover, there was no difference in the CL values between the emulsion-added meat analogues and the OW-added meat analogues.

3.3. Water Holding Capacity

WHC is an important factor as it affects the quality and yield of fresh meat or its products [15]. In meat analogues, WHC represents the ability of protein to hold water and to form the protein gel network. The higher the WHC in meat analogues, the more enhanced the juiciness. Therefore, WHC of meat analogues was investigated depending on the kind of liquid additives before heating and after heating.

Figure 4 shows the effect of liquid additives on the WHC of non-heated and heated-meat analogues. Overall, the WHC of non-heated meat analogues was seen to be lower than that of the heated-meat analogues for all treatments. It was supposed that the heating process causes an enhanced gel network formation of TVP by hydrophobic interactions, which could retain more water when compared to the non-heated TVP system. There are numerous studies that have reported that the heating process can improve the gel network of a soy-based protein [15,23,26]. These studies have justified that once the protein structure of TVP has unfolded and the hydrophobic groups have been exposed by heating, the hydrophobic interaction between proteins is induced. Accordingly, the WHC value of cooked meat analogue is found to be higher than that in the dough condition.

The WHC of the water-added meat analogues, was seen to be affected by SPI under non-heated conditions ($p < 0.05$). Roccia et al. [21] explained the water syneresis of vegetable proteins between SPI and gluten proteins. They reported that the increase of soy addition dramatically decreased free water for syneresis in a mixture of gluten protein and soy protein resulting in high water holding capacity.

The WHC of the oil-added meat analogues was higher than that of the other meat analogues both under non-heated and heated conditions. The high WHC of the former may be explained by an interaction between the hydrocarbon side chains of oil and the hydrophobic amino acids of TVPs. This hydrophobic interaction causes gel matrix formation, which has a greater ability to entrap water. Therefore, the oil addition affects the WHC of the TVP matrix. On the other hand, WHC of lecithin was significantly decreased depending on the concentration of the oil content ($p < 0.05$). When lecithin was added into this system, the hydrophilic head would have bound to water molecules by hydrogen bonds, while the hydrophobic phospholipid fatty acid chains could be bound to the hydrophobic part of TVP or oil. Here, the holding ability of water was weakened with an increasing concentration of oil and lecithin [16]. This could be a result of lecithin blocking the crosslinks between oil and protein. Instead, lecithin strongly attracts the contained water molecules with hydrogen bonds, resulting in water exposure [27]. Therefore, this water may be released after centrifugation for measurement of WHC. This result is consistent with the observed CL. Since hydrogen binding between lecithin and water molecules are destroyed by an increasing temperature [15], WHC of lecithin-oil-TVP was decreased by an increasing concentration of lecithin and oil after heating.

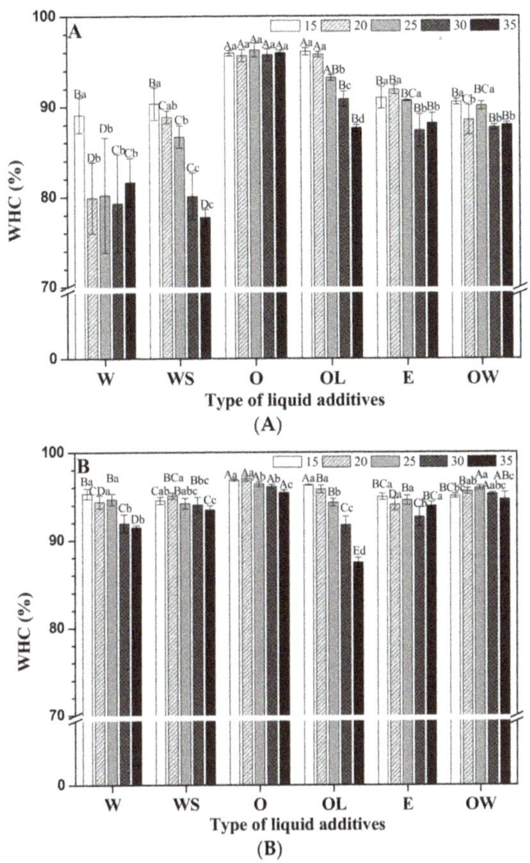

Figure 4. The WHC value of meat analogues based on the type and concentration of liquid additives in dough (**A**) and cooked (**B**) conditions. $^{a–d}$ indicates significant differences in the treatment concentrations ($p < 0.05$); $^{A–E}$ indicates significant differences in the treatment types ($p < 0.05$).

The WHC of emulsion and OW-treated meat analogues was enhanced by the heating process. The WHC values for non-heated meat analogues, however, showed no difference between emulsion and OW treatments. However, after heating, the WHC of OW-treated meat analogues was seen to be slightly higher than that of emulsion-treated meat analogues.

3.4. Texture Profile Analysis

Texture of meat analogues is an important factor to mimic the organoleptic taste of muscle. Figure 5 shows the textural parameters (hardness, cohesiveness, springiness, and chewiness) of meat analogues on treatment with different liquid additives. All TPA parameters were the lowest on water treatment, whereas the highest was on oil treatment ($p < 0.05$). This result is consistent with the rheological properties. Lin et al. [19] reported the hardness, cohesiveness, chewiness, and gumminess decreased as the moisture content increased. This means that the higher water contents formed more softened meat analogues which is the same as our results. On the other hand, Barbut and Marangoni. [11] studied that the smaller size of canola oil can connect protein–protein interaction resulting in a firmer meat product owing to its larger surface area among the protein matrix. These reports were consistent with our results where canola oil formed a compact protein gel network of TVP matrix since the oil globule binds to the hydrophobic amino acids in the protein by a hydrophobic interaction [11].

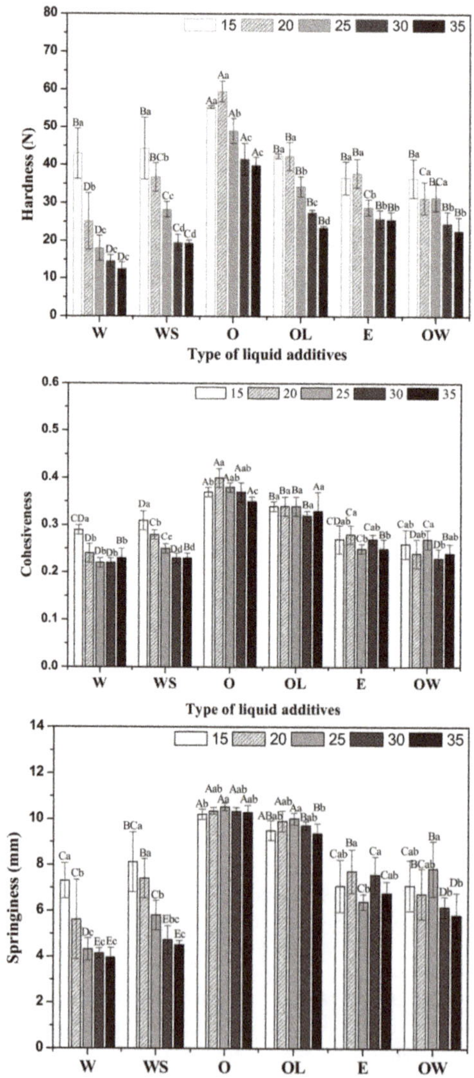

Figure 5. *Cont.*

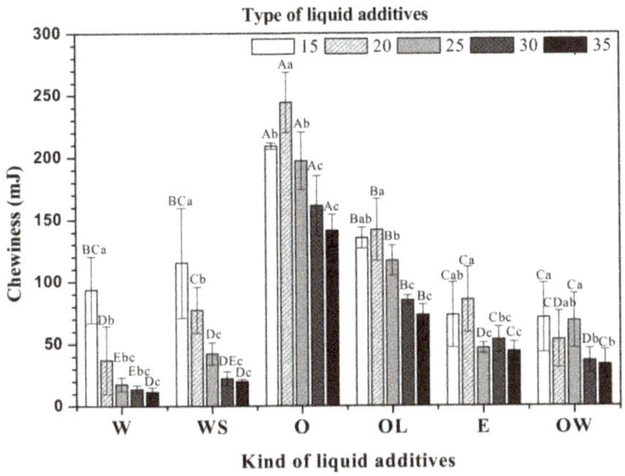

Figure 5. The TPA parameters of meat analogue based on the type and concentration of liquid additives. [a–d] indicates significantly different treatment concentrations ($p < 0.05$); [A–E] indicates significantly different treatment types ($p < 0.05$).

With increasing concentrations of liquid additives, hardness, cohesiveness, springiness, and chewiness decreased in all treatments ($p < 0.05$). This result was inevitable due to the reduction of TVP content in the same weight of meat analogue as the concentration of liquid additive increased [19,28]. All TPA parameters were seen to be significantly increased in meat analogues with water and SPI addition, while there was a significant decrease in meat analogues with oil and lecithin addition ($p < 0.05$). These results indicate that SPI might improve the gel properties of the TVP complex. On the contrary, lecithin may inhibit the gel matrix formation by hydrophobic interactions, thereby enhancing their lubricant effect by interacting between the water molecules and the polar ionic head groups. Wang et al. [15] reported that SPI enhanced the hardness of meat protein due to an increase in hydrophobic interactions. Regarding the role of lecithin, Xia et al. [16] investigated how it inhibits the disulfide binding of the cross-linking of proteins, resulting in a decrease in the hardness of meat protein. Therefore, it is considered that SPI and lecithin may have a similar effect on the TVP protein matrix.

Interestingly, all TPA parameters of the matrix formed by the addition of both oil and water with SPI and lecithin showed intermediate values between WS treatment and OL treatment, but were closer to WS treatment. Accordingly, the TPA parameters seem to be more influenced by the continuous phase (WS) than the dispersed phase (OL) of the added liquid additives. All TPA parameters of the emulsion and OW treatments showed intermediate values between WS and OL treatment because they included a mixture of the continuous phase (WS) and the dispersed phase (OL). However, there was no significant difference between the emulsion and OW treatments.

3.5. Color Analysis

The color measurement according to the type and concentration of liquid additives is shown in Table 2. As the content of liquid additives increased, the lightness (L^*) values of all treatment was seen to significantly increase, except for OL treatment. The small globules such as water or oil cause more light reflection, and they increase the L^* values [11,29]. Youssef and Barbut [30] reported that the L^* values decreased with an increasing protein content in meat products, whereas other researches have reported [29] that L^* values increased with rising canola oil content. Our results showed similar trends where the L^* value was increased during the decrease of TVP contents and an increase in oil content [30]. Especially, the L^* values of emulsion treatment was the highest among all the treatments because of its milky color. Lee et al. [31] reported that the pork patty added with nanoemulsion has the highest L^* values due to its white color. The L^* value of water-SPI-added samples was slightly decreased when compared to only water-added samples due to the increase in protein content [30], also L^* value of oil-lecithin-added samples was greatly decreased when compared to oil treatment due to the yellow-brownish color of the lecithin [16,31] ($p < 0.05$). However, there was no effect from the emulsifiers on the color of the TVP matrix in the emulsion treatment. The redness (a^*) value of all treatments was overall lower compared to the meat products due to a low concentration of myoglobin pigment [4]. As the content of liquid additives increased, the a^* value of all treatments was seen to significantly decrease, except for water treatment. Especially, the a^* value of emulsion treatment showed the lowest value ($p < 0.05$) owing to the milky color of the emulsion [31].

Table 2. Effects of the type and concentration of liquid additives on color of meat analogue.

		Color		
Treatments	**Concentration (g)**	***L****	***a****	***b****
W	15	62.32 ± 4.02 ^{Cc}	2.21 ± 0.23 ^{ABb}	16.19 ± 0.62 ^{Ac}
	20	62.61 ± 1.12 ^{Dc}	1.91 ± 0.07 ^{Bc}	16.25 ± 0.15 ^{Bc}
	25	64.55 ± 2.47 ^{Dbc}	1.93 ± 0.16 ^{Ac}	16.47 ± 0.68 ^{Ac}
	30	65.72 ± 1.71 ^{Cab}	2.62 ± 0.25 ^{Aa}	17.60 ± 0.39 ^{Ab}
	35	67.81 ± 1.04 ^{Ca}	2.62 ± 0.14 ^{Aa}	18.22 ± 0.27 ^{Aa}
WS	15	61.64 ± 0.90 ^{Cb}	2.34 ± 0.03 ^{Aa}	16.19 ± 0.26 ^{Ab}
	20	61.58 ± 0.59 ^{Db}	2.23 ± 0.06 ^{Aab}	16.07 ± 0.17 ^{BCb}
	25	62.41 ± 1.11 ^{Eb}	2.10 ± 0.19 ^{Abc}	16.25 ± 0.20 ^{ABb}
	30	64.54 ± 0.93 ^{Ca}	2.21 ± 0.23 ^{Bab}	16.55 ± 0.77 ^{Bab}
	35	65.38 ± 1.34 ^{Da}	1.98 ± 0.15 ^{Bc}	17.05 ± 0.76 ^{Ba}
O	15	65.26 ± 0.52 ^{Bd}	2.07 ± 0.14 ^{BCa}	15.63 ± 0.30 ^{Ba}
	20	67.64 ± 0.73 ^{Bc}	1.90 ± 0.09 ^{Ba}	15.84 ± 0.34 ^{Ca}
	25	68.94 ± 1.34 ^{Bb}	1.65 ± 0.17 ^{Bb}	15.37 ± 0.44 ^{Ca}
	30	72.12 ± 1.46 ^{Aa}	1.28 ± 0.04 ^{Ec}	15.69 ± 0.96 ^{Ba}
	35	69.41 ± 0.73 ^{Bb}	1.52 ± 0.23 ^{Cb}	16.03 ± 0.52 ^{Ca}
OL	15	65.96 ± 1.40 ^{Ba}	2.26 ± 0.31 ^{ABa}	16.02 ± 0.17 ^{ABbc}
	20	65.87 ± 1.10 ^{Ca}	2.23 ± 0.14 ^{Aa}	16.14 ± 0.20 ^{Bb}
	25	66.80 ± 0.19 ^{Ca}	1.95 ± 0.17 ^{Ab}	15.87 ± 0.29 ^{Bc}
	30	65.77 ± 1.23 ^{Ca}	1.93 ± 0.13 ^{Cb}	16.20 ± 0.03 ^{Bb}
	35	63.89 ± 0.66 ^{Eb}	2.01 ± 0.26 ^{Bab}	16.57 ± 0.07 ^{BCa}
E	15	68.82 ± 0.62 ^{Ac}	1.73 ± 0.04 ^{Da}	16.36 ± 0.20 ^{Aa}
	20	70.09 ± 0.83 ^{Ab}	1.75 ± 0.08 ^{Ca}	16.52 ± 0.23 ^{Aa}
	25	70.75 ± 0.58 ^{Ab}	1.48 ± 0.08 ^{Bb}	16.31 ± 0.04 ^{ABa}
	30	70.08 ± 0.41 ^{Bb}	1.54 ± 0.10 ^{Db}	16.36 ± 0.38 ^{Ba}
	35	71.70 ± 0.54 ^{Aa}	1.32 ± 0.02 ^{Cc}	16.32 ± 0.28 ^{Ca}
OW	15	68.52 ± 0.56 ^{Ab}	1.98 ± 0.04 ^{Ca}	16.19 ± 0.03 ^{Ab}
	20	69.17 ± 0.97 ^{Ab}	1.89 ± 0.05 ^{Ba}	16.17 ± 0.15 ^{Bb}
	25	70.51 ± 0.80 ^{ABa}	1.65 ± 0.01 ^{Bb}	16.17 ± 0.09 ^{ABb}
	30	69.21 ± 0.73 ^{Bb}	1.74 ± 0.05 ^{Cb}	16.52 ± 0.09 ^{Ba}
	35	70.51 ± 0.36 ^{Ba}	1.51 ± 0.14 ^{Cc}	16.13 ± 0.07 ^{Cb}

$^{a–d}$ indicates significant differences in treatment concentrations ($p < 0.05$). $^{A–E}$ indicates significant differences in treatment types ($p < 0.05$).

3.6. Visible Appearance

The external and internal appearance of meat analogues after cooking has been presented in Figure 6. The external appearance showed no difference on observation with water addition and water with SPI treatment. The water-added samples with or without SPI were found to have a rough surface when compared to the other samples due to the high evaporation of moisture on the surface. However, the internal appearance of the water-added samples with SPI obviously appeared more homogeneous and finely structured when compared to the only water-added samples. This therefore proved that the SPI addition might absorb the water molecules to form the fine gel matrix [15]. Furthermore, after cooking, the volume of the meat analogue added with water was slightly reduced due to evaporation of the water [4].

Figure 6. The external (**A–F**) and internal (**A′–F′**) appearance of meat analogue in six different liquid additives. (**A**) and (**A′**): water; (**B**) and (**B′**): water + SPI; (**C**) and (**C′**): canola oil; (**D**) and (**D′**): canola oil + lecithin; (**E**) and (**E′**): O/W emulsion; (**F**) and (**F′**): water + canola oil + SPI + lecithin.

Oil-added samples appeared bright yellow in color with a fine surface texture and volume expansion when compared to lecithin-oil-added samples, which showed cracks. The interior of lecithin-oil-added samples showed void pores when compared to the only oil-added sample. This led to the supposition that the fine gel matrix in oil treatment may hinder the water evaporation and therefore, the captured water molecules during cooking may cause the volume expansion of the meat analogues [4].

The exterior of the meat analogues with emulsion and OW treatment did not show a rough surface or cracks, unlike the ones with WS treatment and OL treatment. On the other hand, the internal appearance of the emulsion and OW-treated meat analogues showed a porous and rough structure like those of the ones treated with water.

3.7. Microstructure

For the structural observation, the amount of liquid additives was fixed at 25 g for all treatments. Microstructure of the six varied TVP matrices after cooking were observed under a scanning electron microscope with a magnification of 100 times (Figure 7) and 300 times (Figure 8). The water-added samples without SPI showed inhomogeneous pore distribution and a large air cell partly (see white arrow in Figure 7A). Meanwhile, SPI-water-added samples displayed more homogeneous and very distinct fibrous pores in the matrix, which exhibited a more sponge-like structure as compared to the others (Figure 8). Wang et al. [15] found a more ordered network between meat protein and SPI by obvious cross-linked strands, which can hold more water to form a fine gel matrix. This report was consistent with our results in meat analogues, and verified that the SPI could absorb water upon hydration and entrap the water molecules inside the gel matrix to form a finer structure of the gel [15], which probably contributes to higher WHC (Figure 4) and hardness (Figure 5) and more elastic properties (Figure 2).

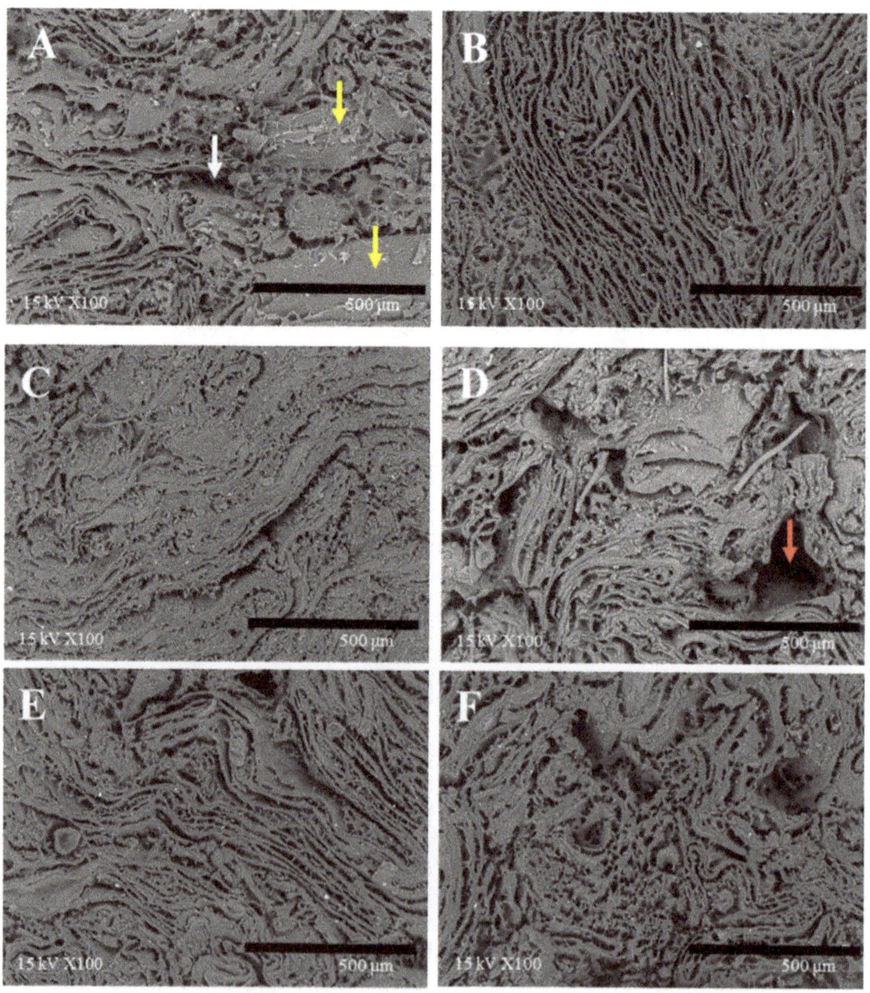

Figure 7. Scanning electron microscope images of samples (magnification 100 times). (**A**): water; (**B**): water + SPI; (**C**): canola oil; (**D**): canola oil + lecithin; (**E**): O/W emulsion; (**F**): water + canola oil + SPI + lecithin.

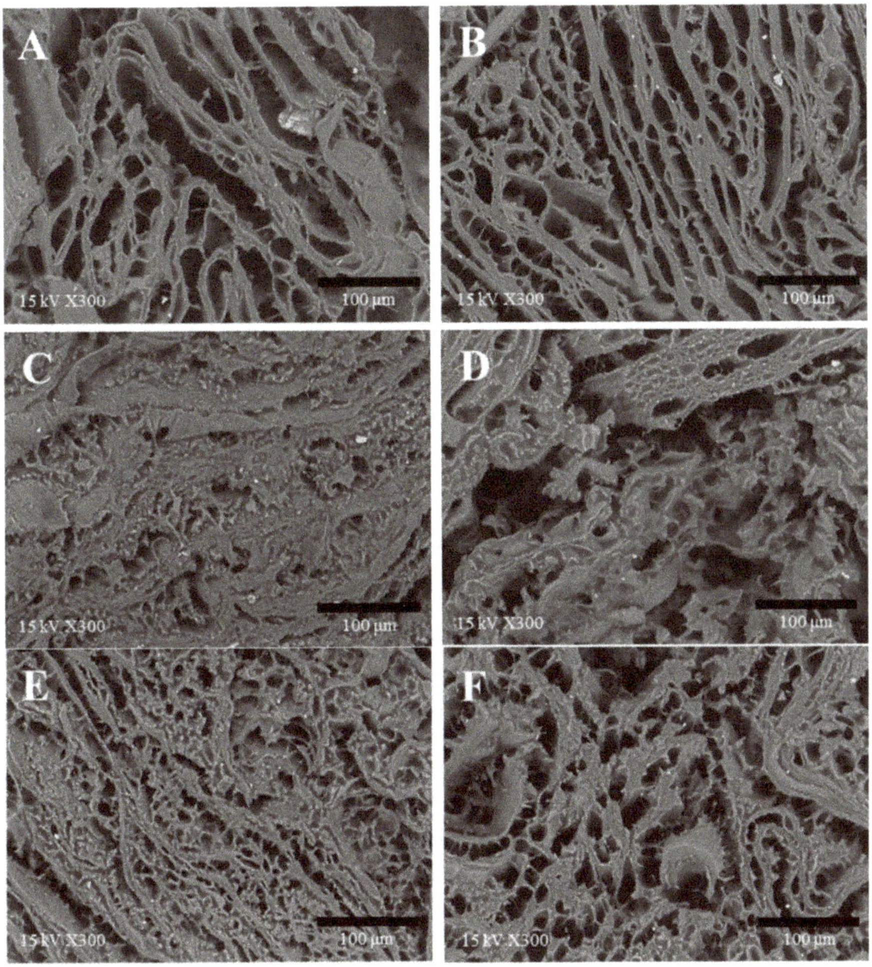

Figure 8. Scanning electron microscope images of samples (magnification 300 times). (**A**): water; (**B**): water + SPI; (**C**): canola oil; (**D**): canola oil + lecithin; (**E**): O/W emulsion; (**F**): water + canola oil + SPI + lecithin.

The oil-added samples without lecithin formed a continuous network comprising of a very fine and tightly connected matrix by hydrophobic interaction between TVPs and oil. On the contrary, the microstructure of the lecithin-oil-added samples showed structural similarities to the water-added samples, having a large air cell (see red arrow in Figure 7D) and was partly cracked, although there was no additional water as a liquid additive. Xia et al. [16] mentioned that the hydrophilic head of lecithin can be bound to water molecules, which indicate that lecithin might attract the water, which was absorbed during the swelling process of TVP preparation before the addition of liquid additives. Thus, the hardness and WHC were decreased, and the cooking loss was increased with an increasing concentration of oil with the lecithin additive [16,32].

The microstructure of the emulsion-added TVP matrix was the most fine and the smallest pore size in the samples. At 100 times magnification, the emulsion-added sample shows an elongated shape of the pore—less than one of the water-SPI-added sample. When observed at 300 times magnification, the emulsion-added samples showed uniform round-shaped pores in the fibrous structure. The emulsions added to the TVPs may improve the homogeneous structure of the composites owing to pre-emulsification of liquid additives. Zhuang et al. [33] reported that the morphology of cooked meat batters applied with pre-emulsified sesame oil showed a homogeneous and compact structure and pores of semi-round shapes. Likewise, Jimenez-Colmenero et al. [34] reported that the microstructure of the frankfurters replaced with emulsion showed small cavities. It is presumed that the emulsifier in an O/W emulsion assists greater dispersion of the oil phase, which was well embedded in the TVP matrix. Comparatively, the OW-addition, which is a non-emulsified liquid additive, presented a coarser and larger air cell than the emulsion-added sample at 100 times magnification. This was also identified in the image obtained at 300 times magnification. Although the microstructures were different between the emulsion-added TVP and the OW-added one, there was no significant difference in the WHC, CL, and TPA parameters. Subsequent studies are needed to determine the difference between emulsion and OW treatment.

3.8. Sensory Evaluation

The sensory evaluation was carried out to analyze the differences in the meat analogues. As shown in Figure 9, intensity and preference score of juiciness were the highest in water treatment and the lowest in oil treatment. According to CL and WHC results, contrary to the expectation that the oil treatment with the lowest CL value and the highest WHC value had the highest juiciness, the juiciness of water treatment was found to be the highest. These results mean that juiciness is more affected by water than oil. Selani et al. [35] reported that a beef burger added with canola oil as a fat replacement enhanced the cohesiveness and springiness in its sensory attributes. According to a report by Kim et al. [25], in a sensory test of vegetable meat, MCT oil showed a positive correlation with compactness and springiness and also showed a negative correlation with juiciness and tenderness. On the other hand, water showed a positive correlation with juiciness and tenderness. In our research, water treatment had the lowest value in all TPA parameters (Figure 5) because of the inhomogeneous structure of the TVP matrix (Figure 7A). However, our results in the sensory test showed that water-treated samples had a high intensity of firmness, elasticity, and compactness in contrast to the TPA results and other research [25]. This reason is supposed to be that the high-density region of the TVP matrix (see yellow arrow in Figure 7A) resulted in a feeling of firmness to chew in terms of mastication.

The preference score of the oil taste at oil treatment was the lowest owing to its excessively oily feeling. The intensity of soy taste did not show any tendency based on oil and water content, and also did not show any tendency between intensity and preference.

Emulsion treatment showed low intensity scores in relation to texture parameters such as firmness, elasticity, stickiness, compactness, and roughness. Therefore, emulsion treatment had a high preference score in relation to texture and overall acceptance. Lee et al. [31] reported that the pork patty added with nanoemulsion showed the highest intensity score for juiciness and tenderness, while it showed the lowest intensity score for compactness and springiness in the sensory test. Kim et al. [25] also reported that meat analogue added with emulsion showed a high juiciness and tenderness, which caused high overall acceptability. Likewise, Pietreasik et al. [6] have proved that most consumers prefer tenderness to be a significant criteria for determining the overall acceptance and quality of meat. These studies are similar results to these results because the emulsion treatment showed the finest structure and the smallest pore size, and a homogeneous structure of the TVP matrix, owing to the emulsification process (Figure 8E), resulted in good texture.

In the sensory evaluation, a difference was seen between the emulsion-added sample and the OW-added sample owing to the differences in pore size (Figure 8E,F). This indicates that the OW-added TVP matrix was less homogeneous than the emulsion-added TVP matrix, resulting in a high intensity of firmness, compactness, and roughness.

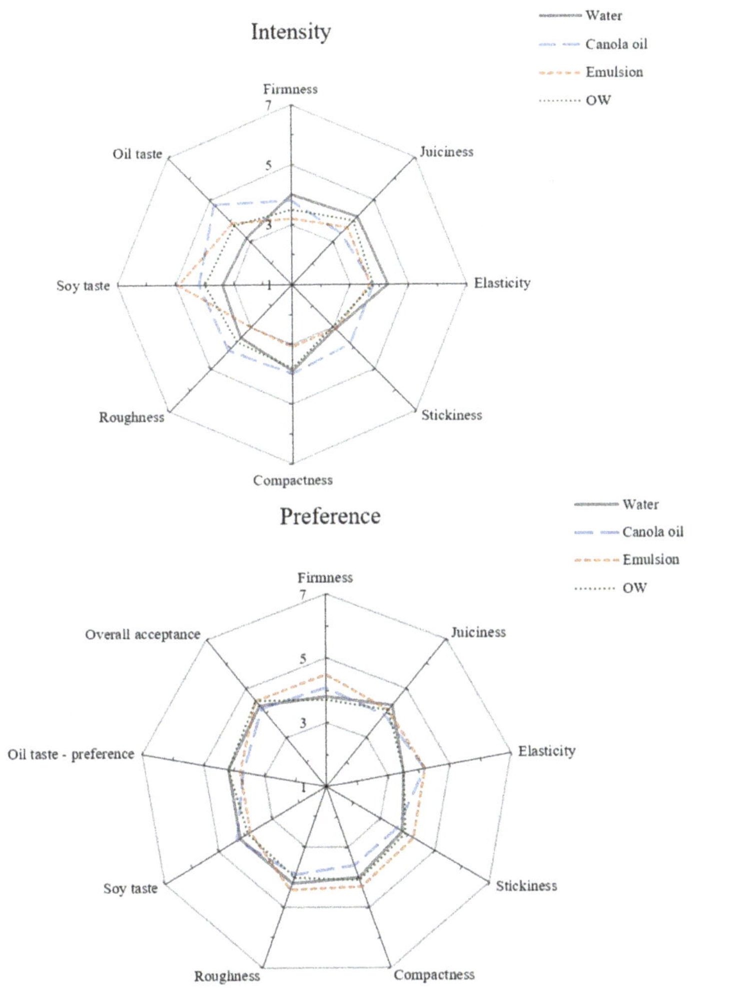

Figure 9. Sensorial profile of meat analogue based on various liquid additives.

4. Conclusions

The present study evaluated the physicochemical, structural, and sensory properties of meat analogues according to the type and concentration of various non-animal-based liquid additives. According to an analysis of CL, WHC, and TPA, SPI had a positive effect on the gel network formation of TVP matrix, whereas lecithin had a negative effect. The values of emulsion treatment are between water treatment added with SPI and oil treatment added with lecithin. There was no significant difference between the pre-emulsified mixture and the separated liquid additives in terms of physicochemical properties, but slight differences were seen in the structural properties and sensory characteristics. Especially, the sensory evaluation showed that juiciness was positively correlated with water, and the preference of texture parameters and overall acceptance were positively correlated with emulsion treatment.

The results of our study have thus effectively analyzed and presented the possibility of using certain liquid additives to improve the acceptability of meat analogues. The substitution of animal fat in meat analogues while maintaining its sensory and physicochemical profile to ensure its acceptability by vegetarians is a present day challenge for the food science industry. This work, if enhanced further, promises to successfully provide answers to this challenge.

Author Contributions: Conceptualization, Y.C. and M.-J.C.; methodology, G.W., J.B., H.K., Y.C. and M.-J.C.; validation, G.W., J.B. and H.K.; writing, G.W. and M.-J.C.; project administration and supervision, and funding acquisition, M.-J.C. All authors have read and agreed to the published version of the manuscript.

Funding: This research was supported by Korea Institute of Planning and Evaluation for Technology in Food, Agriculture, Forestry and Fisheries (IPET) through High Value-added Food Technology Development Program, funded by Ministry of Agriculture, Food and Rural Affairs (MAFRA)(317040).

Conflicts of Interest: The authors declare no conflict of interest.

References

1. Apostolidis, C.; McLeay, F. Should we stop meating like this? Reducing meat consumption through substitution. *Food Policy* **2016**, *65*, 74–89. [CrossRef]
2. Dekkers, B.L.; Boom, R.M.; Goot, A.J. Structuring processes for meat analogues. *Trends Food Sci. Technol.* **2018**, *81*, 25–36. [CrossRef]
3. Hoek, A.C.; Luning, P.A.; Weijzen, P.; Engels, W.; Kok, F.J.; de Graaf, C. Replacement of meat by meat substitutes. A survey on person- and product-related factors in consumer acceptance. *Appetite* **2011**, *56*, 662–673. [CrossRef]
4. Samard, S.; Ryu, G.H. A comparison of physicochemical characteristics, texture, and structure of meat analogue and meats. *J. Sci. Food Agric.* **2018**, *99*, 2708–2715. [CrossRef]
5. Carballo, J.; Barreto, G.; Colmenero, F.J. Starch and Egg White Influence on Properties of Bologna Sausage as Related to Fat Content. *J. Food Sci.* **1995**, *60*, 673–677. [CrossRef]
6. Pietrasik, Z.; Duda, Z. Effect of fat content and soy protein/carrageenan mix on the quality characteristics of comminuted, scalded sausages. *Meat Sci.* **2000**, *56*, 181–188. [CrossRef]
7. Yoo, S.S.; Kook, S.H.; Park, S.Y.; Shim, J.H.; Chin, K.B. Physicochemical characteristics, textural properties and volatile compounds in comminuted sausages as affected by various fat levels and fat replacers. *Int. J. Food Sci. Technol.* **2007**, *42*, 1114–1122. [CrossRef]
8. Baek, K.H.; Utama, D.T.; Lee, S.G.; An, B.K.; Lee, S.K. Effects of Replacing Pork Back Fat with Canola and Flaxseed Oils on Physicochemical Properties of Emulsion Sausages from Spent Layer Meat. *Asian Australas. J. Anim. Sci.* **2016**, *29*, 865–871. [CrossRef]
9. Kyriakopoulu, K.; Dekkers, B.; Van der Goot, A.J. Plant-based meat analogues. In *Sustainable Meat Production and Processing*; Galanakis, C.M., Ed.; Academic Press: Cambridge, MA, USA, 2019; pp. 103–126. [CrossRef]
10. Paneras, E.D.; Bloukas, J.G. Vegetable Oils Replace Pork Backfat for Low-Fat Frankfurters. *J. Food Sci.* **1994**, *59*, 725–728. [CrossRef]
11. Barbut, S.; Marangoni, A. Organogels use in meat processing—Effects of fat/oil type and heating rate. *Meat Sci.* **2019**, *149*, 9–13. [CrossRef]
12. Pietrasik, Z.; Wang, H.; Janz, J.A.M. Effect of canola oil emulsion injection on processing characteristics and consumer acceptability of three muscles from mature beef. *Meat Sci.* **2012**, 93. [CrossRef] [PubMed]
13. Serdaroğlu, M.; Öztürk, B.; Urgu, M. Emulsion characteristics, chemical and textural properties of meat systems produced with double emulsions as beef fat replacers. *Meat Sci.* **2016**, *117*, 187–195. [CrossRef] [PubMed]
14. Freire, M.; Bou, R.; Cofrades, S.; Solas, M.T.; Jimenez-Colmenero, F. Double emulsions to improve frankfurter lipid content: Impact of perilla oil and pork backfat. *J. Sci. Food Agric.* **2016**, *96*, 900–908. [CrossRef] [PubMed]
15. Wang, Z.; Liang, J.; Jiang, L.; Li, Y.; Wang, J.; Zhang, H.; Li, D.; Han, F.; Li, Q.; Wang, R.; et al. Effect of the interaction between myofibrillar protein and heat-induced soy protein isolates on gel properties. *CyTA-J. Food* **2015**, *13*, 527–534. [CrossRef]
16. Xia, W.; Ma, L.; Chen, X.; Li, X.; Zhang, Y. Physicochemical and structural properties of composite gels prepared with myofibrillar protein and lecithin at various ionic strengths. *Food Hydrocoll.* **2018**, *82*, 135–143. [CrossRef]
17. Pathare, P.B.; Roskilly, A.P. Quality and Energy Evaluation in Meat Cooking. *Food Eng. Rev.* **2016**, *8*, 435–447. [CrossRef]

18. Lorenzo, J.M.; Cittadini, A.; Munekata, P.E.; Domínguez, R. Physicochemical properties of foal meat as affected by cooking methods. *Meat Sci.* **2015**, *108*, 50–54. [CrossRef]
19. Lin, S.; Huff, H.E.; Hsieh, F. Texture and Chemical Characteristics of Soy Protein Meat Analog Extruded at High Moisture. *J. Food Sci.* **2000**, *65*, 264–269. [CrossRef]
20. Song, Y.; Zheng, Q. Dynamic rheological properties of wheat flour dough and proteins. *Trends Food Sci. Technol.* **2007**, *18*, 132–138. [CrossRef]
21. Roccia, P.; Ribotta, P.D.; Pérez, G.T.; León, A.E. Influence of soy protein on rheological properties and water retention capacity of wheat gluten. *LWT-Food Sci. Technol.* **2009**, *42*, 358–362. [CrossRef]
22. Azizi, M.H.; Rajabzadeh, N.; Riahi, E. Effect of mono-diglyceride and lecithin on dough rheological characteristics and quality of flat bread. *Lebensm. Wiss. Technol. Food Sci.* **2003**, *36*, 189–193. [CrossRef]
23. Gómez, I.; Ibañez, F.C.; Beriain, M.J. Physicochemical and sensory properties of sous vide meat and meat analog products marinated and cooked at different temperature-time Combinations. *Int. J. Food Prop.* **2019**, *22*, 1693–1708. [CrossRef]
24. Mottola, M.; Vico, R.V.; Villanueva, M.E.; Fanani, M.L. Alkyl esters of l-ascorbic acid: Stability, surface behaviour and interaction with phospholipid monolayers. *J. Colloid Interface Sci.* **2015**, *457*, 232–242. [CrossRef] [PubMed]
25. Kim, H.; Bae, J.; Wi, G.; Kim, H.T.; Cho, Y.; Choi, M.J. Physicochemical properties and sensory evaluation of meat analog mixed with different liquid materials as an animal fat substitute. *Food Eng. Prog.* **2019**, *23*, 62–68. [CrossRef]
26. Huang, X.; Li, C.; Yang, F.; Xie, L.; Xu, X.; Zhou, Y.; Pan, S. Interactions and gel strength of mixed myofibrillar with soy protein, 7S globulin and enzyme-hydrolyzed soy proteins. *Eur. Food Res. Technol.* **2010**, *231*, 751–762. [CrossRef]
27. Gosenca, M.; Bešter-Rogac, M.; Gašperlin, M. Lecithin based lamellar liquid crystals as a physiologically acceptable dermal delivery system for ascorbyl palmitate. *Eur. J. Pharm. Sci.* **2013**, *50*, 114–122. [CrossRef]
28. Kim, H.W.; Setyabrata, D.; Lee, Y.J.; Jones, O.G.; Kim, Y.H.B. Pre-treated mealwork larvae and silkworm pupae as a novel protein ingredient in emulsion sausages. *Innov. Food Sci. Emerg. Technol.* **2016**, *38*, 116–123. [CrossRef]
29. Youssef, M.K.; Barbut, S. Fat reduction in comminuted meat products-effects of beef fat, regular and pre-emulsified canola oil. *Meat Sci.* **2011**, *87*, 356–360. [CrossRef]
30. Youssef, M.K.; Barbut, S. Physicochemical Effects of the Lipid Phase and Protein Level on Meat Emulsion Stability, Texture, and Microstructure. *J. Food Sci.* **2010**, *75*, S108–S114. [CrossRef]
31. Lee, J.; Kim, H.; Choi, M.J.; Cho, Y. Improved Physicochemical Properties of Pork Patty Supplemented with Oil-in-Water Nanoemulsion. *Korean J. Food Sci. Anim. Resour.* **2020**, *40*, 262–273. [CrossRef]
32. Souza, D.N.D.; Mullan, B.P.; Pethick, D.W.; Pluske, J.R.; Dunshea, F.R. Nutritional strategies affect carcass and pork quality but have no effect on intramuscular fat content of pork. *Anim. Prod. Sci.* **2012**, *52*, 276–282. [CrossRef]
33. Zhuang, X.; Han, M.; Kang, Z.; Wang, K.; Bai, Y.; Xu, X.; Zhou, G. Effects of the sugarcane dietary fiber and pre-emulsified sesame oil on low-fat meat batter physicochemical property, texture, and microstructure. *Meat Sci.* **2016**, *113*, 107–115. [CrossRef] [PubMed]
34. Jiménez-Colmenero, F.; Herrero, A.; Pintado, T.; Solas, M.T.; Ruiz-Capillas, C. Influence of emulsified olive oil stabilizing system used for pork backfat replacement in frankfurters. *Food Res. Int.* **2010**, *113*, 107–115. [CrossRef]
35. Selani, M.M.; Shirado, G.A.N.; Margiotta, G.B.; Saldana, E.; Spada, F.P.; Piedade, S.M.S.; Contreras-Castillo, C.J.; Canniatti-Brazaca, S.G. Effects of pineapple byproduct and canola oil as fat replacers on physicochemical and sensory qualities of low-fat beef burger. *Meat Sci.* **2016**, *112*, 69–76. [CrossRef]

© 2020 by the authors. Licensee MDPI, Basel, Switzerland. This article is an open access article distributed under the terms and conditions of the Creative Commons Attribution (CC BY) license (http://creativecommons.org/licenses/by/4.0/).

foods

MDPI

Article

Legume Beverages from Chickpea and Lupin, as New Milk Alternatives

Mariana Lopes [1], Chloé Pierrepont [1], Carla Margarida Duarte [1,*], Alexandra Filipe [2], Bruno Medronho [3,4] and Isabel Sousa [1]

1 LEAF-Linking Landscape, Environment, Agriculture and Food, Instituto Superior de Agronomia, University of Lisbon, Tapada da Ajuda, 1349-017 Lisboa, Portugal; marianacl96@hotmail.com (M.L.); chloe.de-pierrepont@agrosupdijon.fr (C.P.); isabelsousa@isa.ulisboa.pt (I.S.)

2 CIEPQPF, Department of Chemical Engineering, University of Coimbra, Pólo II–R. Silvio Lima, 3030-790 Coimbra, Portugal; xanafilippe@gmail.com

3 MED–Mediterranean Institute for Agriculture, Environment and Development, University of Algarve, Faculty of Sciences and Technology, Campus de Gambelas, Ed. 8, 8005-139 Faro, Portugal; bfmedronho@ualg.pt

4 FSCN, Surface and Colloid Engineering, Mid Sweden University, SE-851 70 Sundsvall, Sweden

* Correspondence: carladuarte@isa.ulisboa.pt

Received: 31 August 2020; Accepted: 9 October 2020; Published: 14 October 2020

Abstract: Recently, milk consumption has been declining and there is a high demand for non-dairy beverages. However, market offers are mainly cereal and nut-based beverages, which are essentially poor in protein (typically, less than 1.5% against the 3.5% in milk) and are not true milk replacers in that sense. In this work, new beverages from different pulses (i.e., pea, chickpea and lupin) were developed using technologies that enable the incorporation of a high level of seed components, with low or no discharge of by-products. Different processing steps were sequentially tested and discussed for the optimization of the sensorial features and stability of the beverage, considering the current commercial non-dairy beverages trends. The lupin beverage protein contents ranged from 1.8% to 2.4% (*w*/*v*) and the chickpea beverage varied between 1.0% and 1.5% (*w*/*v*). The "milk" yield obtained for the optimized procedure B was 1221 g/100 g of dry seed and 1247 g/100 g of dry seed, for chickpea beverage and lupin beverage, respectively. Sensory results show that chickpea beverage with cooking water has the best taste. All pulses-based beverages are typical non-Newtonian fluids, similarly to current non-dairy alternative beverages. In this respect, the sprouted chickpea beverage, without the cooking water, presents the most pronounced shear-thinning behavior of all formulations.

Keywords: non-dairy beverages; pulses; chickpea; lupin; flow behavior

1. Introduction

Milk is an important staple food. According to American market research platforms, although the milk segment is projected to account for the largest market share during the 2018–2023 forecast period, the market for dairy alternatives is projected to grow from USD 17.3 billion in 2018 to USD 29.6 billion by 2023, at a CAGR of 11.4%, whereas the Asia-Pacific region represents the biggest market share [1,2]. Despite milk importance, its consumption has been diminishing. Different reasons are driving the consumers increasing rejection for dairy products, such as: (i) health reasons (e.g., lactose intolerance, cow's milk allergy, hypercholesterolemia, hormones and antibiotic residues); (ii) lifestyle choice (e.g., vegetarian/vegan diet, animal welfare) and (iii) environmental concerns relating livestock breeding with huge environmental impacts (e.g., extensive land-use, water footprint, CO_2 and methane emissions) [3–6], which is estimated to represent 14.5% of all human-induced emissions [7]. The use of native leguminous proteins can help reduce environmental impact; these sources require reduced amounts of fertilizer agents in the soil, due to symbiotic associations with nitrogen-fixing bacteria in the plant roots [8].

Legume beverages are colloidal suspensions of dissolved and disintegrated plant material in water, resembling cow's milk in appearance and consistency [5]. However, the majority of these plant beverages face technological issues, often related to processing or preservation. Legume beverages, present the most balanced composition, rich in proteins and minerals, with low-glycemic index. Its protein content, ca. 3–4%, are similar to cow milk (i.e., 3.3–3.5%), while the other types of cereal and nuts-based beverages typically display values between 0.1% and 1.0% [5]. Soy milk is the most widely consumed legume beverage and contains a similar grade of protein as milk (minimum 3%). Nevertheless, its share has been decreasing because of health concerns related to GMO and allergens, high levels of isoflavones and the CO_2 footprint [9]. Other emerging legume alternatives for milk products, eventually healthier and tastier options than soybean, have been ranked based on sensory results as: pea > lupin > lentil = soybean > chickpea > fababean [10].

The pulse proteins have essential amino acid composition complementary to cereals and are naturally gluten-free, being safe for gluten intolerance/allergic consumers. Thermal processing, pH variation, ionic strength or presence of salts may change protein structure and influence their solubility [11]. Proteins from different pulses generally show higher solubility at alkaline and acidic pH values (pH < 4.0 and pH > 5.0), whereas much less soluble at pH around their isoelectric point (4.6 for lupin proteins and 4.5 for pea and chickpea proteins) [12,13].

Important technological interventions are still needed to improve the quality of legume beverages, such as to increase the product physical stability (e.g., colloidal milling) and to enhance microbial shelf-life [14,15]. Regarding colloidal milling, this technique has been used to reduce the size of dispersed phase particles but microbial spoilage requires further product pasteurization to ensure safe consumption. However, heat treatment, such as pasteurization, may increase legume beverage viscosity, affecting its stability [16]. This is relevant when the legume seed contains a high starch content, such as chickpea and pea. The ultra-high pressure homogenization can be a good thermal processing alternative, to achieve size reduction of colloidal particles and, simultaneously, to destroy microorganisms [14]. Despite that, the off-flavors in legume beverages (considered one of the most challenging barrier to consumer acceptance) can be easily removed by cooking. In fact, the sensory acceptability of legume-based beverages, represents a major limiting factor due to their characteristic "beany" flavor. The "beany" flavor is associated to endogenous lipoxygenases that oxidize unsaturated fatty acids in oil rich pulses, such as soy [17] and peanuts (over 20% fat) and therefore it is expected to be less pronounced in oil poor pulses, such as peas, lupins or chickpeas (1.5% to 5% fat). A promising technique to remove off-flavors in legume-based beverages is the heat inactivation. This needs to be investigated, since high temperatures may cause excessive protein denaturation, aggregation, lower protein solubility and nutrient losses (e.g., vitamins and minerals). In soy beverage the unpleasant "beany" flavor is suppressed by a high temperature (ca. 130 °C) vapor flash (jet cooking) treatment or traditionally by cooking the beans for some time prior to milling [18]. This hydrothermal cooking has the advantage of inactivating protease inhibitors, which increases the digestibility reducing allergen reactions. Despite that, this approach has the disadvantage of denaturing, to a considerable extent, the protein and vitamins present in the seeds [19]. These denatured proteins end up as a solid residue when "milk" is decanted (known as "okara," rich in protein and fiber), produced in large quantities, dramatically reducing the yield of a plant-based beverage, as well as the nutritive potential [5].

Pulse seed bioactives (e.g., phytate, protein inhibitors, phenolics, tannins, lectins and saponins) can have important metabolic effects on a consumer's health [20–24]. Still, some of these bioactives are considered as anti-nutritional factors; lupin and chickpea do not present much of these, being the phytic acid the relevant factor. It has been observed that a soaking step of pulse grains has the ability to reduce polyphenols and eliminate any residual alkaloids present (e.g., in lupin); decrease the cooking time and benefit starch gelatinization (e.g., in pea and chickpea); and also enables protein bioavailability and facilitates peeling [25,26]. The legume seed husk accounts for 75% of the total phenolic content [27], thus the seed's peeling can significantly reduce these compounds by ca. 90% [28]. On the other hand, a germination step is an effective strategy to reduce anti-nutritional factors, since it diminishes the bitterness and "beany" flavor of the grains, due to the presence of phytates [29]. Germination can also reduce the oligosaccharide content [30] and increase the protein bioavailability, thus enhancing the nutritional profile of the legume beverage [31]. Germination has been studied, as a non-chemical, non-thermal processing method, to improve the quality of soy beverage. This approach was shown to increase protein content and also reduce fat, trypsin inhibitors, saponins and phytic acid, inducing proteolysis of the main storage proteins, releasing peptides easier to digest [32]. Additionally, soy beverage from sprouted beans had higher "milk" yield, good color and high sensory acceptability due to the absence of "beany" flavor and odor. The heat treatment also promotes oligosaccharide extraction from legume seeds, as it is strikingly observed during chickpea cooking [28,33].

The main goal of this work was to develop pulse beverages from pea, chickpea and lupin seeds (and their mixtures) focusing on the best technological options to obtain a high-protein beverage (higher than 1.5%) with reduced "beany" flavor, with the least possible discharge of by-products (zero waste). Therefore, a sequence of different processing steps was tested and discussed for the optimization of the sensorial features and beverage's stability. Samples were compared considering the relevant chemical parameters in each process step, such as the total and volatile acidities, protein and carbohydrates content, including starch and glucose. Moreover, the particle size, beverage color and sensory evaluation were also accessed accordingly. The rheological flow behavior of the developed pulse-based beverages were compared with eight commercial non-dairy beverages, selected from a previous study [34], in order to serve as a guide for the preferred consumers mouthfeel and texture.

2. Materials and Methods

2.1. Materials

Four different pulse seeds were used: sweet lupin (*Lupinus albus* L.), chickpea (*Cicer arietinum* L.), green and yellow peas (*Pisum sativum* L.). The seed grains were obtained from a local supplier (Imperial variety of green pea and Branco do Alentejo variety of chickpea) or were kindly offered by the Portuguese research institute INIAV (Elvas) (Estoril variety of sweet lupin, Grisel variety of yellow pea and a mixture of several varieties of chickpea).

2.2. Methods

2.2.1. Pulse Beverage Preparation

Different processing steps were tested for the development of the pulse-based beverages to achieve good sensorial features and remove/mask the "beany" flavor. The beverage's production evolved into the following final optimization (Figure 1a): 150 g of dried seeds was soaked twice in warm tap water (30–35 °C) and once in cold tap water (15–20 °C) for ca. 16 h [26]. All soaking waters were discarded. Then, the soaked seeds were cooked for 30 min after boiling in a pressure pan with 1.5 L of fresh tap water [25]. The cooked seeds were divided into three equal parts and each fraction was processed as follows:

(1) The first fraction was drained and 500 mL of fresh tap water was added. Then, the mixture was milled in the food processor, at 20,500 rpm for 4 min; (2) The second fraction was also drained and the cooking water was replaced with fresh tap water until 500 mL and then milled as (1); (3) The third fraction was divided into two equal parts, one with and the other without cooking water. These parts were later joined with equal shares from similarly processed chickpea pulses, to form mixed beverages (Figure 1b).

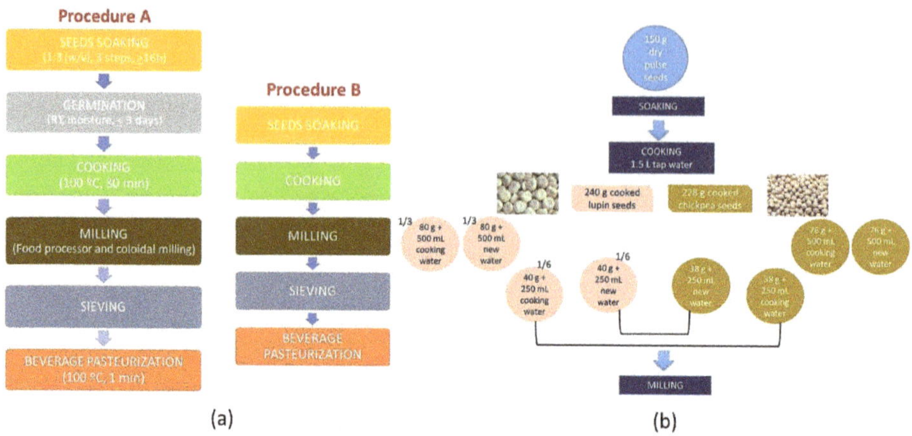

Figure 1. Layout of the production steps used for the different pulse beverages obtained from 10% (*w*/*v*) of dry seeds: (**a**) Procedure A for sprouts; Procedure B for seeds and (**b**) for the pulse beverage mixture formulation.

The milling step adapted from previous studies [19,35], included grinding the seeds (or sprouts) with only 200 mL of the water (cooking or new), in a food processor (Bimby-Worwerk, Wuppertal, Germany) at 20,500 rpm, for 4 min, followed by colloidal milling performed by a mortar grinder, at 70 rpm, for 15 min. (lab scale) using the remaining volume of water (cooking or new). All beverages were sieved with a strainer before being bottled in sterilized flasks (100 °C, 10 min).

In the pasteurization step (adapted from a previous study [35]), the capsulated filled flasks (beverage temperature > 90 °C), were submitted to a thermal shock, inside the pressure cooker for 1 min, in boiling water.

The germination step was adapted from a previous study [32] and included 2 days of incubation where the moist seeds were kept inside an open sterilized flask, at room temperature and protected from light and dust with a cloth. The seeds were washed five times a day with cold water to control moisture content and to avoid mold development.

After production, the pulse beverages were stored at 4 °C for a maximum of 7 days. During this period different analyses were performed as described next.

2.2.2. Color Measurements

The color of the different pulse beverages was measured using a Minolta CR-300 (Tokyo, Japan) tristimulus colorimeter that was calibrated using a white standard porcelain plate (L*96.96; a*0.37; b*2.10). The results were expressed in accordance to the CIELAB uniform color system with reference to standard illuminate D65 (average daylight conditions) and a visual angle of 2°. The color parameters determined were L*, which accounts for the lightness (i.e., 0% for black and 100% for white), a* ranges from green to red and b* from blue to yellow, which corresponds to a numerical variation from −60 to +60. The measurements were conducted at room temperature under similar light conditions (i.e., 50 mm^2 measuring area per measurement) and replicated 6 times on days 1, 3 and 7 after sample processing day.

The total color difference between the samples was calculated according to Equation (1). UHT cow milk and different fat contents were used as references.

$$\Delta E^* = \sqrt{(\Delta L^*)^2 + (\Delta a^*)^2 + (\Delta b^*)^2}. \tag{1}$$

Considering that if $\Delta E^* > 3$, the color difference is detectable by the human eye [36].

2.2.3. Chemical Analysis

The Total and Volatile acidities of the pulse beverages and respective cooking waters were analyzed in accordance with adapted OIV-MA.AS313-01:R2015 and OIV-MA.AS313-02:R2015 international methods, respectively [37], with some modifications. Briefly, for Total Acidity determination, each sample (20 mL) was diluted with 25 mL of boiled water. Then 3 droplets of phenolphthalein were added and the titration was performed with 0.1 N NaOH (aq.). For the Volatile Acidity determination, after the steam distillation of 20 mL of sample with 0.5 g of tartaric acid, the distillate was titrated with 0.1 N NaOH (aq.), again using phenolphthalein as indicator. Both Total and Volatile Acidities were expressed in milliequivalents (mEq) of acid/L.

The protein content of the developed pulse beverages was assessed following the Kjeldahl method [38]. A specific conversion factor for legumes (i.e., 5.4) was used to convert nitrogen into crude protein [39]. Trials were performed in triplicate, and results are expressed in % (*w*/*v*).

The starch content of the developed pulse beverages, and their respective cooking waters was determined with the Total Starch KIT (K-TSTA-100A, Megazyme). Tests were performed in duplicate and data expressed in g/100 mL.

The carbohydrates content of the optimized pulse beverages was carried out according to Dubois et al. [40]. The analysis was performed in triplicate and data expressed as g of carbohydrates per 100 mL of pulse beverage.

2.2.4. High Performance Liquid Chromatography (HPLC) Analysis

The D-glucose content of the optimized pulse beverages (Procedure B) was obtained by HPLC [41]. Briefly, 2 mL of each sample was centrifuged at 12,000 rpm for 10 min and 100 µL of supernatant was collected. After its dilution in H_2SO_4 (≥95%, Fisher Scientific) (50 mM) (1:10 (*v*/*v*)), the samples were centrifuged (12,000 rpm, 10 min) to discard the precipitated protein and filtered through a 0.20 µm-pore-size filter (Whatman, Marlborough, USA). D-glucose was quantified in a high-performance liquid chromatography system (Waters) equipped with a refractive index detector (Waters 2414) and a RezexTM ROA Organic Acid H+ (8%) column (300 mm × 7.8 mm, Phenomenex), at 65 °C. H_2SO_4 (5 mM) was used as mobile phase at 0.5 mL·min^{-1}. Results are expressed as g of glucose per 100 mL of pulse beverage.

2.2.5. Complementary Analyzes

The pH (pH meter CRISON, Barcelona, Spain) of the different pulse beverages was measured at room temperature for the production day and after that, on the third and seventh days.

The sedimentation was measured on the second, fourth and seventh days after production and calculated as the ratio between the height of the sediment and the total height of the beverage in the flask. When considerable sediment was formed, classifications of "puree," "viscous" and "pudding" were applied to describe the consistency of the different beverages.

The "okara" was calculated as the ratio between the weight of the solid residue obtained after sieving the milled beverage and the total weight of the correspondent cooked seeds/sprouts.

The "milk" yield (weight of the beverage (g)/100 g of pulse seed) was also estimated according to the following Equations (2) and (3):

WHOLE SEEDS

$$\frac{weight\ of\ cooked\ seed\ or\ sprout\ +\ weight\ of\ water\ (g)}{weight\ of\ dry\ seed\ (g)} \times 100 \quad (2)$$

DEHULLED SEEDS

$$\frac{weight\ of\ cooked\ seed\ or\ sprout - weight\ of\ husks\ +\ weight\ of\ water\ (g)}{weight\ of\ dry\ seed\ (g)} \times 100 \quad (3)$$

.

2.2.6. Sensory Evaluation

The sensory evaluation was carried out to observe its sensory acceptance. Twenty-nine untrained panelists were asked to score the samples in terms of color, appearance, taste, flavor, consistency, overall appreciation and purchase intent using a hedonic scale from 1 (very unpleasant) to 5 (very pleasant). The samples with 1 to 6 days of cold storage were identified with an alphanumeric code and served to the panelists in a single day.

2.2.7. Characterization of Pulse Beverage Particles

The morphological examination of pulse beverages was conducted by bright field optical microscopy at 20× magnification, using a Zeiss AxioLab (Oberkochen, Germany) A1 equipped with the camera Zeiss Axiocam 105 color with 5 megapixels. A droplet of each beverage was carefully placed in a proper microscope glass slide and covered with a cover slip. Pictures were recorded and analyzed with the software Zen 2.6. The particle size was evaluated by photon correlation spectroscopy using a Zetasizer Nano ZS (Malvern Instruments, Malvern, UK). Each sample was diluted to the appropriate concentration with ultrapure water and placed in a cell cuvette.

2.2.8. Rheological Measurements

The shear viscosity of the pulse-based beverages was measured using a controlled-stress rheometer (Haake MARS III, Germany), at 20 ± 1 °C, with a CCB/CC25 DIN Ti concentric cylinder geometry to avoid phase separation. The steady shear measurements were performed with shear rates from 1.0×10^{-5} to 1.0×10^{3} s^{-1}. Tests took 11 min each and were performed in triplicate with well shaken beverages. The flow curves were fitted to the Carreau model (Equation (4)), since the pulse beverages are non-Newtonian fluids and evidenced shear-thinning behavior, that is, the viscosity decreases as the shear rate increases [42]:

$$\eta = \eta_\infty + \frac{\eta_0 - \eta_\infty}{\left(1 + \left(K\dot{\gamma}\right)^2\right)^{\frac{m}{2}}} \quad (4)$$

where "η_0" is the first limiting ("zero" shear rate) Newtonian viscosity (Pa·s); η_∞ is the second limiting ("infinite" shear rate) Newtonian viscosity (Pa·s); "$\dot{\gamma}$" is the shear rate (s^{-1}); "K" is the relaxation time (s) and the reciprocal, $1/K$ ($\dot{\gamma}c$), is related to the critical shear rate (i.e., onset shear rate for shear-thinning); "m" is the dimensionless constant related to Power Law and accounting for the deviation from the Newtonian behavior.

2.2.9. Statistical Analysis

Analysis of variance (one-way ANOVA) was used to assess significant differences between samples at a significance level of 95% ($p < 0.05$). Multiple comparisons were performed by Tukey HSD (honestly significant difference). All results are presented as mean ± standard deviation.

3. Results

3.1. *Chronological Progress of Beverage Processing Steps*

The initial experimental trials were performed with 5% to 10% (*w*/*v*) of dried seeds in the beverage, to achieve a lower viscosity and a protein content between 1–4% (*w*/*v*). No "okara" was obtained in the following first tests since beverages were not sieved.

At the initial testing, lupin-based beverage always presented phase separation with sediment probably due to high particle size. This is not related to the influence of beverage pH on protein's functional properties. Lupin-based beverages had a pH value of 6.0 ± 0.2, while chickpea and pea-based beverages presented a pH around 6.7. The protein solubility with pH is minimal at pI values (4.6 and

4.5, respectively) but at values of 6 it evidences a good solubility [12,13,43]. In addition and regardless the process step used (i.e., toasting the seeds before soaking or the sprouts before milling; cooking the sprouts/seeds only after milling; testing different concentrations of seeds in the final beverage; beverage thermal shock at 75 °C for 15 min), the fresh chickpea-, green pea- and yellow pea-based beverage samples, were always observed to jellify for the highest concentration of dry seeds used (i.e., 10%). Gelation was also observed for lower contents, ca. 5–6%, when samples aged for 3 days. Therefore, this undesired gelation phenomena affected their viscosity [19] and hampered the formation of homogeneous and liquid-like beverages. The formed pudding-like gel was more robust in green pea- and yellow pea-based beverages than in chickpea. This is related to the higher content of starch, around 45% in pea and chickpea, compared to 6.7% in lupin [44]. This is expected to occur due to the heating-induce breakdown of the amylose and amylopectin intermolecular association, resulting in high viscous solutions and, eventually, changing into a strong gel (retrogradation) upon prolonged storage [45]. This gelation drawback made us retreat the use of peas. Also, at this stage, the toasting step has been no longer considered due to its high energy consuming and lack of sensorial impact in the produced beverages.

Chickpea- and lupin-based beverages still evidenced a slight "beany" flavor at this stage of processing progression. Additional adjustments on the processing steps were considered: the cooking step was performed in pressure pan instead of a food processor; the sieving step was introduced after milling to remove major particles; and the peeling step (before and after cooking) was also included, knowing that the last can contribute to reduce the bitter taste attributed to husk's phenolic compounds [27,28]. The husk removal step of seeds and sprouts from chickpea and lupin (present in procedures A and B) revealed to be unnecessary since similar "milk" yields were obtained as follows—1206 and 1196 g of beverage/100 g of seeds dehulled before cooking; 1203 and 1204 g of beverage/100 g of seeds dehulled after cooking—1206 and 1207 g of beverage/100 g of sprouts dehulled after cooking, for chickpea- and lupin-based beverages, respectively. Nevertheless, the removal of the husks was easier to be performed after cooking the seeds/sprouts. In addition, the cooking, dehulling and the milling steps of seeds/sprouts had no significant effect on the pH of the beverages when stored for 7 days at 4 °C (Table 1). On the other hand, the processing with cooking water or with new water evidenced significant differences in pH values between chickpea pulse-based beverages (Table 1), maybe due to the slightly alkaline tap water (pH 7–8) used. As mentioned, the lupin beverages are usually acidic (pH < 6.0), when compared to chickpea beverages (pH = 6.7–7.2). Adding to that, the low volatile acidity obtained for pulse sprouts-based beverages with 9 days of refrigeration (<6 mEq/L), evidenced the suitability of the heat treatment and bottling, suggesting the absence of bacterial activity.

Table 1. Average of pH values obtained during the 7 days storage at 4 °C of pulse beverages obtained from the processing procedures A and B, when included the dehulling after cooking (where the "1" corresponds to beverages with cooking water and "2" to beverages with new water). The same superscript letter/symbol in beverage-pairs evidences significant difference between them ($p < 0.05$). Values are represented as mean ± standard deviation.

	Lupin Beverage (10% of Lupin)	Chickpea Beverage (10% of Chickpea)	Lupin and Chickpea Beverage (5% of Each)
Dehulled sprouts (A1)	5.8 ± 0.1 [b,l,u,β,Ψ,τ,μ,λ]	6.9 ± 0.1 [a,b,c,d,e,f,g,h,i]	6.3 ± 0.2 [f,p,y,δ,Ψ,Ø,ξ,‡]
Dehulled sprouts (A2)	5.9 ± 0.1 [c,m,v,Ω,Ø,ηγ]	7.2 ± 0.1 [a,j,k,l,m,n,o,p,q,r,s]	6.4 ± 0.1 [g,q,z,θ,τ,η,χ,ƨ,‖]
Dehulled seeds (B1)	5.8 ± 0.1 [d,n,w,π,ξ,χ,σ,ε]	6.7 ± 0.0 [j,t,u,v,w,x,y,z,Σ,α]	6.1 ± 0.0 [h,r,Σ,&,μ,σ,ƨ,ƚ]
Dehulled seeds (B2)	5.9 ± 0.1 [e,o,x,Δ,‡,‖,ɔ]	7.0 ± 0.1 [k,t,β,Ω,π,Δ,δ,θ,&,ω]	6.4 ± 0.0 [i,s,α,ω,λ,γ,ε,ƚ,ɔ]

Separation by sedimentation occurred in all beverages during the cold storage, showing that the developed suspensions were not stable. Moreover, with this processing step, the gel formation was not evidenced in chickpea-based beverages, most likely because the thermal treatments were better controlled, regarding the maximum temperature and time used, when compared to the initial implemented processing steps.

To confirm process suitability (pressure cooking and sieving) to keep the expected protein content, the beverages were analyzed accordingly. The lupin beverages evidenced significant higher values (1.8–2.4%) when compared to the chickpea beverages (1.0–1.5%) (Table 2). Both pulse beverages produced with cooking water presented higher protein values compared to new water, thus confirming the protein solubilization into water during the cooking step.

Table 2. Protein content in pulse beverages obtained from the processing procedures A and B, when included the dehulling after cooking (where the "1" corresponds to beverages with cooking water and "2" to beverages with new water). The same superscript letter/symbol in beverage-pairs evidences significant difference between them ($p < 0.05$). Values are represented as mean ± standard deviation.

	Protein Content (% (*w*/*v*))			
	Sprouts		**Seeds**	
	A1	**A2**	**B1**	**B2**
Lupin	$2.3 \pm 0.1^{a,g,o,s,y,\Sigma,\alpha,\beta}$	$2.0 \pm 0.1^{b,h,p,t,\Omega,\pi,\Delta}$	$2.4 \pm 0.1^{c,I,q,u,\Omega,\delta,\theta,\&,\omega,\Psi}$	$1.8 \pm 0.3^{d,j,v,y,\delta}$
Chickpea	$1.3 \pm 0.1^{a,b,c,d,e,f}$	$1.1 \pm 0.1^{g,h,I,j,k,l,m}$	$1.5 \pm 0.1^{n,o,p,q,r}$	$1.0 \pm 0.1^{n,s,t,u,v,w,x,z}$
Lupin + chickpea	$1.8 \pm 0.1^{e,k,x,\Sigma,\theta}$	$1.6 \pm 0.1^{l,z,\alpha,\pi,\&,\tau}$	$2.0 \pm 0.1^{f,m,r,w,\omega,\tau,\mu}$	$1.4 \pm 0.1^{\beta,\Delta,\Psi,\mu}$

At this point, the pulse beverages developed according to procedure B (i.e., cooked seeds without husks) were selected for sensory analysis. Note that, the germination step (procedure A with dehulling) had not yet been optimized (germination yield less than 50%). The sensory analysis used descriptive and preference tests, which revealed that the beverages were not significantly different from each other (Figure 2). Comparing lupin and chickpea-based beverages, the best color and appearance results were evidenced in both lupin-based beverages. On the other hand, the best flavor was attributed to the chickpea beverage produced with new water and the best appreciation for taste and consistency was obtained for chickpea-based beverage produced with cooking water. The pulse mixture-based beverages (Figure 2) evidenced both chickpea and lupin best sensory characteristics.

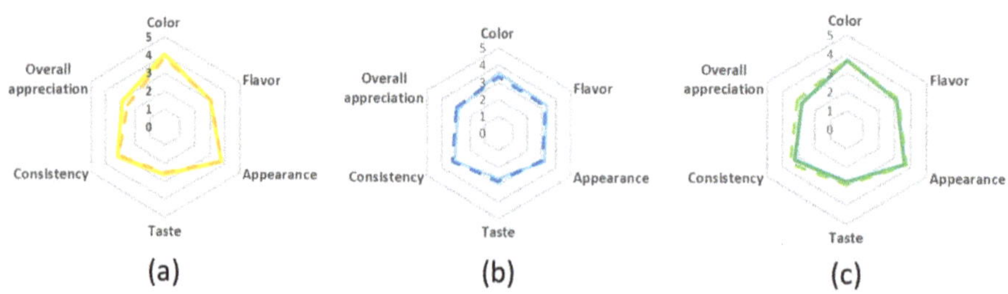

Figure 2. Sensory evaluation of pulse based beverages from procedure B with dehulling step after cooking. (**a**) Lupin-based beverage; (**b**) chickpea-based beverage and (**c**) mixture beverage. The full lines correspond to "new water," while the "dashed lines" correspond to cooking water.

3.2. Achieving the Last Optimization of Beverage Development–Particle Size Reduction

In order to improve the beverage stability and the "sandy mouthfeel," the reduction of the particle dimensions was considered next (see Figure 1 for details). Therefore, the milling step was improved and a prolonged colloidal milling was applied, i.e., about 3 times longer than previously with the Ultraturrax (1 min, 20,500 rpm). The "okara" was further reduced from 16–25% to 1.8–6.5%, evidencing the efficacy of the reduction of the particles dimensions. Comparing chickpea- and lupin-based beverages from both procedures (whole cooked seeds and sprouts, regardless of water used), the "milk" yield values showed no significant difference from each other and also when compared to previous results considering the dehulling step (whole seed: 1221 and 1247 g/100 g of seed; whole sprout: 1222 and 1284 g/100 g of seed, for chickpea- and lupin-based beverages, respectively).

At this stage, the physical stability of the developed pulse-based beverages was also assessed for the final procedures A and B. The chickpea beverage, produced with the cooking water, was the only one showing gel formation, which happened after 3 days of storage at refrigerated conditions. The reason for the observed gelation may rely on the extended duration and heating during colloidal milling, which allowed the gelling of the starch contained in both chickpea seed and cooking water [16]. Accordingly, the lupin, chickpea and their mixture, produced with the cooking waters, displayed the higher starch contents when compared to beverages produced with new water. As expected, the pulse beverages produced from sprouts did not gel due to starch hydrolysis during the germination of seeds. This observation was supported by lower starch contents in lupin sprouts and chickpea sprouts-based beverages when compared to the corresponding non-germinated counterparts.

The protein content of pulse-based beverages developed in final procedures A and B, was not significantly different from the previously determined (Table 2), suggesting that the husks do not contribute meaningfully to the final protein content of the beverages.

The particle size distribution of the pulse-based beverages from procedure B with new water (B2) was analyzed and also their carbohydrates, starch and glucose contents (Table 3). The reason of this selection was due to the fact that the cooking water from non-germinated seeds contain particles and starch that may favor phase separation and hamper beverage's stability. As can be observed in Figure 3, the milling step resulted in beverages with particles of relatively monodisperse sizes and irregular shapes. The particles of lupin beverage are the largest, with a median size of 857 nm and have a narrow size distribution with a "polydispersity index" (PDI) of 0.4. On the other hand, the chickpea particles have a broader size distribution (median size of 469 nm), with an estimated PDI value of 0.7. The lupin and chickpea mixture particles presented a median size of 679 nm, with a PDI value of 0.6.

Table 3. Carbohydrates, including starch and glucose contents in pulse beverages obtained from the processing procedure B-whole seeds, colloidal milling and without the cooking waters.

	Partial Nutritional Composition		
g/100 mL	**Chickpea Beverage**	**Lupin Beverage**	**Lupin + Chickpea Beverage**
Carbohydrates	9.01	3.26	5.36
Starch	0.689	0.006	0.204
Glucose	0.45	0.06	0.28

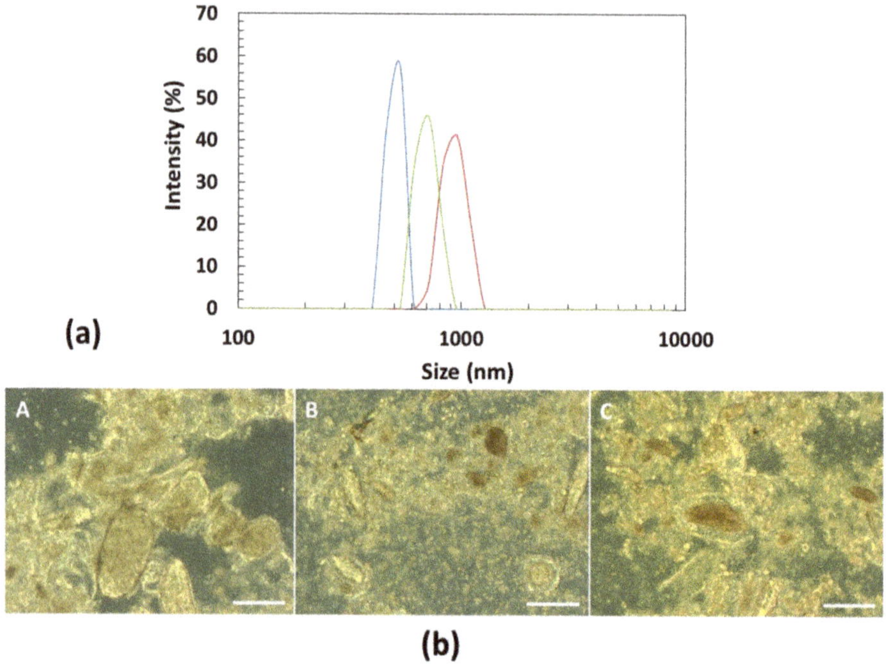

Figure 3. Characterization of the particles from pulse-based beverages with new water (B2): (**a**) particle size distribution of chickpea (blue), lupin (orange) and lupin + chickpea mixture (gray) estimated by dynamic light scattering (Zetasizer); (**b**) optical light microscopy images of chickpea (**A**), lupin+chickpea mixture (**B**) and lupin (**C**). The scale bar represents 100 μm.

The particle sedimentation was analyzed over 24 h and their macroscopic evolution was recorded over time. The onset of particle sedimentation of the chickpea beverage was visible after 15 min of incubation and complete sedimentation occurred after ca. 24 h. Instead, for lupin and lupin + chickpea beverages, the sedimentation started after 24 h and occurred, gradually, during ca. 6 days. The optimization of the milling step enhanced the stability of beverages. Although the chickpea beverage had the smallest particles, its sedimentation was remarkably faster compared to the other beverages, which may be significantly affected by the lower viscosity of the solution and higher bulk density of its particles. As shown in Table 4, the chickpea beverage presents the lowest zero shear-rate viscosity in comparison to the lupin and their mixture-based beverages. Therefore the dispersion viscosity should not be neglected, nor the fact that the bulk density of chickpea is, on average, bigger than lupin [46].

Table 4. Parameters obtained after fitting the flow curves to the Carreau model for all Group X beverages (η_0—zero-shear viscosity; η_∞—infinite-shear viscosity and $\dot{\gamma}$ c—critical shear rate). The superscript (*) indicates the beverage(s) that differ from the largest number of the others (12–13 beverages), per parameter ($p < 0.05$). Values are represented as mean ± standard deviation.

	Group X Beverages		
	η_0 (Pa·s)	η_∞ (Pa·s)	$\dot{\gamma}$
Lupin A1	524.0 ± 95.0	$1.9 \times 10^{-2} \pm 0.0$	$1.8 \times 10^{-4} \pm 0.0 \times 10^{-4}$
Lupin A2	609.5 ± 55.2	$2.4 \times 10^{-2} \pm 0.1$ (*)	$2.5 \times 10^{-4} \pm 0.2 \times 10^{-4}$
Chickpea A1	456.7 ± 54.7	$0.9 \times 10^{-2} \pm 0.0$	$2.6 \times 10^{-4} \pm 0.4 \times 10^{-4}$
Chickpea A2	1243.8 ± 443.4 (*)	$0.9 \times 10^{-2} \pm 0.1$	$2.0 \times 10^{-4} \pm 0.3 \times 10^{-4}$

Table 4. *Cont.*

	Group X Beverages		
	η_0 (Pa·s)	η_∞ (Pa·s)	$\dot{\gamma}$
Lupin + chickpea A1	439.6 ± 200.1	$1.4 \times 10^{-2} \pm 0.0$	$2.5 \times 10^{-4} \pm 0.6 \times 10^{-4}$
Lupin + chickpea A2	233.8 ± 19.2	$1.3 \times 10^{-2} \pm 0.1$	$2.8 \times 10^{-4} \pm 0.3 \times 10^{-4}$
Lupin B1	495.3 ± 4.2	$2.6 \times 10^{-2} \pm 0.3$	$2.0 \times 10^{-4} \pm 0.0 \times 10^{-4}$
Lupin B2	658.1 ± 34.6	$2.9 \times 10^{-2} \pm 0.1$ (*)	$1.9 \times 10^{-4} \pm 0.5 \times 10^{-4}$
Chickpea B1	52.1 ± 12.5	$1.2 \times 10^{-2} \pm 0.0$	$12.7 \times 10^{-4} \pm 4.3 \times 10^{-4}$ (*)
Chickpea B2	176.4 ± 22.4	$1.3 \times 10^{-2} \pm 0.0$	$5.6 \times 10^{-4} \pm 1.1 \times 10^{-4}$
Lupin + chickpea B1	149.3 ± 45.2	$1.5 \times 10^{-2} \pm 0.1$	$3.2 \times 10^{-4} \pm 0.3 \times 10^{-4}$
Lupin + chickpea B2	354.5 ± 34.6	$1.5 \times 10^{-2} \pm 0.3$	$4.0 \times 10^{-4} \pm 1.6 \times 10^{-4}$
Oat beverage	52.6 ± 4.2	$0.7 \times 10^{-2} \pm 0.0$	$18.8 \times 10^{-4} \pm 1.6 \times 10^{-4}$ (*)
Hazelnut beverage	125.4 ± 32.8	$0.8 \times 10^{-2} \pm 0.0$	$15.4 \times 10^{-4} \pm 2.2 \times 10^{-4}$

The carbohydrates, starch and glucose contents of the optimized beverages from Procedure B were also analyzed to infer on how much carbohydrates remained after discarding the cooking water (replaced by new one). The highest values for carbohydrates and starch were evidenced in the chickpea beverage and the lowest in lupin (Table 3), accordingly to their composition. Comparing these results to the corresponding nutritional composition of dried seeds [44], the reference values for carbohydrates in beverages with 10% of dried seeds (10 g/100 mL) are: 5.6 g/100 mL for chickpea and 1 g/100 mL for lupin are lower than those presented in Table 3. For sugars (glucose included), the reference values [44] are 0.3 g/100 mL for chickpea and 0.05 g/100 mL for lupin, compared to the calculated values (Table 3) 0.45 and 0.06 respectively, which are similar and the differences can be attributed to different plant varieties and edapho-climatic conditions. The starch values expected for 10 g of dried seeds per 100 mL, are 4.5 g for chickpea and 0.7 g for lupin [44], the lower values obtained, 0.689 g and 0.006 g, respectively, confirm the elution of starch from beverages during processing/cooking.

3.2.1. Rheological Behavior of Produced Pulse Beverages

The shear flow properties of all developed pulse-based beverages from procedures A and B are depicted in Figure 4. All samples present a typical non-Newtonian shear-thinning behavior, very similar to other current non-dairy alternative beverages and cow milk [39].

The pulse beverages grouped as X displays very similar flow curves, with a well-defined Newtonian plateau at low shear rates followed by the shear-thinning. Their viscosity profile is closer to the oat beverage, one of the consumer's preferences [1]. Nevertheless, when we analyze the obtained fitting parameters from the Carreau model for all pulse-based beverages produced and those from the hazelnut and oat beverages (Table 4), significant differences ($p < 0.05$) can be observed.

The difference between the limiting values of zero-shear viscosity (η_0) and the infinite-shear viscosity ($\eta\infty$) indicates how pronounced is the shear-thinning behavior of a material. Consequently, the chickpea beverage A2 presents the most extended shear-thinning behavior of all, being the most fluid beverage. This is a confirmation of the benefits of the germination step in legume beverages with high starch content and replacement of the cooking water. On the other hand, the chickpea beverage B1 has the less extended shear-thinning behavior (52.1 Pa·s), similar to oat beverage. This fact suggests that its supramolecular structure was more easily destroyed by the applied shear stress than the other pulse beverages and that the micro-scale structures within the fluid rearrange/align to facilitate shearing, reaching sooner the second Newtonian plateau. Nevertheless, the chickpea beverage B1 presents a transitory structural net translated by its higher consistency during cold storage. Most likely, during the resting time some reversible three-dimensional aggregates are formed. Furthermore, the lupin and chickpea beverage B1 is the only sample presenting a beverage shear-thinning behavior similar to the hazelnut beverage. The remaining pulse beverages present a shear-thinning behavior between the extremes chickpea beverage A2 and the chickpea beverage B1.

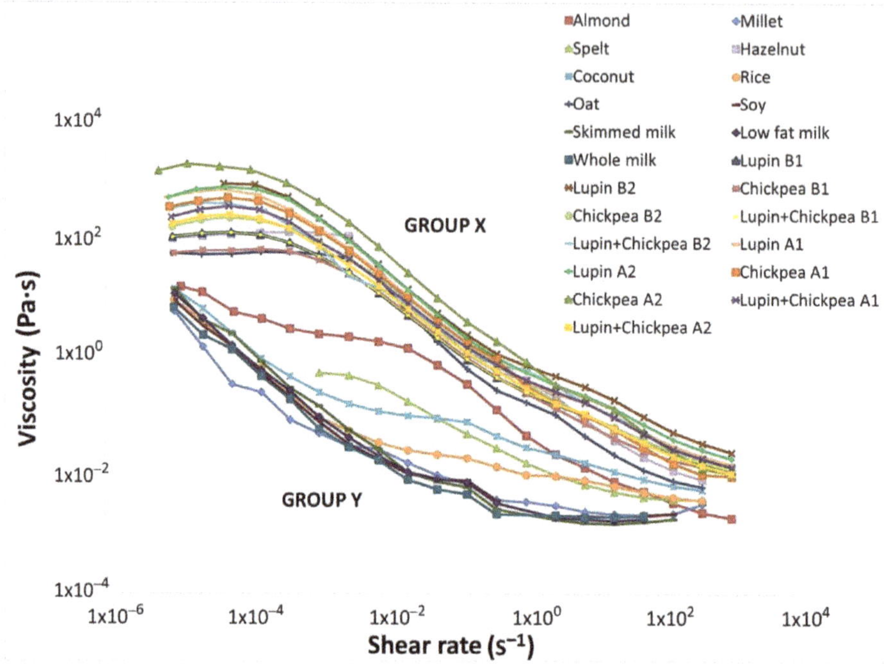

Figure 4. Flow curves showing the shear viscosity evolution for the pulse-based beverages produced by final procedures A and B, skimmed milk, low-fat milk and whole milk. In addition, eight commercial non-dairy beverages were also analyzed. Group X includes the twelve pulse beverages produced (1—with cooking water; 2—with new water) and two commercial non-dairy beverages (hazelnut and oat). Group Y comprises milks of different fat contents (i.e., whole, low and skimmed) and six other commercial non-dairy beverages (i.e., almond, spelt, coconut, millet, rice and soy).

3.2.2. Color Analysis

The pulse beverages prepared in final procedures A and B were also analyzed regarding their color parameters. There are no significant differences between pulse beverages, with or without the germination step. Regarding the lightness parameter L*, the higher values (ca. 79) were observed for the lupin beverages (A1, A2, B1 and B2) while the chickpea beverage A1 presents the lowest one (ca. 68), which may be related to the small particle size. On the other hand, the b* is significantly evidenced in all chickpea beverages (A1, A2, B1, B2) and in lupin beverage B1. The marked yellowness of these samples can be explained by the higher carotenoid content in lupin and chickpea seeds [47,48].

The total color difference was assessed comparing fresh and 7 days aged samples. Only the lupin beverage B1 ($\Delta E^* = 5.6$) and chickpea beverage B1 ($\Delta E^* = 3$) showed a visual color difference, probably due to carbohydrate hydrolysis during seed germination and enhanced dissolution of sugars, reinforced by the cooking water, that can lead to a higher opacity of the beverages. On the other hand, a visual contrast ($\Delta E^* > 3$) is evidenced between all developed pulse beverages and cow milk. Nevertheless, this is not a characteristic that negatively influences the consumer preferences when purchasing non-dairy beverages.

4. Discussion

In this work, novel pulse-based beverages were prepared following two main procedures, each of them consisting of several processing steps, which were systematically optimized. The lupin-, chickpea- and mixture-based beverages protein contents ranged from 1.8% to 2.4%, 1.0% to 1.5% and

1.4% to 2.0%, respectively, and the husks do not contribute meaningfully to the final protein content of the beverages. This observation is supported by the recent work of Niño and coworkers [49], which have determined a protein content in the husks of chickpea of ca. 4.5%. This value corresponds proportionally to the lower content (0.72% to 0.86%) estimated in this work for the entire chickpea seed. Therefore, the final protein contents of these pulse-based beverages are more appealing than the 8 non-dairy beverages mentioned in this work (protein contents below 1.0%) [34].

The "milk" yields obtained for chickpea and lupin beverages during the optimized procedure B, focused on lower discharge of by-products, were around 12.2–12.5 kg of beverage per kg of seed, respectively, which is ca. 4 times higher in comparison to the "milk" yield obtained for oat beverage (2.85 kg/kg of rolled oat) developed by Deswal and co-workers [50] with a technological process using enzymatic hydrolysis to liquefy oat's higher content of starch. On the other hand, the slightly alkaline tap water (pH 7–8) used for soaking and cooking steps may have also contributed to this higher "milk" yield values. A previous study [51] has shown that the alkali soaking has significantly improved the yield of total solids in sesame beverage, which is due to the higher solubility of the plant proteins at these pH values.

The cooking step was observed to be essential for the reduction of the unpleasant "beany" flavor and the sensorial tests performed to lupin-, chickpea- and mixture-based beverages supported that by an acceptable flavor score around 3. However, the first heat treatment conditions tested (extended cooking time) lead to gelled beverages in the green pea and yellow pea-based beverages, due to its high starch content. This would obviously limit their use as homogeneous and a fluid non-dairy alternatives. Despite that, the further optimized lupin, chickpea and mixture-based beverages, showed typical non-Newtonian fluid behavior, displaying a pronounced shear-thinning, which was more pronounced for the sprouted chickpea beverage, without the cooking water. This behavior is common in these kind of fluids with complex composition and particles. The most likely explanation is the alignment of asymmetrical particles to the flow and/or a higher breakage rate of interactions between molecules, rather than the formation of molecules from low energy interactions, like physical entanglements.

5. Conclusions

Legume beverages present the most balanced composition and a protein content similar to cow milk but face technological issues often related to processing or preservation. Heat treatment, such as cooking and pasteurization, is able to remove off-flavors, the most challenging barrier to consumer acceptance. However, high temperatures may cause excessive protein denaturation, lower protein solubility and may increase legume beverage viscosity, affecting its physical stability. The colloidal milling, on the other hand, is one technological intervention capable to increase the beverage physical stability by reducing the size of dispersed particles. In this work, several of these issues have been tackled and, overall, the processing strategies adopted and the optimizations performed led to the development of novel pulse-based beverages with several appealing features (e.g., protein content, rheological behavior, color and appearance) that are highly competitive in the current commercial non-dairy beverages. Therefore, we believe this work strongly contributes to pave the way for the development of novel pulse-based beverages as viable alternatives for cow milk. Nevertheless, further future optimizations must be performed, such as considering the use of enzymes to refine the beverage mouthfeel, the addition of natural flavors for an improved and pleasant sensorial perception and the use of high pressure homogenization to reduce particle's dimensions, improving beverages stabilization and microbial shelf life extension. All these aspects are being considered in ongoing and future studies.

Author Contributions: Conceptualization, I.S.; methodology, B.M. and I.S.; validation, C.M.D., A.F. and B.M.; formal analysis, C.M.D. and A.F.; investigation, M.L., C.P. and A.F.; resources, I.S. and B.M..; data curation, A.F.; writing—original draft preparation, M.L., C.M.D. and A.F.; writing—review and editing, C.M.D., A.F., B.M. and I.S.; visualization, C.M.D., A.F. and B.M.; supervision, C.M.D. and B.M.; project administration, I.S.; funding acquisition, I.S. All authors have read and agreed to the published version of the manuscript.

Funding: This work was supported by the FCT Project PTDC/BAA-AGR/28370/2017: "Legumilk".

Acknowledgments: Authors acknowledge FCT/Portugal for financing the projects PTDC/BAA-AGR/28370/2017 and PTDC/ASP-SIL/30619/2017, the research unit UID/AGR/04129/LEAF, the researcher grant CEECIND/01014/2018 and the Portuguese research institute INIAV (Elvas, Portugal) for the donation of pulse seeds.

Conflicts of Interest: The authors declare no conflict of interest.

References

1. Market and Markets. Dairy Alternatives Market by Source (Soy, Almond, Coconut, Rice, Oats, Hemp), Application (Milk, Cheese, Yogurt, Ice Creams, Creamers), Distribution Channel (Supermarkets, Health Stores, Pharmacies), Formulation and Region—Global Forecast to 2023. Available online: https://www.marketsandmarkets.com/Market-Reports/dairy-alternative-plant-milk-beverages-market-677.html (accessed on 28 February 2019).
2. Mintel Press Office. US Non-Dairy Milk Sales Grow 61% Over the Last Five Years. Available online: https://www.mintel.com/press-centre/food-and-drink/us-non-dairy-milk-sales-grow-61-over-the-last-five-years (accessed on 19 March 2019).
3. Scrimshaw, N.S.; Murray, E.B. The acceptability of milk and milk products in populations with a high prevalence of lactose intolerance. *Am. J. Clin. Nutr.* **1988**, *48*, 1142–1159. [CrossRef] [PubMed]
4. Lifschitz, C.; Szajewska, H. Cow's milk allergy: Evidence-based diagnosis and management for the practitioner. *Eur. J. Pediatr.* **2015**, *174*, 141–150. [CrossRef] [PubMed]
5. Makinen, O.E.; Wanhalinna, V.; Zannini, E.; Arendt, E.K. Foods for special dietary needs: Non-dairy plant-based milk substitutes and fermented dairy-type products. *Crit. Rev. Food Sci. Nutr.* **2016**, *56*, 339–349. [CrossRef]
6. Transparency Market Research. Plant-Based Milk Market. Available online: https://www.transparencymarketresearch.com/plant-based-milk-market.html (accessed on 9 April 2019).
7. Gerber, P.J.; Steinfeld, H.; Henderson, B.; Mottet, A.; Opio, C.; Dijkman, J.; Falcucci, A.; Tempio, G. *Tackling Climate Change Through Livestock—A Global Assessment of Emissions and Mitigation Opportunities*; Food and Agriculture Organization of the United Nations (FAO): Rome, Italy, 2013. Available online: http://www.fao.org/3/a-i3437e.pdf (accessed on 22 March 2019).
8. Franche, C.; Lindström, K.; Elmerich, C. Nitrogen-fixing bacteria associated with leguminous and non-leguminous plants. *Plant Soil* **2009**, *321*, 35–59. [CrossRef]
9. Mintel Press Office. Dairy Milk US 2016: Consumer Market Research Report & Non-dairy Milk US 2016: Consumer Market Research Report. Available online: http://store.mintel.com/dairy-milk-us-march-2016; http://store.mintel.com/non-dairy-milk-us-april-2016 (accessed on 19 March 2019).
10. Sosulski, F.W.; Chakrabotry, P.; Humbert, E.S. Legume-based imitation and blended milk products. *Can. Inst. Food Sci. Technol. J.* **1978**, *11*, 117–123. [CrossRef]
11. Shevkani, K.; Singh, N.; Chen, Y.; Kaur, A.; Yu, L. Pulse proteins: Secondary structure, functionality and applications. *J. Food Sci. Technol.* **2019**, *56*, 2787–2798. [CrossRef]
12. Piornos, J.A.; Burgos-Díaz, C.; Ogura, T.; Morales, E.; Rubilar, M.; Maureira-Butler, I.; Salvo-Garrido, H. Functional and physicochemical properties of a protein isolate from AluProt-CGNA: A novel protein-rich lupin variety (*Lupinus luteus*). *Food Res. Int.* **2015**, *76*, 719–724. [CrossRef]
13. Ladjal-Ettoumi, Y.; Boudries, H.; Chibane, M.; Romero, A. Pea, chickpea and lentil protein isolates: Physicochemical characterization and emulsifying properties. *Food Biophys.* **2016**, *11*, 43–51. [CrossRef]
14. Cruz, N.; Capellas, M.; Hernandez, M.; Trujillo, A.J.; Guamis, B.; Ferragut, V. Ultra high pressure homogenization of soymilk: Microbiological, physicochemical and microstructural characteristics. *Food Res. Int.* **2007**, *40*, 725–732. [CrossRef]
15. Li, Y.Q.; Tian, W.L.; Mo, H.Z.; Zhang, Y.L.; Zhao, X.Z. Effects of pulsed electric field processing on quality characteristics and microbial inactivation of soymilk. *Food Bioprocess. Technol.* **2013**, *6*, 1907–1916. [CrossRef]
16. Fennema, O.R. Chapter 4: Carbohydrates. In *Food Chemistry*, 3rd ed.; Fennema, O.R., Ed.; Marcel Dekker Inc.: New York, NY, USA, 1996; pp. 158–221.
17. Yang, A.; Smyth, H.; Chaliha, M.; James, A. Sensory quality of soymilk and tofu from soybeans lacking lipoxygenases. *Food Sci. Nutr.* **2016**, *4*, 207–215. [CrossRef] [PubMed]

18. Carrão-Panizzi, M.C.; Beléia, A.P.; Prudêncio-Ferreira, S.H.; Oliveira, M.C.N.; Kitamura, K. Effects of isoflavones on beany flavor and astringency of soymilk and cooked whole soybean grains. *Pesquisa Agropecuária Brasileira* **1999**, *34*, 1045–1052. [CrossRef]
19. Sethi, S.; Tyagi, S.K.; Anurag, R.K. Plant-based milk alternatives an emerging segment of functional beverages: A review. *J. Food Sci. Technol.* **2016**, *53*, 3408–3423. [CrossRef]
20. Champ, M. Non-nutrient bioactive substances of pulses. *Br. J. Nutr.* **2010**, *88*, S307–S319. [CrossRef] [PubMed]
21. Campos-Veja, R.; Loarca-Piña, G.; Oomah, B.D. Minor components of pulses and their potential impact on human health. *Food Res. Int.* **2010**, *43*, 461–482. [CrossRef]
22. Clemente, A.; Arques, M.C. Bowman-Birk inhibitors from legumes as colorectal chemopreventive agents. *World J. Gastroenterol.* **2014**, *20*, 10305–10315. [CrossRef] [PubMed]
23. Srikanth, S.; Chen, Z. Plant protease inhibitors in therapeutics-focus on cancer therapy. *Front. Pharmacol.* **2016**, *7*, 470. [CrossRef]
24. Lima, A.I.G.; Mota, J.; Monteiro, S.A.V.S.; Ferreira, R.M.S.B. Legume seeds and colorectal cancer revisited: Protease inhibitors reduce MMP-9 activity and colon cancer cell migration. *Food Chem.* **2016**, *197*, 30–38. [CrossRef]
25. Khandelwal, S.; Udipi, S.A.; Ghugre, P. Polyphenols and tannins in Indian pulses: Effect of soaking, germination and pressure cooking. *Food Res. Int.* **2010**, *43*, 526–530. [CrossRef]
26. Pan, Z.; Tangratanavalee, W. Characteristics of soybeans as affected by soaking conditions. *LWT—Food Sci. Technol.* **2003**, *36*, 143–151. [CrossRef]
27. Singh, U.; Jambunathan, R. Studies on Desi and Kabull Chickpea (*Cicer arietinum* L.) Cultivars: Levels of Protease Inhibitors, Levels of Polyphenolic Compounds and In Vitro Protein Digestibility. *J. Food Sci.* **1981**, *46*, 1364–1367. [CrossRef]
28. Savitri, A.; Desikachar, H.S. A comparative study of flatus production in relation to the oligosaccharides content of some legumes. *Nutr. Rep. Int.* **1985**, *31*, 337–344.
29. Subbulakshmi, G.; Ganeshkumar, K.; Venkataraman, L.V. Effect of germination on the carbohydrates, proteins, tripsin inhibitor, amylase inhibitor and hemagglutinin in horse gram. *Nutr. Rep. Int.* **1976**, *12*, 19–26. Available online: http://ir.cftri.com/id/eprint/6510 (accessed on 8 July 2020).
30. Sampath, S.; Rao, T.M.; Reddy, K.K.; Arun, K.; Reddy, P.V.M. Effect of germination on oligosaccharides in cereals and pulses. *J. Food Sci. Technol. Mysore* **2008**, *45*, 196–198.
31. Kuo, Y.H.; Rozan, P.; Lambein, F.; Frias, J.; Vidal-Valverde, C. Effects of different germination conditions on the contents of free protein and non-proteins amino acids of commercial legumes. *Food Chem.* **2004**, *86*, 537–545. [CrossRef]
32. Murugkar, D.A. Effect of sprouting of soybean on the chemical composition and quality of soymilk and tofu. *J. Food Sci. Technol.* **2014**, *51*, 915–921. [CrossRef] [PubMed]
33. Rao, P.U.; Belavady, B. Oligosaccharides in Pulses: Varietal differences and effects of cooking and germination. *J. Agric. Food Chem.* **1978**, *26*, 316–319. [CrossRef]
34. Lopes, M.; Duarte, C.M.; Nunes, C.; Raymundo, A.; Sousa, I. Flow Behavior of Vegetable Beverages to Replace Milk. In Proceedings of the Iberian Meeting on Rheology, IBEREO, Porto, Portugal, 4–6 September 2019; Galindo-Rosales, F., Campo-Deaño, L., Afonso, A., Alves, M., Pinho, F., Eds.; Springer Proceedings in Materials. Springer: Cham, Switzerland. [CrossRef]
35. Nelson, A.I.; Steinberg, M.P.; Wei, L.S. Illinois process for preparation of soymilk. *J. Food Sci.* **1976**, *41*, 57–61. [CrossRef]
36. Francis, F.J.; Clydesdale, F.M. *Food Colorimetry: Theory and Applications*; AVI Publishing Company Inc.: Westport, CT, USA, 1975; pp. 26–45.
37. International Organization of Vine and Wine. Type I methods: Total Acidity (OIV-MA-AS313-01: R2015); Volatile Acidity (OIV-MA-AS313-02: R2015). In *Compendium of International Methods of Wine and Must Analysis*; IOVW: Paris, France, 2019; Volume 1.
38. International Organization for Standardization. *ISO 20483. Cereals and Pulses—Determination of the Nitrogen Content and Calculation of the Crude Protein Content—Kjeldahl Method*, 1st ed.; ISO: Geneva, Switzerland, 2006; pp. 1–13.
39. Mariotti, F.; Tomé, D.; Mirand, P.P. Converting Nitrogen into Protein—Beyond 6.25 and Jones' Factors'. *Crit. Rev. Food Sci.* **2008**, *48*, 177–184. [CrossRef]

40. Dubois, M.; Gilles, K.; Hamilton, J.K.; Rebers, P.A.; Smith, F. A colorimetric method for the determination of sugars and related substances. *Nature* **1951**, *168*, 167. [CrossRef]
41. Santos, M.V.; Faria, N.T.; Fonseca, C.; Ferreira, F.C. Production of mannosylerythritol lipids from lignocellulose hydrolysates: Tolerance thresholds of *Moesziomyces antarcticus* to inhibitors. *J. Chem. Technol. Biotechnol.* **2019**, *94*, 1064–1107. [CrossRef]
42. Barnes, H.A. Chapter 9: Shear-thinning liquids. In *Handbook of Elementary Rheology*, 1st ed.; Institute of Non-Newtonian Fluid Mechanics, Department of Mathematics, University of Wales: Aberystwyth, Wales, UK, 2000.
43. Sousa, I.M.N.; Morgan, P.J.; Mitchell, J.R.; Harding, S.E.; Hill, S.E. Hydrodynamic characterization of lupin proteins: Solubility, intrinsic viscosity, and molar mass. *J. Agric. Food Chem.* **1996**, *44*, 3018–3021. [CrossRef]
44. PortFIR. Composição de Alimentos. Available online: http://portfir.insa.pt/foodcomp/search (accessed on 21 February 2019).
45. Tako, M.; Tamaki, Y.; Teruya, T.; Takeda, Y. The Principles of Starch Gelatinization and Retrogradation. *Food Nutr. Sci.* **2014**, *5*, 280–291. [CrossRef]
46. Table of Typical Grain Bulk Densities and Angles of Repose. Available online: http://www.leoncooksilos.com.au/Typical%20Grain%20Bulk%20Densities%20and%20Angles%20of%20Repose.pdf (accessed on 17 September 2019).
47. Wang, S.; Errington, S.; Yap, H.H. Studies on Carotenoids from Lupin Seeds. In *Lupins for Health and Wealth, Proceedings of the 12th International Lupin Conference, Fremantle, Australia, 14–18 September 2008*; Palta, J.A., Berger, J.B., Eds.; International Lupin Association: Canterbury, New Zealand, 2008; pp. 198–202.
48. Rezaei, M.K.; Deokar, A.; Tar'an, B. Identification and Expression Analysis of Candidate Genes Involved in Carotenoid Biosynthesis in Chickpea Seeds. *Front. Plant Sci.* **2016**, *7*, 1–12. [CrossRef] [PubMed]
49. Niño-Medina, G.; Muy-Rangel, D.; Garza-Juárez, A.J.; Vázquez-Rodríguez, J.A.; Méndez-Zamora, G.; Urías-Orona, V. Composición nutricional, compuestos fenólicos y capacidad antioxidante de cascarilla de garbanzo (*Cicer arietinum*). *Archivos Latinoamericanos Nutrición* **2017**, *67*, 68–73. Available online: https://www.alanrevista.org/ediciones/2017/1/art-10/ (accessed on 31 January 2020).
50. Deswal, A.; Deora, N.S.; Mishra, H.N. Optimization of enzymatic production process of oat milk using response surface methodology. *Food Bioprocess. Technol.* **2014**, *7*, 610–618. [CrossRef]
51. Prakash, V.; Narasinga Rao, M.S. Physicochemical properties of oilseed protein. *CRC Crit. Rev. Biochem. Mol.* **1986**, *20*, 265–363. [CrossRef]

Publisher's Note: MDPI stays neutral with regard to jurisdictional claims in published maps and institutional affiliations.

© 2020 by the authors. Licensee MDPI, Basel, Switzerland. This article is an open access article distributed under the terms and conditions of the Creative Commons Attribution (CC BY) license (http://creativecommons.org/licenses/by/4.0/).

foods

MDPI

Article

The Effect of Protein Source on the Physicochemical, Nutritional Properties and Microstructure of High-Protein Bars Intended for Physically Active People

Jan Małecki [1,2], Igor Tomasevic [3], Ilija Djekic [4] and Bartosz G. Sołowiej [1,*]

1 Department of Milk Technology and Hydrocolloids, Faculty of Food Sciences and Biotechnology, University of Life Sciences in Lublin, Skromna 8, 20-704 Lublin, Poland; j.malecki@eurohansa.com.pl
2 EUROHANSA Sp. z o.o., ul. Letnia 10-14, 87-100 Toruń, Plant in Puławy, ul. Wiślana 8, 24-100 Puławy, Poland
3 Department of Animal Source Food Technology, Faculty of Agriculture, University of Belgrade, Nemanjina 6, 11080 Belgrade, Serbia; tbigor@agrif.bg.ac.rs
4 Department of Management of Food Safety and Quality, Faculty of Agriculture, University of Belgrade, Nemanjina 6, 11080 Belgrade, Serbia; idjekic@agrif.bg.ac.rs
* Correspondence: bartosz.solowiej@up.lublin.pl; Tel.: +48-81-4623350

Received: 26 September 2020; Accepted: 13 October 2020; Published: 15 October 2020

Abstract: The purpose of this study was to investigate the effect of protein sources (algae, pumpkin, wheat, sunflower, rice, soy, hemp, pea, and whey) on selected physicochemical, nutritional, and structural parameters of high-protein bars. Texture properties, such as hardness, fracturability, cohesiveness, and adhesiveness, have changed depending on the type of protein used. A significant increase, in particular the hardness parameter relating to the control sample (whey protein concentrate—WPC80), was noted for bars containing algae, sunflower, and wheat proteins, with high values of the adhesiveness parameter concurrently. The use of proteins from algae, pea, and wheat resulted in a significant reduction in the water activity of the finished product compared to WPC80. Bars made with the use of wheat, hemp and pumpkin proteins had noticeably higher viscosities than other samples. Color of the tested bars measured by means of Computer Vision System (CVS) was from light cream (soy, pea) to dark green (hemp, pumpkin). Bars prepared of wheat and algae proteins had the highest nutritional value, while the lowest one was recorded in products containing sunflower and hemp proteins. There was a clear differentiation of amino acids (g/100 g) and microstructure in bars depending on the type of protein used. However, a slight similarity can be found between whey and soy proteins (amino acids) and between whey and sunflower proteins (microstructure). Obtained results suggest that selection of the right type of protein for a given application may have a significant impact on the physicochemical features and microstructure of high-protein bars and their nutritional values.

Keywords: animal and plant proteins; computer vision system; nutritional value; texture; water activity; viscosity; microstructure; heavy metals; amino acids

1. Introduction

High-protein products, including bars, have recently become extremely popular. In particular, since products enriched in protein or in which protein is the main ingredient, can be used in products intended for a wide group of consumers [1].

This type of product can be used in the segment of quick snacks (designed to temporarily satisfy hunger), in sports nutrition (muscle tissue growth) or products intended for nutrition of the elderly and sick people who are at risk of developing sarcopenia [2]. As a result of such a large interest in high-protein products on the market, manufacturers meet the consumers' requirements and constantly develop recipes for innovative products that can be part of current trends in healthy and functional nutrition [3]. For this purpose, manufacturers searching for suitable alternatives to commonly used ingredients, such as high fructose and glucose syrups, fats or allergenic proteins, into their alternative components, e.g., polyols, fructo-oligosaccharides, or different protein sources (plant and animal proteins), while maximizing maintaining the technological parameters of the production process. Products resulting from such activities can be of particular interest to people using different types of diets [4,5].

High-protein bars most commonly found on the store shelves contain a small range of proteins of both plant (primarily soy protein concentrates and isolates) and animal origin (especially whey protein concentrates and isolates). It was found that the addition of whey protein hydrolysates used in the application of high-protein bars has a positive effect on maintaining the soft structure of these products, but may affect the slightly bitter aftertaste [6]. Whey derivatives, such as concentrates or isolates, are abundant sources of proteins, in particular alpha-lactalbumin and beta-lactoglobulin. In the food industry, proteins of this type are widely used because of their high nutritional value, desirable sensory properties (milk flavor), and excellent functional properties [7]. For some time, however, there has been a sharp increase in interest in alternative protein sources (especially plant proteins) that could compete with the commonly used WPC protein in terms of physicochemical, textural or nutritional features [8].

Recently, plant proteins have been increasingly used as an economical and versatile alternative replacing animal source in human nutrition, as well as functional ingredients for product formulation. Animal protein presents growing costs and limited supply, which has been highly associated with climate change, freshwater depletion, biodiversity loss, and hazards for human health related to cardiovascular diseases etc. [9]. In addition, the use of plant proteins in food applications (including high-protein bars) may also increase the interest in these products among vegans, vegetarians, and people with an active lifestyle [10].

Examples of proteins that are not currently used on a large scale in Europe are e.g., sunflower, wheat, algae, hemp, and rice proteins. Sunflower seed is one of five major oil sources in the world. The defatted sunflower meals have relatively high content of protein, and have great economic value as a food additive. Due to low amounts of anti-nutritive compounds and no toxic substances found in these raw materials, they can be counted as a promising source for food proteins [11]. Sunflower protein may potentially be a functional protein due to relatively good solubility. It is often a by-product of oil extraction, which is usually denatured during processing and has reduced solubility and functionality. If protein fractions are isolated without being denatured, sunflower proteins may become soluble over a range of ionic strength and pH [10].

The wheat protein isolates are currently of special interest to processors and consumers due to low fat content and as a substitute for egg and dairy proteins. The major functional properties of wheat proteins are: hydration, foaming, improve sheeting properties of dough, light, and floppy texture as well as clean flavor [12]. Algae proteins have been valued around the world since the dawn of time for their nutritional value. Currently, there is an increasing interest in proteins of this origin also due to their functional and health-promoting properties (anti-inflammatory or enriching agents) [13]. Hemp proteins are characterized by a number of pro-health and pharmacological properties e.g., they affect the angiotensin I-converting enzyme (ACE) inhibition, renin inhibition, acetylcholinesterase (AChE) inhibition, metal-binding capacity, antioxidant activity, hypocholesterolemic effect, and serum glucose regulation and have significant amounts of arginine and glutamine [8].

Rice protein is gaining a lot of interest in the food industry due to its unique properties. Moreover, hydrolysis of protein with proteases could produce many potential peptide sequences providing numerous functional and antioxidative properties. It could also enhance the antioxidative properties of native protein by attacking the peptide bonds in the interior of polypeptide chains producing a range of polypeptides that differ in molecular weight or amino acid sequences [14].

Pumpkin proteins are an opportunity to use the large amounts of waste generated during the pumpkin seed oil pressing process. The oil cake obtained as a result of the process has a significant amount of amino acids. Moreover, the nutritional value of the protein preparations obtained from pumpkin is very high, and can be used to improve the nutritional value of food products [15,16].

Soy protein isolates (SPIs) are widely used in the food industry. Owing to the amphipathic (hydrophilic and hydrophobic) nature of soy proteins, SPI possess a favorable capacity to being adsorbed onto the oil-water or air-water interface and maintaining the structure of the corresponding system as stabilizers, i.e., SPI have good foaming and emulsifying abilities. Due to these properties and high nutritional value, low price, and availability, soy proteins are widely used [17].

Pea protein is one such protein that has garnered a great deal of interest based on its low allergenicity, high nutritional value, availability, and low cost. Similar to other plant proteins. However, challenges in utilizing pea protein as a food ingredient exist in terms of limitations in functionality, flavor, and color issues. Pea proteins contain high levels of lysine, but tend to be limiting in methionine and tryptophan. Accordingly, pea proteins are often consumed along with cereal grains, as they have a complementary essential amino acid profile in that cereal proteins are generally deficient in lysine but contain higher levels of sulfur amino acids (methionine, cysteine) [18].

Due to the characteristics of these proteins declared by the producers (nutritional value, content of saturated fatty acids, amino acid composition, fragmentation, etc.), they may be particularly interesting in research on WPC substitutes.

Texture is one of the most important parameters that determines whether a product will be positively assessed by consumers and whether the customer will decide to buy a given product many times [19]. The degree of hardness in protein bars is directly correlated with the concentration of proteins used in the recipe. Too little protein may cause the formation of a liquid and ductile bar mass. On the other hand, overdosing the protein in the application will result in a loose and crumbling structure [20]. However, depending on the type of protein used, these parameters may differ from one other.

The purpose of this article is to discuss the broadly understood physicochemical, textural, microstructural, nutritional, and sensory properties of proteins that may be alternatives to WPC proteins and their possible declarations in accordance with applicable European Union (EU) legislation. According to our current available knowledge, there is no research on ultrasonic viscosity and determining the color differences by Computer Vision System (CVS) for application of high-protein bars. Also, the use of alternative protein sources e.g., algae, pumpkin, hemp, sunflower or wheat in production of high-protein bars is limited. Therefore, the objective of this study was to evaluate the effect of different protein sources (algae, pumpkin, wheat, sunflower, rice, soy, hemp, pea, whey) on selected physicochemical, nutritional, and structural parameters of high-protein bars.

2. Materials and Methods

2.1. Materials

Whey protein concentrate (WPC—80% proteins, 7.4% fat, 4.1% carbohydrates, granule size: <200 μm) was supplied by Polser Sp. z o. o. (Toruń, Poland), soy protein isolate (SPI—87% proteins, 3.1% fat, less than 1% carbohydrates, granule size: <200 μm) was purchased from Solae, pea protein isolate (PAP—82% proteins, 4% fat, 0.8% carbohydrates, granule size: <200 μm) was a product of Cosucra (Warcoing, Belgium), rice protein concentrate (RPC—80% proteins, 1% fat, 6% carbohydrates, granule size: <300 μm) was supplied by Barentz (Warsaw, Poland), wheat protein concentrate (WHP—77% proteins, 4% fat, 4% carbohydrates, granule size: <250 μm) was received from Cargill Polska (Warsaw, Poland), whole algal protein (ALP—60% proteins, 11% fat, 19% carbohydrates, granule size: <600 μm) was a product of TerraVia (San Francisco, CA, USA), sunflower protein (SUP—55% proteins, 2% fat, 9% carbohydrates, granule size: <200 μm), hemp protein (HMP—50% proteins, 10% fat, 5% carbohydrates, granule size: <500 μm) and pumpkin protein (PMP—60% proteins, 13% fat,

3% carbohydrates, granule size: <500 μm) were purchased from All Organic Trading (Wiggensbach, Germany), glucose syrup (Dextrose Equivalent "DE"40) was a product of Amylon (Havlíčkův Brod, Czech Republic), vegetable oil (rapeseed oil) was a product of ZT "Kruszwica" S.A. (Kruszwica, Poland), maltodextrin (Dextrose Equivalent "DE" 15) was purchased from Amylon (Havlíčkův Brod, Czech Republic), powdered barley malt extract was a product of Mountons Ingredients (European Brewery Convention "EBC" color: 5 to 12), soy lecithin (Identity Preserved "IP" 50) was supplied by Brenntag (Kędzierzyn-Koźle, Poland), natural vanilla aroma in powder was received from GBD Aromaty (Warsaw, Poland), chocolate was a product of Barry Callebaut (Łódź, Poland).

2.2. Preparation of High-Protein Bars

Protein concentrates (38.18%, *w*/*w*) with maltodextrin (5.45%, *w*/*w*) and aroma (0.91%, *w*/*w*) were placed in a bowl and mixed using the B10A industrial mixer (Technologies 4ALL Sp. z o. o. Sp. k.; Kępno, Poland) for 1 min at 190 rpm. Barley malt extract (3.64%, *w*/*w*) was dissolved in water (5.45%, *w*/*w*) in a separate laboratory vessel. In another vessel, soy lecithin (0.91%, *w*/*w*) and rapeseed oil (13.64%, *w*/*w*) were combined. Glucose syrup (31.82%, *w*/*w*) was heated to 80 °C and then poured into dry ingredients placed in the mixer bowl. The remaining ingredients prepared earlier were added simultaneously after pouring the syrup. The mass prepared in this way was mixed for 5 min at 365 rpm using an oar end. Finished processed high-protein bars mass were laid onto the conveyor belt using the CONBAR 600 (SOLLICH GmbH & Co. KG, Bad Salzuflen, Germany) and pulled out by forming rollers to a height of 15 mm. The height-adjusted bar mass was then cooled to a temperature of 10 °C for 15 min using the CONBAR 600 cooling tunnel. The chilled bar mass was subjected to longitudinal cutting using the CONBAR 600 longitudinal slitter. The longitudinally cut mass was finally cut transversely into individual bars (95 × 30 mm) using the CONBAR 600 transverse guillotine. The bars prepared in this way, which were to be coated with chocolate, were transferred to the coating machine, and they were covered with chocolate (22%, *w*/*w*) using the DK3520 (A.E. NIELSEN Maskinfabrik ApS., Farum, Denmark). Chocolate had a temper index (TI) oscillating in the range of 5.0–5.5. The parameter was measured using the Temper meter RET-250TMK (ELMI Automatic Systems, Warsaw, Poland). Chocolate coated bars were cooled to a temperature of 10 °C for 15 min using the DK3520 A.E. NIELSEN cooling tunnel. The final high-protein bars were packed in high barrier foil using a manual impulse sealer PFS 200 (NOVUMPACK, Kraków, Poland). The samples were stored at room temperature for 3 weeks in a plastic container. After the storage period, the samples were tested. The bar samples were unpacked from the foil 5 min before measurements. Cylindrical blocks of equal size (height: 15 mm, diameter: 12 mm) were punched out to analyze the texture of bars. To determine water activity and viscosity, the sample was prepared in the same way by weighing 6 g of sample for testing. Every high-protein bar sample was prepared in twenty repetitions. A total of 180 samples were tested considering all tests. The composition of tested high-protein bars without and with chocolate coating are presented in Table 1.

Table 1. Composition of high-protein bars.

Composition of High-Protein Bars without Chocolate Coating	
Ingredient	**Percentage Content in Final Product (% w/w)**
Protein ingredient (WPC, SPI, PAP, RPC, WHP, ALP, SUP, HMP or PMP)	38.18
Glucose syrup	31.82
Rapeseed oil	13.64
Maltodextrin	5.45
Water	5.45
Barley malt extract	3.64
Soy lecithin	0.91
Vanilla flavor (aroma)	0.91
Composition of High-Protein Bars with Chocolate Coating	
Ingredient	**Percentage Content in Final Product (% w/w)**
Protein ingredient (WPC, SPI, PAP, RPC, WHP, ALP, SUP, HMP or PMP)	30.2
Glucose syrup	25.0
Rapeseed oil	10.8
Maltodextrin	4.3
Water	4.3
Barley malt extract	2.9
Soy lecithin	0.7
Vanilla flavor (aroma)	0.7
Chocolate	21.1

2.3. Texture Profile Analysis (TPA)

Texture measurements were carried out on TA-XT2i Texture Analyzer (Stable Micro Systems, Godalming, Surrey, UK) coupled with the Software Texture Expert. The test velocity was 1 mm/s. The high-protein bars were twice compressed by a 36 mm diameter probe (SMS P/36R) to achieve 70% deformation (interval between probe movements: 5 s). High-protein bar samples were evaluated for hardness, fracturability, adhesiveness, and cohesiveness. Analyses were carried out in five replications for each sample. The hardness value was determined as the peak force occurring during the first compression. Fracturability point was occurred where the plot has its first significant peak (where the force falls off) during the probe's first compression of the product. Adhesiveness was calculated using the area over the negative stress–strain curve after the first compression, which represents the work per unit volume. Cohesiveness was defined as the ratio of the area under the second compression curve to the area under the first compression.

2.4. Cutting Test

Cutting strength of high-protein bars was measured using Texture Analyzer (TA-XT2i). The blade set with knife (HDP/BSK) comprising a Warner Bratzler blade (a reversible blade with knife edge) with a slotted blade insert and a blade holder was used for the experiment. In operation, the blade was firmly held employing blade holder, which was screwed directly to the texture analyzer. The slotted blade insert was placed directly onto the heavy-duty platform and acted as a guide for the blade whilst providing support for the product. High-protein bars were placed on the metal plate. Then the blade was lowered at a speed of 2 mm/s. The cutting curve was obtained by recording the maximum force the blade needs to cut the sample completely. Five repetitions were applied for each formulation. The results were based on the maximum peak (maximum force) resulting from the shear stress.

2.5. Water Activity

Water activity (a_w) was measured using the AWMD-10 water activity meter (NAGY, Gäufelden, Germany) with the accuracy of ±0.001 of a_w unit. Before measurement, the apparatus was calibrated with the dedicated humidity standard (95% HR). Measurements were performed at the temperature of 25 °C, in five repetitions. For each sample, two outliers were classified as defective and were excluded from further analysis.

2.6. Computer Vision System (CVS) and Determining Color Differences

Computer Vision System (CVS) was applied according to Tomasevic et al. [21] with the use of Sony Alpha DSLR-A200 digital camera (10.2 Megapixel CCD sensor, SONY, Tokio, Japan). The color was expressed in terms of the International Commission on Illumination (CIELAB) color space with the coordinates being L* (0–100, estimation of lightness), a* (red-green) and b* (yellow-blue) [22]. The noted differences could be described as "marked changes" according to the NBS (National Bureau of Standard) reference scale, which implies that such changes are perceptible to the human eye.

The total color difference was calculated using the formula:

$$\Delta E = \sqrt{(a_1 - a_2)^2 + (b_1 - b_2)^2 + (L_1 - L_2)^2}$$

Moreover, the ΔE* values were converted into National Bureau of Standards (NBS) units by the equation [23]:

$$\text{NBS units} = \Delta E \times 0.92$$

2.7. Ultrasonic Viscosity

The dynamic viscosity of high-protein bars was measured using an ultrasonic viscometer Unipan type 505 (UNIPAN, Warsaw, Poland). Measurements of the viscosity were performed at 25 °C. Prior to each measurement, the ultrasound signal level was checked. The measuring probe was immersed completely in the high-protein bar. The results were read in mPas·g/cm^3. All measurements were performed in three repetitions.

Viscosity tests using ultrasounds rely on the use of a magnetostrictive probe, which produces free vibration [24]. An alternating electric current generates an alternating magnetic field that causes the phenomenon of magnetostriction, i.e., the deformation of ferromagnetic materials. Induced ultrasound waves are damped by the tested material. Ultrasound viscometers display the result as the product of viscosity and density [25]. Ultrasonic viscometer viscosity measurements are performed at high frequency, and for this reason, it is not easy to compare the obtained results with those obtained using other viscometers. In addition, ultrasonic viscometers are used for continuous measurements of viscosity under conditions where measurements can be difficult and it is not possible to use devices such as rotational viscometers [26].

2.8. Nutritional Value

Nutritional value was calculated based on raw material specifications obtained from suppliers of each of the ingredients, which was introduced into the program. Then the recipe was entered into the X-mart (X-mart Group Sp. z o. o., Lublin, Poland) software and the nutritional value of the finished product was calculated per 100 g.

2.9. Sensory Evaluation

A panel of 15 trained consumers was recruited from EUROHANSA Sp. z o. o. The criteria for selection were that the panelists should be between 18 and 60 years old and regular consumers of high-protein bars and not allergic to any raw material used. Panelists were instructed to evaluate the sensorial attributes; color, aroma, consistency and taste. A 5-point hedonic scale (1 = extremely dislike,

5 = extremely like) with significance factors (0.2—color, 0.2—aroma, 0.25—consistency and 0.35—taste) was used [27,28].

2.10. Heavy Metals Analysis

The obtained samples were grounded and about 0.5 g of the sample were weighed from the homogeneous mass on an analytical balance with an accuracy of 0.0001 g. After the tubes were closed, they were transferred to the mineralizer rotor. The mineralization was carried out in a CEM Mars Xpress microwave oven at the temperature of 210 °C and pressure of about 7 atm. The obtained clear mineralizates were quantitatively transferred to 50 cm^3 volumetric flasks and diluted with demineralized water (conductivity 0.055 µS/cm) to the mark. The obtained solutions were analyzed on an inductively coupled plasma mass spectrometer (ICP Mass Spectrometer Varian MS-820, Santa Clara, CA, USA). The gas used to generate the plasma was argon from Messer with a purity of 99.999%. No reaction chamber (CRI) was used in the analysis. The following camera settings were used: Plasma Flow—16 dm^3/min., Flow Nebulizer—0.98 dm^3/min., RF Power—1.38 kW, Sampling Depth—6.5 mm. The following isotopes of the analyzed elements were used: ^{114}Cd, ^{206}Pb, ^{207}Pb, ^{208}Pb.

The determination was made using the standard curve method. Ultra Scientific standards with a purity of 99.999% were used for the analysis. The results obtained are expressed in mg/kg fresh weight. Test quality control during the analysis was applied by measuring blank, duplicate and certified reference material "NIST-1577c Bovine Liver".

2.11. Amino Acids Determination

The sample (approx. 70 mg of pure protein) was hydrolyzed with 6 N HCl at 110 °C for 20 h. After cooling, the solution with the sample was filtered through a G-5 funnel. 4 mL of the hydrolysate was evaporated in a vacuum evaporator. The dry residue from the vacuum flask was dissolved in 5 mL of citrate buffer pH 2.2. The prepared sample was dispensed onto the amino acid analyzer column [29].

Separation of sulfur amino acids was performed as follows: cysteine was oxidized to cysteic acid, and methionine to methionine sulfone using performic acid. The mixture was then flooded with 1 mL of 40% HBr and concentrated in a vacuum evaporator. It was then quenched with 6 N HCl and hydrolyzed at 110 °C for 20 h. Further procedure was the same as for protein amino acids [30].

To determine tryptophan, the sample was subjected to alkaline hydrolysis. The sample weight containing approx. 75 mg of protein was hydrolyzed in $Ba(OH)_2$ solution at 110 °C for 20 h. The sample was then acidified with 6 N HCl and a Na_2SO_4 solution was added. The contents were transferred to centrifuge flasks and centrifuged for 15 min at 3000× *g*. The supernatant, after filtering through a syringe filter, was dosed into the amino acid analyzer.

Amino acids were determined with the AAA 400 amino acid analyzer from Ingos (Prague, Czech Republic). Amino acids were separated through ion exchange chromatography. The column with dimensions of 0.37 × 45 cm was filled with an ion exchanger in the form of a resin. LG ANB ostium was used for the hydrolysates. It is a strong cation exchanger with an average grain size of approx. 12 µm in the form of Na (column temperatures 60 °C and 74 °C). The apparatus detects the amino acids by ninhydrin derivative (this is the detection reagent). The identification of the amino acids was performed by a photometric detector at a wavelength of 570 nm for all amino acids, while for proline—440 nm. Four buffers were used for separation: 1—pH 2.6, 2—pH 3.0, 3—pH 4.25, 4—pH 7.9. After the amino acid separation, the column was regenerated using 0.2 N NaOH.

2.12. Scanning Electron Microscopy (SEM)

The samples were placed in a 4% aqueous glutaraldehyde solution at room temperature for 2 h and then transferred to a refrigerator at ca. 4 °C for 6 h. After this time, the samples were placed in Sörensen's phosphate buffer pH 7.0 and left overnight. After removing from the buffer and washing twice in distilled water, the samples were dehydrated in an acetone series. Concentrations of acetone solutions (p.a.), to which the samples were successively transferred, were: 15, 30, 50, 70, 90, 100%.

The samples were kept for 30 min in each of the solutions. At the end of the dehydration process, the samples were placed in anhydrous acetone dried on silica gel for 30 min. The last stage was carried out twice. For final removal of residual water, the material was subjected to critical point drying with carbon dioxide in an Emitech K-850 dryer (Ashford, UK). Microscope tables were prepared with a carbon substrate placed on them and dried samples of bars were attached to it. The prepared gold was sputtered with an Emitech K-550X sputter (Ashford, UK). After the preparation was completed, the obtained material was placed in a Tescan Vega LMU (Brno, Czech Republic) scanning electron microscope and examined under high vacuum.

2.13. Statistical Analysis

Statistical analysis was carried out with a help of the STATISTICA 13.3 PL software (Stat Soft Polska Sp. z o. o., Kraków, Poland). A one-way ANOVA analysis was performed, and significant differences between samples were determined applying the Tukey *post hoc* test at $p < 0.05$.

3. Results and Discussion

3.1. Texture Profile Analysis (TPA), Cutting Test and Scanning Electron Microscopy (SEM)

The influence of different proteins on hardness, fracturability, adhesiveness, and cohesiveness of the obtained processed high-protein bars with or without chocolate coating is presented in Table 2a,b. Significant differences ($p < 0.05$) were observed. The bar made of sea algae proteins (ALP) had the highest hardness (276.43 N with chocolate and 288.50 N without chocolate coating) in both cases, while the lowest hardness characterized bars made from pea protein (PAP) in the sample without chocolate (13.62 N) and made from rice protein (RPC) in the sample with chocolate coating (20.95 N).

Table 2. Effect of different protein application on high-protein bars texture attributes. (**a**) Data are presented as means ± SD (standard deviation). $^{a–h}$ Means in the same column with different superscripts are significantly different ($p < 0.05$, Tukey's honest significant difference "HSD" test). (**b**) Data are presented as means ± SD (standard deviation). $^{a–i}$ Means in the same column with different superscripts are significantly different ($p < 0.05$, Tukey's HSD test).

(a)

Type of Protein Used in Bars with Chocolate Coating	Texture Attributes			
	Hardness (N)	Fracturability (N)	Adhesiveness (J)	Cohesiveness
WPC	54.66 ^{e} ± 0.303	0.06 ^{a} ± 0.005	382.87 ^{g} ± 4.977	0.14 ^{f} ± 0.001
RPC	20.95 ^{a} ± 0.164	0.06 ^{a} ± 0.004	57.54 ^{c} ± 2.588	0.07 ^{c} ± 0.001
SPI	25.25 ^{c} ± 0.358	20.85 ^{b} ± 0.152	66.85 ^{c} ± 1.773	0.04 ^{a} ± 0.004
SUP	136.61 ^{g} ± 0.406	115.56 ^{f} ± 0.255	225.61 ^{f} ± 4.861	0.10 ^{de} ± 0.001
WHP	88.33 ^{f} ± 0.092	55.38^{e} ± 0.288	123.42 ^{d} ± 3.724	0.11 ^{e} ± 0.004
HEP	27.74 ^{d} ± 0.152	0.03 ^{a} ± 0.005	27.15 ^{b} ± 2.528	0.13 ^{f} ± 0.003
PAP	27.44 ^{d} ± 0.302	25.86 ^{d} ± 0.461	145.78 ^{e} ± 4.853	0.06 ^{b} ± 0.002
PMP	23.52 ^{b} ± 0.338	23.27 ^{c} ± 0.215	7.34 ^{a} ± 0.392	0.09 ^{d} ± 0.004
ALP	276.43 ^{h} ± 0.286	0.13 ^{a} ± 0.012	129.67 ^{d} ± 0.577	0.19 ^{g} ± 0.008

(b)

Type of Protein Used in Bars without Chocolate Coating	Texture Attributes			
	Hardness (N)	Fracturability (N)	Adhesiveness (J)	Cohesiveness
WPC	34.53 ^{f} ± 0.277	0.30 ^{a} ± 0.024	56.23 ^{d} ± 4.336	0.12 ^{d} ± 0.001
RPC	18.64 ^{c} ± 0.327	0.11 ^{a} ± 0.018	34.16 ^{c} ± 2.166	0.05 ^{b} ± 0.001
SPI	16.38 ^{b} ± 0.201	16.39 ^{b} ± 0.306	5.88 ^{a} ± 0.681	0.03 ^{a} ± 0.002
SUP	149.19 ^{h} ± 0.198	122.52 ^{e} ± 0.439	16.23 ^{b} ± 2.171	0.15 ^{e} ± 0.001
WHP	81.31 ^{g} ± 0.172	0.08 ^{a} ± 0.004	130.15 ^{e} ± 2.157	0.22 ^{g} ± 0.010
HEP	21.50 ^{e} ± 0.170	35.59 ^{c} ± 0.450	3.37 ^{a} ± 0.479	0.09 ^{c} ± 0.006
PAP	13.62 ^{a} ± 0.246	36.81 ^{d} ± 0.217	1.69 ^{a} ± 0.246	0.06 ^{b} ± 0.004
PMP	19.67 ^{d} ± 0.167	0.03 ^{a} ± 0.004	322.85 ^{f} ± 2.695	0.21 ^{g} ± 0.006
ALP	288.50 ^{i} ± 0.326	0.07 ^{a} ± 0.004	27.79 ^{c} ± 2.947	0.17 ^{f} ± 0.002

Sensory hardness can be defined as the force necessary to compress a high-protein bar with the molars. Fracturability is the tendency of a material to fracture, crumble, crack, shatter or fail upon the application of a relatively small amount of force or impact. Adhesiveness is the work/force necessary to overcome the attractive forces between the surface of a product and the surface of a material (the probe), with which the product comes in contact. Cohesiveness is the tendency of a product to cohere or stick together [31]. Generally, hardness of high-protein bars is quite high and increases with the addition of protein [32]. The developed high-protein bars are characterized by large variety of parameters. Considering the research of Banach et al. [33], a certain regularity can be noted for bars made of whey proteins. Relatively low value of the hardness parameter translates into high values of the adhesiveness and cohesiveness, and simultaneously low levels of fracturability. A completely different situation can be seen for algae protein. Despite the high hardness of the bar made of this type of protein, high values of the adhesiveness and cohesiveness parameters as well as very low fracturability values were observed in the chocolate-coated bar. The results obtained from the research on a bar made of sunflower proteins (SUP) also deserve attention due to relatively high level of all TPA parameters, in particular in the chocolate-coated sample. The intermolecular attraction, by which the elements of a body or mass of material are held together, determine its cohesiveness [34]. Banach et al. [35] found a connection between the hardness of bars and the size of the protein particles used in the recipe for making those bars. Based on the analysis of results in the Table 2a,b and pictures of the microstructure of bars (Figure 1), it can be assumed that proteins with large particle sizes used in the production of high-protein bars caused a significant increase in the hardness of the final product. According to this lead, fine-grained proteins have much lower tendency to form hard structures during the storage process and allow for the creation of a delicate and soft product structure, which is confirmed by Cho [36]. It was observed that the chocolate-coated bars showed, in most cases, higher hardness than the uncoated samples, but also had increased other TPA parameters. This is most likely related to the greater restriction of air access to these products, which slows down the drying processes of the product (as evidenced by higher adhesiveness results and water activity parameters presented in Figure 2a,b).

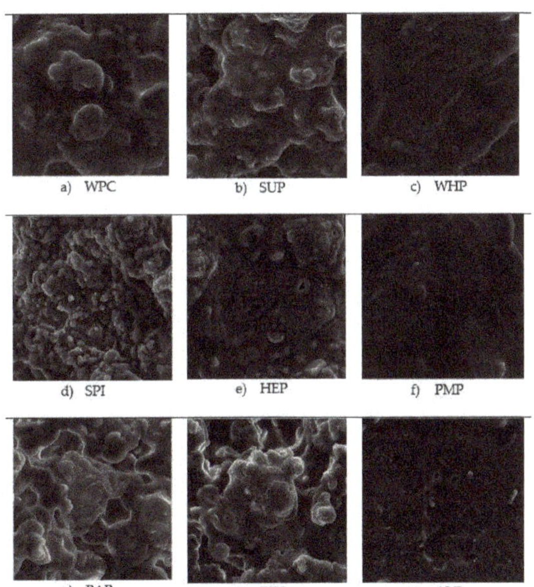

Figure 1. (**a–i**) Microstructure of high-protein bars from the scanning electron microscope (SEM HV: 30 kV, View field: 271 μm, SEM MAG: 800×).

Cutting test indicates the firmness/hardness of a product. If one considers that if the top front teeth were pulled from a curve-shape into a straight line, they would represent a 'knife edge'. Using a knife blade gives a close representation of the biting or cutting action [31].

With reference to Table 3, the bar made of algae proteins (ALP), both without and with chocolate coating, showed the highest cutting resistance (166.82 N without chocolate and 235.45 N with chocolate). Whereas, the most susceptible to this effect turned out to be high-protein bars made of proteins: wheat (WHP: 8.43 N without chocolate and 10.54 N with chocolate), pumpkin (PMP: 7.69 N without chocolate and 22.7 N with chocolate), hemp (HEP: 10.58 N without chocolate and 15.59 N with chocolate) and sunflower (SUP: 14.38 N without chocolate and 22.35 N with chocolate coating). The obtained results correlate with the hardness analysis and confirm the relationship between the hardness of bars and the force needed to cut them. In general, it can be observed that the chocolate-coated bars had higher resistance to the cutting force, except from WPC and PAP. The reason for this phenomenon may result from slight fluctuations in parameters in the degree of chocolate tempering, which slightly change during the chocolate tempering process. Chocolate with a temper index (TI) degree of 5.0–5.5 is characterized by high hardness causing a characteristic crackle when breaking, which is a feature desired by consumers in this type of products. This is also confirmed by the sensory analysis performed. The deviations may be related to the thickness of the chocolate layer in different places of the bars because in production conditions the products are coated with a stream of chocolate, too high layer of chocolate is blown off through a blast of compressed air, which creates a wave on the product characteristic of chocolate-coated products available on the store shelves. A properly tempered chocolate exhibits high gloss, appropriate melting temperature, and fat-bloom stability with desired characteristic crunchiness and hardness during eating depending on the amount of chocolate on the final product [37]. In addition, the flow behavior of tempered chocolate has implications for the processing of chocolate after tempering. Factors such as conching temperature, particle size distribution, fat content, type of emulsifiers, and tempering conditions determine efficiency of mixing, pumping, and transportation of final products during processing [38].

Table 3. Effect of different protein sources on high-protein bars cutting resistance.

Type of Protein Used in Bars with Chocolate Coating	Cutting Resistance Force (N)	Type of Protein Used in Bars without Chocolate Coating	Cutting Resistance Force (N)
WPC	$79.31^{f} \pm 0.298$	WPC	$128.39^{h} \pm 0.393$
RPC	$25.75^{d} \pm 0.198$	RPC	$25.36^{e} \pm 0.084$
SPI	$109.69^{g} \pm 0.112$	SPI	$98.21^{g} \pm 0.162$
SUP	$22.35^{c} \pm 0.298$	SUP	$14.38^{d} \pm 0.149$
WHP	$10.54^{a} \pm 0.073$	WHP	$8.43^{b} \pm 0.020$
HEP	$15.59^{b} \pm 0.271$	HEP	$10.58^{c} \pm 0.126$
PAP	$75.34^{e} \pm 0.222$	PAP	$81.45^{f} \pm 0.194$
PMP	$22.70^{c} \pm 0.143$	PMP	$7.69^{a} \pm 0.148$
ALP	$235.45^{h} \pm 0.366$	ALP	$166.82^{i} \pm 0.138$

Data are presented as means ± SD (standard deviation). $^{a-i}$ Means in the same column with different superscripts are significantly different ($p < 0.05$, Tukey's HSD test).

Based on the electron microscope photos presented in Table 3 and the studies by Labuza and Hyman [39], it can be assumed that large discrepancy in the results of TPA tests may be due to factors related to the structural features, density of protein molecules and their porosity. In addition, Hogan et al. [6,40] proved that the rate of moisture migration in multi-domain foods is slower in foods with smaller pore size, presumably due to more tortuous pathways for moisture diffusion. Air occluded within powder particles may be observed as the proportion of powder volume not subjected to moisture-induced change, and thus may be beneficial to structural stability. A fairly large number of similarities were found between the microstructure of SUP, PAP, WPC and RPC proteins. Pictures of these proteins have the significant number of depressions and embossing. Taking into account the parameters of the texture analysis, such a structure probably has a significant impact on decreasing the hardness and cutting force parameters. Bars with SUP revealing much higher hardness, were the exception. This may be related to the formation of large clusters (agglomerates) of proteins, resulting in the formation of a compact and hard structure, which translates directly into high parameters of hardness, fracturability, and adhesiveness. It was also observed that all the above bars were characterized by quite high susceptibility to the action of shear force. It is worth mentioning that bars made of WHP and PMP proteins, having a wavy structure, without a large number of cavities and air pores, were also characterized by relatively low hardness parameters and high susceptibility to cutting. All bars made of the proteins mentioned above showed similar, technologically acceptable, water activity ($a_w < 0.735$). The bar made of ALP proteins had a very diverse structure. Numerous air pores and unevenly distributed agglomerates of protein particles in the form of bushy protrusions were observed on its surface. Interestingly, the protein protrusions were also interrupted by wavy, relatively smooth protein structures. Referring to the work of Bleakley and Hayes [41], the formation of characteristic agglomerates in the case of algae protein may be related to the presence of lectins in this type of protein. Lectins are glycoproteins known for their aggregation and high specificity binding with carbohydrates without initiating a modification through associated enzymatic activity. Completely different microstructure of ALP bars is probably the reason for the highest hardness. The use of proteins of various botanical origin in the study has a significant impact on the swelling method, reorganization of molecular structures or aggregation of proteins in the product. However, the control of micro- and macrostructures is still very difficult due to the poor knowledge of this issue [42].

3.2. Water Activity

Measurement of water activity of high-protein bars are presented on Figure 2a,b. The water activity of high-protein bars changes during the storage process [43]. Therefore, the bars were stored for 3 weeks in a sealed plastic container (metallized barrier film) at 20 °C. The optimal storage time was adopted due to the results of studies by Banach et al. [35], which showed that a_w increase after this period was not significant. Water activity of the tested samples was shown in Figure 2a,b. The highest a_w value characterized bar with chocolate coating made of sunflower protein (SUP)—0.735 and the lowest (ALP)—0.530 algae protein bar without chocolate coating. Bars made of proteins: sea algae (ALP), pea (PAP) and soybean (SPI) had a_w lower than 0.65, which guarantees the stability of samples during storage (in room temperature) and inhibition of microbial growth [44]. Other bars, made of hemp (HMP), rice (RPC) and sunflower (SUP), had water activity above 0.65. Therefore, it can be suggested that they should be stored in conditions of lower temperature. Increased water activity may indicate movement of water molecules from the intermediate phase, where they act as a plasticizer, to the bulk phase [32]. Proteins included in bars that exceeded 0.65 water activity values, had a fairly high degree of fragmentation, and originated from various plant species, which could also have a significant impact on this parameter.

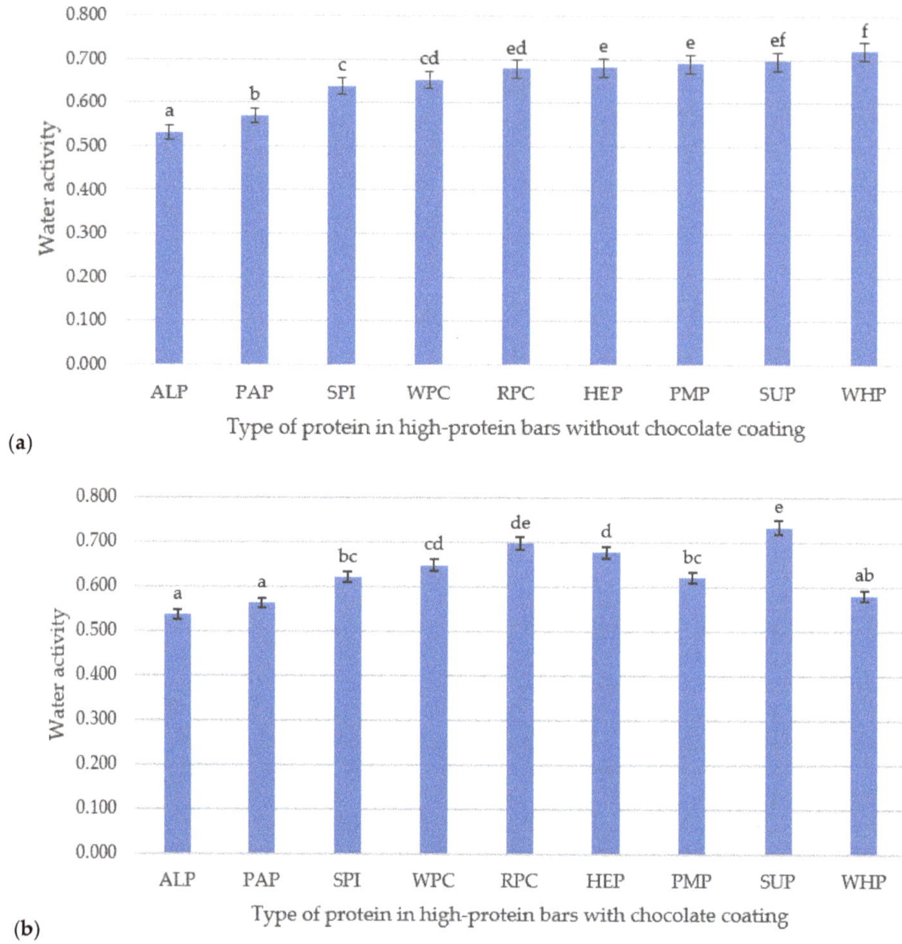

Figure 2. (**a**) Effect of different protein application on water activity of high-protein bars without chocolate coating; (**b**) With chocolate coating. Letters (a–f) indicate significant differences at $p < 0.05$ (Tukey's HSD test).

According to the current knowledge, changes in the structure or organization of proteins in a product may be associated with the formation of disulfide bonds when there are no water molecules associated with the local protein domain. It may also be one of the mechanisms explaining the hardening of high-protein products over time, in particular, if whey proteins have been added to the product [45,46]. High water activity parameters for WHP without chocolate coating may be related to the high water absorption of gluten. On the other hand, the differences in the case of the sample with chocolate coating may result from the restriction of air access to the interior of the product due to the coating with a layer of chocolate. High water activity value of bars made of SUP protein may be related to the high capacity of this type of protein to absorb water and fat, as evidenced by the research of Ren et al. [47]. According to the obtained results, it can be suspected that the addition of other types of proteins, such as ALP, PAP, WHP, or SPI to food products, may have a positive effect on reducing the water activity and modifying the textural parameters of the final product.

3.3. Color Differences Measured with Computer Vision System (CVS)

Typically, the main aspect that reduces the quality of high-protein bars is their hardening over time, but color can also be an important quality indicator for consumers of this type of product. The colors of bars made of various types of proteins after three weeks of storage are presented in Table 4a. The most frequently used methods of color assessment in the analysis of this parameter in food products are colorimetric methods. In turn, the colors generated using CVS closely resemble the real color of the samples being assessed. Moreover, the color is more intense (the colors are more saturated) than for standard colorimetric methods [48]. In this study, the CVS method was used due to the studies by Tomasevic et al. [49], which proved that the CVS method gives much better results in assessing the color of food products.

According to Inami et al. [23], all the color differences expressed by values larger than 6 are considerable. Discrepancies between the >6 prove the high impact of the protein used on the color difference of high-protein bars. Proteins from various sources were characterized by different colors compared to the blank sample made of WPC. The SPI protein was characterized by the highest brightness, which may indicate a slower ability to bind fat than other tested proteins. The fat on the surface of bars probably causes the increased ability to reflect light. The remaining differences in parameters a and b are probably directly related to the origin of the component (plant and animal proteins). Considering the Hasan [50] study, values of the L* parameters were similar for bars made of whey proteins after about 6 weeks of storage. Minor differences in the a* and b* parameters may be related to the analysis of parameters after a longer storage period. On the other hand, comparing the research results with the work of McMahon et al. [51], it can be assumed that bars made with high-fructose and glucose syrups are much more susceptible to darkening processes than in the case of other types of syrups. In this study, only glucose syrup was used in all trials. Therefore, it is probably required to conduct further tests and investigate whether the type of syrup will cause large changes in the L* a* b* parameters for the developed high-protein bars. The color of bars with chocolate coating was not examined, but it can be assumed that they would have brighter colors due to better protection against light and air access, and thus slowing down the Maillard reaction.

3.4. Ultrasonic Viscosity

The obtained ultrasonic viscosity is presented in Figure 3a,b. The ultrasonic viscometer gives the results of measurements in units of dynamic viscosity multiplied by density. The highest viscosity values were recorded for bars made of WHP and HEP proteins. However, the lowest—for ALP, WPC, PAP, and SUP. Low values of the viscosity parameter, in particular for ALP, correlate with high results of the hardness parameters and cutting force. It is also worth paying attention to the microstructure of these proteins, in which numerous clusters of wide pores (probably fat-air), a tightly compact and irregular structure can be seen (Figure 1).

Table 4. (**a**,**b**) Effect of different protein application on color of high-protein bars without chocolate coating measured with Computer Vision System (CVS).

(**a**)

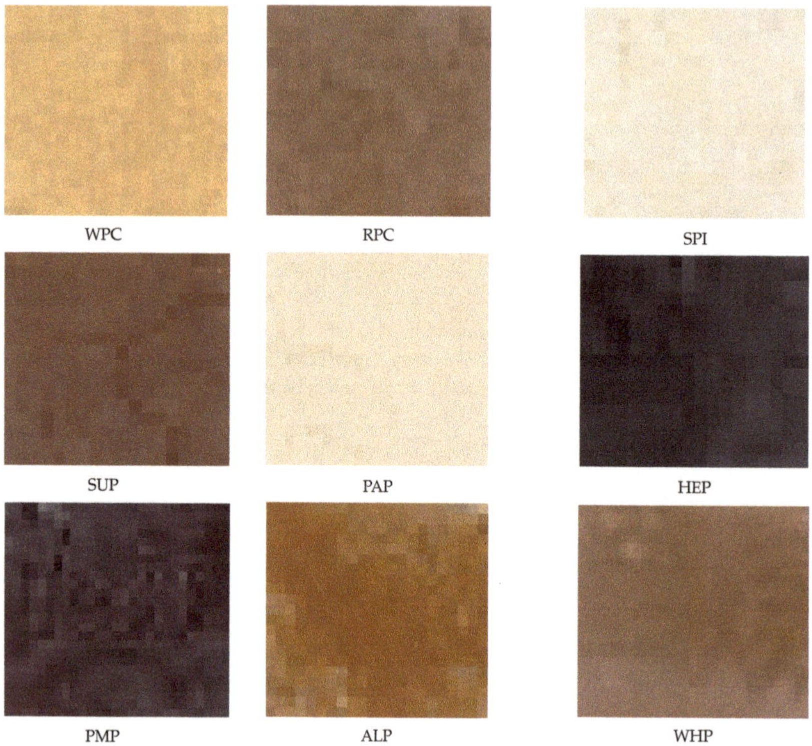

(**b**)

Type of Protein Used in Bars without Chocolate Coating	Attributes			
	L*	a*	b*	NBS Units
WPC	63.86 ^{f} ± 1.069	17.29 ^{d} ± 0.488	44.43 ^{f} ± 0.976	-
RPC	43.71 ^{c} ± 0.488	10.00 ^{c} ± 0.000	21.14 ^{b} ± 0.690	29.12
SPI	79.43 ^{h} ± 0.787	5.29 ^{a} ± 0.488	19.43 ^{a} ± 0.976	29.26
SUP	34.14 ^{a} ± 0.378	11.00 ^{c} ± 0.000	22.29 ^{b} ± 0.756	34.58
WHP	40.57 ^{b} ± 0.976	10.00 ^{c} ± 0.000	31.71 ^{c} ± 1.113	25.32
HEP	50.57 ^{e} ± 0.976	47.43 ^{f} ± 1.272	50.29 ^{g} ± 1.704	30.78
PAP	78.00 ^{g} ± 0.000	8.00 ^{b} ± 0.000	23.00 ^{b} ± 0.000	25.12
PMP	49.43 ^{e} ± 1.134	40.71 ^{e} ± 1.113	34.14 ^{d} ± 2.116	27.02
ALP	47.71 ^{d} ± 0.488	10.86 ^{c} ± 0.378	39.71 ^{e} ± 0.488	16.57

Data are presented as means ± SD (standard deviation). $^{a-h}$ Means in the same column with different superscripts are significantly different ($p < 0.05$, Tukey's HSD test). "-" NBS Units of WPC protein-reference sample, not subject to calculation.

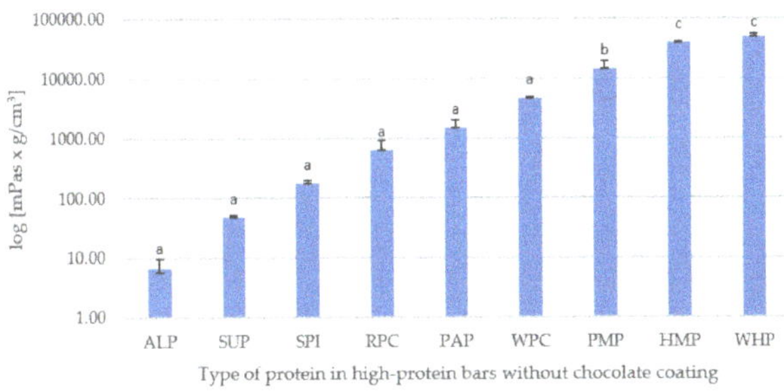

(a)

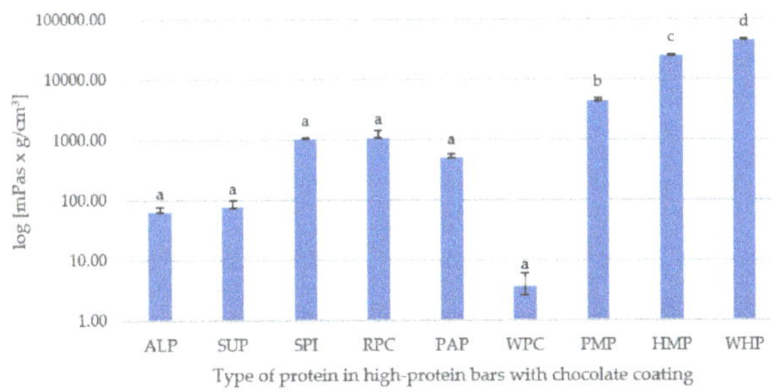

(b)

Figure 3. (**a**) Ultrasonic viscosity measurement results (mPas g/cm^3) of tested high-protein bars without chocolate coating; (**b**) With chocolate coating. Letters (a–d) indicate significant differences at $p < 0.05$ (Tukey's HSD test).

On the other hand, high-protein bars with the highest lightness values had much lower values of parameters related to hardness and cutting force. Their microstructure is free from numerous pores and the surface is much more even compared to other tested samples. Considering the obtained results and comparing them with the research by Tomczyńska-Mleko and Ozimek [52], it can be assumed that the obtained results could be influenced by factors such as degree of aeration in the bar mass and the consistency and protein concentration the product was made of. Ultrasonic viscosity tests are a rare/novel method for the analysis of food products, therefore, according to current knowledge, it is difficult to find publications to compare the obtained results.

3.5. Heavy Metals Analysis

The obtained results of testing the content of heavy metals in the developed high-protein bars are presented in Figure 4. They do not exceed the current permissible concentrations for this type of products according to Commission Regulation (EC) No 1881/2006 of 19 December 2006 [53].

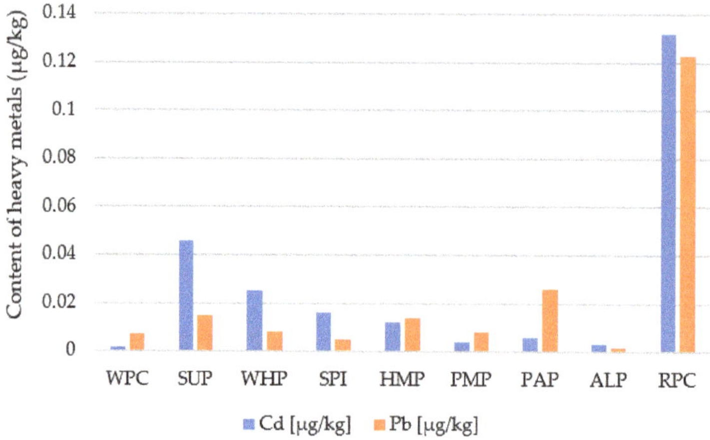

Figure 4. Content of heavy metals (cadmium and lead) in the high-protein bars.

However, a much higher level of cadmium and lead is very visible for the RPC protein. Regarding the research of Huang et al. and Kaneta et al. [54,55], it can be assumed that the increased level of these elements for bars made of RPC may result from the ability of rice, one of the most commonly cultivated plants on earth, to absorb significant amounts of heavy metals from its cultivation sites. Therefore, products made of this type of grain may be characterized by an increased content of not only cadmium and lead, but also arsenic and other heavy metals.

3.6. Amino Acids and Nutritional Value

Often, for people consuming the high-protein products (e.g., athletes, sick people or convalescents), nutritional and pro-health values are important. The obtained results of the content of amino acids in the tested bars presented in Figure 5 show significant differences between the content of individual amino acids in different types of proteins. The largest deviation can be seen in the proline content for bars made of WHP protein.

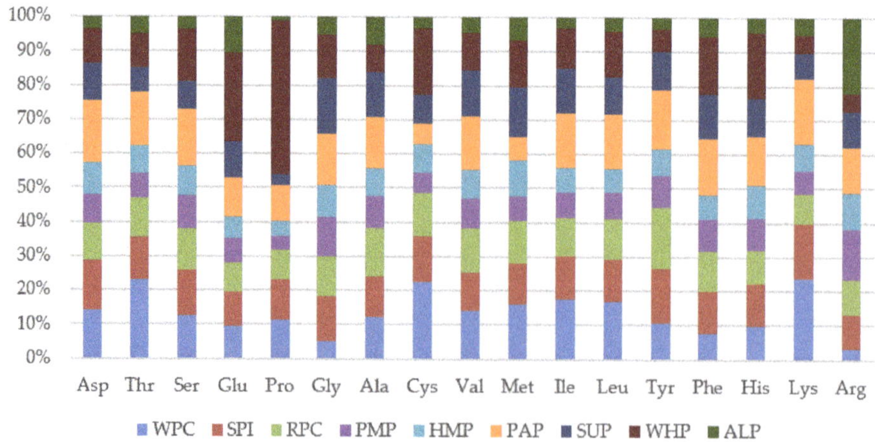

Figure 5. Percentage of individual amino acids in the high-protein bars.

Based on the work of Kowieska et al. [56], it can be assumed that such a high content of endogenous amino acid, i.e., proline, may be caused by a naturally high content of proline in wheat grain. Proline is an important amino acid for physically active people and athletes, as it participates in the formation of secondary structures in collagen. These structures are stabilized by enzymatic hydroxylation (proline hydroxylase) or a substituent having electron withdrawing ability, e.g., fluorine, which significantly increases the stability of the collagen. Deficiencies of proline, vitamin C (being a cofactor of proline hydroxylase), and disorders of enzyme production, can lead to the scurvy [57].

Exogenous amino acids are essential and human body cannot synthesize them from scratch at a rate commensurate with their needs. Therefore, it is necessary to provide them to the body in a properly balanced diet. Out of a total 21 amino acids, nine are considered essential, including phenylalanine, valine, threonine, tryptophan, methionine, leucine, isoleucine, lysine, and histidine. Proteins found in animal sources, such as meat, poultry, fish, eggs, milk, cheese, and yogurt, provide all nine indispensable amino acids and hence they are referred to as "complete proteins." Proteins found in plants, legumes, grains, nuts, seeds, and vegetables tend to be deficient in one or more of the indispensable amino acids and are called "incomplete proteins" [58].

Despite the passage of time and thermal processing during production, the obtained high-protein bars are characterized by a high content of the entire spectrum of amino acids.

Considering the obtained results, it can be assumed that, despite the high content of essential amino acids in the WPC reference sample, alternative sources of plant-derived proteins may be an attractive proposition for people who do not want or cannot consume proteins of animal origin. It is worth paying attention to the high content of arginine in the ALP. As evidenced by the research of Stróżyk et al. [59], arginine is an essential amino acid in the case of increased physical exertion. It is a precursor of nitric oxide, which relaxes the smooth muscles of blood vessel walls, thus improving the blood oxygenation and replenishment of skeletal muscles. It was observed that bars made of SPI, RPC, and PAP proteins showed significant alignment of the entire spectrum of amino acids. Among which, essential amino acids, which have been determined in significant amounts, as for proteins of plant origin, deserve special attention. The high content of essential amino acids in soybean and rice protein isolates was also confirmed by Kalman [60]. Therefore, it can be assumed that these proteins may become more and more popular, especially among vegans and vegetarians and allow the essential amino acids in such diets to be satisfied.

Results obtained from the calculated nutritional value presented in Figure 6a,b and Figure 7a,b indicate slight differences in the protein and fat content as well as nutritional values in individual protein preparations used in the production of high-protein bars. However, these differences are not significant ($p > 0.05$). It is also worth paying attention to the fiber content for bars made of HEP and SUP, the content of which is significantly higher than that of other samples. In addition, this amount of fiber allows the nutrition claim "high fiber content" to be used on the packaging of these products in accordance with Regulation (EC) No 1924/2006 Of The European Parliament And Of The Council of 20 December 2006 [61] on nutrition and health claims made on foods. The consumption of fiber is an important aspect in the diet of every human being due to many positive functions of fiber. Local reactions are related to their presence in the gastrointestinal tract and systemic reactions with an effect on metabolism. Viscosity, the ability to ferment, binding water, binding bile acids, reacting with metal ions, and increased stool weight are just a few of better understood effects of dietary fiber consumption Li and Komarek [62]. Based on Figure 6a,b, it can be concluded that the developed high-protein bars had a low content of saturated fatty acids. This is an important feature from the point of view of athletes and physically active people because saturated and trans fats may also make the lining of blood vessels (the endothelium) less flexible. In addition, trans fats may depress the "good" blood cholesterol (HDL cholesterol) when eaten in large quantities [63].

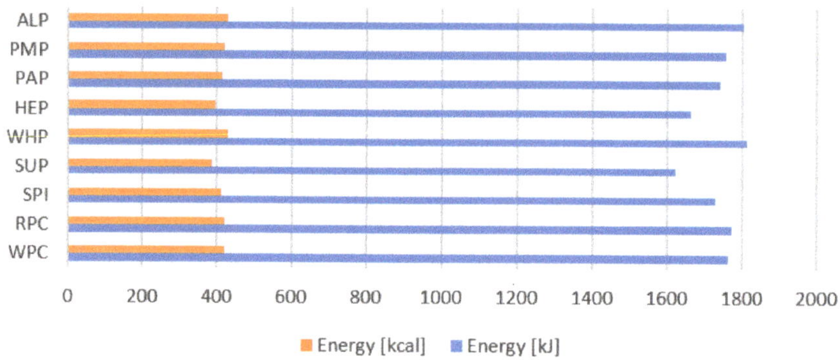

(a)

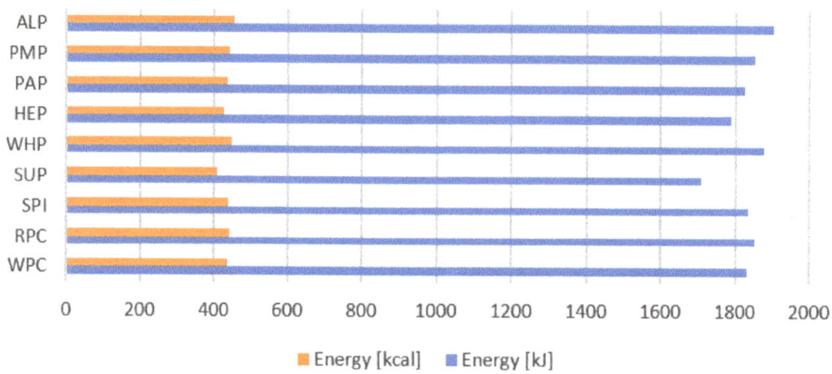

(b)

Figure 6. (**a**) Energy value of high-protein bars without chocolate coating; (**b**) With chocolate coating.

3.7. Sensory Evaluation

Results of the sensory evaluation of the tested high-protein bars are presented in Figure 8a,b. The highest scores during the analysis were obtained for bars made of WPC and PAP proteins. The evaluators appreciated the external appearance, color and taste sensations the most in the highest rated bars. High ratings for these types of proteins are associated with a pleasant consistency, taste and color, according to the judges. The worst rated bar (ALP) had too high hardness, an unpleasant aftertaste and a green-yellow color, which gave it the lowest scores. Negative assessments of the taste of the ALP bar may also be caused by too much dosing of this type of protein for a given application. Based on research by Hall et al. [64], in which the addition of algae protein in bread was 4%, it can be suspected that low taste ratings of bars made of algae proteins were probably caused by too high percentage in the final product (30%, *w/w*). In contrast, the research of Prabhasankar et al. [65] prove that the addition of algae protein in the amount >10%, *w/w* in pasta causes acceptable sensory features for those taking part in the sensory evaluation. Considering the above results, it can be concluded that the chocolate coating of high-protein products significantly increases their palatability. Ratings for individual chocolate bars were clearly higher than their non-chocolate counterparts. The remaining high-protein bars were rated at an average level with a tendency to be more positive. Referring to the research by Usha et al. [66], in which the effect of adding the pumpkin flour to baby food after weaning was examined, the ratings of products, in which addition was 20–30%, *w/w* pumpkin flour, were on the average level (on a nine-point scale) with a tendency to higher one, in particular if dosing at the level of 10%, *w*/*w*. It is worth mentioning that the worst rated parameter by the team was the taste of dishes, which can be correlated with the results obtained from the study of high-protein bars.

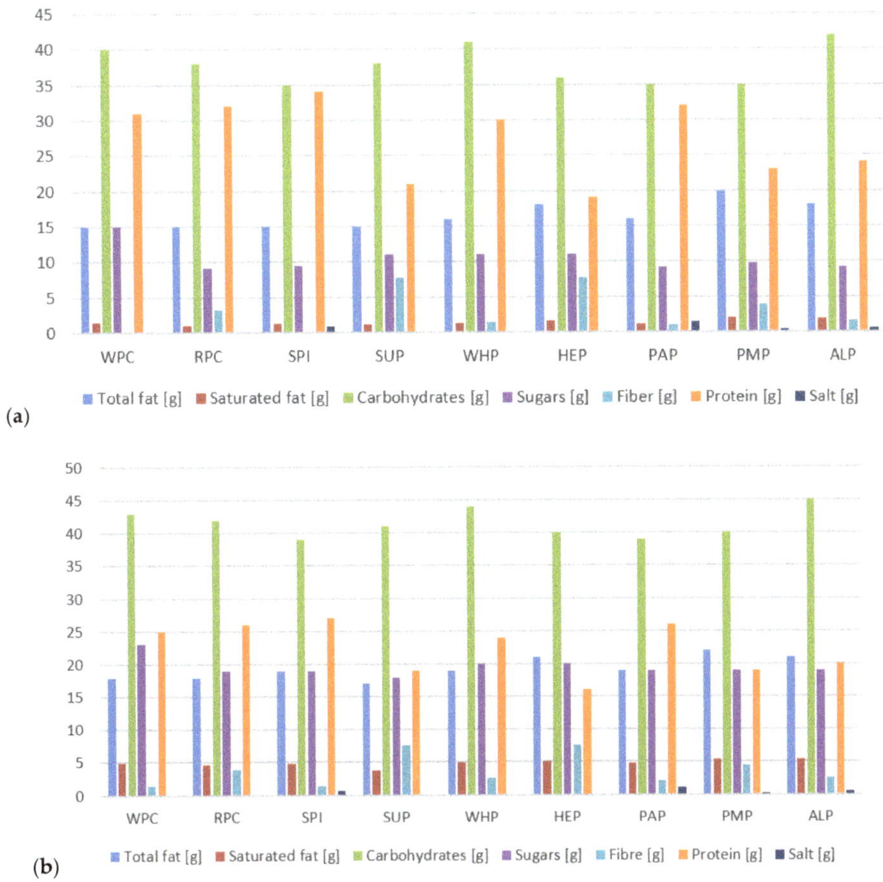

Figure 7. (**a**) Nutritional value of high-protein bars without chocolate coating; (**b**) With chocolate coating.

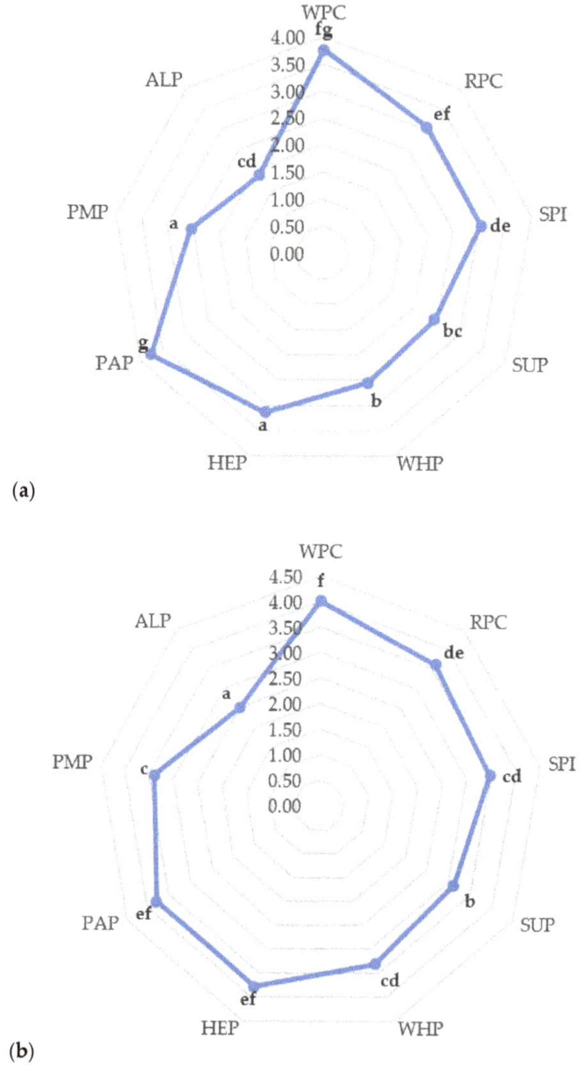

Figure 8. (**a**) Sensory evaluation of high-protein bars without chocolate coating; (**b**) With chocolate coating. [a–g] Means in the same column with different superscripts are significantly different ($p < 0.05$, Tukey's HSD test).

4. Conclusions

Based on the experiment, it can be concluded that changes in textural parameters, nutritional values or physicochemical parameters significantly depend on the type of protein used. Differences in the parameters of texture profile analysis (TPA) and cutting forces show significant differences, of which bars made of RPC, HEP, PAP, and PMP were characterized by the lowest results during the TPA test and ultrasonic viscosity. Differences in the microstructural structure of the tested bars significantly translated into the physical, chemical and textural features presented by proteins. It can be suspected that the microstructure of proteins has a significant effect on the water activity of high-protein bars. In particular, ALP proteins showed the lowest results of this parameter ($a_w < 0.55$), which ensures the microbiological stability of the final product, even during long storage periods. The obtained results of the amino acid content give the possibility that the SPI, RPC, and PAP proteins can compete with the WPC protein in terms of essential amino acid content. Bars made of RPC proteins had an increased content of heavy metals but these values did not exceed the acceptable EU standards. In terms of the nutritional value studied, all types of protein deserve attention due to the low content of saturated fat. Bars made of SUP and HEP proteins, apart from high protein content, also allowed for the declaration of high fiber content (fiber content above 6 g/100 g). Based on sensory analysis and the color assessment using the CVS method of bars without chocolate coating, it can be concluded that the color plays an important role for the consumer. Sensory analysis showed that the coating of high-protein bars with chocolate increases scores of the tested products, masking specific smells, color to a large extent, the aftertaste of some types of proteins, thus contributing to an increase in the overall sensory assessment. Therefore, it can be clearly defined that high-protein bars should be covered with chocolate. The conducted research shows that proteins of plant origin can be successfully used in the food industry as an alternative to WPC proteins, however it is not possible to clearly indicate which type of protein is the best option. Nevertheless, the high content of exogenous amino acids or the technological utility resulting from the texture parameters of bars made of RPC, SPI and PAP proteins speak for these sources. More research is needed on storage trials, microbiological tests, and examining the effect of changing other components in the recipe that may affect the parameters important for the food industry.

Author Contributions: Conceptualization, B.G.S. and J.M.; methodology, B.G.S., J.M., I.T., I.D.; software, B.G.S., J.M., I.T., I.D.; validation, B.G.S., J.M., I.T., I.D.; formal analysis, J.M.; investigation, J.M.; resources, B.G.S., J.M.; data curation, J.M.; writing—original draft preparation, J.M.; writing—review and editing, B.G.S., I.T., I.D.; visualization, J.M.; supervision, B.G.S., I.T., I.D.; final approval of the version to be published, B.G.S., J.M., I.T., I.D.; project administration, B.G.S.; funding acquisition, B.G.S., J.M. All authors have read and agreed to the published version of the manuscript.

Funding: This work was financially supported by the grant of the Ministry of Science and Higher Education in Poland (grant number 0029/DW/2018 "Implementation doctorate").

Acknowledgments: This research was financially supported by EUROHANSA Sp. z o. o. Thanks go to Bożena Czogała—President of EUROHANSA Sp. z o. o. for providing machines and equipment for the production of high-protein bars and to all suppliers for providing raw materials.

Conflicts of Interest: The authors declare no conflict of interest.

References

1. Goodbody, A. Raise the bar. *Min. Mag.* **2013**, 73–78. [CrossRef]
2. Clary, D.; Van de Bussche, T.; Janis, M.; Gossip, M.; Sloan, A.E. The top 10 functional food trends. The next generation. *Food Technol.* **2002**, *56*, 32–57. [CrossRef]
3. Riaz, M.N. *Soy Applications in Food*; CRC Press: Boca Raton, FL, USA, 2005.
4. Walia, A.; Gupta, A.K.; Sharma, V. Role of Bioactive Compounds in Human Health. *Acta Sci. Med. Sci.* **2019**, *3*, 25–33.
5. Potes, N.; Kerry, J.P.; Roos, Y.H. Protein Modifications in High Protein-Oil and Protein-Oil-Sugar Systems at Low Water Activity. *Food Biophys.* **2014**, *9*, 49–60. [CrossRef]
6. Hogan, S.A.; Chaurin, V.; O'Kennedy, B.T.; Kelly, P.M. Influence of dairy proteins on textural changes in high-protein bars. *Int. Dairy J.* **2012**, *26*, 58–65. [CrossRef]
7. Weeks, E.R. Soft jammed materials. *Stat. Phys. Complex Fluids* **2007**, *2*, 87.
8. Wang, Q.; Xiong, Y.L. Processing, Nutrition, and Functionality of Hempseed Protein: A Review. *Compr. Rev. Food Sci. Food Saf.* **2019**, *18*, 936–952. [CrossRef]
9. Sá, A.G.A.; Moreno, Y.M.F.; Carciofi, B.A.M. Plant proteins as high-quality nutritional source for human diet. *Trends Food Sci. Technol.* **2020**, *97*, 170–184. [CrossRef]
10. Ermiş, E.; Karasu, E.N. Püskürtmeli Kurutucu Ile Yaği Alinmiş Ayçiçeği Protein Ekstrakti Tozu Üretimi: Fonksiyonel Özellikleri Ve Toz Karakterizasyonu. *Gida/J. Food* **2020**, *45*, 39–49. [CrossRef]

11. Pihlanto, A.; Mattila, P.; Mäkinen, S.; Pajari, A.M. Bioactivities of alternative protein sources and their potential health benefits. *Food Funct.* **2017**, *8*, 3443–3458. [CrossRef]
12. Hassan, H.M.M.; Afify, A.S.; Basyiony, A.E.; Ahmed, G.T. Nutritional and functional properties of defatted wheat protein isolates. *Aust. J. Basic Appl. Sci.* **2010**, *4*, 348–358.
13. Wells, M.L.; Potin, P.; Craigie, J.S.; Raven, J.A.; Merchant, S.S.; Helliwell, K.E.; Smith, A.G.; Camire, M.E.; Brawley, S.H. Algae as nutritional and functional food sources: Revisiting our understanding. *J. Appl. Phycol.* **2017**, *29*, 949–982. [CrossRef] [PubMed]
14. Phongthai, S.; Homthawornchoo, W.; Rawdkuen, S. Preparation, properties and application of rice bran protein: A review. *Int. Food Res. J.* **2017**, *24*, 25–34.
15. Vinayashree, S.; Vasu, P. Biochemical, nutritional and functional properties of protein isolate and fractions from pumpkin (Cucurbita moschata var. Kashi Harit) seeds. *Food Chem.* **2020**, *340*, 128177. [CrossRef] [PubMed]
16. Rezig, L.; Riaublanc, A.; Chouaibi, M.; Guéguen, J.; Hamdi, S. Functional Properties of Protein Fractions Obtained from Pumpkin (Cucurbita Maxima) Seed. *Int. J. Food Prop.* **2016**, *19*, 172–186. [CrossRef]
17. Xu, Y.T.; Liu, L.L. Structural and Functional Properties of Soy Protein Isolates Modified by Soy Soluble Polysaccharides. *J. Agric. Food Chem.* **2016**, *64*, 7275–7284. [CrossRef]
18. Lam, A.C.Y.; Can Karaca, A.; Tyler, R.T.; Nickerson, M.T. Pea protein isolates: Structure, extraction, and functionality. *Food Rev. Int.* **2018**, *34*, 126–147. [CrossRef]
19. Hogenkamp, P.S.; Stafleu, A.; Mars, M.; Brunstrom, J.M.; de Graaf, C. Texture, not flavor, determines expected satiation of dairy products. *Appetite* **2011**, *57*, 635–641. [CrossRef]
20. De la Fuente, M.A.; Singh, H.; Hemar, Y. Recent advances in the characterisation of heat-induced aggregates and intermediates of whey proteins. *Trends Food Sci. Technol.* **2002**, *13*, 262–274. [CrossRef]
21. Tomasevic, I.; Tomovic, V.; Milovanovic, B.; Lorenzo, J.; Đorđević, V.; Karabasil, N.; Djekic, I. Comparison of a computer vision system vs. traditional colorimeter for color evaluation of meat products with various physical properties. *Meat Sci.* **2019**, *148*, 5–12. [CrossRef]
22. Milovanovic, B.; Djekic, I.; Djordjevic, V.; Tomovic, V.; Barba, F.; Tomasevic, I.; Lorenzo, J.M. Pros and cons of using a computer vision system for color evaluation of meat and meat products. *IOP Conf. Ser. Earth Environ. Sci.* **2019**, *333*, 012008. [CrossRef]
23. Inami, T.; Tanimoto, Y.; Minami, N.; Yamaguchi, M.; Kasai, K. Color stability of laboratory glass-fiber-reinforced plastics for esthetic orthodontic wires. *Korean J. Orthod.* **2015**. [CrossRef]
24. Balter, A.; Szubiakowski, J. Fluorescence probes of viscosity: A comparative study of the fluorescence anisotropy decay of perylene and 3,9-dibromoperylene in glycerol. *J. Fluoresc.* **1993**, *3*, 247–249. [CrossRef]
25. Galant, S. Formation of collagen-like structure of gel after radiation modification of gelatin. *Colloid Polym. Sci.* **1983**, *261*, 445–451. [CrossRef]
26. Sheen, S.H.; Chien, H.-T.; Raptis, A.C. An in-Line Ultrasonic Viscometer. In *Review of Progress in Quantitative Nondestructive Evaluation*; Springer Science & Business Media: Berlin, Germany, 1995.
27. Stone, H. *Sensory Evaluation Practices*; Academic Press: Cambridge, MA, USA, 2012; ISBN 9780123820860.
28. Kowalski, S.; Lukasiewicz, M.; Juszczak, L.; Sikora, M. Charakterystyka teksturalna i sensoryczna mas cukierniczych otrzymanych na bazie miodu naturalnego i wybranych hydrokoloidów polisacharydowych. *Żywność Nauk. Technol. Jakość* **2011**, *18*, 3.
29. Le Roux, L.; Ménard, O.; Chacon, R.; Dupont, D.; Jeantet, R.; Deglaire, A.; Nau, F. Are faba bean and pea proteins potential whey protein substitutes in infant formulas? An in vitro dynamic digestion approach. *Foods* **2020**, *9*, 362. [CrossRef]
30. Badzian, B. Poziom aminokwasów egzogennych w ścianie jelita cienkiego, wątrobie i osoczu krwi kurcząt po zastosowaniu witamin rozpuszczalnych w wodzie. *Acta Sci. Pol. Med. Vet.* **2007**, *6*, 3–14.
31. Novaković, S.; Tomašević, I. A comparison between Warner-Bratzler shear force measurement and texture profile analysis of meat and meat products: A review. In Proceedings of the 59th International Meat Industry Conference MEATCON2017, Zlatibor, Serbia, 1–4 October 2017; Volume 85. [CrossRef]
32. Li, Y.; Szlachetka, K.; Chen, P.; Lin, X.; Ruan, R. Ingredient characterization and hardening of high-protein food bars: An NMR state diagram approach. *Cereal Chem.* **2008**, *85*, 780–786. [CrossRef]
33. Banach, J.C.; Clark, S.; Lamsal, B.P. Particle size of milk protein concentrate powder affects the texture of high-protein nutrition bars during storage. *J. Food Sci.* **2017**, *82*, 913–921. [CrossRef]

34. Trinh, T.; Glasgow, S. On the texture profile analysis test. In Proceedings of the Chemeca 2012: Quality of life through chemical engineering, Wellington, New Zealand, 23–26 September 2012; Available online: https://www.researchgate.net/publication/316093466_On_the_texture_profile_analysis_test (accessed on 9 October 2020).
35. Banach, J.C.; Clark, S.; Lamsal, B.P. Instrumental and Sensory Texture Attributes of High-Protein Nutrition Bars Formulated with Extruded Milk Protein Concentrate. *J. Food Sci.* **2016**, *81*, S1254–S1262. [CrossRef]
36. Cho, M.J. Soy Protein Functionality and Food Bar Texture. In *Chemistry, Texture, and Flavor of Soy*; American Chemical Society: Washington, DC, USA, 2010; ISBN 9780841225619.
37. Debaste, F.; Kegelaers, Y.; Liégeois, S.; Amor, H.B.; Halloin, V. Contribution to the modelling of chocolate tempering process. *J. Food Eng.* **2008**, *88*, 568–575. [CrossRef]
38. Quiñones-Muñoz, T.; Gallegos-Infante, J.A.; Rocha-Guzmán, N.E.; Ochoa-Martinez, L.A.; Morales-Castro, J.; González-Laredo, R.F.; Medina-Torres, L. Mixing and tempering effect on the rheological and particle size properties of dark chocolate coatings. *CYTA J. Food* **2011**, *9*, 109–113. [CrossRef]
39. Labuza, T.P.; Hyman, C.R. Moisture migration and control in multi-domain foods. *Trends Food Sci. Technol.* **1998**, *9*, 47–55. [CrossRef]
40. Hogan, S.A.; Famelart, M.H.; O'Callaghan, D.J.; Schuck, P. A novel technique for determining glass-rubber transition in dairy powders. *J. Food Eng.* **2010**, *99*, 76–82. [CrossRef]
41. Bleakley, S.; Hayes, M. Algal Proteins: Extraction, Application, and Challenges Concerning Production. *Foods* **2017**, *6*, 33. [CrossRef]
42. Purwanti, N.; van der Goot, A.J.; Boom, R.; Vereijken, J. New directions towards structure formation and stability of protein-rich foods from globular proteins. *Trends Food Sci. Technol.* **2010**, *21*, 85–94. [CrossRef]
43. Banach, J.C.; Clark, S.; Lamsal, B.P. Texture and other changes during storage in model high-protein nutrition bars formulated with modified milk protein concentrates. *LWT Food Sci. Technol.* **2014**, *56*, 77–86. [CrossRef]
44. Loveday, S.M.; Hindmarsh, J.P.; Creamer, L.K.; Singh, H. Physicochemical changes in a model protein bar during storage. *Food Res. Int.* **2009**, *42*, 798–806. [CrossRef]
45. Zhou, P.; Liu, X.; Labuza, T.P. Moisture-induced aggregation of whey proteins in a protein/buffer model system. *J. Agric. Food Chem.* **2008**, *56*, 2048–2054. [CrossRef]
46. Zhou, P.; Liu, X.; Labuza, T.P. Effects of moisture-induced whey protein aggregation on protein conformation, the state of water molecules, and the microstructure and texture of high-protein-containing matrix. *J. Agric. Food Chem.* **2008**, *56*, 4534–4540. [CrossRef]
47. Ren, J.; Sun, X.H.; Lin, G.P.; Zheng, X.Q.; Liu, X.L. Isolation and Characterization of Sunflower Protein Isolates and Sunflower Globulins. In *Information Technology and Agricultural Engineering*; Springer: Berlin/Heidelberg, Germany, 2012.
48. Tomasevic, I.; Tomovic, V.; Ikonic, P.; Lorenzo Rodriguez, J.M.; Barba, F.J.; Djekic, I.; Nastasijevic, I.; Stajic, S.; Zivkovic, D. Evaluation of poultry meat colour using computer vision system and colourimeter: Is there a difference? *Br. Food J.* **2019**, *121*, 1078–1087. [CrossRef]
49. Tomasevic, I.; Tomovic, V.; Barba, F.J.; Vasilev, D.; Jokanovic, M.; Sojic, B.; Manuel Lorenzo, J.; Djekic, I. How the color of game meat should be measured. *Fleischwirtschaft* **2019**, *99*, 85–89.
50. Hassan, S.K. Quantitative and qualitative effects of proteins and natural sugars on hardening and color of high-protein nutrition bars during storage. *EurAsian J. Biosci.* **2020**, *14*, 915–932.
51. Mcmahon, D.J.; Adams, S.L.; Mcmanus, W.R. Hardening of high-protein nutrition bars and sugar/polyol-protein phase separation. *J. Food Sci.* **2009**, *74*, E312–E321. [CrossRef]
52. Tomczyńska-Mleko, M.; Ozimek, L. Ultrasound viscosity measurements allow determination of gas volume fraction in foamed gels. *J. Food Process Eng.* **2013**, *36*, 572–578. [CrossRef]
53. The Commission of the European. EU Commision Regulation (EC) No 1881/2006. Setting maximum levels for certain contaminants in foodstuffs. *Off. J. Eur. Union* **2006**, *364*, 5–24.
54. Huang, D.; Liu, L.; Zeng, G.; Xu, P.; Huang, C.; Deng, L.; Wang, R.; Wan, J. The effects of rice straw biochar on indigenous microbial community and enzymes activity in heavy metal-contaminated sediment. *Chemosphere* **2017**, *174*, 545–553. [CrossRef]
55. Kaneta, M.; Hikichi, H.; Endo, S.; Sugiyama, N. Chemical form of cadmium (and other heavy metals) in rice and wheat plants. *Environ. Health Perspect.* **1986**, *65*, 33–37. [CrossRef]
56. Kowieska, A.; Lubowicki, R.; Jaskowska, I. Chemical composition and nutritional characteristics of several cereal grain. *Acta Sci. Pol. Zootech.* **2011**, *10*, 2.

57. Szpak, P. Fish bone chemistry and ultrastructure: Implications for taphonomy and stable isotope analysis. *J. Archaeol. Sci.* **2011**, *38*, 3358–3372. [CrossRef]
58. Jennifer, J.O.; Jennifer Pitzi Hellwig, L.D.M. *Dietary Reference Intakes: The Essential Guide to Nutrient Requirements*; National Academies Press: Washington, DC, USA, 2006; ISBN 9780309100915.
59. Strozyk, A.K.; Pachocka, L. The role of nuts consumption in the primary and secondary prevention of type 2 diabetes. *Clin. Diabetol.* **2017**, *6*, 26–33. [CrossRef]
60. Kalman, D. Amino Acid Composition of an Organic Brown Rice Protein Concentrate and Isolate Compared to Soy and Whey Concentrates and Isolates. *Foods* **2014**, *3*, 394–402. [CrossRef] [PubMed]
61. The European Parliment and the Council of the European Union Regulation (EC) No 1924/2006. *Off. J. Eur. Union* **2011**. Available online: https://esinvesticijos.lt/uploads/main/documents/files/20111018_general_proposal_en.pdf (accessed on 9 October 2020).
62. Li, Y.O.; Komarek, A.R. Dietary fibre basics: Health, nutrition, analysis, and applications. *Food Qual. Saf.* **2017**, *1*, 47–59. [CrossRef]
63. Gershuni, V.M. Saturated Fat: Part of a Healthy Diet. *Curr. Nutr. Rep.* **2018**, *7*, 85–96. [CrossRef]
64. Hall, A.C.; Fairclough, A.C.; Mahadevan, K.; Paxman, J.R. Ascophyllum nodosum enriched bread reduces subsequent energy intake with no effect on post-prandial glucose and cholesterol in healthy, overweight males. A pilot study. *Appetite* **2012**, *58*, 379–386. [CrossRef]
65. Prabhasankar, P.; Ganesan, P.; Bhaskar, N.; Hirose, A.; Stephen, N.; Gowda, L.R.; Hosokawa, M.; Miyashita, K. Edible Japanese seaweed, wakame (Undaria pinnatifida) as an ingredient in pasta: Chemical, functional and structural evaluation. *Food Chem.* **2009**, *115*, 501–508. [CrossRef]
66. Usha, R.; Lakshmi, M.; Ranjani, M. Nutritional, sensory and physical analysis of pumpkin flour incorporated into weaning mix. *Malays. J. Nutr.* **2010**, *16*, 379–387.

Publisher's Note: MDPI stays neutral with regard to jurisdictional claims in published maps and institutional affiliations.

© 2020 by the authors. Licensee MDPI, Basel, Switzerland. This article is an open access article distributed under the terms and conditions of the Creative Commons Attribution (CC BY) license (http://creativecommons.org/licenses/by/4.0/).

foods

Article

Comparative Evaluation of Chemical Composition, Phenolic Compounds, and Antioxidant and Antimicrobial Activities of Tropical Black Bolete Mushroom Using Different Preservation Methods

Jaturong Kumla [1,2,*], Nakarin Suwannarach [1,2], Keerati Tanruean [3] and Saisamorn Lumyong [1,2,4]

1 Department of Biology, Faculty of Science, Chiang Mai University, Chiang Mai 50200, Thailand; suwan.462@gmail.com (N.S.); scboi009@gmail.com (S.L.)
2 Research Center of Microbial Diversity and Sustainable Utilization, Chiang Mai University, Chiang Mai 50200, Thailand
3 Biology Program, Faculty of Science and Technology, Pibulsongkram Rajabhat University, Phitsanulok 65000, Thailand; keerati.t@psru.ac.th
4 Academy of Science, The Royal Society of Thailand, Bangkok 10300, Thailand
* Correspondence: Jaturong_yai@hotmail.com or Jaturong.kumla@gmail.com; Tel.: +66-8-7192-6527

Citation: Kumla, J.; Suwannarach, N.; Tanruean, K.; Lumyong, S. Comparative Evaluation of Chemical Composition, Phenolic Compounds, and Antioxidant and Antimicrobial Activities of Tropical Black Bolete Mushroom Using Different Preservation Methods. *Foods* **2021**, *10*, 781. https://doi.org/10.3390/foods10040781

Academic Editors: Danijela Bursać Kovačević and Predrag Putnik

Received: 28 February 2021
Accepted: 2 April 2021
Published: 5 April 2021

Publisher's Note: MDPI stays neutral with regard to jurisdictional claims in published maps and institutional affiliations.

Copyright: © 2021 by the authors. Licensee MDPI, Basel, Switzerland. This article is an open access article distributed under the terms and conditions of the Creative Commons Attribution (CC BY) license (https://creativecommons.org/licenses/by/4.0/).

Abstract: Tropical black bolete, *Phlebopus portentosus*, provides various nutritional benefits and natural antioxidants to humans. In this study, the chemical composition, phenolic compounds, and antioxidant and antimicrobial activities of fresh mushroom samples and samples stored for a period of one year using different preservation methods (drying, brining, and frozen) were investigated. The results indicated that the brining method significantly reduced the protein and fat contents of the mushrooms. The polyphenol and flavonoid contents of the frozen sample were not significantly different from that of the fresh sample. The results revealed that an inhibition value of 50% (IC_{50}) for the 2,2-diphenyl-1-picrylhydrazyl (DPPH) assay of the extract of the dried and frozen samples was not statistically different from that of the fresh sample. The IC_{50} value of 2,2′-azino-bis(3-ethylbenzothiazoline-6-sulfonic acid (ABTS) assay and ferric reducing antioxidant power (FRAP) value in the extract of the frozen sample were not found to be significantly different from those of the fresh sample. Furthermore, the lowest degree of antioxidant activity was found in the extract of the brined sample. Additionally, the antimicrobial activities of the extracts of the fresh and frozen samples were not significantly different and both extracts could have inhibited the growth of all tested Gram-positive bacteria and *Pseudomonas aeruginosa*.

Keywords: antioxidant properties; edible mushroom; nutrition; phenolic compounds; preservation

1. Introduction

Varieties of certain cultivated edible mushroom genera, such as *Agaricus*, *Auricularia*, *Lentinus*, *Lentinula*, *Pleurotus*, *Tramella*, and *Volvariella*, account for the majority of the mushrooms traditionally consume by humans [1–3]. However, wild mushrooms have recently become increasingly consumed in our diets due to the recognition of their nutritional and pharmacological characteristics [4–6]. Ectomycorrhizal (ECM) mushrooms are a group of wild mushrooms, of which about 2500 recorded species have been defined as edible [7,8]. ECM mushrooms are a good source of essential dietary nutrients, such as carbohydrates, fiber, protein, minerals, and vitamins, all of which make them a valuable food resource for humans [9–11]. Apart from the use of ECM mushrooms as a source of food, they can be important in medicine due to their antibiotic, anticancer, antidiabetic, antioxidant, and increased immune response effects [5,12–15]. Currently, the consumption of edible ECM mushrooms has been increasing and this is mainly attributed to their elevated market value and an increase in consumer demand. Some of them, such as chanterelles (*Cantharellus* spp.), king boletes (*Botetus edulis*), matsutakes (*Tricholoma matsutake*), morels (*Morchella* spp.), and truffles (*Tuber* spp.), are among the world's most expensive foods [7,8,16,17]. Generally, ECM mushrooms are a seasonal and highly perishable crop. After harvesting, various physiological and morphological characteristics of ECM mushrooms change, which makes them unacceptable for human consumption and limits their ability to be effectively marketed. The shelf-life of several ECM mushrooms is only about 2–5 days depending upon the mushroom species [18,19]. Fresh mushrooms are best stored unwashed in brown paper bags in the refrigerator, preferably on the lowest shelf. However, many methods have been employed to improve the preservation of these mushrooms and enhance their shelf-life. The most common methods include brining, canning, drying, freezing, and radiation [20,21]. For the purposes of marketing, popular edible ECM mushrooms, known as chanterelles (*Cantharellus* spp.), king boletes (*B. edulis*), morels (*Morchella* spp.), and truffles (*Tuber* spp.), are commonly preserved through the different processes [22–24].

Thailand is home to a great diversity of wild edible mushroom species that are most abundant during the wet season (mid-May to October) of each year [9,25]. Normally, local farmers collect edible wild ECM mushrooms for both consumption and sale in local, roadside, or city markets. Preliminary investigations at small local markets have revealed the availability of many genera of wild, edible ECM mushrooms such as *Amanita, Astraeus, Boletus, Cantharellus, Lactarius, Phlebopus,* and *Russula* [9,16,26,27]. The tropical black bolete (TBB), *Phlebopus portentous*, is a wild edible ECM mushroom that is known to be distributed throughout Asia, Australasia, and Mexico [9,16,28–31]. This mushroom is widely consumed in southern China, Laos, Myanmar, and northern and northeastern Thailand. In northern Thailand, this mushroom begins producing fruiting bodies at the end of the hot season and does so until the early wet season (May–July). On occasion, a second flush occurs at the end of the wet season (October). The market price for this mushroom is high and it is usually offered at 150–300 Baht/kg (5–10 USD/kg) [9,16]. This bolete produces large fruiting bodies and its texture is similar or even better than that of the king bolete, which is an important edible ECM mushroom in the European market [16,32]. This mushroom variety is known to contain high amounts of protein and is also known to possess a range of medicinal properties [9,13]. The study aimed to investigate the chemical composition, phenolic compounds, and the antioxidant and antimicrobial activities of TBB samples that had been stored for a one-year period using different preservation methods (drying, brining, and freezing). Generally, the preserved mushrooms will have been stored for approximately 2–3 years depending upon the preservation method [18–21]. However, it has been recommended that they should be consumed no later than one year from the sell date on the package. Thus, we selected a storage period of one year for this study. The results of this study will provide valuable information on the nutrient contents, phenolic compounds, and antioxidant and antimicrobial potential of this mushroom through the use of different preservation methods. The results could also be used to further enhance relevant strategies for potential commercial development.

2. Materials and Methods

2.1. Source of Mushroom and Sample Preparation

Fresh fruiting bodies of TBB were bought from local markets at Mae Ta District, Lamphun Province, northern Thailand in 2018 (Figure 1). The fruiting bodies of TBB were characterized by caps: convex to subconvex, centrally depressed, 3.0–15 cm in diameter, olive brown to dark brown; stipes: 3.5–7.0 × 7.0–10 cm, clavate or tapering to apex, concolored with cap; tubes under cap: yellow to yellowish-orange; contexts: soft, yellow (Figure 1). The fresh fruiting bodies were transferred to a laboratory within 24 h of being harvested, and the soil was removed. Mushroom sample was prepared following the method described by Tolera and Abera [33] with some modifications. Whole fruiting bodies were washed under running tap water and drained. The cleaned fruiting bodies were directly sliced into pieces of 4–5 mm thickness using a stainless steel knife. Sliced samples were then used in each preservative treatment.

Figure 1. Local market location and fresh fruiting bodies of the tropical black bolete. Scale bar = 5 cm.

2.2. Preservation Method

2.2.1. Drying

Briefly, the sliced fresh mushroom samples were loaded onto trays and placed in a commercial food dryer at a temperature of 45 °C for 72 h (completely dried). The moisture content was then recorded. The samples were placed in zip-lock plastic bags and kept at room temperature (28 ± 2 °C) in the darkness.

2.2.2. Brining

Sliced fresh mushroom samples were boiled in hot water (95 to 99 °C) for 5 min for the purposes of blanching, according to the method described by Lungu et al. [34]. They were then drained and placed immediately in cold water to cool. Subsequently, 100 g of the boiled product was placed in each 300 mL glass bottle. The brine solution (2.5% salt concentration) was then boiled. After cooling, the brine solution was added to glass bottles that contained the sliced mushrooms until the mushrooms were completely covered. A layer of rice bran oil was added, and they were then steamed for one hour. After cooling, the glass bottles were kept at room temperature in the darkness.

2.2.3. Freezing

The sliced fresh mushroom samples (20 pieces) were placed in each commercial zip-lock plastic bag (25 × 35 cm) and frozen at −20 °C in a freezer.

2.3. Proximate Composition Analysis

The proximate composition of each mushroom sample was identified as carbohydrate, protein, fat, ash, and fiber according to the Association of Official Analytical Chemists (AOAC) method [35] at Central Laboratory (Chiang Mai, Thailand) Company Limited (Chiang Mai, Thailand). The analysis was performed on the samples of each treatment in five replications. The obtained data were compared with the results of the fresh sample.

2.4. Preparation of Mushroom Extracts

Fresh samples were oven-dried in an oven at 45 °C for 72 h before being used. Additionally, all one-year storage samples, except the dried sample, were also oven-dried at 45 °C for 72 h. Each dried sample was ground using a Waring blender (New Hartford, CT, USA). Ground mushroom samples (20 g) were extracted with 200 mL of methanol at

25 °C and at 150 rpm for 24 h following the method described by Kaewnarin et al. [13]. The extracts were then sonicated using a Crest Ultrasonicator (Ewing Township, NJ, USA) and filtered through Whatman's No. 1 filter paper. The residue was then re-extracted twice with methanol as has been described above. The methanol extract was then rotary evaporated at 40 °C to dryness. The extracts were stored at −20 °C for further use.

2.5. *Determination of Phenolic Compounds*

2.5.1. Total Polyphenol Content

The total polyphenol content was estimated using the protocol previously described by Thitilertdecha et al. [36] with some modifications. The mushroom extract of each sample at 0.25 mL was mixed with 2.5 mL deionized water and 0.5 mL of Folin−Ciocalteu reagent. After 5 min, 0.5 mL of Na_2CO_3 (20% *w*/*v*) was added. The reaction mixture was then incubated at room temperature for one hour. Analysis of total polyphenol content was carried out by measurement of the absorbance at 760 nm. The standard curve of gallic acid was used to calculate the total polyphenol content of the samples (25–100 µg/mL; y = 0.006x; R^2 = 0.997). The results were expressed as milligrams of gallic acid equivalent per gram of dry weight (mg GAE/g dw). Five replications were performed for the samples of each treatment.

2.5.2. Total Flavonoid Content

Total flavonoid content was determined according the protocol established by Kaewnarin et al. [37], and quercetin was used as the standard flavonoid. Next, 0.5 mL of the extract was mixed with 2 mL of distilled water, and then 0.15 mL of $NaNO_2$ (50 g/L) was added. After 5 min, 0.15 mL of $AlCl_3$ (100 g/L) was added. The reaction was mixed and incubated at room temperature for 15 min. Absorbance was measured at 415 nm. The standard curve of quercetin was then used to calculate the total flavonoid content in the samples (25–300 µg/mL; y = 0.004x; R^2 = 0.996). The total flavonoid contents were expressed as milligrams quercetin equivalent per gram of dry weight (mg QE/g dw). The data were presented as the average of five replications of each treatment.

2.6. *Antioxidant Assay*

2.6.1. DPPH Scavenging Assay

The DPPH (2,2-diphenyl-1-picrylhydrazyl) scavenging ability was determined according to the method described by Gülçin et al. [38]. Initially, 0.5 mL of different concentrations of each extract sample was mixed with 1.5 mL of the 0.1 mM DPPH solution in methanol. A control mixture was prepared that consisted of 0.5 mL of methanol and 1.5 mL of the DPPH solution. The mixtures were incubated at room temperature for 30 min in the dark. Subsequently, the absorbance was measured at 517 nm. The percentage discoloration of DPPH radical of the samples was then calculated according to the following formula: percentage inhibition = $(A_o - A_s/A_o) \times 100$, where A_o represents the absorbance of the control and A_s represents the absorbance of the mixture containing the extract sample. An inhibition concentration of 50% (IC_{50}) was expressed as a calculation from the plot of percentage inhibition against the extract concentration. Samples of each treatment were analyzed in five replications.

2.6.2. ABTS Scavenging Assay

ABTS (2,2′-azino-bis(3-ethylbenzothiazoline-6-sulfonic acid) scavenging activity was determined according to the method described by Re et al. [39] with some modifications. The solution of ABTS cation chromophore was prepared by facilitating a reaction between 100 mL of the 7.0 mM ABTS solution and 100 mL of 2.45 mM $K_2S_2O_8$. The solution was kept in darkness at 25 °C for 16 h. The ABTS solution was adjusted to an absorbance value of 0.70 ± 0.2 at 734 nm by dilution with the phosphate buffer (50 mM, pH 7.4) before being used. Next, 2.9 mL of the ABTS solution was mixed with 0.1 mL of different concentrations of each extract sample. A mixture composed of the ABTS solution and methanol was used as the control. The mixtures were incubated at room temperature for 30 min in the dark. The absorbance value of the mixture was measured at 734 nm. The percentage of discoloration of the ABTS radical was calculated according to the formula previously described in the DPPH assay, as has been described above. The IC_{50} value was then calculated, and five replications were completed for the samples of each treatment.

2.6.3. FRAP Assay

FRAP (ferric reducing antioxidant power) assay was performed according to the protocol described by Li et al. [40]. The FRAP reagent was prepared using a mixture containing 10 mM 2,4,6-tripyridyl-s-triazine solution in 20 mL of 40 mM HCl, 20 mL of 20 mM ferric (III) chloride, and 5 mL of 300 mM acetate buffer (pH 3.6). Each 0.1 mL of extract was mixed with 1.5 mL of FRAP reagent and 1.4 mL of acetate buffer (300 mM, pH 3.6). The mixture was then incubated at room temperatures for 30 min in the dark. The absorbance at 593 nm was measured. Gallic acid was used to calculate the standard curve (10–100 µg/mL; $y = 0.014x$; $R^2 = 0.995$), and the FRAP value was calculated as mg GAE/g dw. The data were presented as the average of five replications of each treatment.

2.7. Antimicrobial Assay

2.7.1. Microorganisms

The microorganisms used in this study consisted of 8 strains of Gram-positive bacteria (*Bacillus cereus*, *B. subtilis*, *Enterococcus faecalis* ATCC29212, *Listeria monocytogenes*, *Micrococcus luteus*, Methicillin-resistant *Staphylococcus aureus*, *Sta. aureus* ATCC29213, and *Streptococcus pneumoniae* ATCC49699), 10 strains of Gram-negative bacteria (*Escherichia coli* ATCC35218, *E. coli* ATCC25922, *E. coli* O157:H7, *Klebsiella pneumonia*, *Proteus mirabilis*, *Pr. vulgaris*, *Pseudomonas aeruginosa* ATCC27859, *Ps. fluorescens*, *Salmonella typhi*, and *Salmonella* sp. group D), and 2 yeast strains (*Candida albicans* and *Cryptococcus neoformans*). All tested microorganisms were obtained from the Sustainable Development of Biological Resources Laboratory, Department of Biology, Faculty of Science and the Central and Diagnostic Laboratory, Maharaj Nakorn Chiang Mai Hospital, Faculty of Medicine Chiang Mai University, Thailand. Bacteria and yeast samples were grown and maintained on nutrient agar (NA) and yeast extract peptone dextrose agar (YPDA) slants, respectively. The inoculated slants were incubated at 37 °C.

2.7.2. Determination of Antimicrobial Activity

The antimicrobial activity of the mushroom extracts was determined by paper disc diffusion assay [41,42]. Bacteria and yeast samples were cultured in a nutrient broth at 37 °C and yeast extract peptone dextrose broth at 30 °C, respectively. Their cultivation was then achieved on an orbital shaker at 125 rpm for 24 h. The cell density of both test microorganisms was then adjusted to 10^8 CFU/mL that corresponded to the 0.5 McFarland standard. All prepared suspensions of the test microorganisms were swabbed on their respective agar media including NA (for the bacteria test) and YPDA (for the yeast test). Sterile paper discs (8 mm in diameter) were impregnated with 40 µL of each mushroom extract at a concentration of 100 mg/mL. The discs were allowed to dry and then placed on the agar surface of the tested plate. Discs with the methanol were used as the negative control, while standard antibiotics such as ampicillin, streptomycin, and nystatin at concentrations of 10 mg/mL were used as positive controls. The plates were incubated at 37 °C for 24 h. After incubation, the zone of inhibition appeared around the discs and was measured and recorded. Five replications were made for each sample extract obtained from each treatment.

2.8. Statistical Analysis

The data collected were subjected to one-way analysis of variance (ANOVA) by SPSS program version 16.0 for Windows. Tukey's test was used to determine significant differences ($p < 0.05$) between the mean values.

3. Results and Discussion

3.1. Characteristics of Mushroom Samples

Fresh TBB samples contained a moisture content of 82.7 ± 2.0% on a wet-weight basis. Our results agreed with the findings of previous studies that found that the moisture content in fresh bolete mushrooms ranged from 70% to 93% depending upon the bolete species [43–47]. To extend the availability and shelf-life of TBB, a preservation method is recommended. Methods of drying, brining, and freezing were used in this study. Fresh samples and preserved samples after one year of storage were observed, and the pictures depicting the different treatments are shown in Figure 2. Based on a visual observation, the colors of the dried, frozen, and brined samples were cream, light brown, and brown, respectively. It was found that the color of the frozen sample was similar to that of the fresh sample. Additionally, the frozen sample also had a pronounced flavor that was similar to that of the fresh sample. However, both the dried and brined samples revealed a slight change in flavor. These results are in accordance with those of prior studies that reported that different preservation methods can lead to flavor and color changes in mushrooms [20,43,48].

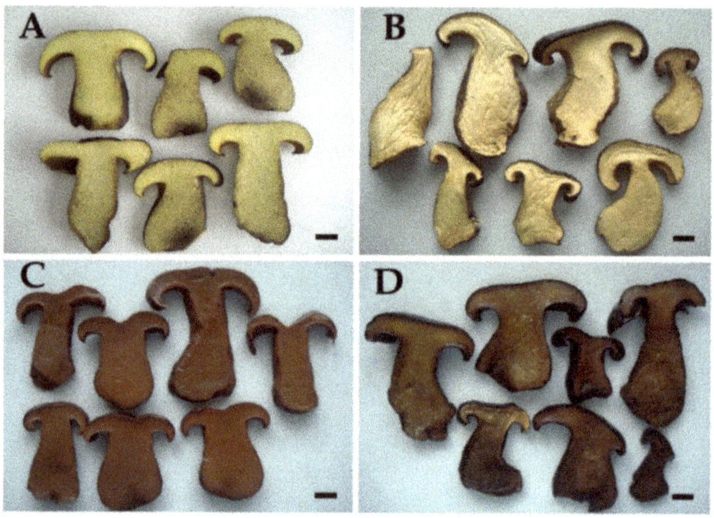

Figure 2. Fresh and one-year storage of tropical black bolete mushroom samples. (**A**) Fresh samples. (**B**) Dried samples. (**C**) Brined samples. (**D**) Frozen samples. Scale bar = 1 cm.

3.2. Proximate Composition

The amounts of ash, carbohydrate, protein, fat, and fiber of the different mushroom samples are shown in Table 1, all of which were within the ranges described in previous reports with regard to the ash (6.7%–27.6%), carbohydrate (33.3%–65.1%), protein (14.0%–36.3%), fat (0.4%–9.5%), and fiber (3.1%–14.7%) contents found in several edible ECM mushrooms including boletes [9–11,42,49–51]. It was found that the drying, brining, and freezing methods did not affect the ash, carbohydrate, and fiber contents of the samples. Notably, the drying and freezing methods did not cause a significant change in protein content when compared with the fresh sample. However, the lowest protein and fat contents were found in the brined sample. This result is in full agreement with the results of previous studies that indicated that the brining method decreased the protein and fat contents of the mushroom [20,34,48]. Pagoń et al. [52] found that the brining method significantly decreased the protein content in boletes (*B. edulis* and *Suillus luteus*) after storage at room temperature for four weeks. Previous studies have reported that lower protein and fat contents in the brined samples can be explained by the much greater amounts of fat, protein, and amino acid solubilizations that exist as a result of employing the brining method [34,48,53,54].

Table 1. Proximate composition on a dry basis (% dry weight) of different tropical black bolete samples.

Sample Type	% Dry Weight					
	Moisture	Ash	Crude Protein	Fat	Fibre	Carbohydrate
Fresh sample	7.2 ± 0.1 a	9.6 ± 0.2 a	19.6 ± 0.4 a	1.0 ± 0.1 a	6.3 ± 0.1 a	54.8 ± 0.7 a
Dried sample	7.1 ± 0.2 a	9.7 ± 0.2 a	19.3 ± 0.4 a	1.0 ± 0.2 a	6.2 ± 0.1 a	54.7 ± 0.6 a
Brined sample	7.0 ± 0.1 a	9.4 ± 0.1 a	17.1 ± 0.8 b	0.7 ± 0.1 b	6.2 ± 0.2 a	53.4 ± 0.5 a
Frozen sample	7.1 ± 0.2 a	9.6 ± 0.1 a	19.4 ± 0.3 a	0.9 ± 0.1 a	6.3 ± 0.2 a	54.6 ± 0.4 a

The results are mean ± standard deviation. Different letters in the same column are considered significantly different according to Tukey's test ($p < 0.05$).

3.3. Determination of Phenolic Compounds

Total polyphenol and flavonoid contents of all methanol extracts of TBB samples are shown in Table 2. The total polyphenol and flavonoid contents in the fresh samples were recorded at 30.10 ± 1.04 mg GAE/g dw and 1.69 ± 0.71 mg QE/g dw, respectively. The polyphenol content obtained in this study was within the range of the polyphenol (1.44–38.44 mg GAE/g dw) and flavonoid (0.03–2.54 mg QE/g dw) contents of other edible ECM mushrooms recorded in previous reports [12–14,55,56], depending upon the mushroom species. It was found that the total polyphenol and flavonoid contents in the preserved samples varied according to different preservation methods. A significant decrease in the polyphenol and flavonoid contents was observed in both the dried and brined samples. The lowest total polyphenol (18.44 ± 0.54 mg GAE/g dw) and flavonoid (0.37 ± 0.31 mg QE/g dw) values were observed in the brined samples. These results were supported by the findings of previous studies that reported that the decrease in the phenolic compound contents in the dried and brined samples could be explained by the application of thermal treatment via boiling, canning, drying, and blanching. These thermal treatments increased the extractability of both the phenolic compounds and the phenolic compounds that were possibly degraded during storage as a result of sensitivity oxidation and solubilization [34,57,58]. Ganguli et al. [59] and Pagoń et al. [52] reported that the loss of total polyphenol contents in white bottom mushrooms (*Agaricus bisporus*) and boletes (*B. edulis* and *Sui. luteus*) were due to the blanching and brining processes, respectively. Nevertheless, the total polyphenol and flavonoid contents in the frozen samples examined in this study were not found to be significantly different when compared with the fresh sample. This result was in full agreement with that the findings presented in previous studies that reported that the freezing process can cause minimal amounts of destruction of the polyphenol and flavonoid compounds in fruits, some vegetables, and mushrooms [57,59]. In addition, the oven-dry process (45 °C for 72 h) used for preparation of the dried samples obtained from the brined and frozen samples may have affected their phenolic compound composition, as well as the total polyphenol and flavonoid compounds. Although the measurement of total polyphenol and flavonoid compounds was carried out using several methods in this study, other techniques such as high-performance liquid chromatography and mass spectrometry should be applied to account for changes to the phenolic profile in each sample that occurred during storage.

Table 2. Total polyphenol and flavonoid contents in the methanolic extracts obtained from different tropical black bolete samples.

Mushroom Extract	Total Polyphenol Content (mg GAE/g dw)	Total Flavonoid Content (mg QE/g dw)
Fresh sample	30.10 ± 1.04 a	1.69 ± 0.17 a
Dried sample	26.70 ± 0.63 b	1.32 ± 0.26 b
Brined sample	18.44 ± 0.54 c	0.37 ± 0.31 c
Frozen sample	29.06 ± 1.20 a	1.63 ± 0.24 a

The results are mean ± standard deviation. Different letters in the same column are considered significantly different according to Tukey's test ($p < 0.05$).

3.4. Antioxidant Assay

Several edible ECM mushroom genera, namely, *Amanita, Astraeus, Boletus, Cantharellus, Lactarius, Leccinum, Phlebopus, Russula, Tricholoma, Suillus,* and *Xerocomus*, have been reported to contain higher amounts of secondary metabolites, which are indicative of multiple biological effects, including the antioxidant activity that can act according to different mechanisms of action for each mushroom species [12–14,47,56,60]. Thus, a single method cannot fully estimate the antioxidant capacity of these mushrooms. Therefore, ABTS, DPPH, and FRAP assays were used to evaluate the possible antioxidant activity of the methanolic extracts of different samples of TBB in this study. ABTS and DPPH values were determined by assessing the scavenging activity on ABTS and DPPH radicals (by measuring the decrease in ABTS and DPPH radical absorption after exposure to radical scavengers), respectively. Furthermore, FRAP assay was used to measure the conversion of the ferric form (Fe^{3+}) to the ferrous form (Fe^{2+}). Results are shown in Table 3.

Table 3. Antioxidant assay of the methanolic extracts obtained from different tropical black bolete samples.

Mushroom Extract	DPPH Assay (IC_{50}, mg/mL)	ABTS Assay (IC_{50}, mg/mL)	FRAP Assay (mg GAE/g dw)
Fresh sample	2.11 ± 0.80 b	1.30 ± 0.20 c	7.86 ± 0.24 a
Dried sample	2.20 ± 0.60 b	1.78 ± 0.27 b	5.53 ± 0.62 b
Brined sample	3.65 ± 0.48 a	2.85 ± 0.31 a	3.81 ± 0.46 c
Frozen sample	2.14 ± 0.72 b	1.72 ± 0.24 b	7.52 ± 0.38 a

The results are mean ± standard deviation. Different letters in the same column are considered significantly different according to Tukey's test ($p < 0.05$).

In the DPPH and ABTS radical scavenging systems, the extract concentrations providing IC50 values were of particular interest. The IC50 value is a comparable parameter widely used to measure the potency of antioxidant activity of test samples. Thus, the lower IC50 value indicated the higher antioxidant activity. In the DPPH scavenging system, the results indicated that the IC50 values of the DPPH assay varied for each sample. The lowest IC50 value of the DPPH activity was observed in the extract of the fresh sample (2.11 ± 0.08 mg/mL), which was not found to be statistically different from that of the dried (2.20 ± 0.60 mg/mL) and frozen samples (2.14 ± 0.72 mg/mL). However, they were significantly lower than that of the brined sample (3.65 ± 0.48 mg/mL). In the ABTS scavenging system, all extracts showed positive results in terms of the ABTS assay and the IC50 values, which varied from 1.30 to 2.85 mg/mL. The lowest IC50 value of ABTS activity was observed in the extract of the fresh sample, followed by the frozen sample and the dried sample.

In the ABTS scavenging system, all extracts showed positive results in terms of the ABTS assay and the IC_{50} values, which varied from 1.30 to 2.85 mg/mL. The lowest IC_{50} value of ABTS activity was observed in the extract of the fresh sample, followed by the frozen sample and the dried sample. However, the highest IC_{50} value of ABTS activity was found in the extract of the brined sample. In the FRAP system, the FRAP values of the extract of the fresh sample (7.86 ± 0.11 mg GAE/g dw) and frozen sample (7.52 ± 0.38 mg

GAE/g dw) were significantly higher than those of the extracts obtained from the other samples. Moreover, the lowest FRAP value was observed from the extract of the brined sample (3.81 ± 0.46 mg GAE/g dw).

Our results are similar to those of previous studies that reported that the methanolic extract of ECM mushrooms, including boletes, revealed notable DPPH, ABTS, and FRAP activities [13,14,47,60,61]. The results indicate that different preservation methods affected the antioxidative activities of the samples. These findings are supported by those of previous studies that reported that the preservation method can influence the antioxidant activity of the mushroom [57,60]. In this study, the brining method led to the highest degree of decreasing value of antioxidant activity. This outcome is in concordance with Ganguli et al. [59] and Muruke [62], who found that the fresh white bottom mushroom and the oyster mushroom (*Pleurotus cystidiosus*) revealed the strongest degrees of antioxidant activity when compared to the brined mushrooms. Moreover, Pagoń et al. [52] reported that the DPPH, ABTS, and FRAP activities in the brined samples of *B. edulis* and *Sui. luteus* after four weeks of storage were lower than in the fresh samples. In this study, the freezing process maintained the antioxidant activity in the TBB sample during storage. This result was supported by the findings of previous studies that indicated that the act of freezing did not induce or even slightly change the degree of antioxidant activity in certain mushrooms (e.g., desert truffles, chanterelles, sweet tooth mushrooms, shitake mushrooms, and chestnut mushrooms) [57] and vegetables (broccoli florets, cauliflower, green asparagus, and kale leaves) [63–66].

3.5. Antimicrobial Assay

Antimicrobial activity of the crude extracts of the fresh and preserved samples after one year of storage was investigated in terms of the inhibition zone against eighteen strains of pathogenic bacteria and yeast. The diameter of the inhibition zone is shown in Table 4. It was found that ampicillin and streptomycin effectively inhibited all tested bacteria. The diameter of the inhibition zone of ampicillin and streptomycin was significantly larger than that of the mushroom extracts. The result revealed that the inhibition zone varied according to the differences in the extract samples. The extracts of the fresh and frozen samples could effectively inhibit the growth of all tested Gram-positive bacteria. However, the extracts of the brined and dried samples could not inhibit the growth of *B. subtilis* and *L. monocytogenes*. The growth of *B. cereus* and *M. luteus* could not be inhibited by the extract of the brined sample. The extract of the fresh sample displayed the largest diameter of the inhibition zone over the other extracts. Most of the extracts showed no antimicrobial activity against all Gram-negative bacteria, except for the extract of the fresh and frozen samples that could inhibit the growth of *Ps. fluorescens*. However, all extracts displayed no antimicrobial activity against all the yeast samples at the concentration values used. Similarly, antimicrobial potential has been observed in the extracts of the other edible ECM mushrooms in that they displayed greater activity against Gram-positive bacteria than against Gram-negative bacteria and yeast specimens [12,14,42,55]. Kosanić et al. [56] and Smolskaitė et al. [61] reported that the intensity of the antimicrobial activity was dependent upon the mushroom species, its concentration value, and the tested microorganisms. Our results indicate that different preservation methods can influence the antimicrobial activity of the mushroom samples.

Table 4. Diameter of inhibition zone of the methanolic extracts from different tropical black bolete samples and antimicrobial compounds.

Microorganism	Diameter of Inhibition Zone (mm)						
	Mushroom Extract (100 mg/mL)				Antimicrobial Compound (10 mg/mL)		
	Fresh Sample	Dried Sample	Brined Sample	Frozen Sample	Ampicillin	Streptomycin	Nystatin
Gram-positive bacteria							
Bacillus cereus	11.2 ± 0.8 c	10.5 ± 0.5 c	–	10.2 ± 0.5 c	28.3 ± 1.5 a	24.5 ± 1.2 b	NT
Bacillus subtilis	10.8 ± 0.7 c	–	–	10.2 ± 0.6 c	29.7 ± 1.2 b	34.1 ± 0.8 a	NT
Enterococcus faecalis ATCC29212	11.5 ± 1.5 c	11.1 ± 0.9 c	9.8 ± 0.2 c	11.1 ± 0.5 c	40.8 ± 1.0 a	22.2 ± 0.5 b	NT
Listeria monocytogenes	11.0 ± 0.5 c	–	–	10.5 ± 0.6 c	34.0 ± 1.7 a	22.7 ± 2.5 b	NT
Micrococcus luteus	11.5 ± 0.9 c	10.5 ± 0.8 cd	–	9.6 ± 0.5 d	17.5 ± 1.3 b	20.8 ± 1.0 a	NT
Methicillin-resistant *Staphylococcus aureus*	11.7 ± 0.6 c	11.0 ± 1.0 cd	9.9 ± 0.6 d	11.2 ± 0.7 c	16.8 ± 1.2 a	14.2 ± 0.8 b	NT
Staphylococcus aureus ATCC29213	11.5 ± 0.5 c	11.2 ± 0.8c	10.5 ± 0.8 c	11.0 ± 0.8 c	17.6 ± 0.5 b	26.5 ± 1.5 a	NT
Streptococcus pneumoniae ATCC49699	16.7 ± 1.2 c	13.2 ± 1.3 d	12.0 ± 1.0 d	14.2 ± 1.3 d	23.8 ± 1.3 b	30.8 ± 1.0 a	NT
Gram-negative bacteria							
Escherichia coli ATCC25922	–	–	–	–	18.0 ± 1.0 a	11.2 ± 1.0 b	NT
Escherichia coli ATCC35218	–	–	–	–	17.6 ± 0.5 a	11.8 ± 1.0 b	NT
Escherichia coli O157:H7	–	–	–	–	17.9 ± 0.2 a	12.5 ± 0.8 b	NT
Klebsiella pneumoniae	–	–	–	–	21.2 ± 1.0 a	11.5 ± 0.5 b	NT
Proteus mirabilis	–	–	–	–	34.2 ± 1.5 a	21.6 ± 1.0 b	NT
Proteus vulgaris	–	–	–	–	12.5 ± 0.7 b	27.3 ± 1.0 a	NT
Pseudomonus fluorescens	–	–	–	–	27.7 ± 2.0 a	10.2 ± 0.5 b	NT
Pseudomonas aeruginosa ATCC27859	9.83 ± 0.8 b	–	–	9.33 ± 0.6 b	20.2 ± 1.0 a	18.2 ± 1.0 a	NT
Salmonella typhi	–	–	–	–	30.7 ± 1.1 b	39.3 ± 1.1 a	NT
Salmonella sp. group D	–	–	–	–	20.7 ± 1.2 a	16.8 ± 0.8 b	NT
Yeasts							
Candida albicans	–	–	–	–	NT	NT	19.0 ± 1.0 a
Cryptococcus neoformans	–	–	–	–	NT	NT	20.7 ± 1.5 a

NT = not tested and (−) = no inhibition zone. The results are mean ± standard deviation. Different letters in the same row are considered significantly different according to Tukey's test ($p < 0.05$).

Additionally, it was found that freezing serves as a suitable preservation method in maintaining the antimicrobial activity in the TBB sample during storage. These results were supported by the findings of previous studies that found that the perseveration process does play a significant role in reducing the antimicrobial activity of mushrooms. Additionally, it was found that the high temperatures reached during the preservation process, for exampl, drying, boiling, and blanching, could partially degrade the antimicrobial active compounds in mushrooms [67–69]. Furthermore, a study conducted by Janeš et al. [70] found that the extracts of several dried mushroom samples displayed lower antibacterial activity than extracts of the frozen samples. Based on our results, the average amount of total polyphenol content of the extracts obtained from fresh, dried, brined, and frozen samples for antibacterial assay was 3.00, 2.66, 1.84, and 2.89 mg GAE at a concentration of 100 mg/mL, respectively. At a concentration extract of 100 mg/mL, the average value of total flavonoid content contained in the extracts obtained from the fresh, dried, brined, and frozen samples was 0.26, 0.20, 0.06, and 0.25 mg QE, respectively. It was found that the high antibacterial potential possessed in the extracts of the fresh and frozen samples could be related to their high total polyphenol and flavonoid contents.

4. Conclusions

TBB is considered a nutritional mushroom with a number of health benefits. Therefore, the ability to extend the shelf-life of this mushroom is crucial to gaining an understanding of how different preservation methods could affect the mushroom's nutrient composition and antioxidant and antimicrobial properties. In this study, three preservation methods were used, namely, brining, drying, and freezing, in order to determine the effect on nutrient composition, phenolic compound content, and antioxidant and antimicrobial properties in

TBB. After one year of storage, the brined sample showed significant reductions in crude protein, fat, polyphenol and flavonoid contents, as well as antioxidant and antimicrobial properties, when compared with the dried, frozen, and fresh samples. The outcomes of this study demonstrate that freezing could serve as a suitable preservation process by resulting in little or no change to the nutrient composition, phenolic compound content, and antioxidant and antimicrobial properties in samples after one year of storage. However, differences in phenolic profiles in terms of the amount of individual phenolic compounds present in the samples during storage should be determined in future studies. Moreover, mineral analysis, sensory evaluation, microbial contamination, and long-term storage of the preserved mushroom samples would be necessary, and to identify a suitable preservation method for potential commercial development.

Author Contributions: Conceptualization: J.K., N.S. and S.L.; methodology: J.K., K.T., and N.S.; software: J.K.; validation: J.K., K.T., N.S. and S.L.; formal analysis: J.K. and K.T.; investigation: J.K. and N.S.; resources: J.K. and N.S.; data curation: J.K., K.T. and N.S.; writing—original draft: J.K. and N.S.; writing—review and editing: J.K., N.S., K.T. and S.L.; supervision: J.K. and S.L. All authors have read and agreed to the published version of the manuscript.

Funding: This work supported by Chiang Mai University, Thailand.

Data Availability Statement: Data sharing not applicable.

Acknowledgments: The authors are grateful to Russell Kirk Hollis for the English correction of this manuscript.

Conflicts of Interest: The authors declare that they have no conflict of interest.

References

1. Marshall, E.; Nair, N.G. *Make Money by Growing Mushrooms*; Rural Infrastructure and Agro-Industries Division Food and Agriculture Organization of the United Nations: Rome, Italy, 2009.
2. Oei, P.; Nieuwenhuijzen, V.B. *Small-scale Mushroom Cultivation: Oyster, Shiitake and Wood Ear Mushroom*; Agrodok: Wageningen, The Netherlands, 2005.
3. Chang, S.T.; Miles, P.G. *Mushrooms: Cultivation, Nutritional Value, Medicinal Effect, and Environmental Impact*, 2nd ed.; CRC Press: Boca Raton, FL, USA, 2008.
4. Barros, L.; Baptista, P.; Correia, D.M.; Casal, S.; Oliveira, B.; Ferreira, I.C.F.R. Fatty acid and sugar compositions, and nutritional value of five wild edible mushrooms from Northeast Portugal. *Food Chem.* **2007**, *105*, 140–145. [CrossRef]
5. Reis, F.S.; Heleno, S.A.; Barros, L.; Sousa, M.J.; Martins, A.; Santos-Buelga, C.; Ferreia, I.C.C.R. Toward the antioxidant and chemical characterization of mycorrhizal mushrooms from Northeast Portugal. *J. Food Sci.* **2001**, *76*, 824–830. [CrossRef] [PubMed]
6. Ferreira, I.C.F.R.; Barros, L.; Abreu, R.M.V. Antioxidants in wild mushrooms. *Curr. Med. Chem.* **2009**, *16*, 1543–1560. [CrossRef]
7. Yun, W.; Hall, I.R. Edible ectomycorrhizal mushrooms: Challenges and achievements. *Can. J. Bot.* **2004**, *82*, 1063–1073. [CrossRef]
8. Hall, I.R.; Yun, W.; Amicucci, A. Cultivation of edible ectomycorrhizal mushrooms. *Trends Biotechnol.* **2003**, *21*, 433–438. [CrossRef]
9. Sanmee, R.; Dell, B.; Lumyong, P.; Izumori, K.; Lumyong, S. Nutritive value of popular wild edible mushroom from Northen Thailand. *Food Chem.* **2003**, *82*, 527–532. [CrossRef]
10. Kalmis, E.; Yildiz, H.; Ergönül, B.; Kalyoncu, F.; Solak, M.H. Chemical composition and nutritional value of a wild edible ectomycorrhizal mushroom, *Tricholoma Anatolicum*. *Turk. J. Bot.* **2011**, *35*, 627–633.
11. Agahar-Murugkar, D.; Subbulakshmi, G. Nutritional value of edible wild mushrooms collected from the Khasi hills of Meghalaya. *Food Chem.* **2005**, *89*, 599–603. [CrossRef]
12. Nowacka, N.; Nowak, R.; Drozd, M.; Olech, M.; Los, R.; Malm, A. Antibacterial, antiradical potential and phenolic compounds of thirty-one polish mushroom. *PLoS ONE* **2015**, *10*, e0140355. [CrossRef]
13. Kaewnarin, K.; Suwannarach, N.; Kumla, J.; Saisamorn, L. Phenolic profile of various wild edible mushroom extracts from Thailand and their antioxidant properties, anti-tyrosinase and hyperglycaemic inhibitory activities. *J. Funct. Foods* **2016**, *27*, 352–364. [CrossRef]
14. Kosanić, M.; Ranković, B.; Dašić, M. Mushrooms as possible antioxidant and antimicrobial agents. *Iran J. Pharm. Res.* **2012**, *11*, 1095–1102.
15. Susanna, B. Medicinal aspects of edible ectomycorrhizal mushrooms. In *Edible Ectomycorrhizal Mushrooms*; Zambonelli, A., Bonito, G., Eds.; Springer Press: Berlin, Germany, 2012; pp. 317–334.
16. Mortimer, P.E.; Karunarathna, S.C.; Li, Q.; Gui, H.; Yang, X.; Yang, X.; He, J.; Ye, L.; Li, H.; Sysouphanthong, P.; et al. Prized edible Asian mushrooms: Ecology, conservation and sustainability. *Fungal Divers.* **2012**, *56*, 31–47. [CrossRef]
17. Zambonelli, A.; Bonito, G.M. *Edible Ectomycorrhizal Mushroom*; Springer Press: Berlin, Germany, 2012.
18. Bernaś, E.; Jaworska, G.; Kmiecik, W. Storage and process of edible mushroom. *ACTA Sci. Pol. Technol. Aliment.* **2006**, *52*, 5–23.

19. Singh, M.; Vijay, B.; Kamal, S.; Wakchure, G.C. *Mushroom Cultivation, Marketing and Consumption*; Indian Council of Agricultural Research: Chambaghat, India, 2011.
20. Rai, R.D.; Arumuganathan, T. *Post Harvest Technology of Mushroom*; National Research Centre for Mushroom: Chambaghat, India, 2008.
21. Apati, G.P.; Furlan, S.A.; Laurindo, J.B. Drying and rehydration of oyster mushroom. *Braz. Arch. Biol. Technol.* **2010**, *53*, 945–952. [CrossRef]
22. Pilz, D.; Norvell, L.; Danell, E.; Molina, R. *Ecology and Management of Commercially Harvested Chanterelle Mushrooms*; US Department of Agriculture, Forest Service, Pacific Northwest Research Station: Portland, OR, USA, 2003.
23. Keefer, M.; Winder, R.; Hobby, T. Commercial development of non-timber forest resources: A case study of morels in the East Kootenay, British Columbia. *BC J. Econ. Manag.* **2010**, *11*, 39–51.
24. Sitta, N.; Davoli., P. Edible ectomycorrhizal mushrooms: International markets and regulations. In *Edible Ectomycorrhizal Mushrooms*; Zambonelli, A., Bonito, G., Eds.; Springer Press: Berlin, Germany, 2012; pp. 355–380.
25. Hyde, K.D.; Norphanphoun, C.; Chen, J.; Dissanayake, A.J.; Doilom, M.; Hongsanan, S.; Jayawardena, R.S.; Jeewon, R.; Perera, R.H.; Thongbai, B.; et al. Thailand's amazing diversity–up to 96% of fungi in northern Thailand are novel. *Fungal Divers.* **2018**, *93*, 215–239. [CrossRef]
26. Butkhup, L.; Samappito, W.; Jorjong, S. Evaluation of bioactivities and phenolic contents of wild edible mushrooms from northeastern Thailand. *Food Sci. Biotechnol.* **2018**, *27*, 193–202. [CrossRef] [PubMed]
27. Ruksawong, P.; Flegel, T.W. *Thai Mushrooms and Other Fungi*; National Center for Genetic Engineering and Biotechnology (BIOTEC), National Science and Technology Development Agency: Bangkok, Thailand, 2001.
28. Kumla, J.; Hobbie, E.A.; Suwannarach, N.; Lumyong, L. The ectomycorrhizal status of a tropical black bolete, *Phlebopus portentosus*, assessed using mycorrhizal synthesis and isotopic analysis. *Mycorrhiza* **2016**, *26*, 333–343. [CrossRef]
29. Kumla, J.; Bussaban, B.; Suwannarach, N.; Lumyong, S.; Danell, E. Basidiome formation od an edible wild, putatively ectomycorrhizal fungus, Phlebopus portentosus without host plant. *Mycologia* **2012**, *104*, 597–603. [CrossRef]
30. Zhang, C.; He, M.; Liu, J.; Xu, X.; Cao, Y.; Gao, F.; Fang, Y.; Wang, W.; Wang, Y. Brief introduction to a unique edible bolete–*Phlebopus portentosus* in southern China. *J. Agric. Sci. Technol. B* **2017**, *7*, 386–394.
31. Bandala, V.M.; Montoya, L.; Jarvio, D. Two interesting records of boletes found in coffee plantations from Eastern Mexico. *Persoonia* **2004**, *18*, 365–380.
32. Lumyong, S.; Sanmee, R.; Lumyong, P. Is large scale cultivation of boletes possible? *Opera Mycol.* **2007**, *1*, 34–37.
33. Tolera, K.D.; Abera, S. Nutritional quality of oyster mushroom (*Pleurotus Ostreatus*) as affected by osmotic pretreatments and drying methods. *Food Sci. Nutr.* **2017**, *5*, 989–996. [CrossRef] [PubMed]
34. Lungu, C.; Baston, O.; Pricop, E.M. The effect of brine preservation on mushroom. *Ann. Food Sci. Technol.* **2015**, *16*, 333–337.
35. AOAC. *Official Methods of Analysis of AOAC International*, 16th ed.; AOAC International: Gaithersburg, MD, USA, 1996.
36. Thitilertdecha, N.; Teerawutgulrag, A.; Rakariyatham, N. Antioxidant and antimicrobial activities of *Nephelium lappacium* L. extracts. *LWT Food Sci. Technol.* **2008**, *41*, 2029–2035. [CrossRef]
37. Kaewnarin, K.; Niamsup, H.; Shank, L.; Rakariyatham, N. Antioxidant and antiglycation activities of some edible and medicinal plants. *Chiang Mai J. Sci.* **2008**, *41*, 105–116.
38. Gülçin, İ.; Oktay, M.; Kireçci, E.; Küfrevioğlu, İ.Ö. Screening of antioxidant and antimicrobial activities of anise (*Pimpinella anisum* L.) seed extracts. *Food Chem.* **2003**, *83*, 371–382. [CrossRef]
39. Re, R.; Pellegrini, N.; Proteggente, A.; Pannala, A.; Yang, M.; Rice-Evans, C. Antioxidant activity applying an improved ABTS radical cation decolorization assay. *Free Radic. Biol. Med.* **1999**, *26*, 1231–1237. [CrossRef]
40. Li, Y.; Guo, C.; Yang, J.; Wei, J.; Xu, J.; Cheng, S. Evaluation of antioxidant properties of pomegranate peel extract in comparison with pomegranate pulp extract. *Food Chem.* **2006**, *96*, 254–260. [CrossRef]
41. Semeniuc, C.A.; Pop, C.R.; Rotar, A.M. Antibacterial activity and interactions of plant essential oil combinations against Gram-positive and Gram-negative bacteria. *J. Food Drug Anal.* **2017**, *25*, 403–408. [CrossRef]
42. Fogarasi, M.; Diaconeasa, Z.M.; Pop, C.R.; Fogarasi, S.; Semeniuc, C.A.; Farcas, A.C.; Tibulcu, D. Elemental composition, antioxidant and antibacterial properties of some wild edible mushrooms from Romania. *Agonomy* **2020**, *10*, 1972.
43. Manzi, P.; Marconi, S.; Aguzzi, A.; Pizzoferrato, L. Commercial mushrooms: Nutritional quality and effect of cooking. *Food Chem.* **2004**, *84*, 201–206. [CrossRef]
44. Gacia-Pascual, P.; Sanjuan, S.; Bon, J.; Carreres, J.E.; Mulet, A. Rehydration process of *Boletus edulis* mushroom: Characteristics and modeling. *J. Sci. Food. Agric.* **2005**, *85*, 1397–1404. [CrossRef]
45. Argyropoulos, D.; Tahir, M.; Muller, J. Effect of air temperature and pre-treatment on color changes and texture of dried *Boletus edulis* mushroom. *Drying Technol.* **2011**, *29*, 1890–1900. [CrossRef]
46. Yuswa, M.H.; On, Y.Y.; Tan, Y.J.; Lai, W.; Zainal, Z.; Sz, J.X.; Daud, F. Nutritional and molecular analysis of wild edible Gelam mushroom (*Boletus* sp.) from Kelantan, Malaysia. *J. Adv. Biol. Biotechnol.* **2017**, *13*, 1–7. [CrossRef]
47. Liaotrakoon, W.; Liaotrakoon, V. Influence of drying process on total phenolics, antioxidative activity and selected physical properties of edible bolete (*Phlebopus colossus* (R. Heim) Singer) and changes during storage. *Food Sci. Technol.* **2018**, *38*, 231–237. [CrossRef]
48. Muyanja, C.; Kyambadde, D.; Namugumya, B. Effect of pretreatments and drying methods on chemical composition and sensory evaluation of oyster mushroom (*Pluerotus oestreatus*) powder and soup. *J. Food Process. Preserv.* **2014**, *38*, 457–465. [CrossRef]

49. Adejumo, T.O.; Awosanya, O.B. Proximate and mineral composition of four edible mushrooms species from south western Nigeria. *Afr. J. Biotechnol.* **2005**, *4*, 1084–1088.
50. Mo, A.; Basu, S.K.; Gyar, S.D.; Goyal, A.; Bhowmik, P.K.; Datta, B.S. Proximate composition and functional properties of mushroom flours from *Ganoderma* spp., *Omphalotus olearius* (DC.) Sing. and *Hebeloma mesophaeum* (Pers.) Quél. used in Nasarawa State, Nigeria. *Malaysian J. Nut.* **2009**, *15*, 233–241.
51. Chittaragi, A.; Naika, R.; Vinayaka, K.S. Nutritive value of few wild mushrooms from the western Ghats of Shivamogga district, Karnataka, India. *Asian J. Pharm. Clin. Res.* **2014**, *7*, 50–53.
52. Pagoń, K.; Gabor, A.; Jaworska, G.; Bernaś, E. Effect of traditional canning in acetic brine on the antioxidants and vitamins in *Boletus edulis* and *Suillus luteus* mushrooms. *J. Food Process. Preserv.* **2016**, *41*, e12826. [CrossRef]
53. Jaworska, G.; Bernaś, E.; Mickowska, B. Effect of production process on the amino acid content of frozen and canned *Pleurotus ostreatus* mushrooms. *Food Chem.* **2011**, *125*, 936–943. [CrossRef]
54. Bernaś, E.; Jaworska, G. Effect of preservation method on amino acid content in selected species of edible mushroom. *LWT Food Sci. Technol.* **2012**, *48*, 242–247. [CrossRef]
55. Rameah, C.; Pattar, M.G. Antimicribial properties, antioxidant activity and bioactive compounds from six wild edible mushrooms of western ghats of Karnataka, India. *Pharm. Res.* **2010**, *2*, 107–112.
56. Kosanić, M.; Rankovic, B.; Daašić, M. Antioxidant and antimicrobial activity of mushrooms. *Bulg. J. Agric. Sci.* **2013**, *19*, 1040–1046.
57. Murcia, M.A.; Martinez-Tome, M.; Jimenez, A.M.; Vera, A.M.; Honrubia, M.; Parras, P.J. Antioxidant activity of edible fungi (truffles and mushrooms): Losses during industrial processing. *J. Food Protect.* **2002**, *65*, 1614–1622. [CrossRef] [PubMed]
58. Rickman, J.C.; Barrett, D.M.; Bruhn, C.M. Nutritional comparison of fresh, frozen and canned fruits and vegetables. Part 1. vitamins C and B and phenolic compounds. *J. Sci. Food Agric.* **2007**, *87*, 930–944. [CrossRef]
59. Ganguli, A.; Ghosh, M.; Singh, S. Antioxidant activities and total phenolics of pickles produced from the edible mushroom, *Agaricus bisporus*. *J. Culin. Sci Technol.* **2006**, *5*, 131–142. [CrossRef]
60. Jaworska, G.; Pagoń, K.; Skrzypczak, A.; Bernaś, E. Composition and antioxidant properties of wild mushrooms *Boletus edulis* and *Xerocomus badius* prepared for consumption. *J. Food Sci. Technol.* **2015**, *52*, 7944–7953. [CrossRef]
61. Smolskaitė, L.; Venskutonis, P.R.; Talou, T. Comprehensive evaluation of antioxidant and antimicrobial properties of different mushroom species. *LWT Food Sci. Technol.* **2015**, *60*, 462–471. [CrossRef]
62. Muruke, M.H. Evaluation of antioxidant and iron chelating activities of a wild edible oyster mushroom *Pleurotus cystidiosus* from Tanzania. *Food Sci. Qual. Manag.* **2014**, *29*, 18–28.
63. Kapusta-Duch, J.; Leszczynska, T.; Borczak, B.; Florkiewicz, A.; Załubska, A. Impact of Different packaging systems on selected antioxidant properties of frozen-stored cauliflower (*Brassica oleracea* L. Var. botrytis). *Polish J. Food Nutr. Sci.* **2017**, *67*, 211–217. [CrossRef]
64. Cai, C.; Miao, H.; Qian, H.; Yao, L.; Wang, B.; Wang, Q. E-ects of industrial pre-freezing processing and freezing handling on glucosinolates and antioxidant attributes in broccoli florets. *Food Chem.* **2016**, *210*, 451–456. [CrossRef] [PubMed]
65. Gebczynski, P. Content of selected antioxidative compounds in green asparagus depending on processing before freezing and on the period and conditions of storage. *Polish J. Food Nutr. Sci.* **2007**, *57*, 209–214.
66. Neri, L.; Faieta, M.; Mattia, C.D.; Sacchetti, G.; Mastrocola, D.; Pittia, P. Antioxidant activity in frozen plant foods: Effect of cryoprotectectants, freezing process and frozen storage. *Foods* **2020**, *9*, 1886. [CrossRef]
67. Shen, H.S.; Shao, S.; Chen, J.C.; Zhou, T. Antimicrobials from mushrooms for assuring food safety. *Compr. Rev. Food Sci. Food Saf.* **2017**, *16*, 316–329. [CrossRef] [PubMed]
68. Diamantopoulou, P.A.; Philippoussis, A.N. Cultivated mushroom: Preservation and processing. In *Handbook of Vegetable Preservation and Processing*, 2nd ed.; CRC Press: Boca Raton, FL, USA, 2016; pp. 495–525.
69. Matser, A.M.; Knott, E.R.; Teunissen, P.G.M.; Bartels, P.V. Effects of high isostatic pressure on mushrooms. *J. Food. Eng.* **2000**, *45*, 11–16. [CrossRef]
70. Janeš, D.; Kreft, S.; Jurc, M.; Seme, K.; Štrukelj, B. Antibacterial activity in higher fungi (mushrooms) and endophytic fungi from Slovenia. *Pharm. Biol.* **2007**, *45*, 700–706. [CrossRef]

foods

MDPI

Review

Safety of Probiotics: Functional Fruit Beverages and Nutraceuticals

Irena Žuntar [1], Zvonimir Petric [2], Danijela Bursać Kovačević [3] and Predrag Putnik [3,*]

1 Faculty of Pharmacy and Biochemistry, University of Zagreb, Ante Kovačića 1, 10000 Zagreb, Croatia; izuntar@pharma.hr

2 Unit of Pharmacokinetics and Drug Metabolism, Department of Pharmacology at the Institute of Neuroscience and Physiology Sahlgrenska Academy at the University of Gothenburg, 40530 Göteborg, Sweden; petric.zvonimir@gmail.com

3 Faculty of Food Technology and Biotechnology, University of Zagreb, Pierottijeva 6, 10000 Zagreb, Croatia; dbursac@pbf.hr

* Correspondence: pputnik@alumni.uconn.edu

Received: 9 June 2020; Accepted: 13 July 2020; Published: 17 July 2020

Abstract: Over the last decade, fruit juice consumption has increased. Their rise in popularity can be attributed to the belief that they are a quick way to consuming a dietary portion of fruit. Probiotics added to fruit juices produce various bioactive compounds, thus probiotic fruit juices can be considered as a new type of functional foods. Such combinations could improve nutritional properties and provide health benefits of fruit juices, due to delivering positive health attributes from both sources (fruit juices and probiotics). However, this review discusses the other side of the same coin, i.e., the one that challenges general beliefs that probiotics are undoubtedly safe. This topic deserves more acknowledgments from the medical and nutritional literature, as it is highly important for health care professionals and nutritionists who must be aware of potential probiotic issues. Still, clinical trials have not adequately questioned the safety of probiotics, as they are generally considered safe. Therefore, this reviews aims to give an evidence-based perspective of probiotic safety, focusing on probiotic fruit beverages and nutraceuticals, by providing documented clinical case reports and studies. Finally, the paper deals with some additional insights from the pharmacological and toxicological point of views, such as pharmacological repercussions of probiotics on health.

Keywords: probiotic safety; toxicity; pathogenicity; functional food industry; pharmacological interactions; functional fruit juices

1. Introduction

Probiotics are commonly defined as viable microorganisms [1]. This refers to both dietary supplements and drugs, as well as microorganisms found in fermented foods as a part of human nutrition. They are generally added to foods in order to improve its nutritional value as increased demand for new probiotic products is constantly growing. Probiotics are conventionally added to dairy products, but in recent times, the food industry is trying to develop other food matrices that are suitable for this purpose. Therefore, the formulation of probiotic beverages based on the fruit juices might be a compromise solution as they provide an excellent carrier for these probiotic bacteria. The probiotic strains produce various bioactive compounds, such as vitamins, antioxidants, amino acids and peptides, and when added to fruit juices may offer a synergy of health benefits from both sources. Such products can be considered as a new type of functional foods [2,3]. However, there are concerns regarding the safety of added probiotics to the foods. Hence, the purpose of this review is to provide the perspective of probiotic safety with focus to probiotic fruit (functional) beverages and nutraceuticals.

2. Safety and Pathogenicity of Probiotics and Their Assessments

The World Health Organization defines probiotics as "live microorganisms that when administered in adequate amounts confer a health benefit on the host" [4], while in the context of a food product, a minimum of 10^6 colony forming units per mL (CFU/mL) must be reached if the food product will be labeled as probiotic [5].

Many probiotics on the market usually contain various bacterial strains from different species, rather than a single strain [6]. This is due to the belief that multiple strains of probiotic products will have a greater success of delivering health benefits and broader efficacy, and additional faith that their effect will be an additive, or even synergistic. However, there is one limitation to this type of reasoning. Namely, it is thought that in some cases there can be antagonistic effects between different probiotic species. Even though, this may sound logical and convincing, studies that compared single vs. multispecies probiotics, concluded that such claims are still not clear and should be further investigated [6].

As many bacteria can have an impact on microbial balance in the intestines, only those bacteria species and strains with confirmed positive effects on the host can be selected as probiotics. Hence, on the global market, probiotics that are mostly found include species of *Lactobacillus*, *Bifidobacterium*, *Lactococcus*, and *Enterococcus*. Some other bacteria, such as *Lactobacillus bulgaricus*, or *Streptococcus thermophilus*, are not normally part of intestinal flora, but still, they are categorized as probiotics because of their use as starters in dairy products. The influence of such bacteria on gut microbial balance is believed to be weak, as they lack the colonization properties [7].

Nowadays, probiotics are found (and regulated) in three categories: (i) Foods (fermented foods), with claimed GRAS (generally recognized as safe) status for *Lactobacillus*, *Bifidobacterium*, and *Lactococcus*; (ii) dietary supplements, which are often sold as over-the-counter (OTC) supplements; and (iii) drugs (pharmaceuticals). Categorization depends on probiotic manufacturers and indications of use, but it is also dependent on demands from different regulatory authorities [8,9]. Safety of foods or pharmaceuticals intended for human consumption, including probiotics, is a paramount factor in order to avoid any health hazards. Some clinical studies confirmed that safety of probiotics is apparent, as due to absence of toxicity in diverse populations including: (i) Healthy adult volunteers [10]; (ii) women during late pregnancy and their children during early infancy [11]; (iii) infants aged 0–2 years [12], and children [13]; (iv) hospitalized children [14]; (v) critically ill children [15]; and (vi) immunocompromised patients. After all, probiotics are *a priori* nonpathogenic, meaning, that they are never supposed to cause, or potentiate any disease in humans, regardless of the source of their intake, i.e., foods or OTC supplements.

2.1. Probiotics Safety

On the contrary and quite recently, papers from prestigious scientific medical journals, such as Lancet, Cell, England Journal of Medicine, and Nature, started questioning if probiotics are truly safe, as some large-scale clinical trials reported unexpected results [16–19]. Some researchers stated that many clinical trials of probiotics have clinical data, which are lacking a proper methodology of collecting and interpreting the results, especially regarding the clinical endpoints, besides the lack of the scientific rigor [19]. This recent paradigm shift of apparent probiotic safety (and efficacy) became even more controversial, as seen from Cochrane reviews of 31 trials, where probiotics are recommended as safe (and effective) when used with antibiotics for the treatment of *Clostridium difficile*. However, a new review from only a year later in 2018 [20] concluded that probiotic use (for some diseases), indeed, lacks sufficient evidence, and more research is required to support probiotic recommendations [21,22]. Additionally, Lerner et al. [23] in recent review article also highlighted the safety concerns of probiotic intake and shared the same suspicion with many other authors [24,25], while stating that the list of probiotics' adverse effects is still underestimated. Therefore, it can be concluded that the safe and effective use of probiotics, from pharmacological and toxicological aspects, seems to be undervalued. On the other hand, it can be often seen how many health professionals, including both, physicians and pharmacists, warmly recommend taking probiotics [26,27]. This is not surprising, considering how much they are advertised, especially to customers and patients, who often, as end-users, have difficulties distinguishing between high and poor-quality products. Namely, probiotic products that contain the same bacterial strains are found to be marketed under various beneficial claims on their labels [28,29].

Probiotics are more frequently categorized as food supplements, and not pharmaceuticals, which implies avoidance of extremely thorough testing which are obligatory for all pharmaceuticals. Unless the probiotic manufacturer makes any specific claims regarding health, probiotics will be classified as food supplements, meaning that the focus on safety c be underestimated. Additionally, if any health claims are found on the packaging of probiotics, regulatory bodies will inspect primarily validity of such claims, and not the safety of the product [30–32]. As the global market of probiotic grows, due to rise of probiotic consumption [33], regulatory questions are becoming increasingly apparent, where a framework which is supposed to be uniformly followed by all manufacturers is still lacking. For example, in the EU, Food Products Directive and Regulation (2000/13/EU, 178/2002/EC) regulates both probiotics and food supplements, while European Food Safety Authority (EFSA) oversees reviewing health claims of probiotics, which are usually stated on the label [34]. The EFSA also issues the Qualified Presumption of Safety (QPS) for various bacterial strains. The word "presumption" is the only criteria linked with the actual true safety of probiotics, meaning that safety valuation is not the primary concern. Furthermore, QPS is focused on a healthy population, i.e., the general population, while those at-risk are not included in the assessment [34]. Interestingly enough, EFSA has excluded all health claims for probiotics with the explanation that amassing a healthy gut flora is not a recognized health benefit. At the same time, the regulation in European Union, states that for safety, traceability, and protection of the intellectual rights, every novel food product should have specified strains. Hence, lawmakers stressed the importance of research in determining the safety and toxicity of probiotics [9].

Post-marketing regulatory vigilance is not performed here either, and the term "health benefit", (often stated on the label) is not a regulated specificity, nor has a clear medical meaning [31,34]. Altogether, it cannot be known if the labels on probiotics are honest, so in the end, it seems justified to have critical thinking about the real efficacy and safety of probiotics. The situation gets even more complex, since there can be so many microorganisms used as probiotics, and there are differences among types of bacterial species, but also among subtypes of the same bacterial species. As a result, probiotics are expected to have different health benefits, but also undesired effects [35–37]. Namely, in the host, the survival of microorganisms is variable, where microorganisms can show different effects. Moreover, probiotic fate in the host can be variable, and affected by the differences in probiotic manufacturing and formulations [38], and the probiotic intake source, e.g., foods or dietary supplements [39]. Hence, one of the most important factors, which contributes to better evaluation of probiotic safety as well as probiotic health risk, is a thorough knowledge of the microbial colonization properties [39]. Therefore, probiotics should never be looked at as a uniform group of viable microorganisms [40], as their properties are strain-dependent, i.e., species-specific, but rather on a case-by-case basis, avoiding the one-size-fits-all approach.

More precisely, potential probiotic health risk can be viewed in two ways [41]. The first way involves the adverse effects of probiotic per se, while the second way involves safety concerns, due to undefined quality standards and manufacturing procedures. Nevertheless, the only standardization of accurate probiotic safety assessment is a retrospective epidemiologic study, accompanied by thorough pharmacological and toxicological post-marketing vigilance [39,42,43] of the product, in order to evaluate further probiotic safety.

We must remain mindful that consumers can respond in different and often unpredicted ways to any medication, dietary supplement or food (allergic reactions), and probiotics are not the exception. Individual differences, such as age, gender, and underlying pathophysiology, are factors that most contribute to such individual response [44]. Moreover, interactions between genes and food (diet) are nowadays becoming the subject of investigations, because many metabolic pathways are found to modulate the development of many diseases. Furthermore, probiotics, as viable microorganisms, grow and colonize the gut, and in theory, under particular circumstances that could be the source of undesired events in the host and represent a serious health concern [39,43].

Fundamental toxicological and pharmacological concepts of how toxic and therapeutic effects of substances are in almost every case dose-dependent are applicable in the assessment of any apparently safe substances [45]. For instance, common sodium salt (NaCl) or even drinking water if taken in large amounts can be lethal. Hence, probiotics (if taken in critical amounts) can be deadly as well with observable toxicological consequences. However, pharmacological and toxicological interplay regarding the probiotic dose-response relationships are not actually studied—not even in animal models.

For a better assessment of probiotic products, many researchers agree that scrutiny of pathogenicity, infectivity, virulence, metabolic activity, and intrinsic properties are required [39,45,46]. However, additional technological characteristics of the manufacturing process and marketing regulation are definitely needed, due to potential unwanted outcomes. More specifically [42], there is a risk of systemic and local infections, but also risks of unwanted and hazardous metabolic activities, or gene transfer, and antibiotic resistance. Moreover, there is a risk of modulating the host's immune response due to inappropriate manufacturing process and health risk with possibly lethal consequences [39,42,45,47]. Finally, there is a concern about pharmacological drug interactions with probiotics, which could have clinical significance that is hardly investigated. Hence, to undoubtedly establish the safety and toxicity of probiotics, human clinical trials of probiotics are indeed justified, despite having a high cost [48].

One could also ask about using animal models for the risk assessment of probiotics. Unfortunately, it is shown that such data gave only limited risk assessment, and the response between species is highly variable. Therefore, extrapolating such results from animals to humans, could be quite unreliable [1]. In addition, human clinical trials, as recommended methodological approach, are not without flaws, as there are health risks for study subjects, especially if they are already having health problems [48].

Regarding the safety concerns of probiotic, let us have a look at a systematic review of 17 studies, which included almost 1530 patients with cancer. This review found five cases of bacteremia, fungemia, and positive blood culture tests associated with probiotics, which confirms that patients, even though immunocompromised, are indeed at risk [49]. Few cases reported bacteremia in patients with HIV/AIDS and Hodgkin's disease after probiotic intake, but such incidences were due to excessive consumption of probiotic-enriched yogurt with *Lactobacillus acidophilus* [50]. Similar observations were also found in animal models [51]. For example, *Lactobacillus gasseri* ATC33323 (purified cell wall fragment), in rats, activated systemic inflammation, and in a higher doses, caused death. Such findings confirmed that the topic of probiotic safety deserves attention, especially if probiotics are intended to be used in critically ill patients, who may have a tendency for unwanted immune modulation and consequently, an inflammatory reaction [52].

2.2. Pathogenicity Studies of Probiotics and Clinical Cases

From the 1990s until today, there are reports and clinical cases that described the invasive fungal infections related to *Saccharomyces cerevisiae* (and *Saccharomyces boulardii*) [53–56]. Even though meta-analysis of probiotics for the prevention of antibiotic-associated diarrhea [56] concluded that *Saccharomyces cerevisiae var. boulardii* is the only effective choice in its management, special caution is still advised in patients with compromised immune function and in those who are critically ill [56].

Besides systemic infections, there are reported cases of local infections as well [42]. The literature describes liver abscess and lung infections. Namely, pneumonia induced by *Lactobacillus*, is noticed even in clinical settings, i.e., under close monitoring of health care professionals. Risk factors that are thought to be responsible for the development of probiotic-induced infections included older age, hepatobiliary disease and diabetes mellitus, patients with a history of malignant diseases, and transplantations [49,50,57,58].

Probiotics containing *Lactobacillus* are related to the cases of bacteremia [59] and endocarditis [60–62] in immunocompromised patients, and in those who have heart defects (with or without prosthetic material). This should not be clinically neglected, as reported mortality from *Lactobacillus* and related endocarditis is 23% [59], and although infections associated with *Lactobacilli* are extremely rare, well-known history of probiotic supplementation is important to exclude probiotics as a cause of endocarditis. Therefore, the use of *Lactobacillus* species, such as *L. rhamnosus*, *L. casei*, *L. acidophilus*, *L. jensenii*, *L. plantarum*, and *L. paracasei*, in immunocompromised patients should be under close monitoring of health care professionals [42,63].

As mentioned earlier, the literature contains conflicting results regarding the positive and negative impacts of probiotics on human health and disease. For instance, one study found that the probiotic intake could not be linked to any negative context of pancreatitis. However, the "PROPATRIA" trial [64] concluded that there were negative impacts and mortality, due to probiotic intake attributable to the bowel ischemia in patients with pancreatitis. In the group of people, who were taking probiotics vs. controls, there was increased mortality due to bowel ischemia. If discussing probiotic induced pathogenicity, this difference was quite significant, i.e., 16% vs. 6%. The proposed mechanism of bowel ischemia is thought to happen due to the increased need for oxygen, after high load of six probiotic strains in these patients. Blood flow was already low, and local inflammation was present; hence, probiotics worsened the clinical picture and led to increased mortality as compared to the control group. As the toxicity of probiotic strain per se, should not be prioritized over the total dose of probiotic, it is clear why it is extremely important not to neglect the dose-response relationship in probiotic combination, as individual toxic responses could be unpredictable.

Sanders et al. speculated about the extent of probiotic colonization, and if there were possible side effects related to a long term of probiotic intake, especially in a population prone to allergies. The consumption of probiotics was related to a higher risk of rhinitis, serious asthma attacks, and atopic dermatitis, allergies and sensitization [65]. On the other hand, other researchers [66,67] demonstrated quite the opposite, where probiotic intake had a positive impact on atopic dermatitis. Nevertheless, in atopic patients, the effects of immunomodulation with probiotics remained to be controversial. However, for such patients, it should be kept in mind that an inadequate immune response can be triggered in some circumstances by any viable microorganisms, including probiotics, as the microenvironment of the host determines the final scenario [59,68].

Some authors reported that the long-term use of probiotics might negatively influence human health and be the cause of antibiotic resistance and higher virulence potential. Probiotics as *Lactobacillus, Lactococcus, and Bifidobacteria* even though as being considered safe and nonpathogenic, in theory, could transfer their antibiotic resistance genes to the opportunistic pathogens, or commensal microflora, with hazardous clinical consequences [69]. Although there are obvious gaps in the current understanding of probiotic resistance to antibiotics, it is demonstrated that *Lactobacillus* species have common intrinsic resistance to tetracycline, vancomycin, and erythromycin. In addition to the reported resistance to streptomycin, clindamycin, gentamicin, oxacillin, and lincosamide. Also, *Bifidobacteria* species showed resistance to tetracycline, streptomycin, erythromycin, gentamicin, and clindamycin, while *Streptococcus* species showed high resistance to tetracycline, ciprofloxacin, and aztreonam [70]. Hence, in one way, probiotics could be considered as a reservoir of resistance while in the case of any probiotic-induced infection, an effective arsenal of antibiotics should be used [71].

Probiotics could be involved in the production of metabolites with toxic potential, as mentioned earlier. One of the possible safety concerns is the production of D-lactate, a compound responsible for the development of D-lactic acidosis [72]. Recently, D-lactic acidosis is related to probiotic consumption, in patients with short bowel syndrome, as well as in infants. *Lactobacillus and Bifidobacterium* are known for fermenting ingested carbohydrates and governing the formation of D-lactate [72,73]. Furthermore, in reported cases of "brain fog" (cognitive impairment), the link between probiotics, D-lactic acidosis (metabolic acidosis), small intestinal bacterial overgrowth (SIBO), and symptoms, such as abdominal bloating, distention, and gas, are not established for sure. However, one study implicated probiotics, as the symptoms of brain fogginess improved when probiotics were discontinued, and when patients received antibiotics. Authors advised caution regarding the excessive use of probiotics, especially in people without any obvious medical reasons, patients suffering from gastrointestinal dysmotility, including the patients who frequently used proton pump inhibitors (PPIs) or opioids [73]. Moreover, there is a case of D-lactic acid encephalopathy, related to the use of probiotics in a 5-year old child (with a history small intestine resection) with a short bowel syndrome [74]. Namely, *Lactomin®* is prescribed (in double amount than regularly) for diarrhea two weeks before the child's neurologic symptoms started to appear. Lactomin® contains *Lactobacillus acidophilis, Lactobacillus bulgaricus, Streptococcus faecalis, and Streptococcus faecium.* In particular, *L. acidophilis* produces D-lactic acid, and it was suggested to be the main cause of D-lactic acid encephalopathy [74].

Some bacteria are able to interfere with amino acids/proteins that can produce potentially toxic substances, such as ammonia, indol, phenols, and biogenic amines [37,45]. This is especially important if such bacteria originated from fermented foods, as it is problematic to estimate the accumulation of these potentially toxic products in the fermentation environment that is difficult to manipulate [75]. To that end, one study reported significant accumulation of cadaverine (a toxic diamine compound, produced by bacterial decarboxylation of lysine), due to the presence of *Lactococcus lactis,* originating from fermented foods [76]. Moreover, biogenic amines, i.e., cadaverine, histamine, or tyramine, from food-fermenting lactic acid bacteria, are known to cause symptoms of severe allergic reactions [77]. Hence, the toxicological significance of consuming fermented foods in larger amounts should be more emphasized [78]. Therefore, current findings regarding the metabolic activity of probiotics and their capacity to produce toxic metabolites, require further clarifications in terms of a real toxicological significance.

Rarely, the use of *S. boulardii* has been related to constipation and increased thirst. Although there are some reports of serious itching rash, fatigue, and pruritus was noticed with some probiotics. It should be noted that fatigue, pruritus, and diarrhea occurred equally in the placebo group as well, so the real toxicological impacts cannot be determined [79,80].

According to some older source of data, probiotics containing *Lactobacillus* were considered as contraindicated in individuals who have a history of hypersensitivity to lactose and milk products [81]. However, recent data provide opposite conclusions, and there are even *Lactobacillus* strains that provide relief for lactose intolerance [82]. Although true that fermented dairy products generally do not contain lactose in the amounts that would be high enough to trigger intolerance reactions in sensitive individuals, still probiotic bacteria can be added to non-fermented dairy products, but their applicability is then limited by the lactose intolerance, milk protein allergies or with diets that require cholesterol restriction [83]. Here fruit juices/beverages are perceived as an alternative because they are healthy and beneficial for all groups of consumers (including vegans and vegetarians), therefore they might be a good nutritional substitute for common dairy foods containing probiotics.

2.3. Drug Interactions of Probiotics

The gut microbiota goes through very vibrant, dynamic changes due to constant variations in nutritional status, disease occurrence, pharmacological modulation, circadian rhythms, and natural environmental influences [47]. Currently, it has been accepted that microbiota is having significant impacts on the bioavailability of many drugs and xenobiotics, their pharmacokinetics (PK), i.e., absorption, distribution, metabolism and elimination (ADME). This is additional to the drug efficacy, response and adverse effects, i.e., drug pharmacodynamics (PD) and toxicology. Therefore, probiotics, as a part of the host's microbiota could affect the "destiny" of many drugs as well [47]. In other words, probiotics could influence the bioavailability of some drugs (defined as *"unchanged drug fraction of an administered dose that enters systemic circulation"*), as well as drug PK/PD (simply defined as "relationship between drug concentration with drug effect") and toxicity [84–86]. Koziolek et al., suggested that changes of the microbiome due to the intake of probiotics need further investigations, as probiotic-drug interactions could be clinically significant and not just speculations.

Earlier, it was thought that a drug absorbed from the gut, cannot interact with the host's microbiota, except in the case when the drug is manufactured as sustained-release dosage form, or when it is a subject to a liver-intestine interplay, i.e., enterohepatic recirculation, which consequently prolongs pharmacological effect [47]. However, new findings regarding the composition of the small intestine, and biotransformation potential of bacteria in the gut, showed that interactions between probiotics and drugs are real, despite lacking the enterohepatic recirculation, and sustained-release dosage form [47,87].

Clear examples can be seen in rats fed with probiotics, who had a significant increase of amiodarone (antiarrhythmic agent) bioavailability [88]. A similar is noted in diabetic rats after they received gliclazide (antihyperglycemic agent) [89] and in rabbits after the administration of amlodipine (antiarrhythmic agent) [90]. Authors that studied amiodarone [88] proposed that the increase of its bioavailability by almost 43% is due to a decrease of pH in the intestine, which consequently, facilitated ionization of the amiodarone and impacted amiodarone transit. Alternatively, it was speculated that increased uptake was caused by the OATP2B1 (influx transporter) upregulation [88]. Surprisingly, up until now, there is no human data available regarding this topic. Thus, only hypothetical relevance regarding the increased bioavailability of amiodarone can be discussed. In brief, as amiodarone already has a risk of causing serious and life-threatening side effects, any additional increase of its bioavailability could be extremely toxic.

Regarding gliclazide [91], when probiotics were given to healthy rats, mucosal efflux drug transporters that control its transport were upregulated. In diabetic rats, the opposite occurred, i.e., mucosal influx drug transporters were upregulated, which could have clinical significance for diabetic patients, as their glucose levels must be maintained at a relatively stable concentration. On the other hand, some other studies reported a decrease in drug bioavailability. For instance, tacrolimus (immunosuppressive agent) required higher doses in patients who had higher amounts of *Faecalibacterium prausnitzii* in their fecal samples [92].

There are numerous other examples of how gut microflora influences the pharmacokinetics of many drugs, such as digoxin, irinotecan, indomethacin, insulin, levodopa, ketoprofen, lovastatin, risperidone, and sulfasalazine, etc. This could have clinical significance regarding the pharmacological response, safety, and toxicity, especially if the drugs have a narrow therapeutic index. Meaning, that even the smallest increase of drug bioavailability will significantly change drug range of concentrations, regarding the effectivity vs. toxicity, towards toxicity and adverse effects [47]. Therefore, it is realistic to expect in the future more exciting research regarding the probiotics and drug pharmacokinetics and interactions.

Some bacteria can interfere with bile acids, due to their bile salt hydrolase (BSH) enzyme. Additionally, there is another fact regarding bile acids and their salts in this context of interfering with probiotics. Namely, bile acids can modulate the absorption of some drugs, especially poorly soluble ones [93]. Pavlovic et al. [94], and Moghimipour et al. [95] indicated that the bile salts increased membrane permeability and fluidity, which is positively correlated with the fraction of the drug absorbed, i.e., drug bioavailability and delivery. Moreover, by forming micelles, bile salts can affect transcellular absorption and increase both, solubility and dissolution of drugs [94,96]. Further, it would be interesting to determine the influence of probiotics on drug PK/PD, as well as to reveal the clinical influence of bile acids on drug PK.

Regarding the well-known pharmacological interactions "probiotic-antibiotic", it is recommended to administer antibiotics for at least two hours before/after probiotic bacteria. Similarly, probiotics containing yeasts, such as *S. boulardii*, interact with antifungals. Hence, antifungal drugs, such as clotrimazole, ketoconazole, griseofulvin, and nystatin, are contraindicated with *S. boulardii* [97].

In the discussion of the general toxicity of probiotics, it is already stated how probiotics should be used with caution in patients who are immunocompromised, or in those who are using chemotherapeutic agents or immunosuppressant drugs (cyclosporine, tacrolimus, azathioprine, etc., as this could induce pathogenic colonization (and infection) in these patients [45]. However, there are conflicting reports in these at-risk populations, and currently, clinical trials are evaluating the safety of a few probiotic strains in cancer patients receiving anticancer therapy. It is speculated that probiotics could lower the occurrence of diarrhea and mucositis—serious adverse reactions of anticancer therapy [98,99].

In the end, the field for exploring the nutritional impacts of probiotics and their effects on the host's health and disease is very broad, unexplored, and interdisciplinary. Thus, very tempting for further scientific investigations by many researchers, coming from different fields of expertise. Many factors are influencing the way of how we look at probiotics, so a "one-size-fits-all" criterion cannot be applied in revealing the missing pieces about their roles, effects, safety, and toxicity. As gut microbial balance can be very easily shifted, so can be the safety paradigm of probiotics.

2.4. Assessment of Probiotic Safety

There are a few ways of assessing the safety of probiotics [68]. Particular focus can be placed on the intrinsic (nonpathogenic) properties of different strains and species, their pharmacokinetic (PK) properties, and strain-host interactions. Intrinsic properties, such as bile salt deconjugation properties, mucin degradation properties, or platelet aggregation properties (which seem to be responsible for cardiac valve colonization and formation of unwanted metabolites in experiments), might be hazardous for human health and can be studied in vitro [38,100]. As the probiotic survival differs for varies bacterial species, to define the specific strain, collecting feces can be used for studying probiotics in vivo. Other approaches include intestinal intubation or performing mucosal biopsies [38,100].

The Food and Agriculture Organization of the United Nations (FAO) and the World Health Organization (WHO) provided guidelines [4] for the evaluation of probiotic safety used in foods. Namely, it is recommended that probiotic strains are characterized by a series of inspections (strain specificity is linked with probiotic effects) which will determine possible health hazard risks. Series of inspections include testing of antibiotic resistance properties, probiotic metabolic activities, and unwanted product formation, e.g., bile salt deconjugation, or D-lactate production. Furthermore, it should be possible to assess adverse effects for consumers by accurate surveillance and epidemiological studies and inspect toxin(s) production and hemolytic activity of probiotics after their intake. Assays should also test probiotic properties in animal models, which will be immunocompromised. Additionally, evaluation of probiotic safety should include tests of anti-mutagenic, anti-carcinogenic, and nonpathogenic probiotic properties [100,101]. Nevertheless, consulting a healthcare expert(s) is always warranted to avoid any issues, regardless of the reasons for probiotic use, especially in cases of serious illness, or hospitalization, which demands close monitoring of patients.

3. Probiotics: Functional Foods

The value of many foods on the market can be often enhanced by the addition of probiotics. It is not surprising that such foods are considered a better choice in the eyes of the consumers [102] who perceive it as the one with health benefits, so its higher cost is simply justified [103]. Nevertheless, when it comes to probiotics as food components, the whole picture must be looked very carefully as consumers are not homogenous groups, so the assessment of probiotic safety should not be generalized. A detailed review on relevant concerns during functional food development was published elsewhere [104]. However, basic steps in manufacturing and development of probiotic (functional) foods are given in Figure 1.

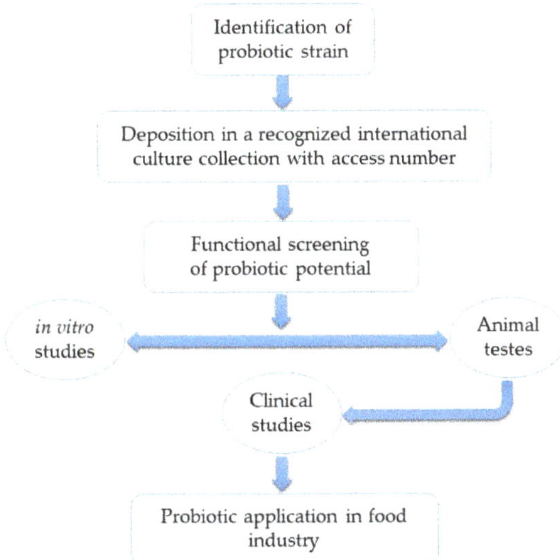

Figure 1. Basic steps for functional probiotic food development.

The most widely used probiotics in the food industry are given in Table 1. Due to the large body of available literature, for further information about the use of various probiotics strains in food industry reader is referred to other sources [79,105,106]. Briefly, strains of *Lactobacillus* and *Bifidobacterium* genera are very heat sensitive [107], therefore *Bacillus coagulans* has attracted the industry interest as this spore-forming bacteria is resistant to heat and possess some characteristics of *Bacillus* and *Lactobacillus* genera [108]. Although probiotics are described as "beneficial" or "friendly" bacteria, however, it should be noted that some types of yeasts, such as *Saccharmyces* (*Saccharomyces cerevisiae* var. *boulardii; S. boulardii*), are also defined as probiotics [37,109]. A crucial point of probiotic stability in foods is their ability to remain in high amounts in the product during processing and storage, together with their viability after ingestion [110,111]. Moreover, various sugars, salt, antimicrobials, compounds used as aroma, water content, oxygen level, pH, temperature, and packaging material impact the probiotic viability in both, positive and negative way [81,112].

Table 1. The most widely used probiotics in the food industry.

Lactobacillus Species	Bifidobacterium Species	Others
L. acidophilus		*Bacillus coagulans*
L. amylovorus		*Bacillus cereus*
L. brevis		*Clostridium botyricum*
L. casei		*Enterococcus faecalis*
L. rhamnosus		*Enterococcus faecium*
L. crispatus	*B. adolescentis*	*Escherichia coli*
L. delbrueckii subsp. Bulgaricus	*B. animalis*	*Lactococcus lactis subsp. Cremoris*
L. fermentum	*B. breve*	*Lactococcus lactis subsp. Lactis*
L. gasseri	*B. bifidum*	*Leuconostoc mesenteroides subsp. Dextranicum*
	B. infantis	
L. helveticus	*B. lactis*	*Pediococcus acidilactici*
L. johnsonii	*B. longum*	*Propionibacterium freudenreichii*
L. lactis		*Saccharomyces boulardii*
L. paracasei		*Streptococcus salivarius subsp. Thermophilus*
L. plantarum		*Sporolactobacillus inulinus*
L. reuteri		
L. salivarius		
L. gallinarum		

In comparison to the pure chemicals, substances, or pharmaceuticals, it is harder to predict the impact of probiotic bacteria in foods and their causal relationship regarding the possible adverse effects. Namely, by exploring the hazardous effects of bacteria from food to date, it is obvious that the risks of toxicity are constantly present [38]. For a better understanding, quantitative risk assessment models should be used. However, "minimal infective dose" for monitoring consumer probiotic safety cannot be straightforwardly determined, as there are just too many parameters. For instance, there is a plethora of microbes and various host's factors, besides the manufacturing process, that influences the probiotic viability. Moreover, there is too large individual variation among consumers, as they can be healthy or with the disease. Additionally, the general cellular mechanism of probiotic effects, safety, and toxicity in humans are still demanding further clarification and studies [38]. In other words, hazards of probiotic intake from food could not be easily toxicologically predictable, especially with inter-individual and intra-individual differences among consumers. Even though theoretical concerns about bacteremia and fungemia are justified, it is unlikely that probiotics from food could show infectivity in a healthy population.

Reported cases of probiotic pathogenicity from foods are related to an immunocompromised people, such as child following bone marrow transplant [113], and 74-year old women with a history of diabetes, who reported a daily intake of 500 mL of dairy drinks containing *L. rhamnosus* GG to relieve her abdominal discomfort [114]. The case of the immunocompromised patient with AIDS, who developed bacteremia from *Lactobacillus acidophilus*, related to excessive consumption of probiotic-enriched yogurts, was already mentioned before [50]. On the positive side, it is interesting to mention that there is some evidence that probiotics can act as potential adsorbents of aflatoxins (ubiquitous contaminants) found in foods [115], and besides, interacting with food components, probiotics recently came in the spotlight, due to the possibility to influence pharmacokinetics/pharmacodynamics of drugs that was previously explained.

Probiotic Fruit Beverages

The production of fruit juices (contain 100% fruit), nectars (up to 25–99% fruit) and juice drinks (up to 25% fruit content) has become one of the largest sectors in the food industry. According to the European Fruit Juice Association (AIJN), global consumption of both fruit juice and nectars was 36,247 million liters in 2017, while all EU countries together consumed 9187 million liters [116]. In particular, fruit juices contain appreciable amounts of dietary fibers, antioxidants, polyphenols, minerals and vitamins, whereas probiotic addition could further enhance benefits of fruit juices consummation. The particular advantage of fruit juices is that they provide a good environment that is capable of stabilizing probiotic strains [83]. Furthermore, the addition of probiotics improves nutritional properties of fruit juices, and enhances native antioxidant properties of beverages. This is additional to the lowering of the pH in the intestines, which has positive repercussions on digestion, absorption of calcium, iron, and magnesium from the native fruit matrix. This is a very desirable property from nutritional aspects [117–119] with supplying ascorbic acid (vitamin C) that has a protective effect on probiotic viability as well [120]. Therefore, such a beneficial combination seems to be an excellent nutritional choice, and it is not surprising that the idea of consumption of such beverages is rapidly growing in the world markets [121,122].

However, there is a question of proper assessment of the bioavailability (fraction of nutrient secreted into circulation and available at the site of action) and bioaccessibility (fraction of bioactive substance that is released from the food matrix) of the health-related bioactive components in the beverages which refers to probiotics as well. Both of these parameters cannot be easily assessed in functional fruit beverages. Secondly, metabolism of health-related bioactive component must also be considered, because sole data of the quantitative input is not sufficient, i.e., the most abundant health-related bioactive compounds from the food matrix do not necessarily imply that it will reach the highest concentrations at the physiological site of the action. For instance, in the case of probiotics, they should be able to survive the exposure to the pepsin [47]. Finally, as seen with medications, absorption of health-related bioactive compounds from food matrix can also differ in the population [48], and it does not help either that there is an evident gap in knowledge about the physicochemical and physiological processes that are involved in the transformation nutrients in the fruit juices [37]. Hence, new techniques and ideas about probiotic functional beverage formulations are more than needed in the near future.

From an industrial perspective, there are always challenges related to adding health-enhancing components, including probiotics, to food matrix. For example, the process of development and formulation of fruit juices as a probiotic carrier is a very complexed task. To design a functional (fruit) beverage with probiotics, it is important that the strains should survive at lower pH [123]. This is additional to resistance to added preservatives and sugars that can negatively influence probiotic viability, and therefore, it is important to examine the stability of probiotic strains in a model juice systems [120]. Usual limitations for the addition of probiotics to fruit juices include: The high acidity, the presence of oxygen, the inadequate amounts of free amino acids, short peptides, and oligosaccharides required for probiotics [124]. Other disadvantages of using probiotics in fruit juices are related to the presence of dyes, flavors, preservatives, antimicrobial components and influence on sensory characteristics [124]. Hence, the proper selection of the right probiotic strain is crucial, as their stability, survival and functionality are more challenging in juices as compared to the addition in common fermented dairy products [125–127].

On the other hand, there are numerous options for fruit juices that could be suitable as carriers of probiotic bacteria [128–131]. Suitable fruits as raw materials include: papaya [132], cranberry, lemon, grapefruit, blackcurrant [130], orange [133], apple [134], acerola [127], apple-carrot juice, and pear juice [121]. Some examples of used probiotics in fruit juices are *Bifidobacterium* and *Lactobacillus* species [119], such as *L. plantarum*, *L. acidophilus*, *L. helveticus*, *L. casei*, *L. paracasei*, *L. rhamnosus*, etc. [135]. To alieve manufacturing limitations, some of the proposed approaches [136] to promote probiotic survival in fruit juices are microencapsulation [127,137], fortification with additional prebiotics [138], probiotic strain exposure to the sub-lethal stress which induces adaptive stress response and survival [125], refrigeration and additional use of antioxidants, such as vitamins [139].

It is also important to mention common authenticity issues regarding the use of fruit juices due to various potential frauds including: Water and sugar addition; partial replacement of fruit juice by juices made from concentrates; added products from undeclared cheaper fruits; addition of undeclared ascorbic acid/vitamin C; addition of undeclared organic acids (e.g., citric acid and malic acid); addition of flavor compounds (natural or synthetic); colorings (e.g., anthocyanin extracts, cochenille red, beetroot); adding the texture influencing agents (e.g., pectin). Moreover, the addition or over-proportional use of fruit extracts, which were produced by unauthorized technology and declaration of false origins or declaration of deceitful fruit varieties [140]. In general, it can be stated that most fraud has no real impact regarding the consumers' safety. However, every food fraud could be a potential health risk, especially in the case of contamination with unexpected agrochemicals [141] or contaminants, even from the probiotic supplementation [142].

Altogether, the recent developments of food processing technologies and constant demands from the consumers regarding the more nutritious and safe food products, fruit juices with added probiotics are soon expected to become a new class of functional foods and important element on general food markets, as well as an integral part of proper nutrition [143,144]. This drive is additionally fostered by an increase in vegetarianism where some companies already have probiotic fruit juice beverages in their portfolio for those seeking a healthy lifestyle [83–145].

4. Conclusions

An enlarged interest of food industry to find new probiotics non-diary vehicles led to increased use of fruit juices as new matrices, representing a new type of functional foods with a great potential for providing even more health benefits for the consumers and for those seeking a healthy lifestyle. Design of functional fruit beverages with probiotics is still a challenging task, but with current and future technological solutions, it should be possible to derive nutritional and economic benefits for consumers and industry from these types of product.

Probiotics do not work in the same way for every individual, and they should be consumed considering the probiotic strain(s) specificity and sources of intake, levels of exposure, manufacturing properties, along with demands from regulatory authorities, pathological states, and general nutritional status with a known history of using medications. From the pharmacological and toxicological aspects, probiotic safety and toxicity, along with their efficacy and observed health benefits are dependent on various factors, and there is not a "one-size-fits-all" criterion for their clinical evaluation and recommendations of intake. Hence, we are suggesting individualized clinical evaluation before any consumption of probiotics. In general, probiotics are considered as safe for a healthy population, but they may pose a threat for at-risk populations, especially if considering documented case reports and theoretical concerns about their safety and toxicity. Regardless of the age, at-risk populations include critically sick patients, patients at intensive care units, postoperative and hospitalized patients, and especially immunocompromised patients. Besides supporting the idea about long-term clinical studies of probiotics, we expect that questions of probiotic efficacy, safety, and toxicity for humans will be the focus of future research focus, and provide the missing pieces of the puzzle needed for defining probiotics as aid or detriment to the health.

Author Contributions: Conceptualization, I.Ž. and P.P.; writing—original draft preparation, I.Ž., Z.P., P.P.; writing—review and editing, I.Ž., P.P., D.B.K.; project administration, P.P., D.B.K. All authors have read and agreed to the published version of the manuscript.

Funding: This research was funded by the Croatian Science Foundation through the funding of the "Hurdle Technology and 3D Printing for Sustainable Fruit Juice Processing and Preservation" project, number "IP-2019-04-2105".

Conflicts of Interest: The authors declare no conflict of interest.

References

1. Hanson, L.A.; Yolken, R.H.; Services, N.N. Probiotics, Other Nutritional Factors, and Intestinal Microflora. In *Nestle Nutrition Workshop Series: Pediatric Program*; Lippincott-Raven: La Tour-de-Peilz, Switzerland, 1999; Volume 42.
2. Mantzourani, I.; Kazakos, S.; Terpou, A.; Alexopoulos, A.; Bezirtzoglou, E.; Bekatorou, A.; Plessas, S. Potential of the probiotic *Lactobacillus Plantarum* ATCC 14917 strain to produce functional fermented pomegranate juice. *Foods* **2018**, *8*, 4. [CrossRef] [PubMed]
3. Chugh, B.; Kamal-Eldin, A. Bioactive compounds produced by probiotics in food products. *Curr. Opin. Food Sci.* **2020**, *32*, 76–82. [CrossRef]
4. Group, J.F.W.W. Report on Drafting Guidelines for the Evaluation of Probiotics in Food. London, Ontario, Canada. 2002. Available online: https://www.who.int/foodsafety/fs_management/en/probiotic_guidelines.pdf (accessed on 15 June 2020).
5. White, J.; Hekmat, S. Development of probiotic fruit juices using *Lactobacillus rhamnosus* GR-1 fortified with short chain and long chain inulin fiber. *Fermentation* **2018**, *4*, 27. [CrossRef]
6. Ouwehand, A.C.; Invernici, M.M.; Furlaneto, F.A.C.; Messora, M.R. Effectiveness of Multistrain Versus Single-strain Probiotics: Current Status and Recommendations for the Future. *J. Clin. Gastroenterol.* **2018**, *52*, S35–S40. [CrossRef] [PubMed]
7. Markowiak, P.; Slizewska, K. Effects of Probiotics, Prebiotics, and Synbiotics on Human Health. *Nutrients* **2017**, *9*, 1021. [CrossRef]
8. Sanders, M.E.; Merenstein, D.; Merrifield, C.A.; Hutkins, R. Probiotics for human use. *Nutr. Bull.* **2018**, *43*, 212–225. [CrossRef]
9. The European Parliament and the Council of the European Union. Regulation (EU) 2015/2283 of the European Parliament and of the Council of 25 November 2015 on novel foods, amending Regulation (EU) No 1169/2011. European Parliament and of the Council and repealing Regulation (EC) No 258/97 of the European Parliament and of the Council and Commission Regulation (EC) No 1852/2001, Ed. *Off. J. Eur. Union* **2015**, *327*, 1–22.

10. Forestier, C.; Mangalat, N.; Liu, Y.; Fatheree, N.Y.; Ferris, M.J.; Van Arsdall, M.R.; Chen, Z.; Rahbar, M.H.; Gleason, W.A.; Norori, J.; et al. Safety and tolerability of *Lactobacillus reuteri* DSM 17938 and effects on biomarkers in healthy adults: Results from a randomized masked trial. *PLoS ONE* **2012**, *7*, e43910. [CrossRef]
11. Allen, S.J.; Jordan, S.; Storey, M.; Thornton, C.A.; Gravenor, M.; Garaiova, I.; Plummer, S.F.; Wang, D.; Morgan, G. Dietary supplementation with Lactobacilli and Bifidobacteria is well tolerated and not associated with adverse events during late pregnancy and early infancy. *J. Nutr.* **2010**, *140*, 483–488. [CrossRef]
12. Dekker, J.W.; Wickens, K.; Black, P.N.; Stanley, T.V.; Mitchell, E.A.; Fitzharris, P.; Tannock, G.W.; Purdie, G.; Crane, J. Safety aspects of probiotic bacterial strains *Lactobacillus rhamnosus* HN001 and *Bifidobacterium animalis* subsp. lactis HN019 in human infants aged 0–2 years. *Int. Dairy J.* **2009**, *19*, 149–154. [CrossRef]
13. Olivares, M.; Castillejo, G.; Varea, V.; Sanz, Y. Double-blind, randomised, placebo-controlled intervention trial to evaluate the effects of *Bifidobacterium longum* CECT 7347 in children with newly diagnosed coeliac disease. *Br. J. Nutr.* **2014**, *112*, 30–40. [CrossRef] [PubMed]
14. Manley, K.J.; Fraenkel, M.B.; Mayall, B.C.; Power, D.A. Probiotic treatment of vancomycin-resistant enterococci: A randomised controlled trial. *Med. J. Aust.* **2007**, *186*, 454–457. [CrossRef] [PubMed]
15. Srinivasan, R.; Meyer, R.; Padmanabhan, R.; Britto, J. Clinical safety of Lactobacillus casei Shirota as a probiotic in critically ill children. *J. Pediatr. Gastroenterol. Nutr.* **2006**, *42*, 171–173. [CrossRef]
16. Kunk, D. Probiotics: Elixir or empty promise? *Lancet Gastroenterol. Hepatol.* **2019**, *4*, 81. [CrossRef]
17. Schnadower, D.; Tarr, P.I.; Casper, T.C.; Gorelick, M.H.; Dean, J.M.; O'Connell, K.J.; Mahajan, P.; Levine, A.C.; Bhatt, S.R.; Roskind, C.G.; et al. *Lactobacillus rhamnosus* GG versus placebo for acute gastroenteritis in children. *N. Engl. J. Med.* **2018**, *379*, 2002–2014. [CrossRef] [PubMed]
18. Freedman, S.B.; Williamson-Urquhart, S.; Farion, K.J.; Gouin, S.; Willan, A.R.; Poonai, N.; Hurley, K.; Sherman, P.M.; Finkelstein, Y.; Lee, B.E.; et al. Multicenter trial of a combination probiotic for children with gastroenteritis. *N. Engl. J. Med.* **2018**, *379*, 2015–2026. [CrossRef]
19. Suez, J.; Zmora, N.; Segal, E.; Elinav, E. The pros, cons, and many unknowns of probiotics. *Nat. Med.* **2019**, *25*, 716–729. [CrossRef]
20. Goldenberg, J.Z.; Yap, C.; Lytvyn, L.; Lo, C.K.-F.; Beardsley, J.; Mertz, D.; Johnston, B.C. Probiotics for the prevention of *Clostridium difficile*-associated diarrhea in adults and children. *Cochrane Database Syst. Rev.* **2017**. [CrossRef]
21. Parker, E.A.; Roy, T.; D'Adamo, C.R.; Wieland, L.S. Probiotics and gastrointestinal conditions: An overview of evidence from the Cochrane Collaboration. *Nutrition* **2018**, *45*, 125–134. [CrossRef] [PubMed]
22. Day, R.L.J.; Harper, A.J.; Woods, R.M.; Davies, O.G.; Heaney, L.M. Probiotics: Current landscape and future horizons. *Future Sci. OA* **2019**, *5*, FSO391. [CrossRef]
23. Lerner, A.; Shoenfeld, Y.; Matthias, T. Probiotics: If it does not help it does not do any harm. Really? *Microorganisms* **2019**, *7*, 104. [CrossRef] [PubMed]
24. Van den Nieuwboer, M.; Brummer, R.J.; Guarner, F.; Morelli, L.; Cabana, M.; Claassen, E. The administration of probiotics and synbiotics in immune compromised adults: Is it safe? *Benefic. Microbes* **2015**, *6*, 3–17. [CrossRef] [PubMed]
25. Doron, S.; Snydman, D.R. Risk and Safety of Probiotics. *Clin. Infect. Dis.* **2015**, *60*, S129–S134. [CrossRef] [PubMed]
26. Van der Geest, A.M.; Flach, J.; Claassen, E.; Sijlmans, A.W.; Van de Burgwal, L.H.M.; Larsen, O.F.A. European General Practitioners perceptions on probiotics: Results of a multinational survey. *PharmaNutrition* **2020**, *11*. [CrossRef]
27. Kokosky, G. Pharmacists Play a Key Role in Counseling Patients about Probiotics. Available online: https://www.pharmacytimes.com/publications/issue/2018/September2018/pharmacists-play-a-key-role-in-counseling-patients-who-may-need-probiotics (accessed on 22 June 2020).
28. Happel, A.-U.; Jaumdally, S.Z.; Pidwell, T.; Cornelius, T.; Jaspan, H.B.; Froissart, R.; Barnabas, S.L.; Passmore, J.-A.S. Probiotics for vaginal health in South Africa: What is on retailers' shelves? *BMC Women Health* **2017**, *17*, 1–10. [CrossRef]
29. Tachedjian, G.; Aldunate, M.; Bradshaw, C.S.; Cone, R.A. The role of lactic acid production by probiotic *Lactobacillus* species in vaginal health. *Res. Microbiol.* **2017**, *168*, 782–792. [CrossRef]
30. Venugopalan, V.; Shriner, K.A.; Wong-Beringer, A. Regulatory oversight and safety of probiotic use. *Emerg. Infect. Dis.* **2010**, *16*, 1661–1665. [CrossRef]

31. United Kingdom National Health Service. Probiotics. Available online: https://www.nhs.uk/conditions/probiotics (accessed on 4 July 2020).
32. De Simone, C. The unregulated probiotic market. *Clin. Gastroenterol. Hepatol.* **2019**, *17*, 809–817. [CrossRef]
33. Gargari, G.; Taverniti, V.; Koirala, R.; Gardana, C.; Guglielmetti, S. Impact of a multistrain probiotic formulation with high bifidobacterial content on the fecal bacterial community and short-chain fatty acid levels of healthy adults. *Microorganisms* **2020**, *8*, 492. [CrossRef]
34. Binnendijk, K.H.; Rijkers, G.T. What is a health benefit? An evaluation of EFSA opinions on health benefits with reference to probiotics. *Benef. Microbes* **2013**, *4*, 223–230. [CrossRef]
35. Betta, P. Not all probiotics are the same: Gut microbiota modulation with a multistrain probiotics. *N. Am. J. Med. Sci.* **2014**, *6*, 58. [CrossRef] [PubMed]
36. Vinderola, G.; Binetti, A.; Burns, P.; Reinheimer, J. Cell viability and functionality of probiotic bacteria in dairy products. *Front. Microbiol.* **2011**, *2*, 70. [CrossRef] [PubMed]
37. California, B.U.o. Probiotics Pros and Cons by Berkeley Wellness. Available online: https://www.berkeleywellness.com/supplements/other-supplements/article/probiotics-pros-and-cons (accessed on 4 July 2020).
38. Marteau, P. Safety aspects of probiotic products. *Näringsforskning* **2016**, *45*, 22–24. [CrossRef]
39. Anadón, A.; Martínez-Larrañaga, M.R.; Ares, I.; Martínez, M.A. Probiotics: Safety and Toxicity Considerations. In *Nutraceuticals*; Gupta, R.C., Ed.; Academic Press: Cambridge, MA, USA, 2016; pp. 777–798.
40. Liong, M.-T. Safety of probiotics: Translocation and infection. *Nutr. Rev.* **2008**, *66*, 192–202. [CrossRef] [PubMed]
41. Sanders, M.E.; Merenstein, D.J.; Ouwehand, A.C.; Reid, G.; Salminen, S.; Cabana, M.D.; Paraskevakos, G.; Leyer, G. Probiotic use in at-risk populations. *J. Am. Pharm. Assoc.* **2016**, *56*, 680–686. [CrossRef] [PubMed]
42. Kothari, D.; Patel, S.; Kim, S.-K. Probiotic supplements might not be universally-effective and safe: A review. *Biomed. Pharmacother.* **2019**, *111*, 537–547. [CrossRef] [PubMed]
43. Redman, M.G.; Ward, E.J.; Phillips, R.S. The efficacy and safety of probiotics in people with cancer: A systematic review. *Ann. Oncol.* **2014**, *25*, 1919–1929. [CrossRef]
44. Nielsen, D.E.; El-Sohemy, A. Applying genomics to nutrition and lifestyle modification. *Personal. Med.* **2012**, *9*, 739–749. [CrossRef]
45. Anukam, K.C. Probiotic toxicity, any evidence? *J. Pharmacol. Toxicol.* **2007**, *2*, 590–598. [CrossRef]
46. Ishibashi, N.; Yamazaki, S. Probiotics and safety. *Am. J. Clin. Nutr.* **2001**, *73*, 465–470. [CrossRef]
47. Enright, E.F.; Gahan, C.G.; Joyce, S.A.; Griffin, B.T. The impact of the gut microbiota on drug metabolism and clinical outcome. *Yale J. Biol. Med.* **2016**, *89*, 375–382. [PubMed]
48. Abdollahi, M.; Abdolghaffari, A.H.; Gooshe, M.; Ghasemi-Niri, F. Safety of Probiotic Bacteria. In *Probiotics, Prebiotics, and Synbiotics: Bioactive Foods in Health Promotion*; Watson, R.R., Victor, R., Preedy, V.R., Eds.; Academic Press: Cambridge, MA, USA, 2016; pp. 227–237. [CrossRef]
49. LeDoux, D.; LaBombardi, V.J.; Karter, D. *Lactobacillus acidophilus* bacteraemia after use of a probiotic in a patient with AIDS and Hodgkin's disease. *Int. J. STD AIDS* **2016**, *17*, 280–282. [CrossRef] [PubMed]
50. Haghighat, L.; Crum-Cianflone, N.F. The potential risks of probiotics among HIV-infected persons: Bacteraemia due to *Lactobacillus acidophilus* and review of the literature. *Int. J. STD AIDS* **2016**, *27*, 1223–1230. [CrossRef] [PubMed]
51. Xu, X.; Hicks, C.; Li, Y.; Su, J.; Shiloach, J.; Kaufman, J.B.; Fitz, Y.; Eichacker, P.Q.; Cui, X. Purified cell wall from the probiotic bacterium *Lactobacillus gasseri* activates systemic inflammation and, at higher doses, produces lethality in a rat model. *Crit. Care* **2014**, *18*, R140. [CrossRef]
52. Venema, K.; Do Carmo, A.P. *Probiotics and Prebiotics, Current Research and Future Trends*; Caister Academic Press: Norfolk, UK, 2015.
53. Enache-Angoulvant, A.; Hennequin, C. Invasive *Saccharomyces* infection: A comprehensive review. *Clin. Infect. Dis.* **2005**, *41*, 1559–1568. [CrossRef]
54. Munoz, P.; Bouza, E.; Cuenca-Estrella, M.; Eiros, J.M.; Perez, M.J.; Sanchez-Somolinos, M.; Rincon, C.; Hortal, J.; Pelaez, T. *Saccharomyces cerevisiae* fungemia: An emerging infectious disease. *Clin. Infect. Dis.* **2005**, *40*, 1625–1634. [CrossRef]
55. Gao, X.; Liu, J.; Li, L.; Liu, W.; Sun, M. A brief review of nutraceutical ingredients in gastrointestinal disorders: Evidence and suggestions. *Int. J. Mol. Sci.* **2020**, *21*, 1822. [CrossRef] [PubMed]
56. Fadhel, M.; Patel, S.; Liu, E.; Levitt, M.; Asif, A. *Saccharomyces cerevisiae* fungemia in a critically ill patient with acute cholangitis and long term probiotic use. *Med. Mycol. Case Rep.* **2019**, *23*, 23–25. [CrossRef]

57. Sherid, M.; Samo, S.; Sulaiman, S.; Husein, H.; Sifuentes, H.; Sridhar, S. Liver abscess and bacteremia caused by lactobacillus: Role of probiotics? Case report and review of the literature. *BMC Gastroenterol.* **2016**, *16*. [CrossRef]
58. Sullivan, Å.; Erik Nord, C. Probiotic lactobacilli and bacteraemia in Stockholm. *Scand. J. Infect. Dis.* **2009**, *38*, 327–331. [CrossRef]
59. Cannon, J.P.; Lee, T.A.; Bolanos, J.T.; Danziger, L.H. Pathogenic relevance of *Lactobacillus*: A retrospective review of over 200 cases. *Eur. J. Clin. Microbiol. Infect. Dis.* **2004**, *24*, 31–40. [CrossRef]
60. Botros, M.; Mukundan, D. *Lactobacillus endocarditis* with prosthetic material: A case report on non-surgical management with corresponding literature review. *Infect. Dis. Rep.* **2014**, *6*. [CrossRef] [PubMed]
61. Salvana, E.M.T.; Frank, M. *Lactobacillus endocarditis*: Case report and review of cases reported since 1992. *J. Infect.* **2006**, *53*, e5–e10. [CrossRef]
62. Ze-Ze, L. Case of aortic endocarditis caused by *Lactobacillus casei*. *J. Med. Microbiol.* **2004**, *53*, 451–453. [CrossRef] [PubMed]
63. Franko, B.; Vaillant, M.; Recule, C.; Vautrin, E.; Brion, J.P.; Pavese, P. *Lactobacillus paracasei endocarditis* in a consumer of probiotics. *Méd. Et Mal. Infect.* **2013**, *43*, 171–173. [CrossRef] [PubMed]
64. Besselink, M.G.H.; van Santvoort, H.C.; Buskens, E.; Boermeester, M.A.; van Goor, H.; Timmerman, H.M.; Nieuwenhuijs, V.B.; Bollen, T.L.; van Ramshorst, B.; Witteman, B.J.M.; et al. Probiotic prophylaxis in predicted severe acute pancreatitis: A randomised, double-blind, placebo-controlled trial. *Lancet* **2008**, *371*, 651–659. [CrossRef]
65. Taylor, A.L.; Dunstan, J.A.; Prescott, S.L. Probiotic supplementation for the first 6 months of life fails to reduce the risk of atopic dermatitis and increases the risk of allergen sensitization in high-risk children: A randomized controlled trial. *J. Allergy Clin. Immunol.* **2007**, *119*, 184–191. [CrossRef]
66. Pelucchi, C.; Chatenoud, L.; Turati, F.; Galeone, C.; Moja, L.; Bach, J.-F.; La Vecchia, C. Probiotics supplementation during pregnancy or infancy for the prevention of atopic dermatitis. *Epidemiology* **2012**, *23*, 402–414. [CrossRef]
67. Garcia-Larsen, V.; Ierodiakonou, D.; Jarrold, K.; Cunha, S.; Chivinge, J.; Robinson, Z.; Geoghegan, N.; Ruparelia, A.; Devani, P.; Trivella, M.; et al. Diet during pregnancy and infancy and risk of allergic or autoimmune disease: A systematic review and meta-analysis. *PLoS Med.* **2018**, *15*. [CrossRef]
68. Olier, M.; Marcq, I.; Salvador-Cartier, C.; Secher, T.; Dobrindt, U.; Boury, M.; Bacquié, V.; Penary, M.; Gaultier, E.; Nougayrède, J.-P.; et al. Genotoxicity of *Escherichia coli* Nissle 1917 strain cannot be dissociated from its probiotic activity. *Gut Microb.* **2014**, *3*, 501–509. [CrossRef]
69. Imperial, I.C.V.J.; Ibana, J.A. Addressing the antibiotic resistance problem with probiotics: Reducing the risk of its double-edged sword effect. *Front. Microbiol.* **2016**, *07*, 1983. [CrossRef]
70. Ashraf, R.; Shah, N.P. Antibiotic resistance of probiotic organisms and safety of probiotic dairy products. *Int. Food Res. J.* **2011**, *18*, 837–853.
71. Gueimonde, M.; Sánchez, B.G.; De los Reyes-Gavilán, C.; Margolles, A. Antibiotic resistance in probiotic bacteria. *Front. Microbiol.* **2013**, *4*, 202. [CrossRef]
72. Connolly, E.; Abrahamsson, T.; Björkstén, B. Safety of D(-)-lactic acid producing bacteria in the human infant. *J. Pediatr. Gastroenterol. Nutr.* **2005**, *41*, 489–492. [CrossRef] [PubMed]
73. Rao, S.S.C.; Rehman, A.; Yu, S.; de Andino, N.M. Brain fogginess, gas and bloating: A link between SIBO, probiotics and metabolic acidosis. *Clin. Transl. Gastroenterol.* **2018**, *9*. [CrossRef] [PubMed]
74. Munakata, S.; Arakawa, C.; Kohira, R.; Fujita, Y.; Fuchigami, T.; Mugishima, H. A case of D-lactic acid encephalopathy associated with use of probiotics. *Brain Dev.* **2010**, *32*, 691–694. [CrossRef]
75. Alvarez, M.A.; Moreno-Arribas, M.V. The problem of biogenic amines in fermented foods and the use of potential biogenic amine-degrading microorganisms as a solution. *Trends Food Sci. Technol.* **2014**, *39*, 146–155. [CrossRef]
76. Kuley, E.; Balıkcı, E.; Özoğul, İ.; Gökdogan, S.; Özoğul, F. Stimulation of cadaverine production by foodborne pathogens in the presence of *Lactobacillus*, *Lactococcus*, and *Streptococcus* spp. *J. Food Sci.* **2012**, *77*, M650–M658. [CrossRef]
77. Jansen, S.C.; van Dusseldorp, M.; Bottema, K.C.; Dubois, A.E.J. Intolerance to dietary biogenic amines: A review. *Ann. Allergy Asthma Immunol.* **2003**, *91*, 233–241. [CrossRef]
78. Linares, D.M.; Martín, M.; Ladero, V.; Alvarez, M.A.; Fernández, M. Biogenic amines in dairy products. *Crit. Rev. Food Sci. Nutr.* **2011**, *51*, 691–703. [CrossRef]

79. Terpou, A.; Papadaki, A.; Lappa, I.K.; Kachrimanidou, V.; Bosnea, L.A.; Kopsahelis, N. Probiotics in food systems: Significance and emerging strategies towards improved viability and delivery of enhanced beneficial value. *Nutrients* **2019**, *11*, 1591. [CrossRef] [PubMed]
80. Hungin, A.P.S.; Mitchell, C.R.; Whorwell, P.; Mulligan, C.; Cole, O.; Agréus, L.; Fracasso, P.; Lionis, C.; Mendive, J.; Philippart de Foy, J.M.; et al. Systematic review: Probiotics in the management of lower gastrointestinal symptoms-An updated evidence-based international consensus. *Aliment. Pharmacol. Ther.* **2018**, *47*, 1054–1070. [CrossRef]
81. Williams, N.T. Probiotics. *Am. J. Health Syst. Pharm.* **2010**, *67*, 449–458. [CrossRef] [PubMed]
82. Pakdaman, M.N.; Udani, J.K.; Molina, J.P.; Shahani, M. The effects of the DDS-1 strain of lactobacillus on symptomatic relief for lactose intolerance-a randomized, double-blind, placebo-controlled, crossover clinical trial. *Nutr. J.* **2016**, *15*, 1–11. [CrossRef] [PubMed]
83. Fernandes Pereira, A.L. Turning Fruit Juice into Probiotic Beverages. In *Fruit Juices Extraction, Composition, Qquality and Analysis*; Rajauria, G., Tiwari, B., Eds.; Academic Press: Cambridge, MA, USA, 2018; pp. 279–285.
84. Koziolek, M.; Alcaro, S.; Augustijns, P.; Basit, A.W.; Grimm, M.; Hens, B.; Hoad, C.L.; Jedamzik, P.; Madla, C.M.; Maliepaard, M.; et al. The mechanisms of pharmacokinetic food-drug interactions—A perspective from the UNGAP group. *Eur. J. Pharm. Sci.* **2019**, *134*, 31–59. [CrossRef]
85. Noh, K.; Kang, Y.R.; Nepal, M.R.; Shakya, R.; Kang, M.J.; Kang, W.; Lee, S.; Jeong, H.G.; Jeong, T.C. Impact of gut microbiota on drug metabolism: An update for safe and effective use of drugs. *Arch. Pharm. Res.* **2017**, *40*, 1345–1355. [CrossRef]
86. Wilson, I.D.; Nicholson, J.K. Gut microbiome interactions with drug metabolism, efficacy, and toxicity. *Transl. Res.* **2017**, *179*, 204–222. [CrossRef]
87. El Aidy, S.; van den Bogert, B.; Kleerebezem, M. The small intestine microbiota, nutritional modulation and relevance for health. *Curr. Opin. Biotechnol.* **2015**, *32*, 14–20. [CrossRef]
88. Deli, M.A.; Matuskova, Z.; Anzenbacherova, E.; Vecera, R.; Tlaskalova-Hogenova, H.; Kolar, M.; Anzenbacher, P. Administration of a probiotic can change drug pharmacokinetics: Effect of *E. coli* Nissle 1917 on amidarone absorption in rats. *PLoS ONE* **2014**, *9*, e87150. [CrossRef]
89. Mikov, M.; Đanić, M.; Pavlović, N.; Stanimirov, B.; Goločorbin-Kon, S.; Stankov, K.; Al-Salami, H. Potential applications of gliclazide in treating Type 1 Diabetes Mellitus: Formulation with bile acids and probiotics. *Eur. J. Drug Metab. Pharmacokinet.* **2017**, *43*, 269–280. [CrossRef]
90. Saputri, F.A.; Kang, D.; Kusuma, A.S.W.; Rusdiana, T.; Hasanah, A.N.; Mutakin; Surono, I.S.; Koyama, H.; Abdulah, R. Lactobacillus plantarum IS-10506 probiotic administration increases amlodipine absorption in a rabbit model. *J. Int. Med. Res.* **2018**, *46*, 5004–5010. [CrossRef] [PubMed]
91. Al-Salami, H.; Butt, G.; Tucker, I.; Skrbic, R.; Golocorbin-Kon, S.; Mikov, M. Probiotic pre-treatment reduces gliclazide permeation (*ex vivo*) in healthy rats but increases it in diabetic rats to the level seen in untreated healthy rats. *Arch. Drug Inf.* **2008**, *1*, 35–41. [CrossRef] [PubMed]
92. Stepkowski, S.; Lee, J.R.; Muthukumar, T.; Dadhania, D.; Taur, Y.; Jenq, R.R.; Toussaint, N.C.; Ling, L.; Pamer, E.; Suthanthiran, M. Gut microbiota and tacrolimus dosing in kidney transplantation. *PLoS ONE* **2015**, *10*, e0122399. [CrossRef]
93. Holm, R.; Müllertz, A.; Mu, H. Bile salts and their importance for drug absorption. *Int. J. Pharm.* **2013**, *453*, 44–55. [CrossRef] [PubMed]
94. Pavlović, N.; Goločorbin-Kon, S.; Đanić, M.; Stanimirov, B.; Al-Salami, H.; Stankov, K.; Mikov, M. Bile acids and their derivatives as potential modifiers of drug release and pharmacokinetic profiles. *Front. Pharmacol.* **2018**, *9*, 1283. [CrossRef] [PubMed]
95. Moghimipour, E.; Ameri, A.; Handali, S. Absorption-enhancing effects of bile salts. *Molecules* **2015**, *20*, 14451–14473. [CrossRef]
96. Pavlović, N.; Stankov, K.; Mikov, M. Probiotics—Interactions with bile acids and impact on cholesterol metabolism. *Appl. Biochem. Biotechnol.* **2012**, *168*, 1880–1895. [CrossRef]
97. Mikawlrawng, K.; Kumar, S.; Bhatnagar, K. Drug interactions, safety and efficacy of probiotics. *Asian J. Med. Health* **2016**, *1*, 1–8. [CrossRef]
98. Vivarelli, S.; Falzone, L.; Basile, M.; Nicolosi, D.; Genovese, C.; Libra, M.; Salmeri, M. Benefits of using probiotics as adjuvants in anticancer therapy (Review). *World Acad. Sci. J.* **2019**. [CrossRef]

99. Mego, M.; Chovanec, J.; Vochyanova-Andrezalova, I.; Konkolovsky, P.; Mikulova, M.; Reckova, M.; Miskovska, V.; Bystricky, B.; Beniak, J.; Medvecova, L.; et al. Prevention of irinotecan induced diarrhea by probiotics: A randomized double blind, placebo controlled pilot study. *Complement. Ther. Med.* **2015**, *23*, 356–362. [CrossRef]
100. Saarela, M.; Mogensen, G.; Fondén, R.; Mättö, J.; Mattila-Sandholm, T. Probiotic bacteria: Safety, functional and technological properties. *J. Biotechnol.* **2000**, *84*, 197–215. [CrossRef]
101. Pradhan, D.; Mallappa, R.H.; Grover, S. Comprehensive approaches for assessing the safety of probiotic bacteria. *Food Control* **2020**, *108*, 106872. [CrossRef]
102. Zielińska, D.; Sionek, B.; Kołożyn-Krajewska, D. Safety of Probiotics. In *Diet, Microbiome and Health*; Mihai Grumezescu, A., Holban, A.M., Eds.; Academic Press: Cambridge, MA, USA, 2018; pp. 131–161. [CrossRef]
103. Khedkar, S.; Carraresi, L.; Bröring, S. Food or pharmaceuticals? Consumers' perception of health-related borderline products. *PharmaNutrition* **2017**, *5*, 133–140. [CrossRef]
104. Granato, D.; Barba, F.J.; Bursać Kovačević, D.; Lorenzo, J.M.; Cruz, A.G.; Putnik, P. Functional Foods: Product Development, Technological Trends, Efficacy Testing, and Safety. *Annu. Rev. Food Sci. Technol.* **2020**, *11*, 93–118. [CrossRef]
105. Zendeboodi, F.; Khorshidian, N.; Mortazavian, A.M.; da Cruz, A.G. Probiotic: Conceptualization from a new approach. *Curr. Opin. Food Sci.* **2020**, *32*, 103–123. [CrossRef]
106. Sarao, L.K.; Arora, M. Probiotics, prebiotics, and microencapsulation: A review. *Crit. Rev. Food Sci. Nutr.* **2017**, *57*, 344–371. [CrossRef]
107. Liu, X.; Champagne, C.P.; Lee, B.H.; Boye, J.I.; Casgrain, M. Thermostability of probiotics and their α-galactosidases and the potential for bean products. *Biotechnol. Res. Int.* **2014**, *2014*, 1–21. [CrossRef]
108. Konuray, G.; Erginkaya, Z. Potential use of *Bacillus coagulans* in the food industry. *Foods* **2018**, *7*, 92. [CrossRef]
109. Banik, A.; Halder, S.K.; Ghosh, C.; Mondal, K.C. Fungal probiotics: Opportunity, challenge, and prospects. In *Recent Advancement in White Biotechnology Through Fungi*; Yadav, A., Singh, S., Mishra, S., Gupta, A., Eds.; Springer: Cham, Switzerland, 2019; pp. 101–117. [CrossRef]
110. Champagne, C.P.; Gomes da Cruz, A.; Daga, M. Strategies to improve the functionality of probiotics in supplements and foods. *Curr. Opin. Food Sci.* **2018**, *22*, 160–166. [CrossRef]
111. Wilkinson, M.G. Flow cytometry as a potential method of measuring bacterial viability in probiotic products: A review. *Trends Food Sci. Technol.* **2018**, *78*, 1–10. [CrossRef]
112. Putta, S.; Yarla, N.S.; Lakkappa, D.B.; Imandi, S.B.; Malla, R.R.; Chaitanya, A.K.; Chari, B.P.V.; Saka, S.; Vechalapu, R.R.; Kamal, M.A.; et al. Probiotics: Supplements, Food, pharmaceutical Industry. In *Therapeutic, Probiotic, and Unconventional Foods*; Grumezescu, A.M., Holban, A.M., Eds.; Academic Press: Cambridge, MA, USA, 2018; pp. 15–25. [CrossRef]
113. Kalima, P.; Masterton, R.G.; Roddie, P.H.; Thomas, A.E. *Lactobacillus rhamnosus* infection in a child following bone marrow transplant. *J. Infect.* **1996**, *32*, 165–167. [CrossRef]
114. Rautio, M.; Jousimies-Somer, H.; Kauma, H.; Pietarinen, I.; Saxelin, M.; Tynkkynen, S.; Koskela, M. Liver abscess due to a *Lactobacillus rhamnosus* strain indistinguishable from *L. rhamnosus* strain GG. *Clin. Infect. Dis.* **1999**, *28*, 1159–1160. [CrossRef] [PubMed]
115. Elsanhoty, R.M.; Salam, S.A.; Ramadan, M.F.; Badr, F.H. Detoxification of aflatoxin M1 in yoghurt using probiotics and lactic acid bacteria. *Food Control* **2014**, *43*, 129–134. [CrossRef]
116. AIJN. AIJF European Fruit Juice Association. Annual Report 2018. Available online: http://viewer.zmags.com/publication/bc62cfea#/bc62cfea/1 (accessed on 12 May 2018).
117. Pimentel, T.C.; Klososki, S.J.; Rosset, M.; Barão, C.E.; Marcolino, V.A. Fruit Juices as Probiotic Foods. In *Sports and Energy Drinks*; Grumezescu, A., Holban, A.M., Eds.; Academic Press: Cambridge, MA, USA, 2019; pp. 483–513. [CrossRef]
118. Šárka, H.; Kristýna, R.; Kristina, B.; Iveta, K.; Marcela, S. Fruit juices with probiotics – new type of functional foods. *Czech J. Food Sci.* **2018**, *36*, 284–288. [CrossRef]
119. Nagpal, R.; Kumar, A.; Kumar, M. Fortification and fermentation of fruit juices with probiotic lactobacilli. *Ann. Microbiol.* **2012**, *62*, 1573–1578. [CrossRef]
120. Shah, N.P.; Ding, W.K.; Fallourd, M.J.; Leyer, G. Improving the stability of probiotic bacteria in model fruit juices using vitamins and antioxidants. *J. Food Sci.* **2010**. [CrossRef]
121. Coman, M.M.; Silvi, S.; Verdenelli, M.C.; Cecchini, C.; Orpianesi, C.; Cresc, A. Fruit and vegetable juices tested as possible probiotic beverage. *Agro Food Ind. Hi Tech* **2010**, *21*, 1–4.

122. Speranza, B.; Petruzzi, L.; Bevilacqua, A.; Gallo, M.; Campaniello, D.; Sinigaglia, M.; Corbo, M.R. Encapsulation of active compounds in fruit and vegetable juice processing: Current state and perspectives. *J. Food Sci.* **2017**, *82*, 1291–1301. [CrossRef]
123. Rodríguez-Roque, M.J.; Rojas-Graü, M.A.; Elez-Martínez, P.; Martín-Belloso, O. In vitro bioaccessibility of health-related compounds as affected by the formulation of fruit juice- and milk-based beverages. *Food Res. Int.* **2014**, *62*, 771–778. [CrossRef]
124. Fenster, K.; Freeburg, B.; Hollard, C.; Wong, C.; Rønhave Laursen, R.; Ouwehand, A. The production and delivery of probiotics: A review of a practical approach. *Microorganisms* **2019**, *7*, 83. [CrossRef] [PubMed]
125. Saarela, M.; Alakomi, H.L.; Mättö, J.; Ahonen, A.M.; Tynkkynen, S. Acid tolerant mutants of *Bifidobacterium animalis* subsp. lactis with improved stability in fruit juice. *LWT Food Sci. Technol.* **2011**, *44*, 1012–1018. [CrossRef]
126. Champagne, C.P.; Gardner, N.J. Effect of storage in a fruit drink on subsequent survival of probiotic lactobacilli to gastro-intestinal stresses. *Food Res. Int.* **2008**, *41*, 539–543. [CrossRef]
127. Antunes, A.E.C.; Liserre, A.M.; Coelho, A.L.A.; Menezes, C.R.; Moreno, I.; Yotsuyanagi, K.; Azambuja, N.C. Acerola nectar with added microencapsulated probiotic. *LWT Food Sci. Technol.* **2013**, *54*, 125–131. [CrossRef]
128. Martins, E.M.F.; Ramos, A.M.; Vanzela, E.S.L.; Stringheta, P.C.; de Oliveira Pinto, C.L.; Martins, J.M. Products of vegetable origin: A new alternative for the consumption of probiotic bacteria. *Food Res. Int.* **2013**, *51*, 764–770. [CrossRef]
129. Nicolesco, C.L.; Buruleanu, L.C. Correlation of some substrate parameters in growing *Lactobacillus acidophilus* on vegetable and fruit cocktail juices. *Bull. Univ. Agric. Sci. Vet. Med. Cluj Napoca Agric.* **2010**, *67*, 352–359.
130. Nualkaekul, S.; Charalampopoulos, D. Survival of *Lactobacillus plantarum* in model solutions and fruit juices. *Int. J. Food Microbiol.* **2011**, *146*, 111–117. [CrossRef]
131. Sheela, T.; Suganya, R.S. Studies on anti-diarrhoeal activity of synbiotic plums juice. *Int. J. Sci. Res. Publ.* **2012**, *2*, 1–5.
132. Priya, N.; Vasudevan, A. Effect of lactic acid bacteria in development of papaya juice using response surface methodology. *Int. J. Biotechnol. Biochem.* **2016**, *12*, 27–32.
133. Da Costa, G.M.; De Carvalho Silva, J.V.; Mingotti, J.D.; Barão, C.E.; Klososki, S.J.; Pimentel, T.C. Effect of ascorbic acid or oligofructose supplementation on *L. paracasei* viability, physicochemical characteristics and acceptance of probiotic orange juice. *LWT Food Sci. Technol.* **2017**, *75*, 195–201. [CrossRef]
134. Pimentel, T.C.; Madrona, G.S.; Garcia, S.; Prudencio, S.H. Probiotic viability, physicochemical characteristics and acceptability during refrigerated storage of clarified apple juice supplemented with *Lactobacillus paracasei* ssp. paracasei and oligofructose in different package type. *LWT Food Sci. Technol.* **2015**, *63*, 415–422. [CrossRef]
135. Patel, A.R. Probiotic fruit and vegetable juices-recent advances and future perspectiv. *Int. Food Res. J.* **2017**, *25*, 1850–1857.
136. Perricone, M.; Bevilacqua, A.; Altieri, C.; Sinigaglia, M.; Corbo, M. Challenges for the production of probiotic fruit juices. *Beverages* **2015**, *1*, 95–103. [CrossRef]
137. Ding, W.K.; Shah, N.P. Survival of free and microencapsulated probiotic bacteria in orange and apple juice. *Int. Food Res. J.* **2008**, *15*, 219–232.
138. Rakin, M.; Vukasinovic, M.; Siler-Marinkovic, S.; Maksimovic, M. Contribution of lactic acid fermentation to improved nutritive quality vegetable juices enriched with brewer's yeast autolysate. *Food Chem.* **2007**, *100*, 599–602. [CrossRef]
139. Gaudreau, H.; Champagne, C.P.; Remondetto, G.E.; Bazinet, L.; Subirade, M. Effect of catechins on the growth of oxygen-sensitive probiotic bacteria. *Food Res. Int.* **2013**, *53*, 751–757. [CrossRef]
140. Granato, D.; Putnik, P.; Bursać Kovačević, D.; Santos, J.S.; Calado, V.; Rocha, R.S.; Cruz, A.G.D.; Jarvis, B.; Rodionova, O.Y.; Pomerantsev, A. Trends in chemometrics: Food authentication, microbiology, and effects of processing. *Compr. Rev. Food Sci. Food Saf.* **2018**, *17*, 663–677. [CrossRef]
141. Rinke, P.; Jamin, E. Fruit Juices. In *Food Integrity Handbook*; 2018; pp. 243–264. Available online: https://secure.fera.defra.gov.uk/foodintegrity/index.cfm?sectionid=83 (accessed on 4 July 2020).
142. Vallabhaneni, S.; Walker, T.A.; Lockhart, S.R.; Ng, D.; Chiller, T.; Melchreit, R.; Brandt, M.E.; Smith, R.M. Notes from the field: Fatal gastrointestinal mucormycosis in a premature infant associated with a contaminated dietary supplement-Connecticut, 2014. *MMWR Morb. Mortal. Wkly. Rep.* **2015**, *64*, 155–156.

143. Santeramo, F.G.; Carlucci, D.; De Devitiis, B.; Seccia, A.; Stasi, A.; Viscecchia, R.; Nardone, G. Emerging trends in European food, diets and food industry. *Food Res. Int.* **2018**, *104*, 39–47. [CrossRef]
144. Mantzourani, I.; Nouska, C.; Terpou, A.; Alexopoulos, A.; Bezirtzoglou, E.; Panayiotidis, M.; Galanis, A.; Plessas, S. Production of a novel functional fruit beverage consisting of cornelian cherry juice and probiotic bacteria. *Antioxidants* **2018**, *7*, 163. [CrossRef]
145. Putnik, P.; Pavlić, B.; Šojić, B.; Zavadlav, S.; Žuntar, I.; Kao, L.; Kitonić, D.; Bursać Kovačević, D. Innovative hurdle technologies for the preservation of functional fruit juices. *Foods* **2020**, *9*, 699. [CrossRef]

© 2020 by the authors. Licensee MDPI, Basel, Switzerland. This article is an open access article distributed under the terms and conditions of the Creative Commons Attribution (CC BY) license (http://creativecommons.org/licenses/by/4.0/).

foods

MDPI

Review

Food–Drug Interactions with Fruit Juices

Zvonimir Petric [1], Irena Žuntar [2,*], Predrag Putnik [3,*] and Danijela Bursać Kovačević [4]

1 Unit of Pharmacokinetics and Drug Metabolism, Department of Pharmacology at the Institute of Neuroscience and Physiology, Sahlgrenska Academy at the University of Gothenburg, 40530 Göteborg, Sweden; petric.zvonimir@gmail.com
2 Faculty of Pharmacy and Biochemistry, University of Zagreb, Ante Kovačića 1, 10000 Zagreb, Croatia
3 Department of Food Technology, University North, Trg dr. Žarka Dolinara 1, 48000 Koprivnica, Croatia
4 Faculty of Food Technology and Biotechnology, University of Zagreb, Pierottijeva 6, 10000 Zagreb, Croatia; dbursac@pbf.hr
* Correspondence: irena.zuntar@pharma.unizg.hr (I.Ž.); pputnik@alumni.uconn.edu (P.P.)

Abstract: Fruit juices contain a large number of phytochemicals that, in combination with certain drugs, can cause food–drug interactions that can be clinically significant and lead to adverse events. The mechanisms behind such interactions are in most cases related to phytochemical interference with the activity of cytochrome P450 metabolizing enzymes (CYPs) or drug transporters. Moreover, alterations in their activity can have a clinical relevance if systemic exposure to the drug is decreased or increased, meaning that the pharmacological drug effects are suboptimal, or the drug will cause toxicity. In general, the common pharmacokinetic parameters found to be altered in food–drug interactions regarding fruit juices are the area under the concentration–time curve, bioavailability, and maximum plasma concentration. In most cases, the results from the drug interaction studies with fruit juices provide only limited information due to the small number of subjects, which are also healthy volunteers. Moreover, drug interactions with fruit juices are challenging to predict due to the unknown amounts of the specific phytochemicals responsible for the interaction, as well as due to the inter-individual variability of drug metabolism, among others. Therefore, this work aims to raise awareness about possible pharmacological interactions with fruit juices.

Keywords: fruit juice; interaction; drug; phytochemical; pharmacokinetics

Citation: Petric, Z.; Žuntar, I.; Putnik, P.; Bursać Kovačević, D. Food–Drug Interactions with Fruit Juices. *Foods* **2021**, *10*, 33. https://dx.doi.org/10.3390/foods10010033

Received: 2 November 2020
Accepted: 21 December 2020
Published: 24 December 2020

Publisher's Note: MDPI stays neutral with regard to jurisdictional claims in published maps and institutional affiliations.

Copyright: © 2020 by the authors. Licensee MDPI, Basel, Switzerland. This article is an open access article distributed under the terms and conditions of the Creative Commons Attribution (CC BY) license (https://creativecommons.org/licenses/by/4.0/).

1. Introduction

By definition, drug interactions (DIs) occur when the pharmacological effect of one drug is altered by the presence of another drug or xenobiotic, which includes herbal medicine, food or drink bioactive components, or any other chemical agents. DIs, if considered to be clinically significant, pose a risk for human health as they can have a direct effect on the therapeutic outcome and even cause life-threatening adverse drug reactions. Traditionally, mechanisms of DIs are classified as pharmacokinetic or pharmacodynamic, depending on the nature of the interactions. Pharmacokinetic DIs involve processes related to drug absorption, distribution, metabolism, and elimination, whereas pharmacodynamic DIs are those in which the drug effects are changed due to the presence of another drug or xenobiotic at its site of action [1]. In other words, depending on the nature of the DIs, different outcomes can be expected, such as decreased drug effectiveness with(out) increased drug toxicity or increased drug effectiveness with(out) increased drug toxicity. Hence, in the context of food–drug interactions, the clinical consequences can be the same, or expected, as in drug–drug interactions [2,3].

As a healthy lifestyle is nowadays becoming an imperative for many people, fruit juices recently came in the spotlight as a novel class of functional beverages, as they are promising carriers of biologically active compounds from many other food sources [4]. Namely, fruit juices, due to the recent advances in food (bio)technology, are now able to provide even more nutritional and health benefits to those seeking a well-balanced diet [5]. Given the fact that many bioactive compounds can be added to fruit juices, it is important to emphasize their clinical, pharmacological, and toxicological aspects regarding the potential of their involvement in DIs [2,3,6].

2. Absorption, Distribution, Metabolism and Elimination

Pharmacokinetics (PK) is a subdiscipline of pharmacology that quantitatively studies how a drug behaves in the body, i.e., what the body does to the drug. More specifically, the acronym ADME (absorption, distribution, metabolism and elimination) is often used to describe the PK processes of many drugs. PK principles can also be applied to patients (clinical pharmacokinetics) in order to provide safe and effective pharmacotherapy. To summarize, PK provides a quantitative relationship between a given dose and observed concentrations of a drug as a function of time, to provide optimization of dosage regimens. Pharmacodynamics (PD), on the other hand, is a subdiscipline of pharmacology, which studies the relationship between the drug concentration (at the site of the action, i.e., receptors) and the drug effects (response); i.e., what the drug does to the body. The pharmacological drug effect (or response) can be therapeutic and/or toxic, depending on the drug exposure. Hence, the PK/PD relationship can be viewed as an exposure–response relationship. Additionally, the relationship between the therapeutic and toxic dose of a drug is expressed by the therapeutic index, which tells about the relative safety of a drug or the narrowness of the therapeutic index, where a relatively small increase in the plasma drug concentration can lead to adverse effects and cause toxicity [7–11].

When the drug is orally taken, it must be absorbed, and this fraction is then carried into the hepatic portal system and liver before reaching the site of action in the unchanged state (systemic circulation). However, there is a possibility of drug metabolism (loss of a drug) along that path in gastrointestinal tissues and the liver. This loss of a drug is called the (presystemic) first-pass effect (metabolism). First-pass metabolism is an enzyme-catalyzed process, where the most common enzymes are of the cytochrome P450 (CYP) type. Besides, various isozymes, such as CYP3A4 (55% of all drugs) and CYP2D6 (30% of all drugs), among many others, are involved in first-pass drug metabolism and show inter-individual variability; i.e., differing enzymatic activity. This activity is of particular interest in determining many DIs, as drugs, bioactive compounds from food (fruit juices), and other xenobiotics, when co-administered, can inhibit or induce CYP activity (CYP inhibitors and CYP inducers), consequently affecting the pharmacodynamic outcome [7–11]. One of the PK parameters often mentioned in this paper is the area under the drug concentration–time curve (AUC), which is used to calculate the bioavailability (F) of a drug. The AUC shows how much of a drug is in the body or describes the total systemic exposure to the drug. Furthermore, F is the fraction (or percentage) of the absorbed drug that will be available at the site of action. In other words, F tells us about the rate and extent of drug absorption. F is a PK parameter that also indicates the extent of the systemic exposure to the drug. Additionally, for the intravenous route of drug administration (i.v.), F is by definition 100% (F = 1). The AUC of the orally administered drug relative to the AUC of the i.v. route gives us the possibility to calculate the absolute oral bioavailability of the drug (Foral). Foral can also be viewed as the product of the fractions of a drug dose that escapes metabolism by the gut (FG) and liver (FH); i.e., Foral = Fabs × FG × FH, where Fabs is the fraction of a dose that is absorbed intact across the enterocytes [7–11].

Many factors can influence the rate of drug absorption, such as P-glycoprotein (drug transporter), and many factors can influence the extent of the drug absorption, such as the first-pass metabolism. Another PK parameter that is mentioned in the paper is the elimination half-life ($t_{1/2}$), which can be obtained from the concentration–time profile of a drug, which gives us information of the time needed to eliminate half of the drug in the body (after drug distribution is over); simply put, it describes the decay of the drug. Moreover, the half-life determines the duration of drug-action for many medications. Half-life is dependent on drug clearance (CL), as CL relates to the rate of drug elimination and drug systemic exposure. The higher the CL, the lower the systemic exposure (AUC). In addition, C_{max} is the peak concentration, or maximum plasma concentration of the drug, which can be easily obtained from the AUC [7–11]. Finally, the number of possible DIs is without a limit, but as a general statement, it can be said that the most pronounced DIs will be those that are affecting oral bioavailability and clearance of drugs, particularly by affecting drug metabolism. DIs are one of the important sources of drug toxicity, as well as responsible for the variable therapeutic responses among individuals [7–11].

3. Importance of Cytochrome P450 Enzymes and Drug Transporters in Drug Interactions

Humans have substrate-specific enzymes (e.g., CYP enzymes/CYP system) that are found in different tissues (e.g., liver, intestine, lungs, etc.), which makes the entire DI concept even more complex, and especially so if the polymorphisms in cytochrome P-450 genes are taken into count. This means that CYP activity shows interindividual differences that is reflected in the drug PK and can make a direct impact on DIs. Moreover, due to various drugs and xenobiotics, which can act as CYP inhibitors or inducers, there also can be intra-individual differences in drug response and toxicity. Additionally, intrinsic factors (such as age, sex, disease, etc.) combined with extrinsic factors (diet, lifestyle, smoking, etc.) further contribute to a variable drug response, also making the prediction of a DI even more challenging for an individual patient.

The CYP system plays an important role in the metabolism/biotransformation of drugs and xenobiotics (as for detoxification), and they are therefore one of the most important factors contributing to DIs, including food–drug interactions [12]. In short, the CYP system is a large family of hemoproteins that catalyze a wide range of reactions (hydroxylation, epoxidation, oxygenation, dealkylation, isomerization, desaturation, reduction, etc.). The common CYPs that are clinically significant for DIs include CYP3A4, CYP2D6, CYP2C9, CYP2C19, CYP2B6, CYP2E1 and CYP1A2 [13]. Hence, any of these chemical reactions in the body can be influenced by food components, including the ones from fruit juices, which consequently can induce, or inhibit, one or more CYP enzymes, with repercussions on drug therapeutic exposure.

Besides the CYP system, drug transporters also have an important role in drug and xenobiotic pharmacokinetics. In general, transporters can be categorized as uptake and efflux transporters, and can be divided into two superfamilies (with more than 500 members), namely, the ATP-binding cassette (ABC) family and the solute carrier (SLC) family. Some of them are ubiquitously expressed, while some are found in tissues such as the liver, brain, small intestine and kidney. Some of the transporters, often being clinically relevant to DIs, include P-glycoprotein (P-gp), also known as multidrug resistance protein 1 (MDR1), breast cancer resistance protein (BCRP), organic anion transporter (OAT1, and OAT3), organic cation transporter (OCT2) and organic anion transporting polypeptide (OATP1B1 and OATP1B3) [14,15]. It is important to mention that drug transporters can also have functional genetic polymorphisms just like CYP system [16].

Altogether, the clinical relevance of the CYP system and drug transporters is recognized in drug–drug interactions, but also in food–drug interactions, because different xenobiotics, including those from fruit juices, e.g., grapefruit juice, can influence their activity and thus have an impact on the systemic drug exposure; i.e., efficacy and toxicity (Figure 1) [17–19]. More importantly, the reader should be aware that drug interactions shown in vitro does not necessarily mean they are also occurring in vivo. Moreover, even if they do occur in vivo (animals), their clinical relevance for humans still could be insignificant, and vice versa. This can be seen in an example of a DI between pomegranate juice and flurbiprofen (a non-steroidal anti-inflammatory agent) [20], which shows discrepancies between in vitro and clinical studies [21].

4. Fruit Juices

Fruits (and vegetables) contain structurally diverse bioactive compounds, such as flavonoids (flavonols, flavones, flavanones, flavanols, anthocyanidins and isoflavones), phenolic acids (hydroxybenzoic acids and hydroxycinnamic acids), carotenoids (β-carotene, lycopene, lutein, zeaxanthin, flavoxanthin, canthaxanthin, capsanthin, capsorubin and β-cryptoxanthin), vitamins (vitamin C–ascorbic acid and vitamin E–tocopherols) and phytoestrogens (isoflavonoids, stilbenes, lignans, coumestans and glucosinolates). Furthermore, it is estimated that almost 5000 phytochemicals have been found in fruits (and vegetables), but a large proportion remains to be discovered. The bioactive compounds in fruits are also greatly variable in amounts due to various parameters, such as the stage of ripeness, cultivar/variety, agricultural practices, environment, harvest and postharvest procedures, processing, storage, etc. [22]. Since most of the fresh fruit is processed, the food industry is increasingly applying new non-thermal processing technologies to preserve the original nutritional and sensory quality of the product, with an emphasis on the preservation of bioactive compounds to the greatest extent [5,23]. Therefore, it can be concluded that fruits (and vegetables) contain a complex combination of constituents, so predicting and determining their interactions with drugs, and the clinical relevance, is very challenging. At this point, it is important to define what is a fruit juice. Even though there are various definitions, the one according to the UK Fruit Juice and Fruit Nectars Regulations seems reasonable, as it gives a broader view of the definition:

> *"Fruit juice is the fermentable but unfermented product obtained from the edible part of the fruit which is sound, ripe and fresh or preserved by chilling or freezing of one or more kinds mixed having the characteristic color, flavor, and taste typical of the juice of the fruit from which it comes."*

However, the specifications and regulations depend on the different types of fruit juices, such as fruit juice from concentrate, concentrated fruit juice, water-extracted fruit juice and dehydrated and powdered fruit juice. Within the EU, fruit juices are regulated by Council Directive 2001/112/EC [24]. Smoothies, on the other hand, do not have a legal definition, but they can contain fruit purees, fruit juice or crushed parts of the fruit. Fruit juices, compared to other foods, contain more beneficial nutrients, especially calcium, iron, vitamin A, thiamin (vitamin B1), riboflavin (vitamin B2) and ascorbic acid (vitamin C), and many other antioxidants, such as tocopherols (vitamin E), beta-carotene, flavonoids, fibers and other minerals [25,26]. It should be kept in mind that fruit juices can also be carriers for many bioactive compounds, which are additionally added to make a novel class of beverages, i.e., functional fruit juices. Hence, concerns regarding interactions between fruit juices and drugs should not sound surprising, especially if we are witnessing clinically significant DIs and unwanted pharmacological outcomes linked to their concomitant consumption [27].

5. Drug Interactions with Common Fruit Juices

The most extensively described DIs with fruit juices are those with grapefruit juice (and grapefruit pulp), for which it is reported to have more than 40 DIs in humans [28]. Therefore, grapefruit juice that was consumed with the medicine lovastatin was used as a representative example for illustration of a DI, shown in Figure 1. The figure shows the CYP3A4 inhibition by the grapefruit juice, leading to development of adverse effects; i.e., myopathy and rhabdomyolysis. Additionally, a human PK study confirmed that C_{max} and the AUC of lovastatin, when taken with grapefruit juice, were increased about 12-fold and 15-fold, respectively [29–31].

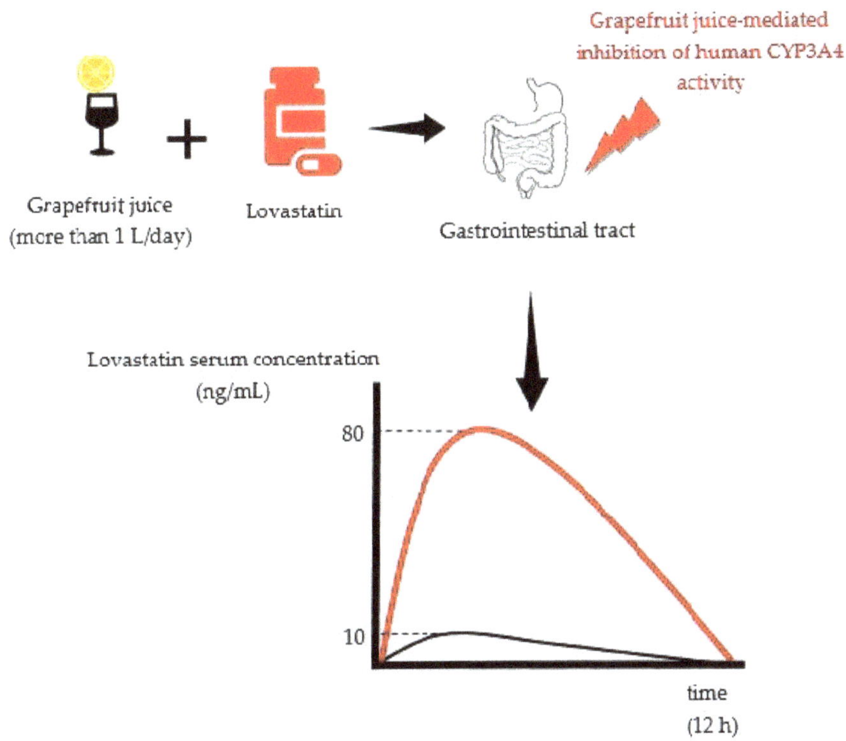

Figure 1. Hypothetical PK profile of lovastatin taken with water (black curve) and with grapefruit juice (red curve). The noticeable increase in the AUC (red curve) implicates an increase in systemic drug exposure, leading to the development of adverse effects; i.e., myopathy and rhabdomyolysis (the C_{max} of the lovastatin taken with water was 7 ± 2.5 ng/mL, while for the grapefruit juice it was 82.4 ± 39.6 ng/mL. The FDA label of lovastatin tablet (40 mg) states that, at the steady-state, the C_{max} of lovastatin should be 7.8 ng/mL) [29,32,33].

The discovery of the grapefruit interaction with drugs was an unexpected result of an interaction study with a focus on ethanol and felodipine (a calcium-channel antagonist) that targeted masking the taste of ethanol with grapefruit juice. The results showed an increase in felodipine bioavailability and C_{max}, as a consequence of irreversible degradation of an intestinal CYP3A by the grapefruit compounds [30,31]. Moreover, grapefruit juice was found to have an impact on the efflux transport via P-glycoprotein in vitro, but also the influx transport via organic anion-transporting polypeptides—OATPs, esterases and sulfotransferases—although the clinical relevance of grapefruit juice on these drug transporters and enzymes was never determined [34]. Other drug classes that are known to have a clinically relevant interaction with grapefruit juice include 3-hydroxy-3-methylglutaryl coenzyme A (HMG-CoA) reductase inhibitors (simvastatin), immunosuppressives (cyclosporine), antiarrhythmics (amiodarone) and anticonvulsants (carbamazepine). However, in some other cases with other drugs, even if the pharmacokinetic parameters were altered by the interaction with grapefruit, the clinical significance was still considered insignificant [34,35]. In addition, it should be kept in mind that, due to the variable amounts of phytochemicals in juices, the results between the different pharmacokinetic studies of the same fruit juices cannot be compared, but the outcome can still be intuitively predicted.

The literature suggests that the specific components in grapefruit juice related to DIs could involve furanocoumarins, namely, bergamottin and 6′7′-dihydroxbergamottin [36], but also naringin, which can inhibit OATP1A2 activity [21]. It is speculated that even micromolar concentrations of naringin were responsible for a DI with fexofenadine, resulting in a decrease of its bioavailability [37], which in this case could be clinically significant but was not confirmed by other authors.

There are some indications that grapefruit juice can reduce the levels of CYP3A4 by 47% only four hours after consumption, and these effects are persistent in the intestinal and liver cells at least 24 h after ingestion, meaning that grapefruit juice, even without co-administration with the drug, can lead to a DI or alter the metabolism of any CYP3A4 substrate for longer periods of time after consumption [38]. Proof for that can be seen in the example of a DI with tacrolimus (immunosuppressant), where a delayed increase in systemic exposure to tacrolimus, e.g., from 4.7 ng/mL to 47.4 ng/mL, happened one week after the last grapefruit juice intake (250 mL, 4 times a day for 3 days). The patient developed severe headache and nausea, luckily without nephrotoxicity [28,39]. Additionally, there is a case report of statin-associated rhabdomyolysis triggered by grapefruit consumption. Rhabdomyolysis is a rare but very serious adverse effect associated with statin therapy, which could cause kidney failure and death. In this case, the metabolism of simvastatin (CYP3A4 substrate) was altered due to CYP3A4 inactivation by the grapefruit, resulting in toxic systemic exposure to simvastatin [40]. What could be expected from this data is that other statins, such as atorvastatin, which are CYP3A4 substrates, will have the same unwanted clinical outcomes. However, statins that are not substrates for the same CYP, such as rosuvastatin, are not expected to have this DI.

Altogether, it is clear that grapefruit juice interactions with orally given drugs cannot be generalized, and clinical relevance cannot be precisely determined. Hence, it is better to avoid grapefruit consumption with drugs that are substrates for an intestinal CYP3A4 and/or P-gp, especially in the cases when a drug has a narrow therapeutic index and a poor oral bioavailability due to the high pre-systemic (first-pass) metabolism mediated via CYP3A4, because even a single consumption of grapefruit juice can lead, in some cases, to drug toxicity. Additionally, the concentration/amount of the grapefruit extracts and as well interindividual variability of intestinal CYP3A4 activity among humans also contribute to the severity of the unwanted outcome [35,41].

Besides grapefruit, other citrus juices, such as orange, lemon, pomelo, and lime, were also reported to cause DIs in some clinical studies. As previous results were contradictory, or the clinical relevance could not be determined, a meta-analysis from Sridharan et al. provides a good summary of the DIs with selected citrus juices and cyclosporine (substrate for CYP3A4). Cyclosporine is an immunosuppressive drug with a narrow therapeutic index, with variable pharmacokinetics from person to person. Pooled results showed that no significant changes in the AUC and C_{max} of cyclosporine were observed with orange juice as compared to the controls, while pomelo juice, on the other hand, increased the AUC and C_{max} and decreased the elimination $t_{1/2}$ of cyclosporine [42]. Hence, orange juice did not have any significant interaction with cyclosporine. Interestingly, an earlier review from different authors [43], regarding orange juice, reported quite the opposite trend. Furthermore, orange juice does not contain bergamottin derivatives that are present in grapefruit juice, which may answer why orange juice did not have any significant interaction with cyclosporine. Pharmacologically, it would be expected that DIs with citrus juices (except orange juice) would have a more significant meaning in the case of liver disease and the concomitant use of other drugs (often seen in elderly patients) that share the same enzymatic pathway.

Orange juice was in one study found to decrease the oral absorption of alendronate (bisphosphonate agent for the treatment of osteoporosis) by approx. 60% compared to water [44]. This is very important, as alendronate, if taken on an empty stomach (after an overnight fast), 2 h before any meal (breakfast), has by default a very poor absolute oral bioavailability—approx. only 0.75% of the total dose [45]. Hence, taking orange juice with alendronate should be avoided.

Seville (sour) orange juice was found to have the same mechanism of CYP3A4 inactivation as grapefruit juice, and thus interacts with felodipine [46]. However, Seville orange juice did not alter the bioavailability of cyclosporine despite the CYP3A4 activity being significantly reduced, in contrast to grapefruit juice [47]. Therefore, it seems that grapefruit juice, besides the inactivation of intestinal CYP3A4, alters some other enzymes or transporters.

Seville orange and lime juices were in one clinical study inspected for a DI with sildenafil, a drug mostly known as Viagra®, which is an agent that improves penile erectile function [48]. Sildenafil has an extensive first-pass metabolism, which results in a relatively low absolute oral bioavailability (F = 40%). Namely, for healthy subjects that received sildenafil (single dose of 50 mg) for three consecutive days and drank 250 mL of juice (or water) just before the drug was taken, the results showed that Seville orange juice increased the AUC and C_{max} of sildenafil by 44%. Although Seville orange juice is considered a moderate CYP3A4 inhibitor, this interaction did not have any adverse effects. The increase in systemic exposure is believed to be due to the intestinal inhibition of CYP3A4 and P-gp. On the other hand, lemon juice did not show any effects on the PK parameters of sildenafil.

Pomelo fruit juice (250 mL) was also inspected for a DI with sildenafil in another clinical study [49]. Surprisingly, pomelo juice decreased the systemic exposure to sildenafil. Namely, the bioavailability of sildenafil with pomelo juice was 60% lower compared with water. This is explained by a possibility that there was some sort of physicochemical interaction with one of the phytochemicals from pomelo juice, or an interaction with some drug transporter. However, the clinical significance of this interaction was not established, but the authors suggested avoiding taking pomelo juice with sildenafil.

Based on the previous information, it can be predicted that sildenafil, as a substrate for CYP3A4, will also have a DI with grapefruit juice, resulting in an increase in the systemic exposure of sildenafil. This was also confirmed in humans, but the adverse effects were not noticed [50]. However, it seems reasonable to avoid this combination. Namely, there are three independent case reports of DI-induced priapism (a persistent and painful penile erection, which is an emergency due to the risk of impotency) due to the concomitant use of pomegranate fruit juice (a high content of flavonoids) and sildenafil [51]. Although evidence of this interaction is based on circumstances, it is still a good example of how DIs with fruit juices are clinically significant, and adverse effects that are generally rare, such as priapism, are believed to be precipitated by a fruit juice. The proposed mechanism behind this interaction is due to CYP3A4 inhibition by phytochemicals from pomegranate juice.

Dresser et al. examined the effect of apple and orange juice on OATP uptake transporters [52]. Namely, in a human volunteer study, subjects were given two fexofenadine tablets (60 mg) (antihistaminic, sold by the trade name Allegra®) with 300 mL of fruit juice, up to a total volume of 1.2 L. The results of the PK study showed that both juices significantly decreased the fexofenadine concentrations in plasma compared with water (AUC (fexofenadine) apple juice = 434 ± 53; AUC (fexofenadine) orange juice = 494 ± 16; and AUC (fexofenadine) water = 1616 ± 120, $p < 0.001$). Among the tested individuals, the extent of the decreased fexofenadine concentrations was variable, or individuals with the highest fexofenadine AUC with water had the greatest decrease with the juices. Fexofenadine is a substrate of P-gp, but as well as of OATP uptake transporters, so Dresser et al. suggest the necessity to determine the individual contribution of each transporter in this case.

Apple juice was also found to have a DI with atenolol, which is antihypertensive drug [53]. In a human volunteer PK study, atenolol (50 mg) was taken with apple juice (600 mL and 1200 mL) and it was shown that atenolol systemic exposure was inversely proportional to the amount of consumed apple juice as compared to water (AUC (atenolol) 600 mL apple juice = 885.3; AUC (atenolol) 1200 mL apple juice = 389.7; and AUC (atenolol) water = 2110). In other words, the AUC of atenolol was decreased by 82% after the ingestion of 1200 mL apple juice, but there were no observed changes in the pharmacodynamic outcome even though there is an evident dose–response relationship. This DI is believed to be due to inhibition of intestinal drug transporter OATP2B1, but it is also believed that the higher acidity in the gastrointestinal tract (a large amount of apple juice) could also have an impact on atenolol absorption.

Other drugs reported as having a decreased systemic exposure with apple juice were montelukast (an anti-asthma agent) and aliskiren (an antihypertensive agent), which are also substrates for uptake transporters [54]. In addition, decreasing the systemic exposure of atenolol and montelukast was also noticed with orange juice [55,56]. However, to our knowledge, there are no documented cases of a DI (from clinical practice) with apple and orange juice to date. One exception could be in the case of calcium-fortified orange juice, where, due to chelation of the drugs with calcium, alterations in the C_{max} and AUC of the fluoroquinolones (antibiotics) could decrease the antibiotic effects, leading to antibiotic resistance [57,58]. The latter statement could be applied to all calcium-fortified fruit juices.

It is worth mentioning that orange juice was found to increase the systemic exposure to aluminum [59] and iron [60] by enhancing their absorption. In the case of aluminum, there is, in theory, a risk of aluminum toxicity in a patient with renal disease, as it was found that urinary excretion of aluminum increased 10 times when it was taken with orange juice.

Fruits, such as oranges, bananas and prunes, along with their fruit juices, or in common combinations with vegetable juices such as carrot juice and tomato juice, contain very high amounts of potassium, which in combination with potassium-sparing drugs, such as angiotensin-converting enzyme inhibitors (ramipril), diuretics (spironolactone, triamterene) and especially in patients with kidney disease (hemodialysis patients) and with hypoaldosteronism, could lead to life-threatening hyperkalemia, so it is advisable to always seek advice from a pharmacist prior to consumption [61–63]. Prune juice is best known for its laxative effect, so it is often sold as an over-the-counter (OTC) dietary supplement in many pharmacies and specialized stores, where consumers/patients might be unaware of the possible side effects [64,65].

Plum and avocado juices were found to contain higher amounts of biogenic amine, tyramine, as compared to other fruit sources [66]. In theory, monoamine oxidase inhibitors (MAOIs) can inhibit tyramine degradation, which leads to its increased systemic exposure with a consequence of developing a hypertensive crisis [67]. However, over the last decades, the use of MAOIs has decreased and documented cases of their toxicity (due to the tyramine overexposure) are very rare, but the caution is still advised, as even 8 mg of tyramine was historically known to cause hypertension with some MAOIs [68,69].

Cranberry fruit juice is nowadays a very popular prophylactic treatment for the infections of the urinary tract. There have been some safety concerns regarding the cranberry fruit juice intake and using warfarin—an anticoagulant—which is known for its narrow therapeutic index and life-threatening side effects in case of increased systemic exposure [70,71]. The main reason for concern was a clinical case report that reported a change in the INR (International Normalized Ratio), a biochemical marker of warfarin's antithrombotic effects, related to cranberry juice intake. INR was in some patients decreased, meaning that the blood coagulates too easily, so the risk for developing blood clots rises [72], or it was increased, meaning that there is increased risk of bleeding [73]. However, in a randomized, double-blind clinical study, this interaction was not confirmed [74]. Ansell et al. commented that cranberry juice was not the one to blame for the results, as the reported cases were not convincing beyond doubt. Namely, patients had a variety of illnesses, the amount of ingested cranberry juice was not known in some cases, as well as the other dietary components, nor was the patients' compliance known, but most importantly, the pharmacogenetic profiling was not done either [75]. Although, it is well-established that polymorphisms of CYP2C9 and vitamin K epoxide reductase (VKORC) affect the INR values and clinical outcome of pharmacotherapy with warfarin. Detailed discussion of this interaction was provided by Zikria et al. [76]. In the literature, there are other described cases of other fruit juices that have in vitro effects on drug-metabolizing enzymes or transporters; however, their clinical significance was never firmly confirmed.

Tangerine fruit juice was reported to upregulate CYP3A4 activity and inhibit P-glycoprotein due to the high content of flavonoid tangeretin [77,78]. Interestingly, some older data reported the opposite effect—tangerine (tangeretin) inhibits CYP3A4 and CYP1A2 in human liver microsomes [79].

Black mulberry was reported to inhibit CYP3A (and OATP-B), but also was wild grape [80]. Black mulberry is traditionally promoted as a fruit with a high content of iron, so it has a beneficial effect on the treatment of anemia. In addition, some studies showed that black mulberry has a beneficial effect on cholesterol levels, liver tissue (hepatoprotective effect) and even anti-obesity potential [81]. In theory, large amounts of black mulberry fruit juice could interact with CYP3A4 substrates, leading to an increase in systemic exposure of such drugs, but clinical relevance was never determined in humans.

Mango stem-bark was found to inhibit CYP1A1, CYP1A2, CYP3A1, CYP2C6, CYP2E1 and P-glycoprotein [82], while grape, due to the resveratrol, was implied in inhibition of CYP1A1/1A2 isoforms and CYP2E1, which are needed for the activation of procarcinogens (polycyclic arylamines, polyaromatic hydrocarbons, aflatoxin B1 and *N*-nitrosamines, respectively). Hence, besides having protective role as antioxidants, it is suggested that their fruit juices could alter the CYP3A4 activity, which is an important biotransformation pathway for many drugs [83]. The same was reported for papaya [84] and black raspberry; however, a recent study with taxane agents (anticancer agents) did not confirm altered CYP3A4 activity [85].

Tropical fruit juices, such as pineapple, papaya, litchi, kiwi, starfruit and passion fruit, were implicated in DIs in vitro due to their inhibitory effects on CYP2C9 or CYP3A4. Pineapple fruit juice, due to a high bromelain content was found to have the most pronounced inhibitory properties on CYP2C9, compared to other fruit juices. The effect of the inhibition was proportionally dependent on the increase in the amount of pineapple juice. Moreover, starfruit juice was found to be a very potent inhibitor of CYP3A4 compared to grapefruit juice. Namely, an assay of midazolam 1-hydroxylase activity of human CYP3A showed that residual activity of midazolam 1-hydroxylase (%) with starfruit juice was only 0.1 ± 0.0, as compared to the 14.7 ± 0.5 for grapefruit juice. It would be interesting to study those inhibitory activities in vivo to determine the clinical relevance of tropical fruit juices on DI.

The next interesting question to address is, could fruit juices be exploited for enhancing the positive pharmacotherapy outcome in some cases? The answer is yes [86].

Grapefruit juice was shown to be a drug-sparing agent (an agent that decreases the therapeutic dose of another drug) in a case of concomitant use with cyclosporine. In other words, patients could avoid the dose-related side effects of cyclosporine [27,87]. In another study, grapefruit enhanced oral bioavailability of artemether, which is antimalarial agent with generally high presystemic metabolism via CYP3A4, indicating more effective treatments of malaria [88]. In other words, grapefruit juice could be useful in maintaining the effectiveness and efficacy of some drugs.

Lime juice and artemisinin combination therapy (antimalarial agents) were given in one study to children (61 males and 50 females) with acute uncomplicated malaria [89]. It was observed that the artemisinin with lime juice caused more rapid clearance of parasites; also, the lime juice is believed to prevent resistance. The proposed explanation for this phenomenon is high amounts of vitamin C and flavonoids present in lime juice, which, besides its low pH, contributed to antioxidant activity. Similar results were shown recently on murine models, but with a lemon decoction resulting in a suppression of parasites by 39% and rapid early parasite clearance, as compared to the controls. However, there is a need to determine the exact effects of lime and lemon juice as supplement treatments in malaria by doing further investigations.

Blueberry juice, prepared from fresh blueberries, was given to 201 children that were receiving etanercept as a treatment for their juvenile idiopathic arthritis [90]. The study showed that the blueberry juice treatment combined with etanercept improved the symptoms of the disease, but as well decreased the side-effects of etanercept.

Regarding the previously mentioned increase of iron absorption with orange juice, it seems that it has a beneficial effect and such a combination could contribute to a better response to iron-deficiency anemia, which is a common problem, especially in children, but also in adults [91]. As DIs precipitated by fruit juices can sometimes be predicted, Table 1 summarizes previously described examples of potentially relevant drug interactions that should be considered in everyday clinical practice.

Table 1. Examples of potentially relevant drug interactions precipitated by fruit juices in humans.

Fruit Juice Type	Examples of Drugs	Suggested Mechanism of an Interaction	Reference
Grapefruit	CYP3A4 substrates HMG-CoA reductase inhibitors (simvastatin, lovastatin) Immunosuppressives (cyclosporine) Antiarrhythmics (amiodarone) Anticonvulsants (carbamazepine)	via CYP3A4, or/and P-gp	[34,35]
	PDE5 inhibitor (sildenafil)		[50,51]
	Antimalarial agent (artemether)		[88]
Orange	Bisphosphonates (alendronate)	physicochemical interaction	[45]
	Antihistamines (fexofenadine)	OATP transporters, or/and P-gp	[52]
	Beta-blocker (atenolol)		[55]
	Anti-asthmatic agent (montelukast)		[56]
Calcium-fortified orange juice	Aluminum and iron supplements	physicochemical interaction	[59,60]
	Antibiotics (fluoroquinolones)		[57,58]
Seville orange	PDE5 inhibitor (sildenafil)	via CYP3A4	[48]
Pomegranate	PDE5 inhibitor (sildenafil)	via CYP3A4	[50,51]
Pomelo	PDE5 inhibitor (sildenafil)	physicochemical interaction	[49]
	Immunosuppressives (cyclosporine)	P-gp	[42,43]
Apple	Antihistamines (fexofenadine)	OATP transporters, or/and P-gp	[52]
	Beta-blocker (atenolol)		[53]
	Anti-asthmatic agent (montelukast)		[54]
	Antihypertensive agent (aliskiren)		[54]
	Antihistamines (fexofenadine)		[52]
Blueberry	TNF-α inhibitor (etanercept)	Beneficial interaction suggested to be due to anti-oxidant/anti-inflammatory properties of blueberries	[90]

HMG-CoA reductase = 3-hydroxy-3-methyl-glutaryl-coenzyme A reductase; PDE5 = phosphodiesterase type 5; OATP = Organic Anion Transporting Polypeptides; P-gp = P-glycoprotein.

6. Conclusions

Fruit juices contain a large number of phytochemicals, but there are not enough clinical studies to evaluate their roles in drug interactions, even though the literature provides evidence that some fruit juices can impact drug disposition and thus interact with drugs. Predicting drug interactions potentiated by fruit juices is challenging due to the unknown number of phytochemicals present in the fruit juice, the unknown doses, and the individual differences among individuals who consume them concomitantly with their medications. Hence, the pharmacodynamic outcome cannot be generalized.

Although many drug interactions with fruit juices are not considered clinically relevant, there still are some that deserve our attention. For the people who prefer avoiding any potential drug interaction precipitated by fruit juices, the best advice is to take the medication with water.

In conclusion, a better understanding of the mechanisms behind drug interactions with fruit juices, and further investigations, are still needed to decrease the adverse drug reactions associated with fruit juice consumption. Patients should always ask specialized healthcare professionals about any concerns regarding their medication and fruit juice interactions, in order to decrease the likelihood of any unwanted effects or unsuccessful pharmacological treatments.

Author Contributions: Conceptualization, Z.P., I.Ž. and P.P.; writing—original draft preparation, Z.P., I.Ž., P.P. and D.B.K.; writing—review and editing, I.Ž., P.P. and D.B.K.; project administration, D.B.K. All authors have read and agreed to the published version of the manuscript.

Funding: This research was funded by the Croatian Science Foundation through the funding of the "Hurdle Technology and 3D Printing for Sustainable Fruit Juice Processing and Preservation" project, number "IP-2019-04-2105".

Conflicts of Interest: The authors declare no conflict of interest. The funders had no role in the design of the study; in the collection, analyses, or interpretation of data; in the writing of the manuscript, or in the decision to publish the results.

References

1. Thanacoody, R. Drug interactions. In *Clinical Pharmacy and Therapeutics*; Whittlesea, C., Hodson, K., Eds.; Elsevier: Amsterdam, The Netherlands, 2019; pp. 53–65.
2. Amadi, C.N.; Mgbahurike, A.A. Selected food/herb–drug interactions. *Am. J. Ther.* **2018**, *25*, 423–433. [CrossRef]
3. Mouly, S.; Lloret-Linares, C.; Sellier, P.-O.; Sene, D.; Bergmann, J.F. Is the clinical relevance of drug-food and drug-herb interactions limited to grapefruit juice and saint-john's wort? *Pharmacol. Res.* **2017**, *118*, 82–92. [CrossRef]
4. Bursać Kovačević, D.; Brdar, D.; Fabečić, P.; Barba, F.J.; Lorenzo, J.M.; Putnik, P. Strategies to achieve a healthy and balanced diet: Fruits and vegetables as a natural source of bioactive compounds. In *Agri-Food Industry Strategies for Healthy Diets and Sustainability*; Barba, F.J., Putnik, P., Bursać Kovačević, D., Eds.; Elsevier: Amsterdam, The Netherlands, 2020; pp. 51–88.
5. Putnik, P.; Pavlić, B.; Šojić, B.; Zavadlav, S.; Žuntar, I.; Kao, L.; Kitonić, D.; Kovačević, D.B. Innovative hurdle technologies for the preservation of functional fruit juices. *Foods* **2020**, *9*, 699. [CrossRef] [PubMed]
6. Zawiah, M.; Yousef, A.-M.; Khan, A.H.; Al-Ashwal, F.Y.; Matar, A.; Alkhawaldeh, B.; Nassar, R.; Abduljabbar, R.; Abdo Ahmed, A.A. Food-drug interactions: Knowledge among pharmacists in jordan. *PLoS ONE* **2020**, *15*, e0234779. [CrossRef] [PubMed]
7. Tozer, T.N.; Rowland, M. *Essentials of Pharmacokinetics and Pharmacodynamics*; Lippincott Williams & Wilkins: Philadelphia, PA, USA, 2016.
8. Tozer, W.K.; Shargel, L.; Yu, A.B.C. *Applied Biopharmaceutics & Pharmacokinetics*; Lippincott Williams & Wilkins Publishers: Philadelphia, PA, USA, 2016.
9. Beringer, P. *Winter's Basic Clinical Pharmacokinetics*; Lippincott Williams & Wilkins: Philadelphia, PA, USA, 2018.
10. Persky, A.M.; Pollack, G.M. *Foundations in Pharmacokinetics*; UNC Press: Chapel Hill, NC, USA, 2017.
11. Derendorf, H.; Schmidt, S. *Rowland and Tozer's Clinical Pharmacokinetics and Pharmacodynamics: Concepts and Applications*, 5th ed.; LWW Lippincott Williams and Wilkins: Philadelphia, PA, USA, 2019.
12. Anderson, K.E. Effects of specific foods and dietary components on drug metabolism. In *Handbook of Drug-Nutrient Interactions*; Humana Press: Totowa, NJ, USA, 2009; pp. 243–265.
13. Rendic, S.; Medic Saric, M. *Metabolism of Drugs and Selected Xenobiotics*; Medicinska Naklada: Zagreb, Croatia, 2013; pp. 154–157.
14. Liang, Y.; Li, S.; Chen, L. The physiological role of drug transporters. *Protein Cell* **2015**, *6*, 334–350. [CrossRef] [PubMed]
15. Mao, Q.; Lai, Y.; Wang, J. Drug transporters in xenobiotic disposition and pharmacokinetic prediction. *Drug Metab. Dispos.* **2018**, *46*, 561–566. [CrossRef]
16. McLean, C.; Wilson, A.; Kim, R.B. Impact of transporter polymorphisms on drug development: Is it clinically significant? *J. Clin. Pharmacol.* **2016**, *56*, S40–S58. [CrossRef]
17. Jargin, S. Grapefruit: Perspectives in nutrition and pharmacology. *J. Intercult. Ethnopharmacol.* **2017**, *6*, 339. [CrossRef]
18. Margină, D.; Ilie, M.; Grădinaru, D.; Androutsopoulos, V.P.; Kouretas, D.; Tsatsakis, A.M. Natural products—Friends or foes? *Toxicol. Lett.* **2015**, *236*, 154–167. [CrossRef]
19. Choi, J.H.; Ko, C.M. Food and drug interactions. *J. Lifestyle Med.* **2017**, *7*, 1–9. [CrossRef]
20. Hanley, M.J.; Masse, G.; Harmatz, J.S.; Court, M.H.; Greenblatt, D.J. Pomegranate juice and pomegranate extract do not impair oral clearance of flurbiprofen in human volunteers: Divergence from in vitro results. *Clin. Pharmacol. Ther.* **2012**, *92*, 651–657. [CrossRef] [PubMed]
21. Farkas, D.; Greenblatt, D.J. Influence of fruit juices on drug disposition: Discrepancies betweenin vitroand clinical studies. *Expert Opin. Drug Metab. Toxicol.* **2008**, *4*, 381–393. [CrossRef] [PubMed]
22. Yalcin, H.; Çapar, T.D. Bioactive compounds of fruits and vegetables. In *Minimally Processed Refrigerated Fruits and Vegetables*; Springer: Berlin, Germany, 2017; pp. 723–745.
23. Koubaa, M.; Barba, F.J.; Bursać Kovačević, D.; Putnik, P.; Santos, M.D.; Queirós, R.P.; Moreira, S.A.; Inácio, R.S.; Fidalgo, L.G.; Saraiva, J.A. Pulsed electric field processing of fruit juices. In *Fruit Juices: Extraction, Composition, Quality and Analysis*; Rajauria, G., Tiwari, B., Eds.; Academic Press: Oxford, UK, 2018; pp. 437–449.

24. The Council of the European Union. Council of the European Union. Council directive 2001/112/ec of 20 december 2001 relating to fruit juices and certain similar products intended for human consumption. In *2001/112/EC*; European Union, Ed.; Official Journal of the European Communities: Brussels, Belgium, 2001; pp. L 10/58–L 10/66.
25. Ashurst, P.R. Introduction. In *Chemistry and Technology of Soft Drinks and Fruit Juices*; Wiley-Blackwell: Hoboken, NJ, USA, 2016; pp. 1–14.
26. Benmeziane, F.; Abdourhamane, A.; Guedaoura, A. Nutritional quality and bioactive compounds of some fruit juices. *Adv. Environ. Biol.* **2016**, *10*, 242–249.
27. Mallhi, T.H.; Sarriff, A.; Adnan, A.S.; Khan, Y.H.; Qadir, M.I.; Hamzah, A.A.; Khan, A.H. Effect of fruit/vegetable-drug interactions on cyp450, oatp and p-glycoprotein: A systematic review. *Trop. J. Pharm. Res.* **2015**, *14*, 1927–1935. [CrossRef]
28. Rodríguez-Fragoso, L.; Martínez-Arismendi, J.L.; Orozco-Bustos, D.; Reyes-Esparza, J.; Torres, E.; Burchiel, S.W. Potential risks resulting from fruit/vegetable-drug interactions: Effects on drug-metabolizing enzymes and drug transporters. *J. Food Sci.* **2011**, *76*, R112–R124. [CrossRef]
29. Kantola, T.; Kivistö, K.T.; Neuvonen, P.J. Grapefruit juice greatly increases serum concentrations of lovastatin and lovastatin acid. *Clin. Pharmacol. Ther.* **1998**, *63*, 397–402. [CrossRef]
30. Bailey, D.G.; Malcolm, J.; Arnold, O.; David Spence, J. Grapefruit juice–drug interactions. *Br. J. Clin. Pharmacol.* **2002**, *46*, 101–110. [CrossRef]
31. Bailey, D.G. Predicting clinical relevance of grapefruit-drug interactions: A complicated process. *J. Clin. Pharm. Ther.* **2017**, *42*, 125–127. [CrossRef]
32. Manson, J.M.; Freyssinges, C.; Ducrocq, M.B.; Stephenson, W.P. Postmarketing surveillance of lovastatin and simvastatin exposure during pregnancy. *Reprod. Toxicol.* **1996**, *10*, 439–446. [CrossRef]
33. ALTOPREV®. Available online: https://www.accessdata.fda.gov/drugsatfda_docs/label/2012/021316s029lbl.pdf (accessed on 1 December 2020).
34. Hanley, M.J.; Cancalon, P.; Widmer, W.W.; Greenblatt, D.J. The effect of grapefruit juice on drug disposition. *Expert Opin. Drug Metab. Toxicol.* **2011**, *7*, 267–286. [CrossRef]
35. Mertens-Talcott, S.U.; Zadezensky, I.; De Castro, W.V.; Derendorf, H.; Butterweck, V. Grapefruit-drug interactions: Can interactions with drugs be avoided? *J. Clin. Pharmacol.* **2006**, *46*, 1390–1416. [CrossRef] [PubMed]
36. Paine, M.F.; Widmer, W.W.; Hart, H.L.; Pusek, S.N.; Beavers, K.L.; Criss, A.B.; Brown, S.S.; Thomas, B.F.; Watkins, P.B. A furanocoumarin-free grapefruit juice establishes furanocoumarins as the mediators of the grapefruit juice–felodipine interaction. *Am. J. Clin. Nutr.* **2006**, *83*, 1097–1105. [CrossRef] [PubMed]
37. Bailey, D.G.; Dresser, G.K.; Leake, B.F.; Kim, R.B. Naringin is a major and selective clinical inhibitor of organic anion-transporting polypeptide 1a2 (oatp1a2) in grapefruit juice. *Clin. Pharmacol. Ther.* **2007**, *81*, 495–502. [CrossRef] [PubMed]
38. Fukuda, K.; Guo, L.; Ohashi, N.; Yoshikawa, M.; Yamazoe, Y. Amounts and variation in grapefruit juice of the main components causing grapefruit–drug interaction. *J. Chromatogr. B* **2000**, *741*, 195–203. [CrossRef]
39. Fukatsu, S.; Fukudo, M.; Masuda, S.; Yano, I.; Katsura, T.; Ogura, Y.; Oike, F.; Takada, Y.; Inui, K.-i. Delayed effect of grapefruit juice on pharmacokinetics and pharmacodynamics of tacrolimus in a living-donor liver transplant recipient. *Drug Metab. Pharmacokinet.* **2006**, *21*, 122–125. [CrossRef]
40. Dreier, J.P.; Endres, M. Statin-associated rhabdomyolysis triggered by grapefruit consumption. *Neurology* **2004**, *62*, 670. [CrossRef]
41. Koziolek, M.; Alcaro, S.; Augustijns, P.; Basit, A.W.; Grimm, M.; Hens, B.; Hoad, C.L.; Jedamzik, P.; Madla, C.M.; Maliepaard, M.; et al. The mechanisms of pharmacokinetic food-drug interactions—A perspective from the ungap group. *Eur. J. Pharm. Sci.* **2019**, *134*, 31–59. [CrossRef]
42. Grenier, J.; Fradette, C.; Morelli, G.; Merritt, G.; Vranderick, M.; Ducharme, M. Pomelo juice, but not cranberry juice, affects the pharmacokinetics of cyclosporine in humans. *Clin. Pharmacol. Ther.* **2006**, *79*, 255–262. [CrossRef]
43. Sermsappasuk, P.; Chaiyakunapruk, N.; Lohitnavy, M.; Kitiyakara, K. Assessing clinical evidence of drug interactions betweencitrus juices and cyclosporine. *Asian Biomed.* **2013**, *7*, 477–489.
44. Gertz, B.J.; Holland, S.D.; Kline, W.F.; Matuszewski, B.K.; Freeman, A.; Quan, H.; Lasseter, K.C.; Mucklow, J.C.; Porras, A.G. Studies of the oral bioavailability of alendronate. *Clin. Pharmacol. Ther.* **1995**, *58*, 288–298. [CrossRef]
45. Gertz, B.J.; Holland, S.D.; Kline, W.F.; Matuszewski, B.K.; Porras, A.G. Clinical pharmacology of alendronate sodium. *Osteoporos. Int.* **1993**, *3*, 13–16. [CrossRef] [PubMed]
46. Malhotra, S. Seville orange juice-felodipine interaction: Comparison with dilute grapefruit juice and involvement of furocoumarins. *Clin. Pharmacol. Ther.* **2001**, *69*, 14–23. [CrossRef] [PubMed]
47. Edwards, D.; Fitzsimmons, M.; Schuetz, E.; Yasuda, K.; Ducharme, M.; Warbasse, L.; Woster, P.; Schuetz, J.; Watkins, P. 6′,7′-dihydroxybergamottin in grapefruit juice and seville orange juice: Effects on cyclosporine disposition, enterocyte cyp3a4, and p-glycoprotein. *Clin. Pharmacol. Ther.* **1999**, *65*, 237–244. [CrossRef]
48. Abdelkawy, K.S.; Donia, A.M.; Turner, R.B.; Elbarbry, F. Effects of lemon and seville orange juices on the pharmacokinetic properties of sildenafil in healthy subjects. *Drugs R D* **2016**, *16*, 271–278. [CrossRef]
49. Al-Ghazawi, M.A.; Tutunji, M.S.; AbuRuz, S.M. The effects of pummelo juice on pharmacokinetics of sildenafil in healthy adult male jordanian volunteers. *Eur. J. Clin. Pharmacol.* **2009**, *66*, 159–163. [CrossRef]

50. Žuntar, I.; Krivohlavek, A.; Kosić-Vukšić, J.; Granato, D.; Bursać Kovačević, D.; Putnik, P. Pharmacological and toxicological health risk of food (herbal) supplements adulterated with erectile dysfunction medications. *Curr. Opin. Food Sci.* **2018**, *24*, 9–15. [CrossRef]
51. Senthilkumaran, S.; Suresh, P.; Thirumalaikolundusubramanian, P.; Balamurugan, N. Priapism, pomegranate juice, and sildenafil: Is there a connection? *Urol. Ann.* **2012**, *4*, 108. [CrossRef]
52. Dresser, G. Fruit juices inhibit organic anion transporting polypeptide–mediated drug uptake to decrease the oral availability of fexofenadine. *Clin. Pharmacol. Ther.* **2002**, *71*, 11–20. [CrossRef]
53. Jeon, H.; Jang, I.-J.; Lee, S.; Ohashi, K.; Kotegawa, T.; Ieiri, I.; Cho, J.-Y.; Yoon, S.H.; Shin, S.-G.; Yu, K.-S.; et al. Apple juice greatly reduces systemic exposure to atenolol. *Br. J. Clin. Pharmacol.* **2013**, *75*, 172–179. [CrossRef]
54. Tapaninen, T.; Neuvonen, P.J.; Niemi, M. Orange and apple juice greatly reduce the plasma concentrations of the oatp2b1 substrate aliskiren. *Br. J. Clin. Pharmacol.* **2011**, *71*, 718–726. [CrossRef]
55. Lilja, J. Orange juice substantially reduces the bioavailability of the β-adrenergic–blocking agent celiprolol. *Clin. Pharmacol. Ther.* **2004**, *75*, 184–190. [CrossRef] [PubMed]
56. Mougey, E.B.; Lang, J.E.; Wen, X.; Lima, J.J. Effect of citrus juice and slco2b1 genotype on the pharmacokinetics of montelukast. *J. Clin. Pharmacol.* **2011**, *51*, 751–760. [CrossRef] [PubMed]
57. Wallace, A.W.; Victory, J.M.; Amsden, G.W. Lack of bioequivalence when levofloxacin and calcium-fortified orange juice are coadministered to healthy volunteers. *J. Clin. Pharmacol.* **2003**, *43*, 539–544. [CrossRef] [PubMed]
58. Neuhofel, A.L.; Wilton, J.H.; Victory, J.M.; Hejmanowski, L.G.; Amsden, G.W. Lack of bioequivalence of ciprofloxacin when administered with calcium-fortified orange juice: A new twist on an old interaction. *J. Clin. Pharmacol.* **2002**, *42*, 461–466. [CrossRef] [PubMed]
59. Fairweather-Tait, S.; Hickson, K.; McGaw, B.; Reid, M. Orange juice enhances aluminium absorption from antacid preparation. *Eur. J. Clin. Nutr.* **1994**, *48*, 71–73. [PubMed]
60. Balay, K.S.; Hawthorne, K.M.; Hicks, P.D.; Griffin, I.J.; Chen, Z.; Westerman, M.; Abrams, S.A. Orange but not apple juice enhances ferrous fumarate absorption in small children. *J. Pediatr. Gastroenterol. Nutr.* **2010**, *50*, 545–550. [CrossRef]
61. Aronson, J.K. *Meyler's Side Effects of Drugs: The International Encyclopedia of Adverse Drug Reactions and Interactions*; Aronson, J.K., Ed.; Elsevier Science: Amsterdam, The Netherlands, 2016; pp. 1030–1053.
62. Weaver, C.M. Potassium and health. *Adv. Nutr.* **2013**, *4*, 368S–377S. [CrossRef]
63. St-Jules, D.E.; Goldfarb, D.S.; Sevick, M.A. Nutrient non-equivalence: Does restricting high-potassium plant foods help to prevent hyperkalemia in hemodialysis patients? *J. Renal Nutr.* **2016**, *26*, 282–287. [CrossRef]
64. Stacewicz-Sapuntzakis, M.; Bowen, P.E.; Hussain, E.A.; Damayanti-Wood, B.I.; Farnsworth, N.R. Chemical composition and potential health effects of prunes: A functional food? *Crit. Rev. Food Sci. Nutr.* **2001**, *41*, 251–286. [CrossRef]
65. Wallace, T. Dried plums, prunes and bone health: A comprehensive review. *Nutrients* **2017**, *9*, 401. [CrossRef]
66. Sánchez-Pérez, S.; Comas-Basté, O.; Rabell-González, J.; Veciana-Nogués, M.; Latorre-Moratalla, M.; Vidal-Carou, M. Biogenic amines in plant-origin foods: Are they frequently underestimated in low-histamine diets? *Foods* **2018**, *7*, 205. [CrossRef]
67. Bushra, R.; Aslam, N.; Khan, A. Food drug interactions. *Oman Med. J.* **2011**, *26*, 77–83. [CrossRef] [PubMed]
68. Garcia, E.; Santos, C. Monoamine Oxidase Inhibitor Toxicity. Available online: https://www.ncbi.nlm.nih.gov/books/NBK459386/ (accessed on 24 October 2020).
69. Chen, J.J. Pharmacologic safety concerns in parkinson's disease: Facts and insights. *Int. J. Neurosci.* **2011**, *121*, 45–52. [CrossRef] [PubMed]
70. Aston, J.L.; Lodolce, A.E.; Shapiro, N.L. Interaction between warfarin and cranberry juice. *Pharmacotherapy* **2006**, *26*, 1314–1319. [CrossRef] [PubMed]
71. Mergenhagen, K.A.; Sherman, O. Elevated international normalized ratio after concurrent ingestion of cranberry sauce and warfarin. *Am. J. Health-Syst. Pharm.* **2008**, *65*, 2113–2116. [CrossRef] [PubMed]
72. Medicines and Healthcare Products Regulatory Agency/Committee on Safety of Medicines. Interaction between warfarin and cranberry juice: New advice. *Curr. Probl. Pharmacovigil.* **2004**, *30*, 10.
73. Suvarna, R. Possible interaction between warfarin and cranberry juice. *BMJ* **2003**, *327*, 1454. [CrossRef]
74. Ansell, J.; McDonough, M.; Zhao, Y.; Harmatz, J.S.; Greenblatt, D.J. The absence of an interaction between warfarin and cranberry juice: A randomized, double-blind trial. *J. Clin. Pharmacol.* **2009**, *49*, 824–830. [CrossRef]
75. Johnson, J.A.; Cavallari, L.H. Warfarin pharmacogenetics. *Trends Cardiovasc. Med.* **2015**, *25*, 33–41. [CrossRef]
76. Zikria, J.; Goldman, R.; Ansell, J. Cranberry juice and warfarin: When bad publicity trumps science. *Am. J. Med.* **2010**, *123*, 384–392. [CrossRef]
77. Backman, J. Lack of correlation between in vitro and in vivo studies on the effects of tangeretin and tangerine juice on midazolam hydroxylation. *Clin. Pharmacol. Ther.* **2000**, *67*, 382–390. [CrossRef] [PubMed]
78. Putnik, P.; Barba, F.J.; Lorenzo, J.M.; Gabrić, D.; Shpigelman, A.; Cravotto, G.; Bursać Kovačević, D. An integrated approach to mandarin processing: Food safety and nutritional quality, consumer preference, and nutrient bioaccessibility. *Compr. Rev. Food Sci. Food Saf.* **2017**, *16*, 1345–1358. [CrossRef]
79. Obermeier, M.T.; White, R.E.; Yang, C.S. Effects of bioflavonoids on hepatic p450 activities. *Xenobiotica* **2009**, *25*, 575–584. [CrossRef]
80. Kim, H.; Yoon, Y.-J.; Shon, J.-H.; Cha, I.-J.; Shin, J.-G.; Liu, K.-H. Inhibitory effects of fruit juices on cyp3a activity. *Drug Metab. Dispos.* **2006**, *34*, 521–523. [CrossRef] [PubMed]

81. Zhang, H.; Ma, Z.; Luo, X.; Li, X. Effects of mulberry fruit (*morus alba* l.) consumption on health outcomes: A mini-review. *Antioxidants* **2018**, *7*, 69. [CrossRef]
82. Rodriguez-Fragoso, L.; Reyes-Esparz, J. Fruit/vegetable-drug interactions: Effects on drug metabolizing enzymes and drug transporters. In *Drug Discovery*; IntechOpen: London, UK, 2013; pp. 1–33.
83. Piver, B.; Berthou, F.; Dreano, Y.; Lucas, D. Inhibition of cyp3a, cyp1a and cyp2e1 activities by resveratrol and other non volatile red wine components. *Toxicol. Lett.* **2001**, *125*, 83–91. [CrossRef]
84. Hidaka, M.; Fujita, K.-i.; Ogikubo, T.; Yamasaki, K.; Iwakiri, T.; Okumura, M.; Kodama, H.; Arimori, K. Potent inhibition by star fruit of human cytochrome p450 3a (cyp3a) activity. *Drug Metab. Dispos.* **2004**, *32*, 581–583. [CrossRef]
85. Eskra, J.N.; Schlicht, M.J.; Bosland, M.C. Effects of black raspberries and their ellagic acid and anthocyanin constituents on taxane chemotherapy of castration-resistant prostate cancer cells. *Sci. Rep.* **2019**, *9*, 1–12. [CrossRef]
86. Gerber, W.; Steyn, J.; Kotzé, A.; Hamman, J. Beneficial pharmacokinetic drug interactions: A tool to improve the bioavailability of poorly permeable drugs. *Pharmaceutics* **2018**, *10*, 106. [CrossRef]
87. Taniguchi, S. Treatment of psoriasis by cyclosporine and grapefruit juice. *Arch. Dermatol.* **1996**, *132*. [CrossRef]
88. van Agtmael, M.A.; Gupta, V.; van der Wösten, T.H.; Rutten, J.P.B.; van Boxtel, C.J. Grapefruit juice increases the bioavailability of artemether. *Eur. J. Clin. Pharmacol.* **1999**, *55*, 405–410. [CrossRef] [PubMed]
89. Adegoke, S.A.; Oyelami, O.A.; Olatunya, O.S.; Adeyemi, L.A. Effects of lime juice on malaria parasite clearance. *Phytother. Res.* **2011**, *25*, 1547–1550. [CrossRef] [PubMed]
90. Zhong, Y.; Wang, Y.; Guo, J.; Chu, H.; Gao, Y.; Pang, L. Blueberry improves the therapeutic effect of etanercept on patients with juvenile idiopathic arthritis: Phase iii study. *Tohoku J. Exp. Med.* **2015**, *237*, 183–191. [CrossRef]
91. Blanco-Rojo, R.; Pérez-Granados, A.M.; Toxqui, L.; González-Vizcayno, C.; Delgado, M.A.; Vaquero, M.P. Efficacy of a microencapsulated iron pyrophosphate-fortified fruit juice: A randomised, double-blind, placebo-controlled study in spanish iron-deficient women. *Br. J. Nutr.* **2011**, *105*, 1652–1659. [CrossRef] [PubMed]

foods MDPI

Review

Optimal Dosing Regimen of Osteoporosis Drugs in Relation to Food Intake as the Key for the Enhancement of the Treatment Effectiveness—A Concise Literature Review

Agnieszka Wiesner [1], Mariusz Szuta [2], Agnieszka Galanty [3] and Paweł Paśko [1,*]

1 Department of Food Chemistry and Nutrition, Faculty of Pharmacy, Jagiellonian University Medical College, 9 Medyczna Str., 30-688 Kraków, Poland; agnieszka.wiesner@doctoral.uj.edu.pl
2 Department of Oral Surgery, Faculty of Medicine, Jagiellonian University Medical College, 4 Montelupich Str., 31-155 Kraków, Poland; mariusz.szuta@uj.edu.pl
3 Department of Pharmacognosy, Faculty of Pharmacy, Jagiellonian University Medical College, 9 Medyczna Str., 30-688 Kraków, Poland; agnieszka.galanty@uj.edu.pl
* Correspondence: p.pasko@uj.edu.pl; Tel.: +48-12-620-5670

Abstract: Bisphosphonates and selective estrogen receptor modulators (SERMs) represent the two most important groups of medications taken orally and employed in osteoporosis treatment. Effectiveness of the therapy may be affected by poor patient adherence, in particular, due to the inconvenient dosing regimen of oral bisphosphonates. With this review we aimed to assess the effects that food, beverages, and dietary supplements consumed during treatment, along with the dosing regimens, may have on pharmacokinetics and pharmacodynamics of oral drugs employed in treating osteoporosis; we also aimed to shape the recommendations valuable for professional patients' counseling and education, to provide appropriate dosing regimens in order to improve adherence to the therapy. Food, beverages such as coffee, juices, and mineral water, as well as dietary supplements containing multivalent cations, e.g., calcium, magnesium, aluminium, iron, showed to have a deleterious effect on the bioavailability of all the investigated oral bisphosphonates, specifically alendronate, risedronate, ibandronate, minodronate, and etidronate. For risedronate, a delayed-release (DR) tablet was designed to solve the malabsorption problem in the presence of food, hence DR risedronate can be ingested following breakfast. For other oral bisphosphonates, the proper interval between drug and food, beverages, and dietary supplements intake should be maintained to minimize the risk of interactions. The effect of food on pharmacokinetic parameters of selective estrogen receptor modulators (SERMs) was found to be clinically irrelevant.

Keywords: bisphosphonates; SERMs; interaction; food; supplements; bioavailability; meal; coffee; juice; mineral water

Citation: Wiesner, A.; Szuta, M.; Galanty, A.; Paśko, P. Optimal Dosing Regimen of Osteoporosis Drugs in Relation to Food Intake as the Key for the Enhancement of the Treatment Effectiveness—A Concise Literature Review. *Foods* **2021**, *10*, 720. https://doi.org/10.3390/foods10040720

Academic Editors: Danijela Bursać Kovačević and Predrag Putnik

Received: 19 February 2021
Accepted: 26 March 2021
Published: 29 March 2021

Publisher's Note: MDPI stays neutral with regard to jurisdictional claims in published maps and institutional affiliations.

Copyright: © 2021 by the authors. Licensee MDPI, Basel, Switzerland. This article is an open access article distributed under the terms and conditions of the Creative Commons Attribution (CC BY) license (https://creativecommons.org/licenses/by/4.0/).

1. Introduction

Being a chronic skeletal disease that leads to a gradual bone density loss and increased bone fragility, as well as susceptibility to fractures, osteoporosis remains a significant reduced life quality risk factor in older patients. At least one in three women and one in five men in their 50s are estimated to experience osteoporotic fracture in their lifetime, often followed by further serious health consequences. Due to the progressive population aging, the problem of treating osteoporosis is more actual than ever. To provide the effective therapy with the two most important groups of drugs used in osteoporosis treatment, namely bisphosphonates and selective estrogen receptor modulators (SERMs), their proper administration in relation to food/meal time should be followed. Poor patient's adherence to drug administration regimens is considered one of the main problems that may affect the efficacy of the therapy, as it depends on the correct medications administration schemes and persistence in applying the treatment [1]. It is estimated that 50% of all patients discontinue therapy with oral bisphosphonates within 1 to 2 years from onset [2], or even earlier, before any clinical effect can be seen [3]. Moreover, even one-third of the patients may administer their medications improperly [4]. It is impossible to indicate one reason for poor adherence, as the problem is multifaceted. Yeam et al. [5] recognized up to 24 factors that can affect medications adherence among patients with osteoporosis. However, in many studies, top two issues reported by patients as the cause of non-adherence to osteoporosis treatment seem to predominate. The first issue is the poor motivation to take the drugs, due to the lack of disease symptoms and hence no clinical signs of drug effectiveness. The second problem includes inconvenient dosing regimen of oral bisphosphonates, especially the need to fast before, and to stay upright after the administration [2–4]. In response to the problem of non-adherence, Cornelissen et al. [6] concluded recently that patient education, supervision, counselling, and shared decision making on changing the dosing regimen, may improve both persistence and compliance, and consequently, also the effectiveness of osteoporosis treatment.

Another general problem is that patients, physicians, pharmacists and dieticians may underestimate the impact of food and dosing regimen on drugs effectiveness [7]. A recent study in a district general hospital suggests that healthcare professionals shall be reminded about proper dosing regimens for oral bisphosphonates. In an observation period of 6 months, Wilcock et al. [8] indicated that among 398 doses of alendronate or risedronate administered to patients in a hospital setting, half of the doses were given less than 30 min before breakfast, that is at the time when food may dramatically reduce bisphosphonates absorption. Although the main limitation of the study was the assumption around the exact time of breakfast, these results should be taken into consideration.

In elderly patients, the greater emphasis should be put on food-drug interactions due to the polypharmacy, and self-medication with dietary and nutritional supplements and OTC drugs (including herbal medicines) [9]. Thus, this review focused on the effects that consumption of food, beverages, and dietary supplements, as well as dosing regimens, may have on pharmacokinetics and pharmacodynamics of oral drugs employed in treating osteoporosis. This is the first such comprehensive review that brings together the information not only about bisphosphonates, but also about selective estrogen modulators, with the emphasis on oral forms of the drugs. Some valuable recommendations, for professional patients' counselling, summarizing the intake of the osteoporosis drugs with relation to food, and presenting how to avoid the potential interactions, are also included in the review.

2. Materials and Methods

To collect information on interactions between food and medications for osteoporosis, the authors, namely AW and PP, performed a literature research in the Medline (via PubMed) and Embase databases, covering reports from 1986 to 2021. The following keywords and phrases were applied in the searching process: drugs names ("alendronate", "risedronate", "ibandronate", "minodronate", "etidronate", "raloxifene", "bazedoxifene") in combinations with "food", "food-drug interaction", "meal", "breakfast", "bioavailability", "juice", "coffee", "tea", "milk", "iron", "aluminum", "calcium", "magnesium", "alcohol", and "mineral water". Furthermore, other resources such as drugs.com, Micromedex, AHFS, and UpToDate, were also researched, as well as characterization charts of particular medicinal products. Additional publications were found by checking the reference lists.

Initially, 155 articles were tracked. After removing 27 duplicates and performing the first evaluation, 78 articles reporting or investigating the effect of meals, beverages, dietary supplements, and dosing regimen on pharmacokinetics and pharmacodynamics of drugs used in oral osteoporosis treatment were considered to be included. No restrictions for study design, participants' characteristics, or sample size were made. 35 reviews, two in vitro studies, and nine studies performed on animals were subsequently excluded. Finally, 32 original studies remained included and are discussed in this review. A flowchart of the searching strategy is presented in Figure 1.

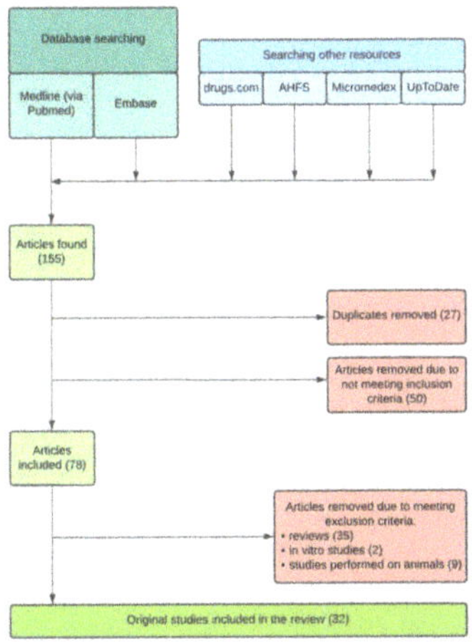

Figure 1. Searching strategy flowchart.

3. Results and Discussion

3.1. Oral Bisphosphonates

Various guidelines on managing osteoporosis recommend oral bisphosphonates as the first-choice drugs [10–13], widely prescribed in postmenopausal osteoporosis, secondary osteoporosis, and male osteoporosis. Bisphosphonates are often referred to as "antiresorptive drugs", due to their mechanism of action, as they inhibit osteoclast-mediated bone resorption [14]. A recent meta-analysis confirmed bisphosphonates' use to decrease the risk of both vertebral and non-vertebral osteoporotic fractures [15]. The treatment becomes effective relatively early, namely within 6–12 months from starting the therapy, and the results tend to prolong [16]. Despite its established effectiveness, therapy with oral bisphosphonates may still be found challenging, due to the overall low bisphosphonates bioavailability, likely gastrointestinal side effects, such as heartburn, irritated esophagus, nausea, gastric ulcers, etc., but also the potential interactions with food and dietary supplements [14,16].

3.1.1. Chemical Characterization of the Bisphosphonates

All bisphosphonates are the analogs of pyrophosphate, containing two phosphate groups linked by carbon atom (P-C-P), instead of an oxygen atom [17]. Additionally, as presented in Figure 2, each bisphosphonate has two side chains bound to the central carbon atom.

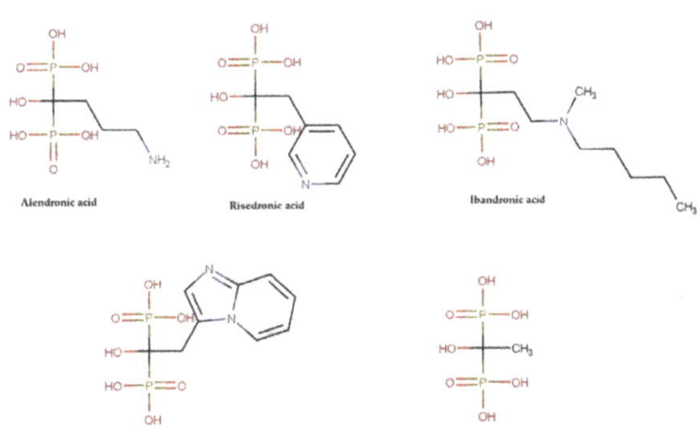

Figure 2. Chemical structures of bisphosphonates discussed in the review.

Of the drugs discussed in this review, only etidronate does not contain nitrogen atom, as it is representative of the oldest, first-generation of bisphosphonates. The addition of nitrogen, either as a constituent of the amine group (alendronate and ibandronate) or heterocyclic ring (in risedronate and minodronate) is known to enhance the antiresorptive activity [17].

The Biopharmaceutical Classification System (BCS) assigns each pharmacologically active compound to one of four classes, based on its water solubility and intestinal membrane permeability, since these parameters determine the rate and extent of drug absorption [18]. Bisphosphonates, as the highly soluble and hardly permeable compounds, belong to the III class of BCS. High water solubility (in other words—very low lipophilicity) of bisphosphonates is due to the presence of hydrophilic phosphate groups. These groups are negatively charged in physiological pH, hence the bisphosphonate compound is completely ionized in the intestine lumen. Ionized and hydrophilic compounds penetrate lipid membranes with great difficulty. That explains both the low permeability and very low oral bioavailability of bisphosphonates, which is known to range from 0.6 to 3%, depending on the drug [19].

3.1.2. Interaction with Di- and Trivalent Cations

Although the overall effect of calcium, or vitamin D alone, on the reduction of fracture risk remains unclear [13,20], the supplementation of both substances is currently recommended for patients at high risk of calcium and vitamin D insufficiency, and in those treated for osteoporosis [21]. In consequence, concomitant use of oral bisphosphonates and calcium preparations is widespread.

Oral bisphosphonates are chelating agents that form salts with multivalent cations present in food, dietary supplements, drugs (e.g., antacids), and mineral water [14]. These salts are insoluble at pH > 5, and as such, bisphosphonate cannot be absorbed in the intestine [22]. In characterization charts of oral bisphosphonates, the recommended minimum time interval between their ingesting and the intake of vitamins, dietary supplements or antacids, containing multivalent cations such as calcium, aluminum, magnesium, and chromium, varies from at least 30 min to 2 h, depending on the drug [23–27], while at least last 2 h for iron preparations [28]. For oral bisphosphonates taken once a week, or once a month, it may be even justified to refrain from administering calcium preparations on the same day as a bisphosphonate.

To improve the patients' understanding of co-treatment rules, but also compliance, fixed-combination packs of bisphosphonates and calcium carbonate were designed and tested. In a cross-over study in 164 postmenopausal women, Ringe et al. [29] revealed that use of the combination pack containing one tablet of risedronate 35 mg and six tablets of calcium carbonate 1250 mg (500 mg of elemental calcium) for one week, significantly improved patients' comprehension of dosing instructions, when compared to separate packaging (80% vs. 70%, respectively). In some countries such fixed-combinations packs are already registered, e.g., for alendronate (*APO-Alendronate Plus D3 and calcium*), or risedronate (*Actonel Combi*).

3.1.3. Problem with Water Type

The results of preclinical studies indicated that alendronate and risedronate absorption may decrease with the increasing concentration of calcium in mineral water used for drug administering [30,31]. As instructed by prescribing guidelines, oral bisphosphonates are recommended to be taken with tap water, to minimize the risk of malabsorption caused by calcium and other cations present in mineral and spring waters. However, tap water can be alkaline as well. Pellegrini et al. [32] reported that in some regions of Italy (e.g., Rome or Milan) calcium content of tap water might be even up to 100 times higher than in some commercially available bottled waters (100–110 mg/L vs. 1–2 mg/L, respectively). Morr et al. [33] obtained similar results after examining tap waters from cities in the USA and Canada, where reportedly, calcium content varied from 1 to 135 mg/L. For comparison, spring waters were found to contain lower calcium concentrations (approximately

21.8 mg/L), and mineral waters in general showed higher calcium content (an average of 208 mg/L). Due to the lack of experimental evidence, neither specific nor approximate maximum calcium intake that would not alter bisphosphonates absorption has been established. For that reason, Azoulay et al. [34] emphasized the need to check the mineral content of both bottled and tap waters before administering bisphosphonates. Moreover, Morr et al. [33] observed that filtering tap water with a home filter system may remove on average 89% of calcium, hence proposed this alternative for patients living in regions with the high calcium content of tap water. Such approach may be thus recommended as one of the elements of preventing the decrease in the effectiveness of bisphosphonates therapy.

3.1.4. Alendronate

Alendronate (syn. alendronic acid, alendronate sodium) is available in tablet (*Fosamax*, *Ostemax*), effervescent tablet (*Binosto*), and oral solution (*Fosamax*) forms, taken daily or weekly. The same doses of different alendronate formulations are bioequivalent [23,35]. Additionally, in 2012, the oral jelly formulation (*Bonalon*) has been approved in Japan, to reduce the risk of choking and to prevent gastrointestinal side effects [36]. In some countries combination of alendronate in the same tablet with vitamin D (*Fosamax Plus*) is registered as well.

Oral bioavailability of alendronate under fasting conditions is 0.64% in women (for dosage range 5–70 mg), and 0.59% in men (for dose 10 mg); in the presence of food and beverages further decrease is observed [23,37].

Gertz et al. [22] conducted series of open-label, randomized, cross-over trials to establish the most optimal regimen for alendronate administration regarding food and beverages. In the first study of 10 post-menopausal women, a single dose of 20 mg alendronate tablet was administered after an overnight fast, in three dosing regimens: (1) 2 h before breakfast—reference regimen, (2) 1 h before breakfast, or (3) 30 min before breakfast; with or without calcium carbonate supplement (containing 1000 mg of elemental calcium). The detailed composition of the breakfast was described in Table 1—Set No 1. In the second study of 49 post-menopausal women, a single dose of 10 mg alendronate tablet was taken after an overnight fast, in five dosing regimens: (1) 2 h before breakfast—reference regimen, (2) 1 h before breakfast, (3) 30 min before breakfast, (4) immediately after breakfast, or (5) 2 h after breakfast (Table 1, Set No 2). Results of both studies were comparable and found that the administration of alendronate 30 min and 1 h before breakfast reduced its bioavailability approximately by 40%, relative to the reference regimen. Moreover, the second study revealed that ingesting alendronate concomitantly with breakfast, or even 2 h after the meal, may drastically impair drug absorption (by 85–90%). Adding calcium supplement to breakfast in the first study did not cause an additional decrease in absorption when compared to ingesting the meal alone. In the third study of 40 post-menopausal women, a single dose of 10 mg alendronate tablet was taken, followed an overnight fast, 2 h before breakfast, with (1) 240 mL of tap water—reference regimen, (2) black unsweetened coffee, or (3) orange juice. Coffee and orange juice were found to significantly reduce alendronate absorption (by 60% for both beverages) relative to administration with water. What is worth noting, the study is probably the only example, describing the interactions between any bisphosphonate and coffee and fruit juice, thus the important results should be further continued to draw more reliable conclusions.

Wagener et al. [38] assessed whether the ingestion of alendronate during the day, with 1 h fast before a meal, could be an effective dosing regimen. After 8 weeks of study, significant changes in biochemical markers of bone turnover, namely osteocalcin, alkaline phosphatase, and dipeptidyl peptidase, were observed, suggesting that despite a non-optimal dosing regimen, the efficacy of treatment was maintained.

Table 1. Detailed composition of the meals used in the studies described in this review.

Type of Meal	Food Items	Nutrition Facts	Energy	References
		Alendronate		
Breakfast Set No 1	2 pieces of white toast, jelly/marmalade (20 g), 1 fried egg, 2 strips of bacon, and orange juice (250 mL)	70.6 g of carbohydrates, 6.3 g of fat, and 22.4 g of protein	Approx. 350 kcal	[22]
Breakfast Set No 2	1 piece of white toast, 2 pats of butter, 2 strips of bacon, 2 fried eggs, hash brown potatoes (2–4 oz), and whole milk (240 mL).	40.3 g of carbohydrates, 29 g of fat, and 89.4 g of protein	Approx. 665 kcal	[22]
		Risedronate		
Lunch Set No 3	Smoked turkey, vegetable and beef soup with crackers, and whole wheat bread with lettuce (283 g), tossed salad (142 g) with light salad dressing (12 g), mayonnaise (15 mL), 2 canned peach halves, and skimmed milk (283 g)	104 g of carbohydrates, 19 g of fat, and 38 g of protein	Approx. 716 kcal	[39]
Breakfast Set No 4	2 slices of white toast, 2 pats of butter, 2 slices of bacon, hash brown potatoes (57 g), 2 eggs fried in butter, and whole milk (226 g)	50 g of carbohydrates, 46 g of fat, and 30 g of protein	Approx. 730 kcal	[39]
Dinner Set No 5	baked chicken breast (113 g), 1 baked potato, light gravy (28 g), 1 pat of margarine, 0.5 cup of apple sauce, 0.5 cup of carrot rounds, 1 peanut butter cookie, and lemonade (283 g)	103 g of carbohydrates, 16 g of fat, and 40 g of protein	Approx. 700 kcal	[39]
		Minodronate		
Breakfast Set No 6	chicken drumstick, fried egg, hamburger, and milk	30% of carbohydrates, 55% of fat, 15% of protein	Approx. 900 kcal	[40]

The impact of food on the bioavailability of alendronate was investigated for effervescent tablet formulation as well. A study of 119 healthy women revealed that taking 70 mg alendronate 15 min before standard breakfast may decrease its bioavailability by approximately 50%, when compared to administering 4 h before a meal [35].

Interestingly, a recent prospective case-control study [2] suggested that administering alendronate after breakfast, despite the existing evidence of the impaired drug absorption for this regimen, may have positive influence on a patient's persistence and adherence, and hence improve the efficacy of treatment. Park et al. [2] compared 1-year persistence and therapy effectiveness in patients taking 5 mg alendronate tablet in two regimens: (1) according to the guidelines. i.e., 30 min before breakfast, and (2) after breakfast. After 1 year, the medication possession ratio (MPR), measured as the sum of administration days divided by 365 days, was significantly higher in patients taking alendronate after breakfast, relative to the recommended regimen (0.71 vs. 0.66, respectively), which indicate enhanced persistence and adherence. Additionally, significant improvement in lumbar and hip T-scores was observed in both groups, together with no significant differences between the groups, which suggests the therapy effectiveness to be similar for both treatment regimens.

3.1.5. Risedronate

Risedronate (syn. risedronic acid, risedronate sodium) is available in two formulations: immediate-release (IR) tablets (*Actonel*, *Risendros*) taken daily, once a week or once a month, and delayed-release (DR) tablets (*Atelvia*) taken once a week.

Immediate-Release (IR) Tablets

The mean oral bioavailability of the 30 mg IR tablet of risedronate is 0.63%, when administered while fasting, whereas the concomitant intake of food decreases it further [24]. Mitchell et al. [39] conducted a randomized, parallel study in 127 healthy volunteers, to compare four different timing regimens for risedronate administration: (1) after night fasting, 4 h before lunch (Table 1, Set No 3) that was the reference regimen, (2) after night fasting, 1 h before a high-fat breakfast (Table 1, Set No 4), (3) after night fasting, 30 min before a high-fat breakfast (Table 1, Set No 4), (4) 2 h after a standard dinner (Table 1, Set No 5). After administering a single dose of 30 mg risedronate IR tablets in each group, pharmacokinetic parameters such as $AUC_{0-\infty}$ and C_{max} were measured. Risedronate, both taken 30 min before breakfast (group 3), and 2 h after dinner (group 4), showed the lowest extend of absorption, when compared to the reference regimen; AUC and C_{max} were reduced by 56 and 32% in group 3; and by 52 and 75% in group 4, respectively. To compare, when risedronate was taken 1 h before breakfast (group 2), AUC and C_{max} were reduced by 32 and 14%, respectively. Mitchell et al. concluded that although the most optimal dosing regimen for risedronate in IR tablets is 1 h before breakfast, a flexible-dosing approach may be considered, since dosing either 30 min before breakfast or 2 h after dinner resulted in a similar extend of risedronate absorption.

Ogura et al. [12] performed a similar study in 12 healthy Japanese volunteers, randomly assigned to administer 5 mg of IR risedronate (1) under fasting conditions, without breakfast—a reference regimen, (2) 30 min before breakfast, (3) 30 min after breakfast, and (4) 3 h after breakfast. Breakfast contained 200 mg of elemental calcium. To compare the level of risedronate absorption in different regimens, such pharmacokinetic parameters as AUC_{0-24}, C_{max}, t_{max}, and $t_{1/2}$ were measured. Administering risedronate 30 min after breakfast (3) caused the highest decrease in AUC_{0-24} and C_{max} (both by 94%), and the highest increase in t_{max} and $t_{1/2}$ (by 62 and 39%, respectively), when compared to the reference regimen. Contrastingly, administration 30 min before breakfast decreased AUC_{0-24} by 63%, and C_{max} by 26%, hence Ogura et al. recommended this regimen as the most optimal for Japanese patients. However, it should be emphasized that in this study administration 60 min before breakfast was not investigated.

In two studies of a total number of 1582 postmenopausal women with osteoporosis, Kendler et al. [41] assessed the influence of dosing regimen on risedronate efficacy measured by lumbar spine bone mineral density (BMD). Risedronate IR tablets in a daily dose of 5 mg were administered (1) under fasted conditions, at least 30 min before the first meal of the day, or (2) between meals, at least 2 h from a meal or 30 min before bedtime. Both studies provided similar observations. Both dosing regimens resulted in a significant increase from the baseline in lumbar spine BMD, however response to risedronate was smaller for flexible dosing. In the first study, an increase in lumbar spine BMD after 6 months was 2.9% for (1) regimen and 1.5% for (2) regimen, and in the second study after 12 months, 4.4 and 3.2% respectively. Kendler et al. [41] suggested that minor response to treatment in the flexible dosing group may be due to less than optimal risedronate absorption, when taken between meals, or poorer compliance to this regimen, especially in certain geographic regions.

Agrawal et al. [42] also studied the efficacy of the flexible-dosing approach in a randomized, double-blind trial in 60 residents of the nursing home (NH). 31 of participants were weekly administered 30 mg of risedronate in IR tablets, between meals but at least 2 h from a meal, and 29 received placebo. The effect of the treatment was measured by the changes in the serum levels of bone turnover markers (e.g., bone-specific alkaline phosphatise—BSAP). Risedronate administered in-between-meal schedule significantly reduced serum BSAP after 6 weeks of study relative to the control group, but this effect was not maintained after 12 weeks. Agrawal et al. [42] provided possible reasons for the ineffectiveness of flexible-dosing regimen in NH residents, such as delayed gastric emptying in the elderly, leading to impaired absorption of risedronate, when taken between meals, or high prevalence of vitamin D insufficiency that may contribute to the increase of bone turnover.

Delayed-Release (DR) Tablets

Enteric-coated DR risedronate formulation was developed with the addition of ethylenediaminetetraacetic acid (EDTA), the chelator of metal ions. As a result, di- and trivalent cations in food may be preferentially bound by EDTA instead of risedronate, which eliminates the need for fasting and improves risedronate bioavailability, when compared with IR formulation [14]. For example, 35 mg of risedronate in DR tablet taken after a high-fat breakfast was found to have similar bioavailability as the same dose in IR tablet administered 4 h before a meal, and 2–4-fold greater bioavailability than the IR tablet administered 30 min before breakfast [25].

In a recent randomized, double-blind, placebo-controlled study of 68 healthy postmenopausal women, Fukase et al. [43] investigated the influence of dosing regimens on pharmacokinetic and pharmacodynamic parameters of risedronate in DR tablets. A single dose of 37.5 mg was administered (1) under fasting conditions—a reference regimen, (2) 30 min before a meal, (3) immediately after a meal, or (4) 30 min after a meal. The meal consisted of 492 kcal and contained 4% of fat. When compared to fasting conditions, taking risedronate DR tablets immediately after a meal resulted in a significant decrease of $AUC_{0-\infty}$ and C_{max} (by 58 and 69%, respectively). However, pharmacokinetic and pharmacodynamic profiles for regimen (2), (3), and (4) were comparable, hence Fukase et al. suggested that DR formulation can be taken independently from meals.

Similar results were obtained in a cross-over pharmacokinetic study, where administration of DR tablet with high-fat breakfast resulted in a 30% decrease of bioavailability, compared to fasting state [25]. However, it should be noted that when DR tablets are taken on an empty stomach, an 8–9% increase in gastrointestinal adverse events may occur [25]. A possible solution to this problem could comprise administering risedronate after dinner. In a separate study, taking DR tablets in that regimen improved bioavailability by 87%, compared to the administration with breakfast. Nevertheless, the data for the safety and efficacy of taking risedronate DR tablets at bedtime is scarce [25].

In a randomized 2-year study of postmenopausal women with osteoporosis, McClung et al. [44,45] compared the efficacy of weekly administered 35 mg risedronate in DR tablets in two dosing regimens: at least 30 min before breakfast (BB), or following breakfast (FB) with 5 mg IR tablets taken daily while fasting. The effectiveness of the treatment was measured by changes in lumbar spine BMD. After 1 year of the study [44], the mean percent changes in lumbar spine BMD for 767 women were 3.4% in the 35 mg DR BB group, 3.3% in the 35 mg DR FB group, and 3.1% in the 5 mg IR daily group. After 2 years of treatment [45], the mean percent changes in lumbar spine BMD for 722 women were higher in all groups: 5.4% in the DR BB weekly group, 5.5% in the DR FB weekly group, and 4.4% in the IR daily group. McClung et al. concluded that weekly treatment with risedronate 35 mg DR is at least as effective as daily administered risedronate 5 mg IR, and DR formulation can be taken either before or after breakfast with similar efficacy.

On the contrary, a recent randomized phase II/III double-blind study [46] in Japanese patients with involutional osteoporosis revealed that the effectiveness of risedronate DR therapy may significantly differ, depending upon dosing regimen and formulation. Two doses of DR risedronate—25 and 37.5 mg were administered monthly for 1 year: (1) under fasting conditions, (2) concomitantly with breakfast, or (3) 30 min after breakfast. The efficacy of treatment was measured by changes in lumbar spine BMD and compared with daily administered IR 2.5 mg risedronate. At the end of the study, for DR 25 mg, percentage changes in lumbar spine BMD for regimens (1), (2), (3) were as follows: 3.82, 3.36, 3.93%, for DR 37.5 mg: 4.81, 4.11, 4.36%, respectively, and for IR 2.5 mg: 5.07%. Results indicated that risedronate ingestion immediately with breakfast is the least effective, and the non-inferiority of treatment with IR and DR formulations cannot be declared.

Although most of the concerned studies indicate that DR formulation can be administered while eating, without significant changes in bioavailability and efficacy, still the intake of supplements containing di- and trivalent cations should be delayed. In a cross-over study on 101 postmenopausal women, the bioavailability of a single 35 mg dose of risedronate DR tablets was evaluated, when taken following breakfast, with or without the supplement containing 600 mg of elemental calcium and 400 IU of vitamin D. Concomitant intake of risedronate and calcium-containing supplement reduced drug absorption by 38% [25].

3.1.6. Ibandronate

Ibandronate (syn. ibandronic acid, ibandronate sodium) for oral use is available in tablets (*Bonviva, Bondronat*), taken at monthly intervals. The mean oral bioavailability is approximately 0.6% for a dose of 2.5 mg [26].

In a pharmacokinetic study, the bioavailability of ibandronate was found to be dramatically reduced (by 90%), when 150 mg tablet was ingested concomitantly with a standard breakfast, relative to the fasting state [26]. Ingesting ibandronate 2 h after a standard meal or 30 min before breakfast significantly reduced the bioavailability as well (by 75% and 30%, respectively) [47]. No meaningful reduction in ibandronate bioavailability was observed for administration at least 60 min before breakfast [26,47].

In single-centre clinical trials, Nakai et al. [48] investigated the influence of fasting intervals on ibandronate bioavailability measured as $AUC_{0-\infty}$. In the first study, 24 healthy postmenopausal women were administered 2.5 mg ibandronate tablets 30 or 60 min before breakfast. Measured $AUC_{0-\infty}$ were 1.12 ± 0.95 and 1.40 ± 0.77 ng·h/mL, respectively. In the second study, 24 healthy men and postmenopausal women were given 50 mg ibandronate tablets. Measured $AUC_{0-\infty}$ values were 11.1 ± 23.5 ng·h/mL for 30-min fasting interval, and 16.0 ± 15.6 ng·h/mL for 60-min fasting interval. Nakai et al. concluded that shortening the fasting period from 60 to 30 min may result in lower ibandronate bioavailability, and hence recommended delaying food intake for 1 h after drug administration.

In a 48-week, multi-centre, open-label study, Tanko et al. [49] examined whether different dosing regimens may influence the effectiveness of ibandronate treatment, measured by percentage changes in lumbar spine BMD from baseline. 184 postmenopausal women were randomly assigned to administer 2.5 mg ibandronate tablets 30 or 60 min before their standard breakfast. At the end of the study, significant increases in lumbar spine BMD were observed in both groups, however, in a group where a 60-min fasting interval was maintained, the mean change in the lumbar spine BMD was higher than in a group with a 30-min fasting interval (4.95 vs. 3.07%, respectively). Taking it into consideration, Tanko et al. suggested that fasting interval after ingesting ibandronate can be reduced to 30 min, however increasing drug dose may be required to maintain the efficacy of treatment.

3.1.7. Minodronate

Minodronate (syn. minodronic acid) is approved for the treatment of osteoporosis in Japan. Among available oral bisphosphonates, it is the strongest inhibitor of bone resorption, hence it can be administered once a month [50]. Bioavailability of minodronate after oral administration was calculated to be 1.21% [51].

In an open-label study of 12 healthy volunteers, Zhou et al. [40] investigated the influence of meals on minodronate absorption. A single dose of 4 mg was administered while fasting or 30 min before a high-fat breakfast (Table 1, Set No 6) consisted of 400 mg of elemental calcium. Taking minodronate 30 min before breakfast reduced $AUC_{0-\infty}$ and C_{max} by 72 and 55%, respectively, and increased t_{max} from 0.75 to 2.15 h relative to fasted conditions. These results indicate a necessity to administer minodronate while fasting.

3.1.8. Etidronate

Etidronate (syn. etidronic acid, etidronate disodium) is available in tablets (*Didronel*) taken in 90-days cycles: daily for 14 days, followed by calcium supplementation for the remaining 76 days. Although etidronate is no longer recommended in American College of Physicians guidelines for the treatment of osteoporosis [13], and it is not FDA-approved for this indication as well, in some countries etidronate use is still considered as the second-line therapy for osteoporosis [52].

The oral bioavailability of etidronate is approximately 3% [27]. A single-dose study on 10 healthy volunteers revealed that administering 400 mg etidronate during breakfast may reduce drug absorption to zero [53].

In a retrospective study of 110 patients with osteoporosis, Cook et al. [54] assessed the influence of etidronate dosing regimen on treatment effectiveness, measured by changes in BMD. Patients were administered etidronate (1) after waking up, (2) at bedtime, or (3) during the day, and maintained 2 h fasting intervals before and after drug ingestion. After approximately 2.6 years of study, no significant differences in percentage change in BMD were found between the three groups, indicating that the dosing regimen of etidronate does not affect the efficacy of therapy.

Nevertheless, in another study of 70 patients treated with etidronate, Ryan et al. [55] revealed that drug ingestion during the day, even with 2 h fast before and after dosing, may still result in significantly impaired treatment response (lower percentage changes in BMD), when compared to administration in the early morning or late evening.

3.2. Selective Estrogen Receptor Modulators (SERMs)

Selective estrogen receptor modulators (SERMs) act as agonists on estrogen receptors in bone—modulate skeletal homeostasis by diminishing the activity of osteoclasts and maintaining the function of osteoblasts [56]. SERMs decrease bone remodelling, prevent bone loss by increasing bone mineral density [56] and are recommended as the most beneficial in younger postmenopausal women with mild and moderate osteoporosis [57]. The most common side effects of treatment with SERMs are hot flashes, leg cramps, and an increased risk of deep vein thrombosis [56,57].

The core structure of SERMs is based on 17β-estradiol template. Chemical structures of the discussed SERMs representatives are presented in Figure 3.

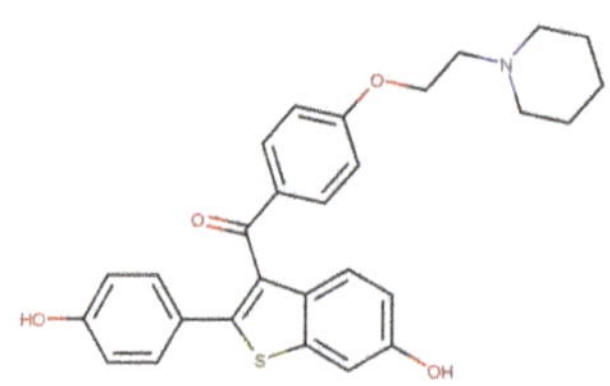

Raloxifene

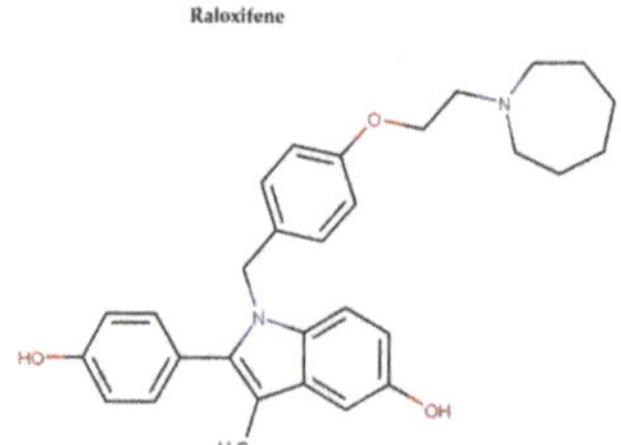

Bazedoxifene

Figure 3. Chemical structures of SERMs discussed in the review.

Both raloxifene and bazedoxifene belong to the II class of BCS—as lipophilic compounds they are hardly soluble in water and easily permeable by membrane cells. Low water solubility, together with the first-pass metabolism are the main factors negatively affecting oral bioavailability of these drugs [58].

3.2.1. Raloxifene

Raloxifene is available in tablets (*Evista*), taken once daily. On average 60% of an oral dose is rapidly absorbed from the gastrointestinal tract, however, due to the extensive first-pass hepatic metabolism, absolute oral bioavailability is only 2% [59]. The presence of food does not influence raloxifene pharmacokinetics. Concomitant ingestion with a high-fat meal increased AUC and C_{max} by 16 and 28%, respectively, though these changes were considered insignificant [60].

3.2.2. Bazedoxifene

Bazedoxifene is available in tablets taken daily, both alone (*Conbriza*) and in combination with conjugated estrogens (*Duavee*). The absolute oral bioavailability of bazedoxifene is on average 6% [61].

To examine the impact of food on bazedoxifene pharmacokinetic parameters, McKeand et al. [62] administered a single 10 mg bazedoxifene tablet to 84 healthy postmenopausal women, either while fasting, or 10 min before a high-fat breakfast. Concomitant intake with a meal reduced bazedoxifene C_{max} by 35%, but this change was considered insignificant; AUC for both dosing regimens were comparable.

Other studies revealed that the administration of bazedoxifene with food may even slightly enhance drug absorption. After taking a 20 mg tablet with a high-fat meal, AUC and C_{max} increased by 28 and 22%, respectively, and after ingestion with a medium-fat meal, by 35 and 42%, correspondingly. However, these changes were found clinically irrelevant, and it was concluded that bazedoxifene may be taken regardless of food [61].

The influence of food was also investigated for bazedoxifene combined with conjugated estrogens. In a crossover study, 23 postmenopausal women were administered a single 20 mg/0.625 mg tablet either under fasting conditions, or with a high-fat, high-calorie meal. Ingestion with a meal increased $AUC_{0-\infty}$ by 25%, while C_{max} remained unchanged, suggesting that combined preparation can be administered without regard to meals as well [63].

3.3. *Recomendations*

Table 2 below presents the summary of the most important data and recommendations for appropriate intake of drugs employed in treating osteoporosis with regard to food. Knowledge on the subject, with special emphasis on oral bisphosphonates, is critical to educate patients professionally and effectively, which is the key intervention to enhance adherence to oral osteoporosis treatment.

Table 2. The summary of recommendations for appropriate intake of drugs employed in treating osteoporosis with regard to food.

Drug	Available Oral Formulations	Dosing Frequency	Recommendations with Regard to Food	Other Recommendations	References
Alendronate	tablets effervescent tablets oral solution jelly	daily or weekly	should be taken at least 30 min before breakfast should be taken at least 30 min before vitamins, mineral supplements, or antacids high in calcium, aluminum, or magnesium should be taken at least 2 h before iron preparations swallowing with coffee and juice should be avoided	tablets should be swallowed with a full glass of tap water effervescent tablets should be dissolved in a half a glass of tap water the oral solution should be taken with at least a quarter a glass of tap water jelly can be administered without water should be taken while standing or sitting in an upright position lying down for at least 30 min after administration should be avoided	[2,22,23,35,37,38]
Risedronate	immediate-release (IR) tablets	daily, weekly or monthly	should be taken at least 30 min before breakfast the flexible-dosing approach might be considered calcium, magnesium, and aluminum preparations should be administered at a different time of the day should be taken at least 2 h before iron preparations	should be swallowed with a full glass of tap water should be taken while standing or sitting in an upright position lying down for at least 30 min after administration should be avoided	[12,24,39,41,42]
	delayed-release (DR) tablets	Weekly	should be taken immediately following breakfast administration at bedtime might be considered calcium, magnesium, and aluminum preparations should be administered at a different time of the day should be taken at least 2 h before iron preparations	should be swallowed with at least half a glass of tap water should be taken while standing or sitting in an upright position lying down for at least 30 min after administration should be avoided	[14,25,43–46]
Ibandronate	tablets	Monthly	should be taken at least 1 h before breakfast should be taken at least 1 h before vitamins, mineral supplements, or antacids high in calcium, aluminum, or magnesium should be taken at least 2 h before iron preparations	should be swallowed with a full glass of tap water should be taken while standing or sitting in an upright position lying down for at least 60 min after administration should be avoided	[26,47–49]
Minodronate	tablets	Monthly	should be taken at least 30 min before breakfast	should be swallowed with a tap water should be taken while standing or sitting in an upright position lying down for at least 30 min after administration should be avoided	[40]
Etidronate	tablets	daily for 14 days, in 90-days cycles	should be taken at least 2 h before food (especially high-calcium products), early in the morning, or at bedtime should be taken at least 2 h before vitamins, mineral supplements, or antacids high in calcium, magnesium, aluminum, iron	should be swallowed with a full glass of tap water should be taken while standing or sitting in an upright position lying down immediately after administration should be avoided	[53–55]
Raloxifene	tablets	Daily	can be taken irrespectively of food		[60]
Bazedoxifene	tablets	Daily	can be taken irrespectively of food		[61–63]

4. Summary

A proper regimen of administration for medications used in oral osteoporosis therapy is crucial not only for the treatment effectiveness but also for minimizing the risk of adverse events and hence improving the patient's quality of life. For the first-choice drugs, namely oral bisphosphonates, their combined administration with food, beverages, and dietary supplements containing multivalent cations, showed to have a deleterious effect on the bioavailability, and treatment efficacy. Changing drug formulation to delayed-release tablets may partially solve this problem, however more studies are needed. On the contrary, the effect of food on pharmacokinetic parameters of selective estrogen receptor modulators (SERMs) is clinically irrelevant. It is also worth noting that the data on the interactions of osteoporosis drugs with the individual food components is scarce, while the results of the existing studies suggests the importance of the problem. Thus, in our opinion, further in-depth studies are highly recommended, the results of which may complete the current strategy in enhancing the effectiveness of osteoporosis treatment.

Author Contributions: Conceptualization P.P. and A.W.; methodology P.P. and A.W.; investigation, P.P., A.W., A.G., and M.S.; resources, P.P.; writing—original draft preparation, A.W. and P.P.; writing—review and editing, P.P., M.S. and A.G.; visualization, A.W.; supervision, P.P.; project administration, P.P.; funding acquisition, P.P. All authors have read and agreed to the published version of the manuscript.

Funding: This work was partially financed and supported by grant Social Responsibility of Science (SONP/SP/461418/2020) Polish Ministry of Education and Science "Interaction between drugs and food—knowledge, awareness, effectiveness and safety".

Institutional Review Board Statement: Not applicable.

Informed Consent Statement: Not applicable.

Data Availability Statement: Data availability on the request.

Conflicts of Interest: The authors declare no conflict of interest.

References

1. Cramer, J.A.; Roy, A.; Burrell, A.; Fairchild, C.J.; Fuldeore, M.J.; Ollendorf, D.A.; Wong, P.K. Medication compliance and persistence: Terminology and definitions. *Value Health* **2008**, *11*, 44–47. [CrossRef] [PubMed]
2. Park, C.H.; Jung, K.J.; Nho, J.H.; Kim, J.H.; Won, S.H.; Chun, D.I.; Byun, D.W. Impact on bisphosphonate persistence and compliance: Daily postprandial administration. *J. Bone Metab.* **2019**, *26*, 39–44. [CrossRef] [PubMed]
3. Donato, P.; Pepe, J.; Colangelo, L.; Danese, V.; Cecchetti, V.; Minisola, S.; Cipriani, C. Adherence to bisphosphonates in the general population: A study in patients referred to a primary care service. *Arch. Osteoporos.* **2019**, *14*, 10–14. [CrossRef] [PubMed]
4. Kothawala, P.; Badamgarav, E.; Ryu, S.; Miller, R.M.; Halbert, R.J. Systematic Review and Meta-analysis of Real-World Adherence to Drug Therapy for Osteoporosis. *Mayo Clin. Proc.* **2007**, *82*, 1493–1501. [CrossRef]
5. Yeam, C.T.; Chia, S.; Tan, H.C.C.; Kwan, Y.H.; Fong, W.; Seng, J.J.B. A systematic review of factors affecting medication adherence among patients with osteoporosis. *Osteoporos. Int.* **2018**, *29*, 2623–2637. [CrossRef]
6. Cornelissen, D.; de Kunder, S.; Si, L.; Reginster, J.Y.; Evers, S.; Boonen, A.; Hiligsmann, M. Interventions to improve adherence to anti-osteoporosis medications: An updated systematic review. *Osteoporos. Int.* **2020**, *31*, 1645–1669. [CrossRef] [PubMed]
7. Wiesner, A.; Gajewska, D.; Paśko, P. Levothyroxine Interactions with Food and Dietary Supplements—A Systematic Review. *Pharmaceuticals* **2021**, *14*, 206. [CrossRef]
8. Wilcock, M.; Pothecary, A.; Roberson, C. What do we know about the administration of oral bisphosphonates in a hospital setting? *Pharmacoepidemiol. Drug Saf.* **2018**, *27*, 3–14. [CrossRef]
9. Paśko, P. Food-drug interactions among the elderly: Risk assessment and recommendations for patients. *Encycl. Biomed. Gerontol.* **2019**, 107–114. [CrossRef]
10. Eastell, R.; Rosen, C.J.; Black, D.M.; Cheung, A.M.; Murad, M.H.; Shoback, D. Pharmacological management of osteoporosis in postmenopausal women: An endocrine society clinical practice guideline. *J. Clin. Endocrinol. Metab.* **2019**, *104*, 1595–1622. [CrossRef] [PubMed]
11. National Institute for Health and Care Excellence (NICE). *Bisphosphonates for Treating Osteoporosis*; National Institute for Health and Care Excellence (NICE): London, UK, 2017; pp. 1–25.
12. Ogura, Y.; Gonsho, A.; Cyong, J.C.; Orimo, H. Clinical trial of risedronate in Japanese volunteers: A study on the effects of timing of dosing on absorption. *J. Bone Miner. Metab.* **2004**, *22*, 120–126. [CrossRef]
13. Qaseem, A.; Forciea, M.A.; McLean, R.M.; Denberg, T.D. Treatment of Low Bone Density or Osteoporosis to Prevent Fractures in Men and Women: A Clinical Practice Guideline Update From the American College of Physicians. *Ann. Intern. Med.* **2017**, *166*, 818. [CrossRef] [PubMed]
14. Pazianas, M.; Abrahamsen, B.; Ferrari, S.; Russell, R.G.G. Eliminating the need for fasting with oral administration of bisphosphonates. *Ther. Clin. Risk Manag.* **2013**, *9*, 395–402. [CrossRef]
15. Byun, J.-H.; Jang, S.; Lee, S.; Park, S.; Yoon, H.K.; Yoon, B.-H.; Ha, Y.-C. The Efficacy of Bisphosphonates for Prevention of Osteoporotic Fracture: An Update Meta-analysis. *J. Bone Metab.* **2017**, *24*, 37. [CrossRef]
16. Kinov, P.; Boyanov, M. Clinical utility of risedronate in postmenopausal osteoporosis: Patient considerations with delayed-release formulation. *Int. J. Womens Health* **2012**, *4*, 167–174. [CrossRef] [PubMed]

17. Soares, A.P.; do Espírito Santo, R.F.; Line, S.R.P.; Pinto, M.d.G.F.; de Moura Santos, P.; Toralles, M.B.P.; do Espírito Santo, A.R. Bisphosphonates: Pharmacokinetics, bioavailability, mechanisms of action, clinical applications in children, and effects on tooth development. *Environ. Toxicol. Pharmacol.* **2016**, *42*, 212–217. [CrossRef] [PubMed]
18. Cheng, L.; Wong, H. Food effects on oral drug absorption: Application of physiologically-based pharmacokinetic modeling as a predictive tool. *Pharmaceutics* **2020**, *12*, 672. [CrossRef] [PubMed]
19. Biernacka, J.; Betlejewska-Kielak, K.; Kniosíska-Szmurło, E.; Pluciński, F.A.; Mazurek, A.P. Prediction of bioavailability of selected bisphosphonates using in silico methods towards categorization into biopharmaceutical classification system. *Acta Pol. Pharm. Drug Res.* **2013**, *70*, 877–882.
20. Reid, I.R.; Bolland, M.J. Calcium and/or vitamin D supplementation for the prevention of fragility fractures: Who needs it? *Nutrients* **2020**, *12*, 1011. [CrossRef] [PubMed]
21. Harvey, N.C.; Biver, E.; Kaufman, J.M.; Bauer, J.; Branco, J.; Brandi, M.L.; Bruyère, O.; Coxam, V.; Cruz-Jentoft, A.; Czerwinski, E.; et al. The role of calcium supplementation in healthy musculoskeletal ageing: An expert consensus meeting of the European Society for Clinical and Economic Aspects of Osteoporosis, Osteoarthritis and Musculoskeletal Diseases (ESCEO) and the International Foundat. *Osteoporos. Int.* **2017**, *28*, 447–462. [CrossRef]
22. Gertz, B.J.; Holland, S.D.; Kline, W.F.; Matuszewski, B.K.; Freeman, A.; Quan, H.; Lasseter, K.C.; Mucklow, J.C.; Porras, A.G. Studies of the oral bioavailability of alendronate. *Clin. Pharmacol. Ther.* **1995**, *58*, 288–298. [CrossRef]
23. Fosamax (Merck & Co., Inc.). Prescribing Information. Available online: https://www.accessdata.fda.gov/drugsatfda_docs/label/2012/021575s017lbl.pdf (accessed on 5 February 2021).
24. Actonel (Procter & Gamble Pharmaceuticals, Inc.). Prescribing Information. Available online: https://www.accessdata.fda.gov/drugsatfda_docs/label/2009/020835s035lbl.pdf (accessed on 5 February 2021).
25. Atelvia (Warner Chilcott (US), LLC). Prescribing Information. Available online: https://www.accessdata.fda.gov/drugsatfda_docs/label/2015/022560s007lbl.pdf (accessed on 5 February 2021).
26. Bonviva (Roche Therapeutics Inc.). Prescribing Information. Available online: https://www.accessdata.fda.gov/drugsatfda_docs/label/2010/021455s009lbl.pdf (accessed on 6 February 2021).
27. Didronel (Procter & Gamble Pharmaceuticals, Inc.). Prescribing Information. Available online: https://www.accessdata.fda.gov/drugsatfda_docs/label/2009/017831s055lbl.pdf (accessed on 6 February 2021).
28. Ferracru (Norgine B.V.). Product Characteristics. Available online: https://www.ema.europa.eu/en/documents/product-information/feraccru-epar-product-information_en.pdf (accessed on 6 February 2021).
29. Ringe, J.D.; Van Der Geest, S.A.P.; Möller, G. Importance of calcium co-medication in bisphosphonate therapy of osteoporosis: An approach to improving correct intake and drug adherence. *Drugs Aging* **2006**, *23*, 569–578. [CrossRef] [PubMed]
30. Akagi, Y.; Sakaue, T.; Yoneyama, E.; Aoyama, T. Influence of mineral water on absorption of oral alendronate in rats. *Yakugaku Zasshi* **2011**, *131*, 801–807. [CrossRef] [PubMed]
31. Itoh, A.; Akagi, Y.; Shimomura, H.; Aoyama, T. Interaction between bisphosphonates and mineral water: Study of oral risedronate absorption in rats. *Biol. Pharm. Bull.* **2016**, *39*, 323–328. [CrossRef] [PubMed]
32. Pellegrini, R. Which water for alendronate administration? *Osteoporos. Int.* **2009**, *20*, 1451. [CrossRef]
33. Morr, S.; Cuartas, E.; Alwattar, B.; Lane, J.M. How much calcium is in your drinking water? A survey of calcium concentrations in bottled and tap water and their significance for medical treatment and drug administration. *HSS J.* **2006**, *2*, 130–135. [CrossRef]
34. Azoulay, A.; Garzon, P.; Eisenberg, M.J. Comparison of the mineral content of tap water and bottled waters. *J. Gen. Intern. Med.* **2001**, *16*, 168–175. [CrossRef]
35. Binosto (Mission Pharmacal Company). Prescribing Information. Available online: https://www.accessdata.fda.gov/drugsatfda_docs/label/2016/202344s011lbl.pdf (accessed on 7 February 2021).
36. Imai, K. Alendronate sodium hydrate (oral jelly) for the treatment of osteoporosis: Review of a novel, easy to swallow formulation. *Clin. Interv. Aging* **2013**, *8*, 681–688. [CrossRef]
37. Gertz, B.J.; Holland, S.D.; Kline, W.F.; Matuszewski, B.K.; Porras, A.G. Clinical pharmacology of alendronate sodium. *Osteoporos. Int.* **1993**, *3*, 13–16. [CrossRef]
38. Wagener, P. Safety and efficacy of alendronate when not taken in the morning. Results of an 8-week prospective study. *Eur. J. Geriatr.* **2000**, *2*, 35–37.
39. Mitchell, D.Y.; Heise, M.A.; Pallone, K.A.; Clay, M.E.; Nesbitt, J.D.; Russell, D.A.; Melson, C.W. The effect of dosing regimen on the pharmacokinetics of risedronate. *Br. J. Clin. Pharmacol.* **1999**, *48*, 536–542. [CrossRef]
40. Zhou, Y.; He, X.; Li, H.; Ni, Y.; Xu, M.; Sattar, H.; Chen, H.; Li, W. Pharmacokinetics and Tolerability of Minodronic Acid Tablets in Healthy Chinese Subjects and Food and Age Effects on the Pharmacokinetics. *Clin. Ther.* **2015**, *37*, 869–876. [CrossRef]
41. Kendler, D.L.; Ringe, J.D.; Ste-Marie, L.G.; Vrijens, B.; Taylor, E.B.; Delmas, P.D. Risedronate dosing before breakfast compared with dosing later in the day in women with postmenopausal osteoporosis. *Osteoporos. Int.* **2009**, *20*, 1895–1902. [CrossRef] [PubMed]
42. Agrawal, S.; Krueger, D.C.; Engelke, J.A.; Nest, L.J.; Krause, P.F.; Drinka, P.J.; Binkley, N.C. Between-meal risedronate does not alter bone turnover in nursing home residents. *J. Am. Geriatr. Soc.* **2006**, *54*, 790–795. [CrossRef]
43. Fukase, H.; Tanaka, S.; Ohishi, H.; Oikawa, I.; Furuie, H. Phase I clinical trial of NE-58095 DR, a risedronate delayed-release tablet: Dose escalation and food effect. *Jpn. J. Clin. Pharmacol. Ther.* **2018**, *49*, 113–124. [CrossRef]

44. McClung, M.R.; Miller, P.D.; Brown, J.P.; Zanchetta, J.; Bolognese, M.A.; Benhamou, C.L.; Balskc, A.; Burgio, D.E.; Sarley, J.; McCullough, L.K.; et al. Efficacy and safety of a novel delayed-release risedronate 35 mg once-a-week tablet. *Osteoporos. Int.* **2012**, *23*, 267–276. [CrossRef] [PubMed]
45. McClung, M.R.; Balske, A.; Burgio, D.E.; Wenderoth, D.; Recker, R.R. Treatment of postmenopausal osteoporosis with delayed-release risedronate 35 mg weekly for 2 years. *Osteoporos. Int.* **2013**, *24*, 301–310. [CrossRef]
46. Soen, S.; Kishimoto, H.; Hagino, H.; Sone, T.; Ohishi, H.; Fujimoto, T.; Sasaki, E.; Tanaka, S.; Sugimoto, T. Phase II/III, randomized, double-blind, parallel-group study of monthly delayed-release versus daily immediate-release risedronate tablets in Japanese patients with involutional osteoporosis. *J. Bone Miner. Metab.* **2020**, *38*, 86–98. [CrossRef]
47. Bondronat (Roche Therapeutic Inc.). Product Characteristics. Available online: https://www.ema.europa.eu/en/documents/product-information/bondronat-epar-product-information_en.pdf (accessed on 7 February 2021).
48. Nakai, K. The effect of fasting interval on oral bioavailability of ibandronate. *Jpn. Pharmacol. Ther.* **2015**, *43*, 1553–1560.
49. Tankó, L.B.; McClung, M.R.; Schimmer, R.C.; Mahoney, P.; Christiansen, C. The efficacy of 48-week oral ibandronate treatment in postmenopausal osteoporosis when taken 30 versus 60 min before breakfast. *Bone* **2003**, *32*, 421–426. [CrossRef]
50. Ohishi, T.; Matsuyama, Y. Minodronate for the treatment of osteoporosis. *Ther. Clin. Risk Manag.* **2018**, *14*, 729–739. [CrossRef] [PubMed]
51. Matsumoto, T.; Endo, I. Minodronate. *Bone* **2020**, *137*, 115432. [CrossRef]
52. Papaioannou, A.; Morin, S.; Cheung, A.M.; Atkinson, S.; Brown, J.P.; Feldman, S.; Hanley, D.A.; Hodsman, A.; Jamal, S.A.; Kaiser, S.M.; et al. 2010 clinical practice guidelines for the diagnosis and management of osteoporosis in Canada: Summary. *CMAJ* **2010**, *182*, 1864–1873. [CrossRef] [PubMed]
53. Fogelman, I.; Smith, L.; Mazesss, R.; Wilson, M.A.; Bevan, J.A. Absorption of Oral Diphosphonate in Normal Subjects. *Clin. Endocrinol.* **1986**, *24*, 57–62. [CrossRef]
54. Cook, G.J.R.; Blake, G.M.; Fogelman, I. The time of day that Etidronate is ingested does not influence its therapeutic effect in osteoporosis. *Scand. J. Rheumatol.* **2000**, *29*, 62–64. [CrossRef] [PubMed]
55. Ryan, P.J. Timing of cyclical etidronate. *Clin. Exp. Rheumatol.* **2000**, *18*, 609–612. [PubMed]
56. Rosen, H. Selective Estrogen Receptor Modulators for Prevention and Treatment of Osteoporosis. Available online: https://www.uptodate.com/contents/selective-estrogen-receptor-modulators-for-prevention-and-treatment-of-osteoporosis (accessed on 23 January 2021).
57. Whitaker Elam, R. What Is the Role of Raloxifene (Evista) in the Treatment of Osteoporosis? Available online: https://www.medscape.com/answers/330598-83093/what-is-the-role-of-raloxifene-evista-in-the-treatment-of-osteoporosis#:~{}:text=Raloxifene-may-be-most-useful-the-endometrium-was-not-stimulated (accessed on 23 January 2021).
58. Jha, R.K.; Tiwari, S.; Mishra, B. Bioadhesive microspheres for bioavailability enhancement of raloxifene hydrochloride: Formulation and pharmacokinetic evaluation. *AAPS PharmSciTech* **2011**, *12*, 650–657. [CrossRef] [PubMed]
59. Morello, K.C.; Wurz, G.T.; DeGregorio, M.W. Pharmacokinetics of Selective Estrogen Receptor Modulators. *Clin. Pharmacokinet.* **2003**, *42*, 361–372. [CrossRef]
60. Evista (SUBSTIPHARM). Prescribing Information. Available online: https://www.accessdata.fda.gov/drugsatfda_docs/label/2007/022042lbl.pdf (accessed on 9 February 2021).
61. Conbriza (Pfizer Europe). Product Information. Available online: https://www.ema.europa.eu/en/documents/product-information/conbriza-epar-product-information_en.pdf (accessed on 9 February 2021).
62. McKeand, W. Pharmacokinetics, Dose Proportionality, and Bioavailability of Bazedoxifene in Healthy Postmenopausal Women. *Clin. Ther.* **2017**, *39*, 1769–1779. [CrossRef]
63. Duavee (Pfeizer Inc.). Prescribing Information. Available online: https://www.accessdata.fda.gov/drugsatfda_docs/label/2015/022247s002lbl.pdf (accessed on 9 February 2021).

foods

MDPI

Article

The Effect of UV Irradiation on Vitamin D_2 Content and Antioxidant and Antiglycation Activities of Mushrooms

Francesca Gallottiand Vera Lavelli *

Department of Food, Environmental and Nutritional Sciences (DeFENS), University of Milan, 20133 Milan, Italy; francesca.gallotti@unimi.it

* Correspondence: vera.lavelli@unimi.it; Tel.: +39-02-5031-9172

Received: 24 July 2020; Accepted: 4 August 2020; Published: 10 August 2020

Abstract: Mushroom irradiation has been considered a sustainable process to generate high amounts of vitamin D_2 due to the role of this vitamin for human health and of the global concerns regarding its deficient or inadequate intake. Mushrooms are also receiving increasing interest due to their nutritional and medicinal properties. However, there is still a knowledge gap regarding the effect of UV irradiation on mushroom bioactive compounds. In this study, two of the most cultivated mushroom species worldwide, *Agaricus bisporus* and *Pleurotus ostreatus*, were irradiated with UV-B, and the effect of processing was investigated on the contents of vitamin D_2 as well as on antioxidant and antiglycation activities. UV irradiation increased vitamin D_2 up to 57 μg/g d.w, which is an adequate level for the fortification of a number of target foods. UV irradiation decreased the antioxidant activity when measured by the Folin–Ciocalteu reagent, the 2,2-diphenyl-1-(2,4,6 trinitrophenyl) hydrazyl radical assay and the ferric ion-reducing antioxidant power assay, but did not decrease the mushroom's ability to inhibit glycation of a target protein. These results open up a new area of investigation aimed at selecting mushroom species with high nutraceutical benefits for irradiation in order to maintain their potential properties to inhibit oxidative and glycation processes responsible for human diseases.

Keywords: mushroom; vitamin D; antioxidant activity; reducing capacity; glycation

1. Introduction

Deficient and inadequate vitamin D intake (vitamin D represents both D_2 and D_3 forms) is a worldwide public health issue causing both skeletal diseases and increased risk of various chronic diseases [1]. Sources of vitamin D for humans are sunlight exposure, yielding vitamin D_3; a few foods of animal origin providing vitamin D_3; and yeasts and mushrooms, providing vitamin D_2. Both vitamin D_2 and D_3 are metabolized in the liver to 25-hydroxyvitamin D (25(OH)D) and then in the kidneys to 1,25-dihydroxyvitamin D (1,25(OH)2D). According to the serum concentration of 25(OH)D, vitamin D status is defined as deficient when serum 25(OH)D is lower than 25 nmol/L (10 ng/mL) and insufficient when serum 25(OH)D is between 25 and 50 nmol/L (10–20 ng/mL). Vitamin D status can also be assessed by validated food questionnaires [2]. Populations at risk of vitamin D deficiency are young children, the elderly, pregnant women and non-western immigrants in Europe [1]. The occurrence of vitamin D deficiency was found to be 48.1% in a Chinese population of preschool-aged children [3]. A large survey throughout Europe has shown that the intake of vitamin D is inadequate for 77–100% of adults (19–64 years old) and for 55–100% of elderly adults (>64 years old) [4]. From the 2005–2016 National Health and Nutrition Examination Surveys (NHANES), the prevalence of inadequacy for vitamin D intake among US population was found to be 95% [5].

Despite the fact that vitamin D_3 can be formed by the conversion of its precursors in the skin upon sunlight exposure, the above-reported studies document that this process is not sufficient to ensure appropriate levels of this vitamin. Moreover, the intake of vitamin D_3 from animal-based foods is not a proper solution because the current food systems, which are highly dependent on animal-based food sources, are not sustainable from an environmental point of view but also from a health and food security perspective [6]. Hence, one strategy to overcome the global demand of vitamin D in a sustainable way is to produce this vitamin from mushroom irradiation. This way, vitamin D can be delivered through fortified foods [7]. The amount of vitamin D_2 in mushrooms is generally low. However, mushrooms have high amounts of the vitamin D_2 precursor, i.e., ergosterol. Processes that generate vitamin D_2 from ergosterol have been developed by UV irradiation on either the fresh or dried fruit body [8]. Recent trends have tested the efficiency of vitamin D_2 generation by combining supercritical CO_2 extraction with dissolution of the extract in ethanol or methanol followed by UV irradiation [9]. *Pleurotus* spp. are among the basidiomycetes that can be used for vitamin D_2 generation because their cultivation can be performed on various agri-food waste at a low cost [10,11]. Another basidiomycete of interest is *Agaricus bisporus*, which is the most cultivated mushroom worldwide [12].

Mushrooms are known to have medicinal properties and interest in this food source is increasingly expanding due to many studies that demonstrated their potential roles on human health due to antioxidant, antitumor, antimicrobial, anti-inflammatory, immunomodulator, antiatherogenic and hypoglycemic activities [10,11].

There is still a knowledge gap regarding the effect of UV irradiation on mushrooms' bioactive properties. Nevertheless, it is known that UV radiation can promote photo-oxidation via two major routes. The first of these involves direct photo-oxidation arising from the absorption of UV radiation by chromophores, thereby generating excited state species (singlet or triplets) or radicals as a result of photo-ionization. The second major process involves indirect oxidation of targets via the formation and subsequent reactions of singlet oxygen ($^1O_2$). Antioxidant enzymes that eliminate $^1O_2$ have not evolved. Instead, the highly reactive $^1O_2$ can interact with potential targets by either physical quenching or a chemical reaction. The former results in energy transfer and de-excitation of the singlet state but no chemical change in the energy acceptor. The latter causes modification of the target and in the initiation of radical type reactions [13]. Hence, it may be expected that UV irradiation affects the potential ability of mushrooms to prevent cell damage.

Oxidative stress has been found to mediate cell damage, thus triggering a number of human diseases [14]. The term is used to describe the condition of oxidative damage resulting when the critical balance between reactive oxygen species (ROS) enzymatic and non-enzymatic generation and antioxidant defenses (low molecular weight antioxidants, antioxidant enzymes and repair enzymes) is unfavorable.

Moreover, carbonyl stress is another route leading to cell damage that is associated with a number of human diseases [15]. It involves highly reactive dicarbonyl compounds that are formed mainly through non-enzymatic protein glycation and in turn are involved in the formation of various harmful cross-linked adducts, which are collectively called advanced glycation end-products (AGEs).

The bioactive properties of mushrooms are due to the polysaccharides, proteins, lipids and molecules from their secondary metabolism, such as terpenoids, eritadenine, ergothioneine and phenolic compounds [10,11,16–18]. The antioxidant activity of edible mushrooms was studied in mice and has been associated mainly with the polysaccharide fraction, ergothioneine, and phenolic compounds [9,10]. There is little information on the antiglycation activity of mushrooms. In a previous study, polysaccharides isolated from *Ganoderma lucidum* were supplemented in high-fat-diet/streptozotocin diabetic rats and found to decrease the level of AGEs and augment the activities of antioxidant enzymes [19].

In this study, *P. ostreatus* and *A. bisporus* were submitted to UV irradiation and the effects of the process on vitamin D_2 content, as well as antioxidant and antiglycation properties were investigated to obtain overall knowledge regarding the impact of this process on the potential health benefit of mushrooms.

2. Materials and Methods

2.1. Chemicals

Chemicals were purchased from Sigma Aldrich (Milan, Italy).

2.2. Mushrooms

P. ostreatus and *A. bisporus* were purchased from the market. The irradiation treatment was performed as described previously [20]. In brief, about 10 kg of body fruits were used for each mushroom type, cut into 4 mm slices, spread on an oven rack in a single layer and air-dried at 37 °C for 48 h. During the first 24 h of air-drying, the mushrooms were irradiated using two fluorescent lamps installed at 30 cm above the mushroom layer. In accordance to most of the previous studies so far performed [8,12], a UV-B source was chosen for the irradiation. The lamps delivered a UV-B (280–315 nm) irradiance of 0.4 W/m^2. At the end of the process, about 1 kg of dried mushroom was obtained from each mushroom type and the product was grinded to a fine powder using a Thermomix TM 31 (Vorwerk Contempora S.r.l., Milan, Italy). Control *P. ostreatus* and *A. bisporus* powders were obtained by processing mushrooms with the same drying and grinding procedure without UV irradiation.

2.3. Determination of Vitamin D_2

Vitamin D_2 and ergosterol were extracted in triplicate from non-irradiated and irradiated mushroom powders after saponification as described previously [21]. To check the recovery, preliminary samples of mushroom powder were added with 0.1 mL of vitamin D_3 (400 mg/L in methanol). The observed recovery was always higher than 90%. Then, samples were analyzed without vitamin D_3 in order to check the possible presence of vitamin D_2 photoproducts with elution times close to that of vitamin D_3 [22]. Vitamin D_2 and ergosterol were identified by a previously published procedure [23]. A Shimadzu HPLC system (Kyoto, Japan), including an LC-20 AD pump and an SPD-M20A photodiode array detector operated by Labsolution Software, was used. The column was a C18 Sunfire (4.6 mm × 250 mm × 5 mm, Waters, Milan, Italy) and the mobile phase was methanol:water (95:5, *v*/*v*), at a flow rate of 1.0 mL/min. UV detection was performed at 254 nm. The retention times for vitamin D_2, vitamin D_3 and ergosterol were 19.5, 20.5 and 23 min, respectively. The content of vitamin D_2, vitamin D_3 and ergosterol were calculated on the basis of the calibration curve of pure compounds.

2.4. Recovery of Bioactive Fractions

Mushroom powders of *P. ostreatus* and *A. bisporus* were extracted in triplicate with cold water (0.25 g into 5 mL) for 24 h at 4 °C under magnetic stirring. Then, the mixtures were centrifuged (Centrikon T-42K, Kontron, Instruments, Milan, Italy) at 12,000 rpm for 30 min at 20 °C and the supernatant was used for further characterization [24]. The ethanol-insoluble fractions were obtained by adding 8.3 mL of 96% ethanol to 2 mL of the water extracts. The precipitate was recovered by centrifugation (Centrikon T-42K) at 12,000 rpm for 30 min at 20 °C and redissolved in water for further characterization [25]. The yield of water-soluble extracts and ethanol-insoluble precipitates was determined by measuring the weight of dry solids after drying in a vacuum oven at 70 °C and 50 Torr for 18 h.

2.5. Antioxidant Activity

The Folin–Ciocalteu (FC) assay was performed on the water-soluble fractions and the ethanol-insoluble fractions redissolved in water, as described previously [26]. A calibration curve was built using gallic acid. FC-reducing compounds were expressed as milligram of gallic acid equivalents (GAE) per gram of dry fraction.

The free radical scavenging capacity of the water-soluble fractions and the ethanol-insoluble fractions redissolved in water was evaluated using the stable 2,2-diphenyl-1-(2,4,6 trinitrophenyl) hydrazyl radical (DPPH) as described previously [27]. Additionally, 6-hydroxy-2,5,7,8-tetramethylchroman-2-carboxylic acid (Trolox) was used as a reference antioxidant and the results were expressed as micromoles of Trolox equivalents (TE) per gram of dry fractions.

The ferric ion-reducing antioxidant power (FRAP) assay was performed on of the water-soluble fractions and the ethanol-insoluble fractions redissolved in water as described previously [26]. $FeSO_4$ was used for calibration and the results were expressed as micromoles of Fe II equivalents per gram of dry fractions.

2.6. *Antiglycation Activity Using Bovine Serum Albumin (BSA)/Fructose Model Systems*

The antiglycation activity was determined on the water-soluble fractions and the ethanol-insoluble fractions redissolved in water, as described previously [28]. The reaction mixture was prepared by adding 100 μL of sample extracts or standard diluted in water, 900 μL of 200 mM potassium phosphate buffer, pH 7.4 with 0.02% sodium azide, 300 μL of bovine serum albumin (BSA) solution (50 mg/mL of BSA in phosphate buffer) and 300 μL of fructose solution (1.25 M fructose in phosphate buffer). A BSA solution (blank sample) and control reaction without sample addition were prepared in parallel. The mixtures were incubated at 37 °C for 3 days in the dark and then analyzed for fluorescence on a Perkin-Elmer LS 55 Luminescence Spectrometer (Perkin-Elmer, Milan, Italy) with an excitation/emission wavelength pair of 350/420 nm and 5 nm slit width, read against phosphate buffer. Aminoguanidine was used as a positive control. For each sample extract, 3–4 dilutions were assessed in duplicate. Dose–response curves were built reporting % inhibition of BSA glycation as a function of sample or standard concentration. The results were expressed as milligrams of aminoguanidine equivalents (AG) per gram of dry fractions.

2.7. *Statistical Analysis of Data*

Experimental data were analyzed using one-way ANOVA with the least significant difference (LSD) as a multiple range test using Statgraphics 5.1 (STCC Inc., Rockville, MD, USA). These results are reported as the average ± standard error (SE).

3. Results and Discussion

3.1. *The Effect of Mushroom Irradiation on Vitamin D_2 Content*

Initial content of vitamin D_2 was low in both commercial mushrooms considered. This result is consistent with previous studies and could be associated with the conditions used for industrial cultivation of mushrooms that generally occurs with low exposure to sunlight [8,12]. UV irradiation led to an increase in vitamin D_2 content from 3.1 to 37 μg/g d.w. and from 0.3 to 57 μg/g d.w. in *P. ostreatus* and *A. bisporus*, respectively (Table 1). The initial contents of ergosterol were 3600 and 5700 μg/g d.w. in *P. ostreatus* and *A. bisporus*, respectively (Table 1). These values fall in the range previously reported [29] and remained high after the generation of significant amounts of vitamin D_2. This result may suggest that upon UV irradiation. the residual amount of ergosterol is still relevant for drug development and could contribute to the health-promoting effects of mushrooms [30].

Table 1. Vitamin D_2 and ergosterol content in non-irradiated and UV-irradiated mushrooms [a].

Mushroom	Vitamin D_2 µg/g d.w	Ergosterol µg/g d.w.
P. ostreatus	$3.1^{c} \pm 0.1$	$3600^{c} \pm 400$
P. ostreatus$_{UV}$	$37^{b} \pm 0.1$	$3200^{c} \pm 300$
A. bisporus	$0.31^{d} \pm 0.1$	$5700^{a} \pm 300$
A. bisporus$_{UV}$	$57^{a} \pm 0.1$	$4800^{b} \pm 300$

[a] Values are average ± SE. Different letters in the same column (a–d) indicate significant differences among samples (LSD, $p < 0.05$).

The contents of vitamin D_2 obtained were in the recovery range described by previous studies, i.e., from 3.9 to 741 µg/g d.w. for *A. bisporus* and from 27.89 to 239.67 µg/g d.w. for *P. ostreatus*, respectively [8]. In the present study, no vitamin D_2 photoproducts were detected (Figure 1). According to [22], irradiation of *P. ostreatus* led to significant formation of vitamin D_2, tachysterol$_2$ and lumysterol$_2$, but the conditions applied were different from those used in the present study and the amount of vitamin D_2 produced was 141.2 mg/kg, i.e., higher than those observed in the present study.

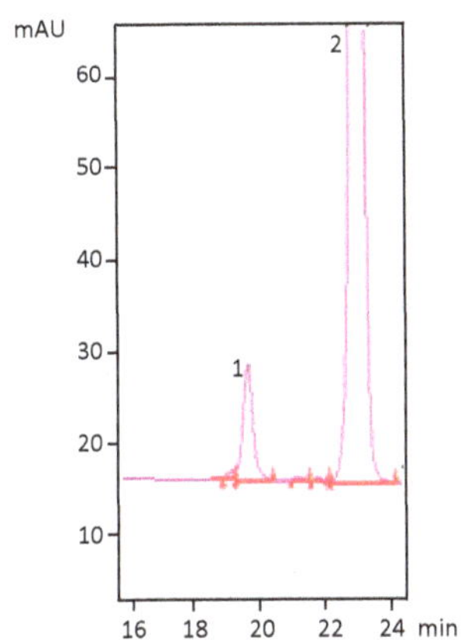

Figure 1. HPLC profile of the *n*-hexane extract of irradiated *P. ostreatus* powder. 1. Vitamin D_2; 2. ergosterol.

The yield of vitamin D_2 formation upon irradiation was previously found to depend on conditions that can be controlled during processing, such as the light source, the distance between the lamp and the mushroom, temperature and the duration of the irradiation. However, the yield also depends on factors that cannot be controlled, such as the shape of mushroom, which conditions the surface exposed to light and the moisture levels, that varies during drying [8]. The final powder obtained after irradiation can be standardized for vitamin D_2 content by mixing the treated powder with untreated powder. Hence, considering a target final value for vitamin D_2 of 40 µg/g d.w., a simulation was conducted to calculate the amount of irradiated powder that could be necessary to fortify vehicle foods in order to achieve health benefits (Table 2).

To perform this simulation, three studies were considered to define the most appropriate target foods and fortification levels, namely, a study in line with the regulation dealing with vitamin D fortification in the United States and Canada [31], a study considering the European regulation scenario on vitamin D fortification [32] and a study aimed at exploring food matrices for vitamin D fortification in low/lower-middle income countries [33]. These latter studies defined somewhat different fortification strategies due to different dietary habits and needs of the populations involved and to different regulations of the countries of interest. In any case, it can be observed that very low fortification levels of mushroom powder, in the range 0.03 to 0.38 g/100 g of food, could lead to the target fortification level, thus confirming the potential role of irradiation technology in increasing the intake of vitamin D (Table 2).

Table 2. Modelled formulations of foods with irradiated mushroom powder to achieve a target level of vitamin D_2 [a].

Food Category	Target Level for Vitamin D µg/100 g Food	Reference	Amount of UV-Irradiated Mushroom Powder g/100 Food
Ready-to-eat breakfast cereals	8.75	[31]	0.22
Milk	1.05	[31]	0.03
Yogurt	2.225	[31]	0.06
Margarine	8.275	[31]	0.21
Edible oil	15	[32]	0.38
Milk	2	[32]	0.05
Wheat flour	2.8	[32]	0.07
Milk	1	[33]	0.03
Orange juice	10.5	[33]	0.26

[a] The mushroom powder considered was obtained from *P. ostreatus* and *A. bisporus* through UV irradiation as described in the Material and Methods Section and contained 40 µg/g of vitamin D_2.

3.2. The Effect of Mushroom Irradiation on Antioxidant Activity

Considering the variety of mushroom compounds that can act as antioxidants, various extraction procedures have been proposed to study the antioxidant activity of mushrooms. Comparing the yields of ethanol and hot water extractions for various mushroom species, it was observed that ethanol only accounted for 5.89–18.89% of solids, while hot water fraction accounted for 38.3–57.2% of total solids [34]. Moreover, information obtained by using hot water is considered more valuable because it corresponds to the procedure used to recover bioactive compounds from medicinal mushrooms by Chinese traditional medicine [34]. However, studies conducted on *P. citrinopileatus* have shown a greater bioactivity in vitro of the cold aqueous extract compared to the hot water extract due to protein denaturation and degradation of phenolic compounds at high temperatures [24]. Moreover, the ethanol precipitation of the aqueous extract has been proposed for the isolation of proteoglycans with anticancer properties [25]. Hence, in the current study, both the cold water extract and ethanol precipitate fractions of *P. ostreatus* and *A. bisporus* were studied to assess the effects of UV irradiation. The water-soluble fraction of *P. ostreatus* accounted for 54% of solids, while the ethanol-insoluble fraction was 7.2% of solids (Table 3). Higher percentage of the water extracts was observed for *A. bisporus*, i.e., 69%, but the percentage of ethanol-insoluble fraction was lower, i.e., 3.3% (Table 3). Due to the high extraction yield, the cold water extracts can be considered representative of the mushroom matrix to investigate the effect of irradiation on potential bioactive components.

The antioxidant activity in both non-irradiated and UV-irradiated mushrooms was evaluated by the FC, FRAP and DPPH assays (Table 3).

The levels of FC-reducing compounds for the aqueous extracts of *P. ostreatus* and *A. bisporus* before irradiation were 24.6 and 16.2 mg GAE/g of fraction, respectively. These values can be considered to be high according to the ranking defined previously in a screening study on 23 species of mushrooms, which reported levels of FC-reducing compounds of the aqueous extract in the range of from 2 to 37 mg GAE/g [35]. The ethanol-insoluble fractions of both mushrooms showed a higher specific content of FC-reducing compounds than those of the respective water-soluble fraction for both mushrooms. This result may suggest that bioactive compounds partitioned mostly into the ethanol-insoluble fraction. UV irradiation decreased the FC-reducing compounds of *P. ostreatus* by 20%, while the decrease was lower for *A. bisporus* (7%).

The DPPH values for the water extracts of *P. ostreatus* and *A. bisporus* before irradiation were 56 and 48 µmol TE/g fraction. As observed for the FC-reducing compounds, the specific DPPH values for the ethanol-insoluble fractions of both mushrooms were higher than those of the corresponding water-soluble fractions. The genus *Agaricus* has been previously found to have the highest DPPH radical scavenging properties among the most widely appreciated cultivated mushrooms [36]. UV irradiation decreased the DPPH values of all fractions, except for the ethanol-insoluble fraction of *A. bisporus*.

Table 3. Fraction yields and antioxidant activity evaluated by the FC, DPPH and FRAP assays of the water extracts and ethanol-insoluble fractions of *P. ostreatus* and *A. bisporus* before and after UV irradiation.[a].

Mushroom and Treatment	Fraction Yield g/100 g	Antioxidant Activity		
		FC mg GAE/g fraction	DPPH μmol TE/g fraction	FRAP μmol FeII/g fraction
P. ostreatus WE	54 ^{b} ± 3	24.6 ^{c} ± 1.3	56 ^{c} ± 4	82 ^{d} ± 5
P. ostreatus $_{UV}$ WE	54 ^{b} ± 3	19.6 ^{d} ± 1.5 (20%)	48 ^{d} ± 1 (14%)	58 ^{e} ± 14 (29%)
A. bisporus WE	67 ^{a} ± 1	16.2 ^{e} ± 0.5	48 ^{d} ± 6	208 ^{b} ± 7
A. bisporus $_{UV}$ WE	70 ^{a} ± 1	15.0 ^{f} ± 0.2 (7%)	37 ^{e} ± 7 (23%)	131 ^{c} ± 28 (37%)
P. ostreatus EP	7.2 ^{c} ± 0.1	32.1 ^{b} ± 0.6	71 ^{b} ± 1	222 ^{b} ± 1
P. ostreatus $_{UV}$ EP	7.3 ^{c} ± 0.1	25.8 ^{c} ± 0.7 (20%)	49 ^{d} ± 3 (31%)	127 ^{c} ± 14 (43%)
A. bisporus EP	3.4 ^{d} ± 0.1	53.4 ^{a} ± 1.6	119 ^{a} ± 14	473 ^{a} ± 43
A. bisporus $_{UV}$ EP	3.2 ^{d} ± 0.1	53.4 ^{a} ± 0.8	103 ^{a} ± 17	505 ^{a} ± 14

[a]Values are average ± SE; values in brackets are the percent decreases for the UV irradiated samples with respect to non-irradiated samples. WE = water extract, EP = ethanol precipitate, FC = Folin–Ciocalteu, DPPH = 2,2-diphenyl-1-(2,4,6 trinitrophenyl)hydrazyl radical, FRAP = ferric ion-reducing antioxidant power, GAE = gallic acid equivalents, TE = Trolox equivalents. Different letters in the same column (a–e) indicate significant differences among samples (LSD, $p < 0.05$).

The FRAP values for the water extracts of *P. ostreatus* and *A. bisporus* before irradiation were 82 and 208 μmol FeII/g fraction, respectively. As observed for FC and DPPH, higher specific FRAP values were present in the ethanol-insoluble fractions than in the water-soluble fractions for both mushroom species, with values of 222 and 473 μmol FeII/g fraction. This result may be related to the preferential partition of the bioactive compounds in the ethanol-insoluble fractions. A screening of 1943 plant-based foods revealed that the average FRAP value is 115.7 μmol FeII/g [37]. Hence, mushrooms can be considered important as a dietary source of reducing compounds. On the other hand, irradiation caused a marked decrease in the FRAP values, except for the ethanol-insoluble fraction of *A. bisporus*.

3.3. The Effect of Mushroom Irradiation on Antiglycation Activity

The antiglycation activity in both non-irradiated and UV-irradiated mushrooms was evaluated by the BSA/fructose assay (Table 4).

Table 4. Antiglycation activity of the water extracts and ethanol-insoluble fractions of *P. ostreatus* and *A. bisporus* before and after UV irradiation.

Mushroom and Treatment	Antiglycation Agents mg AG/g fraction
P. ostreatus WE	113 ^{c} ± 22
P. ostreatus $_{UV}$ WE	131 ^{c} ± 25
A. bisporus WE	83 ^{d} ± 9
A. bisporus $_{UV}$ WE	89 ^{d} ± 10
P. ostreatus EP	177 ^{b} ± 10
P. ostreatus $_{UV}$ EP	170 ^{b} ± 6
A. bisporus EP	555 ^{a} ± 80
A. bisporus $_{UV}$ EP	500 ^{a} ± 89

Values are average ± SE. WE = water extract, EP = ethanol precipitate, AG = aminoguanidine. Fraction yield is reported in Table 3. Different letters in the same column (a–d) indicate significant differences among samples (LSD, $p < 0.05$).

The water extracts of *P. ostreatus* and *A. bisporus* were able to inhibit protein glycation, exhibiting 113 and 83 mg AG/g fraction. As observed for the antioxidant activity, which was higher in the ethanol-insoluble fractions, higher levels of antiglycation activity was observed in the ethanol-insoluble fractions with respect to the water-soluble fractions, corresponding to 177 and 555 mg AG/g fraction for *P. ostreatus* and *A. bisporus*, respectively. There is no information on the antiglycation activity of these mushroom species. In previous studies, the antiglycation activities of a crude water-soluble fraction extracted from the sclerotia of *Inonotus obliquus* [38] and *Lignosus rhinocerus* [39] were found to be 23 and 133 mg AG/g fraction. Compared to these latter studies, the antiglycation activities of the ethanol-insoluble fractions of *P. ostreatus* and *A. bisporus* were found to be much higher. However, even higher antiglycation activity was observed for a polysaccharide purified from the fruiting body of *Boletus snicus*, which was found to have the same antiglycation activity of aminoguanidine in the BSA/glucose model system, i.e., 1000 mg AG/g [40].

The UV irradiation process did not cause any significant change to the antiglycation activity of *P. ostreatus* and *A. bisporus* (Table 4). Despite redox-active compounds generally possessing antiglycation activity, this activity is also dependent on metal chelation and carbonyl-trapping abilities [16].

4. Conclusions

P. ostreatus and *A. bisporus*, which are among the most cultivated mushroom species in the world, were able generate relevant amounts of vitamin D_2 through UV irradiation. The concept that this latter technology can address the global need of vitamin D_2 was highlighted by providing a scenario of possible future application of the vitamin D_2 enriched mushroom in target foods. *P. ostreatus* and *A. bisporus* were also found to possess high antioxidant and antiglycation activities. Irradiation caused a decrease in antioxidant activity in both mushroom species but did not affect antiglycation activity. The ethanol-insoluble fraction of *A. bisporus* was not affected by UV irradiation. Results from this research open up a new area of investigation. In fact, while the relevance of UV irradiation of mushrooms to address the global need of vitamin D_2 has attracted a lot of research interest, the identification of the effects of irradiation on the mushroom matrices is still lacking and deserves particular attention.

One limitation of the current study is that it included only two mushroom species. Considering the wide biodiversity among mushrooms, future studies should be directed to extend knowledge on the effects of irradiation to other mushroom species in order to assist a more efficient design of the process for the generation of vitamin D_2 with a major focus on the retention of bioactive properties.

Author Contributions: Conceptualization, V.L.; investigation, F.G.; writing—review and editing, V.L., F.G.; project administration, V.L.; funding acquisition, V.L. All authors have read and agreed to the published version of the manuscript.

Funding: This research was funded by Cariplo foundation, grant number 2016-0740.

Conflicts of Interest: The authors declare no conflict of interest.

References

1. Holick, M.F.; Chen, T.C. Vitamin D deficiency: A worldwide problem with health consequences. *Am. J. Clin. Nutr.* **2008**, *87*, 1080S–1086S. [CrossRef] [PubMed]
2. Ganji, V.; Abu-Dbaa, R.; Othman, H.; Zewein, M.; Al-Abdi, T.; Zumin, S. Validation of vitamin D-specific food frequency questionnaire against food records for Qatari women. *Foods* **2020**, *9*, 195. [CrossRef] [PubMed]
3. Zhang, H.; Li, Z.; Wei, Y.; Fu, J.; Feng, Y.; Chen, D.; Xu, D. Status and influential factors of vitamin D among children aged 0 to 6 years in a Chinese population. *BMC Public Health* **2020**, *20*, 1–6. [CrossRef] [PubMed]
4. Roman Viñas, B.; Ribas Barba, L.; Ngo, J.; Gurinovic, M.; Novakovic, R.; Cavelaars, A.; de Groot, L.C.P.G.M.; Veer, P.v.; Matthys, C.; Majem, L.S. Projected prevalence of inadequate nutrient intakes in Europe. *Ann. Nutr. Metab.* **2011**, *59*, 84–95. [CrossRef] [PubMed]
5. Reider, C.A.; Chung, R.-Y.; Devarshi, P.P.; Grant, R.W.; Mitmesser, S.H. Inadequacy of immune health nutrients: Intakes in US adults, the 2005–2016 NHANES. *Nutrients* **2020**, *12*, 1735. [CrossRef]

6. Galanakis, C.M. The food systems in the era of the coronavirus (COVID-19) pandemic crisis. *Foods* **2020**, *9*, 523. [CrossRef]
7. McClements, D.J. Development of next-generation nutritionally fortified plant-based milk substitutes: Structural design principles. *Foods* **2020**, *9*, 421. [CrossRef]
8. Taofiq, O.; Fernandes, A.; Barrosa, L.; Barreiro, M.F.; Ferreira, I.C.F.R. UV-irradiated mushrooms as a source of vitamin D_2: A review. *Trends Food Sci. Technol.* **2017**, *70*, 82–94. [CrossRef]
9. Morales, D.; Gil-Ramirez, A.; Smiderle, F.R.; Piris, A.J.; Ruiz-Rodriguez, A.; Soler-Rivas, C. Vitamin D-enriched extracts obtained from shiitake mushrooms (lentinula edodes) by supercritical fluid extraction and UV-irradiation. *Inn. Food Sci. Emerg. Technol.* **2017**, *41*, 330–333. [CrossRef]
10. Carrasco-González, J.A.; Serna-Saldívar, S.O.; Gutiérrez-Uribe, J.A. Nutritional composition and nutraceutical properties of the *Pleurotus* fruiting bodies: Potential use as food ingredient. *J. Food Compos. Anal.* **2017**, *58*, 69–81. [CrossRef]
11. Lavelli, V.; Proserpio, C.; Gallotti, F.; Laureati, M.; Pagliarini, E. Circular reuse of bio-resources: The role of *Pleurotus* spp. in the development of functional foods. *Food Funct.* **2018**, *9*, 1353–1372. [CrossRef] [PubMed]
12. Cardwell, G.; Bornman, J.F.; James, A.P.; Black, L.J. A review of mushrooms as a potential source of dietary vitamin D. *Nutrients* **2018**, *10*, 1498. [CrossRef] [PubMed]
13. Redmond, R.W.; Kocheva, I.E. Symposium-in-print: Singlet oxygen invited review. Spatially resolved cellular responses to singlet oxygen. *Photochem. Photobiol.* **2006**, *82*, 1178–1186. [CrossRef] [PubMed]
14. Lobo, V.; Patil, A.; Phatak, A.; Chandra, N. Free radicals, antioxidants and functional foods: Impact on human health. *Pharmacogn. Rev.* **2010**, *4*, 118–126. [CrossRef]
15. Sri Harsha, P.S.C.; Lavelli, V. Use of grape pomace phenolics to counteract endogenous and exogenous advanced glycation end-product formation. *Nutrients* **2019**, *11*, 1917. [CrossRef]
16. Ritota, M.; Manzi, P. Pleurotus spp. cultivation on different agri-food by-products: Example of biotechnological application. *Sustainability* **2019**, *11*, 5049. [CrossRef]
17. Bekiaris, G.; Tagkouli, D.; Koutrotsios, G.; Kalogeropoulos, N.; Zervakis, G.I. *Pleurotus* mushrooms content in glucans and ergosterol assessed by ATR-FTIR spectroscopy and multivariate analysis. *Foods* **2020**, *9*, 535. [CrossRef]
18. Morales, D.; Tabernero, M.; Largo, C.; Polo, G.; Piris, A.J.; Soler-Rivas, C. Effect of traditional and modern culinary processing, bioaccessibility, biosafety and bioavailability of eritadenine, a hypocholesterolemic compound from edible mushrooms. *Food Funct.* **2018**, *9*, 6360–6368. [CrossRef]
19. Menga, G.; Zhua, H.; Yangb, S.; Wua, F.; Zheng, H.; Chen, E.; Xua, J. Attenuating effects of *Ganoderma lucidum* polysaccharides on myocardial collagen cross-linking relates to advanced glycation end product and antioxidant enzymes in high-fat-diet and streptozotocin-induced diabetic rats. *Carbohydr. Polym.* **2011**, *84*, 180–185. [CrossRef]
20. Pedrali, D.; Gallotti, F.; Proserpio, C.; Pagliarini, E.; Lavelli, V. Kinetic study of vitamin D_2 degradation in mushroom powder to improve its applications in fortified foods. *LWT Food Sci. Technol.* **2020**, *125*, 109248. [CrossRef]
21. Sławińska, A.; Fornal, E.; Radzki, W.; Skrzypczak, K.; Zalewska-Korona, M.; Michalak-Majewska, M.; Parfieniuk, E.; Stachniuk, A. Study on vitamin D_2 stability in dried mushrooms during drying and storage. *Food Chem.* **2016**, *199*, 203–209. [CrossRef] [PubMed]
22. Wittig, M.; Krings, U.; Berger, R.G. Single-run analysis of vitamin D photoproducts in oyster mushroom (*Pleurotus ostreatus*) after UV-B treatment. *J. Food Compos. Anal.* **2013**, *31*, 266–274. [CrossRef]
23. Huang, S.J.; Lin, C.P.; Tsai, S.Y. Vitamin D_2 content and antioxidant properties of fruit body and mycelia of edible mushrooms by UV-B irradiation. *J. Food Compos. Anal.* **2015**, *42*, 38–45. [CrossRef]
24. Chen, P.; Weng, Y.; Yu, Z.; Koo, M.; Wang, B. Extraction temperature affects the activities of antioxidation, carbohydrate-digestion enzymes, and angiotensin-converting enzyme of *Pleurotus citrinopileatus* extract. *J. Food Drug Anal.* **2016**, *24*, 548–555. [CrossRef]
25. Sarangi, I.; Gosh, D.; Sanjaya, S.; Maiti, K. Anti-tumor and immunomodulating effects of *Pleurotus ostreatus* mycelia-derived proteoglycans. *Int. Immunopharmacol.* **2008**, *6*, 1287–1297. [CrossRef] [PubMed]
26. Lavelli, V.; Sri Harsha, P.S.C.; Spigno, G. Modelling the stability of maltodextrin-encapsulated grape skins phenolics used as a new ingredient in apple puree. *Food Chem.* **2016**, *209*, 323–331. [CrossRef]

27. Lavelli, V.; Hidalgo, A.; Pompei, C.; Brandolini, A. Radical scavenging activity of einkorn (*Triticum monococcum* L. subsp. *monococcum*) wholemeal flour and its relationship to soluble phenolic and lipophilic antioxidant content. *J. Cereal Sci.* **2009**, *49*, 319–321. [CrossRef]
28. Lavelli, V.; Sri Harsha, P.S.C. Microencapsulation of grape skin phenolics for pH controlled release of antiglycation agents. *Food Res. Int.* **2019**, *119*, 822–828. [CrossRef]
29. Jasinghe, V.J.; Perera, C.O. Distribution of ergosterol in different tissues of mushrooms and its effect on the conversion of ergosterol to vitamin D_2 by UV irradiation. *Food Chem.* **2005**, *92*, 541–546. [CrossRef]
30. Merdivan, S.; Lindequist, U. Ergosterol peroxide: A mushroom-derived compound with promising biological activities-A review. *Int. J. Med. Mushrooms* **2017**, *19*, 93–105. [CrossRef]
31. Calvo, M.S.; Whiting, S.J.; Barton, C.N. Vitamin D fortification in the United States and Canada: Current status and data needs. *Am. J. Clin. Nutr.* **2004**, *80*, 1710S–1716S. [CrossRef] [PubMed]
32. O'Mahony, L.; Stepien, M.; Gibney, M.J.; Nugent, A.P.; Brennan, L. The potential role of vitamin D enhanced foods in improving vitamin D status. *Nutrients* **2011**, *3*, 1023–1041. [CrossRef] [PubMed]
33. Cashman, K.D.; O'Dea, R. Exploration of strategic food vehicles for vitamin D fortification in low/lower-middle income countries. *J. Steroid Biochem. Mol. Biol.* **2019**, *195*, 105479. [CrossRef] [PubMed]
34. Tsai, S.; Huang, S.; Lo, S.H.; Wu, T.; Lian, P.; Mau, J. Flavor compounds and antioxidant properties of several cultivated mushrooms. *Food Chem.* **2009**, *113*, 578–584. [CrossRef]
35. Puttaraju, N.G.; Yenkateshaiah, S.U.; Dharmesh, S.M.; Nanjarajurs, S.M.; Somasundaram, R. Antioxidant activity of indigenous edible mushrooms. *J. Agric. Food Chem.* **2006**, *54*, 9764–9772. [CrossRef]
36. Reis, F.S.; Martins, A.; Barros, L.; Ferreira, I.C.F.R. Antioxidant properties and phenolic profile of the most widely appreciated cultivated mushrooms: A comparative study between *in vivo* and *in vitro* samples. *Food Chem. Toxicol.* **2012**, *50*, 1201–1207. [CrossRef]
37. Carlsen, M.H.; Halvorsen, B.L.; Holte, K.; Bøhn, S.K.; Dragland, S.; Sampson, L. The total antioxidant content of more than 3100 foods, beverages, spices, herbs and supplements used worldwide. *Nutr. J.* **2010**, *9*, 3. [CrossRef]
38. Wang, C.; Chen, Z.; Pan, Y.; Gao, X.; Chen, H. Anti-diabetic effects of *Inonotus obliquus* polysaccharides-chromium (III) complex in type 2 diabetic mice and its sub-acute toxicity evaluation in normal mice. *Food Chem. Toxicol.* **2017**, *108*, 498–509. [CrossRef]
39. Yap, H.-Y.Y.; Tan, N.-H.; Ng, S.-T.; Tan, C.-S.; Fung, S.-Y. Inhibition of protein glycation by tiger milk mushroom [*Lignosus rhinocerus* (cooke) ryvarden] and search for potential anti-diabetic activity-related metabolic pathways by genomic and transcriptomic data mining. *Front. Pharmacol.* **2018**, *9*, 103. [CrossRef]
40. Liping, S.; Xuejiao, S.; Yongliang, Z. Preparation, characterization and antiglycation activities of the novel polysaccharides from *Boletus snicus*. *Int. J. Biol. Macromol.* **2016**, *92*, 607–614. [CrossRef]

© 2020 by the authors. Licensee MDPI, Basel, Switzerland. This article is an open access article distributed under the terms and conditions of the Creative Commons Attribution (CC BY) license (http://creativecommons.org/licenses/by/4.0/).

MDPI
St. Alban-Anlage 66
4052 Basel
Switzerland
Tel. +41 61 683 77 34
Fax +41 61 302 89 18
www.mdpi.com

Foods Editorial Office
E-mail: foods@mdpi.com
www.mdpi.com/journal/foods

www.ingramcontent.com/pod-product-compliance
Lightning Source LLC
Chambersburg PA
CBHW041015270926
51850CB00024B/209
* 9 7 8 3 0 3 6 5 5 3 0 1 6 *